Theory and Decision Library C

Game Theory, Social Choice, Decision Theory, and Optimization

Volume 47

More information about this series at http://www.springer.com/series/6618

David W.K. Yeung • Leon A. Petrosyan

Subgame Consistent Cooperation

A Comprehensive Treatise

Springer

David W.K. Yeung
Center of Game Theory
Saint Petersburg State University
St Petersburg, Russia

SRS Consortium for Advanced Study
in Cooperative Dynamic Games
Hong Kong Shue Yan University
Hong Kong

Leon A. Petrosyan
Faculty of Applied Mathematics-Control
Processes
St Petersburg State University
St Petersburg, Russia

ISSN 0924-6126 ISSN 2194-3044 (electronic)
Theory and Decision Library C
ISBN 978-981-10-9382-1 ISBN 978-981-10-1545-8 (eBook)
DOI 10.1007/978-981-10-1545-8

Mathematics Subject Classification: 91A12, 91A25, 49N70, 91A35, 91A80

Printed on acid-free paper

This Springer imprint is published by Springer Nature
The registered company is Springer Science+Business Media Singapore Pte Ltd.

Preface

It is well known that noncooperative behaviors among participants would lead to an outcome which is not Pareto optimal and it could even be highly undesirable. There exist a large number of problems – like transboundary pollution, overpopulation, water and food shortage, international trade, provision of public goods, infectious diseases, military conflicts, and nuclear proliferation – that could not be solved effectively by noncooperative individual actions. Cooperation suggests the best possibility of obtaining socially optimal and group-efficient solutions to problems like these. Though collaborative schemes like global cooperation in environmental control hold out the best promise of effective action, limited success has been observed. In particular, one can hardly be convinced that multinational joint initiatives like the Kyoto Protocol, the Copenhagen Agreement, or the Cancun Agreements can offer a long-term solution because there is no guarantee that participants will always be better off throughout the entire duration of the agreement. To create a cooperative solution that every party would commit to from beginning to end, the proposed arrangement must always remain acceptable to all the participants at any time instance within the period of cooperation. This is a "classic" game-theoretic problem.

Formulation of optimal behaviors for players is a fundamental element in the theory of cooperative games. The players' behaviors satisfying some specific optimality principles constitute a solution of the game. Dynamic cooperation is one of the most intriguing forms of optimization analysis, and its complexity leads to great difficulties in the derivation of satisfactory solutions. To ensure sustainability of dynamic cooperative schemes, a stringent condition on the cooperative solution is required – that is *subgame consistency*. A cooperative solution is subgame consistent if the specific optimality principle agreed upon at the outset remains effective at any subgame with a later starting time and a state brought about by prior optimal behaviors. The notion of subgame consistency is crucial to the success of cooperation in a dynamic framework. This book provides a comprehensive treatise on subgame consistent cooperation emanated from our works on the topic in the past two decades.

We are very grateful to George Leitmann and (the late) John Nash for inspiration from their classic work in game theory on which many results in the book are based. We are also very grateful to our friends and colleagues whom we have learned a lot from through communication and collaboration, particularly Georges Zaccour, Steffen Jørgensen, Alain Haurie, Vladimir Mazalov, Jerzy Filar, Nikolay Zenkevich, Michelle Breton, and Joseph Shinar. Our families have been an enormous and continuing source of inspiration throughout our careers. We thank Nina and Ovanes (LP) and Stella and Patricia (DY) for their love and patience during this projects which on occasion might have diverted our attention away from them. We thank Cynthia Yingxuan Zhang for her outstanding research assistance and manuscript formatting. Research supports from Saint Petersburg State University (research project 9.38.245.2014) and Shue Yan University are gratefully acknowledged.

Finally, we would dedicate this book to the memory of our dear friend and legendary game theorist, John Forbes Nash Jr.

St Perersburg, Russia — David W.K. Yeung
Hong Kong
St Perersburg, Russia — Leon A. Petrosyan

Contents

Chapter 1
Introduction

Strategic behavior in the human and social world has been increasingly recognized in theory and practice. From a decision-maker's perspective, it becomes important to consider and accommodate the interdependencies and interactions of human decisions. As a result, game theory has emerged as a fundamental instrument in pure and applied research. In addition, since human beings live in time and decisions generally lead to effects over time, most of the strategic interactions are dynamic rather than static. One particularly complex and fruitful branch of game theory is dynamic games, which investigates interactive decision making over time. Differential (continuous-time dynamic) games were originated by Rufus Isaacs (1965). Discrete-time dynamic games (usually referred to as dynamic games) are multi-stage counterparts of differential games using Bellman's (1957) discrete-time dynamic programming technique to obtain their solutions. Since then research involving continuous-time and discrete-time dynamic games continue to grow in a large number of fields and studies including economics, engineering, business, biology, mathematics, environmental studies, and social and political sciences. Rather exhaustive collections of differential and dynamic game applications in economics and business can be found in Dockner et al. (2000), Jørgensen and Zaccour (2004) and Long (2010).

It is well known that non-cooperative behaviours among participants would, in general, lead to an outcome which is not Pareto optimal. Worse still, highly undesirable outcomes (like the prisoner's dilemma) and even devastating results (like the tragedy of the commons) could appear when the involved parties only care about their individual self interests in a non-cooperative situation. In a dynamic world, non-cooperative behaviours guided by short-sighted individual rationality could be a source for series of disastrous consequences in the future. The phenomenon of the 'inter-temporal tragedy of temporal individual rationality' becomes rather common in many real world dynamic interactive activities. Cooperation suggests the possibility of obtaining socially optimal and group efficient solutions to decision problems involving strategic actions. The calls for cooperation had not been scarce:

D.W.K. Yeung, L.A. Petrosyan, *Subgame Consistent Cooperation*,
Theory and Decision Library C 47, DOI 10.1007/978-981-10-1545-8_1

> You look at the large problems that we face – that would be overpopulation, water shortages, global warming and AIDS, I suppose – all of that needs international cooperation to be solved. – (Molly Ivins).

> Competition has been shown to be useful up to a certain point and no further, but cooperation, which is the thing we must strive for today, begins where competition leaves off. – (Franklin D. Roosevelt).

> When times are tough and people are frustrated and angry and hurting and uncertain, the politics of constant conflict may be good, but what is good politics does not necessarily work in the real world. What works in the real world is cooperation. – (William J. Clinton).

> The keystone of successful business is cooperation. Friction retards progress. – (James Cash Penney).

> The only thing that will redeem mankind is cooperation. – (Bertrand Russell).

> Nature is based on harmony. So it says if we want to survive and become more like nature, then we actually have to understand that it's cooperation versus competition. – (Bruce Lipton).

> When it comes to the fundamental issues that humanity faces, I think that solutions involve shifting consciousness towards cooperation. – (Jeremy Gilley).

> With closely knitted transnational interests, zero-sum games and conflict con-frontations had long been outdated; helping each other on the same boat, co-operation and mutual victory are the need of our time. – (Xi Jinping).

However dynamic cooperation cannot be sustainable if there is no guarantee that the participants will always be better off within the entire duration of the cooperation. More than anything else, it is due to the lack of this kind of guarantees that cooperative schemes fail to last till its end. Dynamic cooperation represents one of the most intriguing forms of optimization analysis. The complexity of the problem leads to great difficulties in the derivation of satisfactory solutions. Similar to the case of static (one-shot) cooperation two fundamental factors – individual rationality and group optimal – must be maintained in dynamic cooperation. Group optimality ensures that all potential gains from cooperation are captured. Failure to fulfil group optimality leads to the condition where the participants prefer to deviate from the agreed upon solution plan in order to extract the unexploited gains. Individual rationality is required to hold so that the payoff allocated to any participant under cooperation will be no less than his noncooperative payoff. Failure to guarantee individual rationality leads to the condition where the concerned participants would reject the agreed upon solution plan and act noncooperatively. Yet for dynamic cooperation group optimality and individual rationality have to be satisfied at all time instants during the entire cooperation duration.

On top of individual rationality and group optimality being satisfied throughout the cooperation duration sustainability of dynamic cooperation requires the satisfaction of a stringent condition – *subgame consistency*. A cooperative solution is subgame consistent if an extension of the solution policy to a subgame with a later starting time and a state brought about by prior optimal behaviors would remain optimal. In particular, it implies that the specific optimality principle agreed upon at the outset must remain effective at any instant of time throughout the game along the optimal state trajectory. Brexit – the attempted exit of UK from the European Union is a clear example of subgame inconsistency. Demonstration of the notion of subgame consistency in a numerical example is provided in the Appendix of this Chapter.

Petroysan (1997) introduced the notion of agreeable solutions in cooperative differential games in which subgame consistent (then referred to as time consistent solutions in the deterministic framework) solution was developed. Crucial to the derivation of a subgame consistent cooperative solution is the formulation of a payment distribution mechanism that would lead to the realization of the solution. Yeung and Petrosyan (2004) developed a generalized method for the derivation of analytically tractable subgame consistent solutions in stochastic differential games. This has made possible the rigorous study of subgame consistent solutions in continuous-time dynamic cooperation. Applications and exegeses of subgame consistent solutions in differential games and stochastic differential games can be found in Yeung (2005, 2006b, 2010) and Yeung and Petrosyan (2006a, 2007a, b, c, 2008, 2013b, 2014a).

The analysis on subgame consistent solution was further extended to randomly furcating stochastic differential games in which both the state dynamics and future payoffs are stochastic in Petrosyan and Yeung (2007). Applications of subgame consistent solutions in randomly furcating stochastic differential games are found in Petrosyan and Yeung (2006), Yeung (2008) and Yeung and Petrosyan (2012c). Yeung (2011) analyzed subgame consistent solutions in differential games with asynchronous players' horizons. In developing the analysis for games with nontransferable payoffs Yeung (2014) derived the nontransferable individual payoff functions under cooperation in stochastic differential games with nontransferable payoffs. Yeung and Petrosyan (2005) and Yeung et al. (2007) analyzed subgame consistent solutions in cooperative stochastic differential games with nontransferable payoffs.

For discrete-time analyses, Yeung and Petrosyan (2010) developed a generalized method for the derivation of analytically tractable subgame consistent solutions in stochastic dynamic games. This has made possible the rigorous study of subgame consistent solutions in discrete-time dynamic cooperation. Yeung (2014) analyzed subgame consistent solutions in a dynamically cooperative game of environmental management with the possibility of switching the choice of control. To accommodate the possibility of uncertain game duration Yeung and Petrosyan (2011) developed subgame consistent solution mechanisms for cooperative dynamic games with random horizon. The analysis was extended to the case where the state dynamics and the game horizon are stochastic in Yeung and Petrosyan (2012b).

The notion of subgame consistency and solution mechanisms for randomly-furcating cooperative stochastic dynamic games were developed by Yeung and Petrosyan (2013a). Applications of subgame consistent cooperative solution in randomly furcating stochastic dynamic games in collaborative provision of public goods was given in Yeung and Petrosyan (2014b). To analyze subgame consistency under randomly furcating payoffs, stochastic dynamics and uncertain horizon Yeung and Petrosyan (2014c) developed subgame consistent cooperative solution mechanisms for randomly furcating stochastic dynamic games with uncertain horizon.

Yeung and Petrosyan (2015a) developed subgame consistent solution mechanisms in cooperative dynamic games with non-transferable payoffs/utility (NTU) using a variable payoffs weights scheme. A theorem for characterizing subgame consistent solutions is derived. The use of a variable payoff weights scheme allows the derivation of subgame consistent solutions under a wide range of optimality principles. A stochastic version was provided in Yeung and Petrosyan (2015b). Yeung (2013) derived individual player's payoff functions in NTU cooperative stochastic dynamic games which are used in formulating subgame consistent solutions.

The book Cooperative Stochastic Differential Games by Yeung and Petrosyan (2006b) was the first text focusing on subgame consistent solutions in the field of cooperative stochastic differential games. The book Subgame Consistent Economic Optimization by Yeung and Petrosyan (2012a) was the first text on the treatment of subgame consistent solutions in economic optimization.

This book presents a comprehensive treatise on subgame consistent cooperation emanated from the works of its authors in the field of cooperative subgame consistency and cooperative dynamic games. Some novel extensions and elaborated expositions on the existing work are also provided. The text is organized into three parts. Part 1, which includes Chaps. 2, 3, 4, 5, and 6, consists of continuous-time analyses. Part II, which includes Chaps. 7, 8, 9, 10 and 11, consists of discrete-time analyses. Part III, which includes Chaps. 12, 13, 14, and 15, presents applications of subgame consistent solutions in various areas. Each chapter is designed to be self contained and slightly repeated technical preliminary settings may appear.

Chapter 2 considers subgame consistent cooperative solutions in differential games. It integrates the works of Chapter 2 of Yeung and Petrosyan (2006b), Chapter 4 of Yeung and Petrosyan (2012a) and the deterministic version of Yeung and Petrosyan (2004). The basic formulations of cooperative differential games, group optimality, individual rationality under cooperation, and the notion of subgame consistency are presented. An analysis on subgame consistent dynamic cooperation and the derivation of a subgame consistent payoff distribution procedure are provided. Subgame consistency in infinite horizon cooperative differential games is also examined.

Chapter 3 introduces stochastic elements in the state dynamics and considers subgame consistent cooperative solutions in stochastic differential games. It provides an integrated exposition the works of Yeung and Petrosyan (2004), Chapter 4 of Yeung and Petrosyan (2006b), and Chapter 8 of Yeung and Petrosyan (2012a). An analysis on cooperative subgame consistency under uncertainty, derivation of a subgame consistent payoff distribution procedure and illustrations in cooperative fishery are presented.

Chapter 4 considers subgame consistency in randomly-furcating cooperative stochastic differential games. This class of games allow random shocks in the state dynamics and stochastic changes in the players' payoff structures. The Chapter presents an $n-$ player counterpart of the Petrosyan and Yeung's (2007) 2-player analysis on subgame-consistent cooperative solutions in randomly-furcating stochastic differential games. The basic game formulation, an analysis on subgame consistent dynamic cooperation and derivation of a subgame consistent payoff distribution procedure are provided.

In many game situations, the players' time horizons differ. This may arise from different life spans, different entry and exit times, and different durations of leases and contracts. Chapter 5 considers subgame consistency under asynchronous players' horizons. In particular, it is an integrated disquisition of the analysis in Yeung (2011) with an extension incorporating stochastic state dynamics. Dynamic cooperation among players coexisting in the same duration is examined and an analysis on payoff distribution procedures leading to subgame consistent solutions in this asynchronous horizons scenario is provided. An illustration in cooperative resource extraction is shown.

Chapter 6 considers subgame consistent cooperative solutions in non-transferable utility/payoff (NTU) stochastic differential games. In the case when payoffs are nontransferable, transfer of payoffs cannot be made and subgame consistent solution mechanism becomes extremely complicated. The Chapter is an integrated exposition of the works in Yeung and Petrsoyan (2005) and Yeung et al. (2007). The notion of subgame consistency in NTU cooperative stochastic differential games under time invariant payoff weights is examined and a class of cooperative stochastic differential games with nontransferable payoffs is developed to illustrate the derivation of subgame consistent solutions.

Part II considers subgame consistent cooperative solutions in a discrete-time dynamic framework. In fact, in many game situations, the evolutionary process is in discrete time rather than in continuous time. Chapter 7 considers subgame consistent cooperative solutions in dynamic games. It integrates the works of Yeung and Petrosyan (2010) and Chapters 12 and 13 of Yeung and Petrosyan (2012a). The notions of group optimality and individual rationality, subgame consistent cooperative solutions and corresponding payoff distribution procedures are derived. A heuristic approach of obtaining subgame consistent solutions is provided to widen the application to a wide range of cooperative game problems in which only estimates of the expected cooperative payoffs and individual non-cooperative payoffs with acceptable degrees of accuracy are available.

Chapter 8 considers subgame consistent cooperative solutions in random horizon dynamic games. Examples of this kind of problems arise from uncertainties in the renewal of lease, the terms of offices of elected authorities, contract renewal and continuation of agreements subjected to periodic negotiations. The analysis is based on the work in Yeung and Petrosyan (2011). A dynamic programming technique for solving inter-temporal problems with random horizon is developed to serve as the foundation of solving the game problem. The noncooperative equilibrium is characterized with a set of random duration discrete-time Isaacs-Bellman equations. The issues of dynamic cooperation under random horizon, group optimality and individual rationality are analyzed.

Chapter 9 considers subgame consistent cooperative solutions in randomly furcating stochastic dynamic games. In this type of games, the evolution of the state is stochastic and future payoff structures are not known with certainty. The analysis is based on Yeung and Petrosyan (2013a). Non-cooperative Nash equilibria of the games are characterized and subgame-consistent solutions are derived for cooperative paradigms. A discrete-time analytically tractable payoff distribution procedure contingent upon specific random realizations of the state and payoff structure is derived.

Chapter 10 investigates the class of randomly furcating stochastic dynamic games with uncertain game horizon. In particular, there exist uncertainties in the state dynamics, future payoff structures and game horizon. The non-cooperative Nash outcome, subgame-consistent cooperative solutions and discrete-time analytically tractable payoff distribution procedures contingent upon specific random realizations of the state and payoff structure are derived. Corresponding Bellman equations for solving inter-temporal problems with randomly furcating payoffs and random horizon are derived. A set of random duration discrete-time Hamilton-Jacobi-Bellman equations for a non-cooperative equilibrium are presented. The analysis is developed along the work of Yeung and Petrosyan (2014c).

Chapter 11 considers subgame consistent solutions in NTU cooperative dynamic games via the use of variable payoff weights. It is based on an elaborated exposition of the analysis in Yeung and Petrosyan (2015a, b). The notion of subgame consistency in NTU dynamic games under a variable weights scheme is presented. Derivations of subgame consistent cooperative strategies via variable weights are shown and an illustration in public capital build-up is given. An extension the analysis to NTU cooperative stochastic dynamic games is also provided. The use of variable payoff weights provides an effective way in achieving subgame consistency for non-transferrable payoffs games under a wide range of optimality principles.

In Part III, various applications of subgame consistent solutions are presented. Chapter 12 provides subgame consistent cooperative schemes to resolve the classic problem of cooperative public goods provisions. The first application is from Yeung and Petrosyan (2013b) in which the analysis is in a cooperative stochastic differential game framework with multiple asymmetric agents in public capital build-up. The second is from Yeung and Petrosyan (2014b) in which the analysis is conducted in a randomly-furcating stochastic dynamic game framework which allows uncertainties in the capital accumulation dynamics and payoff structures.

Chapter 13 presents collaborative schemes for environmental management in a cooperative differential game framework and derives subgame consistent solutions for the schemes. Due to the geographical diffusion of pollutants, unilateral response on the part of one country or region is often ineffective. This Chapter gives an integrated exposition of the work of Yeung and Petrosyan (2008) on a subgame consistent scheme of dynamic cooperation in transboundary industrial pollution management. An extension of the Yeung and Petrosyan (2008) analysis to incorporate uncertainties in future payoffs is also presented.

Under the current situation of environmental degradation, even substantial reduction in industrial production using conventional production technique would only slow down the rate of increase and not be able to reverse the trend of continual pollution accumulation. Adoption of environment-preserving technique plays a central role to solving the problem effectively. Chapter 14 presents a cooperative dynamic model of collaborative environmental management with production technique choices and derives a subgame consistent solution. The analysis is based on Yeung's (2014) work on subgame consistent collaborative environmental management with the availability of environment-preserving production techniques. An extension of Yeung's (2014) analysis to multiple types of environment-preserving techniques is also provided.

Chapter 15 presents two applications in business collaboration. The first one is on corporate joint venture and the second one is on cartel. The joint venture analysis is based on Yeung and Petrosyan (2006a), Yeung (2010) and Chapter 9 of Yeung and Petrosyan (2012a). An analysis of a dynamic corporate joint venture in which gains can be obtained from cost saving cooperation is provided. Subgame consistent solutions with optimality principles requiring the sharing of cooperative payoff proportionally to the firms' expected noncooperative payoffs and the sharing of cooperative payoff according to the Shapley value are derived. The Cartel analysis is extracted from Yeung (2005) and Chapter 11 of Yeung and Petrosyan (2012a). It presents a stochastic dynamic dormant-firm cartel. The basic settings, market outcome, optimal cartel output and subgame-consistent cartel profit sharing are investigated.

Dynamic optimization techniques are provided in the Technical Appendices at the end of the book. Finally, worth noting is that the text does not only provide rigorous solution mechanisms for subgame consistent cooperation it also presents a heuristic approach in Chap. 7 to derive subgame consistent solutions in situations where it may not be possible or practical to obtain all the information needed. The heuristic approach allows the application of subgame consistent solution in dynamic games if estimates of the expected cooperative payoffs and individual non-cooperative payoffs with acceptable degrees of accuracy are available. This approach would be helpful to resolving the unstable elements in cooperative schemes for a wide range of game theoretic real-world problems.

Appendix: Numerical Demonstration of Subgame Consistency

To demonstrate the notion of subgame consistency in cooperative dynamic games in a clear way with minimal technical requirement, we consider a simple numerical example with two players in a 4-stage game horizon. The players derive incomes in each stage and there is a state variable x_t, for $t \in \{1, 2, 3, 4\}$. The players' incomes and the values of the state variable are affected by the actions of these players.

Non-cooperative Outcome

In a non-cooperative scenario, the players' incomes in each stage, the values of the state variable and the players' payoffs are summarized in Table 1.1 below.

Note that the payoff of a player at stage t refers to the sum of stage incomes that he will receive from stage t to the last stage of the game (that is stage 4). Now consider the case when the players agree to cooperate and enhance their joint incomes.

Cooperation and Optimality Principle

Consider the case where the players agree to act cooperative under an optimality principle: "Maximize the joint payoff and share the cooperative payoff proportional to their non-cooperative payoffs". If any side decides to opt out the cooperative plan will be cancelled and the players will revert to playing non-cooperatively. Acting

Table 1.1 Players' payoffs and stage incomes and the state path under non-cooperation

PLAYER 1 (PI)					
PI Non-coop Stage Income:	100	95	85	70	350
Stage:	1	2	3	4	TOTAL
State Path:	x_1 10	x_2 8.5	x_3 7	x_4 5.5	x_5 4
PI Stages 1-4 Payoff $V^1(1,x_1)$:	100	95	85	70	350
PI Stages 2-4 Payoff $V^1(2,x_2)$:		95	85	70	250
PI Stages 3-4 Payoff $V^1(3,x_3)$:			85	70	155
PI Stage 4 Payoff $V^1(4,x_4)$:				70	70
PLAYER 2 (PII)					
PII Non-coop Stage Income:	60	53	47	40	200
Stage:	1	2	3	4	TOTAL
State Path:	x_1 10	x_2 8.5	x_3 7	x_4 5.5	x_5 4
PII Stages 1-4 Payoff $V^2(1,x_1)$:	60	53	47	40	200
PII Stages 2-4 Payoff $V^2(2,x_2)$:		53	47	40	140
PII Stages 3-4 Payoff $V^2(3,x_3)$:			47	40	87
PII Stage 4 Payoff $V^2(4,x_4)$:				40	40

cooperatively in maximizing the joint payoff, the players' stage cooperative incomes and the values of the state variable are given in Table 1.2 below.

Summing the stage incomes of player 1 and that of player 2 yields the cooperative joint stage income. The cooperative joint incomes in each stage, the values of the state variable and the maximized joint payoffs are given in Table 1.3 below.

Let $\xi^i(1,x_1^*)$ denote the payoff that player i will receive in stage 1 under cooperation for $i \in \{1,2\}$. At initial stage 1, according to the agreed-upon optimality principle the players would share the cooperative payoff proportional to their non-cooperative payoffs the cooperative payoff, that is:

$$\xi^i(1,x_1^*) = \frac{V^i(1,x_1^*)}{V^1(1,x_1^*) + V^2(1,x_1^*)} W(1,x_1^*) \quad \text{for } i \in \{1,2\}.$$

Using Tables 1.1 and 1.3, the cooperative payoffs of players 1 and 2 at stage 1 are respectively:

Table 1.2 Players' stage cooperative incomes and the cooperative state path

PLAYER 1 (PI)					
PI Coop Stage Income:	115	140	130	125	510
Stage:	1	2	3	4	TOTAL
Co-op State Path:	x_1^*	x_2^*	x_3^*	x_4^*	x_5^*
	10	12	15	18	15
PLAYER 2 (PII)					
PII Coop Stage Income:	65	60	60	55	240
Stage:	1	2	3	4	TOTAL

Table 1.3 Cooperative joint incomes in each stage, the cooperative state path and maximized joint payoffs

Co-op Joint Stage Income:	180	200	190	180	750
Stage:	1	2	3	4	TOTAL
Co-op State Path:	x_1^*	x_2^*	x_3^*	x_4^*	x_5^*
	10	12	15	18	15
Co-op Joint Payoff $W(1,x_1^*)$:	180	200	190	180	750
Co-op Joint Payoff $W(2,x_2^*)$:		200	190	180	570
Co-op Joint Payoff $W(3,x_3^*)$:			190	180	370
Co-op Joint Payoff $W(4,x_4^*)$:				180	180

$$\xi^1(1,x_1^*) = \frac{350}{350+200}750 = 477.273 \text{ and}$$
$$\xi^2(1,x_1^*) = \frac{200}{350+200}750 = 272.727. \tag{1.1}$$

Hence under the optimality principle agreed-upon in the initial stage, player 1 is expected to realize a payoff $\xi^1(1,x_1^*) = 477.273$ and player 2 is expected to realize a payoff $\xi^2(1,x_1^*) = 272.727$. However, according to Table 1.2, the joint payoff maximization scheme would yield a payoff of 510 to player 1 and a payoff of 240 to player 2. A payoff distribution procedure (PDP) has to be designed so that the payoffs according to the optimality principle can be realized.

Notion of Subgame Consistency

In a multi-stage game, a stringent condition for the sustainability of cooperation is the notion of subgame consistency. The idea of subgame consistency is that the specific agreed-upon optimality principle at the initial time must be maintained at subsequent time throughout the game horizon along the optimal state trajectory. Let $\xi^i(t,x_t^*)$ denote the payoff that player i will receive in stage t under cooperation for $i\in\{1,2\}$ and $t\in\{1,2,3,4\}$. At subsequent stage $t\in\{2,3,4\}$ after the initial stage, according to the agreed-upon optimality principle: "Maximize the joint payoff and share the cooperative payoff proportional to their non-cooperative payoffs", player i's payoff under cooperation in stage t should be

$$\xi^i(t,x_t^*) = \frac{V^i(t,x_t^*)}{V^1(t,x_t^*)+V^2(t,x_t^*)}W(t,x_t^*) \quad \text{for } i\in\{1,2\} \quad \text{and } t\in\{2,3,4\}\ .$$

The non-cooperative payoffs of the players along the cooperative path, that is $V^i(t,x_t^*)$, for $i\in\{1,2\}$ and $t\in\{2,3,4\}$, are given in Table 1.4 below.

Therefore in stage 2 the players would adopt the optimality principle: "Maximize the joint payoff and share the cooperative payoff proportional to their non-cooperative payoffs". Hence the payoffs to the players in stage 2 have to satisfy:

$$\xi^i(2,x_2^*) = \frac{V^i(2,x_2^*)}{V^1(2,x_2^*)+V^2(2,x_2^*)}W(2,x_2^*) \quad \text{for } i\in\{1,2\}\ .$$

Table 1.4 Non-cooperative payoffs of the players along the cooperative state path

PLAYER 1 (PI)					
PI Stage 4 Payoff $V^1(4,x_4^*)$:				75	75
PI Stages 3-4 Payoff $V^1(3,x_3^*)$:			92	78	170
PI Stages 2-4 Payoff $V^1(2,x_2^*)$:		105	95	80	280
PI Stages 1-4 Payoff $V^1(1,x_1^*)$:	100	95	85	70	350
PLAYER 2 (PII)					
PII Stage 4 Payoff $V^2(4,x_4^*)$:				45	45
PII Stages 3-4 Payoff $V^2(3,x_3^*)$:			52	48	100
PII Stages 2-4 Payoff $V^2(2,x_2^*)$:		60	55	50	165
PII Stages 1-4 Payoff $V^2(1,x_1^*)$:	60	53	47	40	200

Using Tables 1.2 and 1.3, the cooperative payoffs of players 1 and 2 at stage 2 are respectively:

$$\xi^1(2,x_2^*) = \frac{280}{280+165}570 = 358.65 \text{ and}$$
$$\xi^2(2,x_2^*) = \frac{165}{280+165}570 = 211.35. \tag{1.2}$$

Similarly, at stage 3 using the agreed-upon optimality principle the cooperative payoffs of players 1 and 2 are respectively:

$$\xi^1(3,x_3^*) = \frac{170}{170+100}370 = 232.96 \text{ and}$$
$$\xi^2(3,x_3^*) = \frac{100}{170+100}370 = 137.04. \tag{1.3}$$

Finally at stage 4, using the agreed-upon optimality principle the cooperative payoffs of players 1 and 2 are respectively:

$$\xi^1(4,x_4^*) = \frac{75}{75+45}180 = 112.5 \text{ and}$$
$$\xi^2(4,x_4^*) = \frac{45}{75+45}180 = 67.5. \tag{1.4}$$

A system of payoffs as in (1.1, 1.2, 1.3 and 1.4) leads a subgame consistent solution. A payoff distribution procedure (PDP) has to be designed so that the

Table 1.5 Payoff Distribution Procedure (PDP) of a subgame consistent solution

PLAYER 1 (PI)					
PI Stage Income:	118.623	125.69	120.46	112.5	477.273
Stage:	1	2	3	4	TOTAL
State Path:	x_1	x_2	x_3	x_4	x_5
	10	8.5	7	5.5	4
PLAYER 2 (PII)					
PII Stage Income:	61.377	74.31	69.54	67.5	272.727
Stage:	1	2	3	4	TOTAL

Table 1.6 Transfer payments for a subgame consistent PDP

PI Stage Transfer:	3.623	-14.31	-9.54	-12.5	-32.727
Stage:	1	2	3	4	TOTAL
PII Stage Transfer:	-3..623	14.31	9.54	12.5	32.727

payoffs of players 1 and 2 in stage 1 to stage 4 will be realized as (1.1, 1.2, 1.3 and 1.4). Leaving the general theorem for the derivation of PDP to be explained in Chap. 7 we proceed to verify that a PDP with the following cooperative stage incomes would lead to (1.1, 1.2, 1.3 and 1.4):

To verify that the PDP in Table 1.5 would lead to (1.1, 1.2, 1.3 and 1.4) we can readily obtain:

$$\begin{aligned}\xi^1(1,x_1^*) &= 118.623 + 125.69 + 120.46 + 112.5 = 477.273,\\ \xi^1(2,x_2^*) &= 125.69 + 120.46 + 112.5 = 358.65,\\ \xi^1(3,x_3^*) &= 120.46 + 112.5 = 232.96,\\ \xi^1(4,x_4^*) &= 112.5.\end{aligned}$$

Similarly, from Table 1.5, we obtain $\xi^2(1,x_1^*) = 272.727, \xi^2(2,x_2^*) = 211.35$, $\xi^2(3,x_3^*) = 137.04$ and $\xi^2(4,x_4^*) = 67.5$.

In order to achieve the PDP in Table 1.5 a transfer payment scheme with stage income transfers received/paid (+/−) in Table 1.6 below has to be adopted.

Adding the stage transfer payments in Table 1.6 to the players' cooperative stage incomes in Table 1.2 yields the PDP in Table 1.5 (which leads a subgame consistent solution). Finally, note that player i's cooperative payoff $\xi^i(t,x_t^*)$ is always higher than his non-cooperative payoff $V^i(t,x_t^*)$, for $i \in \{1,2\}$ and $t \in \{1,2,3,4\}$, and therefore the players would have no incentive to deviate from the above subgame consistent solution at any stage of the game.

Part I
Continuous-Time Analysis

Chapter 2
Subgame Consistent Cooperative Solution in Differential Games

In game theory, strategic behavior and decision making are modeled in terms of the characteristics of players, the objective or payoff function of each individual, the actions open to each player throughout the game, the order of such actions, and the information available at each stage of play. Optimal decisions are then determined under different assumptions regarding the availability and transmission of information, and the opportunities and possibilities for individuals to communicate, negotiate, collude, offer inducements, and enter into agreements which are binding or enforceable to varying degrees and at varying costs. Cooperative games suggest the possibility of socially optimal and group efficient solutions to decision problems involving strategic action. As discussed in Chap. 1, individual rationality, group optimality and subgame consistency are crucial elements of a cooperative game solution. This chapter presents an analysis on subgame consistent solutions which entail group optimality and individual rationality for cooperative differential games. It integrates the works of Chapter 2 of Yeung and Petrosyan (2006b), Chapter 4 of Yeung and Petrosyan (2012a) and the deterministic version of Yeung and Petrosyan (2004).

The organization of the Chapter is as follows. Section 2.1 presents the basic formulation of cooperative differential games. Section 2.2 presents an analysis on subgame consistent dynamic cooperation. Derivation of a subgame consistent payoff distribution procedure is provided in Sect. 2.3. An illustration of the solution mechanism is given in a cooperative fishery game in Sect. 2.4. Subgame consistency in infinite horizon cooperative differential games is examined in Sect. 2.5. In Sect. 2.6, a subgame consistent solution of an infinite horizon cooperative resource extraction scheme is derived. Chapter notes are given in Sect. 2.7 and problems in Sect. 2.8.

D.W.K. Yeung, L.A. Petrosyan, *Subgame Consistent Cooperation*,
Theory and Decision Library C 47, DOI 10.1007/978-981-10-1545-8_2

2.1 Basic Formulation of Cooperative Differential Games

Consider the general form of n-person differential games in which player i seeks to maximize its objective:

$$\int_{t_0}^{T} g^i[s, x(s), u_1(s), u_2(s), \cdots, u_n(s)] \exp\left[-\int_{t_0}^{s} r(y)dy\right] ds \\ + \exp\left[-\int_{t_0}^{T} r(y)dy\right] q^i(x(T)), \tag{1.1}$$

for $i \in N = \{1, 2, \cdots, n\}$, where $r(y)$ is the discount rate, $x(s) \in X \subset R^m$ denotes the state variables of game, $q^i(x(T))$ is player i's valuation of the state at terminal time T and $u_i \in U^i$ is the control of player i, for $i \in N$. The payoffs of the players are transferrable.

The state variable evolves according to the dynamics

$$\dot{x}(s) = f[s, x(s), u_1(s), u_2(s), \cdots, u_n(s)], \; x(t_0) = x_0, \tag{1.2}$$

where $x(s) \in X \subset R^m$ denotes the state variables of game, and $u_i \in U^i$ is the control of player i, for $i \in N$. The functions $f[s, x, u_1, u_2, \cdots, u_n]$, $g^i[s, \cdot, u_1, u_2, \cdots, u_n]$ and $q^i(\cdot)$, for $i \in N$, and $s \in [t_0, T]$ are differentiable functions.

2.1.1 Non-cooperative Feedback Equilibria

To analyze the cooperative outcome we first characterize the non-cooperative equilibria as a benchmark for negotiation in a cooperative scheme. Since in a non-cooperative situation it is difficult to prevent the players from revising their strategies during the game duration, therefore they would consider adopting feedback strategies which are decision rules that are dependent upon the current state $x(t)$ and current time t, for $t_0 \le t \le s$.

For the n-person differential game (1.1 and 1.2), an n -tuple of feedback strategies $\{u_i^*(s) = \phi_i^*(s, x) \in U^i$, for $i \in N\}$ constitutes a *Nash equilibrium solution* if the following relations for each $i \in N$ are satisfied:

$$V^{(t_0)i}(t, x) = \int_{t}^{T} g^i[s, x^*(s), \phi_1^*(s, x^*(s)), \phi_2^*(s, x^*(s)), \ldots, \phi_n^*(s, x^*(s))] \exp\left[-\int_{t_0}^{s} r(y)dy\right] ds \\ + q^i(x^*(T)) \exp\left[-\int_{t_0}^{T} r(y)dy\right]$$

$$\geq \int_t^T g^i\left[s, x^i(s), \phi_1^*\left(s, x^i(s)\right), \phi_2^*\left(s, x^i(s)\right), \ldots, \phi_{i-1}^*\left(s, x^i(s)\right), \phi_i\left(s, x^i(s)\right),\right.$$

$$\left.\phi_{i+1}^*\left(s, x^i(s)\right), \ldots\ldots, \phi_n^*\left(s, x^i(s)\right)\right] \exp\left[-\int_{t_0}^s r(y)dy\right] ds$$

$$+ q^i\left(x^i(T)\right) \exp\left[-\int_{t_0}^T r(y)dy\right], \forall \phi_i^*(s,x) \in U^i, x \in R^m, \tag{1.3}$$

where on the interval $[t_0, T]$,

$$\dot{x}^*(s) = f\left[s, x^*(s), \phi_1^*\left(s, x^*(s)\right), \phi_2^*\left(s, x^*(s)\right), \ldots, \phi_n^*\left(s, x^*(s)\right)\right], \quad x^*(t) = x;$$

and

$$\dot{x}^i(s) = f\left[s, x^i(s), \phi_1^*(s, x^i(s)), \phi_2^*(s, x^i(s)), \ldots, \phi_{i-1}^*(s, x^i(s)), \phi_i(s, x^i(s)), \phi_{i+1}^*(s, x^i(s)), \ldots\right.$$
$$\left.\ldots, \phi_n^*(s, x^i(s))\right], \quad x^i(t) = x, \text{ for } i \in N.$$

A feedback Nash equilibrium solution of the game (1.1 and 1.2) satisfying (1.3) can be characterized by the following Theorem.

Theorem 1.1 An n-tuple of strategies $\{u_i^*(t) = \phi_i^*(t,x) \in U^i, \text{ for } i \in N\}$ provides a feedback Nash equilibrium solution to the game (1.1 and 1.2) if there exist continuously differentiable functions $V^{(t_0)i}(t,x) : [t_0, T] \times R^m \to R, i \in N$, satisfying the following set of partial differential equations:

$$-V_t^{(t_0)i}(t,x) = \max_{u_i} \left\{ g^i\left[t, x, \phi_1^*(t,x), \phi_2^*(t,x), \Lambda, \phi_{i-1}^*(t,x), u_i(t,x), \phi_{i+1}^*(t,x), \Lambda\right.\right.$$

$$\left.\Lambda, \phi_n^*(t,x)\right] \exp\left[-\int_{t_0}^t r(y)dy\right]$$

$$\left.+V_x^{(t_0)i}(t,x) f\left[t, x, \phi_1^*(t,x), \phi_2^*(t,x), \Lambda, \phi_{i-1}^*(t,x), u_i(t,x), \phi_{i+1}^*(t,x), \Lambda, \phi_n^*(t,x)\right]\right\}$$

$$= g^i\left[t, x, \phi_1^*(t,x), \phi_2^*(t,x), \Lambda, \phi_n^*(t,x) \exp\left[-\int_{t_0}^t r(y)dy\right]\right.$$

$$+V_x^{(t_0)i}(t,x) f\left[t, x, \phi_1^*(t,x), \phi_2^*(t,x), \Lambda, \phi_n^*(t,x)\right],$$

$$V^{(t_0)i}(T,x) = q^i(x) \exp\left[-\int_{t_0}^T r(y)dy\right], \quad i \in N.$$

Proof Invoking the dynamic programming technique in Theorem A.1 of the Technical Appendices, $V^{(t_0)i}(t,x)$ is the maximized payoff of player i for given strategies $\left\{u_j^*(s) = \phi_j^*(t,x) \in U^j, \text{ for } j \in N \text{ and } j \neq i\right\}$ of the other $n-1$ players. Hence a Nash equilibrium appears. ■

A remark that will be utilized in subsequent analysis is given below.

Remark 1.1 Let $V^{(\tau)i}(t,x)$ denote the feedback Nash equilibrium payoff of player i at time t given the state x in a game with payoffs (1.1) and dynamics (1.2) which starts at time τ for $\tau \in [t_0, T)$. Note that the equilibrium feedback strategies depend on current time and current state. One can readily verify that

$$
\begin{aligned}
&\exp\left[\int_{t_0}^{\tau} r(y)dy\right] V^{(t_0)i}(t,x) = \exp\left[\int_{t_0}^{\tau} r(y)dy\right] \\
&\times \int_t^T g^i\left[s, x^*(s), \phi_1^*(s, x^*(s)), \phi_2^*(s, x^*(s)), \Lambda, \phi_n^*(s, x^*(s))\right] \exp\left[-\int_{t_0}^{s} r(y)dy\right] ds \\
&= \int_t^T g^i\left[s, x^*(s), \phi_1^*(s, x^*(s)), \phi_2^*(s, x^*(s)), \Lambda, \phi_n^*(s, x^*(s))\right] \exp\left[-\int_{\tau}^{s} r(y)dy\right] ds \\
&= V^{(\tau)i}(t,x),
\end{aligned}
$$

for $\tau \in [t_0, T)$. ■

While non-cooperative outcomes are (in general) not Pareto optimal the players would consider cooperation to enhance their payoffs. This will be analyzed in the following section.

2.1.2 Dynamic Cooperation

Cooperative games suggest the possibility of socially optimal and group efficient solutions to decision problems involving strategic action. Now consider the case when the players agree to cooperate and distribute the payoff among themselves according to an optimality principle. Two essential properties that a cooperative scheme has to satisfy are group optimality and individual rationality. Group optimality ensures that the joint payoff of all the players under cooperation is maximized. Failure to fulfill group optimality leads to the condition where the participants prefer to deviate from the agreed-upon solution plan in order to extract the unexploited gains. Individual rationality is required to hold so that the payoff allocated to any player under cooperation will be no less than his noncooperative payoff. Failure to guarantee individual rationality leads to the condition where the concerned participants would deviate from the agreed upon solution plan and play noncooperatively.

2.1.2.1 Group Optimality Under Cooperation

Since payoffs are transferable, group optimality requires the players to maximize their joint payoff. The players must then solve the following optimal control problem:

$$\max_{u_1, u_2, \cdots, u_n} \left\{ \int_{t_0}^{T} \sum_{j=1}^{n} g^j[s, x(s), u_1(s), u_2(s), \cdots, u_n(s)] \exp\left[-\int_{t_0}^{s} r(y)dy \right] ds \right. \\ \left. + \exp\left[-\int_{t_0}^{T} r(y)dy \right] \sum_{j=1}^{n} q^j(x(T)) \right\} \tag{1.4}$$

subject to (1.2).

An optimal solution to the control problem (1.2) and (1.4) characterizing the set of group optimal control strategies is provided by the theorem below.

Theorem 1.2 A set of controls $\{\psi_i^*(t,x), \text{for } i \in N \text{ and } t \in [t_0, T]\}$ provides an optimal solution to the control problem (1.2) and (1.4) if there exists continuously differentiable function $W^{(t_0)}(t,x) : [t_0, T] \times R^m \to R$ satisfying the following Bellman equation:

$$-W_t^{(t_0)}(t,x) \\ = \max_{u_1, u_2, \cdots, u_n} \left\{ \sum_{j=1}^{n} g^j[t, x, u_1, u_2, \cdots, u_n] \exp\left[-\int_{t_0}^{t} r(y)dy \right] + W_x^{(t_0)} f[t, x, u_1, u_2, \cdots, u_n] \right\},$$

$$W^{(t_0)}(T,x) = \exp\left[-\int_{t_0}^{T} r(y)dy \right] \sum_{j=1}^{n} q^j(x).$$

Proof Follow the proof of Theorem A.1 in the Technical Appendices. ■

Hence the players will adopt the cooperative control $\{\psi_i^*(t,x)$, for $i \in N$ and $t \in [t_0, T]\}$ to obtain the maximized level of joint profit. Substituting this set of control into (1.2) yields the dynamics of the optimal (cooperative) trajectory as:

$$\dot{x}(s) = f\left[s, x(s), \psi_1^*(s, x(s)), \psi_2^*(s, x(s)), \cdots, \psi_n^*(s, x(s))\right], \quad x(t_0) = x_0. \tag{1.5}$$

Let $x^*(t)$ denote the solution to (1.5). The optimal trajectory $\{x^*(t)\}_{t=t_0}^{T}$ can be expressed as:

$$x^*(t) = x_0 + \int_{t_0}^{t} f\left[s, x^*(s), \psi_1^*(s, x^*(s)), \psi_2^*(s, x^*(s)), \cdots, \psi_n^*(s, x^*(s))\right] ds. \tag{1.6}$$

For notational convenience, we use the terms $x^*(t)$ and x_t^* interchangeably.

Note that for group optimality to be achievable, the cooperative controls $\{\psi_i^*(t, x^*(t))$, for $i \in N$ and $t \in [t_0, T]\}$ must be exercised throughout time interval $[t_0, T]$.

The maximized cooperative payoff over the interval $[t, T]$, for $t \in [t_0, T)$, can be expressed as:

$$W^{(t_0)}(t, x_t^*) =$$
$$\int_t^T \sum_{j=1}^n g^j[s, x^*(s), \psi_1^*(s, x^*(s)), \psi_2^*(s, x^*(s)), \cdots, \psi_n^*(s, x^*(s))] \exp\left[-\int_{t_0}^s r(y)dy\right] ds$$
$$+ \exp\left[-\int_{t_0}^T r(y)dy\right] \sum_{j=1}^n q^j(x^*(T))$$

A remark that will be utilized in subsequent analysis is given below.

Remark 1.2 Let $W^{(\tau)}(t, x_t^*)$ denote the maximized cooperative payoff of the control problem

$$\max_{u_1, u_2, \cdots, u_n} \left\{ \int_t^T \sum_{j=1}^n g^j[s, x(s), u_1(s), u_2(s), \cdots, u_n(s)] \exp\left[-\int_\tau^s r(y)dy\right] ds \right.$$
$$\left. + \exp\left[-\int_\tau^T r(y)dy\right] \sum_{j=1}^n q^j(x(T)) \right\},$$

subject to

$$\dot{x}(s) = f[s, x(s), u_1(s), u_2(s), \cdots, u_n(s)], \; x(t) = x_t^*.$$

One can readily verify that

$$\exp\left[\int_{t_0}^\tau r(y)dy\right] W^{(t_0)}(t, x_t^*) = \exp\left[\int_{t_0}^\tau r(y)dy\right] \times$$
$$\left\{ \int_t^T \sum_{j=1}^n g^j[s, x^*(s), \psi_1^*(s, x^*(s)), \psi_2^*(s, x^*(s)), \cdots, \psi_n^*(s, x^*(s))] \exp\left[-\int_{t_0}^s r(y)dy\right] ds \right.$$
$$\left. + \exp\left[-\int_{t_0}^T r(y)dy\right] \sum_{j=1}^n q^j(x^*(T)) \right\} =$$
$$\int_t^T \sum_{j=1}^n g^j[s, x^*(s), \psi_1^*(s, x^*(s)), \psi_2^*(s, x^*(s)), \cdots, \psi_n^*(s, x^*(s))] \exp\left[-\int_\tau^s r(y)dy\right] ds$$
$$+ \exp\left[-\int_\tau^T r(y)dy\right] \sum_{j=1}^n q^j(x^*(T)) = W^{(\tau)}(t, x_t^*),$$

for $\tau \in [t_0, T]$ and $t \in [\tau, T)$. ■

2.1.2.2 Individual Rationality

After the players agree to cooperate and maximize their joint payoff, they have to distribute the cooperative payoff among themselves. At time t_0, with the state being x_0, the term $\xi^{(t_0)i}(t_0, x_0)$ is used to denote the imputation of payoff (received over the time interval $[t_0, T]$) to player i. A necessary condition for group optimality and individual rationality to be upheld is:

$$\text{(i)}\ \sum_{j=1}^{n} \xi^{(t_0)j}(t_0, x_0) = W^{(t_0)}(t_0, x_0), \text{ and}$$

$$\text{(ii)}\ \xi^{(t_0)i}(t_0, x_0) \geq V^{(t_0)i}(t_0, x_0), \quad \text{for } i \in N \tag{1.7}$$

Condition (i) of (1.7) ensures group optimality and condition (ii) guarantees individual rationality at time t_0.

For the optimization scheme to be upheld throughout the game horizon both group rationality and individual rationality are required to be satisfied throughout the cooperation period $[t_0, T]$. At time $\tau \in [t_0, T]$, let $\xi^{(\tau)i}(\tau, x_\tau^*)$ denote the imputation of payoff to player i over the time interval $[\tau, T]$. Therefore the conditions

$$\text{(i)}\ \sum_{j=1}^{n} \xi^{(\tau)j}(\tau, x_\tau^*) = W^{(\tau)}(\tau, x_\tau^*), \text{ and}$$

$$\text{(ii)}\ \xi^{(\tau)i}(\tau, x_\tau^*) \geq V^{(\tau)i}(\tau, x_\tau^*); \text{ for } i \in N \text{ and } \tau \in [t_0, T], \tag{1.8}$$

have to be fulfilled.

In particular, condition (i) ensures Pareto optimality and condition (ii) guarantees individual rationality, throughout the cooperation period $[t_0, T]$. Failure to guarantee individual rationality leads to the condition where the concerned participants would reject the agreed upon solution plan and play noncooperatively.

Dockner and Jørgensen (1984); Dockner and Long (1993), Tahvonen (1994); Mäler and de Zeeuw (1998) and Rubio and Casino (2002) examines group optimal solutions in cooperative differential games. Haurie and Zaccour (1986, 1991); Kaitala and Pohjola (1988, 1990, 1995); Kaitala et al. (1995) and Jørgensen and Zaccour (2001) presented classes of transferable-payoff cooperative differential games with solutions which satisfy group optimality and individual rationality.

2.1.3 Distribution of Cooperative Payoffs

With the players using the cooperative strategies $\{\psi_i^*(s, x_s^*), \text{ for } s \in [t_0, T] \text{ and } i \in N\}$, player i would derive a direct payoff :

$$W^{(t_0)i}(t_0,x_0) = \int_{t_0}^{T} g^i\left[s,x^*(s),\psi_1^*(s,x^*(s)),\psi_2^*(s,x^*(s)),\cdots,\psi_n^*(s,x^*(s))\right]\exp\left[-\int_{t_0}^{s} r(y)dy\right]ds + \exp\left[-\int_{t_0}^{T} r(y)dy\right]q^i(x^*(T)),\ \text{for}\, i\in N. \tag{1.9}$$

At initial time t_0, for cooperation to begin the cooperative payoff to player i $W^{(t_0)i}(t_0,x_0)$ must be no less than the non-cooperative $V^{(t_0)i}(t_0,x_0)$ for all player $i\in N$. However as time proceeds there is no guarantee that adopting the cooperative strategies would lead to $W^{(t)i}(t,x_t^*)\geq V^{(t)i}(t,x_t^*)$ for all player $i\in N$. In case there exists some player i such that $V^{(t)i}(t,x_t^*) > W^{(t)i}(t,x_t^*)$, then player i would have an incentive to deviate from the cooperation plan. Hence the cooperation scheme has to include transfer payments to overcome this problem. Let $\chi^i(s)$ denote the instantaneous transfer payment allocated to agent i at time $s\in[t_0,T]$. With players using the cooperative strategies $\{\psi_i^*(s,x_s^*)$, for $s\in[t_0,T]$ and $i\in N\}$, the payoff that player i's payoff under cooperation at time t_0 becomes:

$$\xi^{(t_0)i}(t_0,x_0) = \int_{t_0}^{T}\{g^i\left[s,x^*(s),\psi_1^*(s,x^*(s)),\psi_2^*(s,x^*(s)),\cdots,\psi_n^*(s,x^*(s))\right]+\chi^i(s)\}\exp\left[-\int_{t_0}^{s} r(y)dy\right]ds + \exp\left[-\int_{t_0}^{T} r(y)dy\right]q^i(x^*(T)),$$

$$\text{for } i\in N;$$

$$\text{and} \sum_{j=1}^{n}\int_{t_0}^{T}\chi^j(s)\exp\left[-\int_{t_0}^{s} r(y)dy\right]ds = 0. \tag{1.10}$$

In order to uphold individual rationality one has to device a time path of instantaneous transfer payments $\chi^i(s)$ for $s\in[t_0,T]$ satisfying:

$$\int_{\tau}^{T}\{g^i\left[s,x^*(s),\psi_1^*(s,x^*(s)),\psi_2^*(s,x^*(s)),\cdots,\psi_n^*(s,x^*(s))\right]+\chi^i(s)\}\exp\left[-\int_{\tau}^{s} r(y)dy\right]ds + \exp\left[-\int_{\tau}^{T} r(y)dy\right]q^i(x^*(T)) \geq V^{(\tau)i}(\tau,x_\tau^*),\ \text{for } i\in N; \tag{1.11}$$

and

$$\sum_{j=1}^{n}\int_{\tau}^{T}\chi^j(s)\exp\left[-\int_{\tau}^{s} r(y)dy\right]ds = 0,\ \text{for } \tau\in[t_0,T]. \tag{1.12}$$

There exist a large number of $\chi^i(s)$ for $s\in[t_0,T]$ paths which leads to the satisfaction of individual rationality for all players to be selected. Nevertheless, just satisfying individual rationality may not be acceptable. For instance, players with larger non-cooperative payoffs (or sizes) would demand a larger share proportionally. In the next Section we will consider the derivation of $\chi^i(s)$ for $s\in[t_0,T]$ paths which would keep the original agree-upon imputation throughout the cooperation duration.

2.2 Subgame Consistent Dynamic Cooperation

Though group optimality and individual rationality constitute two essential properties for cooperation, their fulfillment does not necessarily guarantee a dynamically stable solution in cooperation because there is no guarantee that the agreed-upon optimality principle is fulfilled throughout the cooperative duration. The question of dynamic stability in differential games has been explored rigorously in the past four decades. Haurie (1976) discussed the problem of instability in extending the Nash bargaining solution to differential games. Petrosyan (1977) formalized mathematically the notion of dynamic stability in solutions of differential games. Petrosyan and Danilov (1979, 1982) introduced the notion of "imputation distribution procedure" for cooperative solution.

To ensure stability in dynamic cooperation over time, a stringent condition is required: the specific agreed-upon optimality principle must be maintained at any instant of time throughout the game along the optimal state trajectory. This condition is the notion of *subgame consistency*.

Let $\Gamma_c(x_0, T-t_0)$ denote a cooperative game in which player i's payoff is (1.1) and the state dynamics is (1.2). The players agree to act according to an agreed-upon optimality principle. The agreement on how to act cooperatively and allocate cooperative payoff constitutes the solution optimality principle of a cooperative scheme. In particular, the solution optimality principle includes

(i) an agreement on a set of cooperative strategies/controls,
(ii) an imputation vector stating the allocation of the cooperative payoff to individual players, and
(iii) a mechanism to distribute total payoff among players.

2.2.1 *Optimality Principle*

Let there be an optimality principle agreed upon by all players in the cooperative game $\Gamma_c(x_0, T-t_0)$. Based on the agreed upon optimality principle the solution of the game $\Gamma_c(x_0, T-t_0)$ at time t_0 includes

(i) a set of cooperative strategies $u^{(t_0)*}(s) = \left[u_1^{(t_0)*}(s), u_2^{(t_0)*}(s), \cdots, u_n^{(t_0)*}(s)\right]$, for $s \in [t_0, T]$;

(ii) an imputation vector $\xi^{(t_0)}(t_0, x_0) = \left[\xi^{(t_0)1}(t_0, x_0), \xi^{(t_0)2}(t_0, x_0), \cdots, \xi^{(t_0)n}(t_0, x_0)\right]$ to allot the cooperative payoff to the players; and

(iii) a payoff distribution procedure $B^{t_0}(s) = \left[B_1^{t_0}(s), B_2^{t_0}(s), \cdots, B_n^{t_0}(s)\right]$ for $s \in [t_0, T]$, where $B_i^{t_0}(s)$ is the instantaneous payments for player i at time s. In particular,

$$\xi^{(t_0)i}(t_0, x_0) = \int_{t_0}^{T} B_i^{t_0}(s) \exp\left[-\int_{t_0}^{s} r(y)dy\right] ds + q^i(x_T) \exp\left[-\int_{t_0}^{T} r(y)dy\right],$$

for $i \in N$.

This means that the players agree at the outset on a set of cooperative strategies $u^{(t_0)*}(s)$, an imputation $\xi^{(t_0)i}(t_0, x_0)$ of the gains to the i th player covering the time interval $[t_0, T]$, and a payoff distribution procedure $\{B^{t_0}(s)\}_{s=t_0}^{T}$ to allocate payments to the players over the game interval.

Using the agreed-upon cooperative strategies the state evolves according to the state dynamics:

$$\dot{x}(s) = f\left[s, x(s), u_1^{(t_0)*}(s), u_2^{(t_0)*}(s), \cdots, u_n^{(t_0)*}(s)\right], \quad x(t_0) = x_0. \tag{2.1}$$

The solution to (2.1) yields the optimal cooperative trajectory which is denoted by $\{x^c(s)\}_{s=t_0}^{T}$. For notational convenience we use $x^c(s)$ and x_s^c interchangeably.

When time $t \in (t_0, T]$ has arrived, the situation becomes a cooperative game in which player i's payoff is:

$$\begin{aligned} &\int_t^T g^i[s, x(s), u_1(s), u_2(s), \cdots, u_n(s)] \exp\left[-\int_t^s r(y)dy\right] ds \\ &\quad + \exp\left[-\int_t^T r(y)dy\right] q^i(x(T)), \quad \text{for } i \in N, \end{aligned} \tag{2.2}$$

and the evolutionary dynamics of the state is

$$\dot{x}(s) = f[s, x(s), u_1(s), u_2(s), \cdots, u_n(s)], \; x(t) = x_t^c. \tag{2.3}$$

We use $\Gamma_c(x_t^c, T-t)$ to denote a cooperative game in which player i's objective is (2.2) with state dynamics (2.3). At time $t \in (t_0, T]$ when the state is x_t^c, according to the agreed-upon principle the solution of the game $\Gamma_c(x_t^c, T-t)$ includes:

(i) a set of cooperative strategies $u^{(t)*}(s) = \left[u_1^{(t)*}(s), u_2^{(t)*}(s), \cdots, u_n^{(t)*}(s)\right]$, for $s \in [t, T]$;

(ii) an imputation vector $\xi^{(t)}(t, x_t^c) = \left[\xi^{(t)1}(t, x_t^c), \xi^{(t)2}(t, x_t^c), \cdots, \xi^{(t)n}(t, x_t^c)\right]$ to allot the cooperative payoff to the players; and

(iii) a payoff distribution procedure $B^t(s) = \left[B_1^t(s), B_2^t(s), \cdots, B_n^t(s)\right]$ for $s \in [t, T]$, where $B_i^t(s)$ is the instantaneous payments for player i at time s. In particular,

$$\xi^{(t)i}(t, x_t^c) = \int_t^T B_i^t(s) \exp\left[-\int_t^s r(y)dy\right] ds + q^i(x_T^c) \exp\left[-\int_t^T r(y)dy\right], \quad (2.4)$$

for $i \in N$ and $t \in [t_0, T]$.

This means that under the agreed-upon optimality principle, the players agree on a set of cooperative strategies $u^{(t)*}(s)$, an imputation of the gains in such a way that the gain under cooperation of the i th player over the time interval $[t, T]$ is equal to $\xi^{(t)i}(t, x_t^c)$ and a payoff distribution procedure $\{B^t(s)\}_{s=t}^T$ to allocate payments to the players over the game interval $[t, T]$.

Examples of optimality principles include:

(i) joint payoff maximization and equal sharing of gains from cooperation,
(ii) joint payoff maximization and sharing gains proportional to non-cooperative payoffs,
(iii) joint payoff maximization and time varying sharing weights,
(iv) different combinations of (i), (ii) and (iii),
(v) joint payoff maximization and sharing gains according to the Shapley value,
(vi) joint payoff maximization and sharing gains according to the von Neumann-Morgenstern solution, or
(vii) joint payoff maximization and sharing gains according to the nucleolus.

2.2.2 *Cooperative Subgame Consistency*

To satisfy subgame consistency, the cooperative strategies, imputation and payoff distribution procedure $\{ u^{(t_0)*}(s)$ and $B^{t_0}(s)$ for $s \in [t_0, T]$; $\xi^{(t_0)}(t_0, x_0) \}$ generated by the agreed-upon optimality principle in the cooperative game $\Gamma_c(x_0, T - t_0)$ must be consistent with the cooperative strategies, imputation and payoff distribution procedure $\{ u^{(t)*}(s)$ and $B^t(s)$ for $s \in [t, T]$; $\xi^{(t)}(t, x_t^c) \}$ generated by the same optimality principle in the cooperative game $\Gamma_c(x_t^c, T - t)$ along the optimal cooperative trajectory $\{x_s^c\}_{s=t_0}^T$.

If this consistency does not appear, there is no guarantee that the players would not abandon the cooperative scheme and switch to other plans including the non-cooperative scheme. Dynamical instability would arise as participants found that their agreed upon optimality principle could not be maintained after cooperation has gone on for some time.

2.2.2.1 Subgame Consistent Cooperative Strategies

First we consider the cooperative strategies adopted under the agreed-upon optimality principle in the game $\Gamma_c(x_0, T - t_0)$. At time t_0 when the initial state is x_0, the set of cooperative strategies is

$$u^{(t_0)*}(s) = \left[u_1^{(t_0)*}(s), u_2^{(t_0)*}(s), \cdots, u_n^{(t_0)*}(s)\right], \text{ for } s \in [t_0, T].$$

Consider the case when the game has proceeded to time t and the state variable became x_t^c. Then one has a cooperative game $\Gamma_c(x_t^c, T - t)$ which starts at time t with initial state x_t^c. According to the agreed upon optimality principle a set of cooperative strategies

$$u^{(t)*}(s) = \left[u_1^{(t)*}(s), u_2^{(t)*}(s), \cdots, u_n^{(t)*}(s)\right], \text{ for } s \in [t, T],$$

will be adopted.

Definition 2.1 The set of cooperative strategies $u^{(t_0)*}(s) = \left[u_1^{(t_0)*}(s), u_2^{(t_0)*}(s), \cdots, u_n^{(t_0)*}(s)\right]$ in the game $\Gamma_c(x_0, T - t_0)$ is subgame consistent if $\left[u_1^{(t_0)*}(s), u_2^{(t_0)*}(s), \cdots, u_n^{(t_0)*}(s)\right] = \left[u_1^{(t)*}(s), u_2^{(t)*}(s), \cdots, u_n^{(t)*}(s)\right]$ in the game $\Gamma_c(x_t^c, T - t)$ under the agreed-upon optimality principle, for $s \in [t, T]$ and $t \in [t_0, T]$. ■

If the condition in Definition 2.1 is satisfied at each instant of time $t \in [t_0, T]$ along the optimal trajectory $\{x^c(t)\}_{t=t_0}^T$, the continuation of the original cooperative strategies $u^{(t_0)*}(s)$ coincides with the cooperative strategies $u^{(t)*}(s)$ in the cooperative game $\Gamma_c(x_t^c, T - t)$. Hence the set of cooperative strategies $u^{(t_0)*}(s)$ is subgame consistent. Recall that to ensure group optimality the players have to maximize the players' joint payoffs. An optimality principle which requires group optimality would yield a set of cooperative controls that solves the problem:

$$\max_{u_1, u_2, \cdots, u_n} \left\{ \int_{t_0}^{T} \sum_{j=1}^{n} g^j[s, x(s), u_1(s), u_2(s), \cdots, u_n(s)] \exp\left[-\int_{t_0}^{s} r(y)dy\right] ds \right.$$
$$\left. + \exp\left[-\int_{t_0}^{T} r(y)dy\right] \sum_{j=1}^{n} q^j(x(T)) \right\}, \tag{2.5}$$

$$\text{subject to } \dot{x}(s) = f[s, x(s), u_1(s), u_2(s), \cdots, u_n(s)], \quad x(t_0) = x_0. \tag{2.6}$$

A set of group optimal cooperative strategies $\{\psi_i^*(s, x^*(s))$, for $i \in N$ and $s \in [t_0, T]\}$ which solves the problem (2.5 and 2.6) could be characterized by Theorem 1.2. In particular, $\{x^*(t)\}_{t=t_0}^T$ is the solution path of the optimal cooperative trajectory:

$$\dot{x}(s) = f\left[s, x(s), \psi_1^*(s, x(s)), \psi_2^*(s, x(s)), \cdots, \psi_n^*(s, x(s))\right], \; x(t_0) = x_0.$$

Invoking Remark 1.2 in Sect. 2.1, one can show that the joint payoff maximizing controls for the cooperative game $\Gamma_c(x_t^*, T-t)$ over the time interval $[t, T]$ is identical to the joint payoff maximizing controls for the cooperative game $\Gamma_c(x_0, T-t_0)$ over the same time interval.

Therefore the solution to an optimality principle which requires group optimality yields a system of subgame consistent cooperative strategies. Given that group optimality is an essential element in dynamic cooperation, a valid optimality principle would require the maximization of joint payoff and the cooperative strategies $u^{(t_0)*}(s) = u_1^{(t)*}(s) = \psi_i^*(s, x^*(s))$, for $s \in [t, T]$ and $t \in [t_0, T]$.

2.2.2.2 Subgame Consistent Imputation

Now, we consider subgame consistency in imputation and payoff distribution procedure. In the cooperative game $\Gamma_c(x_0, T-t_0)$, according to the agreed-upon optimality principle the players would use the payoff distribution procedure $\{B^{t_0}(s)\}_{s=t_0}^{T}$ to bring about an imputation to player i as:

$$\xi^{(t_0)i}(t_0, x_0) = \int_{t_0}^{T} B_i^{t_0}(s) \exp\left[-\int_{t_0}^{s} r(y)dy\right] ds + q^i(x_T) \exp\left[-\int_{t_0}^{T} r(y)dy\right], \quad (2.7)$$

for $i \in N$.

When the game proceeds to time $t \in (t_0, T]$, the current state is x_t^c. According to the same optimality principle player i will receive an imputation (in present value viewed at time t_0) equaling

$$\xi^{(t_0)i}(t, x_t^c) = \int_{t}^{T} B_i^{t_0}(s) \exp\left[-\int_{t_0}^{s} r(y)dy\right] ds + q^i(x_T^c) \exp\left[-\int_{t_0}^{T} r(y)dy\right], \quad (2.8)$$

over the time interval $[t, T]$.

At time $t \in (t_0, T]$ when the current state is x_t^c, we have a cooperative game $\Gamma_c(x_t^c, T-t)$. According to the agreed-upon optimality principle the players would use the payoff distribution procedure $\{B^t(s)\}_{s=t}^{T}$ to bring about an imputation to player i as:

$$\xi^{(t)i}(t, x_t^c) = \int_{t}^{T} B_i^{t}(s) \exp\left[-\int_{t}^{s} r(y)dy\right] ds + q^i(x_T^c) \exp\left[-\int_{t}^{T} r(y)dy\right], \quad (2.9)$$

for $i \in N$.

For the imputation and payoff distribution procedure in the game $\Gamma_c(x_0, T - t_0)$ to be consistent with those from $\Gamma_c(x_t^c, T - t)$, it is essential that

$$\exp\left[\int_{t_0}^{t} r(y)dy\right]\xi^{(t_0)}\left(t, x_t^c\right) = \xi^{(t)}\left(t, x_t^c\right), \text{ for } t \in [t_0, T].$$

In addition, in the game $\Gamma_c(x_0, T - t_0)$ according to the agreed-upon optimality principle the payoff distribution procedure is

$$B^{t_0}(s) = \left[B_1^{t_0}(s), B_2^{t_0}(s), \cdots, B_n^{t_0}(s)\right], \text{ for } s \in [t_0, T].$$

Consider the case when the game has proceeded to time t and the state variable became x_t^c. Then one has a cooperative game $\Gamma_c\left(x_t^c, T - t\right)$ which starts at time t with initial state x_t^c. According to the agreed-upon optimality principle the payoff distribution procedure

$$B^t(s) = \left[B_1^t(s), B_2^t(s), \cdots, B_n^t(s)\right], \text{ for } s \in [t, T]$$

will be adopted.

For the continuation of the payoff distribution procedure $B^{t_0}(s)$ for $s \in [t, T]$ to be consistent with $B^t(s)$ in the game $\Gamma_c\left(x_t^c, T - t\right)$, it is required that

$$B^{t_0}(s) = B^t(s), \text{ for } s \in [t, T] \text{ and } t \in [t_0, T].$$

Therefore a formal definition can be presented as below.

Definition 2.2 The imputation and payoff distribution procedure $\left\{\xi^{(t_0)}(t_0, x_0) \text{ and } B^{t_0}(s) \text{ for } s \in [t_0, T]\right\}$ are subgame consistent if

(i) $$\exp\left[\int_{t_0}^{t} r(y)dy\right]\xi^{(t_0)i}\left(t, x_t^c\right)$$
$$\equiv \exp\left[\int_{t_0}^{t} r(y)dy\right]\left\{\int_t^T B_i^{t_0}(s)\exp\left[-\int_t^s r(y)dy\right]ds + q^i\left(x_T^c\right)\exp\left[-\int_{t_0}^{T} r(y)dy\right]\right\}$$
$$= \xi^{(t)i}\left(t, x_t^c\right) \equiv \int_t^T B_i^t(s)\exp\left[-\int_t^s r(y)dy\right]ds + q^i\left(x_T^c\right)\exp\left[-\int_t^T r(y)dy\right],$$
for $i \in N$ and $t \in [t_0, T]$, and (2.10)

(ii) the payoff distribution procedure $B^{t_0}(s) = \left[B_1^{t_0}(s), B_2^{t_0}(s), \cdots, B_n^{t_0}(s)\right]$ for $s \in [t, T]$ is identical to

$$B^t(s) = \left[B_1^t(s), B_2^t(s), \cdots, B_n^t(s)\right], \text{ for } t \in [t_0, T] \quad (2.11)$$

■

Thus cooperative strategies, payoff distribution procedures and imputations satisfying the conditions in Definitions 2.1 and 2.2 are subgame consistent.

2.3 Subgame Consistent Payoff Distribution Procedure

Crucial to obtaining a subgame consistent solution is the derivation of a payoff distribution procedure satisfying Definition 2.2 in Sect. 2.2.

2.3.1 Derivation of Payoff Distribution Procedures

Invoking part (ii) of Definition 2.2, we have $B^{t_0}(s) = B^t(s)$ for $t \in [t_0, T]$ and $s \in [t, T]$. We use $B(s) = \{B_1(s), B_2(s), \cdots, B_n(s)\}$ to denote $B^t(s)$ for all $t \in [t_0, T]$. Along the optimal trajectory $\{x^c(s)\}_{s=t_0}^T$ we then have:

$$\xi^{(\tau)i}(\tau, x_\tau^c) = \int_\tau^T B_i(s) \exp\left[-\int_\tau^s r(y)dy\right] ds + q^i(x_T^c) \exp\left[-\int_\tau^T r(y)dy\right], \quad (3.1)$$

for $i \in N$ and $\tau \in [t_0, T]$; and

$$\sum_{j=1}^n B_j(s) = \sum_{j=1}^n g^j\left[s, x_s^c, u_1^{(\tau)*}(s), u_2^{(\tau)*}(s), \cdots, u_n^{(\tau)*}(s)\right].$$

Moreover, for $t \in [\tau, T]$, we use the term

$$\xi^{(\tau)i}(t, x_t^c) = \int_t^T B_i(s) \exp\left[-\int_\tau^s r(y)dy\right] ds + q^i(x_T^c) \exp\left[-\int_\tau^T r(y)dy\right], \quad (3.2)$$

to denote the present value (with initial time being τ) of player i's payoff under cooperation over the time interval $[t, T]$ according to the agreed-upon optimality principle along the cooperative state trajectory.

Invoking (3.1) and (3.2) we have

$$\xi^{(\tau)i}(t, x_t^c) = \exp\left[-\int_\tau^t r(y)dy\right] \xi^{(t)i}(t, x_t^c),$$
$$\text{for } i \in N \text{ and } \tau \in [t_0, T] \text{ and } t \in [\tau, T]. \quad (3.3)$$

One can readily verify that a payoff distribution procedure $\{B(s)\}_{s=t_0}^T$ which satisfies (3.3) would give rise to subgame consistent imputations satisfying part (ii) of Definition 2.2. The next task is the derivation of a payoff distribution procedure $\{B(s)\}_{s=t_0}^T$ that leads to the realization of (3.1, 3.2 and 3.3).

We first consider the following condition concerning the imputation $\xi^{(\tau)}(t, x_t^c)$, for $\tau \in [t_0, T]$ and $t \in [\tau, T]$.

Condition 3.1 For $i \in N$ and $t \in [\tau, T]$ and $\tau \in [t_0, T]$, the imputation $\xi^{(\tau)i}(t, x_t^c)$, for $i \in N$, is a function that is continuously differentiable in t and x_t^c. ■

A theorem characterizing a formula for $B_i(s)$, for $s \in [t_0, T]$ and $i \in N$, which yields (3.1, 3.2 and 3.3) is provided as follows.

Theorem 3.1 If Condition 3.1 is satisfied, a PDP with a terminal payment $q^i(x_T^c)$ at time T and an instantaneous payment at time $s \in [\tau, T]$:

$$\begin{aligned} B_i(s) = & -\left[\xi_t^{(s)i}\left(t, x_t^c\right)\Big|_{t=s}\right] \\ & - \left[\xi_{x_s^c}^{(s)i}\left(s, x_s^c\right)\right] f\left[s, x_s^c, \psi_1^*\left(s, x_s^c\right), \psi_2^*\left(s, x_s^c\right), \cdots, \psi_n^*\left(s, x_s^c\right)\right], \text{for } i \in N, \end{aligned} \tag{3.4}$$

yields imputation vector $\xi^{(\tau)}(\tau, x_\tau^c)$, for $\tau \in [t_0, T]$ which satisfy (3.1, 3.2 and3.3).

Proof Invoking (3.1, 3.2 and 3.3), one can obtain

$$\begin{aligned} \xi^{(\upsilon)i}\left(\upsilon, x_\upsilon^c\right) = & \int_\upsilon^{\upsilon+\Delta t} B_i(s) \exp\left[-\int_\upsilon^s r(y)dy\right] ds + \\ & \exp\left[-\int_\upsilon^{\upsilon+\Delta t} r(y)dy\right] \xi^{(\upsilon+\Delta t)i}\left(\upsilon + \Delta t, x_\upsilon^c + \Delta x_\upsilon^c\right), \\ & \text{for } \upsilon \in [\tau, T] \text{ and } i \in N; \end{aligned} \tag{3.5}$$

where $\Delta x_\upsilon^c = f\left[\upsilon, x_\upsilon^c, \psi_1^*\left(\upsilon, x_\upsilon^c\right), \psi_2^*\left(\upsilon, x_\upsilon^c\right), \cdots, \psi_n^*\left(\upsilon, x_\upsilon^c\right)\right] \Delta t + o(\Delta t)$, and $o(\Delta t)/\Delta t \to 0$ as $\Delta t \to 0$.

From (3.2) and (3.5), one obtains

$$\begin{aligned} & \int_\upsilon^{\upsilon+\Delta t} B_i(s) \exp\left[-\int_\upsilon^s r(y)dy\right] ds \\ & = \xi^{(\upsilon)i}\left(\upsilon, x_\upsilon^c\right) - \exp\left[-\int_\upsilon^{\upsilon+\Delta t} r(y)dy\right] \xi^{(\upsilon+\Delta t)i}\left(\upsilon + \Delta t, x_\upsilon^c + \Delta x_\upsilon^c\right) \\ & = \xi^{(\upsilon)i}\left(\upsilon, x_\upsilon^c\right) - \xi^{(\upsilon)i}\left(\upsilon + \Delta t, x_\upsilon^c + \Delta x_\upsilon^c\right), \\ & \text{for all } \upsilon \in [t_0, T] \text{ and } i \in N \end{aligned} \tag{3.6}$$

If the imputations $\xi^{(\upsilon)}(t, x_t^c)$, for $\upsilon \in [t_0, T]$, satisfy Condition 3.1, as $\Delta t \to 0$, one can express condition (3.6) as:

$$\begin{aligned} B_i(\upsilon)\Delta t = & -\left[\xi_t^{(\upsilon)i}\left(t, x_t^c\right)\Big|_{t=\upsilon}\right] \Delta t \\ & - \left[\xi_{x_\upsilon^c}^{(\upsilon)i}\left(\upsilon, x_\upsilon^c\right)\right] f\left[\upsilon, x_\upsilon^c, \psi_1^*\left(\upsilon, x_\upsilon^c\right), \psi_2^*\left(\upsilon, x_\upsilon^c\right), \cdots, \psi_n^*\left(\upsilon, x_\upsilon^c\right)\right] \Delta t - o(\Delta t). \end{aligned} \tag{3.7}$$

Dividing (3.7) throughout by Δt, with $\Delta t \to 0$, yields (3.4).

$$B_i(v) = -\left[\xi_t^{(v)i}(t, x_t^c)\Big|_{t=v}\right] - \left[\xi_{x_v^c}^{(v)i}(v, x_v^c)\right] f\left[v, x_v^c, \psi_1^*(v, x_v^c), \psi_2^*(v, x_v^c), \cdots, \psi_n^*(v, x_v^c)\right]$$

Thus the payoff distribution procedure in $B_i(s)$ in (3.4) would lead to the realization of $\xi^{(\tau)i}(\tau, x_\tau^c)$, for $\tau \in [t_0, T]$ which satisfy (3.1, 3.2 and 3.3). ■

Assigning the instantaneous payments according to the payoff distribution procedure in (3.4) leads to the realization of the imputation $\xi^{(\tau)}(\tau, x_\tau^c)$ governed by the agreed-upon optimality principle in the game $\Gamma_c(x_\tau^c, T-\tau)$ for $\tau \in [t_0, T]$. Therefore the payoff distribution procedure in $B_i(s)$ in (3.4) yields a subgame consistent solution.

With players using the cooperative strategies $\{\psi_i^*(\tau, x_\tau^*)$, for $\tau \in [t_0, T]$ and $i \in N\}$, the instantaneous payment received by player i at time instant τ is:

$$\zeta_i(\tau) = g^i\left[\tau, x_\tau^*, \psi_1^*(\tau, x_\tau^*), \psi_2^*(\tau, x_\tau^*), \cdots, \psi_n^*(\tau, x_\tau^*))\right],$$
$$\text{for } \tau \in [t_0, T] \text{ and } i \in N. \tag{3.8}$$

According to Theorem 3.1, the instantaneous payment that player i should receive under the agreed-upon optimality principle is $B_i(\tau)$ as stated in (3.2). Hence an instantaneous transfer payment

$$\chi^i(\tau) = B_i(\tau) - \zeta_i(\tau) \tag{3.9}$$

has to be given to player i at time τ, for $i \in N$ and $\tau \in [t_0, T]$.

2.3.2 Subgame Consistent Solution under Specific Optimality Principle

In this section we present examples of subgame consistent solutions under various optimality principles.

Case I Consider the cooperative differential game $\Gamma_c(x_0, T-t_0)$. In particular, the players agree with an optimality principle which entails

(i) group optimality and
(ii) the division of the excess of the total cooperative payoff over the sum of individual noncooperative payoffs equally.

According to the optimality principle the imputation to player j in $\Gamma_c(x_0, T - t_0)$ is:

$$\xi^{(\tau)i}\left(\tau, x_\tau^c\right) = V^{(\tau)i}\left(\tau, x_\tau^c\right) + \frac{1}{n}\left[W^{(\tau)}\left(\tau, x_\tau^c\right) - \sum_{j=1}^{n} V^{(\tau)j}\left(\tau, x_\tau^c\right)\right], \tag{3.10}$$

for $i \in N$ and $\tau \in [t_0, T]$.

The imputation in (3.10) yields

(i) $\xi^{(\tau)i}\left(\tau, x_\tau^c\right) \geq V^{(\tau)i}\left(\tau, x_\tau^c\right)$, for $i \in N$ and $\tau \in [t_0, T]$; and

(ii) $\sum_{j=1}^{n} \xi^{(\tau)j}\left(\tau, x_\tau^c\right) = W^{(\tau)}\left(\tau, x_\tau^c\right)$ for $\tau \in [t_0, T]$.

Hence the imputation vector $\xi^{(\tau)i}(\tau, x_\tau^*)$ satisfies individual rationality and group optimality.

Applying Theorem 3.1 a subgame consistent solution under the optimal principle can be characterized by $\left\{u(s) \text{ and } B(s) \text{ for } s \in [t_0, T] \text{ and } \xi^{(t_0)}(t_0, x_0)\right\}$ in which

(i) $u(s)$ for $s \in [t_0, T]$ is the set of group optimal strategies $\psi^*(s, x_s^*)$ in the game $\Gamma_c(x_0, T - t_0)$, and

(ii) the imputation distribution procedure

$$B(s) = \{B_1(s), B_2(s), \cdots, B_n(s)\} \text{ for } s \in [t_0, T] \text{ where}$$

$$B_i(s) = -\frac{\partial}{\partial t}\left[V^{(s)i}\left(t, x_t^*\right) + \frac{1}{n}\left(W^{(s)}\left(t, x_t^*\right) - \sum_{j=1}^{n} V^{(s)j}\left(t, x_t^*\right)\right)\Bigg|_{t=s}\right]$$

$$- \frac{\partial}{\partial x_s^*}\left[V^{(s)i}\left(s, x_s^*\right) + \frac{1}{n}\left(W^{(s)}\left(s, x_s^*\right) - \sum_{j=1}^{n} V^{(s)j}\left(s, x_s^*\right)\right)\right]$$

$$\times f\left[s, x_s^*, \psi_1^*\left(s, x_s^*\right), \psi_2^*\left(s, x_s^*\right), \cdots, \psi_n^*\left(s, x_s^*\right)\right], \tag{3.11}$$

for $i \in N$.

Case II Consider the cooperative differential game $\Gamma_c(x_0, T - t_0)$. In particular, the players agree with an optimality principle which entails

(i) group optimality and

(ii) the sharing of the excess of the total cooperative payoff over the sum of individual noncooperative payoffs proportional to the players' noncooperative payoffs.

$$\xi^{(\tau)i}\left(\tau, x_\tau^c\right) = \frac{V^{(\tau)i}\left(\tau, x_\tau^c\right)}{\sum_{j=1}^{n} V^{(\tau)j}\left(\tau, x_\tau^c\right)} + W^{(\tau)}\left(\tau, x_\tau^c\right), \tag{3.12}$$

for $i \in N$ and $\tau \in [t_0, T]$.

Applying Theorem 3.1 a subgame consistent solution under the optimal principle will yield the imputation distribution procedure
$B(s) = \{B_1(s), B_2(s), \cdots, B_n(s)\}$ for $s \in [t_0, T]$ where

$$B_i(s) = -\frac{\partial}{\partial t}\left[\frac{V^{(s)i}(t, x_t^*)}{\sum_{j=1}^{n} V^{(s)j}(t, x_t^*)} W^{(s)}(t, x_t^*)\Bigg|_{t=s}\right]$$

$$-\frac{\partial}{\partial x_s^*}\left[\frac{V^{(s)i}(s, x_s^*)}{\sum_{j=1}^{n} V^{(s)j}(s, x_s^*)} W^{(s)}(s, x_s^*)\right]$$

$$\times f\left[s, x_s^*, \psi_1^*(s, x_s^*), \psi_2^*(s, x_s^*), \cdots, \psi_n^*(s, x_s^*)\right] \text{ for } i \in N \tag{3.13}$$

Case III Consider the cooperative differential game $\Gamma_c(x_0, T - t_0)$ with two players. In particular, the players agree with an optimality principle which entails

(i) group optimality and
(ii) the division of the excess of the total cooperative payoff over the sum of individual noncooperative payoffs by the time-varying weights $-\frac{\tau}{T+\alpha}$ for player 1 and $\frac{T+\alpha-\tau}{T+\alpha}$ for player 2 at time $\tau \in [t_0, T]$.

According to optimality principle the imputations to player 1 and player 2 in Γ_c $(x_0, T - t_0)$ are:

$$\xi^{(\tau)1}(\tau, x_\tau^c) = V^{(\tau)1}(\tau, x_\tau^c) + \frac{\tau}{T+\alpha}\left[W^{(\tau)}(\tau, x_\tau^c) - \sum_{j=1}^{2} V^{(\tau)j}(\tau, x_\tau^c)\right]$$

for player 1, and

$$\xi^{(\tau)2}(\tau, x_\tau^c) = V^{(\tau)2}(\tau, x_\tau^c) + \frac{T+\alpha-\tau}{T+\alpha}\left[W^{(\tau)}(\tau, x_\tau^c) - \sum_{j=1}^{2} V^{(\tau)j}(\tau, x_\tau^c)\right] \tag{3.14}$$

for player 2; $\tau \in [t_0, T]$.

Applying Theorem 3.1 a subgame consistent solution under the optimal principle will yield the imputation distribution procedure
$B(s) = \{B_1(s), B_2(s), \cdots, B_n(s)\}$ for $s \in [t_0, T]$ where

$$B_1(s) = -\frac{\partial}{\partial t}\left[V^{(s)1}(t,x_t^*) + \frac{t}{T+\alpha}\left(W^{(s)}(t,x_t^*) - \sum_{j=1}^{2} V^{(s)j}(t,x_t^*)\right)\Big|_{t=s}\right]$$

$$-\frac{\partial}{\partial x_s^*}\left[V^{(s)1}(s,x_s^*) + \frac{s}{T+\alpha}\left(W^{(s)}(s,x_s^*) - \sum_{j=1}^{2} V^{(s)j}(s,x_s^*)\right)\right]$$

$$\times f\left[s, x_s^*, \psi_1^*(s,x_s^*), \psi_2^*(s,x_s^*)\right]$$

$$B_2(s) = -\frac{\partial}{\partial t}\left[V^{(s)2}(t,x_t^*) + \frac{T-t+\alpha}{T+\alpha}\left(W^{(s)}(t,x_t^*) - \sum_{j=1}^{2} V^{(s)j}(t,x_t^*)\right)\Big|_{t=s}\right]$$

$$-\frac{\partial}{\partial x_s^*}\left[V^{(s)1}(s,x_s^*) + \frac{T-s+\varepsilon}{T+\alpha}\left(W^{(s)}(s,x_s^*) - \sum_{j=1}^{2} V^{(s)j}(s,x_s^*)\right)\right]$$

$$\times f\left[s, x_s^*, \psi_1^*(s,x_s^*), \psi_2^*(s,x_s^*)\right]. \tag{3.15}$$

A variety of optimality principles with various imputation schemes can be constructed.

2.4 An Illustration in Cooperative Fishery

Consider a deterministic version of an example in Yeung and Petrosyan (2004) in which two nations are harvesting fish in common waters. The growth rate of the fish stock is characterized by the differential equation:

$$\dot{x}(s) = ax(s)^{1/2} - bx(s) - u_1(s) - u_2(s), \quad x(t_0) = x_0 \in X, \tag{4.1}$$

where $u_i \in U_i$ is the (nonnegative) amount of fish harvested by nation i, for $i \in \{1, 2\}$, a and b are positive constants.

The harvesting cost for nation $i \in \{1, 2\}$ depends on the quantity of resource extracted $u_i(s)$, the resource stock size $x(s)$, and a parameter c_i. In particular, nation i's extraction cost can be specified as $c_i u_i(s) x(s)^{-1/2}$. The fish harvested by nation i at time s will generate a net benefit of the amount $[u_i(s)]^{1/2}$. The horizon in concern is $[t_0, T]$. At time T, nation i will receive a termination bonus $q_i x(T)^{1/2}$, where q_i is nonnegative. There exists a positive discount rate r.

At time t_0 the payoff of nation $i \in [1, 2]$ is:

$$\int_{t_0}^{T}\left[[u_i(s)]^{1/2} - \frac{c_i}{x(s)^{1/2}} u_i(s)\right] \exp[-r(s-t_0)]\, ds$$

$$+ \exp[-r(T-t_0)] q_i x(T)^{\frac{1}{2}}. \tag{4.2}$$

Following the above analysis a set of feedback strategies $\{u_i^*(t) = \phi_i^*(t,x)$, for $i \in \{1,2\}\}$ provides a feedback Nash equilibrium solution to the game (4.1 and 4.2), if there exist continuously differentiable functions $V^{(\tau)i}(t,x) : [\tau, T] \times R \to R$, $i \in \{1,2\}$, satisfying the following partial differential equations:

$$-V_t^{(\tau)i}(t,x) = \max_{u_i} \left\{ \left[u_i^{1/2} - \frac{c_i}{x^{1/2}}u_i\right] \exp[-r(t-\tau)] \right.$$
$$\left. + V_x^{(\tau)i}(t,x)\left[ax^{1/2} - bx - u_i - \phi_j^*(t,x)\right] \right\}, \text{ and}$$
$$V^{(\tau)i}(T,x) = q_i x^{1/2} \exp[-r(T-\tau)] \text{ for } i \in \{1,2\},\ j \in \{1,2\} \text{ and } j \neq i. \tag{4.3}$$

Performing the indicated maximization yields:

$$\phi_i^*(t,x) = \frac{x}{4\left[c_i + V_x^{(\tau)i}\exp[r(t-\tau)]x^{1/2}\right]^2}, \quad \text{for } i \in \{1,2\} \tag{4.4}$$

Substituting $\phi_1^*(t,x)$ and $\phi_2^*(t,x)$ into (4.3) and upon solving (4.43 one obtains can obtain the feedback Nash equilibrium payoff of nation i in the game (4.1 and 4.2) as:

$$V^{(\tau)i}(t,x) = \exp[-r(t-\tau)]\left[A_i(t)x^{1/2} + C_i(t)\right],$$
$$\text{for } i \in \{1,2\} \text{ and } t \in [\tau, T] \text{ and } \tau \in [t_0, T], \tag{4.5}$$

where $A_i(t), C_i(t), A_j(t)$ and $C_j(t)$, for $i \in \{1,2\}$ and $j \in \{1,2\}$ and $i \neq j$, satisfy:

$$\dot{A}_i(t) = \left[r + \frac{b}{2}\right]A_i(t) - \frac{1}{2[c_i + A_i(t)/2]} + \frac{c_i}{4[c_i + A_i(t)/2]^2}$$
$$+ \frac{A_i(t)}{8[c_i + A_i(t)/2]^2} + \frac{A_i(t)}{8\left[c_j + A_j(t)/2\right]^2}$$
$$\dot{C}_i(t) = rC_i(t) - \frac{a}{2}A_i(t) \text{ and } A_i(T) = q, \text{ and } C_i(T) = 0. \tag{4.6}$$

Now consider the case when the nations agree to cooperate in harvesting the fishery. Let $\Gamma_c(x_0, T - t_0)$ denote a cooperative game with the game structure of $\Gamma(x_0, T - t_0)$ in which the players agree to act according to the optimality principle that they would

(i) maximize the sum of their payoffs and
(ii) divide the excess of the total cooperative payoff over the sum of individual noncooperative payoffs equally.

To maximize the joint payoffs, the nations would consider the optimal control problem:

$$\int_{t_0}^{T}\left(\left[u_1(s)^{1/2}-\frac{c_1}{x(s)^{1/2}}u_1(s)\right]+\left[u_2(s)^{1/2}-\frac{c_2}{x(s)^{1/2}}u_2(s)\right]\right)\exp[-r(t-t_0)]\,ds$$
$$+\,2\exp[-r(T-t_0)]qx(T)^{\frac{1}{2}}, \tag{4.7}$$

subject to (4.1).

Let $[\psi_1^*(t,x),\psi_2^*(t,x)]$ denote a set of controls that provides a solution to the optimal control problem (4.1) and (4.7) and $W^{(t_0)}(t,x):[t_0,T]\times R^n\to R$ denote the value function that satisfies the equations:

$$\begin{aligned}&-W_t^{(t_0)}(t,x)\\&=\max_{u_1,u_2}\left\{\left(\left[u_1{}^{1/2}-\frac{c_1}{x^{1/2}}u_1\right]+\left[u_2{}^{1/2}-\frac{c_2}{x^{1/2}}u_2\right]\right)\exp[-r(t-t_0)]\right.\\&\left.+\,W_x^{(t_0)}(t,x)\left[ax^{1/2}-bx-u_1-u_2\right]\right\},\ \text{and}\\&W^{(t_0)}(T,x)=2\exp[-r(T-t_0)]qx^{\frac{1}{2}}.\end{aligned} \tag{4.8}$$

Performing the indicated maximization we obtain:

$$\psi_1^*(t,x)=\frac{x}{4\left[c_1+W_x^{(t_0)}\exp[r(t-t_0)]x^{1/2}\right]^2},\ \text{and}$$
$$\psi_2^*(t,x)=\frac{x}{4\left[c_2+W_x^{(t_0)}\exp[r(t-t_0)]x^{1/2}\right]^2}.$$

Substituting $\psi_1^*(t,x)$ and $\psi_2^*(t,x)$ above into (4.8) yields the value function

$$W^{(t_0)}(t,x)=\exp[-r(t-t_0)]\left[\hat{A}(t)x^{1/2}+\hat{C}(t)\right],$$

where $\dot{\hat{A}}(t)=\left[r+\frac{b}{2}\right]\hat{A}(t)-\frac{1}{2\left[c_1+\hat{A}(t)/2\right]}-\frac{1}{2\left[c_2+\hat{A}(t)/2\right]}$

$$+\frac{c_1}{4\left[c_1+\hat{A}(t)/2\right]^2}+\frac{c_2}{4\left[c_2+\hat{A}(t)/2\right]^2}+\frac{\hat{A}(t)}{8\left[c_1+\hat{A}(t)/2\right]^2}+\frac{\hat{A}(t)}{8\left[c_2+\hat{A}(t)/2\right]^2},$$

$$\dot{\hat{C}}(t)=r\hat{C}(t)-\frac{a}{2}\hat{A}(t),\ \ \hat{A}(T)=2q,\ \text{and}\ \hat{B}(T)=0. \tag{4.9}$$

The optimal cooperative controls can then be obtained as:

$$\psi_1^*(t,x) = \frac{x}{4\left[c_1 + \hat{A}(t)/2\right]^2} \text{ and } \psi_2^*(t,x) = \frac{x}{4\left[c_2 + \hat{A}(t)/2\right]^2}. \tag{4.10}$$

Substituting these control strategies into (4.1) yields the dynamics of the state trajectory under cooperation:

$$\dot{x}(s) = ax(s)^{1/2} - bx(s) - \frac{x(s)}{4\left[c_1 + \hat{A}(s)/2\right]^2} - \frac{x(s)}{4\left[c_2 + \hat{A}(s)/2\right]^2},\ x(t_0) = x_0. \tag{4.11}$$

Solving (4.11) yields the optimal cooperative state trajectory for $\Gamma_c(x_0, T - t_0)$ as:

$$x^*(s) = \varpi(t_0, s)^2\left[x_0^{1/2} + \int_{t_0}^{s} \varpi^{-1}(t_0, t) H_1 dt\right]^2, \text{ for } s \in [t_0, T], \tag{4.12}$$

where $\varpi(t_0, s) = \exp\left[\int_{t_0}^{s} H_2(\tau) d\tau\right], H_1 = \frac{1}{2}a$, and $H_2(s) = -\left[\frac{1}{2}b + \frac{1}{8\left[c_1 + \hat{A}(s)/2\right]^2} + \frac{1}{8\left[c_2 + \hat{A}(s)/2\right]^2}\right]$.

The cooperative control for the game $\Gamma_c(x_0, T - t_0)$ over the time interval $[t_0, T]$ along the optimal trajectory $\{x^*(t)\}_{t=t_0}^{T}$ can be expressed precisely as:

$$\psi_1^*(t, x_t^*) = \frac{x_t^*}{4\left[c_1 + \hat{A}(t)/2\right]^2}, \text{ and } \psi_2^*(t, x_t^*) = \frac{x_t^*}{4\left[c_2 + \hat{A}(t)/2\right]^2}. \tag{4.13}$$

Following the above analysis, the value function of the optimal control problem with dynamics structure (4.1) and payoff structure (4.7) which starts at time τ with initial state x_τ^* can be obtained as $W^{(\tau)}(t,x) = \exp[-r(t-\tau)]\left[\hat{A}(t)x^{1/2} + \hat{B}(t)\right]$, and the corresponding optimal controls as

$$\psi_1^*(t, x_t^*) = \frac{x_t^*}{4\left[c_1 + \hat{A}(t)/2\right]^2}, \text{ and } \psi_2^*(t, x_t^*) = \frac{x_t^*}{4\left[c_2 + \hat{A}(t)/2\right]^2},$$

over the time interval $[\tau, T]$.

The agreed-upon optimality principle entails an imputation

$$\xi^{(\tau)i}(\tau, x_\tau^*) = V^{(\tau)i}(\tau, x_\tau^*) + \frac{1}{n}\left[W^{(\tau)}(\tau, x_\tau^*) - \sum_{j=1}^{n} V^{(\tau)j}(\tau, x_\tau^*)\right], \ i \in \{1, 2\}, \tag{4.14}$$

in the cooperative game $\Gamma_c(x_\tau^*, T - \tau)$ for $\tau \in \{t_0, T\}$.

Applying Theorem 3.1 a subgame consistent solution under the above optimal principle for the cooperative game $\Gamma_c(x_0, T - t_0)$ can be obtained as:

$\{u(s)$ and $B(s)$ for $s \in [t_0, T]$ and $\xi^{(t_0)}(t_0, x_0)\}$ in which

(i) $u(s)$ for $s \in [t_0, T]$ is the set of group optimal strategies $\psi_1^*(s, x_s^*) = \frac{x_s^*}{4[c_1 + \hat{A}(s)/2]^2}$, and $\psi_2^*(s, x_s^*) = \frac{x_s^*}{4[c_2 + \hat{A}(s)/2]^2}$; and

(ii) the imputation distribution procedure

$B(s) = \{B_1(s), B_2(s)\}$ for $s \in [t_0, T]$ where

$$
\begin{aligned}
B_i(s) = \frac{-1}{2}\Bigg\{ & \left(\left[\dot{A}_i(s)\left(x_s^*\right)^{1/2} + \dot{C}_i(s)\right] + r\left[A_i(s)\left(x_s^*\right)^{1/2} + C_i(s)\right]\right) \\
& + \left[\frac{1}{2}A_i(s)\left(x_s^*\right)^{-1/2}\right]\left[a\left(x_s^*\right)^{1/2} - bx_s^* - \frac{x_s^*}{4\left[c_i + \hat{A}(s)/2\right]^2} - \frac{x_s^*}{4\left[c_j + \hat{A}(s)/2\right]^2}\right]\Bigg\} \\
& -\frac{1}{2}\Bigg\{ \left(\left[\dot{\hat{A}}(s)\left(x_s^*\right)^{1/2} + \dot{\hat{C}}(s)\right] + r\left[\hat{A}(s)\left(x_s^*\right)^{1/2} + \hat{C}(s)\right]\right) \\
& + \left[\frac{1}{2}\hat{A}(s)\left(x_s^*\right)^{-1/2}\right] \\
& \left[a\left(x_s^*\right)^{1/2} - bx_s^* - \frac{x_s^*}{4\left[c_i + \hat{A}(s)/2\right]^2} - \frac{x_s^*}{4\left[c_j + \hat{A}(s)/2\right]^2}\right]\Bigg\} \\
& +\frac{1}{2}\Bigg\{ \left(\left[\dot{A}_j(s)\left(x_s^*\right)^{1/2} + \dot{C}_j(s)\right] + r\left[A_j(s)\left(x_s^*\right)^{1/2} + C_j(s)\right]\right) \\
& + \left[\frac{1}{2}A_j(s)\left(x_s^*\right)^{-1/2}\right] \\
& \left[a\left(x_s^*\right)^{1/2} - bx_s^* - \frac{x_s^*}{4\left[c_i + \hat{A}(s)/2\right]^2} - \frac{x_s^*}{4\left[c_j + \hat{A}(s)/2\right]^2}\right]\Bigg\},
\end{aligned}
$$

$$\text{for } i, j \in \{1, 2\} \text{ and } i \neq j, \tag{4.15}$$

where $\dot{A}_i(s)$ and $\dot{C}_i(s)$ are given in (4.6); and $\dot{\hat{A}}(s)$ and $\dot{\hat{C}}(s)$ are given in (4.9).

With players using the cooperative strategies, the instantaneous receipt of player i at time instant τ is:

$$\zeta_i(\tau) = \frac{\left(x_\tau^*\right)^{1/2}}{2[c_i + A(\tau)/2]} - \frac{c_i\left(x_\tau^*\right)^{1/2}}{4[c_i + A(\tau)/2]^2}, \tag{4.16}$$

Under cooperation the instantaneous payment that player i should receive is $B_i(\tau)$ as stated in (4.15). Hence an instantaneous transfer payment

$$\chi^i(\tau) = B_i(\tau) - \zeta_i(\tau) \tag{4.17}$$

has to be given to player i at time τ, for $i \in \{1, 2\}$ and $\tau \in [t_0, T]$.

2.5 Infinite Horizon Analysis

In this section we consider infinite horizon cooperative differential games in which player i's payoff is:

$$\int_\tau^\infty g^i[x(s), u_1(s), u_2(s), \cdots, u_n(s)] \exp[-r(s - \tau)] ds, \text{ for } i \in N. \tag{5.1}$$

The state dynamics is

$$\dot{x}(s) = f[x(s), u_1(s), u_2(s), \cdots, u_n(s)], x(\tau) = x_\tau. \tag{5.2}$$

Since s does not appear in $g^i[x(s), u_1(s), u_2(s)]$ and the state dynamics, the game (5.1 and 5.2) is an autonomous problem. Consider the alternative game $\Gamma(x)$ which starts at time $t \in [t_0, \infty)$ with initial state $x(t) = x$:

$$\max_{u_i} \int_t^\infty g^i[x(s), u_1(s), u_2(s), \cdots, u_n(s)] \exp[-r(s - t)] ds, \text{ for } i \in N, \tag{5.3}$$

subject to the state dynamics

$$\dot{x}(s) = f[x(s), u_1(s), u_2(s), \cdots, u_n(s)], \; x(t) = x. \tag{5.4}$$

The infinite-horizon autonomous game $\Gamma(x)$ is independent of the choice of t and dependent only upon the state at the starting time, that is x.

A feedback Nash equilibrium solution for the infinite-horizon autonomous game (5.3) and (5.4) can be characterized as follows:

Theorem 5.1 An n-tuple of strategies $\{u_i^* = \phi_i^*(\cdot) \in U^i, \text{ for } i \in N\}$ provides a feedback Nash equilibrium solution to the infinite-horizon game (5.3) and (5.4) if there exist continuously differentiable functions $\hat{V}^i(x) : R^m \to R, i \in N$, satisfying the following set of partial differential equations:

$$r\hat{V}^i(x) = \max_{u_i} \left\{ g^i\left[x, \phi_1^*(x), \phi_2^*(x), \cdots, \phi_{i-1}^*(x), u_i, \phi_{i+1}^*(x), \cdots, \phi_n^*(x)\right] \right.$$
$$\left. + \hat{V}_x^i(x) f\left[x, \phi_1^*(x), \phi_2^*(x), \cdots, \phi_{i-1}^*(x), u_i, \phi_{i+1}^*(x), \cdots, \phi_n^*(x)\right] \right\}, \text{ for } i \in N.$$

Proof By Theorem A.2 in the Technical Appendices, $\hat{V}^{i}(x)$ is the value function associated with the optimal control problem of player i, $i \in N$. Hence the conditions in Theorem 5.1 imply a Nash equilibrium. ■

Now consider the case when the players agree to act cooperatively. Let $\Gamma_c(\tau, x_\tau)$ denote a cooperative game in which player i's payoff is (5.1) and the state dynamics is (5.2). The players agree to act according to an agreed upon optimality principle $P(\tau, x_\tau)$ which entails

(i) group optimality and
(ii) the distribution of the total cooperative payoff according to an imputation vector $\xi^{(\upsilon)}(\upsilon, x_\upsilon^*)$ for $\upsilon \in [\tau, \infty)$ over the game duration. Moreover, the function $\xi^{(\upsilon)i}(\upsilon, x_\upsilon^*) \in \xi^{(\upsilon)}(\upsilon, x_\upsilon^*)$, for $i \in N$, is continuously differentiable in υ and x_υ^*.

The solution of the cooperative game $\Gamma_c(\tau, x_\tau)$ includes

(i) a set of group optimal cooperative strategies $u^{(\tau)*}(s) = \left[u_1^{(\tau)*}(s), u_2^{(\tau)*}(s), \cdots, u_n^{(\tau)*}(s)\right]$, for $s \in [\tau, \infty)$;
(ii) an imputation vector $\xi^{(\tau)}(\tau, x_\tau) = \left[\xi^{(\tau)1}(\tau, x_\tau), \xi^{(\tau)2}(\tau, x_\tau), \cdots, \xi^{(\tau)n}(\tau, x_\tau)\right]$ to allot the cooperative payoff to the players; and
(iii) a payoff distribution procedure $B^\tau(s) = \left[B_1^\tau(s), B_2^\tau(s), \cdots, B_n^\tau(s)\right]$ for $s \in [\tau, \infty)$, where $B_i^\tau(s)$ is the instantaneous payments for player i at time s. In particular,

$$\xi^{(\tau)i}(\tau, x_\tau) = \int_\tau^\infty B_i^\tau(s) \exp[-r(s-\tau)]\, ds, \quad \text{for } i \in N \tag{5.5}$$

In thc following sub-sections, we characterize the cooperative strategies and payoff distribution procedure of the cooperative game $\Gamma_c(\tau, x_\tau)$ under the agreed-upon optimality principle.

2.5.1 Group Optimal Cooperative Strategies

To ensure group rationality the players maximize the sum of their payoffs, the players solve the problem:

$$\max_{u_1, u_2, \cdots, u_n} \left\{ \int_\tau^\infty \sum_{j=1}^n g^j[x(s), u_1(s), u_2(s), \cdots, u_n(s)] \exp[-r(s-\tau)] ds \right\}, \tag{5.6}$$

subject to (5.2).

Following Theorem A.2 in the Technical Appendices, we note that a set of controls $\{\psi_1^*(x), \text{for } i \in N\}$ provides a solution to the optimal control problem (5.6) if there exists continuously differentiable function $W(x): R^m \to R$ satisfying the infinite-horizon Bellman equation:

$$rW(x) = \max_{u_1, u_2, \cdots, u_n} \left\{ \sum_{j=1}^{2} g^j[x, u_1, u_2, \cdots, u_n] + W_x f[x, u_1, u_2, \cdots, u_n] \right\}. \quad (5.7)$$

The players will adopt the cooperative control $\{\psi_1^*(x)$, for $i \in N\}$ characterized in (5.7). Note that these controls are functions of the current state x only. Substituting this set of control into the state dynamics yields the optimal (cooperative) trajectory as;

$$\dot{x}(s) = f\left[x(s), \psi_1^*(x(s)), \psi_2^*(x(s)), \cdots, \psi_n^*(x(s))\right],\ x(\tau) = x_\tau. \quad (5.8)$$

Let $x^*(s)$ denote the solution to (5.8). The optimal trajectory $\{x^*(s)\}_{s=\tau}^{\infty}$ can be expressed as:

$$x^*(s) = x_\tau + \int_\tau^s f\left[x^*(v), \psi_1^*(x^*(v)), \psi_2^*(x^*(v)), \cdots, \psi_n^*(x^*(v))\right] dv.$$

For notational convenience, we use the terms $x^*(s)$ and x_s^* interchangeably. The cooperative control for the game can be expressed more precisely as:

$$\{\psi_i^*(x_s^*),\ \text{for } i \in N \text{ and } s \in [\tau, \infty)\},$$

which are functions of the current state x_s^* only. The term

$$W(x_\tau^*) = \int_\tau^\infty \sum_{j=1}^{n} g^j\left[x^*(s), \psi_1^*(x^*(s)), \psi_2^*(x^*(s)), \cdots, \psi_n^*(x^*(s))\right] \exp[-r(s-\tau)]\, ds$$

yields the maximized cooperative payoff at current time τ, given that the state is x_τ^*

2.5.2 *Subgame Consistent Imputation and Payoff Distribution Procedure*

According to the agreed-upon optimality principle $P(\tau, x_\tau)$ the players would use the Payoff Distribution Procedure $\{B^\tau(s)\}_{s=\tau}^{\infty}$ to bring about an imputation to player i as:

$$\xi^{(\tau)i}(\tau, x_\tau) = \int_\tau^\infty B_i^\tau(s) \exp[-r(s-\tau)]\ ds,\ \text{for } i \in N. \quad (5.9)$$

At time τ, we define the present value of player i's payoff over the time interval $[t, \infty)$ as:

$$\xi^{(\tau)i}\left(t, x_t^*\right) = \int_t^\infty B_i^\tau(s)\exp[-r(s-\tau)]\ ds,\ \text{for}\, i \in N, \tag{5.10}$$

where $t > \tau$ and $x_t^* \in \{x^*(s)\}_{s=\tau}^\infty$.

Consider the case when the game has proceeded to time t and the state variable became x_t^*. Then one has a cooperative game $\Gamma_c(t, x_t^*)$ which starts at time t with initial state x_t^*. According to the agreed-upon optimality principle, an imputation

$$\xi^{(t)i}\left(t, x_t^*\right) = \int_t^\infty B_i^t(s)\exp[-r(s-t)]\ ds,$$

will be allotted to player i, for $i \in N$.

However, according to the optimality principle, the imputation (in present value viewed at time τ) to player i over the period $[t, \infty)$ is

$$\xi^{(\tau)i}\left(t, x_t^*\right) = \int_t^\infty B_i^\tau(s)\exp[-r(s-\tau)]\ ds,\ \text{for}\ i \in N, \tag{5.11}$$

For the imputations from the optimality principle to be consistent throughout the cooperation duration, it is essential that

$$\exp[r(t-\tau)]\xi^{(\tau)i}\left(t, x_t^*\right) = \xi^{(t)i}\left(t, x_t^*\right),\ \text{for}\, t \in (\tau, \infty).$$

In addition, at time τ when the initial state is x_τ, according to the optimality principle the payoff distribution procedure is

$$B^\tau(s) = \left[B_1^\tau(s), B_2^\tau(s), \cdots, B_n^\tau(s)\right],\ \text{for}\, s \in [\tau, \infty).$$

When the game has proceeded to time t and the state variable became x_t^*. According to the optimality principle the payoff distribution procedure

$$B^t(s) = \left[B_1^t(s), B_2^t(s), \cdots, B_n^t(s)\right],\ \text{for}\, s \in [t, \infty),$$

will be adopted.

For the continuation of the payoff distribution procedure to be consistent it is required that

$$B^{t_0}(s) = B^t(s),\ \text{for}\ s \in [t, \infty)\ \text{and}\, t \in [\tau, \infty).$$

Definition 5.1 The imputation and payoff distribution procedure $\{\xi^{(\tau)}(\tau, x_\tau)$ and $B^\tau(s)$ for $s \in [\tau, \infty)\}$ are subgame consistent if

$$\text{(i) } \exp[r(t-\tau)]\xi^{(\tau)i}(t, x_t^*) \equiv \exp[r(t-\tau)] \int_t^\infty B_i^\tau(s)\exp[-r(s-\tau)]\, ds \\ = \xi^{(t)i}(t, x_t^*), \text{ for } t \in (\tau, \infty) \text{ and } i \in N; \text{ and} \tag{5.12}$$

(ii) the payoff distribution procedure $B^\tau(s) = [B_1^\tau(s), B_2^\tau(s), \cdots, B_n^\tau(s)]$ for $s \in [t, \infty)$ is identical to $B^t(s) = [B_1^t(s), B_2^t(s), \cdots, B_n^t(s)] \in (t, x_t^*)$. ■

Definition 5.1 yields the infinite horizon subgame consistent imputation and payoff distribution procedure.

2.5.3 *Derivation of Subgame Consistent Payoff Distribution Procedure*

A payoff distribution procedure leading to subgame consistent imputation has to satisfy Definition 5.1. Invoking Definition 5.1, we have $B_i^\tau(s) = B_i^t(s) = B_i(s)$, for $s \in [\tau, \infty)$ and $t \in [\tau, \infty)$ and $i \in N$.

Therefore along the cooperative trajectory $\{x^*(t)\}_{t \ge t_0}$,

$$\xi^{(\tau)i}(\tau, x_\tau^*) = \int_\tau^\infty B_i(s)\exp[-r(s-\tau)]ds, \text{ for } i \in N, \text{ and}$$

$$\xi^{(\upsilon)i}(\upsilon, x_\upsilon^*) = \int_\upsilon^\infty B_i(s)\exp[-r(s-\upsilon)]ds, \text{ for } i \in N, \text{ and}$$

$$\xi^{(t)i}(t, x_t^*) = \int_t^\infty B_i(s)\exp[-r(s-t)]ds, \text{ for } i \in N \text{ and } t \ge \upsilon \ge \tau \tag{5.13}$$

Moreover, for $i \in N$ and $t \in [\tau, \infty)$, we define the term

$$\xi^{(\upsilon)i}(t, x_t^*) = \left\{ \left(\int_t^\infty B_i(s)\exp[-r(s-\upsilon)]ds \right) \middle| \, x(t) = x_t^* \right\}, \tag{5.14}$$

to denote the present value of player i's cooperative payoff over the time interval $[t, \infty)$, given that the state is x_t^* at time $t \in [\upsilon, \infty)$, under the solution $P(\upsilon, x_\upsilon^*)$.

Invoking (5.13) and (5.14) one can readily verify that $\exp[r(t-\tau)]\xi^{(\tau)i}(t, x_t^*) = \xi^{(t)i}(t, x_t^*)$, for $i \in N$ and $\tau \in [t_0, T]$ and $t \in [\tau, T]$.

The next task is to derive $B_i(s)$, for $s \in [\tau, \infty)$ and $t \in [\tau, \infty)$ so that (5.13) can be realized. Consider again the following condition.

Condition 5.1 For $i \in N$ and $t \geq \upsilon$ and $\upsilon \in [\tau, T]$, the term $\xi^{(\upsilon)i}(t, x_t^*)$ is a function that is continuously differentiable in t and x_t^*.

A theorem characterizing a formula for $B_i(s)$, for $i \in N$ and $s \in [\upsilon, \infty)$, which yields (5.14) is provided as follows.

Theorem 5.2 If Condition 5.1 is satisfied, a PDP with instantaneous payments at time s equaling

$$B_i(s) = -\left[\xi_t^{(s)i}(t, x_t^*)\Big|_{t=s}\right] - \xi_{x_s^*}^{(s)i}(s, x_s^*) f\left[x_s^*, \psi_1^*(x_s^*), \psi_2^*(x_s^*), \cdots, \psi_n^*(x_s^*)\right], \quad (5.15)$$

for $i \in N$ and $s \in [\upsilon, \infty)$,

yields imputation $\xi^{(\upsilon)i}(\upsilon, x_\upsilon^c)$, for $\upsilon \in [\tau, \infty)$ which satisfy (5.13).

Proof Note that along the cooperative trajectory $\{x^*(t)\}_{t \geq \tau}$

$$\xi^{(\upsilon)i}(t, x_t^*) = \int_t^\infty B_i(s) \exp[-r(s-\upsilon)] ds = \exp[-r(t-\upsilon)]\, \xi^{(t)i}(t, x_t^*),$$

for $i \in N$ and $t \in [\upsilon, \infty)$. (5.16)

For $\Delta t \to 0$, Eq. (5.13) can be expressed as

$$\begin{aligned} \xi^{(\upsilon)i}(\tau, x_\tau^*) &= \int_\upsilon^\infty B_i(s) \exp[-r(s-\upsilon)] ds \\ &= \int_\upsilon^{\upsilon+\Delta t} B_i(s) \exp[-r(s-\upsilon)]\ ds + \xi^{(\upsilon)i}(\upsilon + \Delta t, x_\upsilon^* + \Delta x_\upsilon^*), \end{aligned} \quad (5.17)$$

where

$\Delta x_\upsilon^* = f\left[x_\upsilon^*, \psi_1^*(x_\upsilon^*), \psi_2^*(x_\upsilon^*), \cdots, \psi_n^*(x_\upsilon^*)\right] \Delta t + o(\Delta t)$, and $o(\Delta t)/\Delta t \to 0$ as $\Delta t \to 0$.

Replacing the term $x_\upsilon^* + \Delta x_\upsilon^*$ with $x_{\upsilon+\Delta t}^*$ and rearranging (5.17) yields:

$$\begin{aligned} &\int_\upsilon^{\upsilon+\Delta t} B_i(s) \exp[-r(s-\upsilon)]\ ds \\ &= \xi^{(\upsilon)i}(\upsilon, x_\upsilon^*) - \xi^{(\upsilon)i}(\upsilon + \Delta t, x_{\upsilon+\Delta t}^*), \text{ for all } \upsilon \in [\tau, \infty) \text{ and } i \in N. \end{aligned} \quad (5.18)$$

Consider the following condition concerning $\xi^{(\upsilon)i}(t, x_t^*)$, for $\upsilon \in [\tau, \infty)$ and $t \in [\upsilon, \infty)$:

With Condition 5.1 holding and $\Delta t \to 0$, (5.18) can be expressed as:

$$\begin{aligned} B_i(\upsilon)\Delta t = &-\left[\xi_t^{(\upsilon)i}(t, x_t^*)\Big|_{t=\tau}\right] \Delta t \\ &- \xi_{x_\upsilon^*}^{(\upsilon)i}(\upsilon, x_\upsilon^*) f\left[x_\upsilon^*, \psi_1^*(x_\upsilon^*), \psi_2^*(x_\upsilon^*), \cdots, \psi_n^*(x_\upsilon^*)\right] \Delta t - o(\Delta t). \end{aligned} \quad (5.19)$$

Dividing (5.19) throughout by Δt, with $\Delta t \to 0$, yields (5.15). Thus the payoff distribution procedure in $B_i(\upsilon)$ in (5.15) would lead to the realization of the imputations which satisfy (5.15). ■

Since the payoff distribution procedure in $B_i(\tau)$ in (5.15) leads to the realization of (5.13), it would yields subgame consistent imputations satisfying Definition 5.1.

A more succinct form of Theorem 5.2 can be derived as follows. Note that, a PDP with instantaneous payments at time s equaling

$$B_i(s) = r\,\xi^{(s)i}(s, x_s^*) - \xi_{x_s^*}^{(s)i}(s, x_s^*) f\left[x_s^*, \psi_1^*(x_s^*), \psi_2^*(x_s^*), \cdots, \psi_n^*(x_s^*)\right],$$
$$\text{for } i \in N \text{ and } s \in [\upsilon, \infty), \tag{5.20}$$

yields imputation $\xi^{(\upsilon)i}(\upsilon, x_\upsilon^c)$, for $\upsilon \in [\tau, \infty)$ which satisfy (5.13).

To demonstrate that (5.20) is an alternative form for (5.15) in Theorem 5.2, we define

$$\hat{\xi}^i(x_\upsilon^*) = \left\{ \int_\upsilon^\infty B_i(s) \exp[-r(s-\upsilon)]ds \,\middle|\, x(\upsilon) = x_\upsilon^* \right\} = \xi^{(\upsilon)i}(\tau, x_\upsilon^*), \text{ and}$$
$$\hat{\xi}^i(x_t^*) = \left\{ \int_t^\infty B_i(s) \exp[-r(s-t)]ds \,\middle|\, x(t) = x_t^* \right\} = \xi^{(t)i}(t, x_t^*),$$

for $i \in N$ and $\upsilon \in [\tau, \infty)$ and $t \in [\upsilon, \infty)$ along the optimal cooperative trajectory $\{x_s^*\}_{s=\tau}^{\infty}$.

We then have:

$$\xi^{(\upsilon)i}(t, x_t^*) = \exp[-r(t-\upsilon)]\,\hat{\xi}^i(x_t^*).$$

Differentiating $\xi^{(\upsilon)i}(t, x_t^*)$ with respect to t yields:

$$\left[\xi_t^{(\upsilon)i}(t, x_t^*)\Big|_{t=\upsilon}\right] = -r \exp[-r(t-\upsilon)]\,\hat{\xi}^i(x_t^*) = -r\,\xi^{(\upsilon)i}(t, x_t^*).$$

At $t = \upsilon$, $\xi^{(\upsilon)i}(t, x_t^*) = \xi^{(\upsilon)i}(\upsilon, x_\upsilon^*)$, therefore

$$\left[\xi_t^{(\upsilon)i}(t, x_t^*)\Big|_{t=\upsilon}\right] = r\xi^{(\upsilon)i}(t, x_t^*) = r\xi^{(\upsilon)i}(\upsilon, x_\upsilon^*). \tag{5.21}$$

Substituting (5.21) into (5.15) yields (5.20). Since the infinite-horizon autonomous game $\Gamma(x)$ is independent of the choice of time s and dependent only upon the state, Eq. (5.20) can be expressed as:

$$B_i(x_s^*) = r\,\hat{\xi}^i(x_s^*) - \hat{\xi}_{x_s^*}^i(x_s^*) f\left[x_s^*, \psi_1^*(x_s^*), \psi_2^*(x_s^*), \cdots, \psi_n^*(x_s^*)\right], \text{ for } i \in N. \tag{5.22}$$

Therefore a subgame consistent solution for the cooperative game $\Gamma_c(\tau, x_\tau)$ with optimality principle $P(\tau, x_\tau)$ includes the cooperative strategies and Payoff Distribution Procedure:

$\{u(s)$ and $B(x_s^*)$ for $s \in [\tau, \infty)\}$ in which

(i) $u(s)$ is the set of group optimal strategies $\psi^*(x_s^*)$ for the game $\Gamma_c(\tau, x_\tau)$, and
(ii) the payoff distribution procedure

$$B(x_s^*) = \{B_1(x_s^*), B_2(x_s^*), \cdots, B_n(x_s^*)\} \text{ where}$$
$$B_i(x_s^*) = r\hat{\xi}^i(x_s^*) - \hat{\xi}^i_{x_s^*}(x_s^*) f[x_s^*, \psi_1^*(x_s^*), \psi_2^*(x_s^*), \cdots, \psi_n^*(x_s^*)], \tag{5.23}$$

for $i \in N$.

With players using the cooperative strategies $\{\psi_i^*(x_\upsilon^*)$, for $i \in N$ and $\upsilon \in [\tau, \infty)\}$, the instantaneous receipt of player i at time instant υ is:

$$\zeta_i(x_\upsilon^*) = g^i[x_\upsilon^*, \psi_1^*(x_\upsilon^*), \psi_2^*(x_\upsilon^*), \cdots, \psi_n^*(x_\upsilon^*)],$$
$$\text{for } i \in N. \tag{5.24}$$

According to Theorem 5.2, the instantaneous payment that player i should receive under the agreed-upon optimality principle is $B_i(\upsilon)$ in (5.15) or equivalently $B_i(x_\upsilon^*)$ in (5.23). Hence an instantaneous transfer payment

$$\chi^i(x_\upsilon^*) = B_i(x_\upsilon^*) - \zeta_i(x_\upsilon^*) \tag{5.25}$$

has to be given to player i at time υ, for $i \in N$.

2.6 Infinite Horizon Resource Extraction

Consider an infinite horizon version of the cooperative fishery game in Sect. 2.5. At initial time τ, the payoff of nation 1 and that of nation 2 are respectively:

$$\int_\tau^\infty \left[u_1(s)^{1/2} - \frac{c_1}{x(s)^{1/2}} u_1(s) \right] \exp[-r(t-\tau)]\, ds$$

and

$$\int_\tau^\infty \left[u_2(s)^{1/2} - \frac{c_2}{x(s)^{1/2}} u_2(s) \right] \exp[-r(t-\tau)]\, ds. \tag{6.1}$$

The resource stock $x(s) \in X \subset R$ follows the dynamics

$$\dot{x}(s) = ax(s)^{1/2} - bx(s) - u_1(s) - u_2(s), \;\; x(\tau) = x_\tau \in X, \tag{6.2}$$

Using Theorem 5.1, the value function $\hat{V}^i(t, x)$ reflecting the payoff of nation i in a noncooperative feedback Nash equilibrium can be obtained as:

$$\hat{V}^i(t, x) = \left[A_i x^{1/2} + C_i\right], \tag{6.3}$$

where for $i, j \in \{1, 2\}$ and $i \neq j, A_i, C_i, A_j$ and C_j satisfy:

$$\begin{aligned}
&\left[r + \frac{b}{2}\right] A_i - \frac{1}{2[c_i + A_i/2]} + \frac{c_i}{4[c_i + A_i/2]^2} \\
&+ \frac{A_i}{8[c_i + A_i/2]^2} + \frac{A_i}{8\left[c_j + A_j/2\right]^2} = 0, \text{ and} \\
&C_i = \frac{a}{2} A_i.
\end{aligned}$$

The game equilibrium strategies can be obtained as:

$$\phi_1^*(x) = \frac{x}{4[c_1 + A_1/2]^2}, \text{ and } \phi_2^*(x) = \frac{x}{4[c_2 + A_2/2]^2}. \tag{6.4}$$

Consider the case when these two nations agree to act according to an agreed upon optimality principle which entails

(i) group optimality, and
(ii) the distribution of the cooperative payoff according to the imputation that divides the excess of the total cooperative payoff over the sum of individual noncooperative payoffs equally.

To maximize their joint payoff for group optimality, the nations have to solve the control problem of maximizing

$$\int_\tau^\infty \left(\left[u_1(s)^{1/2} - \frac{c_1}{x(s)^{1/2}} u_1(s) \right] + \left[u_2(s)^{1/2} - \frac{c_2}{x(s)^{1/2}} u_2(s) \right] \right) \exp[-r(t - \tau)]\, ds \tag{6.5}$$

subject to (6.2).

Invoking Eq. (5.7), we obtain:

$$\begin{aligned}
rW(x) = \max_{u_1, u_2} \Bigg\{ & \left(\left[u_1{}^{1/2} - \frac{c_1}{x^{1/2}} u_1 \right] + \left[u_2{}^{1/2} - \frac{c_2}{x^{1/2}} u_2 \right] \right) \\
& + W_x(x) \left[ax^{1/2} - bx - u_1 - u_2\right] \Bigg\}.
\end{aligned}$$

The value function $W(x)$ which reflects the maximized joint payoff can be obtained as:

$$W(x) = \left[Ax^{1/2} + C\right],$$

where $\left[r + \frac{b}{2}\right]A - \frac{1}{2[c_1 + A/2]} - \frac{1}{2[c_2 + A/2]}$

$$+\frac{c_1}{4[c_1 + A/2]^2} + \frac{c_2}{4[c_2 + A/2]^2} + \frac{A}{8[c_1 + A/2]^2} + \frac{A}{8[c_2 + A/2]^2} = 0, \text{ and}$$

$$C = \frac{a}{2r}A$$

The optimal cooperative controls can then be obtained as:

$$\psi_1^*(x) = \frac{x}{4[c_1 + A/2]^2} \text{ and } \psi_2^*(x) = \frac{x}{4[c_2 + A/2]^2}. \tag{6.6}$$

Substituting these control strategies into (6.2) yields the dynamics of the state trajectory under cooperation:

$$\dot{x}(s) = ax(s)^{1/2} - bx(s) - \frac{x(s)}{4[c_1 + A/2]^2} - \frac{x(s)}{4[c_2 + A/2]^2}, \ x(\tau) = x_\tau. \tag{6.7}$$

Solving (6.7) yields the optimal cooperative state trajectory $\{x^*(s)\}_{\tau=t_0}^{\infty}$ for the cooperative game (6.1 and 6.2) as:

$$x^*(s) = \left[\frac{a}{2H} + \left((x_\tau)^{\frac{1}{2}} - \frac{a}{2H}\right)\exp[-H(s-\tau)]\right]^2, \tag{6.8}$$

where $H = -\left[\frac{b}{2} + \frac{1}{8[c_1 + A/2]^2} + \frac{1}{8[c_2 + A/2]^2}\right]$.

According to the agreed-upon optimality principle these nations will distribute the cooperative payoff according to the imputation which divides the excess of the total cooperative payoff over the sum of individual noncooperative payoffs equally. Hence the imputation $\xi(\upsilon, x_\upsilon^*) = \left[\hat{\xi}^1(x_\upsilon^*), \hat{\xi}^2(x_\upsilon^*)\right]$ has to satisfy:

Condition 6.1

$$\hat{\xi}^i(x_\upsilon^*) = \hat{V}^i(x_\upsilon^*) + \frac{1}{2}\left[W(x_\upsilon^*) - \sum_{j=1}^{2}\hat{V}^j(x_\upsilon^*)\right], \tag{6.9}$$

for $i \in \{1, 2\}$ and $\upsilon \in [\tau, \infty)$. ■

Applying Theorem 5.2 and Eq. (5.23) a subgame consistent solution payoff distribution procedure $B(x_s^*) = \{B_1(x_s^*), B_2(x_s^*)\}$ for $s \in [\tau, \infty)$ can be obtained as:

$$
\begin{aligned}
B_i(x_s^*) &= \frac{1}{2}\left\{ r\left[A_i(x_s^*)^{1/2} + C_i\right] + r\left[A(x_s^*)^{1/2} + C\right] - r\left[A_j(x_s^*)^{1/2} + C_j\right] \right\} \\
&\quad - \frac{1}{4}\left\{ A_i(x_s^*)^{-1/2} + A(x_s^*)^{-1/2} - A_j(x_s^*)^{-1/2} \right\} \\
&\quad \times \left[a(x_s^*)^{1/2} - bx_s^* - \frac{x_s^*}{4[c_1 + A/2]^2} - \frac{x_s^*}{4[c_2 + A/2]^2} \right],
\end{aligned}
\tag{6.10}
$$

for $i, j \in \{1, 2\}$ and $i \neq j$.

With players using the cooperative strategies $\{\psi_i^*(x_\upsilon^*), i \in \{1, 2\}\}$ along the cooperative trajectory, the instantaneous receipt of player i at time instant υ becomes:

$$
\zeta_i(x_\upsilon^*) = \frac{(x_\upsilon^*)^{1/2}}{2[c_i + A/2]} - \frac{c_i(x_\upsilon^*)^{1/2}}{4[c_i + A/2]^2}, \tag{6.11}
$$

According to (6.10), the instantaneous payment that player i should receive under the agreed-upon optimality principle is $B_i(x_\upsilon^*)$. Hence an instantaneous transfer payment

$$
\chi^i(x_\upsilon^*) = B_i(x_\upsilon^*) - \zeta_i(x_\upsilon^*) \tag{6.12}
$$

has to be given to player i at time $\upsilon \in [\tau, \infty)$, for $i \in \{1, 2\}$.

2.7 Chapter Notes

Significant contributions to general game theory include von Neumann and Morgenstern (1944); Nash (1950, 1953); Vorob'ev (1972); Shapley (1953) and Shubik (1959a, b). Dynamic optimization techniques are essential in the derivation of solutions to differential games. The origin of differential games was established by Rufus Isaacs in the late 1940s (the complete work was published in Isaacs (1965)). In the meantime, control theory reached its maturity in the *Optimal Control Theory* of Pontryagin et al. (1962) and Bellman's *Dynamic Programming* (1957). Berkovitz (1964) developed a variational approach to differential games, and Leitmann and Mon (1967) investigated the geometry of differential games. Pontryagin (1966) solved differential games in open-loop solution in terms of the maximum principle. Cooperative games suggest the possibility of socially optimal and group efficient solutions to decision problems involving strategic action. As discussed above, Individual rationality and group optimality are essential element of a cooperative game solution. Dockner and Jørgensen (1984); Dockner and Long (1993); Tahvonen (1994); Mäler and de Zeeuw (1998) and Rubio and Casino (2002) presented cooperative solutions satisfying group optimality in differential games. The majority of cooperative differential games adopt solutions satisfying the essential criteria for dynamic stability – group optimality and individual rationality. Haurie and Zaccour (1986, 1991), Kaitala and Pohjola (1988, 1990, 1995), Kaitala et al. (1995) and Jørgensen and Zaccour (2001) presented classes of transferable-payoff cooperative differential games with solutions which satisfy group optimality and individual rationality. Miao et al. (2010) studied a cooperative differential game on transmission rate in wireless networks. Lin et al. (2014) presented a cooperative differential game for model energy-bandwidth efficiency tradeoff in the Internet. Huang et al. (2016) presented a cooperative differential game of transboundary industrial pollution with a Stackelberg game between firms and local governments while the governments cooperate in pollution reduction. Tolwinski et al. (1986) considered cooperative equilibria in differential games in which memory-dependent strategies and threats are introduced to maintain the agreed-upon control path. Petrosyan and Danilov (1982); Petrosyan and Zenkevich (1996) and Petrosyan (1997) provided a detailed analysis of subgame consistent (then referred to as time consistent solutions in the deterministic framework) imputation distribution schemes in cooperative differential games. Filar and Petrosyan (2000) considered dynamic cooperative games in characteristic functions which evolve over time in a dynamic equation that is influenced by the current (instantaneous) characteristic function and cooperative solution concept adopted.

Yeung and Petrosyan (2004) presented subgame consistent solution in stochastic differential games and Yeung and Petrosyan (2012a) gave a comprehensive account of the topic. Application of subgame consistent solutions in differential games in cost-saving joint venture, collaborative environmental management and dormant firm cartel can be found in Yeung and Petrosyan (2012a). Other examples of cooperative differential games with solutions satisfying subgame consistency can be found in Petrosyan (1997), Jørgensen and Zaccour (2001). A note concerning the notations used in Petrosyan (1997) and Yeung and Petrosyan (2004) is given in Yeung and Petrosyan (2012d). A non-cooperative-equivalent imputation formula in cooperative differential games is provided by Yeung (2007b) and an irrational-behaviour proof condition in cooperative differential games is given in Yeung (2006a). A study on the tragedy of the commons in a dynamic game framework can be found in Hartwick and Yeung (1997).

2.8 Problems

1. Consider the case of three nations harvesting fish in common waters. The growth rate of the fish biomass is characterized by the differential equation:

$$\dot{x}(s) = 4x(s)^{1/2} - 0.5x(s) - u_1(s) - u_2(s),\ x(0) = 50,$$

where $u_i \in U_i$ is the (nonnegative) amount of fish harvested by nation i, for $i \in \{1, 2\}$. The horizon of the game is $[0, 5]$.

The harvesting cost for nation $i \in \{1, 2\}$ depends on the quantity of resource extracted $u_i(s)$ and the resource stock size $x(s)$. In particular, nation 1's extraction cost is $u_1(s)x(s)^{-1/2}$ and nation 2's is $2u_2(s)x(s)^{-1/2}$. The revenue of fish harvested by nation 1 at time s is $2[u_1(s)]^{1/2}$ and that by nation 2 is $[u_2(s)]^{1/2}$. The interest rate is 0.05.

Characterize a feedback Nash equilibrium solution for this fishery game.

2. If these nations agree to cooperate and maximize their joint payoff, obtain a group optimal cooperative solution.
3. Furthermore, if these nations agree to share the excess of their gain from cooperation equally along the optimal trajectory, derive a subgame consistent cooperative solution.
4. If the game horizon of the above problems is extended to infinity, what would be the answers to Problems 1, 2 and 3?

Chapter 3
Subgame Consistent Cooperation in Stochastic Differential Games

An essential characteristic of time – and hence decision making over time – is that though the individual may, through the expenditure of resources, gather past and present information, the future is inherently unknown and therefore (in the mathematical sense) uncertain. An empirically meaningful theory must therefore incorporate time-uncertainty in an appropriate manner. This Chapter considers subgame consistent cooperation in stochastic differential games. It provides an integrated exposition the works of Yeung and Petrosyan (2004), Chapter 4 of Yeung and Petrosyan (2006b), and Chapter 8 of Yeung and Petrosyan (2012a).

The organization of the Chapter is as follows. Section 3.1 presents the basic formulation of cooperative stochastic differential games. Section 3.2 presents an analysis on cooperative subgame consistency under uncertainty. Derivation of a subgame consistent payoff distribution procedure is provided in Sect. 3.3. An illustration in cooperative fishery under uncertainty is given in Sect. 3.4. Infinite horizon subgame consistency under uncertainty is examined in Sect. 3.5. In Sect. 3.6, a subgame consistent solution for infinite horizon cooperative fishery under uncertainty is presented. Chapter notes are provided in Sect. 3.7 and problems in Sect. 3.8.

3.1 Cooperative Stochastic Differential Games

Consider the general form of n-person stochastic differential games in which player i seeks to maximize its expected payoffs:

$$E_{t_0}\left\{ \int_{t_0}^{T} g^i[s, x(s), u_1(s), u_2(s), \cdots, u_n(s)] \exp\left[-\int_{t_0}^{s} r(y)dy\right] ds \right.$$
$$\left. + \exp\left[-\int_{t_0}^{T} r(y)dy\right] q^i(x(T)) \right\}, \quad \text{for } i \in N, \tag{1.1}$$

D.W.K. Yeung, L.A. Petrosyan, *Subgame Consistent Cooperation*, Theory and Decision Library C 47, DOI 10.1007/978-981-10-1545-8_3

with $E_{t_0}\{\cdot\}$ denoting the expectation operation taken at time t_0, and the dynamics of the state is

$$dx(s) = f[s, x(s), u_1(s), u_2(s), \cdots, u_n(s)]ds + \sigma[s, x(s)]dz(s), \qquad x(t_0) = x_0, \quad (1.2)$$

where $\sigma[s, x(s)]$ is a $m \times \Theta$ matrix and $z(s)$ is a Θ-dimensional Wiener process and the initial state x_0 is given. Let $\Omega[s, x(s)] = \sigma[s, x(s)]\ \sigma[s, x(s)]'$ denote the covariance matrix with its element in row h and column ζ denoted by $\Omega^{h\zeta}[s, x(s)]$. Moreover, $E[dz_\varpi] = 0$ and $E[dz_\varpi dt] = 0$ and $E\left[(dz_\varpi)^2\right] = dt$, for $\varpi \in [1, 2, \cdots, \Theta]$; and $E[dz_\varpi dz_\omega] = 0$, for $\varpi \in [1, 2, \cdots, \Theta]$, $\varpi \in [1, 2, \cdots, \Theta]$ and $\varpi \neq \omega$.

3.1.1 Non-cooperative Equilibria

Again, we first characterize the non-cooperative equilibria of the game as a benchmark for negotiation in the cooperative scheme. A feedback Nash equilibrium solution of the stochastic differential game (1.1) and (1.2) can be characterized by the following Theorem.

Theorem 1.1 An N-tuple of feedback strategies $\{\phi_i^*(t, x) \in U^i; i \in N\}$ provides a Nash equilibrium solution to the game (1.1) and (1.2) if there exist suitably smooth functions $V^{(t_0)i}(t, x) : [t_0, T] \times R^m \to R, i \in N$, satisfying the partial differential equations

$$-V_t^{(t_0)i}(t, x) - \frac{1}{2}\sum_{h,\zeta=1}^{m} \Omega^{h\zeta}(t, x) V_{x^h x^\zeta}^{(t_0)i}(t, x) =$$

$$\max_{u_i} \{g^i[t, x, \phi_1^*(t, x), \phi_2^*(t, x), \cdots, \phi_{i-1}^*(t, x), u_i(t), \phi_{i+1}^*(t, x), \cdots, \phi_n^*(t, x)]$$

$$\exp\left[-\int_{t_0}^{s} r(y)dy\right]$$

$$+ V_x^{(t_0)i}(t, x) f[t, x, \phi_1^*(t, x), \phi_2^*(t, x), \cdots, \phi_{i-1}^*(t, x), u_i(t), \phi_{i+1}^*(t, x), \cdots, \phi_n^*(t, x)]\},$$

$$V^{(t_0)i}(T, x) = q^i(x)\exp\left[-\int_{t_0}^{T} r(y)dy\right], \ i \in N.$$

Proof This result follows readily from the definition of Nash equilibrium and from the stochastic control result in Theorem A.3 of the Technical Appendices. ■

In particular, $V^{(t_0)i}(t, x)$ represents the expected game equilibrium payoff of player i at time $t \in [t_0, T]$ with the state being x, that is

$$V^{(t_0)i}(t,x)$$
$$= E_{t_0}\left\{ \int_t^T g^i\left[s, x^*(s), \phi_1^*(s,x^*(s)), \phi_2^*(s,x^*(s)), \cdots, \phi_n^*(s,x^*(s))\right] \right.$$
$$\exp\left[-\int_{t_0}^{s} r(y)dy\right] ds$$
$$\left. + q^i(x^*(T))\exp\left[-\int_{t_0}^{T} r(y)dy\right] \right\}.$$

A remark that will be utilized in subsequent analysis is given below.

Remark 1.1 Let $V^{(\tau)i}(t,x)$ denote the feedback Nash equilibrium payoff of nation i in the game with stochastic dynamics (1.1) and expected payoffs (1.2) which starts at time τ for $\tau \in [t_0, T)$. Note that the equilibrium feedback strategies are Markovian in the sense that they depend on current time and current state. One can readily verify that

$$\exp\left[\int_{t_0}^{\tau} r(y)dy\right] V^{(t_0)i}(t,x) = \exp\left[\int_{t_0}^{\tau} r(y)dy\right]$$
$$\times E_{t_0}\left\{ \int_t^T g^i\left[s, x^*(s), \phi_1^*(s,x^*(s)), \phi_2^*(s,x^*(s)), \cdots, \phi_n^*(s,x^*(s))\right] \right.$$
$$\left. \exp\left[-\int_{t_0}^{s} r(y)dy\right] ds \right\}$$
$$\times E_{t_0}\left\{ \int_t^T g^i\left[s, x^*(s), \phi_1^*(s,x^*(s)), \phi_2^*(s,x^*(s)), \cdots, \phi_n^*(s,x^*(s))\right] \right.$$
$$\left. \exp\left[-\int_{t_0}^{s} r(y)dy\right] ds \right\}$$
$$= E_t\left\{ \int_t^T g^i\left[s, x^*(s), \phi_1^*(s,x^*(s)), \phi_2^*(s,x^*(s)), \cdots, \phi_n^*(s,x^*(s))\right] \right.$$
$$\left. \exp\left[-\int_{\tau}^{s} r(y)dy\right] ds \right\}$$
$$= V^{(\tau)i}(t,x), \text{ for } \tau \in [t_0, T).$$

■

3.1.2 Dynamic Cooperation Under Uncertainty

The participating players agree to act according to an agreed-upon optimality principle. Based on this optimality principle, the solution of the cooperative differential game $\Gamma_c(x_0, T - t_0)$ at time t_0 includes

(i) a set of cooperative strategies $u^{(t_0)*}(s, x_s) = \left[u_1^{(t_0)*}(s, x_s), u_2^{(t_0)*}(s, x_s), \cdots, u_n^{(t_0)*}(s, x_s)\right]$, for $s \in [t_0, T]$ given that the state is x_s at time s;

(ii) an imputation vector $\xi^{(t_0)}(t_0,x_0)=\left[\xi^{(t_0)1}(t_0,x_0),\xi^{(t_0)2}(t_0,x_0),\cdots,\xi^{(t_0)n}(t_0,x_0)\right]$ to allot the cooperative payoff to the players; and

(iii) a payoff distribution procedure $B^{t_0}(s,x_s)=\left[B_1^{t_0}(s,x_s),B_2^{t_0}(s,x_s),\cdots,B_n^{t_0}(s,x_s)\right]$ for $s\in[t_0,T]$, where $B_i^{t_0}(s,x_s)$ is the instantaneous payments for player i at time s given that the state is x_s. In particular,

$$\xi^{(t_0)i}(t_0,x_0)=E_{t_0}\left\{\int_{t_0}^{T}B_i^{t_0}(s,x_s)\exp\left[-\int_{t_0}^{s}r(y)dy\right]ds+q^i(x_T)\exp\left[-\int_{t_0}^{T}r(y)dy\right]\right\},$$

for $i\in N$.

$$(1.3)$$

This means that the players agree at the outset on a set of cooperative strategies $u^{(t_0)*}(s,x_s)$, an imputation $\xi^{(t_0)i}(t_0,x_0)$ of the gains to the ith player covering the time interval $[t_0,T]$, and a payoff distribution procedure $\{B^{t_0}(s,x_s)\}_{s=t_0}^{T}$ to allocate payments to the players over the game interval.

Recall that group optimality is an essential element in dynamic cooperation, an optimality principle has to require the players have to maximize their expected joint payoff:

$$E_{t_0}\left\{\sum_{j=1}^{n}\int_{t_0}^{T}g^j[s,x(s),u_1(s),u_2(s),\cdots,u_n(s)]\exp\left[-\int_{t_0}^{s}r(y)dy\right]ds\right.$$
$$\left.+\sum_{j=1}^{n}\exp\left[-\int_{t_0}^{T}r(y)dy\right]q^i(x(T))\right\},\tag{1.4}$$

subject to (1.2).

Let $W^{(t_0)}(t,x)$ denote maximized expected payoff of the stochastic control problem at time t given that the state is x, that is:

$$W^{(t_0)}(t,x)$$
$$=\max_{\substack{u_1(s),u_2(s),\cdots,u_n(s);\\ for\ s\in[t,T]}}E_{t_0}\left\{\right.$$
$$\left.\sum_{j=1}^{n}\int_{t}^{T}g^j[s,x(s),u_1(s),u_2(s),\cdots,u_n(s)]\exp\left[-\int_{t_0}^{s}r(y)dy\right]ds+\sum_{j=1}^{n}\exp\left[-\int_{t_0}^{T}r(y)dy\right]q^i(x(T))\right\}.$$

An optimal solution to the stochastic dynamic programming control problem (1.2) and (1.4) is provided by the theorem below.

Theorem 1.2 A set of controls $\left\{u_i^*(t)=\psi_i^*(t,x),\text{ for } i\in N\right\}$ constitutes an optimal solution to the stochastic control problem (1.2) and (1.4), if there exist continuously twice differentiable functions $W^{(t_0)}(t,x):[t_0,T]\times R_m\to R$, satisfying the following partial differential equation:

$$-W_t^{(t_0)}(t,x) - \frac{1}{2}\sum_{h,\zeta=1}^{m}\Omega^{h\zeta}(t,x)W_{x^h x^\zeta}^{(t_0)}(t,x) =$$
$$\max_{u_1,u_2,\cdots,u_n}\left\{\sum_{j=1}^{n} g^j[t,x,u_1,u_2,\cdots,u_n]\exp\left[-\int_{t_0}^{t} r(y)dy\right]\right.$$
$$\left. + W_x^{(t_0)}(t,x)f[t,x,u_1,u_2,\cdots,u_n]\right\}, \text{ and}$$
$$W^{(t_0)}(T,x) = \sum_{j=1}^{n} q^j(x)\exp\left[-\int_{t_0}^{T} r(y)dy\right]. \tag{1.5}$$

Proof Follow the proof of Theorem A.3 in the Technical Appendices. ■

Hence the players will adopt the cooperative control $\{\psi_i^*(t,x)$, for $i \in N$ and $t \in [t_0,T]\}$ to obtain the maximized level of expected joint profit. Substituting this set of control into (1.1) yields the dynamics of the optimal (cooperative) trajectory as:

$$\begin{aligned} dx(s) = & f\left[s,x(s),\psi_1^*(s,x(s)),\psi_2^*(s,x(s)),\cdots,\psi_n^*(s,x(s))\right] ds \\ & + \sigma[s,x(s)]dz(s), \;\; x(t_0) = x_0 \end{aligned} \tag{1.6}$$

The solution to (1.6) can be expressed as:

$$\begin{aligned} x^*(t) = x_0 & + \int_{t_0}^{t} f\left[s,x^*(s),\psi_1^*\left(s,x^*(s)\right),\psi_2^*\left(s,x^*(s)\right),\cdots,\psi_n^*\left(s,x^*(s)\right)\right] ds \\ & + \int_{t_0}^{t} \sigma\left[s,x^*(s)\right]dz(s) \end{aligned} \tag{1.7}$$

We use X_t^* to denote the set of realizable values of $x^*(t)$ at time t generated by (1.7). The term $x_t^* \in X_t^*$ is used to denote an element in X_t^*. We use the terms $x^*(t)$ and x_t^* interchangeably in case where there is no ambiguity.

The cooperative control for the game (1.2) and (1.4) over the time interval $[t_0,T]$ can be expressed more precisely as

$$\{\psi_i^*\left(t,x_t^*\right), for\, i \in N \,\text{and}\, t \in [t_0,T] \,\text{when}\, x_t^* \in X_t^* \,\text{is realized}\}. \tag{1.8}$$

The expected cooperative payoff over the interval $[t,T]$, for $t \in \left[t_0,T\right)$, can be expressed as:

$$\begin{aligned} & W^{(t_0)}\left(t,x_t^*\right) = E_{t_0}\Bigg\{ \\ & \int_t^T \sum_{j=1}^{n} g^j\left[s,x^*(s),\psi_1^*\left(s,x^*(s)\right),\psi_2^*\left(s,x^*(s)\right),\cdots,\psi_n^*\left(s,x^*(s)\right)\right]\exp\left[-\int_{t_0}^{s} r(y)dy\right] ds \\ & + \exp\left[-\int_{t_0}^{T} r(y)dy\right]\sum_{j=1}^{n} q^j\left(x^*(T)\right)\Bigg|\, x^*(t) = x_t^* \in X_t^* \Bigg\}. \end{aligned} \tag{1.9}$$

To verify whether the player would find it optimal to adopt the cooperative controls (1.8) throughout the cooperative duration, we consider a stochastic control problem with dynamics (1.2) and payoff (1.4) which begins at time $\tau \in [t_0, T]$ with initial state $x_\tau^* \in X_t^*$. At time τ, the optimality principle ensuring group optimality requires the players to solve the problem:

$$\max_{u_1, u_2, \cdots, u_n} E_\tau \left\{ \int_\tau^T \sum_{j=1}^n g^j[s, x(s), u_1(s), u_2(s), \cdots, u_n(s)] \exp\left[-\int_\tau^s r(y)dy\right] ds \right.$$
$$\left. + \exp\left[-\int_\tau^T r(y)dy\right] \sum_{j=1}^n q^j(x(T)) \right\}, \tag{1.10}$$

subject to

$$dx(s) = f[s, x(s), u_1(s), u_2(s), \cdots, u_n(s)]\, ds + \sigma[s, x(s)]dz(s),\ x(\tau) = x_\tau^* \in X_t^*. \tag{1.11}$$

Note that

$$\max_{u_1, u_2, \cdots, u_n} E_{t_0} \left\{ \int_\tau^T \sum_{j=1}^n g^j[s, x(s), u_1(s), u_2(s), \cdots, u_n(s)] \exp\left[-\int_{t_0}^s r(y)dy\right] ds \right.$$
$$\left. + \exp\left[-\int_{t_0}^T r(y)dy\right] \sum_{j=1}^n q^j(x(T)) \middle|\, x(\tau) = x_\tau^* \in X_\tau^* \right\}$$
$$= \max_{u_1, u_2, \cdots, u_n} E_{t_0} \left\{ \exp\left[-\int_{t_0}^\tau r(y)dy\right] \times \right.$$
$$\left(\int_\tau^T \sum_{j=1}^n g^j[s, x(s), u_1(s), u_2(s), \cdots, u_n(s)] \exp\left[-\int_\tau^s r(y)dy\right] ds \right.$$
$$\left.\left. + \exp\left[-\int_\tau^T r(y)dy\right] \sum_{j=1}^n q^j(x(T)) \right) \middle|\, x(\tau) = x_\tau^* \in X_\tau^* \right\}.$$
$$= \exp\left[-\int_{t_0}^\tau r(y)dy\right] \times$$
$$\max_{u_1, u_2, \cdots, u_n} E_\tau \left\{ \right.$$
$$\left(\int_\tau^T \sum_{j=1}^n g^j[s, x(s), u_1(s), u_2(s), \cdots, u_n(s)] \exp\left[-\int_\tau^s r(y)dy\right] ds \right.$$
$$\left.\left. + \exp\left[-\int_\tau^T r(y)dy\right] \sum_{j=1}^n q^j(x(T)) \right) \middle|\, x(\tau) = x_\tau^* \right\}. \tag{1.12}$$

Hence the stochastic optimal controls strategies for problem (1.10) and (1.11) are analogous to the controls strategies for problem (1.2) and (1.4) in the time interval $[t, T]$.

A remark that will be utilized in subsequent analysis is given below.

Remark 1.2 Let $W^{(\tau)}(t,x_t^*)$ denote the expected cooperative payoff of control problem (1.10) and (1.11). One can readily verify that

$$\exp\left[\int_{t_0}^{\tau} r(y)dy\right] W^{(t_0)}\left(t,x_t^*\right) = \exp\left[\int_{t_0}^{\tau} r(y)dy\right] W^{(\tau)}\left(t,x_t^*\right),$$

for $\tau \in [t_0,T]$ and $t \in \left[\tau,T\right)$ and $x_\tau^* \in X_t^*$. ■

Again, we use $\Gamma_c\left(x_t^*, T-t\right)$ to denote the cooperative game with player payoffs (1.1) and dynamics (1.2) which starts at time $t \in \left[t_0,T\right)$ given the state $x(\tau) = x_\tau^* \in X_\tau^*$. Let there exist a solution under the agreed-upon optimality principle, $t_0 \le t \le T$ along the optimal trajectory $\left\{x^*(t)\right\}_{t=t_0}^{T}$. If this condition is not satisfied it is impossible for the players to adhere to the chosen principle of optimality.

For $\xi^{(t)}\left(t,x_t^*\right)$, $t \in [t_0,T]$, to be valid imputations, it is required that both group optimality and individual rationality have to be satisfied. Hence a valid optimality principle $P\left(x_t^*, T-t\right)$ would yield a solution which contains

(i)

$$\sum_{j=1}^{n} \xi^{(t)j}\left(t,x_t^*\right) = W^{(t)}\left(t,x_t^*\right), \text{ for } t \in [t_0,T]; \text{ and}$$

(ii)

$$\xi^{(t)i}\left(t,x_t^*\right) \ge V^{(t)i}\left(t,x_t^*\right), \text{ for } i \in N \text{ and } t \in [t_0,T].$$

3.2 Cooperative Subgame Consistency Under Uncertainty

In this Section we examine the properties of subgame consistency in cooperative stochastic differential games.

3.2.1 Principle of Subgame Consistency

In a stochastic environment, the condition of subgame consistency requires the optimality principle agreed upon at the outset to remain effective in a subgame with a later starting time and any realizable state brought about by prior optimal behavior. Assume that at the start of the game the players execute the solution under an agreed-upon optimality principle (which includes a set of cooperative strategies, an imputation to distribute the cooperative payoff and a payoff distribution procedure). When the game proceeds to time t and the state becomes $x_t^* \in X_t^*$, the continuation of the scheme for the game $\Gamma_c(x_0, T - t_0)$ has to be consistent with the solution to the game $\Gamma_c(x_t^*, T - t)$ under the same optimality principle. If this consistency condition is violated, some of the players will have an incentive to deviate from the initial agreement and instability arises.

To verify whether the solution is indeed subgame consistent, one has to verify whether the agreed upon cooperative strategies, payoff distribution procedures and imputations are all subgame consistent. Using Remark 1.2, one can show that joint expected payoff maximizing strategies are subgame consistent. In the next subsection, subgame consistent imputation and payoff distribution procedure are examined.

3.2.2 Subgame-Consistency in Imputation and Payoff Distribution Procedure

In this Section, we consider subgame consistency in imputation and payoff distribution procedure. In the cooperative game $\Gamma_c(x_0, T - t_0)$ according to the solution generated by the agreed-upon optimality principle, the players would use the payoff distribution procedure $\{B^{t_0}(s, x_s^*)\}_{s=t_0}^{T}$ to bring about an imputation to player i as:

$$\xi^{(t_0)i}(t_0, x_0) = E_{t_0}\left\{ \int_{t_0}^{T} B_i^{t_0}(s, x_s^*) \exp\left[-\int_{t_0}^{s} r(y)dy\right] ds + q^i(x^*(T)) \exp\left[-\int_{t_0}^{T} r(y)dy\right] \right\}, \tag{2.1}$$

for $i \in N$.

When the game proceeds to time $t \in (t_0, T]$, the current state is $x_t^* \in X_t^*$. According to the solution of the game $\Gamma_c(x_0, T - t_0)$ generated by the agreed-upon optimality principle player i will receive an imputation (in present value viewed at time t_0) equaling

$$\xi^{(t_0)i}(t, x_t^*) = E_{t_0}\left\{ \int_{t}^{T} B_i^{t_0}(s, x_s^*) \exp\left[-\int_{t_0}^{s} r(y)dy\right] ds + q^i(x^*(T)) \exp\left[-\int_{t_0}^{T} r(y)dy\right] \middle| x(t) = x_t^* \right\}, \tag{2.2}$$

over the time interval $[t, T]$.

Note that at time $t \in (t_0, T]$ when the current state is $x_t^* \in X_t^*$, we have a cooperative game $\Gamma_c(x_t^*, T-t)$. According to the solution generated by the same optimality principle, the players would use the payoff distribution procedure $\{B^t(s, x_s^*)\}_{s=t}^{T}$ to bring about an imputation to player i as:

$$\xi^{(t)i}(t, x_t^*) = E_t\left\{ \int_t^T B_i^t(s, x_s^*) \exp\left[-\int_t^s r(y)dy \right] ds \right.$$
$$\left. + q^i(x^*(T)) \exp\left[-\int_t^T r(y)dy \right] \right\}, \text{ for } i \in N. \tag{2.3}$$

For the imputation and payoff distribution procedure of the game $\Gamma_c(x_0, T-t_0)$ to be consistent with those of the game $\Gamma_c(x_t^c, T-t)$ under the same agreed-upon optimality principle, it is essential that

$$\exp\left[\int_{t_0}^t r(y)dy \right] \xi^{(t_0)}(t, x_t^*) = \xi^{(t)}(t, x_t^*), \text{ for } t \in [t_0, T].$$

In addition, the payoff distribution procedure of the game $\Gamma_c(x_0, T-t_0)$ generated by the agreed upon optimality principle is

$$B^{t_0}(s, x_s^*) = \left[B_1^{t_0}(s, x_s^*), B_2^{t_0}(s, x_s^*), \cdots, B_n^{t_0}(s, x_s^*) \right], \text{ for } s \in [t_0, T].$$

Consider the case when the game has proceeded to time t and the state variable became $x_t^* \in X_t^*$. Then one has a cooperative game $\Gamma_c(x_t^*, T-t)$ which starts at time t with initial state x_t^*. According to the same optimality principle, the payoff distribution procedure

$$B^t(s, x_s^*) = \left[B_1^t(s, x_s^*), B_2^t(s, x_s^*), \cdots, B_n^t(s, x_s^*) \right], \text{ for } s \in [t, T],$$

will be adopted.

For the continuation of the payoff distribution procedure $B^{t_0}(s, x_s^*)$ of the game $\Gamma_c(x_0, T-t_0)$ to be consistent with $B^t(s, x_s^*)$ of the game $\Gamma_c(x_t^*, T-t)$, it is required that

$$B^{t_0}(s, x_s^*) = B^t(s, x_s^*), \text{ for } s \in [t, T] \text{ and } t \in [t_0, T].$$

Therefore a formal definition can be presented as below.

Definition 2.1 The imputation and payoff distribution procedure $\{\xi^{(t_0)}(t_0, x_0)$ and $B^{t_0}(s, x_s^*)$ for $s \in [t_0, T]\}$ under the agreed-upon optimality principle are subgame consistent if

(i)
$$\exp\left[\int_{t_0}^{t} r(y)dy\right] \xi^{(t_0)i}(t, x_t^*)$$
$$\equiv \exp\left[\int_{t_0}^{t} r(y)dy\right] E\left\{ \int_t^T B_i^{t_0}(s, x_s^*) \exp\left[-\int_t^s r(y)dy\right] ds \right.$$
$$\left. + q^i(x^*(T)) \exp\left[-\int_{t_0}^{T} r(y)dy\right] \middle| x(t) = x_t^* \right\} = \xi^{(t)i}(t, x_t^*) \equiv$$
$$E_t\left\{ \int_t^T B_i^t(s, x_s^*) \exp\left[-\int_t^s r(y)dy\right] ds + q^i(x^*(T)) \exp\left[-\int_t^T r(y)dy\right] \right\}$$

under $P(x_t^*, T-t)$, for $i \in N$ and $t \in [t_0, T]$; and

(ii) the payoff distribution procedure $B^{t_0}(s, x_s^*) = [B_1^{t_0}(s, x_s^*), B_2^{t_0}(s, x_s^*), \cdots, B_n^{t_0}(s, x_s^*)]$ for $s \in [t, T]$ is identical to $B^t(s, x_s^*) = [B_1^t(s, x_s^*), B_2^t(s, x_s^*), \cdots, B_n^t(s, x_s^*)]$ of the game $\Gamma_c(x_t^*, T-t)$. ■

3.3 Subgame Consistent Payoff Distribution Procedure

Crucial to obtaining a subgame consistent solution is the derivation of a payoff distribution procedure satisfying Definition 2.1 in Sect. 3.2. Invoking part (ii) of Definition 2.1, we have $B^{t_0}(s, x_s^*) = B^t(s, x_s^*)$ for $t \in [t_0, T]$ and $s \in [t, T]$. We use $B(s, x_s^*) = \{B_1(s, x_s^*), B_2(s, x_s^*), \cdots, B_n(s, x_s^*)\}$ to denote $B^t(s, x_s^*)$ for all $t \in [t_0, T]$. Along the optimal trajectory $\{x^*(s)\}_{s=t_0}^{T}$ we then have:

$$\xi^{(\tau)i}(\tau, x_\tau^*) = E_\tau\left\{ \int_\tau^T B_i^\tau(s, x^*(s)) \exp\left[-\int_\tau^s r(y)dy\right] ds \right.$$
$$\left. + q^i(x_T^*) \exp\left[-\int_\tau^T r(y)dy\right] \middle| x^*(\tau) = x_\tau^* \in X_\tau^* \right\}, \tag{3.1}$$

for $i \in N$ and $\tau \in [t_0, T]$.

Moreover, for $t \in [\tau, T]$, we use the term

$$\xi^{(\tau)i}(t, x_t^*) = E_\tau\left\{ \int_t^T B_i^\tau(s, x^*(s)) \exp\left[-\int_\tau^s r(y)dy\right] ds \right.$$
$$\left. + q^i(x_T^*) \exp\left[-\int_\tau^T r(y)dy\right] \middle| x^*(t) = x_t^* \in X_t^* \right\}, \tag{3.2}$$

to denote the expected present value (with initial time being τ) of player i's expected payoff under cooperation over the time interval $[t, T]$ according to the optimality principle $P(x_\tau^*, T-\tau)$ along the cooperative state trajectory.

Invoking (3.1) and (3.2) we have

$$\xi^{(\tau)i}\left(t,x_t^c\right)=\exp\left[-\int_\tau^t r(y)dy\right]\xi^{(t)i}\left(t,x_t^*\right),$$
$$\text{for } i\in N \text{ and } \tau\in[t_0,T] \text{ and } t\in[\tau,T] \tag{3.3}$$

One can readily verify that a payoff distribution procedure $\left\{B\left(s,x_s^*\right)\right\}_{s=t_0}^T$ which satisfies (3.3) would give rise to time-consistent imputations satisfying part (i) of Definition 2.1. The next task is the derivation of a payoff distribution procedure $\left\{B\left(s,x_s^*\right)\right\}_{s=t_0}^T$ that leads to the realization of (3.1), (3.2), and (3.3).

We first consider the following condition concerning the imputation $\xi^{(\tau)}(t,x_t^*)$, for $\tau\in[t_0,T]$ and $t\in[\tau,T]$.

Condition 3.1 For $i\in N$ and $t\in[\tau,T]$ and $\tau\in[t_0,T]$, the imputation $\xi^{(\tau)i}(t,x_t^*)$, for $i\in N$, is a function that is twice continuously differentiable in t and $x_t^*\in X_t^*$. ■

A theorem characterizing a formula for $B_i(s,x_s^*)$, for $s\in[t_0,T]$, $x_s^*\in X_s^*$ and $i\in N$, which yields (3.1), (3.2), and (3.3) can be provided as follows.

Theorem 3.1 If Condition 3.1 is satisfied, a PDP with a terminal payment $q^i(x_T^*)$ at time T and an instantaneous payment at time $s\in[\tau,T]$:

$$\begin{aligned}B_i\left(s,x_s^*\right)=&-\left[\xi_t^{(s)i}\left(t,x_t^*\right)\Big|_{t=s}\right]\\&-\left[\xi_{x_t^*}^{(s)i}\left(t,x_t^*\right)\Big|_{t=s}\right]f\left[s,x_s^*,\psi_1^*\left(s,x_s^*\right),\psi_2^*\left(s,x_s^*\right),\cdots,\psi_n^*\left(s,x_s^*\right)\right]\\&-\frac{1}{2}\sum_{h,\zeta=1}^m\Omega^{h\zeta}\left(s,x_s^*\right)\left[\xi_{x_t^hx_t^\zeta}^{(s)i}\left(t,x_t^*\right)\Big|_{t=s}\right],\ \text{for } i\in N \text{ and } x_s^*\in X_s^*,\end{aligned}\tag{3.4}$$

yields imputation vector $\xi^{(\tau)}(\tau,x_\tau^*)$, for $\tau\in[t_0,T]$ which satisfy (3.1), (3.2), and (3.3).

Proof Invoking (3.1), (3.2) and (3.3), one can obtain

$$\begin{aligned}\xi^{(v)i}\left(v,x_v^*\right)=E_v\Bigg\{&\int_v^{v+\Delta t}B_i\left(s,x_s^*\right)\exp\left[-\int_v^s r(y)dy\right]ds+\\ \exp&\left[-\int_v^{v+\Delta t}r(y)dy\right]\xi^{(v+\Delta t)i}\left(v+\Delta t,x_v^*+\Delta x_v^*\right)\Bigg|\,x(v)=x_v^*\in X_v^*\Bigg\},\\ \text{for } &v\in[\tau,T] \text{ and } i\in N;\end{aligned}\tag{3.5}$$

where

$\Delta x_v^*=f\left[v,x_v^c,\psi_1^*\left(v,x_v^*\right),\psi_2^*\left(v,x_v^*\right),\cdots,\psi_n^*\left(v,x_v^*\right)\right]\Delta t+\sigma\left[v,x_v^*\right]\Delta z_v+o(\Delta t)$,
$\Delta z_v=Z(v+\Delta t)-z(v)$, and $E_v[o(\Delta t)]/\Delta t\rightarrow 0$ as $\Delta t\rightarrow 0$.

From (3.2) and (3.5), one obtains

$$E_\upsilon\left\{ \int_\upsilon^{\upsilon+\Delta t} B_i(s, x_s^*) \exp\left[-\int_\upsilon^s r(y)dy\right] ds \middle| x(\upsilon) = x_\upsilon^* \right\}$$
$$= E_\upsilon\left\{ \xi^{(\upsilon)i}(\upsilon, x_\upsilon^*) - \exp\left[-\int_\upsilon^{\upsilon+\Delta t} r(y)dy\right] \xi^{(\upsilon+\Delta t)i}(\upsilon + \Delta t, x_\upsilon^* + \Delta x_\upsilon^*) \right\}$$
$$= E_\upsilon\left\{ \xi^{(\upsilon)i}(\upsilon, x_\upsilon^*) - \xi^{(\upsilon)i}(\upsilon + \Delta t, x_\upsilon^* + \Delta x_\upsilon^*) \right\},$$
$$\text{for all } \upsilon \in [t_0, T] \text{ and } i \in N. \tag{3.6}$$

If the imputations $\xi^{(\upsilon)}(t, x_t^*)$, for $\upsilon \in [t_0, T]$, satisfy Condition 3.1, as $\Delta t \to 0$, one can express condition (3.6) as:

$$E_\upsilon\left\{ B_i(\upsilon, x_\upsilon^*)\Delta t + o(\Delta t) \right\} = E_\upsilon\left\{ -\left[\xi_t^{(\upsilon)i}(t, x_t^c)\Big|_{t=\upsilon}\right] \Delta t \right.$$
$$- \left[\xi_{x_\upsilon^c}^{(\upsilon)i}(\upsilon, x_\upsilon^c)\right] f\left[\upsilon, x_\upsilon^c, \psi_1^*(\upsilon, x_\upsilon^c), \psi_2^*(\upsilon, x_\upsilon^c), \cdots, \psi_n^*(\upsilon, x_\upsilon^c)\right] \Delta t$$
$$- \frac{1}{2} \sum_{h,\zeta=1}^m \Omega^{h\zeta}(\upsilon, x_\upsilon^*) \left[\xi_{x_t^h x_t^\zeta}^{(\upsilon)i}(t, x_t^*)\Big|_{t=\upsilon}\right] - \left[\xi_{x_\upsilon^c}^{(\upsilon)i}(\upsilon, x_\upsilon^c)\right] \sigma[\upsilon, x_\upsilon^*] \Delta z_\upsilon$$
$$\left. - o(\Delta t). \right\} \tag{3.7}$$

Dividing (3.7) throughout by Δt, with $\Delta t \to 0$, and taking expectation yield (3.4). Thus the payoff distribution procedure in $B_i(s, x_s^*)$ in (3.4) would lead to the realization of $\xi^{(\tau)i}(\tau, x_\tau^c)$, for $\tau \in [t_0, T]$ which satisfy (3.1)–(3.3). ■

Assigning the instantaneous payments according to the payoff distribution procedure in (3.4) leads to the realization of the imputation $\xi^{(\tau)}(\tau, x_\tau^*) \in P(x_\tau^*, T - \tau)$ for $\tau \in [t_0, T]$ and $x_\tau^* \in X_\tau^*$.

With players using the cooperative strategies $\{\psi_i^*(\tau, x_\tau^*)$, for $\tau \in [t_0, T]$ and $i \in N\}$, the instantaneous payment received by player i at time instant τ is:

$$\zeta_i(\tau, x_\tau^*) = g^i\left[\tau, x_\tau^*, \psi_1^*(\tau, x_\tau^*), \psi_2^*(\tau, x_\tau^*), \cdots, \psi_n^*(\tau, x_\tau^*))\right],$$
$$\text{for } \tau \in [t_0, T],\ x_\tau^* \in X_\tau^* \text{ and } i \in N. \tag{3.8}$$

According to Theorem 3.1, the instantaneous payment that player i should receive under the agreed-upon optimality principle is $B_i(\tau, x_\tau^*)$ as stated in (3.4). Hence an instantaneous transfer payment

$$\chi^i(\tau, x_\tau^*) = B_i(\tau, x_\tau^*) - \zeta_i(\tau, x_\tau^*) \tag{3.9}$$

has to be given to player i at time τ, for $i \in N$ and $\tau \in [t_0, T]$ when the state is $x_\tau^* \in X_\tau^*$.

3.4 An Illustration in Cooperative Fishery Under Uncertainty

Consider the stochastic resource extraction game with two asymmetric extractors.

The resource stock $x(s) \in X \subset R$ follows the stochastic dynamics:

$$dx(s) = \left[ax(s)^{1/2} - bx(s) - u_1(s) - u_2(s)\right]ds + \sigma x(s)dz(s), \quad x(t_0) = x_0 \in X \tag{4.1}$$

where $u_i(s)$ is the harvest rate of extractor $i \in \{1, 2\}$. The instantaneous payoffs at time $s \in [t_0, T]$ for extractor 1 and extractor 2 are, respectively, $\left[u_1(s)^{1/2} - \frac{c_1}{x(s)^{1/2}} u_1(s)\right]$ and $\left[u_2(s)^{1/2} - \frac{c_2}{x(s)^{1/2}} u_2(s)\right]$, where c_1 and c_2 are constants and $c_1 \neq c_2$. At time T, each extractor will receive a termination bonus $qx(T)^{1/2}$. Payoffs are transferable between extractors and over time. Given the constant discount rate r, values received at time t are discounted by the factor $\exp[-r(t - t_0)]$.

At time t_0, the expected payoff of extractor i is:

$$E_{t_0}\left\{ \int_{t_0}^{T} \left[u_i(s)^{1/2} - \frac{c_i}{x(s)^{1/2}} u_i(s)\right] \exp[-r(t - t_0)]\, ds \right.$$
$$\left. + \exp[-r(T - t_0)] qx(T)^{\frac{1}{2}} \right\}, \text{ for } i \in \{1, 2\}. \tag{4.2}$$

Let $[\phi_1^*(t, x), \phi_2^*(t, x)]$ for $t \in [t_0, T]$ denote a set of strategies that provides a feedback Nash equilibrium solution to the game (4.1) and (4.2), and $V^{(t_0)i}(t, x) : [t_0, T] \times R^n \to R$ denote the feedback Nash equilibrium payoff of extractor $i \in \{1, 2\}$ that satisfies the equations:

$$-V_t^{(t_0)i}(t, x) - \frac{1}{2}\sigma^2 x^2 V_{xx}^{(t_0)i}(t, x)$$
$$= \max_{u_i} \left\{ \left[(u_i)^{1/2} - \frac{c_i}{x^{1/2}} u_i\right] \exp[-r(t - t_0)] \right.$$
$$\left. + V_x^{(t_0)i}(t, x)\left[ax^{1/2} - bx - u_i - \phi_j^{(t_0)*}(t, x)\right] \right\}, \text{ and}$$
$$V^{(t_0)i}(T, x) = \exp[-r(T - t_0)] qx(T)^{\frac{1}{2}}, \text{ for } i \in \{1, 2\} \text{ and } j \in \{1, 2\} \text{ and } j \neq i. \tag{4.3}$$

Performing the indicated maximization in (4.3) yields:

$$\phi_i^*(t,x) = \frac{x}{4\left[c_i + V_x^{(t_0)i}\exp[r(t-t_0)]x^{1/2}\right]^2}, \text{ for } i \in \{1,2\}.$$

To completely characterize a feedback solution, we derive the feedback Nash equilibrium payoffs of the extractors in the game (4.1) and (4.2) as:

Proposition 4.1 The feedback Nash equilibrium payoff of extractor $i \in \{1,2\}$ in the game (4.1) and (4.2) is:

$$V^{(t_0)i}(t,x) = \exp[-r(t-t_0)]\left[A_i(t)x^{1/2} + C_i(t)\right], \tag{4.4}$$

where for $i,j \in \{1,2\}$ and $i \neq j, A_i(t), B_i(t), A_j(t)$ and $B_j(t)$ satisfy:

$$\begin{aligned}
\dot{A}_i(t) &= \left[r + \frac{1}{8}\sigma^2 + \frac{b}{2}\right]A_i(t) - \frac{1}{2[c_i + A_i(t)/2]} + \frac{c_i}{4[c_i + A_i(t)/2]^2} \\
&\quad + \frac{A_i(t)}{8[c_i + A_i(t)/2]^2} + \frac{A_i(t)}{8\left[c_j + A_j(t)/2\right]^2}. \\
\dot{C}_i(t) &= rC_i(t) - \frac{a}{2}A_i(t), \\
A_i(T) &= q \text{ and } C_i(T) = 0
\end{aligned}$$

Proof First substitute $\phi_1^*(t,x)$ and $\phi_2^*(t,x), V^{(t_0)i}(t,x)$ from (4.4) and the corresponding derivatives $V_t^{(t_0)i}(t,x), V_x^{(t_0)i}(t,x)$ and $V_{xx}^{(t_0)i}(t,x)$ into (4.3). Upon solving (4.3) one obtains Proposition 4.1. ■

Invoking Remark 4.1 in Chap. 2, we can obtain the feedback Nash equilibrium payoff of player i in the game with dynamics (4.1) and expected payoffs (4.2) which starts at time τ for $\tau \in [t_0, T)$ as:

$$V^{(\tau)i}(t,x) = \exp[-r(t-\tau)]\left[A_i(t)x^{1/2} + B_i(t)\right], \text{ for } i \in \{1,2\}.$$

3.4.1 Cooperative Extraction Under Uncertainty

Now consider the case when the resource extractors agree to act cooperatively and follow the optimality principle under which they would

(i) maximize their joint expected payoffs and
(ii) share the excess of the total expected cooperative payoff over the sum of expected individual noncooperative payoffs proportional to the extractors' expected noncooperative payoffs.

Hence the extractors maximize the sum of their expected profits:

$$E_{t_0}\Bigg\{ \int_{t_0}^{T}\left(\left[u_1(s)^{1/2}-\frac{c_1}{x(s)^{1/2}}u_1(s)\right]+\left[u_2(s)^{1/2}-\frac{c_2}{x(s)^{1/2}}u_2(s)\right]\right)\exp[-r(t-t_0)]\,ds + 2\exp[-r(T-t_0)]qx(T)^{\frac{1}{2}}\Bigg\}, \tag{4.5}$$

subject to the stochastic dynamics (4.1).

Invoking Theorem A.3 in the Technical Appendices yields the characterization of solution of the problem (4.1) and (4.5) as a set of controls $\{u_i^*(t)=\psi_i^*(t,x)$, for $i\in\{1,2\}\}$ which satisfies the following partial differential equation:

$$\begin{aligned}
&-W_t^{(t_0)}(t,x)-\frac{1}{2}\sigma^2x^2W_{xx}^{(t_0)}(t,x)\\
&\quad=\max_{u_1,u_2}\Bigg\{\left(\left[u_1{}^{1/2}-\frac{c_1}{x^{1/2}}u_1\right]+\left[u_2{}^{1/2}-\frac{c_2}{x^{1/2}}u_2\right]\right)\exp[-r(t-t_0)]\\
&\qquad+W_x^{(t_0)}(t,x)\left[ax^{1/2}-bx-u_1-u_2\right]\Bigg\},\text{ and}\\
&W^{(t_0)}(T,x)=2\exp[-r(T-t_0)]qx^{\frac{1}{2}}.
\end{aligned} \tag{4.6}$$

Performing the indicated maximization we obtain:

$$\begin{aligned}
\psi_1^{(t_0)*}(t,x)&=\frac{x}{4\left[c_1+W_x^{(t_0)}\exp[r(t-t_0)]x^{1/2}\right]^2},\text{ and}\\
\psi_2^{(t_0)*}(t,x)&=\frac{x}{4\left[c_2+W_x^{(t_0)}\exp[r(t-t_0)]x^{1/2}\right]^2}.
\end{aligned} \tag{4.7}$$

The maximized expected joint profit of the extractors can be obtained as:

Proposition 4.2

$$W^{(t_0)}(t,x)=\exp[-r(t-t_0)]\left[A(t)x^{1/2}+C(t)\right], \tag{4.8}$$

where

$$\begin{aligned}
\dot{A}(t)&=\left[r+\frac{\sigma^2}{8}+\frac{b}{2}\right]A(t)-\frac{1}{2[c_1+A(t)/2]}-\frac{1}{2[c_2+A(t)/2]}+\frac{c_1}{4[c_1+A(t)/2]^2}\\
&\quad+\frac{c_2}{4[c_2+A(t)/2]^2}+\frac{A(t)}{8[c_1+A(t)/2]^2}+\frac{A(t)}{8[c_2+A(t)/2]^2},\\
\dot{C}(t)&=rC(t)-\frac{a}{2}A(t),\ A(T)=2q,\text{ and }C(T)=0.
\end{aligned}$$

Proof Upon substituting the optimal strategies in (4.7), $W^{(t_0)}(t,x)$ in (4.8), and the relevant derivatives $W_t^{(t_0)}(t,x), W_x^{(t_0)}(t,x)$ and $W_{xx}^{(t_0)}(t,x)$ into (4.6) yields the results in Proposition 4.2. ■

The optimal cooperative controls can then be obtained as:

$$\psi_1^*(t,x) = \frac{x}{4[c_1 + A(t)/2]^2} \text{ and } \psi_2^*(t,x) = \frac{x}{4[c_2 + A(t)/2]^2}. \tag{4.9}$$

Substituting these control strategies into (4.1) yields the dynamics of the state trajectory under cooperation:

$$dx(s) = \left[ax(s)^{1/2} - bx(s) - \frac{x(s)}{4[c_1 + A(s)/2]^2} - \frac{x(s)}{4[c_2 + A(s)/2]^2} \right] ds$$
$$+ \sigma x(s)dz(s),\ x(t_0) = x_0. \tag{4.10}$$

Solving (4.11) yields the optimal cooperative state trajectory as:

$$x^*(s) = \varpi(t_0,s)^2 \left[x_0^{1/2} + \int_{t_0}^{s} \varpi^{-1}(t_0,t) H_1 dt \right]^2, \text{ for } s \in [t_0, T] \tag{4.11}$$

Where $\varpi(t_0,s) = \exp\left[\int_{t_0}^{s} \left[H_2(\tau) - \frac{\sigma^2}{8}\right] d\upsilon + \int_{t_0}^{s} \frac{\sigma}{2} dz(\upsilon)\right], H_1 = \frac{1}{2}a,$

and $H_2(s) = -\left[\frac{1}{2}b + \frac{1}{8[c_1+A(s)/2]^2} + \frac{1}{8[c_2+A(s)/2]^2} + \frac{\sigma^2}{8}\right].$

The cooperative control for the game $\Gamma_c(x_0, T - t_0)$ over the time interval $[t_0, T]$ along the optimal trajectory can be expressed as:

$$\psi_1^*(t,x_t^*) = \frac{x_t^*}{4[c_1 + A(t)/2]^2}, \text{ and } \psi_2^*(t,x_t^*) = \frac{x_t^*}{4[c_2 + A(t)/2]^2},$$
$$\text{for } t \in [t_0, T] \text{ and } x_t^* \in X_t^*. \tag{4.12}$$

3.4.2 Subgame Consistent Cooperative Extraction

The agreed-upon optimality principle requires the extractors to share the excess of the total expected cooperative payoff over the sum of individual noncooperative payoffs proportional to the extractors' expected noncooperative payoffs. Therefore the following imputation has to be satisfied.

Condition 4.1 An imputation

$$\xi^{(\tau)i}(\tau, x_\tau^*) = \frac{V^{(\tau)i}(\tau, x_\tau^*)}{\sum_{j=1}^{2} V^{(\tau)j}(\tau, x_\tau^*)} W^{(\tau)}(\tau, x_\tau^*)$$

$$= \frac{\left[A_i(\tau)(x_\tau^*)^{1/2} + C_i(\tau)\right]}{\sum_{j=1}^{2}\left[A_j(\tau)(x_\tau^*)^{1/2} + C_j(\tau)\right]} \left[A(\tau)(x_\tau^*)^{1/2} + C(\tau)\right] \tag{4.13}$$

is assigned to extractor i, for $i \in \{1,2\}$ if $x_\tau^* \in X_\tau^*$ occurs at time $\tau \in [t_0, T]$. ■

Applying Theorem 3.1 a subgame-consistent solution under the optimal principle $P(x_0, T - t_0)$ for the cooperative game $\Gamma_c(x_0, T - t_0)$ can be obtained as: $\{\ u(s, x_s^*)$ and $B(s, x_s^*)$ for $s \in [t_0, T]$ and $\xi^{(t_0)}(t_0, x_0)\}$ in which

(i) $u(s, x_s^*)$ for $s \in [t_0, T]$ is the set of group optimal strategies $\psi_1^*(s, x_s^*) = \frac{x_s^*}{4[c_1 + A(s)/2]^2}$, and $\psi_2^*(s, x_s^*) = \frac{x_s^*}{4[c_2 + A(s)/2]^2}$; and

(ii) $B(s, x_s^*) = \{B_1(s, x_s^*), B_2(s, x_2^*)\}$ for $s \in [t_0, T]$ where

$$B_i(s, x_s^*) = -\left[\xi_t^{(s)i}(t, x_t^*)\Big|_{t=s}\right]$$
$$-\left[\xi_{x_s^*}^{(s)i}(s, x_s^*)\right]\left[a(x_s^*)^{1/2} - bx_s^* - \frac{x_s^*}{4[c_1 + A(s)/2]^2} - \frac{x_s^*}{4[c_2 + A(s)/2]^2}\right]$$
$$-\frac{1}{2}\sigma^2(x_s^*)^2\left[\xi_{x_s^* x_s^*}^{(s)i}(s, x_s^*)\right], \text{ for } i \in \{1,2\} \tag{4.14}$$

where

$$\left[\xi_t^{(s)i}(t, x_t^*)\Big|_{t=s}\right]$$
$$= \frac{\left[A_i(s)(x_s^*)^{1/2} + C_i(s)\right]}{\left(\sum_{j=1}^{2}\left[A_j(s)(x_s^*)^{1/2} + C_j(s)\right]\right)} \left\{\left[\dot{A}(s)(x_s^*)^{1/2} + \dot{C}(s)\right] - r\left[A(s)(x_s^*)^{1/2} + C(s)\right]\right\}$$
$$+ \frac{\left[A(s)(x_s^*)^{1/2} + B(s)\right]}{\left(\sum_{j=1}^{2}\left[A_j(s)(x_s^*)^{1/2} + B_j(s)\right]\right)}$$
$$\left\{\left[\dot{A}_i(s)(x_s^*)^{1/2} + \dot{B}_i(s)\right] - r\left[A_i(s)(x_s^*)^{1/2} + B_i(s)\right]\right\}$$

$$
-\frac{\left[A_i(s)\left(x_s^*\right)^{1/2}+B_i(s)\right]\left[A(s)\left(x_s^*\right)^{1/2}+B(s)\right]}{\left(\sum_{j=1}^{2}\left[A_j(s)\left(x_s^*\right)^{1/2}+B_j(s)\right]\right)^2}
$$

$$
\times\sum_{j=1}^{2}\left\{\left[\dot{A}_j(s)\left(x_s^*\right)^{1/2}+\dot{C}_j(s)\right]-r\left[A_j(s)\left(x_s^*\right)^{1/2}+C_j(s)\right]\right\}
$$

$$
\left[\xi_{x_s^*}^{(s)i}\left(s,x_s^*\right)\right]=
$$

$$
\frac{\left[A_i(s)\left(x_s^*\right)^{1/2}+C_i(s)\right]A(s)\left(x_s^*\right)^{-1/2}+\left[A(s)\left(x_s^*\right)^{1/2}+C(s)\right]A_i(s)\left(x_s^*\right)^{-1/2}}{2\sum_{j=1}^{2}\left[A_j(s)\left(x_s^*\right)^{1/2}+C_j(s)\right]}
$$

$$
-\frac{\left[A_i(s)\left(x_s^*\right)^{1/2}+C_i(s)\right]\left[A(s)\left(x_s^*\right)^{1/2}+C(s)\right]}{\left(\sum_{j=1}^{2}\left[A_j(s)\left(x_s^*\right)^{1/2}+C_j(s)\right]\right)^2}\left(\frac{1}{2}\sum_{j=1}^{2}A_j(s)\left(x_s^*\right)^{-1/2}\right);
$$

and

$$
\left[\xi_{x_s^*x_s^*}^{(s)i}\left(s,x_s^*\right)\right]=-\frac{C_i(s)A(s)\left(x_s^*\right)^{-3/2}+C(s)A_i(s)\left(x_s^*\right)^{-3/2}}{4\sum_{j=1}^{2}\left[A_j(s)\left(x_s^*\right)^{1/2}+C_j(s)\right]}
$$

$$
-\frac{\left[A_i(s)\left(x_s^*\right)^{1/2}+C_i(s)\right]A(s)\left(x_s^*\right)^{-1/2}+\left[A(s)\left(x_s^*\right)^{1/2}+C(s)\right]A_i(s)\left(x_s^*\right)^{-1/2}}{\left(2\sum_{j=1}^{2}\left[A_j(s)\left(x_s^*\right)^{1/2}+C_j(s)\right]\right)^2}
$$

$$
\times\sum_{j=1}^{2}\left[A_j(s)\left(x_s^*\right)^{-1/2}\right]
$$

$$
+\frac{\left[A_i(s)\left(x_s^*\right)^{1/2}+C_i(s)\right]\left[A(s)\left(x_s^*\right)^{1/2}+C(s)\right]}{\left(\sum_{j=1}^{2}\left[A_j(s)\left(x_s^*\right)^{1/2}+C_j(s)\right]\right)^2}\left(\frac{1}{4}\sum_{j=1}^{2}A_j(s)\left(x_s^*\right)^{-3/2}\right)
$$

$$
-\left(\frac{1}{2}\sum_{j=1}^{2}A_j(s)\left(x_s^*\right)^{-1/2}\right)
$$

$$
\times\left[\frac{A_i(s)A(s)+\frac{1}{2}\left[A_i(s)C(s)+A(s)C_i(\tau)\right]\left(x_s^*\right)^{-1/2}}{\left(\sum_{j=1}^{2}\left[A_j(s)\left(x_s^*\right)^{1/2}+C_j(s)\right]\right)^2}\right.
$$

$$
\left.-\frac{\left[A_i(s)\left(x_s^*\right)^{1/2}+C_i(s)\right]\left[A(s)\left(x_s^*\right)^{1/2}+C(s)\right]}{\left(\sum_{j=1}^{2}\left[A_j(s)\left(x_s^*\right)^{1/2}+C_j(s)\right]\right)^3}\sum_{j=1}^{2}A_j(s)\left(x_s^*\right)^{-1/2}\right].
$$

With extractors using the cooperative strategies in (4.13), the instantaneous receipt of extractor i at time instant τ is:

$$\zeta_i(\tau, x_\tau^*) = \frac{(x_\tau^*)^{1/2}}{2[c_i + A(\tau)/2]} - \frac{c_i (x_\tau^*)^{1/2}}{4[c_i + A(\tau)/2]^2},$$
$$\text{for } \tau \in [t_0, T], x_\tau^* \in X_\tau^* \text{ and } i \in \{1, 2\}. \quad (4.15)$$

Under cooperation the instantaneous payment that extractor $i \in \{1, 2\}$ should receive $B_i(\tau, x_\tau^*)$ in (4.15). Hence an instantaneous transfer payment

$$\chi^i(\tau, x_\tau^*) = B_i(\tau, x_\tau^*) - \zeta_i(\tau, x_\tau^*) \quad (4.16)$$

has to be given to extractor i at time τ, for $i \in \{1, 2\}$ and $\tau \in [t_0, T]$ when the state is $x_\tau^* \in X_\tau^*$.

3.5 Infinite Horizon Subgame Consistency Under Uncertainty

Consider the infinite stochastic differential game in which player i seeks to

$$\max_{u_i} E_\tau \left\{ \int_\tau^\infty g^i[x(s), u_1(s), u_2(s), \cdots, u_n(s)] \exp[-r(s - \tau)]\, ds \right\},$$
$$\text{for } i \in N, \quad (5.1)$$

subject to the stochastic dynamics

$$dx(s) = f[x(s), u_1(s), u_2(s), \cdots, u_n(s)]ds + \sigma[x(s)]dz(s), \qquad x(\tau) = x_\tau. \quad (5.2)$$

Consider the alternative game which starts at time $t \in [t_0, \infty)$ with initial state $x(t) = x$:

$$\max_{u_i} E_t \left\{ \int_t^\infty g^i[x(s), u_1(s), u_2(s), \cdots, u_n(s)] \exp[-r(s - t)] ds \right\},$$
$$\text{for } i \in N, \quad (5.3)$$

subject to the stochastic dynamics

$$dx(s) = f[x(s), u_1(s), u_2(s), \cdots, u_n(s)]ds + \sigma[x(s)]dz(s), \qquad x(t) = x_t. \quad (5.4)$$

Let $\Omega[x(s)] = \sigma[x(s)]\sigma[x(s)]^T$ denote the covariance matrix with its element in row h and column ζ denoted by $\Omega^{h\zeta}[x(s)]$.

The infinite horizon autonomous game (5.4) and (5.5) is independent of the choice of t and dependent only upon the state at the starting time, that is x. A Nash equilibrium solution for the infinite-horizon stochastic differential game (5.4) and (5.5) can be characterized by the following theorem.

Theorem 5.1 An n-tuple of strategies $\{u_i^* = \phi_i^*(\cdot) \in U^i$, for $i \in N\}$ provides a Nash equilibrium solution to the game (5.4) and (5.5) if there exist continuously twice differentiable functions $\hat{V}^i(x) : R^m \to R$, $i \in N$, satisfying the following set of partial differential equations:

$$r\hat{V}^i(x) - \frac{1}{2}\sum_{h,\zeta=1}^{m} \Omega^{h\zeta}(x)\hat{V}^i_{x^h x^\zeta}(x)$$
$$= \max_{u_i} \{g^i[x, \phi_1^*(x), \phi_2^*(x), \cdots, \phi_{i-1}^*(x), u_i(x), \phi_{i+1}^*(x), \cdots, \phi_n^*(x)]$$
$$+ \hat{V}^i_x(x) f[x, \phi_1^*(x), \phi_2^*(x), \cdots, \phi_{i-1}^*(x), u_i(x), \phi_{i+1}^*(x), \cdots, \phi_n^*(x)]\}$$
$$= \{g^i[x, \phi_1^*(x), \phi_2^*(x), \cdots, \phi_n^*(x)] + \hat{V}^i_x(x) f[x, \phi_1^*(x), \phi_2^*(x), \cdots, \phi_n^*(x)]\},$$

for $i \in N$.

Proof This result follows readily from the definition of Nash equilibrium and from the infinite horizon stochastic control Theorem A.4 in the Technical Appendices. ■

Now consider the case when the players agree to act cooperatively. Let $\Gamma_c(\tau, x_\tau)$ denote a cooperative game in which player i's payoff is (5.2) and the state dynamics is (5.3). The players agree to act according to an agreed upon optimality principle which entails

(i) group optimality and
(ii) the distribution of the total cooperative payoff according to an imputation which equals $\xi^{(\upsilon)}(\upsilon, x_\upsilon^*)$ for $\upsilon \in [\tau, \infty)$ over the game duration. Moreover, the function $\xi^{(\upsilon)i}(\upsilon, x_\upsilon^*)$, for $i \in N$, is continuously differentiable in υ and x_υ^*.

The solution of the cooperative game $\Gamma_c(\tau, x_\tau)$ under the agreed-upon optimality principle includes

(i) a set of cooperative strategies

$$u^{(\tau)*}(s, x_s^*) = \left[u_1^{(\tau)*}(s, x_s^*), u_2^{(\tau)*}(s, x_s^*), \cdots, u_n^{(\tau)*}(s, x_s^*)\right], \text{ for } s \in [\tau, \infty);$$

(ii) an imputation vector $\xi^{(\tau)}(\tau, x_\tau) = [\xi^{(\tau)1}(\tau, x_\tau), \xi^{(\tau)2}(\tau, x_\tau), \cdots, \xi^{(\tau)n}(\tau, x_\tau)]$ to allot the cooperative payoff to the players; and
(iii) a payoff distribution procedure $B^\tau(s, x_s^*) = [B_1^\tau(s, x_s^*), B_2^\tau(s, x_s^*), \cdots, B_n^\tau(s, x_s^*)]$ for $s \in [\tau, \infty)$, where $B_i^\tau(s, x_s^*)$ is the i at time s when the state is $x_s^* \in X_s^*$. In particular,

$$\xi^{(\tau)i}(\tau, x_\tau) = E_\tau\left\{\int_\tau^\infty B_i^\tau(s, x_s^*)\exp[-r(s-\tau)]\, ds\right\}, \text{ for } i \in N. \quad (5.5)$$

3.5.1 Group Optimal Cooperative Strategies

To ensure group rationality the players maximize the sum of their expected payoffs, the players solve the problem:

$$\max_{u_1, u_2, \cdots, u_n} E_\tau \left\{ \int_\tau^\infty \sum_{j=1}^n g^j[x(s), u_1(s), u_2(s), \cdots, u_n(s)] \exp[-r(s-\tau)] ds \right\}, \tag{5.6}$$

subject to (5.3).

Invoking Theorem A.4 in the Technical Appendices, a set of controls $\{\psi_i^*(x) \in U^i; i \in N\}$ constitutes an optimal solution to the infinite horizon stochastic control problem (5.3) and (5.7) if there exists continuously twice differentiable function $W(x)$ defined on $R^m \to R$ which satisfies the following equation:

$$rW(x) - \frac{1}{2} \sum_{h,\zeta=1}^m \Omega^{h\zeta}(x) W_{x^h x^\zeta}(x)$$
$$= \max_{u_1, u_2, \cdots, u_n} \left\{ \sum_{j=1}^n g^j[x, u_1, u_2, \cdots, u_n] + W_x(x) f[x, u_1, u_2, \cdots, u_n] \right\}. \tag{5.7}$$

Hence the players will adopt the cooperative control $\{\psi_i^*(x)$, for $i \in N\}$ to obtain the maximized level of expected joint profit. Substituting this set of control into (6.5) yields the dynamics of the optimal (cooperative) trajectory as:

$$dx(s) = f\left[x(s), \psi_1^*(x(s)), \psi_2^*(x(s)), \cdots, \psi_n^*(x(s))\right] ds + \sigma[x(s)] dz(s), \; x(\tau) = x_\tau. \tag{5.8}$$

The solution to (5.9) can be expressed as:

$$x^*(s) = x_\tau + \int_\tau^s f\left[x^*(\upsilon), \psi_1^*(x^*(\upsilon)), \psi_2^*(x^*(\upsilon)), \cdots, \psi_n^*(x^*(\upsilon))\right] d\upsilon$$
$$+ \int_\tau^s \sigma\left[x^*(\upsilon)\right] dz(\upsilon). \tag{5.9}$$

We use X_s^* to denote the set of realizable values of $x^*(s)$ at time s generated by (5.9). The term $x_s^* \in X_s^*$ is used to denote an element in X_s^*. The terms $x^*(s)$ and x_s^* will be used interchangeably in case where there is no ambiguity.

The expected cooperative payoff can be expressed as:

$$W\left(x_\tau^*\right) = E_\tau\Bigg\{$$
$$\int_\tau^\infty \sum_{j=1}^n g^j\left[x^*(s), \psi_1^*\left(x^*(s)\right), \psi_2^*\left(x^*(s)\right), \cdots, \psi_n^*\left(x^*(s)\right)\right]\exp[-r(s-\tau)]ds$$
$$\left| x^*(\tau) = x_\tau^* \right\}.$$

Moreover, one can easily verify that the joint payoff maximizing controls for the cooperative game $\Gamma_c(\tau, x_\tau)$ over the time interval $[t, \infty)$ is identical to the joint payoff maximizing controls for the cooperative game $\Gamma_c(t, x_t^*)$ over the time interval $[t, \infty)$.

3.5.2 Subgame Consistent Imputation and Payoff Distribution Procedure

In the game $\Gamma_c(t, x_t^*)$, according to optimality principle the players would use the Payoff Distribution Procedure $\left\{B^\tau\left(s, x_s^*\right)\right\}_{s=\tau}^\infty$ to bring about an imputation to player i such that:

$$\xi^{(\tau)i}(\tau, x_\tau) = E_\tau\left\{\int_\tau^\infty B_i^\tau\left(s, x_s^*\right)\exp[-r(s-\tau)]\ ds\right\},\ \text{for } i \in N.$$

We define

$$\xi^{(\tau)i}\left(t, x_t^*\right) = E_\tau\left\{\int_t^\infty B_i^\tau\left(s, x_s^*\right)\exp[-r(s-\tau)]\ ds \middle| x(t) = x_t^* \in X_t^*\right\},$$
$$\text{for } i \in N, \tag{5.10}$$

where $t > \tau$ and $x_t^* \in \left\{x^*(s)\right\}_{s=\tau}^\infty$.

At time τ, according to $P(\tau, x_\tau)$ player i is supposed to receive a payoff $\xi^{(\tau)i}(t, x_t^*)$ over the remaining time interval $[t, \infty)$ if the state is $x_t^* \in X_t^*$.

Consider the case when the game has proceeded to time t and the state variable becames $x_t^* \in X_t^*$. Then one has a cooperative game $\Gamma_c(t, x_t^*)$ which starts at time t with initial state x_t^*. According to the agreed-upon optimality principle, an imputation

$$\xi^{(t)i}\left(t, x_t^*\right) = E_t\left\{\int_t^\infty B_i^t\left(s, x_s^*\right)\exp[-r(s-t)]\ ds \middle| x(t) = x_t^* \in X_t^*\right\},$$

will be allotted to player i, for $i \in N$.

However, according to the solution to the game $\Gamma_c(\tau, x_\tau)$, the imputation (in present value viewed at time τ) to player i over the period $[t, \infty)$ is $\xi^{(\tau)i}(t, x_t^*)$

in (5.11). For the imputation from $\Gamma_c(\tau, x_\tau)$ to be consistent with those from $\Gamma_c(t, x_t^*)$, it is required that

$$\begin{aligned}&\exp[r(t-\tau)]\xi^{(\tau)i}(t, x_t^*) = \xi^{(t)i}(t, x_t^*) \text{ from the game } \Gamma_c(t, x_t^*)\\&\text{under the same optimality principle,} \quad \text{for } t \in (\tau, \infty).\end{aligned} \tag{5.11}$$

The payoff distribution procedure of the game $\Gamma_c(\tau, x_\tau)$ according to the agreed-upon optimality principle is

$$B^\tau(s, x_s^*) = [B_1^\tau(s, x_s^*), B_2^\tau(s, x_s^*), \cdots, B_n^\tau(s, x_s^*)], \text{ for } s \in [\tau, \infty) \text{ and } x_s^* \in X_s^*.$$

When the game has proceeded to time t and the state variable became $x_t^* \in X_t^*$, we have the game $\Gamma_c(t, x_t^*)$. According to the agreed-upon optimality principle the payoff distribution procedure of the game $\Gamma_c(t, x_t^*)$ is

$$B^t(s, x_s^*) = [B_1^t(s, x_s^*), B_2^t(s, x_s^*), \cdots, B_n^t(s, x_s^*)], \text{ for } s \in [t, \infty) \text{ and } x_s^* \in X_s^*.$$

For the continuation of the payoff distribution procedure $B^\tau(s, x_s^*)$ to be consistent with $B^t(s, x_s^*)$, it is required that

$$B^{t_0}(s, x_s^*) = B^t(s, x_s^*), \text{ for } s \in [t, \infty) \text{ and } t \in [\tau, \infty) \text{ and } x_s^* \in X_s^*.$$

Definition 5.1 The imputation and payoff distribution procedure $\{\xi^{(\tau)}(\tau, x_\tau)$ and $B^\tau(s, x_s^*)$ for $s \in [\tau, \infty)\}$ are subgame consistent if

(i)

$$\begin{aligned}&\exp[r(t-\tau)]\,\xi^{(\tau)i}(t, x_t^*)\\&\equiv \exp[r(t-\tau)]E_\tau\left\{\int_t^\infty B_i^\tau(s, x_s^*)\exp[-r(s-\tau)]\,ds \,\middle|\, x(t) = x_t^* \in X_t^*\right\}\\&= \xi^{(t)i}(t, x_t^*), \text{ for } t \in (\tau, \infty) \text{ and } i \in N; \text{ and}\end{aligned} \tag{5.12}$$

(ii) the payoff distribution procedure $B^\tau(s, x_s^*)$ for $s \in [t, \infty)$ is identical to $B^t(s, x_s^*)$. ■

3.5.3 *Payoff Distribution Procedure Leading to Subgame Consistency*

A payoff distribution procedure leading to subgame consistent imputation has to satisfy Definition 5.1. Invoking Definition 5.1, we have $B_i^\tau(s, x_s^*) = B_i^t(s, x_s^*) = B_i(s, x_s^*)$, for $s \in [\tau, \infty)$, $x_s^* \in X_s^*$ and $t \in [\tau, \infty)$ and $i \in N$.

Therefore along the cooperative trajectory,

$$\xi^{(\tau)i}(\tau,x_\tau)=E_\tau\left\{\int_\tau^\infty B_i(s,x_s^*)\exp[-r(s-\tau)]ds\right\},\ \text{for } i\in N,\ \text{and}$$
$$\xi^{(\upsilon)i}(\upsilon,x_\upsilon^*)=E_\upsilon\left\{\int_\upsilon^\infty B_i(s,x_s^*)\exp[-r(s-\upsilon)]ds\,\middle|\,x(\upsilon)=x_\upsilon^*\in X_\upsilon^*\right\},\ \text{for } i\in N,\ \text{and}$$
$$\xi^{(t)i}(t,x_t^*)=E_t\left\{\int_t^\infty B_i(s,x_s^*)\exp[-r(s-t)]ds\,\middle|\,x(t)=x_t^*\in X_t^*\right\},$$
$$\text{for } i\in N \text{ and } t\geq\upsilon\geq\tau. \tag{5.13}$$

Moreover, for $i\in N$ and $t\in[\tau,\infty)$, we define the term

$$\xi^{(\upsilon)i}(t,x_t^*)=E_\upsilon\left\{\left(\int_t^\infty B_i(s,x_s^*)\exp[-r(s-\upsilon)]ds\right)\middle|\,x(t)=x_t^*\right\}, \tag{5.14}$$

to denote the present value of player i's cooperative payoff over the time interval $[t,\infty)$, given that the state is x_t^* at time $t\in[\upsilon,\infty)$, under the optimality principle $P(\upsilon,x_\upsilon^*)$.

Invoking (5.14) and (5.15) one can readily verify that $\exp[r(t-\tau)]\xi^{(\tau)i}(t,x_t^*) = \xi^{(t)i}(t,x_t^*)$, for $i\in N$ and $\tau\in[t_0,T]$ and $t\in[\tau,T]$.

The next task is to derive $B_i(s,x_s^*)$, for $s\in[\tau,\infty)$ and $t\in[\tau,\infty)$ so that (5.14) can be realized. Consider again the following condition.

Condition 5.1 For $i\in N$ and $t\geq\upsilon$ and $\upsilon\in[\tau,T]$, the term $\xi^{(\upsilon)i}(t,x_t^*)$ is a function that is continuously differentiable in t and x_t^*.

A theorem characterizing a formula for $B_i(s,x_s^*)$, for $i\in N$ and $s\in[\upsilon,\infty)$, which yields (5.15) is provided as follows.

Theorem 5.2 If Condition 5.1 is satisfied, a PDP with instantaneous payments at time s with the state being $x_s^*\in X_s^*$ equaling

$$\begin{aligned}B_i(s,x_s^*)=&-\left[\xi_t^{(s)i}(t,x_t^*)\Big|_{t=s}\right]\\&-\left[\xi_{x_t^*}^{(s)i}(t,x_t^*)\Big|_{t=s}\right]f\left[x_s^*,\psi_1^*(s,x_s^*),\psi_2^*(s,x_s^*),\cdots,\psi_n^*(s,x_s^*)\right]\\&-\frac{1}{2}\sum_{h,\zeta=1}^m\Omega^{h\zeta}(x_s^*)\left[\xi_{x_t^hx_t^\zeta}^{(s)i}(t,x_t^*)\Big|_{t=s}\right],\ \text{for } i\in N \text{ and } s\in[\upsilon,\infty),\end{aligned} \tag{5.15}$$

yields imputation $\xi^{(\upsilon)i}(\upsilon,x_\upsilon^*)$ for $\upsilon\in[\tau,\infty)$ and $x_\upsilon^*\in X_\upsilon^*$ which satisfy (5.14).

Proof Note that along the cooperative trajectory

$$\begin{aligned}\xi^{(\upsilon)i}(t,x_t^*)&=E_\upsilon\left\{\int_t^\infty B_i(s,x_s^*)\exp[-r(s-\upsilon)]ds\,\middle|\,x(t)=x_t^*\in X_t^*\right\}\\&=\exp[-r(t-\upsilon)]\xi^{(t)i}(t,x_t^*),\ \text{for } i\in N \text{ and } t\in[\upsilon,\infty).\end{aligned} \tag{5.16}$$

For $\Delta t \to 0$, equation (5.14) can be expressed as

$$\xi^{(v)i}(v, x_v^*) = E_v\left\{ \int_v^\infty B_i(s, x_s^*)\exp[-r(s-v)]ds \right\}$$
$$= E_v\left\{ \int_v^{v+\Delta t} B_i(s, x_s^*)\exp[-r(s-v)]\ ds + \xi^{(v)i}(v+\Delta t, x_v^* + \Delta x_v^*) \right\}, \quad (5.17)$$

where

$$\Delta x_v^* = f\left[x_v^*, \psi_1^*(x_v^*), \psi_2^*(x_v^*), \cdots, \psi_n^*(x_v^*)\right]\Delta t + \sigma(x_v^*)\,\Delta z_v + o(\Delta t),$$
$$\Delta z_v = Z(v+\Delta t) - z(v),\ \text{and}\, E_v[o(\Delta t)]/\Delta t \to 0 \text{ as } \Delta t \to 0.$$

Replacing the term $x_v^* + \Delta x_v^*$ with $x_{v+\Delta t}^*$ and rearranging (5.18) yields:

$$E_v\left\{ \int_v^{v+\Delta t} B_i(s)\exp[-r(s-v)]\ ds \right\} = E_v\left\{ \xi^{(v)i}(v, x_v^*) - \xi^{(v)i}(v+\Delta t, x_{v+\Delta t}^*) \right\},$$
$$\text{for all } v \in [\tau, \infty) \text{ and } i \in N. \quad (5.18)$$

With Condition 5.1 holding and $\Delta t \to 0$, (5.19) can be expressed as:

$$E_v\left\{ B_i(s, x_s^*)\Delta t + o(\Delta t) \right\} = E_v\left\{ -\left[\xi_t^{(s)i}(t, x_t^*)\Big|_{t=s}\right]\Delta t \right.$$
$$-\left[\xi_{x_t^*}^{(s)i}(t, x_t^*)\Big|_{t=s}\right] f\left[x_s^*, \psi_1^*(s, x_s^*), \psi_2^*(s, x_s^*), \cdots, \psi_n^*(s, x_s^*)\right]\Delta t$$
$$\left. -\frac{1}{2}\sum_{h,\zeta=1}^{m} \Omega^{h\zeta}(x_s^*)\left[\xi_{x_t^h x_t^\zeta}^{(s)i}(t, x_t^*)\Big|_{t=s}\right]\Delta t - \left[\xi_{x_t^*}^{(s)i}(t, x_t^*)\Big|_{t=s}\right]\sigma(x_v^*)\,\Delta z_v - o(\Delta t) \right\}. \quad (5.19)$$

Dividing (5.20) throughout by Δt, with $\Delta t \to 0$ and taking expectation yields (5.16). Thus the payoff distribution procedure in $B_i(v, x_v^*)$ in (5.16) would lead to the realization of the imputations which satisfy (5.14). ■

Since the payoff distribution procedure in $B_i(\tau)$ in (5.16) leads to the realization of (5.14), it would yields subgame consistent imputations satisfying Definition 5.1.

A more succinct form of the PDP instantaneous payment in (5.14) can be derived as follows. First we define

$$\hat{\xi}^i(x_v^*) = E_v\left\{ \int_v^\infty B_i(s)\exp[-r(s-v)]ds \middle| x(v) = x_v^* \right\}\xi^{(v)i}(\tau, x_v^*), \text{ and}$$
$$\hat{\xi}^i(x_t^*) = E_t\left\{ \int_t^\infty B_i(s)\exp[-r(s-t)]ds \middle| x(t) = x_t^* \right\} = \xi^{(t)i}(t, x_t^*),$$

for $i \in N$ and $v \in [\tau, \infty)$ and $t \in [v, \infty)$ along the optimal cooperative trajectory $\{x_s^*\}_{s=\tau}^\infty$.

We then have:

$$\xi^{(v)i}\left(t,x_t^*\right) = \exp[-r(t-v)]\hat{\xi}^i\left(x_t^*\right).$$

Differentiating the above condition with respect to t yields:

$$\left[\xi_t^{(v)i}\left(t,x_t^*\right)\Big|_{t=v}\right] = -r\exp[-r(t-v)]\hat{\xi}^i\left(x_t^*\right) = -r\xi^{(v)i}\left(t,x_t^*\right).$$

At $t=v$, $\xi^{(v)i}\left(t,x_t^*\right) = \xi^{(v)i}\left(v,x_v^*\right)$, therefore

$$\left[\xi_t^{(v)i}\left(t,x_t^*\right)\Big|_{t=v}\right] = r\xi^{(v)i}\left(t,x_t^*\right) = r\xi^{(v)i}\left(v,x_v^*\right). \tag{5.20}$$

Substituting (5.21) into (5.16) yields,

$$B_i\left(s,x_s^*\right) = r\xi^{(s)i}\left(s,x_s^*\right) - \xi_{x_s^*}^{(s)i}\left(s,x_s^*\right)f\left[x_s^*,\psi_1^*\left(x_s^*\right),\psi_2^*\left(x_s^*\right),\cdots,\psi_n^*\left(x_s^*\right)\right]$$
$$-\frac{1}{2}\sum_{h,\zeta=1}^{m}\Omega^{h\zeta}\left(x_s^*\right)\left[\xi_{x_t^h x_t^\zeta}^{(s)i}\left(t,x_t^*\right)\Big|_{t=s}\right],\ \text{for } i\in N,\ x_s^*\in X_s^* \text{ and } s\in[v,\infty). \tag{5.21}$$

An alternative form of Theorem 5.2 can be expressed as:

Theorem 5.3 A PDP with instantaneous payments with the state being x^* equaling

$$B_i\left(x^*\right) = r\hat{\xi}^i\left(x^*\right) - \xi_{x^*}^i\left(x^*\right)f\left[x^*,\psi_1^*\left(x^*\right),\psi_2^*\left(x^*\right),\cdots,\psi_n^*\left(x^*\right)\right]$$
$$-\frac{1}{2}\sum_{h,\zeta=1}^{m}\Omega^{h\zeta}\left(x^*\right)\hat{\xi}_{x^h x^\zeta}^i\left(x^*\right),\ \text{for } i\subset N. \tag{5.22}$$

yields imputation $\hat{\xi}^i\left(x^*\right)$.

Proof Multiplying (5.22) throughout by $\exp[r(t-v)]$ yields

$$B_i\left(x_s^*\right) = r\hat{\xi}^i\left(x_s^*\right) - \hat{\xi}_{x_s^*}^i\left(x_s^*\right)f\left[x_s^*,\psi_1^*\left(x_s^*\right),\psi_2^*\left(x_s^*\right),\cdots,\psi_n^*\left(x_s^*\right)\right]$$
$$-\frac{1}{2}\sum_{h,\zeta=1}^{m}\Omega^{h\zeta}\left(x_s^*\right)\hat{\xi}_{x_s^h x_s^\zeta}^i\left(x_s^*\right),\ \text{for } i\in N,\ x_s^*\in X_s^* \text{ and } s\in[v,\infty).$$

Recall that the infinite-horizon autonomous game $\Gamma(x)$ is independent of the choice of time s and dependent only upon the state, equation (5.22) can be expressed as (5.23). ■

With agents using the cooperative strategies, when the state is $x^*\in X^*$ the instantaneous receipt of agent i is:

$$\zeta_i\left(x^*\right) = g^i\left[x^*,\psi_1^*\left(x^*\right),\psi_2^*\left(x^*\right),\cdots,\psi_n^*\left(x^*\right)\right],\ \text{for } i\in N. \tag{5.23}$$

According to Theorem 5.2 and (5.23), the instantaneous payment that player i should receive under the agreed-upon optimality principle is $B_i(x^*)$ as stated in (5.23). Hence an instantaneous transfer payment

$$\chi^i(x^*) = B_i(x^*) - \zeta_i(x^*), \text{ for } i \in N \tag{5.24}$$

has to be given to player i when the state is $x^* \in X^*$.

3.6 Infinite Horizon Cooperative Fishery Under Uncertainty

Consider an infinite horizon version of the cooperative fishery in Sect. 3.5. At time τ, the expected payoff of extractor 1 and that of extractor 2 are respectively:

$$E_\tau\left\{ \int_\tau^\infty \left[u_1(s)^{1/2} - \frac{c_1}{x(s)^{1/2}} u_1(s) \right] \exp[-r(t-\tau)]\, ds \right\} \text{ and}$$
$$E_\tau\left\{ \int_\tau^\infty \left[u_2(s)^{1/2} - \frac{c_2}{x(s)^{1/2}} u_2(s) \right] \exp[-r(t-\tau)]\, ds \right\}. \tag{6.1}$$

The fish resource stock $x(s) \in X \subset R$ follows the stochastic dynamics:

$$dx(s) = \left[ax(s)^{1/2} - bx(s) - u_1(s) - u_2(s) \right] ds + \sigma x(s) dz(s), \; x(\tau) = x_\tau, \tag{6.2}$$

Invoking Theorem 5.1, the set of strategies $[\phi_1^*(x), \phi_2^*(x)]$ for $t \in [t_0, T]$ that provides a feedback Nash equilibrium solution to the game (6.2) and (6.3) can be characterized by:

$$r\hat{V}^i(x) - \frac{1}{2}\sigma^2 x^2 \hat{V}^i_{xx}(x) = \max_{u_i} \left\{ u_i^{1/2} - \frac{c_i}{x^{1/2}} u_i + \hat{V}^i_x(x) \left[ax^{1/2} - bx - u_i - \phi_j^*(x) \right] \right\}$$
for $i, j \in \{1, 2\}$ and $i \neq j$. (6.3)

Performing the indicated maximization in (6.4) and using the derived game equilibrium strategies one obtains the value function of extractor $i \in \{1, 2\}$ as:

$$\hat{V}^i(t, x) = \left[A_i x^{1/2} + C_i \right], \tag{6.4}$$

where for $i, j \in \{1, 2\}$ and $i \neq j, A_i, C_i, A_j$ and C_j satisfy:

$$\left[r+\frac{\sigma^2}{8}+\frac{b}{2}\right]A_i-\frac{1}{2[c_i+A_i/2]}+\frac{c_i}{4[c_i+A_i/2]^2}$$
$$+\frac{A_i}{8[c_i+A_i/2]^2}+\frac{A_i}{8\left[c_j+A_j/2\right]^2}=0;\text{ and}$$
$$C_i=\frac{a}{2}A_i.$$

3.6.1 Cooperative Extraction

Consider the case when these two nations agree to act according to an agreed upon optimality principle which entails

(i) group optimality, and
(ii) the distribution of the excess of the total expected cooperative payoff over the sum of expected individual noncooperative payoffs proportional to the extractors' expected noncooperative payoffs.

To maximize their joint expected payoff for group optimality, the nations have to solve the stochastic control problem of maximizing

$$E_t\Bigg\{\int_t^\infty\left(\left[u_1(s)^{1/2}-\frac{c_1}{x(s)^{1/2}}u_1(s)\right]+\left[u_2(s)^{1/2}-\frac{c_2}{x(s)^{1/2}}u_2(s)\right]\right)\exp[-r(t-t)]\,ds\Bigg\}. \quad (6.5)$$

subject to (6.3).

Invoking Theorem A.4 in the Technical Appendices yields the characterization of solution of the problem (6.3) and (6.6) as:

Corollary 6.1 A set of controls $\{\psi_i^*(x)$, for $i \in \{1, 2\}\}$ constitutes an optimal solution to the stochastic control problem (6.3) and (6.6), if there exist continuously twice differentiable functions $W(x): R^m \to R$,, satisfying the following partial differential equation:

$$rW(x)-\frac{1}{2}\sigma^2x^2W_{xx}(x)=\max_{u_1,u_2}\Bigg\{\left(\left[u_1{}^{1/2}-\frac{c_1}{x^{1/2}}u_1\right]+\left[u_2{}^{1/2}-\frac{c_2}{x^{1/2}}u_2\right]\right)$$
$$+W_x(x)\left[ax^{1/2}-bx-u_1-u_2\right]\Bigg\}. \quad (6.6)\blacksquare$$

Performing the indicated maximization and solving (6.7) one obtains the maximized expected joint profit can be derived as:

$$W(x) = \left[Ax^{1/2} + C\right], \tag{6.7}$$

where

$$\begin{aligned}
&\left[r + \frac{\sigma^2}{8} + \frac{b}{2}\right]A - \frac{1}{2[c_1 + A/2]} - \frac{1}{2[c_2 + A/2]} \\
&+ \frac{c_1}{4[c_1 + A/2]^2} + \frac{c_2}{4[c_2 + A/2]^2} + \frac{A}{8[c_1 + A/2]^2} + \frac{A}{8[c_2 + A/2]^2} = 0, \text{ and} \\
&C = \frac{a}{2r}A.
\end{aligned} \tag{6.8}$$

The optimal cooperative controls can then be obtained as:

$$\psi_1^*(x) = \frac{x}{4[c_1 + A/2]^2}, \text{ and } \psi_2^*(x) = \frac{x}{4[c_2 + A/2]^2}. \tag{6.9}$$

Substituting these control strategies into (6.3) yields the dynamics of the state trajectory under cooperation:

$$\begin{aligned}
dx(s) = &\left[ax(s)^{1/2} - bx(s) - \frac{x(s)}{4[c_1 + A/2]^2} - \frac{x(s)}{4[c_2 + A/2]^2} \right] ds \\
&+ \sigma x(s) dz(s), \; x(t_0) = x_0.
\end{aligned} \tag{6.10}$$

Solving (6.11) yields the optimal cooperative state trajectory as:

$$x^*(s) = \varpi(t_0, s)^2 \left[x_0^{1/2} + \int_{t_0}^{s} \varpi^{-1}(t_0, t) H_1 dt \right]^2, \text{ for } s \in [t_0, T], \tag{6.11}$$

where

$$\varpi(t_0, s) = \exp\left[\int_{t_0}^{s} \left[H_2(\tau) - \frac{\sigma^2}{8}\right] d\upsilon + \int_{t_0}^{s} \frac{\sigma}{2} dz(\upsilon)\right], H_1 = \frac{1}{2}a,$$

$$\text{and } H_2(s) = -\left[\frac{1}{2}b + \frac{1}{8[c_1 + A(s)/2]^2} + \frac{1}{8[c_2 + A(s)/2]^2} + \frac{\sigma^2}{8}\right].$$

3.6.2 *Subgame Consistent Payoff Distribution*

With the extractors using the cooperative strategies (6.10) along the stochastic cooperative path, they agree to share the excess of the total expected cooperative payoff over the sum of individual noncooperative payoffs proportional to the extractors' expected noncooperative payoffs. Therefore the following imputation has to be satisfied.

Condition 6.1 An imputation

$$\xi^{(\upsilon)i}\left(\upsilon, x_\upsilon^*\right) = \frac{\hat{V}^i\left(x_\upsilon^*\right)}{\sum_{j=1}^{2} \hat{V}^j\left(x_\upsilon^*\right)} W\left(x_\upsilon^*\right) = \frac{\left[A_i\left(x_\upsilon^*\right)^{1/2} + C_i\right]}{\sum_{j=1}^{2}\left[A_j\left(x_\upsilon^*\right)^{1/2} + C_j\right]} \left[A\left(x_\upsilon^*\right)^{1/2} + C\right] \quad (6.12)$$

is assigned to extractor i, for $i \in \{1, 2\}$ if $x_\upsilon^* \in X_\upsilon^*$ occurs at time $\upsilon \in [\tau, \infty)$. ■

Applying Theorem 5.3 a subgame-consistent solution for the cooperative game $\Gamma_c(\tau, x_\tau)$ includes:

(i) a set of group optimal strategies

$$\psi_1^*\left(x_s^*\right) = \frac{x_s^*}{4[c_1 + A/2]^2} \text{ and } \psi_2^*\left(x_s^*\right) = \frac{x_s^*}{4[c_2 + A/2]^2}; \text{ and}$$

(ii) a Payoff Distribution Procedure
$B\left(s, x_s^*\right) = \left\{B_1\left(s, x_s^*\right), B_2\left(s, x_s^*\right), \cdots, B_n\left(s, x_s^*\right)\right\}$for $s \in [\tau, \infty)$ with

$$\begin{aligned} B_i\left(s, x_s^*\right) &= r\,\xi^{(s)i}\left(s, x_s^*\right) \\ &\quad - \xi_{x_s^*}^{(s)i}\left(s, x_s^*\right)\left[a\left(x_s^*\right)^{1/2} - bx_s^* - \frac{x_s^*}{4[c_1 + A/2]^2} - \frac{x_s^*}{4[c_2 + A/2]^2} \right] \\ &\quad - \frac{1}{2}\sigma^2\left(x_s^*\right)^2 \xi_{x_s^h x_s^\zeta}^{(\tau)i}\left(s, x_s^*\right), \text{ for } i \in \{1, 2\}, \end{aligned}$$

where

$$\xi_{x_s^*\ xs*}^{(s)i\ si}\left(s, x_s^*\right) = \frac{\left[A_i\left(x_s^*\right)^{1/2} + C_i\right] A\left(x_s^*\right)^{-1/2} + \left[A\left(x_s^*\right)^{1/2} + C\right] A_i\left(x_s^*\right)^{-1/2}}{2\sum_{j=1}^{2}\left[A_j\left(x_s^*\right)^{1/2} + C_j\right]}$$
$$- \frac{\left[A_i\left(x_s^*\right)^{1/2} + C_i\right]\left[A\left(x_s^*\right)^{1/2} + C\right]}{\left(\sum_{j=1}^{2}\left[A_j\left(x_s^*\right)^{1/2} + C_j\right]\right)^2} \left(\frac{1}{2}\sum_{j=1}^{2} A_j\left(x_s^*\right)^{-1/2}\right);$$

$$\text{and } \xi_{x_s^* x_s^*}^{(\tau)i}\left(s, x_s^*\right) = -\frac{C_i A\left(x_s^*\right)^{-3/2} + C A_i\left(x_s^*\right)^{-3/2}}{4\sum_{j=1}^{2}\left[A_j\left(x_s^*\right)^{1/2} + C_j\right]}$$
$$- \frac{\left[A_i\left(x_s^*\right)^{1/2} + C_i\right] A\left(x_s^*\right)^{-1/2} + \left[A\left(x_s^*\right)^{1/2} + C\right] A_i\left(x_s^*\right)^{-1/2}}{\left(2\sum_{j=1}^{2}\left[A_j\left(x_s^*\right)^{1/2} + C_j\right]\right)^2}$$

$$\sum_{j=1}^{2}\left[A_j\left(x_s^*\right)^{-1/2}\right]$$
$$+\frac{\left[A_i\left(x_s^*\right)^{1/2}+C_i\right]\left[A\left(x_s^*\right)^{1/2}+C\right]}{\left(\sum_{j=1}^{2}\left[A_j\left(x_s^*\right)^{1/2}+C_j\right]\right)^2}\left(\frac{1}{4}\sum_{j=1}^{2}A_j\left(x_s^*\right)^{-3/2}\right)$$
$$-\left(\frac{1}{2}\sum_{j=1}^{2}A_j\left(x_s^*\right)^{-1/2}\right)\times\left[\frac{A_iA+\frac{1}{2}\left[A_iC+AC_i\right]\left(x_s^*\right)^{-1/2}}{\left(\sum_{j=1}^{2}\left[A_j\left(x_s^*\right)^{1/2}+C_j\right]\right)^2}\right.$$
$$\left.-\frac{\left[A_i\left(x_\tau^*\right)^{1/2}+C_i\right]\left[A\left(x_\tau^*\right)^{1/2}+C\right]}{\left(\sum_{j=1}^{2}\left[A_j\left(x_\tau^*\right)^{1/2}+C_j\right]\right)^3}\sum_{j=1}^{2}A_j\left(x_\tau^*\right)^{-1/2}\right]. \tag{6.13}$$

With extractors using the cooperative strategies in (6.13), the instantaneous receipt of extractor i at time instant $\upsilon \in [\tau, \infty)$ with the state being x_υ^* is:

$$\zeta_i\left(\upsilon, x_\upsilon^*\right)=\frac{\left(x_\upsilon^*\right)^{1/2}}{2\left[c_i+A/2\right]}-\frac{c_i\left(x_\upsilon^*\right)^{1/2}}{4\left[c_i+A/2\right]^2}, \text{ for } i \in \{1,2\}, \tag{6.14}$$

Under the cooperative agreement, the instantaneous payment that extractor $i \in \{1,2\}$ should receive under the agreed-upon optimality principle is $B_i(\upsilon, x_\upsilon^*)$ in (6.14). Hence an instantaneous transfer payment

$$\chi^i\left(\upsilon, x_\upsilon^*\right)=B_i\left(\upsilon, x_\upsilon^*\right)-\zeta_i\left(\upsilon, x_\upsilon^*\right) \tag{6.15}$$

has to be given to extractor i at time υ, for $i \in \{1,2\}$ and $x_\upsilon^* \in X_\upsilon^*$.

3.7 Chapter Notes

The analysis on subgame consistent solution in stochastic differential games was presented in Yeung and Petrosyan (2004). In particular, a generalized theorem for the derivation of an analytically tractable "payoff distribution procedure" which would lead to subgame-consistent solutions was developed. Examples of cooperative stochastic differential games with solutions satisfying subgame consistency can be found in Yeung (2005, 2007a, 2008, 2010) and Yeung and Petrosyan (2004, 2006a, b, 2007a, b, c, 2008, 2012c, 2014a). Theorem 3.1 could be applied to obtain subgame consistent cooperative solution for existing differential games in economic analysis. Solution mechanisms for cooperative stochastic differential games can be found in Yeung (2006b).

3.8 Problems

1. Consider the case of two nations harvesting fish in common waters. The growth rate of the fish biomass is subject to stochastic shocks and follows the differential equation:

$$dx(s) = \left[12x(s)^{1/2} - x(s) - u_1(s) - u_2(s)\right]ds + 0.1x(s)dz(s),\ x(0) = 100,$$

where $z(s)$ is a Wiener process, $x(s)$ is the fish stock and $u_i(s)$ is the amount of fish harvested by nation i, for $i \in \{1, 2\}$. The horizon of the game is $[0, 3]$.

The harvesting cost for nation $i \in \{1, 2\}$ depends on the quantity of resource extracted $u_i(s)$ and the resource stock size $x(s)$. In particular, nation 1's extraction cost is $2u_1(s)x(s)^{-1/2}$ and nation 2's is $u_2(s)x(s)^{-1/2}$. The fish harvested by nation 1 at time s will generate a net benefit of the amount $3[u_1(s)]^{1/2}$ and the fish harvested by nation 2 at time s will generate a net benefit of the amount $2[u_2(s)]^{1/2}$. At terminal time 5, nations 1 and 2 will receive termination bonuses $8x(3)^{1/2}$ and $6x(3)^{1/2}$ while the interest rate is 0.05.

Characterize a feedback Nash equilibrium solution for this stochastic fishery game.

2. If these nations agree to cooperate and maximize their expected joint payoff, obtain a group optimal cooperative solution.
3. Furthermore, if these nations agree to share the expected gain proportional to their non-cooperative payoffs, derive a subgame consistent solution.
4. Consider the case when the game horizon in exercise 1 is extended to infinity.

   (i) Characterize a feedback Nash equilibrium solution for this stochastic dynamic game.

   (ii) If these nations agree to cooperate and maximize their expected joint payoff and share the excess of their expected gain equally, derive a subgame consistent solution.

Chapter 4
Subgame Consistency in Randomly-Furcating Cooperative Stochastic Differential Games

An essential characteristic of time – and hence decision making over time – is that though an individual may, through the expenditure of resources, gather past and present information, the future is inherently unknown and therefore (in the mathematical sense) uncertain. There is no escape from this fact, regardless of what resources the individual should choose to devote to obtaining data, information, and to forecasting. An empirically meaningful theory must therefore incorporate time-uncertainty in an appropriate manner. Important forms of structure uncertainty follow from uncertainty of payoffs and perturbing stochastic state dynamics. Causes of structure uncertainty include (a) Imprecise or incomplete knowledge about the game's payoffs over time – the benefits and costs from playing are generally known only probabilistically, and (b) imperfect knowledge regarding the behavior of the game's state variables – generally, how the game evolves over time is only known probabilistically. To meet the challenges following from structure-uncertainty, randomly-furcating stochastic differential games allows random shocks in the stock dynamics and stochastic changes in payoffs. Since future payoff are not known with certainty, the term "randomly-furcating" is introduced to highlight the fact that a particularly useful way to analyze the situation is to assume that payoffs change at any future time instant according to (known) probability distributions defined in terms of multiple-branching stochastic processes (see Yeung (2001) and Yeung (2003)).

This Chapter presents an $n-$ player counterpart of the Petrosyan and Yeung's (2007) 2-player analysis on subgame-consistent cooperative solutions in randomly-furcating stochastic differential games. The organization of the Chapter is as follows. Section 4.1 presents the basic formulation of randomly-furcating cooperative differential games. Section 4.2 presents an analysis on subgame consistent dynamic cooperation of this class of games. Derivation of a subgame consistent payoff distribution procedure is provided in Sect. 4.3. An illustration of the solution mechanism is given in a cooperative fishery game in Sect. 4.4. Subgame consistency in infinite horizon randomly-furcating cooperative

D.W.K. Yeung, L.A. Petrosyan, *Subgame Consistent Cooperation*,
Theory and Decision Library C 47, DOI 10.1007/978-981-10-1545-8_4

differential games is examined in Sect. 4.4. Chapter notes are given in Sect. 4.5 and problems in Sect. 4.6.

4.1 Game Formulation and Noncooperative Outcomes

Consider a class of randomly furcating stochastic differential game in which there are n players. The game interval is $[t_0, T]$. When the game commences at t_0, the payoff structures of the players in the interval $[t_0, t_1)$ are known. In future instants of time t_k $(k = 1, 2, \cdots, m)$, where $t_0 < t_m < T \equiv t_{m+1}$, the payoff structures in the time interval $[t_k, t_{k+1})$ are affected by a series of random events Θ^k. In particular, Θ^k for $k \in \{1, 2, \cdots, m\}$, are independent and identically distributed random variables with range $\{\theta_1, \theta_2, \ldots, \theta_\eta\}$ and corresponding probabilities $\{\lambda_1, \lambda_2, \ldots, \lambda_\eta\}$. Changes in preference, technology, legal arrangements and the physical environments are examples of factors which constitute the change in payoff structures. At time T a terminal value $q^i(x(T))$ will be given to player i. Specifically player i seeks to maximize the expected payoff:

$$E_{t_0}\left\{\int_{t_0}^{t_1} g^{[i,\theta_0^0]}[s,x(s),u_1(s),u_2(s),\cdots,u_n(s)]\; e^{-r(s-t_0)}ds\right.$$
$$\left.+\sum_{h=1}^{m}\sum_{a_h=1}^{\eta}\lambda_{a_h}\int_{t_h}^{t_{h+1}} g^{\left[i,\theta_{a_h}^h\right]}[s,x(s),u_1(s),u_2(s),\cdots,u_n(s)]e^{-r(s-t_0)}+e^{-r(T-t_0)}q^i(x(T))\right\},$$
$$\text{for } i\in\{1,2,\cdots,n\}\equiv N, \tag{1.1}$$

where $x(s) \in X \subset R^\kappa$ is a vector of state variables, $\theta_{a_k}^h \in \{\theta_1, \theta_2, \ldots, \theta_\eta\}$ for $k \in \{1, 2, \cdots, m\}$, $\theta_{a_0} = \theta_0^0$ is known at time t_0, r is the discount rate, $u_i \in U^i$ is the control of player i, and E_{t_0} denotes the expectation operator performed at time t_0. The payoffs of the players are transferable.

The state dynamics of the game is characterized by the vector-valued stochastic differential equations:

$$dx(s) = f[s,x(s),u_1(s),u_2(s),\cdots,u_n(s)]ds + \sigma[s,x(s)]dz(s),$$
$$x(t_0) = x_0, \tag{1.2}$$

where $\sigma[s, x(s)]$ is a $\kappa \times \upsilon$ matrix and $z(s)$ is a υ -dimensional Wiener process and the initial state x_0 is given. Let $\Omega[s, x(s)] = \sigma[s, x(s)]\sigma[s, x(s)]^T$ denote the covariance matrix with its element in row h and column ζ denoted by $\Omega^{h\zeta}[s, x(s)]$. $u_i \in U_i \subset compR^\ell$ is the control vector of player i, for $i \in N$.

To obtain a Nash equilibrium solution for the game (1.1 and 1.2), we first consider the solution for the subgame in the last time interval, that is $[t_m, T]$. For the case where $\theta^m_{a_m} \in \{\theta_1, \theta_2, \ldots, \theta_\eta\}$ has occurred at time instant t_m and $x(t_m) = x_{t_m} \in X$, player i maximizes the payoff:

$$E_{t_m}\left\{ \int_{t_m}^{T} g^{\left[i, \theta^m_{\alpha m}\right]}[s, x(s), u_1(s), u_2(s), \cdots, u_n(s)] \; e^{-r(s-t_0)} ds \right.$$
$$\left. + q^i(x(T)) e^{-r(T-t_0)} \middle| \, x(t_m) = x_{t_m} \right\}, \tag{1.3}$$

subject to

$$dx(s) = f[s, x(s), u_1(s), u_2(s), \cdots, u_n(s)]ds + \sigma[s, x(s)]dz(s), x(t_m) = x_{t_m}. \tag{1.4}$$

The conditions characterizing a Nash equilibrium solution of the game (1.3 and 1.4) is provided in the lemma below.

Lemma 1.1 A set of feedback strategies $\{u_i^{(m)\theta^m_{\alpha m}}(t) = \phi_i^{(m)\theta^m_{\alpha m}}(t, x); i \in \{1, 2\}$ and $t \in [t_m, T]\}$ constitutes a Nash equilibrium solution for the game (1.3 and 1.4), if there exist continuously differentiable functions $V^{i\left[\theta^m_{\alpha m}\right](m)}(t, x) : [t_m, T] \times R^\kappa \to R$, for $i \in \{1, 2\}$, which satisfy the following partial differential equations:

$$-V_t^{i\left[\theta^m_{\alpha m}\right](m)}(t, x) - \frac{1}{2}\sum_{h,\zeta=1}^{n} \Omega^{h\zeta}(t, x) V_{x^h x^\zeta}^{i\left[\theta^m_{\alpha m}\right](m)}(t, x)$$
$$= \max_{u_i^{\theta_{\alpha m}} \in U^i} \left\{ g^{\left[i, \theta^m_{\alpha m}\right]}\left[t, x, u_i^{(m)\theta^m_{\alpha m}}, \underline{\phi}_{N \backslash i}^{(m)\theta^m_{\alpha m}}(t, x)\right] e^{-r(t-t_0)} \right.$$
$$\left. + V_x^{i\left[\theta^m_{\alpha m}\right](m)}(t, x) f\left[t, x, u_i^{(m)\theta^m_{\alpha m}}, \underline{\phi}_{N \backslash i}^{(m)\theta^m_{\alpha m}}(t, x)\right] \right\}, \text{ and}$$
$$V^{i\left[\theta^m_{\alpha m}\right](m)}(T, x) = e^{-r(T-t_m)} q^i(x), \quad \text{for } i \in N, \ j \in N \text{ and } j \neq i, \tag{1.5}$$

where

$$\underline{\phi}_{N \backslash i}^{(m)\theta^m_{\alpha m}}(t, x) =$$
$$\left[\phi_1^{(m)\theta^m_{\alpha m}}(t, x), \phi_2^{(m)\theta^m_{\alpha m}}(t, x), \cdots, \phi_{i-1}^{(m)\theta^m_{\alpha m}}(t, x), \phi_{i+1}^{(m)\theta^m_{\alpha m}}(t, x), \cdots, \phi_n^{(m)\theta^m_{\alpha m}}(t, x)\right].$$

Proof System (1.5) satisfies the optimal conditions in stochastic dynamic programming in Theorem A.3 in the Technical Appendices for each player and the Nash equilibrium condition (1951). Hence Lemma 1.1 follows. ■

For ease of exposition and sidestepping the issue of multiple equilibria, we assume that a particular noncooperative Nash equilibrium is adopted in the entire subgame. In order to formulate the subgame in the second last time interval $[t_{m-1}, t_m)$, it is necessary to identify the expected terminal payoffs at time t_m.

If $\theta^m_{a_m}$ occurs at time t_m, one can invoke Lemma 1.1 and obtain player i's payoffs at time t_m as $V^{i[\theta^m_{a_m}](m)}(t_m, x_{t_m})$. Note that $V^{i[\theta^m_{a_m}](m)}(t_m, x_{t_m})$ gives the expected payoff to player i for playing the subgame in the last interval if $\theta^m_{a_m}$ occurs at time t_m. Taking into consideration of all the possibilities of $\theta^m_{a_m} \in \{\theta_1, \theta_2, \ldots, \theta_\eta\}$, the expected payoff to player i for playing the subgame in the last interval payoff can be obtained as:

$$\sum_{a=1}^{\eta} \lambda_a V^{i[\theta^m_a](m)}(t_m, x_{t_m}). \tag{1.6}$$

The expected terminal payoff of player i, for $i \in N$, in the subgame over the time interval $[t_{m-1}, t_m]$ is reflected by (1.6) under the assumption that a particular Nash equilibrium is adopted in each of the possible subgame scenarios in the time interval $[t_m, T]$. If $\theta^{m-1}_{a_{m-1}} \in \{\theta_1, \theta_2, \ldots, \theta_\eta\}$ occurs at time t_{m-1}, the subgame in the time interval $[t_{m-1}, t_m]$ can be formally set up as:

$$\max_{u_i} E_{t_{m-1}} \left\{ \int_{t_{\tau-1}}^{t_\tau} g^{[i,\theta^{m-1}_{a_{m-1}}]}[s, x(s), u_1(s), u_2(s), \cdots, u_n(s)]\ e^{-r(s-t_0)} ds + \sum_{a=1}^{\eta} \lambda_a V^{i[\theta^m_a](m)}(t_m, x(t_m)) \middle| x(t_{m-1}) = x_{t_{m-1}} \right\}, \text{ for } i \in N \tag{1.7}$$

subject to

$$\begin{aligned} &dx(s) = f[s, x(s), u_1(s), u_2(s), \cdots, u_n(s)]ds + \sigma[s, x(s)]dz(s), \\ &x(t_{m-1}) - x_{t_{m-1}} \subset X \end{aligned} \tag{1.8}$$

Similarly, if $\theta^k_{a_k} \in \{\theta_1, \theta_2, \ldots, \theta_\eta\}$ occurs at time t_k the subgame in the time interval $[t_k, t_{k+1})$, for $k \in \{0, 1, 2, \cdots, m-2\}$ can be set up as:

$$\max_{u_i} \left\{ \int_{t_k}^{t_{k+1}} g^{[i,\theta^k_{a_k}]}[s, x(s), u_1(s), u_2(s), \cdots, u_n(s)]\ e^{-r(s-t_0)} ds + \sum_{a=1}^{\eta} \lambda_a V^{i[\theta^{k+1}_a](k+1)}(t_{k+1}, x(t_{k+1})) \middle| x(t_k) = x_{t_k} \right\}, \text{ for } i \in N, \tag{1.9}$$

subject to

$$\begin{aligned} &dx(s) = f[s, x(s), u_1(s), u_2(s), \cdots, u_n(s)]ds + \sigma[s, x(s)]dz(s), \\ &x(t_k) = x_{t_k} \in X. \end{aligned} \tag{1.10}$$

Following Lemma 1.1 a Nash equilibrium solution of game (1.1 and 1.2) can be characterized by the following theorem.

Theorem 1.1 A set of feedback strategies $\{u_i^{(m)\theta_{\alpha_m}^m}(t) = \phi_i^{(m)\theta_{\alpha_m}^m}(t,x), \text{for } t \in [t_m, T];$ $u_i^{(k)\theta_{\alpha_k}^k}(t) = \phi_i^{(k)\theta_{\alpha_k}^k}(t,x)$, for $t \in [t_k, t_{k+1})$, $k \in \{0,1,2,\cdots,m-1\}$ and $i \in N\}$, contingent upon the events $\theta_{\alpha_m}^m \in \{\theta_1, \theta_2, \ldots, \theta_\eta\}$ and $\theta_{\alpha_k}^k \in \{\theta_1, \theta_2, \ldots, \theta_\eta\}$ for $k \in \{1,2,\cdots,m-1\}$ constitutes a Nash equilibrium solution for the game (1.1 and 1.2), if there exist continuously differentiable functions $V^{i[\theta_{\alpha_m}^m](m)}(t,x)$: $[t_m, T] \times R^\kappa \to R$ and $V^{i[\theta_{\alpha_k}^k](k)}(t,x)$: $[t_k, t_{k+1}] \times R^\kappa \to R$, for $k \in \{0,1,2,\cdots,m-1\}$ and $i \in N$, which satisfy the following partial differential equations:

$$-V_t^{i[\theta_{\alpha_m}^m](m)}(t,x) - \frac{1}{2}\sum_{h,\zeta=1}^{n} \Omega^{h\zeta}(t,x) V_{x^h x^\zeta}^{i[\theta_{\alpha_m}^m](m)}(t,x)$$

$$= \max_{u_i^{\theta_{\alpha_m}} \in U^i} \Bigg\{ g^{[i,\theta_{\alpha_m}^m]}\left[t,x,u_i^{(m)\theta_{\alpha_m}^m}, \underline{\phi}_{N\backslash i}^{(m)\theta_{\alpha_m}^m}(t,x)\right] e^{-r(t-t_0)}$$

$$+V_x^{i[\theta_{\alpha_m}^m](m)}(t,x) f\left[t,x,u_i^{(m)\theta_{\alpha_m}^m}, \underline{\phi}_{N\backslash i}^{(m)\theta_{\alpha_m}^m}(t,x)\right]\Bigg\}, \text{ and}$$

$$V^{i[\theta_{\alpha_m}^m](m)}(T,x) = e^{-r(T-t_0)} q^i(x);$$

$$-V_t^{i[\theta_{\alpha_k}^k](k)}(t,x) - \frac{1}{2}\sum_{h,\zeta=1}^{n} \Omega^{h\zeta}(t,x) V_{x^h x^\zeta}^{i[\theta_{\alpha_k}^k](k)}(t,x)$$

$$= \max_{u_i^{\theta_{\alpha_k}} \in U^i} \Bigg\{ g^{[i,\theta_{\alpha_k}^k]}\left[t,x,u_i^{(k)\theta_{\alpha_k}^k}, \underline{\phi}_{N\backslash i}^{(k)\theta_{\alpha_k}^k}(t,x)\right] e^{-r(t-t_0)}$$

$$+V_x^{i[\theta_{\alpha_k}^k](k)}(t,x) f\left[t,x,u_i^{(k)\theta_{\alpha_k}^k}, \underline{\phi}_{N\backslash i}^{(k)\theta_{\alpha_k}^k}(t,x)\right]\Bigg\}, \text{ and}$$

$$V^{i[\theta_{\alpha_k}^k](k)}(t_{k+1},x) = \sum_{a=1}^{\eta} \lambda_a V^{i[\theta_a^{k+1}](k+1)}(t_{k+1},x),$$

for $i \in N$ and $k \in \{0,1,2,\cdots,m-1\}$.

Proof The results in Theorem 1.1 satisfy the optimal conditions in stochastic dynamic programming in Technical Appendix A.3 for each player and the Nash equilibrium condition (1951). Hence Theorem 1.1 follows. ■

Two remarks given below will be utilized in subsequent analysis.

Remark 1.1 One can readily verify that $\overline{V}^{i[\theta_{\alpha_k}^k](k)}(t_k, x_{t_k}) = V^{i[\theta_{\alpha_k}^k](k)}(t_k, x_{t_k}) e^{r(t_k - t_0)}$ is the expected feedback Nash equilibrium payoff of player i in the game

$$\max_{u_i} \Bigg\{ \int_{t_k}^{t_{k+1}} g^{[i,\theta_{\alpha_k}^k]}[s, x(s), u_1(s), u_2(s), \cdots, u_n(s)]\; e^{-r(s-t_k)} ds$$

$$+ e^{-r(t_{k+1}-t_k)} \sum_{a=1}^{\eta} \lambda_a \overline{V}^{i[\theta_a^{k+1}](k+1)}(t_{k+1}, x(t_{k+1})) \Bigg|\; x(t_k) = x_{t_k} \Bigg\},$$

for $i \in N$,

subject to

$$dx(s) = f[s, x(s), u_1(s), u_2(s), \cdots, u_n(s)]ds + \sigma[s, x(s)]dz(s),$$
$$x(t_k) = x_{t_k} \in X.$$

Remark 1.2 One can also readily verify that $\overline{V}^{i\left[\theta^k_{a_k}\right](k)\tau}(\tau, x_\tau) = V^{i\left[\theta^k_{a_k}\right](k)}(\tau, x_\tau)e^{r(\tau - t_0)}$, for $\tau \in [t_k, t_{k+1})$, is the expected feedback Nash equilibrium payoff of player i in the game

$$\max_{u_i} \left\{ \int_\tau^{t_{k+1}} g^{\left[i, \theta^k_{a_k}\right]}[s, x(s), u_1(s), u_2(s), \cdots, u_n(s)] \; e^{-r(s-\tau)} ds \right.$$
$$\left. + e^{-r(t_{k+1} - \tau)} \sum_{a=1}^{\eta} \lambda_a \overline{V}^{i\left[\theta^{k+1}_a\right](k+1)}(t_{k+1}, x(t_{k+1})) \middle| \; x(\tau) = x_\tau \right\}, \quad \text{for } i \in N,$$

subject to

$$dx(s) = f[s, x(s), u_1(s), u_2(s), \cdots, u_n(s)]ds + \sigma[s, x(s)]dz(s),$$
$$x(\tau) = x_\tau \in X.$$

4.2 Dynamic Cooperation

Now consider the case when the players want to cooperate and agree to act and allocate the cooperative payoff according to a set of agreed upon optimality principles. The agreement on how to act cooperatively and allocate cooperative payoff constitutes the solution optimality principle of a cooperative scheme. In particular, the optimality principle includes:

(i) an agreement on a set of cooperative strategies/controls, and
(ii) a mechanism to distribute total payoff between players.

Both group rationality and individual rationality are required in a cooperative plan. Group rationality requires the players to seek a set of cooperative strategies/controls that yields a Pareto optimal solution. The allocation principle has to satisfy individual rationality in the sense that no player would be worse off than before under cooperation.

4.2.1 Group Rationality

Since payoffs are transferable, group rationality requires the players to maximize their expected joint payoff

$$E_{t_0}\left\{\sum_{j=1}^{n}\int_{t_0}^{t_1} g^{[j,\theta_0^0]}[s,x(s),u_1(s),u_2(s),\cdots,u_n(s)]\ e^{-r(s-t_0)}ds\right.$$
$$+\sum_{j=1}^{n}\sum_{h=1}^{m}\sum_{a_h=1}^{\eta}\lambda_{a_h}\int_{t_h}^{t_{h+1}} g^{\left[j,\theta_{a_h}^h\right]}[s,x(s),u_1(s),u_2(s),\cdots,u_n(s)]\,e^{-r(s-t_0)}ds$$
$$\left.+e^{-r(T-t_0)}\sum_{j=1}^{n}q^j(x(T))\right\} \tag{2.1}$$

subject to (1.2).

We solve the control problem (2.1) and (1.1) in a manner similar to that we used to solve the game (1.1 and 1.2). In particular, an optimal solution of the problem (1.2) and (2.1) is characterized by the theorem below.

Theorem 2.1 A set of controls $\{\ u_i^{(m)\theta_{\alpha_m}^m}(t)=\psi_i^{(m)\theta_{\alpha_m}^m}(t,x),$ for $t\in[t_m,T]$; $u_i^{(k)\theta_{\alpha_k}^k}(t)=\psi_i^{(k)\theta_{\alpha_k}^k}(t,x),$ for $t\in[t_k,t_{k+1}),$ $k\in\{0,1,2,\cdots,m-1\}$ and $i\in N\ \}$, contingent upon the events $\theta_{\alpha_m}^m$ and $\theta_{\alpha_k}^k$ constitutes an optimal solution for the stochastic control problem (2.1 and 1.2), if there exist continuously differentiable functions $W^{[\theta_{\alpha_m}^m](m)}(t,x):[t_m,T]\times R^\kappa\to R$ and $W^{\left[\theta_{\alpha_k}^k\right](k)}(t,x):[t_k,t_{k+1}]\times R^\kappa\to R$ for $k\in\{0,1,2,\cdots,m-1\}$ which satisfy the following partial differential equations:

$$-W_t^{[\theta_{\alpha_m}^m](m)}(t,x)-\frac{1}{2}\sum_{h,\zeta=1}^{n}\Omega^{h\zeta}(t,x)W_{x^hx^\zeta}^{[\theta_{\alpha_m}^m](m)}(t,x)$$
$$=\max_{u_1^{\theta_{\alpha_m}},u_2^{\theta_{\alpha_m}},\cdots,u_n^{\theta_{\alpha_m}}}\left\{\sum_{j=1}^{n}g^{[j,\theta_{\alpha_m}^m]}\left[t,x(t),u_1^{(m)\theta_{\alpha_m}^m},u_2^{(m)\theta_{\alpha_m}^m},\cdots,u_n^{(m)\theta_{\alpha_m}^m}\right]e^{-r(t-t_\tau)}\right.$$
$$\left.+W_x^{[\theta_{\alpha_m}^m](m)}(t,x)f\left[t,x,u_1^{(m)\theta_{\alpha_m}^m},u_2^{(m)\theta_{\alpha_m}^m},\cdots,u_n^{(m)\theta_{\alpha_m}^m}\right]\right\},\text{ and}$$
$$W^{[\theta_{\alpha_m}^m](m)}(T,x)=e^{-r(T-t_0)}\sum_{j=1}^{n}q^j(x);$$
$$-W_t^{\left[\theta_{\alpha_k}^k\right](k)}(t,x)-\frac{1}{2}\sum_{h,\zeta=1}^{n}\Omega^{h\zeta}(t,x)W_{x^hx^\zeta}^{\left[\theta_{\alpha_k}^k\right](k)}(t,x)$$
$$=\max_{u_1^{\theta_{\alpha_k}},u_2^{\theta_{\alpha_k}},\cdots,u_2^{\theta_{\alpha_k}}}\left\{\sum_{j=1}^{n}g^{\left[j,\theta_{\alpha_k}^k\right]}\left[t,x,u_1^{(k)\theta_{\alpha_k}^k},u_2^{(k)\theta_{\alpha_k}^k},\cdots,u_n^{(k)\theta_{\alpha_k}^k}\right]e^{-r(t-t_k)}\right.$$
$$\left.+W_x^{\left[\theta_{\alpha_k}^k\right](k)}(t,x)f\left[t,x,u_1^{(k)\theta_{\alpha_k}^k},u_2^{(k)\theta_{\alpha_k}^k},\cdots,u_n^{(k)\theta_{\alpha_k}^k}\right]\right\},\text{ and}$$
$$W^{\left[\theta_{\alpha_k}^k\right](k)}(t_{k+1},x)=\sum_{a=1}^{\eta}\lambda_aW^{[\theta_a^{k+1}](k)}(t_{k+1},x),\ \text{for}\,k\in\{0,1,2,\cdots,m-1\}.$$

Proof Following the argument in the analysis in Sect. 4.1 we obtain $\sum_{a=1}^{\eta} \lambda_a W^{[\theta_a^m](k+1)}(t_{k+1}, x_{t_{k+1}})$ as the expected terminal value for the stochastic control problem in the time interval $[t_k, t_{k+1}]$, for $k \in \{0, 1, 2, \cdots, m\}$. Then direct application of the stochastic control technique in Theorem A.3 in the Technical Appendices and the Nash equilibrium condition yields Theorem 2.1. ■

Hence under cooperation the players will adopt the cooperative strategy $\left[\psi_i^{(h)\theta_{a_h}^h}(t,x), \psi_2^{(h)\theta_{a_h}^h}(t,x), \cdots, \psi_n^{(h)\theta_{a_h}^h}(t,x)\right]$ in the time interval $[t_h, t_{h+1})$ if $\theta_{a_h} \in \{\theta_1, \theta_2, \ldots, \theta_\eta\}$ occurs at time t_h, for $h \in \{0, 1, 2, \cdots, m\}$. In a cooperative framework, the issue of non-uniqueness of the optimal controls can be resolved by agreement between the players on a particular set of controls. Substituting the set of cooperative strategy into (1.2) yields the dynamics of the cooperative state trajectory in the time interval $[t_k, t_{k+1})$ for $k \in \{0, 1, 2, \cdots, m\}$ as

$$
\begin{aligned}
dx(s) = f&\left[s, x(s), \psi_1^{(k)\theta_{a_k}^k}(s, x(s)), \psi_2^{(k)\theta_{a_k}^k}(s, x(s)), \cdots, \psi_n^{(k)\theta_{a_k}^k}(s, x(s))\right] ds \\
&+ \sigma[s, x(s)]\, dz(s),
\end{aligned}
\tag{2.2}
$$

$x(t_k) = x_{t_k}$, for $s \in [t_k, t_{k+1})$, if $\theta_{a_k}^k \in \{\theta_1, \theta_2, \ldots, \theta_\eta\}$ occurs at time t_k.

For simplicity in exposition we denote the set of state variable realizable at time t according to (2.2) by X_t^*, and use x_t^* to denote an element in X_t^* that would occur.

Finally, similar to Remarks 1.1 and 1.2 we have two results that will be utilized in subsequent analysis:

Remark 2.1 One can readily verify that $\overline{W}^{[\theta_{a_k}^k](k)}(t_k, x_k) = W^{[\theta_{a_k}^k](k)}(t_k, x_k) e^{r(t_k - t_0)}$ is the maximized value of the stochastic control problem

$$
\begin{aligned}
\max_{u_1, u_2, \cdots, u_n} E_{t_k} \Bigg\{ & \sum_{j=1}^{n} \int_{t_k}^{t_{k+1}} g^{[j, \theta_{a_k}^k]}[s, x(s), u_1(s), u_2(s), \cdots, u_n(s)]\; e^{-r(s-t_k)} ds \\
& + \sum_{j=1}^{n} \sum_{h=k+1}^{m} \sum_{a_h=1}^{\eta} \lambda_{a_h} \int_{t_h}^{t_{h+1}} g^{[j, \theta_{a_h}^h]}[s, x(s), u_1(s), u_2(s), \cdots, u_n(s)] e^{-r(s-t_k)} ds \\
& + e^{-r(T-t_k)} \sum_{j=1}^{n} q^j(x(T)) \Bigg\}
\end{aligned}
$$

subject to

$$
\begin{aligned}
& dx(s) = f[s, x(s), u_1(s), u_2(s), \cdots, u_n(s)]ds + \sigma[s, x(s)]dz(s), \\
& x(t_k) = x_{t_k} \in X.
\end{aligned}
$$

Remark 2.2 One can readily verify that

$$\overline{W}^{\left[\theta_{a_k}^k\right](k)\tau}(\tau, x_\tau) = W^{\left[\theta_{a_k}^k\right](k)}(\tau, x_\tau)e^{r(\tau - t_0)}, \text{ for } \tau \in \left[t_k, t_{k+1}\right),$$

is the maximized value of the stochastic control problem

$$\begin{aligned}
\max_{u_1, u_2, \cdots, u_n} \; & E_\tau\left\{ \sum_{j=1}^{n} \int_\tau^{t_{k+1}} g^{\left[j, \theta_{a_k}^k\right]}[s, x(s), u_1(s), u_2(s), \cdots, u_n(s)] \; e^{-r(s-\tau)} ds \right. \\
& + \sum_{j=1}^{2} \sum_{h=k+1}^{m} \sum_{a_h=1}^{\eta} \lambda_{a_h} \int_{t_h}^{t_{h+1}} g^{\left[j, \theta_{a_h}^h\right]}[s, x(s), u_1(s), u_2(s), \cdots, u_n(s)] \, e^{-r(s-\tau)} ds \\
& \left. + \, e^{-r(T-\tau)} \sum_{j=1}^{n} q^j(x(T)) \right\}
\end{aligned}$$

subject to

$$dx(s) = f[s, x(s), u_1(s), u_2(s), \cdots, u_n(s)]ds + \sigma[s, x(s)]dz(s), x(\tau) = x_\tau \in X.$$

4.2.2 *Individual Rationality*

Assume that at time t_0 when the initial state is x_0 the agreed upon optimality principle assigns a set of imputation vectors contingent upon the events θ_0^0 and $\theta_{a_h}^h$ for $\theta_{a_h}^h \in \{\theta_1, \theta_2, \quad \ldots, \theta_\eta\}$ and $h \in \{1, 2, \cdots, m\}$. We use

$$\left[\xi^{1\left[\theta_0^0\right](0)}(t_0, x_0), \xi^{2\left[\theta_0^0\right](0)}(t_0, x_0), \cdots, \xi^{n\left[\theta_0^0\right](0)}(t_0, x_0)\right]$$

to denote an imputation vector of the gains in such a way that the share of the ith player over the time interval $[t_0, T]$ is equal to $\xi^{i\left[\theta_0^0\right](0)t_0}(t_0, x_0)$.

Individual rationality requires that

$$\xi^{i\left[\theta_0^0\right](0)t_0}(t_0, x_0) \geq V^{i\left[\theta_0^0\right](0)}(t_0, x_0), \text{ for } i \in N.$$

In a dynamic framework, individual rationality has to be maintained at every instant of time $t \in [t_0, T]$ along the cooperative trajectory. At time t, for $t \in \left[t_0, t_1\right)$, if the players are allowed to reconsider their cooperative plan, they will compare their expected cooperative payoff to their expected noncooperative payoff at that time. Using the same optimality principle, at time t, for $t \in \left[t_0, t_1\right)$, an imputation vector will assign the shares of the players over the time interval $[t, T]$ as

$\left[\xi^{1[\theta_0^0](0)t}(t,x_t^*),\xi^{2[\theta_0^0](0)t}(t,x_t^*),\cdots,\xi^{n[\theta_0^0](0)t}(t,x_t^*)\right]$ (in current value at time t). Individual rationality requires that

$$\xi^{i[\theta_0^0](0)t}(t,x_t^*)\geq\overline{V}^{i[\theta_0^0](0)t}(t,x_t^*),\ \text{for}\, i\in N\,\text{and}\, t\in[t_0,t_1).$$

At time t_h, for $h\in\{1,2,\cdots,m\}$, if $\theta_{a_h}^h\in\{\theta_1,\theta_2,\ \ldots,\theta_\eta\}$ has occurred and the state is $x_{t_h}^*$, the same optimality principle assigns an imputation vector $\left[\xi^{1\left[\theta_{a_h}^h\right](h)t_h}\left(t_h,x_{t_h}^*\right),\xi^{2\left[\theta_{a_h}^h\right](h)t_h}\left(t_h,x_{t_h}^*\right),\cdots,\xi^{n\left[\theta_{a_h}^h\right](h)t_h}\left(t_h,x_{t_h}^*\right)\right]$ (in current value at time t_h). Individual rationality is satisfied if:

$$\xi^{i\left[\theta_{a_h}^h\right](h)t_h}\left(t_h,x_{t_h}^*\right)\geq\overline{V}^{i\left[\theta_{a_h}^h\right](h)t_h}\left(t_h,x_{t_h}^*\right).\qquad \text{for}\, i\in N.$$

Using the same optimality principle, at time t, for $t\in[t_h,t_{h+1})$, an imputation vector will assign the shares of the players over the time interval $[t,T]$ as $\left[\xi^{1\left[\theta_{a_h}^h\right](h)t}(t,x_t^*),\xi^{2\left[\theta_{a_h}^h\right](h)t}(t,x_t^*),\cdots,\xi^{n\left[\theta_{a_h}^h\right](h)t}(t,x_t^*)\right]$ (in terms of current value at time t). Individual rationality requires that

$$\xi^{i\left[\theta_{a_h}^h\right](h)t}(t,x_t^*)\geq\overline{V}^{i\left[\theta_{a_h}^h\right](h)t}(t,x_t^*),\ \text{for}\ i\in N, t\in[t_h,t_{h+1})\ \text{and}\ h\in\{1,2,\cdots,m\}.$$

4.3 Subgame Consistent Solution and Payoff Distribution

A stringent requirement for solutions of cooperative stochastic differential games to be dynamically stable is the property of subgame consistency. Under subgame consistency, an extension of the solution policy to a situation with a later starting time and any feasible state brought about by prior optimal behaviors would remain optimal. In particular, when the game proceeds, at each instant of time the players are guided by the same optimality principles, and hence do not have any ground for deviation from the previously adopted optimal behavior throughout the game. A dynamically stable solution to the randomly furcating game (1.1 and 1.2) is sought in this section.

4.3.1 Solution Imputation Vector

According to the solution optimality principle the players agree to share their cooperative payoff according to the following set of imputation vectors

$$\left[\xi^{1[\theta_0^0](0)}(t_0,x_0),\xi^{2[\theta_0^0](0)}(t_0,x_0),\cdots,\xi^{n[\theta_0^0](0)}(t_0,x_0)\right] \text{ at time } t_0,$$

$$\left[\xi^{1[\theta_0^0](0)t}(t,x_t^*),\xi^{2[\theta_0^0](0)t}(t,x_t^*),\cdots,\xi^{n[\theta_0^0](0)t}(t,x_t^*)\right] \text{ for } t\in[t_0,t_1),$$

$$\left[\xi^{1\left[\theta_{a_h}^h\right](h)}\left(t_h,x_{t_h}^*\right),\xi^{2\left[\theta_{a_h}^h\right](h)}\left(t_h,x_{t_h}^*\right),\cdots,\xi^{n\left[\theta_{a_h}^h\right](h)}\left(t_h,x_{t_h}^*\right)\right] \text{ at time } t_h,$$

for $\theta_{a_h}^h\in\{\theta_1,\theta_2,\ \ldots,\theta_\eta\}$ and $h\in\{1,2,\cdots,m\}$,

$$\left[\xi^{1\left[\theta_{a_h}^h\right](h)t}(t,x_t^*),\xi^{1\left[\theta_{a_h}^h\right](h)t}(t,x_t^*),\cdots,\xi^{n\left[\theta_{a_h}^h\right](h)t}(t,x_t^*)\right]$$

for $t\in[t_h,t_{h+1})$ and $\theta_{a_h}^h\in\{\theta_1,\theta_2,\ \ldots,\theta_\eta\}$ and $h\in\{1,2,\cdots,m\}$. (3.1)

Since (3.1) is guided by a solution optimality principle group optimality and individual rationality are satisfied.

The solution imputation $\xi^{i\left[\theta_{a_k}^k\right](k)\tau}(t,x_t^*)$ may be governed by many specific principles. For instance, the players agree to maximize the sum of their payoffs and equally divide the excess of the cooperative payoff over the noncooperative payoff. The imputation scheme has to satisfy:

Scheme 3.1

$$\xi^{i\left[\theta_{a_k}^k\right](k)}\left(t_k,x_{t_k}^*\right)=\overline{V}^{i\left[\theta_{a_k}^k\right](k)}\left(t_k,x_{t_k}^*\right)+\frac{1}{n}\left[\overline{W}^{\left[\theta_{a_k}^k\right](k)}\left(t_k,x_{t_k}^*\right)\right.$$
$$\left.-\sum_{j=1}^{n}\overline{V}^{j\left[\theta_{a_k}^k\right](k)}\left(t_k,x_{t_k}^*\right)\right], \text{ and}$$

$$\xi^{i\left[\theta_{a_k}^k\right](k)t}(t,x_t^*)=\overline{V}^{i\left[\theta_{a_k}^k\right](k)t}(t,x_t^*)+\frac{1}{n}\left[\overline{W}^{\left[\theta_{a_k}^k\right](k)}\left(t_k,x_{t_k}^*\right)\right.$$
$$\left.-\sum_{j=1}^{n}\overline{V}^{j\left[\theta_{a_k}^k\right](k)t}\left(t_k,x_{t_k}^*\right)\right],$$

for $i\in N$ and $t\in(t_k,t_{k+1})$.

As another example, the solution imputation $\xi^{i\left[\theta_{a_k}^k\right](k)\tau}(t,x_t^*)$ may be an allocation principle in which the players allocate the total joint payoff according to the relative sizes of the firms' noncooperative profits. Hence the imputation scheme has to satisfy

Scheme 3.2

$$\xi^{i\left[\theta_{a_k}^k\right](k)}\left(t_k,x_{t_k}^*\right)=\frac{\overline{V}^{i\left[\theta_{a_k}^k\right](k)}\left(t_k,x_{t_k}^*\right)}{\sum_{j=1}^{n}\overline{V}^{j\left[\theta_{a_k}^k\right](k)}\left(t_k,x_{t_k}^*\right)}\overline{W}^{\left[\theta_{a_k}^k\right](k)}\left(t_k,x_{t_k}^*\right), \text{ and}$$

$$\xi^{i\left[\theta_{a_k}^k\right](k)t}(t,x_t^*)=\frac{\overline{V}^{i\left[\theta_{a_k}^k\right](k)t}(t,x_t^*)}{\sum_{j=1}^{n}\overline{V}^{i\left[\theta_{a_k}^k\right](k)t}(t,x_t^*)}\overline{W}^{\left[\theta_{a_k}^k\right](k)t}(t,x_t^*), \text{ for } i\in N \text{ and } t\in(t_k,t_{k+1}).$$

Crucial to the analysis is the formulation of a payoff distribution mechanism that would lead to the realization of Condition (3.1). This will be done in the next subsection.

4.3.2 Subgame-Consistent Payoff Distribution Procedure

First consider the cooperative subgame in the last time interval, that is $[t_m, T]$ in which $\theta_{a_m}^m \in \{\theta_1, \theta_2, \ldots, \theta_\eta\}$ has occurred at time t_m. To maximize expected joint payoff the players

$$\max_{u_1, u_2, \cdots, u_n} E_{t_m}\left\{\sum_{j=1}^{n}\int_{t_m}^{T} g^{[j,\theta_{a_m}^m]}[s, x(s), u_1(s), u_2(s), \cdots, u_n(s)]\, e^{-r(s-t_m)} ds + e^{-r(T-t_m)}\sum_{j=1}^{n} q^j(x(T))\right\} \tag{3.2}$$

subject to

$$dx(s) = f[s, x(s), u_1(s), u_2(s), \cdots, u_n(s)]ds + \sigma[s, x(s)]dz(s),$$
$$x(t_m) = x_{t_m}^*. \tag{3.3}$$

According to (3.1) the players agree to share their cooperative payoff according to the imputation

$$\left[\xi^{1[\theta_{a_m}^m](m)t_m}\left(t_m, x_{t_m}^*\right), \xi^{2[\theta_{a_m}^m](m)t_m}\left(t_m, x_{t_m}^*\right), \cdots, \xi^{n[\theta_{a_m}^m](m)t_m}\left(t_m, x_{t_m}^*\right)\right].$$

Following Yeung and Petrosyan (2004), we formulate a payoff distribution over time so that the agreed imputations can be realized. Let the vectors $\left[B_1^{(\theta_{a_m}^m)m}(s), B_2^{(\theta_{a_m}^m)m}(s), \cdots, B_n^{(\theta_{a_m}^m)m}(s)\right]$ denote the instantaneous payoff at time $s \in [t_m, T]$ for the cooperative subgame (3.2 and 3.3). In other words, player i, for $i \in N$, obtains an instantaneous payment $B_i^{(\theta_{a_m}^m)m}(s)$ at time instant s. A terminal value of $q^i(x_T^*)$ is received by player i at time T.

In particular, $B_i^{(\theta_{a_m}^m)m}(s)$ and $q^i(x_T^*)$ constitute a payoff distribution for the subgame in the sense that

$$\xi^{i[\theta_{a_m}^m](m)t_m}\left(t_m, x_{t_m}^*\right) = E_{t_m}\left\{\left(\int_{t_m}^{T} B_i^{(\theta_{a_m}^m)m}(s)\, e^{-r(s-t_m)} ds + e^{-r(T-t_m)} q^i\left(x_T^*\right)\right) \middle| \, x(t_m) = x_{t_m}^*\right\},$$

for $i \in N$. (3.4)

As the game proceed to at time t, for $t \in [t_m, T)$, using the same optimality principle an imputation vector will assign the shares of the players over the time interval $[t, T]$ as $\xi^{i[\theta_{a_m}^m](m)t}(t, x_t^*)$. For consistency reasons, it is required that

$$\begin{aligned}
&\xi^{i[\theta_{a_m}^m](m)t}(t, x_t^*) \\
&= E_t\left\{\left(\int_t^T B_i^{(\theta_{a_m}^m)m}(s)\, e^{-r(s-t)} ds + e^{-r(T-t)} q^i(x_T^*) \right) \middle|\, x(t) = x_t^* \right\}, \\
&\text{for } t \in [t_m, T].
\end{aligned} \tag{3.5}$$

To fulfill group optimality, it is required that

$$\begin{aligned}
&\sum_{j=1}^{n} \xi^{j[\theta_{a_m}^m](m)t}(t, x_t^*) = \overline{W}^{[\theta_{a_m}^m](m)t}(t, x_t^*) \text{ for } t \in [t_m, T], \text{ and} \\
&\sum_{j=1}^{n} B^{j[\theta_{a_m}^m](m)t}(t) \\
&= \sum_{j=1}^{n} g^{[j, \theta_{a_m}^m]}\left[t, x_t^*, \psi_1^{(m)\theta_{a_m}^m}(t, x_t^*), \psi_2^{(m)\theta_{a_m}^m}(t, x_t^*), \cdots, \psi_n^{(m)\theta_{a_m}^m}(t, x_t^*)\right].
\end{aligned} \tag{3.6}$$

If the conditions from (3.4) to (3.6) are satisfied, one can say that the solution imputations are time-consistent in the sense that (3.1) can be realized.

Now we consider

$$\begin{aligned}
&\xi^{i[\theta_{a_m}^m](m)t_m}(t, x_t^*) = E_{t_m}\left\{\left(\int_t^T B_i^{(\theta_{a_m}^m)m}(s)\, e^{-r(s-t_m)} ds \right.\right. \\
&\left.\left. + e^{-r(T-t_m)} q^i(x_T^*) \right) \middle|\, x(t) = x_t^* \right\}, \\
&\text{for } t \in [t_m, T] \text{ and } i \in N.
\end{aligned} \tag{3.7}$$

Using (3.4), (3.5) and (3.7), we have

$$\xi^{i[\theta_{a_m}^m](m)t_m}(t, x_t^*) = e^{-r(t-t_m)} \xi^{i[\theta_{a_m}^m](m)t}(t, x_t^*), \quad \text{for } t \in [t_m, T]. \tag{3.8}$$

Moreover, we can write

$$\begin{aligned}
&\xi^{i[\theta_{a_m}^m](m)\tau}(\tau, x_\tau^*) = E_\tau\left\{ \int_\tau^{\tau+\Delta t} B_i^{(\theta_{a_m}^m)m}(s)\, e^{-r(s-\tau)} ds \right. \\
&\left. + e^{-r(\Delta t)} \xi^{i[\theta_{a_m}^m](m)\tau+\Delta t}(\tau + \Delta t, x_\tau^* + \Delta x_\tau^*) \middle|\, x(\tau) = x_\tau^* \right\} \\
&\text{for } \tau \in [t_m, T] \text{ and } i \in N;
\end{aligned} \tag{3.9}$$

where

$$\Delta x_\tau^* = f\left[\tau, x_\tau^*, \psi_1^{(m)\theta_{am}^m}(\tau, x_\tau^*), \psi_2^{(m)\theta_{am}^m}(\tau, x_\tau^*), \cdots, \psi_n^{(m)\theta_{am}^m}(\tau, x_\tau^*)\right]\Delta t$$
$$+\sigma\left[\tau, x_\tau^*\right]\Delta z_\tau + o(\Delta t),$$

and

$\Delta z_\tau = z(\tau + \Delta t) - z(\tau)$, and $E_t[o(\Delta t)]/\Delta t \to 0$ as $\Delta t \to 0$.

From (3.9) we obtain

$$E_\tau\left\{\int_\tau^{\tau+\Delta t} B_i^{(\theta_{am}^m)m}(s)\, e^{-r(s-\tau)} ds \,\middle|\, x(\tau) = x_\tau^*\right\}$$
$$= \xi^{i[\theta_{am}^m](m)t}(t, x_t^*) - e^{-r(\Delta t)} \xi^{i[\theta_{am}^m](m)t+\Delta t}(t + \Delta t, x_t^* + \Delta x_t^*). \tag{3.10}$$

Invoking (3.8) yields

$$E_\tau\left\{\int_t^{t+\Delta t} B_i^{(\theta_{am}^m)m}(s)\, e^{-r(s-t)} ds \,\middle|\, x(t) = x_t^*\right\}$$
$$= \xi^{i[\theta_{am}^m](m)\tau}(\tau, x_\tau^*) - \xi^{i[\theta_{am}^m](m)\tau}(\tau + \Delta t, x_\tau^* + \Delta x_\tau^*), \tag{3.11}$$

For imputations $\xi^{i[\theta_{am}^m](m)\tau}(t, x_t^*)$, for $\tau \in [t_m, T]$ and $t \in [\tau, T]$ being functions that are continuously twice differentiable in t and x_t^*, one can express (3.11), with $\Delta t \to 0$, as:

$$E_\tau\left\{B_i^{(\theta_{am}^m)m}(\tau)\Delta t + o(\Delta t)\right\} = E_\tau\left\{-\left[\xi_t^{i[\theta_{am}^m](m)\tau}(t, x_t^*)|_{t=\tau}\right]\Delta t\right.$$
$$-\left[\xi_{x_t^*}^{i[\theta_{am}^m](m)\tau}(t, x_t^*)|_{t=\tau}\right]$$
$$f\left[\tau, x_\tau^*, \psi_1^{(m)\theta_{am}^m}(\tau, x_\tau^*), \psi_2^{(m)\theta_{am}^m}(\tau, x_\tau^*), \cdots, \psi_n^{(m)\theta_{am}^m}(\tau, x_\tau^*)\right]\Delta t$$
$$-\frac{1}{2}\sum_{h,\zeta=1}^{n} \Omega^{h\zeta}(\tau, x_\tau^*)\left[\xi_{x_t^h x_t^\zeta}^{i[\theta_{am}^m](m)\tau}(t, x_t^*)\Big|_{t=\tau}\right]\Delta t$$
$$\left.-\left[\xi_{x_t}^{i[\theta_{am}^m](m)\tau}(t, x_t^*)\Big|_{t=\tau}\right]\sigma\left[\tau, x_\tau^*\right], \Delta z_\tau, -, o(\Delta t)\right\}. \tag{3.12}$$

Dividing (3.12) throughout by Δt, with $\Delta t \to 0$, and taking expectation yield

$$B_i^{(\theta_{am}^m)m}(\tau) = -\left[\xi_t^{i[\theta_{am}^m](m)\tau}(t, x_t^*)|_{t=\tau}\right]$$
$$-\left[\xi_{x_t^*}^{i[\theta_{am}^m](m)\tau}(t, x_t^*)|_{t=\tau}\right]$$
$$f\left[\tau, x_\tau^*, \psi_1^{(m)\theta_{am}^m}(\tau, x_\tau^*), \psi_2^{(m)\theta_{am}^m}(\tau, x_\tau^*), \cdots, \psi_n^{(m)\theta_{am}^m}(\tau, x_\tau^*)\right]$$
$$-\frac{1}{2}\sum_{h,\zeta=1}^{n} \Omega^{h\zeta}(\tau, x_\tau^*)\left[\xi_{x_t^h x_t^\zeta}^{i[\theta_{am}^m](m)\tau}(t, x_t^*)\Big|_{t=\tau}\right], \text{ for } i \in N, \tag{3.13}$$

One can repeat the analysis from (3.4) to (3.13) for all $\xi^{i[\theta^m_{a_m}](m)\tau}(\tau, x^*_\tau)$ each associated with an $\theta^m_{a_m} \in \{\theta_1, \theta_2, \ldots, \theta_\eta\}$ and obtain the corresponding $B_i^{(\theta^m_{a_m})m}(\tau)$ for $\tau \in [t_m, T]$.

In order to formulate the cooperative subgame in the second last time interval $[t_{m-1}, t_m]$, it is necessary to identify the expected terminal payoffs at time t_m. Using Theorem 2.1, one can obtain $W^{[\theta^m_{a_m}](m)}(t_m, x^*_m)$ if $\theta^m_{a_m} \in \{\theta_1, \theta_2, \ldots, \theta_\eta\}$ occurs at time t_m. The term $\sum_{a=1}^{\eta} W^{[\theta^m_a](m)}(t_m, x^*_m)$ gives the expected joint payoff of the cooperative game over the duration $[t_m, T]$ and hence is the expected terminal joint payoff for the cooperative subgame in the time interval $[t_{m-1}, t_m]$. In a similar manner, the term $\sum_{a=1}^{\eta} W^{[\theta^{k+1}_a](k+1)}(t_{k+1}, x^*_{k+1})$ gives the expected terminal joint payoff for the cooperative subgame in the time interval $[t_k, t_{k+1}]$ for $k \in \{0, 1, 2, \cdots, m-1\}$. In general, the cooperative subgame in the time interval $[t_k, t_{k+1}]$ if $\theta^k_{a_k} \in \{\theta_1, \theta_2, \ldots, \theta_\eta\}$ occurs at time t_k for $k \in \{0, 1, 2, \cdots, m-1\}$ can be expressed as:

$$\max_{u_1, u_2} E_{t_k}\left\{ \sum_{j=1}^{2} \int_{t_k}^{t_{k+1}} g^{[j, \theta^k_{a_k}]}[s, x(s), u_1(s), u_2(s), \cdots, u_n(s)]\, e^{-r(s-t_k)} ds \right.$$
$$\left. + e^{-r(t_{k+1}-t_k)} \sum_{j=1}^{2} \sum_{a=1}^{\eta} \overline{W}^{[\theta^{k+1}_a](k+1)}(t_{k+1}, x(t_{k+1})) \right\} \tag{3.14}$$

subject to

$$dx(s) = f[s, x(s), u_1(s), u_2(s), \cdots, u_n(s)]ds + \sigma[s, x(s)]dz(s),$$
$$x(t_k) = x^*_k. \tag{3.15}$$

One can repeat the analysis from (3.4) to (3.13) for all $\xi^{i[\theta^k_{a_k}](k)\tau}(\tau, x^*_\tau)$ each associated with an $\theta^k_{a_k} \in \{\theta_1, \theta_2, \ldots, \theta_\eta\}$ for $k \in \{0, 1, 2, \cdots, m-1\}$ and derive the corresponding $B_i^{(\theta^k_{a_k})k}(\tau)$ for $\tau \in [t_k, t_{k+1})$.

A theorem characterizing a subgame consistent PDP is provided below.

Theorem 3.1 If the solution imputations $\xi^{i[\theta^k_{a_k}](k)\tau}(t, x^*_t)$, for $i \in N$ and $\tau \in [t_k, t_{k+1}]$ and $t \in [\tau, t_{k+1}]$ and $k \in \{0, 1, 2, \cdots, m-1\}$, satisfy group optimality, individual rationality and are differentiable in t and x^*_t, a PDP with a terminal payment $q^i(x^*_T))$ at time T and an instantaneous payment at time $\tau \in [t_k, t_{k+1}]$:

$$B_i^{(\theta_{a_k}^k)k}(\tau) = -\left[\xi_t^{i\left[\theta_{a_k}^k\right](k)\tau}(t, x_t^*)|_{t=\tau}\right]$$
$$-\left[\xi_{x_t^*}^{i\left[\theta_{a_k}^k\right](k)\tau}(t, x_t^*)|_{t=\tau}\right]$$
$$f\left[\tau, x_\tau^*, \psi_1^{(k)\theta_{a_k}^k}(\tau, x_\tau^*), \psi_2^{(k)\theta_{a_k}^k}(\tau, x_\tau^*), \cdots, \psi_n^{(k)\theta_{a_k}^k}(\tau, x_\tau^*)\right]$$
$$-\frac{1}{2}\sum_{h,\zeta=1}^{n}\Omega^{h\zeta}(\tau, x_\tau^*)\left[\xi_{x_t^h x_t^\zeta}^{i\left[\theta_{a_k}^k\right](k)\tau}(t, x_t^*)\bigg|_{t=\tau}\right], \tag{3.16}$$

for $i \in N$ and $k \in \{1, 2, \cdots, m\}$,

contingent upon $\theta_{a_k}^k \in \{\theta_1, \theta_2, \ \ldots, \theta_\eta\}$ has occurred at time t_k,

yields a subgame-consistent cooperative solution to the randomly furcating stochastic differential game (1.1 and 1.2).

Proof Theorem 3.1 can be proved by following the analysis from (3.4) to (3.15). ■

4.4 An Illustration in Cooperative Resource Extraction

Consider a resource extraction game, in which two extractors are awarded leases to extract a renewable resource over the time interval $[t_0, T]$. The resource stock $x(s) \in X \subset R$ follows the dynamics:

$$dx(s) = \left[ax(s)^{1/2} - bx(s) - u_1(s) - u_2(s)\right]ds + \sigma x(s)dz(s),\ x(t_0) = x_0 \in X, \tag{4.1}$$

where $u_1(s)$ is the harvest rate of extractor 1 and $u_2(s)$ is the harvest rate of extractor 2. The dynamics is adopted from Jørgensen and Yeung (1996).

The instantaneous payoff at time $s \in [t_0, T]$ for player 1 and player 2 are respectively:

$$\left[u_1(s)^{1/2} - \frac{\varepsilon_1^{[a]}c_1}{x(s)^{1/2}}u_1(s)\right] \text{ and } \left[u_2(s)^{1/2} - \frac{\varepsilon_2^{[a]}c_2}{x(s)^{1/2}}u_2(s)\right],$$

if the event θ_a happens for $a \in \{1, 2, 3\}$, where $\varepsilon_1^{[a]}$, $\varepsilon_2^{[a]}$, c_1 and c_2 are constants.

At time t_0, it is known that θ_1 has occurred. θ_1 will remain in effect until time $t_1 \in (t_0, T)$. At time t_1, the corresponding probabilities for the events $\{\theta_1, \theta_2, \theta_3\}$ to occur are $\{\lambda_1, \lambda_2, \lambda_3\} = \{1/4,\ 1/2,\ 1/4\}$. The occurred event will remain until the end of the game, that is time T. At time T, each extractor will receive a termination bonus $qx(T)^{1/2}$, which depends on the resource remaining at the terminal time. Payoffs are transferable between player 1 and player 2 and over time. There is a constant discount rate r.

Applying Theorem **1.1**, we obtain the following value functions for the associating noncooperative games.

$$V^{i[\theta_a^1](1)}(t,x) = \exp[-r(t-t_0)]\left[A_i^{\theta_a^1(1)}(t)x^{1/2} + C_i^{\theta_a^1(1)}(t)\right],$$
$$\text{for}\, i \in \{1,2\}, a \in \{1,2,3\}\,\text{and}\, t \in [t_1,T]; \tag{4.2}$$

$$V^{i[\theta_1](0)}(t,x) = \exp[-r(t-t_0)]\left[A_i^{\theta_1(0)}(t)x^{1/2} + C_i^{\theta_1(0)}(t)\right],$$
$$\text{for}\, i \in \{1,2\}\,\text{and}\, t \in [t_0,t_1], \tag{4.3}$$

where $A_i^{\theta_a^1(1)}(t)$, $C_i^{\theta_a^1(1)}(t)$, $A_i^{\theta_1(0)}(t)$ and $C_i^{\theta_1(0)}(t)$ satisfy:

$$\dot{A}_i^{\theta_a^1(1)}(t) = \left[r + \frac{\sigma^2}{8} + \frac{b}{2}\right]A_i^{\theta_a^1(1)}(t)$$
$$-\frac{1}{2\left[\varepsilon_i^{[a]}c_i + A_i^{\theta_a^1(1)}(t)/2\right]} + \frac{\varepsilon_i^{[a]}c_i}{4\left[\varepsilon_i^{[a]}c_i + A_i^{\theta_a^1(1)}(t)/2\right]^2}$$
$$+\frac{A_i^{\theta_a^1(1)}(t)}{8\left[\varepsilon_i^{[a]}c_i + A_i^{\theta_a^1(1)}(t)/2\right]^2} + \frac{A_{\theta_a^1(1)i}(t)}{8\left[\varepsilon_j^{[a]}c_j + A_j^{\theta_a^1(1)}(t)/2\right]^2},$$
$$\dot{C}_i^{\theta_a^1(1)}(t) = rC_i^{\theta_a^1(1)}(t) - \frac{\alpha}{2}A_i^{\theta_a^1(1)}(t),$$
$$A_i^{\theta_a^1(1)}(T) = q,\ \text{and}\, C_i^{\theta_a^1(1)}(T) = 0;\quad \text{for}\, i \in \{1,2\}\,\text{and}\, a \in \{1,2,3\};$$

$$\dot{A}_i^{\theta_1(0)}(t) = \left[r + \frac{\sigma^2}{8} + \frac{b}{2}\right]A_i^{\theta_1(0)}(t)$$
$$-\frac{1}{2\left[\varepsilon_i^{[1]}c_i + A_i^{\theta_1(0)}(t)/2\right]} + \frac{\varepsilon_i^{[1]}c_i}{4\left[\varepsilon_i^{[1]}c_i + A_i^{\theta_1(0)}(t)/2\right]^2}$$
$$+\frac{A_i^{\theta_1(0)}(t)}{8\left[\varepsilon_i^{[1]}c_i + A_i^{\theta_1(0)}(t)/2\right]^2} + \frac{A_i^{\theta_1(0)}(t)}{8\left[\varepsilon_j^{[1]}c_j + A_j^{\theta_1(0)}(t)/2\right]^2},$$
$$\dot{C}_i^{\theta_1(0)}(t) = rC_i^{\theta_1(0)}(t) - \frac{\alpha}{2}A_i^{\theta_1(0)}(t),$$
$$A_i^{\theta_1(0)}(t_1) = \sum_{h=1}^{3}\lambda_h A_i^{\theta_h^1(1)}(t_1),\ \text{and}\, C_i^{\theta_1(0)}(t_1) = \sum_{h=1}^{3}\lambda_h C_i^{\theta_h^1(1)}(t_1).$$

Applying Theorem 2.1, we obtain

$$W^{[\theta_a^1](1)}(t,x) = \exp[-r(t-t_0)]\left[\hat{A}^{\theta_a^1(1)}(t)x^{1/2} + \hat{B}^{\theta_a^1(1)}(t)\right],$$
$$\text{for}\, a \in \{1,2,3\}\,\text{and}\, t \in [t_1,T]; \tag{4.4}$$

$$W^{[\theta_1](0)}(t,x) = \exp[-r(t-t_0)]\left[\hat{A}^{\theta_1(0)}(t)x^{1/2} + \hat{B}^{\theta_1(0)}(t)\right],$$
$$\text{for}\, t \in [t_0,t_1]. \tag{4.5}$$

where $\hat{A}^{\theta_a^1(1)}(t)$, $\hat{B}^{\theta_a^1(1)}(t)$, $\hat{A}^{\theta_1(0)}(t)$ and $\hat{B}^{\theta_1(0)}(t)$ satisfy:

$$\dot{\hat{A}}^{\theta_a^1(1)}(t) = \left[r + \frac{\sigma^2}{8} + \frac{b}{2}\right]\hat{A}^{\theta_a^1(1)}(t) - \sum_{j=1}^{2} \frac{1}{2\left[\varepsilon_j^{[a]} c_j + \hat{A}^{\theta_a^1(1)}(t)/2\right]}$$

$$+ \sum_{j=1}^{2} \frac{\varepsilon_j^{[a]} c_j}{4\left[\varepsilon_j^{[a]} c_j + \hat{A}^{\theta_a^1(1)}(t)/2\right]^2} + \sum_{j=1}^{2} \frac{\hat{A}^{\theta_a^1(1)}(t)}{8\left[\varepsilon_j^{[a]} c_j + \hat{A}^{\theta_a^1(1)}(t)/2\right]^2},$$

$$\dot{\hat{B}}^{\theta_a^1(1)}(t) = r\hat{B}^{\theta_a^1(1)}(t) - \frac{a}{2}\hat{A}^{\theta_a^1(1)}(t),$$

$$\hat{A}^{\theta_a^1(1)}(T) = 2q, \text{ and } \hat{B}^{\theta_a^1(1)}(T) = 0;$$

$$\dot{\hat{A}}^{\theta_1(0)}(t) = \left[r + \frac{\sigma^2}{8} + \frac{b}{2}\right]\hat{A}^{\theta_1(0)}(t) - \sum_{j=1}^{2} \frac{1}{2\left[\varepsilon_j^{[1]} c_j + \hat{A}^{\theta_1(0)}(t)/2\right]}$$

$$+ \sum_{j=1}^{2} \frac{\varepsilon_j^{[1]} c_j}{4\left[\varepsilon_j^{[1]} c_j + \hat{A}^{\theta_1(0)}(t)/2\right]^2} + \sum_{j=1}^{2} \frac{\hat{A}^{\theta_1(0)}(t)}{8\left[\varepsilon_j^{[1]} c_j + \hat{A}^{\theta_1(0)}(t)/2\right]^2},$$

$$\dot{\hat{B}}^{\theta_1(0)}(t) = r\hat{B}^{\theta_1(0)}(t) - \frac{a}{2}\hat{A}^{\theta_1(0)}(t),$$

$$\hat{A}^{\theta_1(0)}(t_1) = \sum_{h=1}^{3} \lambda_h \hat{A}^{\theta_h^1(1)}(t_1), \text{ and } \hat{B}^{\theta_1(0)}(T) = \sum_{h=1}^{3} \lambda_h \hat{B}^{\theta_h^1(1)}(t_1).$$

Using (4.4) and (4.5) the optimal cooperative controls can then be obtained as:

$$\psi_i^{(0)\theta_1}(t,x) = \frac{x}{4\left[\varepsilon_i^{[1]} c_i + \hat{A}^{\theta_1(0)}(t)/2\right]^2}, \text{ for } i \in \{1,2\} \text{ and } t \in [t_0, t_1); \tag{4.6}$$

$$\psi_i^{(1)\theta_a^1}(t,x) = \frac{x}{4\left[\varepsilon_i^{[a]} c_i + \hat{A}^{\theta_a^1(1)}(t)/2\right]^2}, \text{ for } i \in \{1,2\} \text{ and } t \in [t_1, T], \tag{4.7}$$

if $\theta_a^1 \in \{\theta_1, \theta_2, \theta_3\}$ occurs at time t_1.

Substituting these control strategies into (2.2) yields the dynamics of the state trajectory under cooperation. The optimal cooperative state trajectory in the time interval $[t_0, t_1)$ can be obtained as:

$$x^*(t) = \varpi(t_0, t, \theta_1)^2 \left[x_0^{1/2} + \int_{t_0}^{t} \varpi^{-1}(t_0, s) \frac{\alpha}{2} ds \right]^2, \text{ for } t \in [t_0, t_1), \tag{4.8}$$

where $\varpi(t_0, t, \theta_1) = \exp\left[\int_{t_0}^{t} \left[H_0(\theta_1, \upsilon) - \frac{\sigma^2}{8}\right] d\upsilon + \int_{t_0}^{t} \frac{\sigma}{2} dz(\upsilon)\right]$, and

$$H_0(\theta_1, s) = -\left[\frac{b}{2} + \sum_{j=1}^{2} \frac{1}{8\left[\varepsilon_j^{[1]} c_j + \hat{A}^{\theta_1(0)}(s)/2\right]^2} + \frac{\sigma^2}{8}\right].$$

If $\theta_a^1 \in \{\theta_1, \theta_2, \theta_3\}$ occurs at time t_1, the optimal cooperative state trajectory in the interval $[t_1, T]$ becomes

$$x^*(t) = \varpi\left(t_0, t, \theta_a^1\right)^2 \left[\left(x_{t_1}^*\right)^{1/2} + \int_{t_0}^{t} \varpi^{-1}\left(t_1, s, \theta_a^1\right) \frac{\alpha}{2} ds \right]^2, \text{ for } t \in [t_1, T], \quad (4.9)$$

where $\varpi\left(t_1, t, \theta_a^1\right) = \exp\left[\int_{t_1}^{t} \left[H_1\left(\theta_a^1, \upsilon\right) - \frac{\sigma^2}{8}\right] d\upsilon + \int_{t_1}^{t} \frac{\sigma}{2} dz(\upsilon)\right]$, and

$$H_1\left(\theta_a^1, s\right) = -\left[\frac{b}{2} + \sum_{j=1}^{2} \frac{1}{8\left[\varepsilon_j^{[1]} c_j + \hat{A}^{\theta_a^1(0)}(s)/2\right]^2} + \frac{\sigma^2}{8}\right].$$

Now suppose that the players agree to divide their cooperative gains according to scheme 3.1 in the time interval $\left[t_0, t_1\right)$, according scheme 3.1 if θ_1 occurs at time t_1 and according scheme 3.2 if θ_2 or θ_3 occurs at time t_1.

Using Schemes 3.1 and 3.2, Theorem 3.1 and the results derived in section, an instantaneous payment at time $\tau \in \left[t_k, t_{k+1}\right]$:

$$B_i^{\left(\theta_{a_k}^k\right)k}(\tau) = -\left[\xi_t^{i\left[\theta_{a_k}^k\right](k)\tau}\left(t, x_t^*\right)|_{t=\tau}\right]$$

$$-\left[\xi_{x_t^*}^{i\left[\theta_{a_k}^k\right](k)\tau}\left(t, x_t^*\right)|_{t=\tau}\right] f\left[\tau, x_\tau^*, \psi_1^{(k)\theta_{a_k}^k}\left(\tau, x_\tau^*\right), \psi_2^{(k)\theta_{a_k}^k}\left(\tau, x_\tau^*\right)\right]$$

$$-\frac{1}{2}\sum_{h,\zeta=1}^{n} \Omega^{h\zeta}\left(\tau, x_\tau^*\right)\left[\xi_{x_t^h x_t^\zeta}^{i\left[\theta_{a_k}^k\right](k)\tau}\left(t, x_t^*\right)\bigg|_{t=\tau}\right]$$

for $i \in \{1, 2\}$, $k \in \{0, 1\}$, $\theta_{a_0}^0 = \theta_1$ and $\theta_a^1 \in \{\theta_1, \theta_2, \theta_3\}$ can be obtained explicitly using the results derived in (4.2) to (4.8).

4.5 Chapter Notes

This chapter considers subgame-consistent cooperative solutions in randomly furcating stochastic differential games. This approach widens the application of cooperative stochastic differential game theory to problems where future environments are not known with certainty. If the state dynamics is deterministic the above analysis yields subgame consistent cooperative solutions for randomly-furcating differential games. Yeung (2008) considered subgame consistent solutions for a pollution management differential game in collaborative abatement under uncertain future payoffs.

Finally, the random event Θ^k, for $k \in \{1, 2, \cdots, m\}$, affecting the payoffs may be more complex stochastic processes, like a branching process with a series of random events Θ^k, for $k \in \{1, 2, \cdots, m\}$, which is a random variable stemming from the branching process as described below.

Given that $\theta^1_{a_1}$ is realized in time interval $[t_1, t_2)$, for $a_1 = 1, 2, \ldots, \eta_1$, the process Θ^2 in time interval $[t_2, t_3)$ has a range $\theta^2 = \left\{\theta_1^{2[(1,a_1)]}, \theta_2^{2[(1,a_1)]}, \ldots, \theta_{\eta_{2[(1,a_1)]}}^{2[(1,a_1)]}\right\}$ with the corresponding probabilities $\left\{\lambda_1^{2[(1,a_1)]}, \lambda_2^{2[(1,a_1)]}, \ldots, \lambda_{\eta_{2[(1,a_1)]}}^{2[(1,a_1)]}\right\}$.

Given that $\theta^1_{a_1}$ is realized in time interval $[t_1, t_2)$ and $\theta_{a_2}^{2[(1,a_1)]}$ is realized in time interval $[t_2, t_3)$, for $a_1 = 1, 2, \ldots, \eta_1$ and $a_2 = 1, 2, \ldots, \eta_{2[(1,a_1)]}$, $\theta^3 = \left\{\theta_1^{3[(1,a_1)\,(2,a_2)]}, \theta_2^{3[(1,a_1)\,(2,a_2)]}, \ldots, \theta_{\eta_{3[(1,a_1)\,(2,a_2)]}}^{3[(1,a_1)\,(2,a_2)]}\right\}$ would be realized with the corresponding probabilities $\left\{\lambda_1^{3[(1,a_1)\,(2,a_2)]}, \lambda_2^{3[(1,a_1)\,(2,a_2)]}, \ldots, \lambda_{\eta_{3[(1,a_1)\,(2,a_2)]}}^{3[(1,a_1)\,(2,a_2)]}\right\}$.

In general, given that $\theta^1_{a_1}$ is realized in time interval $[t_1, t_2)$, $\theta_{a_2}^{2[(1,a_1)]}$ is realized in time interval $[t_2, t_3)$, ..., and $\theta_{a_{k-1}}^{k-1[(1,a_1)(2,a_2)\ldots(k-2,a_{k-2})]}$ is realized in time interval $[t_{k-1}, t_k)$, for $a_1 = 1, 2, \ldots, \eta_1$, $a_2 = 1, 2, \ldots, \eta_{2[(1,a_1)]}$, ..., $a_{k-1} = 1, 2, \ldots, \eta_{k-1[(1,a_1)(2,a_2)\ldots(k-1,a_{k-1})]}$, $\theta^k = \left\{\theta_1^{k[(1,a_1)\,(2,a_2)\ldots(k-1,a_{k-1})]}, \theta_2^{k[(1,a_1)\,(2,a_2)\ldots(k-1,a_{k-1})]}, \ldots, \theta_{\eta_{k[(1,a_1)(2,a_2)\ldots(k-1,a_{k-1})]}}^{k[(1,a_1)\,(2,a_2)\ldots(k-1,a_{k-1})]}\right\}$

would be realized with the corresponding probabilities

$$\left\{\lambda_1^{k[(1,a_1)\,(2,a_2)\ldots(k-1,a_{k-1})]}, \lambda_2^{k[(1,a_1)\,(2,a_2)\ldots(k-1,a_{k-1})]}, \ldots, \lambda_{\eta_{k[(1,a_1)(2,a_2)\ldots(k-1,a_{k-1})]}}^{k[(1,a_1)\,(2,a_2)\ldots(k-1,a_{k-1})]}\right\}$$

for $k = 1, 2, \ldots, \tau$.

4.6 Problems

1. Consider a resource extraction game, in which two extractors are awarded leases to extract a renewable resource over the time interval $[0, 4]$. The resource stock $x(s) \in X \subset R$ follows the dynamics:

$$dx(s) = \left[10x(s)^{1/2} - x(s) - u_1(s) - u_2(s)\right]ds + 0.05x(s)dz(s), x(0) = 80,$$

where $u_1(s)$ is the harvest rate of extractor 1 and $u_2(s)$ is the harvest rate of extractor 2.

The instantaneous payoff at time $s \in [0,2)$ for player 1 and player 2 are known to be respectively:

$$\left[2u_1(s)^{1/2} - \frac{2}{x(s)^{1/2}}u_1(s)\right] \text{ and } \left[u_2(s)^{1/2} - \frac{2}{x(s)^{1/2}}u_2(s)\right].$$

The instantaneous payoff at time $s \in [2,4]$ for player 1 and player 2 are known to be respectively:

$$\left[2u_1(s)^{1/2} - \frac{2}{x(s)^{1/2}}u_1(s)\right] \text{ and } \left[3u_2(s)^{1/2} - \frac{2}{x(s)^{1/2}}u_2(s)\right] \text{ with probability 0.3,}$$

$$\left[2u_1(s)^{1/2} - \frac{1}{x(s)^{1/2}}u_1(s)\right] \text{ and } \left[2u_2(s)^{1/2} - \frac{1}{x(s)^{1/2}}u_2(s)\right] \text{ with probability 0.4,}$$

$$\text{and } \left[3u_1(s)^{1/2} - \frac{0.5}{x(s)^{1/2}}u_1(s)\right] \text{ and } \left[4u_2(s)^{1/2} - \frac{2}{x(s)^{1/2}}u_2(s)\right] \text{ with probability 0.3.}$$

At terminal time 4, extractor 1 will receive a termination bonus $2x(4)^{1/2}$ and extractor 2 will receive a termination bonus $x(4)^{1/2}$. The discount rate is 0.05. Characterize a feedback Nash equilibrium.

2. Obtain a group optimal solution which maximizes the joint expected payoff of the extractors.
3. Derive a subgame consistent solution in which the players share the excess gain from cooperation equally.

Chapter 5
Subgame Consistency Under Asynchronous Players' Horizons

In many game situations, the players' time horizons differ. This may arise from different life spans, different entry and exit times, and the different duration for leases and contracts. Asynchronous horizon game situations occur frequently in economic and social activities. In this Chapter, subgame consistent cooperative solutions are derived for differential games with asynchronous players' horizons and uncertain types of future players. Analytically tractable payoff distribution mechanisms which lead to the realization of these solutions are derived. This analysis extends the application of cooperative differential game theory to problems where the players' game horizons are asynchronous and the types of future players are uncertain. In particular, the Chapter is an integrated disquisition of the analysis in Yeung (2011) with an extension to incorporate stochastic state dynamics.

The organization of the chapter is as follows. Section 5.1 presents the game formulation and characterizes noncooperative outcomes. Dynamic cooperation among players coexisting in the same duration is examined in Sect. 5.2. Section 5.3 provides an analysis on payoff distribution procedures leading to dynamically consistent solutions in this asynchronous horizons scenario. An illustration in cooperative resource extraction is given in Sect. 5.4. An extension to stochastic dynamics is provided in Sect. 5.5. Chapter notes are given in Sect. 5.6 and problems in Sect. 5.7.

5.1 Game Formulation and Noncooperative Outcome

In this section we present an analytical framework of differential games with asynchronous players' horizons and characterize the noncooperative outcome.

D.W.K. Yeung, L.A. Petrosyan, *Subgame Consistent Cooperation*,
Theory and Decision Library C 47, DOI 10.1007/978-981-10-1545-8_5

5.1.1 Game Formulation

For clarity in exposition and without loss of generality, we consider a general class of differential games, in which there are $\upsilon + 1$ overlapping cohorts or generations of players. The game begins at time t_1 and terminates at time $t_{\upsilon+1}$. In the time interval $[t_1, t_2)$, there coexist a generation 0 player whose game horizon is $[t_1, t_2)$ and a generation 1 player whose game horizon is $[t_1, t_3)$. In the time interval $[t_k, t_{k+1})$ for $k \in \{2, 3, \cdots, \upsilon - 1\}$, there coexist a generation $k - 1$ player whose game horizon is $[t_{k-1}, t_{k+1})$ and a generation k player whose game horizon is $[t_k, t_{k+2})$. In the last time interval $[t_\upsilon, t_{\upsilon+1}]$, there coexist a generation $\upsilon - 1$ player and a generation υ player whose game horizon is just $[t_\upsilon, t_{\upsilon+1}]$.

When the game starts at initial time t_1, it is known that in the time interval $[t_1, t_2)$, there coexist a type ω_0^1 generation 0 player and a type ω_1^1 generation 1 player. At time t_1, it is also known that the probability of the generation k player being type $\omega_k^{a_k} \in \{\omega_k^1, \omega_k^2, \cdots, \omega_k^{\varsigma_k}\}$ is $\lambda_k^{a_k} \in \{\lambda_k^1, \lambda_k^2, \cdots, \lambda_k^{\varsigma_k}\}$, for $k \in \{2, 3, \cdots, \upsilon\}$. The type of generation k player will become known with certainty at time t_k.

The instantaneous payoffs and terminal rewards of the type $\omega_k^{a_k}$ generation k player and the type $\omega_{k-1}^{a_{k-1}}$ generation $k - 1$ player coexisting in the time interval $[t_k, t_{k+1})$ are respectively:

$$g^{k-1(\omega_{k-1}^{a_{k-1}})}\left[s, x(s), u_k^{(\omega_{k-1}^{a_{k-1}}, O)\omega_k^{a_k}}(s), u_k^{(\omega_k^{a_k}, Y)\omega_{k-1}^{a_{k-1}}}(s)\right] \text{ and } q^{k-1(\omega_{k-1}^{a_{k-1}})}[t_{k+1}, x(t_{k+1})] \text{ and}$$
$$g^{k(\omega_k^{a_k})}\left[s, x(s), u_k^{(\omega_{k-1}^{a_{k-1}}, O)\omega_k^{a_k}}(s), u_k^{(\omega_k^{a_k}, Y)\omega_{k-1}^{a_{k-1}}}(s)\right] \text{ and } q^{k(\omega_k^{a_k})}[t_{k+2}, x(t_{k+2})], \tag{1.1}$$

for $k \in \{1, 2, 3, \cdots, \upsilon\}$,

where $u_k^{(\omega_{k-1}^{a_{k-1}}, O)\omega_k^{a_k}}(s)$ is the vector of controls of the type $\omega_{k-1}^{a_{k-1}}$ generation $k - 1$ player when he is in his second (old) life stage while the type $\omega_k^{a_k}$ generation k player is coexisting;

and $u_k^{(\omega_k^{a_k}, Y)\omega_{k-1}^{a_{k-1}}}(s)$ is that of the type $\omega_k^{a_k}$ generation k player when he is in his first (young) life stage while the type $\omega_{k-1}^{a_{k-1}}$ generation $k - 1$ player is coexisting.

Note that the superindex "O" in $u_k^{(\omega_{k-1}^{a_{k-1}}, O)\omega_k^{a_k}}(s)$ denote "Old" and the superindex "Y" in $u_k^{(\omega_k^{a_k}, Y)\omega_{k-1}^{a_{k-1}}}(s)$ denote "Young". The state dynamics of the game is characterized by the vector-valued differential equations:

$$\dot{x}(s) = f\left[s, x(s), u_k^{(\omega_{k-1}^{a_{k-1}}, O)\omega_k^{a_k}}(s), u_k^{(\omega_k^{a_k}, Y)\omega_{k-1}^{a_{k-1}}}(s)\right], \text{ for } s \in [t_k, t_{k+1}), \tag{1.2}$$

if type $\omega_k^{a_k}$ generation k player and type $\omega_{k-1}^{a_{k-1}}$ generation $k - 1$ player co-exist in the time interval $[t_k, t_{k+1})$ for $k \in \{1, 2, 3, \cdots, \upsilon\}$, and $x(t_1) = x_0 \in X$.

In the game interval $\left[t_k, t_{k+1}\right)$ for $k \in \{1, 2, 3, \cdots, \upsilon - 1\}$ with type $\omega_{k-1}^{a_{k-1}}$ generation $k - 1$ player and type $\omega_k^{a_k}$ generation k player, the type $\omega_{k-1}^{a_{k-1}}$ generation $k - 1$ player seeks to maximize:

$$\int_{t_k}^{t_{k+1}} g^{k-1\left(\omega_{k-1}^{a_{k-1}}\right)}\left[s, x(s), u_k^{\left(\omega_{k-1}^{a_{k-1}}, O\right)\omega_k^{a_k}}(s), u_k^{\left(\omega_k^{a_k}, Y\right)\omega_{k-1}^{a_{k-1}}}(s)\right] e^{-r(s-t_k)} ds$$
$$+ e^{-r(t_{k+1}-t_k)} q^{k-1\left(\omega_{k-1}^{a_{k-1}}\right)}\left[t_{k+1}, x(t_{k+1})\right], \tag{1.3}$$

and the type ω_k generation k player seeks to maximize:

$$\int_{t_k}^{t_{k+1}} g^{k\left(\omega_k^{a_k}\right)}\left[s, x(s), u_k^{\left(\omega_{k-1}^{a_{k-1}}, O\right)\omega_k^{a_k}}(s), u_k^{\left(\omega_k^{a_k}, Y\right)\omega_{k-1}^{a_{k-1}}}(s)\right] e^{-r(s-t_k)} ds$$
$$+ \sum_{\ell=1}^{\varsigma_{k+1}} \lambda_{k+1}^{\ell} \int_{t_{k+1}}^{t_{k+2}} g^{k\left(\omega_k^{a_k}\right)}\left[s, x(s), u_{k+1}^{\left(\omega_k^{a_k}, O\right)\omega_{k+1}^{\ell}}(s), u_{k+1}^{\left(\omega_{k+1}^{\ell}, Y\right)\omega_k^{a_k}}(s)\right] e^{-r(s-t_k)} ds$$
$$+ e^{-r(t_{k+2}-t_k)} q^{k\left(\omega_k^{a_k}\right)}\left[t_{k+2}, x(t_{k+2})\right] \tag{1.4}$$

subject to dynamics (1.2), where r is the discount rate.

In the last time interval $[t_\upsilon, t_{\upsilon+1}]$ where the generation $\upsilon - 1$ player is of type $\omega_{\upsilon-1}^{a_{\upsilon-1}}$ and the generation υ player is of type $\omega_\upsilon^{a_\upsilon}$, the type $\omega_{\upsilon-1}^{a_{\upsilon-1}}$ generation $\upsilon - 1$ player seeks to maximize:

$$\int_{t_\upsilon}^{t_{\upsilon+1}} g^{\upsilon-1\left(\omega_{\upsilon-1}^{a_{\upsilon-1}}\right)}\left[s, x(s), u_\upsilon^{\left(\omega_{\upsilon-1}^{a_{\upsilon-1}}, O\right)\omega_\upsilon^{a_\upsilon}}(s), u_\upsilon^{(\omega_\upsilon^{a_\upsilon}, Y)\omega_{\upsilon-1}^{a_{\upsilon-1}}}(s)\right] e^{-r(s-t_\upsilon)} ds$$
$$+ e^{-r(t_{\upsilon+1}-t_\upsilon)} q^{\upsilon-1\left(\omega_{\upsilon-1}^{a_{\upsilon-1}}\right)}\left[t_{\upsilon+1}, x(t_{\upsilon+1})\right], \tag{1.5}$$

and the type $\omega_\upsilon^{a_\upsilon}$ generation υ player seeks to maximize:

$$\int_{t_\upsilon}^{t_{\upsilon+1}} g^{\upsilon(\omega_\upsilon^{a_\upsilon})}\left[s, x(s), u_\upsilon^{\left(\omega_{\upsilon-1}^{a_{\upsilon-1}}, O\right)\omega_\upsilon^{a_\upsilon}}(s), u_\upsilon^{(\omega_\upsilon^{a_\upsilon}, Y)\omega_{\upsilon-1}^{a_{\upsilon-1}}}(s)\right] e^{-r(s-t_\upsilon)} ds$$
$$+ e^{-r(t_{\upsilon+1}-t_\upsilon)} q^{\upsilon(\omega_\upsilon^{a_\upsilon})}\left[t_{\upsilon+1}, x(t_{\upsilon+1})\right], \tag{1.6}$$

subject to dynamics (1.2).

The game formulated is a finite overlapping generations version of Jørgensen and Yeung's (2005) infinite generations game.

5.1.2 Noncooperative Outcomes

To obtain a characterization of a noncooperative solution to the asynchronous horizons game mentioned above we first consider the solutions of the game in the last time interval $[t_\upsilon, t_{\upsilon+1}]$, that is the game (1.5 and 1.6). One way to characterize and derive a feedback solution to the game in $[t_\upsilon, t_{\upsilon+1}]$ is provided in the lemma below.

Lemma 1.1 If the generation $\upsilon-1$ player is of type $\omega_{\upsilon-1}^{a_{\upsilon-1}} \in \{\omega_{\upsilon-1}^1, \omega_{\upsilon-1}^2, \cdots, \omega_{\upsilon-1}^{\varsigma_{\upsilon-1}}\}$ and the generation υ player is of type $\omega_\upsilon^{a_\upsilon} \in \{\omega_\upsilon^1, \omega_\upsilon^2, \cdots, \omega_\upsilon^{\varsigma_\upsilon}\}$ in the time interval $[t_\upsilon, t_{\upsilon+1}]$, a set of feedback strategies $\left\{\phi_\upsilon^{(\omega_{\upsilon-1}^{a_{\upsilon-1}},O)\omega_\upsilon^{a_\upsilon}}(t,x);\ \phi_\upsilon^{(\omega_\upsilon^{a_\upsilon},Y)\omega_{\upsilon-1}^{a_{\upsilon-1}}}(t,x)\right\}$ constitutes a Nash equilibrium solution for the game (1.5 and 1.6), if there exist continuously differentiable functions $V^{\upsilon-1(\omega_{\upsilon-1}^{a_{\upsilon-1}},O)\omega_\upsilon^{a_\upsilon}}(t,x) : [t_\upsilon, t_{\upsilon+1}] \times R^m \to R$ and $V^{\upsilon(\omega_\upsilon^{a_\upsilon},Y)\omega_{\upsilon-1}^{a_{\upsilon-1}}}(t,x) : [t_\upsilon, t_{\upsilon+1}] \times R^m \to R$ satisfying the following partial differential equations:

$$
\begin{aligned}
-V_t^{\upsilon-1(\omega_{\upsilon-1}^{a_{\upsilon-1}},O)\omega_\upsilon^{a_\upsilon}}(t,x) &= \max_{u_\upsilon^O}\Big\{g^{\upsilon-1(\omega_{\upsilon-1}^{a_{\upsilon-1}})}\Big[t,x,u_\upsilon^O,\phi_\upsilon^{(\omega_\upsilon^{a_\upsilon},Y)\omega_{\upsilon-1}^{a_{\upsilon-1}}}(t,x)\Big]e^{-r(t-t_\upsilon)} \\
&\quad + V_x^{\upsilon-1(\omega_{\upsilon-1}^{a_{\upsilon-1}},O)\omega_\upsilon^{a_\upsilon}}(t,x) f\Big[t,x,u_\upsilon^O,\phi_\upsilon^{(\omega_\upsilon^{a_\upsilon},Y)\omega_{\upsilon-1}^{a_{\upsilon-1}}}(t,x)\Big]\Big\}, \\
V^{\upsilon-1(\omega_{\upsilon-1}^{a_{\upsilon-1}},O)\omega_\upsilon^{a_\upsilon}}(t_{\upsilon+1},x) &= e^{-r(t_{\upsilon+1}-t_\upsilon)}q^{\upsilon-1(\omega_{\upsilon-1}^{a_{\upsilon-1}})}(t_{\upsilon+1},x), \text{ and} \\
-V_t^{\upsilon(\omega_\upsilon^{a_\upsilon},Y)\omega_{\upsilon-1}^{a_{\upsilon-1}}}(t,x) &= \max_{u_\upsilon^Y}\Big\{g^{\upsilon(\omega_\upsilon^{a_\upsilon})}\Big[t,x,\phi_{\upsilon-1}^{(\omega_{\upsilon-1}^{a_{\upsilon-1}},O)\omega_\upsilon^{a_\upsilon}}(t,x),u_\upsilon^Y\Big]e^{-r(t-t_\upsilon)} \\
&\quad + V_x^{\upsilon(\omega_\upsilon^{a_\upsilon},Y)\omega_{\upsilon-1}^{a_{\upsilon-1}}}(t,x) f\Big[t,x,\phi_{\upsilon-1}^{(\omega_{\upsilon-1}^{a_{\upsilon-1}},O)\omega_\upsilon^{a_\upsilon}}(t,x),u_\upsilon^Y\Big]\Big\}, \\
V^{\upsilon(\omega_\upsilon^{a_\upsilon},Y)\omega_{\upsilon-1}^{a_{\upsilon-1}}}(t_{\upsilon+1},x) &= e^{-r(t_{\upsilon+1}-t_\upsilon)}q^{\upsilon(\omega_\upsilon^{a_\upsilon})}[t_{\upsilon+1},x(t_{\upsilon+1})]
\end{aligned} \tag{1.7}
$$

Proof Follow the proof of Theorem 1.1 in Chap. 2. ■

For ease of exposition and sidestepping the issue of multiple equilibria, the analysis focuses on solvable games in which a particular noncooperative Nash equilibrium is chosen by the players in the entire subgame.

We proceed to examine the game in the second last interval $[t_{\upsilon-1}, t_\upsilon)$. If the generation $\upsilon-2$ player is of type $\omega_{\upsilon-2}^{a_{\upsilon-2}} \in \{\omega_{\upsilon-2}^1, \omega_{\upsilon-2}^2, \cdots, \omega_{\upsilon-2}^{\varsigma_{\upsilon-2}}\}$ and the generation $\upsilon-1$ player is of type $\omega_{\upsilon-1}^{a_{\upsilon-1}} \in \{\omega_{\upsilon-1}^1, \omega_{\upsilon-1}^2, \cdots, \omega_{\upsilon-1}^{\varsigma_{\upsilon-1}}\}$. The type $\omega_{\upsilon-2}^{a_{\upsilon-2}}$ generation $\upsilon-2$ player seeks to maximize:

$$
\begin{aligned}
&\int_{t_{\upsilon-1}}^{t_\upsilon} g^{\upsilon-2(\omega_{\upsilon-2}^{a_{\upsilon-2}})}\Big[s,x(s),u_{\upsilon-2}^{(\omega_{\upsilon-2}^{a_{\upsilon-2}},O)\omega_{\upsilon-1}^{a_{\upsilon-1}}}(s),u_{\upsilon-1}^{(\omega_{\upsilon-1}^{a_{\upsilon-1}},Y)\omega_{\upsilon-2}^{a_{\upsilon-2}}}(s)\Big]e^{-r(s-t_{\upsilon-1})}ds \\
&\quad + e^{-r(t_\upsilon-t_{\upsilon-1})}q^{\upsilon-2(\omega_{\upsilon-2}^{a_{\upsilon-2}})}[t_\upsilon,x(t_\upsilon)].
\end{aligned} \tag{1.8}
$$

In the subgame in the time interval $[t_{\upsilon-1}, t_\upsilon)$ the expected payoff of the type $\omega_{\upsilon-1}^{a_{\upsilon-1}}$ generation $\upsilon-1$ player at time t_υ can be expressed as:

$$
\sum_{\ell=1}^{\varsigma_\upsilon} \lambda_\upsilon^\ell V^{\upsilon-1(\omega_{\upsilon-1}^{a_{\upsilon-1}},O)\omega_\upsilon^\ell}(t_\upsilon,x). \tag{1.9}
$$

Therefore the type $\omega_{\upsilon-1}^{a_{\upsilon-1}}$ generation $\upsilon-1$ player then seeks to maximize:

$$\int_{t_{\upsilon-1}}^{t_\upsilon} g^{\upsilon-1(\omega_{\upsilon-1}^{a_{\upsilon-1}})}\left[s,x(s),u_{\upsilon-2}^{(\omega_{\upsilon-2}^{a_{\upsilon-2}},O)\omega_{\upsilon-1}^{a_{\upsilon-1}}}(s),u_{\upsilon-1}^{(\omega_{\upsilon-1}^{a_{\upsilon-1}},Y)\omega_{\upsilon-2}^{a_{\upsilon-1}}}(s)\right]e^{-r(s-t_{\upsilon-1})}ds$$
$$+e^{-r(t_\upsilon-t_{\upsilon-1})}\sum_{\ell=1}^{\varsigma_\upsilon}\lambda_\upsilon^\ell V^{\upsilon-1(\omega_{\upsilon-1}^{a_{\upsilon-1}},O)\omega_\upsilon^\ell}(t_\upsilon,x(t_\upsilon)).$$

Similarly, in the subgame in the interval $[t_k,t_{k+1})$ the expected payoff of the type ω_k generation k player at time t_{k+1} can be expressed as:

$$\sum_{\ell=1}^{\varsigma_{k+1}}\lambda_{k+1}^\ell V^{k(\omega_k^{a_k},O)\omega_{k+1}^\ell}(t_{k+1},x),\qquad \text{for}\, k\in\{1,2,\cdots,\upsilon-3\}. \tag{1.10}$$

Consider the game in the time interval $[t_k,t_{k+1})$ involving the type $\omega_k^{a_k}$ generation k player and the type $\omega_{k-1}^{a_{k-1}}$ generation $k-1$ player, for $k\in\{1,2,\cdots,\upsilon-3\}$. The type $\omega_{k-1}^{a_{k-1}}$ generation $k-1$ player will maximize the payoff

$$\int_{t_k}^{t_{k+1}} g^{k-1(\omega_{k-1}^{a_{k-1}})}\left[s,x(s),u_{k-1}^{(\omega_{k-1}^{a_{k-1}},O)\omega_k^{a_k}}(s),u_k^{(\omega_k^{a_k},Y)\omega_{k-1}^{a_{k-1}}}(s)\right]\ e^{-r(s-t_k)}ds$$
$$+e^{-r(t_{k+1}-t_k)}q^{k-1(\omega_{k-1}^{a_{k-1}})}[t_{k+1},x(t_{k+1})], \tag{1.11}$$

and the type $\omega_k^{a_k}$ generation k player will maximize the payoff:

$$\int_{t_k}^{t_{k+1}} g^{k(\omega_k^{a_k})}\left[s,x(s),u_{k-1}^{(\omega_{k-1}^{a_{k-1}},O)\omega_k^{a_k}}(s),u_k^{(\omega_k^{a_k},Y)\omega_{k-1}^{a_{k-1}}}(s)\right]\ e^{-r(s-t_k)}ds$$
$$+e^{-r(t_{k+1}-t_k)}\sum_{\ell=1}^{\varsigma_{k+1}}\lambda_{k+1}^\ell V^{k(\omega_k^{a_k},O)\omega_{k+1}^\ell}(t_{k+1},x(t_{k+1})), \tag{1.12}$$

subject to (1.2).

A feedback solution to the game (1.5, 1.6) and (1.11, 1.12) can be characterized by the lemma below.

Lemma 1.2 A set of feedback strategies $\left\{\phi_{k-1}^{(\omega_{k-1}^{a_{k-1}},O)\omega_k^{a_k}}(t,x);\ \phi_k^{(\omega_k^{a_k},Y)\omega_{k-1}^{a_{k-1}}}(t,x)\right\}$ constitutes a Nash equilibrium solution for the game (1.5, 1.6) and (1.11, 1.12), if there exist continuously differentiable functions $V^{k-1(\omega_{k-1}^{a_{k-1}},O)\omega_k^{a_k}}(t,x):[t_k,t_{k+1}]\times R^m\to R$ and $V^{k(\omega_k^{a_k},Y)\omega_{k-1}^{a_{k-1}}}(t,x):[t_k,t_{k+1}]\times R^m\to R$ satisfying the following partial differential equations:

$$-V_t^{\upsilon(\omega_\upsilon^{a_\upsilon},Y)\omega_{\upsilon-1}^{a_{\upsilon-1}}}(t,x)=\max_{u_\upsilon^Y}\left\{g^{\upsilon(\omega_\upsilon^{a_\upsilon})}\left[t,x,\phi_{\upsilon-1}^{(\omega_{\upsilon-1}^{a_{\upsilon-1}},O)\omega_\upsilon^{a_\upsilon}}(t,x),u_\upsilon^Y\right]e^{-r(t-t_\upsilon)}\right.$$
$$\left.+V_x^{\upsilon(\omega_\upsilon^{a_\upsilon},Y)\omega_{\upsilon-1}^{a_{\upsilon-1}}}(t,x)f\left[t,x,\phi_{\upsilon-1}^{(\omega_{\upsilon-1}^{a_{\upsilon-1}},O)\omega_\upsilon^{a_\upsilon}}(t,x),u_\upsilon^Y\right]\right\}$$
$$V^{\upsilon(\omega_\upsilon^{a_\upsilon},Y)\omega_{\upsilon-1}^{a_{\upsilon-1}}}(t_{\upsilon+1},x)=e^{-r(t_{\upsilon+1}-t_\upsilon)}q^{\upsilon(\omega_\upsilon^{a_\upsilon})}[t_{\upsilon+1},x(t_{\upsilon+1})];\text{ and}$$
$$-V_t^{k-1(\omega_{k-1}^{a_{k-1}},O)\omega_k^{a_k}}(t,x)=\max_{u_k^O}\left\{g^{k-1(\omega_{k-1})}\left[t,x,u_k^O,\phi_k^{(\omega_k^{a_k},Y)\omega_{k-1}^{a_{k-1}}}(t,x)\right]e^{-r(t-t_k)}\right.$$
$$\left.+V_x^{k-1(\omega_{k-1}^{a_{k-1}},O)\omega_k^{a_k}}(t,x)f\left[t,x,u_k^O,\phi_k^{(\omega_k^{a_k},Y)\omega_{k-1}^{a_{k-1}}}(t,x)\right]\right\},$$
$$V^{k-1(\omega_{k-1}^{a_{k-1}},O)\omega_k^{a_k}}(t_{k+1},x)=e^{-r(t_{k+1}-t_k)}q^{k-1(\omega_{k-1}^{a_{k-1}})}(t_{k+1},x),\text{ and}$$
$$-V_t^{k(\omega_k^{a_k},Y)\omega_{k-1}^{a_{k-1}}}(t,x)=\max_{u_k^Y}\left\{g^{(k,\omega_k)}\left[t,x,\phi_{k-1}^{(\omega_{k-1}^{a_{k-1}},O)\omega_k^{a_k}},u_k^Y\right]e^{-r(t-t_k)}\right.$$
$$\left.+V_x^{k(\omega_k^{a_k},Y)\omega_{k-1}^{a_{k-1}}}(t,x)f\left[t,x,\phi_{k-1}^{(\omega_{k-1}^{a_{k-1}},O)\omega_k^{a_k}},u_k^Y\right]\right\}$$
$$V^{k(\omega_k^{a_k},Y)\omega_{k-1}^{a_{k-1}}}(t_{k+1},x)=e^{-r(t_{k+1}-t_k)}\sum_{\ell=1}^{\varsigma_{k+1}}\lambda_{k+1}^\ell V^{k(\omega_k^{a_k},O)\omega_{k+1}^\ell}(t_{k+1},x),\tag{1.13}$$

for $k\in\{1,2,\cdots,\upsilon-1\}$.

Proof Again follow the proof of Theorem 1.1 in Chap. 2. ■

5.2 Dynamic Cooperation Among Coexisting Players

Now consider the case when coexisting players want to cooperate and agree to act and allocate the cooperative payoff according to a set of agreed upon optimality principles. The agreement on how to act cooperatively and allocate cooperative payoff constitutes the solution optimality principle of a cooperative scheme. In particular, the solution optimality principle for the cooperative game includes (i) an agreement on a set of cooperative strategies/controls, and (ii) an imputation of their payoffs.

Consider the game in the time interval $[t_k,t_{k+1})$ involving the type $\omega_k^{a_k}$ generation k player and the type $\omega_{k-1}^{a_{k-1}}$ generation $k-1$ player. Let $\varpi_h^{(\omega_{k-1}^{a_{k-1}},\omega_k^{a_k})}$ denote the probability that the type $\omega_k^{a_k}$ generation k player and the type $\omega_{k-1}^{a_{k-1}}$ generation $k-1$ player would agree to the solution imputation

$\left[\xi^{k-1(\omega_{k-1}^{a_{k-1}},O)\omega_k^{a_k}[h]}(t,x),\ \xi^{k(\omega_k^{a_k},Y)\omega_{k-1}^{a_{k-1}}[h]}(t,x)\right]$, over the time interval $[t_k,t_{k+1})$,

$$\text{where}\quad\sum_{h=1}^{\varsigma(\omega_{k-1}^{a_{k-1}},\omega_k^{a_k})}\varpi_h^{(\omega_{k-1}^{a_{k-1}},\omega_k^{a_k})}=1.$$

At time t_1, the agreed-upon imputation for the type ω_0^1 generation 0 player and the type ω_1^1 generation 1 player are known to be $\left[\xi^{0(\omega_0^1,O)\omega_1^1[1]}(t,x),\ \xi^{1(\omega_1^1,Y)\omega_0^1[1]}(t,x)\right]$, over the time interval $[t_1, t_2)$.

The solution imputation may be governed by many specific principles. For instance, the players may agree to maximize the sum of their expected payoffs and equally divide the excess of the cooperative payoff over the noncooperative payoff. As another example, the solution imputation may be an allocation principle in which the players allocate the total joint payoff according to the relative sizes of the players' noncooperative payoffs. Finally, it is also possible that the players refuse to cooperate. In that case, the imputation vector becomes $\left[V^{k-1(\omega_{k-1}^{a_{k-1}},O)\omega_k^{a_k}}(t,x),\ V^{k(\omega_k^{a_k},Y)\omega_{k-1}^{a_{k-1}}}(t,x)\right]$.

Both group optimality and individual rationality are required in a cooperative plan. Group optimality requires the players to seek a set of cooperative strategies/controls that yields a Pareto optimal solution. The allocation principle has to satisfy individual rationality in the sense that neither player would be no worse off than before under cooperation.

5.2.1 Group Optimality

Since payoffs are transferable, group optimality requires the players coexisting in the same time interval to maximize their expected joint payoff. Consider the last time interval $[t_\upsilon, t_{\upsilon+1}]$, in which the generation $\upsilon - 1$ player is of type $\omega_{\upsilon-1}^{a_{\upsilon-1}} \in \{\omega_{\upsilon-1}^1, \omega_{\upsilon-1}^2, \cdots, \omega_{\upsilon-1}^{\varsigma_{\upsilon-1}}\}$ and the generation υ player is of type $\omega_\upsilon^{a_\upsilon} \in \{\omega_\upsilon^1, \omega_\upsilon^2, \cdots, \omega_\upsilon^{\varsigma_\upsilon}\}$. The players maximize their joint payoff

$$\int_{t_\upsilon}^{t_{\upsilon+1}} \left(g^{\upsilon-1(\omega_{\upsilon-1}^{a_{\upsilon-1}})}\left[s, x(s), u_\upsilon^{(\omega_{\upsilon-1}^{a_{\upsilon-1}},O)\omega_\upsilon^{a_\upsilon}}(s), u_\upsilon^{(\omega_\upsilon^{a_\upsilon},Y)\omega_{\upsilon-1}^{a_{\upsilon-1}}}(s)\right] \right.$$
$$\left. + g^{\upsilon(\omega_\upsilon^{a_\upsilon})}\left[s, x(s), u_\upsilon^{(\omega_{\upsilon-1}^{a_{\upsilon-1}},O)\omega_\upsilon^{a_\upsilon}}(s), u_\upsilon^{(\omega_\upsilon^{a_\upsilon},Y)\omega_{\upsilon-1}^{a_{\upsilon-1}}}(s)\right]\right) e^{-r(s-t_\upsilon)} ds$$
$$+ e^{-r(t_{\upsilon+1}-t_\upsilon)}\left(q^{\upsilon-1(\omega_{\upsilon-1}^{a_{\upsilon-1}})}[t_{\upsilon+1}, x(t_{\upsilon+1})] + q^{\upsilon(\omega_\upsilon^{a_\upsilon})}[t_{\upsilon+1}, x(t_{\upsilon+1})]\right), \tag{2.1}$$

subject to (1.2).

An optimal solution of the problem (2.1 and 1.2) can be characterized by the following lemma.

Lemma 2.1 A set of Controls $\left\{\psi_{\upsilon-1}^{(\omega_{\upsilon-1}^{a_{\upsilon-1}},O)\omega_\upsilon^{a_\upsilon}}(t,x);\ \psi_\upsilon^{(\omega_\upsilon^{a_\upsilon},Y)\omega_{\upsilon-1}^{a_{\upsilon-1}}}(t,x)\right\}$ constitutes an optimal solution for the control problem (2.1 and 1.2), if there exist continuously differentiable functions $W^{[t_\upsilon,t_{\upsilon+1}](\omega_{\upsilon-1}^{a_{\upsilon-1}},\omega_\upsilon^{a_\upsilon})}(t,x) : [t_\upsilon, t_{\upsilon+1}] \times R^m \to R$ satisfying the following partial differential equations:

$$-W_t^{[t_\upsilon,t_{\upsilon+1}]\left(\omega_{\upsilon-1}^{a_{\upsilon-1}},\omega_\upsilon^{a_\upsilon}\right)}(t,x) = \max_{u_\upsilon^O,u_\upsilon^O}\left\{g^{\upsilon-1\left(\omega_{\upsilon-1}^{a_{\upsilon-1}}\right)}\left[t,x,u_\upsilon^O,u_\upsilon^Y\right]e^{-r(t-t_\upsilon)}\right.$$
$$\left.+\,g^{\upsilon\left(\omega_\upsilon^{a_\upsilon}\right)}\left[t,x,u_\upsilon^O,u_\upsilon^Y\right]e^{-r(t-t_\upsilon)} + W_x^{[t_\upsilon,t_{\upsilon+1}]\left(\omega_{\upsilon-1}^{a_{\upsilon-1}},\omega_\upsilon^{a_\upsilon}\right)}(t,x)f\left[t,x,u_\upsilon^O,u_\upsilon^Y\right]\right\},$$
$$W^{[t_\upsilon,t_{\upsilon+1}]\left(\omega_{\upsilon-1}^{a_{\upsilon-1}},\omega_\upsilon^{a_\upsilon}\right)}(t_{\upsilon+1},x) = e^{-r(t_{\upsilon+1}-t_\upsilon)}\left[q^{\upsilon-1\left(\omega_{\upsilon-1}^{a_{\upsilon-1}}\right)}(t_{\upsilon+1},x) + q^{\upsilon\left(\omega_\upsilon^{a_\upsilon}\right]}(t_{\upsilon+1},x)\right]. \tag{2.2}$$

Proof Invoking Bellman's techniques of dynamic programming stated in Theorem A.1 of the Technical Appendices an optimal solution of the problem (2.1 and 1.2) can be characterized as (2.2). ■

We proceed to examine joint payoff maximization problem in the time interval $\left[t_{\upsilon-1},t_\upsilon\right)$ involving the type $\omega_{\upsilon-1}^{a_{\upsilon-1}}$ generation $\upsilon-1$ player and type $\omega_{\upsilon-2}^{a_{\upsilon-2}}$ generation $\upsilon-2$ player. A critical problem is to determent type $\omega_{\upsilon-1}^{a_{\upsilon-1}}$ generation $\upsilon-1$ player's expected valuation of his optimization problem in the time interval $\left[t_{\upsilon-1},t_\upsilon\right)$ at time t_υ. At time t_υ, the $\omega_{\upsilon-1}^{a_{\upsilon-1}}$ generation $\upsilon-1$ player may co-exist with the type $\omega_\upsilon^{a_\upsilon}$ $\in\left\{\omega_\upsilon^1,\omega_\upsilon^2,\cdots,\omega_\upsilon^{\varsigma_\upsilon}\right\}$ generation υ player with probabilities $\left\{\lambda_\upsilon^1,\lambda_\upsilon^2,\cdots,\lambda_\upsilon^{\varsigma_\upsilon}\right\}$. Consider the case in the time interval $\left[t_\upsilon,t_{\upsilon+1}\right)$ in which the type $\omega_{\upsilon-1}^{a_{\upsilon-1}}$ generation $\upsilon-1$ player and the type $\omega_\upsilon^{a_\upsilon}$ generation υ player co-exist. The probability that the type $\omega_{\upsilon-1}^{a_{\upsilon-1}}$ generation player and the type $\omega_\upsilon^{a_\upsilon}$ generation player would agree to the solution imputation

$$\left[\xi^{\upsilon-1\left(\omega_{\upsilon-1}^{a_{\upsilon-1}},O\right)\omega_\upsilon^{a_\upsilon}[h]}(t,x),\ \xi^{\upsilon\left(\omega_\upsilon^{a_\upsilon},Y\right)\omega_{\upsilon-1}^{a_{\upsilon-1}}[h]}(t,x)\right] \text{ is } \varpi_h^{\left(\omega_{\upsilon-1}^{a_{\upsilon-1}},\omega_\upsilon^{a_\upsilon}\right)},$$
$$\text{where } \sum_{h=1}^{\zeta\left(\omega_{\upsilon-1}^{a_{\upsilon-1}},\omega_\upsilon^{a_\upsilon}\right)} \varpi_h^{\left(\omega_{\upsilon-1}^{a_{\upsilon-1}},\omega_\upsilon^{a_\upsilon}\right)} = 1. \tag{2.3}$$

In the optimization problem within the time interval $\left[t_{\upsilon-1},t_\upsilon\right)$, the expected reward to the $\omega_{\upsilon-1}$ generation $\upsilon-1$ player at time t_υ can be expressed as:

$$\sum_{\ell=1}^{\varsigma_\upsilon}\lambda_\upsilon^\ell\sum_{h=1}^{\zeta\left(\omega_{\upsilon-1}^{a_{\upsilon-1}},\omega_\upsilon^\ell\right)}\varpi_h^{\left(\omega_{\upsilon-1}^{a_{\upsilon-1}},\omega_\upsilon^\ell\right)}\xi^{\upsilon-1\left(\omega_{\upsilon-1}^{a_{\upsilon-1}},O\right)\omega_\upsilon^\ell[h]}(t_\upsilon,x). \tag{2.4}$$

Similarly for the optimization problem within the time interval $\left[t_k,t_{k+1}\right)$, the expected reward to the type $\omega_k^{a_k}$ generation k player at time t_{k+1} can be expressed as:

$$\sum_{\ell=1}^{\varsigma_{k+1}} \lambda_{k+1}^{\ell} \sum_{h=1}^{\zeta\left(\omega_k^{a_k},\omega_{k+1}^{\ell}\right)} \varpi_h^{\left(\omega_k^{a_k},\omega_{k+1}^{\ell}\right)} \xi^{k\left(\omega_k^{a_k},O\right)\omega_{k+1}^{\ell}[h]}(t_{k+1},x), \qquad \text{for} \, k\in\{1,2,\cdots,\upsilon-2\}. \tag{2.5}$$

The joint maximization problem in the time interval $\left[t_k, t_{k+1}\right)$, for $k\in\{1,2,\cdots,\upsilon-2\}$, involving the type $\omega_k^{a_k}$ generation k player and type $\omega_{k-1}^{a_{k-1}}$ generation $k-1$ player can be expressed as the maximization of joint payoff

$$\begin{aligned}
&\Bigg\{\int_{t_k}^{t_{k+1}} \Bigg(g^{k-1\left(\omega_{k-1}^{a_{k-1}}\right)}\left[s, x(s), u_k^{\left(\omega_{k-1}^{a_{k-1}},O\right)\omega_k^{a_k}}(s), u_k^{\left(\omega_k^{a_k},Y\right)\omega_{k-1}^{a_{k-1}}}(s)\right] \\
&\quad + g^{k\left(\omega_k^{a_k}\right)}\left[s, x(s), u_k^{\left(\omega_{k-1}^{a_{k-1}},O\right)\omega_k^{a_k}}(s), u_k^{\left(\omega_k^{a_k},Y\right)\omega_{k-1}^{a_{k-1}}}(s)\right]\Bigg) e^{-r(s-t_k)}ds \\
&\quad + e^{-r(t_{k+1}-t_k)}\Bigg(q^{k-1\left(\omega_{k-1}^{a_{k+1}}\right)}[t_{k+1}, x(t_{k+1})] \\
&\quad + \sum_{\ell=1}^{\varsigma_{k+1}} \lambda_{k+1}^{\ell} \sum_{h=1}^{\zeta\left(\omega_k^{a_k},\omega_{k+1}^{\ell}\right)} \varpi_h^{\left(\omega_k^{a_k},\omega_{k+1}^{\ell}\right)} \xi^{k\left(\omega_k^{a_k},O\right)\omega_{k+1}^{\ell}[h]}(t_{k+1}, x(t_{k+1}))\Bigg)\Bigg\},
\end{aligned} \tag{2.6}$$

subject to (1.2).

The conditions characterizing an optimal solution of the problem of maximizing (2.6) subject to (1.2) are given in the following theorem.

Theorem 2.1 A set of controls $\left\{\psi_{k-1}^{\left(\omega_{k-1}^{a_{k-1}},O\right)\omega_k^{a_k}}(t,x);\ \psi_k^{\left(\omega_k^{a_k},Y\right)\omega_{k-1}^{a_{k-1}}}(t,x)\right\}$ constitutes an optimal solution for the control problem (1.2 and 2.6), if there exist continuously differentiable functions $W^{[t_k,t_{k+1}]\left(\omega_{k-1}^{a_{k-1}},\omega_k^{a_k}\right)}(t,x) : \left[t_k, t_{k+1}\right) \times R^m \to R$ satisfying the following partial differential equations:

$$\begin{aligned}
&-W_t^{[t_\upsilon,t_{\upsilon+1}]\left(\omega_{\upsilon-1}^{a_{\upsilon-1}},\omega_\upsilon^{a_\upsilon}\right)}(t,x) = \max_{u_\upsilon^O,\, u_\upsilon^Y} \Big\{ g^{\upsilon-1\left(\omega_{\upsilon-1}^{a_{\upsilon-1}}\right)}\left[t, x, u_\upsilon^O, u_\upsilon^Y\right] e^{-r(t-t_\upsilon)} \\
&+ g^{\upsilon\left(\omega_\upsilon^{a_\upsilon}\right)}\left[t, x, u_\upsilon^O, u_\upsilon^Y\right] e^{-r(t-t_\upsilon)} + W_x^{[t_\upsilon,t_{\upsilon+1}]\left(\omega_{\upsilon-1}^{a_{\upsilon-1}},\omega_\upsilon^{a_\upsilon}\right)}(t,x) f\left[t, x, u_\upsilon^O, u_\upsilon^Y\right]\Big\} \\
&W^{[t_\upsilon,t_{\upsilon+1}]\left(\omega_{\upsilon-1}^{a_{\upsilon-1}},\omega_\upsilon^{a_\upsilon}\right)}(t_{\upsilon+1},x) = e^{-r(t_{\upsilon+1}-t_\upsilon)} \\
&\left[q^{\upsilon-1\left(\omega_{\upsilon-1}^{a_{\upsilon-1}}\right)}(t_{\upsilon+1},x) + q^{\upsilon\left(\omega_\upsilon^{a_\upsilon}\right]}(t_{\upsilon+1},x)\right] \text{and} \\
&-W_t^{[t_k,t_{k+1}]\left(\omega_{k-1}^{a_{k-1}},\omega_k^{a_k}\right)}(t,x) = \max_{u_k^O,\, u_\upsilon^Y} \Big\{ g^{k-1\left(\omega_{k-1}^{a_{k-1}}\right)}\left[t, x, u_k^O, u_k^Y\right] e^{-r(t-t_k)} \\
&+ g^{k\left(\omega_k^{a_k}\right)}\left[t, x, u_k^O, u_k^Y\right] e^{-r(t-t_k)} + W_x^{[t_k,t_{k+1}]\left(\omega_{k-1}^{a_{k-1}},\omega_k^{a_k}\right)}(t,x) f\left[t, x, u_k^O, u_k^Y\right]\Big\} \\
&W^{[t_k,t_{k+1}]\left(\omega_{k-1}^{a_{k-1}},\omega_k^{a_k}\right)}(t_{k+1},x) = e^{-r(t_{k+1}-t_k)}\Bigg(q^{k-1\left(\omega_{k-1}^{a_{k-1}}\right)}(t_{k+1},x) \\
&+\sum_{\ell=1}^{\varsigma_{k+1}} \lambda_{k+1}^{\ell} \sum_{h=1}^{\zeta\left(\omega_k^{a_k},\omega_{k+1}^{\ell}\right)} \varpi_h^{\left(\omega_k^{a_k},\omega_{k+1}^{\ell}\right)} \xi^{k\left(\omega_k^{a_k},O\right)\omega_{k+1}^{\ell}[h]}(t_{k+1}, x(t_{k+1}))\Bigg), \text{for } k\in\{1,2,\cdots,\upsilon-1\}.
\end{aligned} \tag{2.7}$$

Proof Invoking Bellman's (1957) technique of dynamic programming stated in Theorem A.1 of the Technical Appendices we obtain the conditions characterizing an optimal solution of the problem (1.2) and (2.6) as in (2.7). ■

In particular, $W^{[t_k,t_{k+1}](\omega_{k-1}^{a_{k-1}},\omega_k^{a_k})}(t,x)$ gives the maximized joint payoff of the type $\omega_k^{a_k}$ generation k player and type $\omega_{k-1}^{a_{k-1}}$ generation $k-1$ player at time $t\in[t_k,t_{k+1}]$ with the state x in the control problem

$$\max_{u_k^{(\omega_{k-1}^{a_{k-1}},O)\omega_k^{a_k}},\,u_k^{(\omega_k^{a_k},Y)\omega_{k-1}^{a_{k-1}}}}\left\{\int_t^{t_{k+1}}\left(g^{k-1(\omega_{k-1}^{a_{k-1}})}\left[s,x(s),u_k^{(\omega_{k-1}^{a_{k-1}},O)\omega_k^{a_k}}(s),u_k^{(\omega_k^{a_k},Y)\omega_{k-1}^{a_{k-1}}}(s)\right]\right.\right.$$
$$\left.+g^{k(\omega_k^{a_k})}\left[s,x(s),u_k^{(\omega_{k-1}^{a_{k-1}},O)\omega_k^{a_k}}(s),u_k^{(\omega_k^{a_k},Y)\omega_{k-1}^{a_{k-1}}}(s)\right]\right)e^{-r(s-t_k)}ds$$
$$+e^{-r(t_{k+1}-t_k)}\left(q^{k-1(\omega_{k-1}^{a_{k+1}})}[t_{k+1},x(t_{k+1})]\right.$$
$$\left.\left.+\sum_{\ell=1}^{\varsigma_{k+1}}\lambda_{k+1}^{\ell}\sum_{h=1}^{\zeta_{(\omega_k^{a_k},\omega_{k+1}^{\ell})}}\varpi_h^{(\omega_k^{a_k},\omega_{k+1}^{\ell})}\xi^{k(\omega_k^{a_k},O)\omega_{k+1}^{\ell}[h]}(t_{k+1},x(t_{k+1}))\right)\right\}$$

subject to

$$\dot{x}(s)=f\left[s,x(s),u_k^{(\omega_{k-1}^{a_{k-1}},O)\omega_k^{a_k}}(s),u_k^{(\omega_k^{a_k},Y)\omega_{k-1}^{a_{k-1}}}(s)\right],\quad \text{for } s\in[t_k,t_{k+1})$$
$$x(t)=x.$$

Substituting the set of cooperative strategies into (1.2) yields the dynamics of the cooperative state trajectory in the time interval $[t_k,t_{k+1})$

$$\dot{x}(s)=f\left[s,x(s),\psi_{k-1}^{(\omega_{k-1}^{a_{k-1}},O)\omega_k^{a_k}}(s,x(s)),\psi_k^{(\omega_k^{a_k},Y)\omega_{k-1}^{a_{k-1}}}(s,x(s))\right],\tag{2.8}$$

if type $\omega_k^{a_k}$ generation k player and type $\omega_{k-1}^{a_{k-1}}$ generation $k-1$ player coexist in $[t_k,t_{k+1})$, for $s\in[t_k,t_{k+1})$, $k\in\{1,2,\cdots,\upsilon\}$ and $x(t_k)=x_{t_k}\in X$.

Let $\left\{x^{(\omega_{k-1}^{a_{k-1}},\omega_k^{a_k})*}(t)\right\}_{t=t_k}^{t_{k+1}}$ denote the cooperative solution path governed by (2.8).

For simplicity in exposition we denote $x^{(\omega_{k-1}^{a_{k-1}},\omega_k^{a_k k})*}(t)$ by $x_t^{(\omega_{k-1}^{a_{k-1}},\omega_k^{a_k})*}$.

To fulfill group optimality, the imputation vectors have to satisfy:

$$\xi^{k-1(\omega_{k-1}^{a_{k-1}},O)\omega_k^{a_k}[h]}(t,x^*)+\xi^{k(\omega_k^{a_k},Y)\omega_{k-1}^{a_{k-1}}[h]}(t,x^*)=W^{[t_k,t_{k+1}](\omega_{k-1}^{a_{k-1}},\omega_k^{a_k})}(t,x^*),\tag{2.9}$$

for $t\in[t_k,t_{k+1})$, $\omega_k^{a_k}\in\{\omega_k^1,\omega_k^2,\cdots,\omega_k^{\varsigma_k}\}$, $\omega_{k-1}^{a_{k-1}}\in\{\omega_{k-1}^1,\omega_{k-1}^2,\cdots,\omega_{k-1}^{\varsigma_{k-1}}\}$, $h\in\left\{1,2,\cdots,\zeta_{(\omega_k^{a_k},\omega_{k+1}^{a_{k-1}})}\right\}$ and $k\in\{0,1,2,\cdots,\upsilon\}$.

5.2.2 *Individual Rationality*

In a dynamic framework, individual rationality requires that the imputation received by a player has to be no less than his noncooperative payoff throughout the time interval in concern. Hence for individual rationality to hold along the cooperative trajectory $\left\{x^{\left(\omega_{k-1}^{a_{k-1}},\omega_k^{a_k}\right)*}(t)\right\}_{t=t_k}^{t_{k+1}}$,

$$\xi^{k-1\left(\omega_{k-1}^{a_{k-1}},O\right)\omega_k^{a_k}[h]}\left(t,x_t^*\right) \geq V^{k-1\left(\omega_{k-1}^{a_{k-1}},O\right)\omega_k^{a_k}}\left(t,x_t^*\right) \text{ and}$$
$$\xi^{k\left(\omega_k^{a_k},Y\right)\omega_{k-1}^{a_{k-1}}[h]}\left(t,x_t^*\right) \geq V^{k\left(\omega_k^{a_k},Y\right)\omega_{k-1}^{a_{k-1}}}\left(t,x_t^*\right), \tag{2.10}$$

for $t\in\left[t_k,t_{k+1}\right)$, $\omega_k^{a_k}\in\left\{\omega_k^1,\omega_k^2,\cdots,\omega_k^{\varsigma_k}\right\}$, $\omega_{k-1}^{a_{k-1}}\in\left\{\omega_{k-1}^1,\omega_{k-1}^2,\cdots,\omega_{k-1}^{\varsigma_{k-1}}\right\}$, $h\in\left\{1,2,\cdots,\zeta_{\left(\omega_k^{a_k},\omega_{k+1}^{a_{k-1}}\right)}\right\}$ and $k\in\{0,1,2,\cdots,v\}$,

where x_t^* is the short form for $x_t^{\left(\omega_{k-1}^{a_{k-1}},\omega_k^{a_k}\right)*}$.

For instance, using the results derived, an imputation vector equally dividing the excess of the cooperative payoff over the noncooperative payoff can be expressed as:

$$\xi^{k-1\left(\omega_{k-1}^{a_{k-1}},O\right)\omega_k^{a_k}[h]}\left(t,x_t^*\right) = V^{k-1\left(\omega_{k-1}^{a_{k-1}},O\right)\omega_k^{a_k}}\left(t,x_t^*\right) + 0.5\left[W^{[t_k,t_{k+1}]\left(\omega_{k-1}^{a_{k-1}},\omega_k^{a_k}\right)}\left(t,x_t^*\right)\right.$$
$$\left. - V^{k-1\left(\omega_{k-1}^{a_{k-1}},O\right)\omega_k^{a_k}}\left(t,x_t^*\right) - V^{k\left(\omega_k^{a_k},Y\right)\omega_{k-1}^{a_{k-1}}}\left(t,x_t^*\right)\right], \text{ and}$$
$$\xi^{k\left(\omega_k^{a_k},Y\right)\omega_{k-1}^{a_{k-1}}[h]}\left(t,x_t^*\right) = V^{k\left(\omega_k^{a_k},Y\right)\omega_{k-1}^{a_{k-1}}}\left(t,x_t^*\right) + 0.5\left[W^{[t_k,t_{k+1}]\left(\omega_{k-1}^{a_{k-1}},\omega_k^{a_k}\right)}\left(t,x_t^*\right)\right.$$
$$\left. - V^{k-1\left(\omega_{k-1}^{a_{k-1}},O\right)\omega_k^{a_k}}\left(t,x_t^*\right) - V^{k\left(\omega_k^{a_k},Y\right)\omega_{k-1}^{a_{k-1}}}\left(t,x_t^*\right)\right]. \tag{2.11}$$

One can readily see that the imputations in (2.11) satisfy individual rationality and group optimality.

5.3 Subgame Consistent Solutions and Payoff Distribution

A stringent requirement for solutions of cooperative differential games to be dynamically stable is the property of subgame consistency. Under subgame consistency, an extension of the solution policy to a situation with a later starting time and any feasible state brought about by prior optimal behaviors would remain optimal. In particular, when the game proceeds, at each instant of time the players are guided by the same optimality principles. According to the solution optimality principle the players agree to share their cooperative payoff according to the imputations

$$\left[\xi^{k-1\left(\omega_{k-1}^{a_{k-1}},O\right)\omega_k^{a_k}[h]}\left(t,x_t^*\right),\ \xi^{k\left(\omega_k^{a_k},Y\right)\omega_{k-1}^{a_{k-1}}[h]}\left(t,x_t^*\right)\right] \tag{3.1}$$

over the time interval $\left[t_k,t_{k+1}\right)$.

To achieve dynamic consistency, a payment scheme has to be derived so that imputation (3.1) will be maintained throughout the time interval $\left[t_k,t_{k+1}\right)$. Following the analysis in Chap. 3, we formulate a payoff distribution procedure (PDP) over time so that the agreed imputations (3.1) can be realized. Let $B_{k-1}^{\left(\omega_{k-1}^{a_{k-1}},O\right)\omega_k^{a_k}[h]}(s)$ and $B_k^{\left(\omega_k^{a_k},Y\right)\omega_{k-1}^{a_{k-1}}[h]}(s)$ denote the instantaneous payments at time $s\in\left[t_k,t_{k+1}\right)$ allocated to the type $\omega_{k-1}^{a_{k-1}}$ generation $k-1$ (old) player and type $\omega_k^{a_k}$ generation k (young) player.

In particular, the imputation vector can be expressed as:

$$\begin{aligned}
&\xi^{k-1\left(\omega_{k-1}^{a_{k-1}},O\right)\omega_k^{a_k}[h]}\left(t,x_t^*\right)\\
&\quad=\int_{t_k}^{t_{k+1}}B_{k-1}^{\left(\omega_{k-1}^{a_{k-1}},O\right)\omega_k^{a_k}[h]}(s)\,e^{-r(s-t_k)}ds+e^{-r(t_{k+1}-t_k)}q^{k-1\left(\omega_{k-1}^{a_{k-1}}\right)}\left[t_{k+1},x^*\left(t_{k+1}\right)\right]\\
&\xi^{k\left(\omega_k^{a_k},Y\right)\omega_{k-1}^{a_{k-1}}[h]}\left(t,x_t^*\right)=\int_{t_k}^{t_{k+1}}B_k^{\left(\omega_k^{a_k},Y\right)\omega_{k-1}^{a_{k-1}}[h]}(s)\,e^{-r(s-t_k)}ds\\
&\quad+e^{-r(t_{k+1}-t_k)}\sum_{\ell=1}^{\varsigma_{k+1}}\lambda_{k+1}^{\ell}\sum_{h=1}^{\varsigma\left(\omega_k^{a_k},\omega_{k+1}^{\ell}\right)}\varpi_h^{\left(\omega_k^{a_k},\omega_{k+1}^{\ell}\right)}\xi^{k\left(\omega_k^{a_k},O\right)\omega_{k+1}^{\ell}[h]}\left(t_{k+1},x^*\left(t_{k+1}\right)\right),
\end{aligned} \tag{3.2}$$

for $k\in\{1,2,\cdots,\upsilon-1\}$, and

$$\begin{aligned}
&\xi^{\upsilon-1\left(\omega_{\upsilon-1}^{a_{\upsilon-1}},O\right)\omega_\upsilon^{a_\upsilon}[h]}\left(t,x_t^*\right)=\int_{t_\upsilon}^{t_{\upsilon+1}}B_{\upsilon-1}^{\left(\omega_{\upsilon-1}^{a_\upsilon 1},O\right)\omega_\upsilon^{a_\upsilon}[h]}(s)\,e^{-r(s-t_\upsilon)}ds\\
&\quad+e^{-r(t_{\upsilon+1}-t_\upsilon)}q^{\upsilon-1\left(\omega_{\upsilon-1}^{a_{\upsilon-1}}\right)}\left[t_{\upsilon+1},x^*\left(t_{\upsilon+1}\right)\right]\\
&\xi^{\upsilon(\omega_\upsilon^{a_\upsilon},Y)\omega_{\upsilon-1}^{a_{\upsilon-1}}[h]}\left(t,x_t^*\right)=\int_{t_\upsilon}^{t_{\upsilon+1}}B_\upsilon^{(\omega_\upsilon^{a_\upsilon},Y)\omega_{\upsilon-1}^{a_{\upsilon-1}}[h]}(s)\,e^{-r(s-t_\upsilon)}ds\\
&+e^{-r(t_{\upsilon+1}-t_\upsilon)}q^{\upsilon(\omega_\upsilon^{a_\upsilon})}\left[t_{\upsilon+1},x^*\left(t_{\upsilon+1}\right)\right].
\end{aligned} \tag{3.3}$$

Using the analysis in Chap. 2 we obtain a PDP leading to the realization of the imputation vectors in (3.2 and 3.3) in the following theorem.

Theorem 3.1 If the imputation vector $\left[\xi^{k-1\left(\omega_{k-1}^{a_{k-1}},O\right)\omega_k^{a_k}[h]}\left(t,x_t^*\right),\right.$ $\left.\xi^{k\left(\omega_k^{a_k},O\right)\omega_{k-1}^{a_{k-1}}[h]}\left(t,x_t^*\right)\right]$ are functions that are continuously differentiable in t and x_t^*, a PDP with an instantaneous payment at time $t\in\left[t_k,t_{k+1}\right)$:

$$B_{k-1}^{(\omega_{k-1}^{a_{k-1}},O)\omega_k^{a_k}[h]}(t) = -\xi_t^{k-1(\omega_{k-1}^{a_{k-1}},O)\omega_k^{a_k}[h]}(t,x_t^*)$$
$$-\xi_x^{k-1(\omega_{k-1}^{a_{k-1}},O)\omega_k^{a_k}[h]}(t,x_t^*)f\left[t,x_t^*,\psi_{k-1}^{(\omega_{k-1}^{a_{k-1}},O)\omega_k^{a_k}}(t,x_t^*),\psi_k^{(\omega_k^{a_k},Y)\omega_{k-1}^{a_{k-1}}}(t,x_t^*)\right] \quad (3.4)$$

allocated to the type $\omega_{k-1}^{a_{k-1}}$ generation $k-1$ player;

and an instantaneous payment at time $t\in[t_k,t_{k+1})$:

$$B_k^{(\omega_k^{a_k},Y)\omega_{k-1}^{a_{k-1}}[h]}(t) = -\xi_t^{k(\omega_k^{a_k},Y)\omega_{k-1}^{a_{k-1}}[h]}(t,x_t^*)$$
$$-\xi_x^{k(\omega_k^{a_k},Y)\omega_{k-1}^{a_{k-1}}[h]}(t,x_t^*)f\left[t,x_t^*,\psi_{k-1}^{(\omega_{k-1}^{a_{k-1}},O)\omega_k^{a_k}}(t,x_t^*),\psi_k^{(\omega_k^{a_k},Y)\omega_{k-1}^{a_{k-1}}}(t,x_t^*)\right]$$

allocated to the type $\omega_k^{a_k}$ generation k player,

yields a mechanism leading to the realization of the imputation vector

$$\left[\xi^{k-1(\omega_{k-1}^{a_{k-1}},O)\omega_k^{a_k}[h]}(t,x_t^*),\ \xi^{k(\omega_k^{a_k},Y)\omega_{k-1}^{a_{k-1}}[h]}(t,x_t^*)\right],\ \text{for}\, k\in\{1,2,\cdots,\upsilon\}.$$

Proof Follow the proof leading to Theorem 3.1 in Chap. 2 with the imputation vector in present value (rather than in current value). ■

5.4 An Illustration in Resource Extraction

Consider the game in which there are 4 overlapping generations of players with generation 0 and generation 1 players in $[t_1,t_2)$, generation 1 and generation 2 players in $[t_2,t_3)$, generation 2 and generation 3 players in $[t_3,t_4]$. Players are of either type 1 or type 2. The instantaneous payoffs and terminal rewards of the type 1 generation k player and the type 2 generation k player are respectively:

$$\left[(u_k)^{1/2}-\frac{c_1}{x^{1/2}}u_k\right]\ \text{and}\, q_1x^{1/2},\quad \text{and}\ \left[(u_k)^{1/2}-\frac{c_2}{x^{1/2}}u_k\right]\ \text{and}\, q_2x^{1/2}, \quad (4.1)$$

where the state variable $x(s)$ is the biomass of a renewable resource. $u_k(s)$ is the harvest of the generation k extraction firm. The type $i\in\{1,2\}$ generation k extraction firm's extraction cost is $c_iu_k(s)x(s)^{-1/2}$.

At initial time t_1, it is known that the generation 0 player is of type 1 and the generation 1 player is also of type 1. It is also known that the generation 2 and generation 3 players may be of type 1 with probability $\lambda_k^1=0.4$ and of type 2 with probability $\lambda_k^2=0.6$ in time interval $[t_k,t_{k+1})$ for $k\in\{2,3\}$.

The state dynamics of the game is characterized by:

$$\dot{x}(s) = ax(s)^{1/2} - bx(s) - u_k^{(i,O)j}(s) - u_k^{(j,Y)i}(s), \tag{4.2}$$

if the old generation $k-1$ extractor is of type i and the young generation k extractor is of type j, for $s \in [t_k, t_{k+1})$ and $k \in \{1,2,3\}$;

$$x(t_1) = x_0 \in X \subset R,$$

where $u_k^{(i,O)j}(s)$ denote the harvest of the type i generation $k-1$ old extractor and $u_k^{(j,Y)i}(s)$ denote the harvest of the type j generation k young extractor.

The death rate of the resource is b. The rate of growth is $a/x^{1/2}$ which reflects the decline in the growth rate as the biomass increases. The game is an asynchronous horizons version of the synchronous-horizon resource extraction game in Yeung and Petrosyan (2006b).

This asynchronous horizon game can be expressed as follows. In the time interval $[t_3, t_4]$, consider the case with a type $i \in \{1,2\}$ generation 2 firm and a type $j \in \{1,2\}$ generation 3 firm, the game becomes

$$\begin{aligned}
&\max_{u_3^{(i,O)j}} \left\{ \int_{t_3}^{t_4} \left[\left[u_3^{(i,O)j}(s) \right]^{1/2} - \frac{c_i}{x(s)^{1/2}} u_3^{(i,O)j}(s) \right] \exp[-r(s-t_3)]\, ds \right. \\
&\qquad \left. + \exp[-r(t_4 - t_3)] q_i x(t_4)^{\frac{1}{2}} \right\}, \\
&\max_{u_3^{(j,Y)i}} \left\{ \int_{t_3}^{t_4} \left[\left[u_3^{(j,Y)i}(s) \right]^{1/2} - \frac{c_j}{x(s)^{1/2}} u_2^{(j,Y)i}(s) \right] \exp[-r(s-t_3)]\, ds \right. \\
&\qquad \left. + \exp[-r(t_4 - t_3)] q_j x(t_4)^{\frac{1}{2}} \right\},
\end{aligned} \tag{4.3}$$

subject to (4.2).

In the time interval $[t_k, t_{k+1})$, for $k \in \{1,2\}$, consider the case with a type $i \in \{1,2\}$ generation $k-1$ firm and a type $j \in \{1,2\}$ generation k firm, the game becomes

$$\begin{aligned}
&\max_{u_k^{(i,O)j}} \left\{ \int_{t_k}^{t_{k+1}} \left[\left[u_k^{(i,O)j}(s) \right]^{1/2} - \frac{c_i}{x(s)^{1/2}} u_k^{(i,O)j}(s) \right] \exp[-r(s-t_k)]\, ds \right. \\
&\left. + \exp[-r(t_{k+1} - t_k)] q_i x(t_{k+1})^{\frac{1}{2}} \right\}, \\
&\max_{u_k^{(j,Y)i},\, u_{k+1}^{(j,O)\ell}} \left\{ \int_{t_k}^{t_{k+1}} \left[\left[u_k^{(j,Y)i}(s) \right]^{1/2} - \frac{c_j}{x(s)^{1/2}} u_k^{(j,Y)i}(s) \right] \exp[-r(s-t_k)]\, ds \right. \\
&+ \sum_{\ell=1}^{2} \lambda_{k+1}^{\ell} \int_{t_{k+1}}^{t_{k+2}} \left[\left[u_{k+1}^{(j,O)\ell}(s) \right]^{1/2} - \frac{c_j}{x(s)^{1/2}} u_{k+1}^{(j,O)\ell}(s) \right] \exp[-r(s-t_k)]\, ds \\
&\left. + \exp[-r(t_{k+2} - t_k)] q_j x(t_{k+2})^{\frac{1}{2}} \right\},
\end{aligned} \tag{4.4}$$

subject to (4.2).

5.4.1 Noncooperative Outcomes

In this section we first characterize the noncooperative outcome of the asynchronous horizons game (4.2, 4.3 and 4.4) as follows.

Proposition 4.1 The feedback Nash equilibrium payoffs for the type $i\in\{1,2\}$ generation $k-1$ firm and the type $j\in\{1,2\}$ generation k firm coexisting in the game interval $[t_k, t_{k+1})$ can be obtained as:

$$\begin{aligned}
&V^{k-1(i,O)j}(t,x)=\exp[-r(t-t_k)]\left[A_{k-1}^{(i,O)j}(t)x^{1/2}+C_{k-1}^{(i,O)j}(t)\right], \text{ and}\\
&V^{k(j,Y)i}(t,x)=\exp[-r(t-t_k)]\left[A_{k}^{(j,Y)i}(t)x^{1/2}+C_{k}^{(j,Y)i}(t)\right],
\end{aligned}\tag{4.5}$$

for $k\in\{1,2,3\}$ and $i,j\in\{1,2\}$,

where
$A_{k-1}^{(i,O)j}(t)$, $C_{k-1}^{(i,O)j}(t)$, $A_{k}^{(j,Y)i}(t)$ and $C_{k}^{(j,Y)i}(t)$ satisfy:

$$\begin{aligned}
&\dot{A}_{k-1}^{(i,O)j}(t)=\left[r+\frac{b}{2}\right]A_{k-1}^{(i,O)j}(t)-\frac{1}{2\left[c_i+A_{k-1}^{(i,O)j}(t)/2\right]}\\
&+\frac{c_i}{4\left[c_i+A_{k-1}^{(i,O)j}(t)/2\right]^2}+\frac{A_{k-1}^{(i,O)j}(t)}{8\left[c_i+A_{k-1}^{(i,O)j}(t)/2\right]^2}+\frac{A_{k-1}^{(i,O)j}(t)}{8\left[c_j+A_{k}^{(j,Y)i}(t)/2\right]^2},\\
&\dot{C}_{k-1}^{(i,O)j}(t)=rC_{k-1}^{(i,O)j}(t)-\frac{a}{2}A_{k-1}^{(i,O)j}(t);\\
&A_{k-1}^{(i,O)j}(t_{k+1})=q_i \text{ and } C_{k-1}^{(i,O)j}(t_{k+1})=0,\quad \text{for } k\in\{1,2,3\};
\end{aligned}\tag{4.6}$$

$$\begin{aligned}
&\dot{A}_{k}^{(j,Y)i}(t)=\left[r+\frac{b}{2}\right]A_{k}^{(j,Y)i}(t)-\frac{1}{2\left[c_j+A_{k}^{(j,Y)i}(t)/2\right]}+\frac{c_j}{4\left[c_j+A_{k}^{(j,Y)i}(t)/2\right]^2}\\
&+\frac{A_{k}^{(j,Y)i}(t)}{8\left[c_j+A_{k}^{(j,Y)i}(t)/2\right]^2}+\frac{A_{k}^{(j,Y)i}(t)}{8\left[c_i+A_{k-1}^{(i,O)j}(t)/2\right]^2}\\
&\dot{C}_{k}^{(j,Y)i}(t)=rC_{k}^{(j,Y)i}(t)-\frac{a}{2}A_{k}^{(j,Y)i}(t),\quad \text{for } k\in\{1,2,3\};\\
&A_{k}^{(j,Y)i}(t_{k+1})=e^{-r(t_{k+1}-t_k)}\sum_{\ell=1}^{2}\lambda_{k+1}^{\ell}A_{k+1}^{(j,O)\ell}(t_{k+1}) \text{ and}\\
&C_{k}^{(j,Y)i}(t_{k+1})=e^{-r(t_{k+1}-t_k)}\sum_{\ell=1}^{2}\lambda_{k+1}^{\ell}C_{k+1}^{(j,O)\ell}(t_{k+1}),\\
&\text{for } k\in\{1,2\}, \text{ and } A_{3}^{(j,Y)i}(t_4)=q_j \text{ and } C_{3}^{(j,Y)i}(t_4)=0.
\end{aligned}\tag{4.7}$$

Proof Performing the indicated maximization in (4.4) and solving the system yield (4.5). Hence Proposition 4.1 follows. ∎

The solution time paths $A_{k-1}^{(i,O)j}(t)$, $C_{k-1}^{(i,O)j}(t)$, $A_k^{(j,Y)i}(t)$ and $C_k^{(j,Y)i}(t)$ for the system of first order differential equations in (4.6) and (4.7) can be computed numerically for given values of the model parameters $r, q_1, q_2, c_1, c_2, a, b, \lambda_k^1$ and λ_k^2.

The game equilibrium strategies then can be expressed as:

$$\phi_k^{(i,O)j}(t,x) = \frac{x}{4\left[c_i + A_k^{(i,O)j}(t)/2\right]^2} \text{ and } \phi_k^{(j,Y)i}(t,x) = \frac{x}{4\left[c_j + A_k^{(j,Y)i}(t)/2\right]^2}.$$

5.4.2 Dynamic Cooperation

Now consider the case when coexisting firms want to cooperate and agree to act and allocate the cooperative payoff according to a set of agreed upon optimality principles. Let there be three acceptable imputations for the extractor firms.

Imputation I: the firms would share the excess gain from cooperation equally with weights $w_k^{o(1)} = 0.5$ for the generation $k-1$ firm and $w_k^{Y(1)} = 0.5$ for the generation k firm.

Imputation II: the generation $k-1$ firm acquires $w_k^{o(2)} = 0.6$ of the excess gain from cooperation and the generation k firm acquires $w_k^{Y(2)} = 0.4$ of the excess gain.

Imputation III: the generation $k-1$ firm acquires $w_k^{o(3)} = 0.4$ of the excess gain from cooperation and the generation k firm acquires $w_k^{Y(3)} = 0.6$ of the excess gain.

In time interval $[t_k, t_{k+1})$, if both the generation $k-1$ firm and the generation k firm are of type 1, the probabilities that the firms would agree to Imputations I, II and III are respectively $\varpi_k^{(1,1)1} = 0.8$, $\varpi_k^{(1,1)2} = 0.1$ and $\varpi_k^{(1,1)3} = 0.1$, for $k \in \{2,3\}$.

If both the generation $k-1$ firm and the generation k firm are of type 2, the probabilities that the firms would agree to Imputations I, II and III are respectively $\varpi_k^{(2,2)1} = 0.7$, $\varpi_k^{(2,2)2} = 0.15$ and $\varpi_k^{(2,2)3} = 0.15$, for $k \in \{2,3\}$.

If the generation $k-1$ firm is of type 1 and the generation k firm are of type 2, the probabilities that the firms would agree to Imputations I, II and III are respectively $\varpi_k^{(1,2)1} = 0.15$, $\varpi_k^{(1,2)2} = 0.75$ and $\varpi_k^{(1,2)3} = 0.1$, for $k \in \{2,3\}$.

If the generation $k-1$ firm is of type 2 and the generation k firm are of type 1, the probabilities that the firms would agree to Imputations I, II and III are respectively $\varpi_k^{(2,1)1} = 0.15$, $\varpi_k^{(2,1)2} = 0.1$ and $\varpi_k^{(2,1)3} = 0.75$, for $k \in \{2,3\}$.

At initial time t_1, the type 1 generation 0 firm and the type 1 generation 1 firm are assumed to have agreed to Imputation II.

Since payoffs are transferable, group optimality requires the firms coexisting in the same time interval to maximize their joint payoff. Consider the last time interval $[t_3, t_4]$, in which the generation 2 firm is of type $i \in \{1, 2\}$ and the generation 3 firm is of type $j \in \{1, 2\}$. The firms maximize their joint profit

$$\left\{ \int_{t_3}^{t_4} \left[\left[u_3^{(i,O)j}(s) \right]^{1/2} - \frac{c_i}{x(s)^{1/2}} u_3^{(i,O)j}(s) \right] \exp[-r(s - t_3)]\, ds \right.$$
$$+ \int_{t_3}^{t_4} \left[\left[u_3^{(j,O)i}(s) \right]^{1/2} - \frac{c_j}{x(s)^{1/2}} u_3^{(j,O)i}(s) \right] \exp[-r(s - t_3)]\, ds$$
$$\left. + \exp[-r(t_4 - t_3)] q_i x(t_4)^{\frac{1}{2}} + \exp[-r(t_4 - t_3)] q_j x(t_4)^{\frac{1}{2}} \right\}, \tag{4.8}$$

subject to (4.2).

The maximized joint payoffs of the players in the last subgame interval can be characterized by the proposition below.

Proposition 4.2 The maximized joint payoff with type $i \in \{1, 2\}$ generation 2 firm and the type $j \in \{1, 2\}$ generation 3 firm coexisting in the game interval $[t_3, t_4)$ can be obtained as:

$$W^{[t_3,t_4](i,j)}(t, x) = \exp[-r(t - t_3)] \left[A^{[t_3,t_4](i,j)}(t) x^{1/2} + C^{[t_3,t_4](i,j)}(t) \right], \tag{4.9}$$

where $A^{[t_3,t_4](i,j)}(t)$ and $C^{[t_3,t_4](i,j)}(t)$ satisfy:

$$\dot{A}^{[t_3,t_4](i,j)}(t) = \left[r + \frac{b}{2} \right] A^{[t_3,t_4](i,j)}(t) - \frac{1}{2\left[c_i + A^{[t_3,t_4](i,j)}(t)/2\right]}$$
$$- \frac{1}{2\left[c_j + A^{[t_3,t_4](i,j)}(t)/2\right]} + \frac{c_i}{4\left[c_i + A^{[t_3,t_4](i,j)}(t)/2\right]^2}$$
$$+ \frac{c_j}{4\left[c_j + A^{[t_3,t_4](i,j)}(t)/2\right]^2}$$
$$+ \frac{A^{[t_3,t_4](i,j)}(t)}{8\left[c_i + A^{[t_3,t_4](i,j)}(t)/2\right]^2} + \frac{A^{[t_3,t_4](i,j)}(t)}{8\left[c_j + A^{[t_3,t_4](i,j)}(t)/2\right]^2}$$
$$\dot{C}^{[t_3,t_4](i,j)}(t) = rC^{[t_3,t_4](i,j)}(t) - \frac{a}{2} A^{[t_3,t_4](i,j)}(t),$$
$$A^{[t_3,t_4](i,j)}(t_4) = q_i + q_j \text{ and } C^{[t_3,t_4](i,j)}(t_4) = 0. \tag{4.10}$$

Proof Invoking the dynamic programming techniques in Theorem A.1 in the Technical Appendices one can obtain (4.9 and 4.10). ■

The solution time paths $A^{[t_3,t_4](i,j)}(t)$ and $C^{[t_3,t_4](i,j)}(t)$ for the system of first order differential equations in (4.9 and 4.10) can be computed numerically for given values of the model parameters r, q_1, q_2, c_1, c_2, a and b.

In the game interval $[t_3, t_4)$ with type $i \in \{1,2\}$ generation 2 firm and the type $j \in \{1,2\}$ generation 3 firm coexisting, if imputation $h \in \{1,2,3\}$ is chosen the imputations of the firms under cooperation can be expressed as:

$$\begin{aligned}
&\xi^{2(i,O)j[h]}(t,x) = V^{2(i,O)j}(t,x) + w_3^{o(h)}\left[W^{[t_3,t_4](i,j)}(t,x) - V^{2(i,O)j}(t,x)\right.\\
&\left.- V^{3(j,Y)i}(t,x)\right],\\
&\xi^{3(j,Y)i[h]}(t,x) = V^{3(j,Y)i}(t,x) + w_3^{Y(h)}\left[W^{[t_3,t_4](i,j)}(t,x) - V^{2(i,O)j}(t,x)\right.\\
&\left.-V^{3(j,Y)i}(t,x)\right].
\end{aligned} \tag{4.11}$$

Now we proceed to the second last interval $[t_k, t_{k+1})$ for $k = 2$. Consider the case in which the generation k firm is of type $j \in \{1,2\}$ and the generation $k-1$ firm is known to be of type $i = 2$. Following the analysis in (2.4) and (2.5), the expected terminal reward to the type j generation k firm at time t_{k+1} can be expressed as:

$$\sum_{\ell=1}^{2} \lambda_k^{\ell} \sum_{h=1}^{3} \varpi_h^{(j,\ell)} \xi^{k(j,O)\ell[h]}(t_{k+1}, x), \quad \text{for } k = 2. \tag{4.12}$$

A review of Proposition 4.1, Proposition 4.2 and (4.11) shows the term in (4.12) can be written as:

$$A_k^{\zeta(j,O)} x^{1/2} + C_k^{\zeta(j,O)}, \tag{4.13}$$

where $A_k^{\zeta(j,O)}$ and $C_k^{\zeta(j,O)}$ are constant terms.

The joint maximization problem in the time interval $[t_k, t_{k+1})$, for $k \in \{1,2\}$, involving the type j generation k player and type i generation $k-1$ player can be expressed as:

$$\begin{aligned}
&\max_{u_k^{(i,O)j},\, u_k^{(j,Y)i}} \left\{ \int_{t_k}^{t_{k+1}} \left[\left[u_k^{(i,O)j}(s)\right]^{1/2} - \frac{c_i}{x(s)^{1/2}} u_k^{(i,O)j}(s) \right] \exp[-r(s-t_k)]\, ds \right.\\
&+ \int_{t_k}^{t_{k+1}} \left[\left[u_k^{(j,Y)i}(s)\right]^{1/2} - \frac{c_j}{x(s)^{1/2}} u_k^{(j,Y)i}(s) \right] \exp[-r(s-t_k)]\, ds\\
&\left. + \exp[-r(t_{k+1}-t_k)] \left[q_i x(t_{k+1})^{\frac{1}{2}} + \sum_{\ell=1}^{2} \lambda_k^{\ell} \sum_{h=1}^{3} \varpi_h^{(j,\ell)} \xi^{k(j,O)\ell[h]}(t_{k+1}, x) \right] \right\},
\end{aligned} \tag{4.14}$$

subject to (4.2).

The maximized joint payoff of the players in the first two subgame intervals can be characterized by the proposition below.

Proposition 4.3 The maximized joint payoff with type $i \in \{1,2\}$ generation $k-1$ firm and the type $j \in \{1,2\}$ generation k firm coexisting in the game interval $[t_k, t_{k+1})$, for $k \in \{1,2\}$, can be obtained as:

$$W^{[t_k,t_{k+1}](i,j)}(t,x) = \exp[-r(t-t_k)]\left[A^{[t_k,t_{k+1}](i,j)}(t)x^{1/2} + C^{[t_k,t_{k+1}](i,j)}(t)\right], \tag{4.15}$$

where $A^{[t_k,t_{k+1}](i,j)}(t)$ and $C^{[t_k,t_{k+1}](i,j)}(t)$ satisfy:

$$\begin{aligned}
\dot{A}^{[t_k,t_{k+1}](i,j)}(t) &= \left[r+\frac{b}{2}\right]A^{[t_k,t_{k+1}](i,j)}(t) - \frac{1}{2\left[c_i + A^{[t_k,t_{k+1}](i,j)}(t)/2\right]} \\
&\quad - \frac{1}{2\left[c_j + A^{[t_k,t_{k+1}](i,j)}(t)/2\right]} + \frac{c_i}{4\left[c_i + A^{[t_k,t_{k+1}](i,j)}(t)/2\right]^2} \\
&\quad + \frac{c_j}{4\left[c_j + A^{[t_k,t_{k+1}](i,j)}(t)/2\right]^2} \\
&\quad + \frac{A^{[t_k,t_{k+1}](i,j)}(t)}{8\left[c_i + A^{[t_k,t_{k+1}](i,j)}(t)/2\right]^2} + \frac{A^{[t_k,t_{k+1}](i,j)}(t)}{8\left[c_j + A^{[t_k,t_{k+1}](i,j)}(t)/2\right]^2} \\
\dot{C}^{[t_k,t_{k+1}](i,j)}(t) &= rC^{[t_k,t_{k+1}](i,j)}(t) - \frac{a}{2}A^{[t_k,t_{k+1}](i,j)}(t), \\
A^{[t_k,t_{k+1}](i,j)}(t_{k+1}) &= q_i + A_k^{\zeta(j,O)} \text{ and } C^{[t_k,t_{k+1}](i,j)}(t_{k+1}) = C_k^{\zeta(j,O)}.
\end{aligned} \tag{4.16}$$

Proof Performing the maximization operator in (4.14) and invoking (4.13) one can obtain the results in (4.15) and (4.16). ■

The solution time paths $A^{[t_k,t_{k+1}](i,j)}(t)$ and $C^{[t_k,t_{k+1}](i,j)}(t)$ for the system of first order differential equations in (4.16) can be computed numerically for given values of the model parameters r, q_1, q_2, c_1, c_2, a, b, λ_k^1, λ_k^2, and $\varpi_h^{(j,\ell)}$ for $h \in \{1,2,3\}$ and $j,\ell \in \{1,2\}$.

The optimal cooperative controls can then be obtained as:

$$\begin{aligned}
\psi_{k-1}^{(i,O)j}(t,x) &= \frac{x}{4\left[c_i + A^{[t_k,t_{k+1}](i,j)}(t)/2\right]^2}, \text{ and} \\
\psi_k^{(j,Y)i}(t,x) &= \frac{x}{4\left[c_j + A^{[t_k,t_{k+1}](i,j)}(t)/2\right]^2}.
\end{aligned} \tag{4.17}$$

Substituting these control strategies into (4.2) yields the dynamics of the state trajectory under cooperation. The optimal cooperative state trajectory in the time interval $[t_k, t_{k+1})$ can be obtained as:

$$x^{(i,j)*}(t) = \left[\Omega_{(i,j)}(t_k,t)\right]^2 \left[(x_{t_k})^{1/2} + \int_{t_k}^{t} \Omega_{(i,j)}^{-1}(t_k,s)\frac{a}{2}ds \right]^2, \tag{4.18}$$

where $\Omega_{(i,j)}(t_k,t) = \exp\left[\int_{t_k}^{t} H_{(i,j)}(\upsilon)d\upsilon\right]$ and

$$H_{(i,j)}(s) = -\left[\frac{b}{2} + \frac{1}{8\left[c_i + A^{[t_k,t_{k+1}](i,j)}(s)/2\right]^2} + \frac{1}{8\left[c_j + A^{[t_k,t_{k+1}](i,j)}(s)/2\right]^2}\right]$$

The term x_t^* is used to denote $x^{(i,j)*}(t)$ whenever there is no ambiguity.

5.4.3 *Dynamically Consistent Payoff Distribution*

According to the solution optimality principle the players agree to share their cooperative payoff according to the solution imputations:

$$\xi^{k-1(i,O)j[h]}(t,x) = V^{k-1(i,O)j}(t,x) + w_{k-1}^h\left[W^{[t_k,t_{k+1}](i,j)}(t,x) - V^{k-1(i,O)j}(t,x) - V^{k(j,Y)i}(t,x)\right],$$
$$\xi^{k(j,Y)i[h]}(t,x) = V^{k(j,Y)i}(t,x) + w_k^h\left[W^{[t_k,t_{k+1}](i,j)}(t,x) - V^{k-1(i,O)j}(t,x) - V^{k(j,Y)i}(t,x)\right],$$

for $h \in \{1,2,3\}$, $i,j \in \{1,2\}$ and $k \in \{1,2,3\}$.

These imputations are continuous differentiable in x and t. If an imputation vector $\left[\xi^{k-1(i,O)j[h]}(t,x),\ \xi^{k(j,Y)i[h]}(t,x)\right]$ is chosen, a crucial process is to derive a payoff distribution procedure (PDP) so that this imputation could be realized for $t \in [t_k, t_{k+1})$ along the cooperative trajectory $\{x_t^*\}_{t=t_k}^{t_{k+1}}$.

Following Theorem 3.1, a PDP leading to the realization of the imputation vector $\left[\xi^{k-1(i,O)j[h]}(t,x),\ \xi^{k(j,Y)i[h]}(t,x)\right]$ can be obtained as:

Corollary 4.1 A PDP with an instantaneous payment at time $t \in [t_k, t_{k+1})$:

$$B_k^{(i,O)j[h]}(t) = -\xi_t^{k-1(i,O)j[h]}(t,x_t^*) - \xi_x^{k-1(i,O)j[h]}(t,x_t^*)\left[a(x_t^*)^{1/2} - bx_t^* - \frac{x_t^*}{4\left[c_i + A^{[t_k,t_{k+1}](i,j)}(t)/2\right]^2} - \frac{x_t^*}{4\left[c_j + A^{[t_k,t_{k+1}](i,j)}(t)/2\right]^2}\right], \tag{4.19}$$

allocated to the type i generation $k-1$ player;

and an instantaneous payment at time $t \in [t_k, t_{k+1})$:

$$B_k^{(j,Y)i[h]}(t) = -\xi_t^{k(j,Y)i[h]}(t, x_t^*) - \xi_x^{k(j,Y)i[h]}(t, x_t^*)\Bigg[a(x_t^*)^{1/2} - bx_t^* \\ - \frac{x_t^*}{4\left[c_i + A^{[t_k,t_{k+1}](i,j)}(t)/2\right]^2} - \frac{x_t^*}{4\left[c_j + A^{[t_k,t_{k+1}](i,j)}(t)/2\right]^2}\Bigg] \tag{4.20}$$

allocated to the type j generation k player,
yields a mechanism leading to the realization of the imputation vector

$\left[\xi^{k-1(i,O)j[h]}(t,x),\ \xi^{k(j,Y)i[h]}(t,x)\right]$, for $k \in \{1,2,3\}$, $h \in \{1,2,3\}$ and $i,j \in \{1,2\}$. ■

Since the imputations $\xi^{k-1(i,O)j[h]}(t,x)$ and $\xi^{k(j,Y)i[h]}(t,x)$ are in terms of explicit differentiable functions, the relevant derivatives can be derived using the results in Propositions 4.1, 4.2 and 4.3. Hence, the PDP $B_k^{(i,O)j[h]}(t)$ and $B_k^{(j,Y)i[h]}(t)$ in (4.19) and (4.20) can be obtained explicitly.

5.5 Extension to Stochastic Dynamics

In this Section we extend the analysis to the case where the state dynamics is stochastic and governed by the stochastic differential equations:

$$dx(s) = f\left[s, x(s), u_{k-1}^{(\omega_{k-1},O)\omega_k}(s), u_k^{(\omega_k,Y)\omega_{k-1}}(s)\right]ds + \sigma[s,x(s)]dz(s), \\ x(t_1) = x_0 \in X, \tag{5.1}$$

for $s \in [t_k, t_{k+1})$, if the type ω_{a_k} generation k player and the type $\omega_{a_{k-1}}$ generation a_{k-1} player coexist in the time interval $[t_k, t_{k+1})$ for $k \in \{1,2,3,\cdots,v\}$, and where σ[s, x (s)] is a n × Θ matrix and $z(s)$ is a Θ-dimensional Wiener process. Let $\Omega[s,x(s)] = \sigma[s,x(s)]\,\sigma[s,x(s)]'$ denote the covariance matrix with its element in row h and column ζ denoted by $\Omega^{h\zeta}[s,x(s)]$.

5.5.1 Noncooperative Outcomes and Joint Maximization

Following the analysis in Sect. 5.1 of this Chapter and Sect. 3.1 of Chap. 3 a counterpart of Lemma 1.2 characterizing the noncooperative outcomes of the game the stochastic dynamic problem (1.3, 1.4, 1.5, 1.6 and 5.1) can be obtained as Lemma 5.1 below.

Lemma 5.1 A set of feedback strategies $\left\{\phi_{k-1}^{(\omega_{k-1}^{a_{k-1}},O)\omega_k^{a_k}}(t,x);\ \phi_k^{(\omega_k^{a_k},Y)\omega_{k-1}^{a_{k-1}}}(t,x)\right\}$ constitutes a Nash equilibrium solution for the game (1.3, 1.4, 1.5, 1.6 and 5.1), if there exist continuously differentiable functions $V^{k-1(\omega_{k-1}^{a_{k-1}},O)\omega_k^{a_k}}(t,x):[t_k,t_{k+1}]\times R^m\to R$ and $V^{k(\omega_k^{a_k},Y)\omega_{k-1}^{a_{k-1}}}(t,x):[t_k,t_{k+1}]\times R^m\to R$ satisfying the following partial differential equations:

$$-V_t^{\upsilon(\omega_\upsilon^{a_\upsilon},Y)\omega_{\upsilon-1}^{a_{\upsilon-1}}}(t,x)-\frac{1}{2}\sum_{h,\zeta=1}^{m}\Omega^{h\zeta}(t,x)\,V_{x^hx^\zeta}^{\upsilon(\omega_\upsilon^{a_\upsilon},O)\omega_{\upsilon-1}^{a_{\upsilon-1}}}(t,x)$$
$$=\max_{u_\upsilon^Y}\left\{g^{\upsilon(\omega_\upsilon^{a_\upsilon})}\left[t,x,\phi_{\upsilon-1}^{(\omega_{\upsilon-1}^{a_{\upsilon-1}},O)\omega_\upsilon^{a_\upsilon}}(t,x),u_\upsilon^Y\right]e^{-r(t-t_\upsilon)}\right.$$
$$\left.+V_x^{\upsilon(\omega_\upsilon^{a_\upsilon},Y)\omega_{\upsilon-1}^{a_{\upsilon-1}}}(t,x)f\left[t,x,\phi_{\upsilon-1}^{(\omega_{\upsilon-1}^{a_{\upsilon-1}},O)\omega_\upsilon^{a_\upsilon}}(t,x),u_\upsilon^Y\right]\right\},$$
$$V^{\upsilon(\omega_\upsilon^{a_\upsilon},Y)\omega_{\upsilon-1}^{a_{\upsilon-1}}}(t_{\upsilon+1},x)=e^{-r(t_{\upsilon+1}-t_\upsilon)}q^{\upsilon(\omega_\upsilon^{a_\upsilon})}[t_{\upsilon+1},x(t_{\upsilon+1})];\ \text{and}$$
$$-V_t^{k-1(\omega_{k-1}^{a_{k-1}},O)\omega_k^{a_k}}(t,x)-\frac{1}{2}\sum_{h,\zeta=1}^{m}\Omega^{h\zeta}(t,x)\,V_{x^hx^\zeta}^{k-1(\omega_{k-1}^{a_{k-1}},O)\omega_k^{a_k}}(t,x)$$
$$=\max_{u_k^O}\left\{g^{k-1(\omega_{k-1})}\left[t,x,u_k^O,\phi_k^{(\omega_k^{a_k},Y)\omega_{k-1}^{a_{k-1}}}(t,x)\right]e^{-r(t-t_k)}\right.$$
$$\left.+V_x^{k-1(\omega_{k-1}^{a_{k-1}},O)\omega_k^{a_k}}(t,x)f\left[t,x,u_k^O,\phi_k^{(\omega_k^{a_k},Y)\omega_{k-1}^{a_{k-1}}}(t,x)\right]\right\}$$
$$V^{k-1(\omega_{k-1}^{a_{k-1}},O)\omega_k^{a_k}}(t_{k+1},x)=e^{-r(t_{k+1}-t_k)}q^{k-1(\omega_{k-1}^{a_{k-1}})}(t_{k+1},x),\ \text{and}$$
$$-V_t^{k(\omega_k^{a_k},Y)\omega_{k-1}^{a_{k-1}}}(t,x)-\frac{1}{2}\sum_{h,\zeta=1}^{m}\Omega^{h\zeta}(t,x)\,V_{x^hx^\zeta}^{k(\omega_k^{a_k},Y)\omega_{k-1}^{a_{k-1}}}(t,x)$$
$$=\max_{u_k^Y}\left\{g^{(k,\omega_k)}\left[t,x,\phi_{k-1}^{(\omega_{k-1}^{a_{k-1}},O)\omega_k^{a_k}},u_k^Y\right]e^{-r(t-t_k)}\right.$$
$$\left.+V_x^{k(\omega_k^{a_k},Y)\omega_{k-1}^{a_{k-1}}}(t,x)\,f\left[t,x,\phi_{k-1}^{(\omega_{k-1}^{a_{k-1}},O)\omega_k^{a_k}},u_k^Y\right]\right\}$$
$$V^{k(\omega_k^{a_k},Y)\omega_{k-1}^{a_{k-1}}}(t_{k+1},x)=e^{-r(t_{k+1}-t_k)}\sum_{\ell=1}^{\varsigma_{k+1}}\lambda_{k+1}^{\ell}\,V^{k(\omega_k^{a_k},O)\omega_{k+1}^{\ell}}(t_{k+1},x),\tag{5.2}$$

for $k\in\{1,2,\cdots,\upsilon-1\}$.

Proof Follow the proof of Theorem 1.1 in Chap. 3. ■

Now consider the case when coexisting players want to cooperate and maximize their joint expected payoff. Following the analysis in Sect. 5.2, the joint maximization problem in the time interval $[t_\upsilon, t_{\upsilon+1})$ involving type $\omega_\upsilon^{a_\upsilon}$ generation υ player and type $\omega_{\upsilon-1}^{a_{\upsilon-1}}$ generation $\upsilon - 1$ player can be expressed as the expected joint payoff

$$E_{t_\upsilon}\Bigg\{\int_{t_\upsilon}^{t_{\upsilon+1}}\Bigg(g^{\upsilon-1(\omega_{\upsilon-1}^{a_{\upsilon-1}})}\Big[s, x(s), u_{\upsilon-1}^{(\omega_{\upsilon-1}^{a_{\upsilon-1}}, O)\omega_\upsilon^{a_\upsilon}}(s), u_\upsilon^{(\omega_\upsilon^{a_\upsilon}, Y)\omega_{\upsilon-1}^{a_{\upsilon-1}}}(s)\Big]$$
$$+ g^{\upsilon(\omega_\upsilon^{a_\upsilon})}\Big[s, x(s), u_{\upsilon-1}^{(\omega_{\upsilon-1}^{a_{\upsilon-1}}, O)\omega_\upsilon^{a_\upsilon}}(s), u_\upsilon^{(\omega_\upsilon^{a_\upsilon}, Y)\omega_{\upsilon-1}^{a_{\upsilon-1}}}(s)\Big]\Bigg)e^{-r(s-t_\upsilon)}ds$$
$$+ e^{-r(t_{\upsilon+1}-t_\upsilon)}\Big(q^{\upsilon-1(\omega_{\upsilon-1}^{a_{\upsilon-1}})}[t_{\upsilon+1}, x(t_{\upsilon+1})] + q^{\upsilon(\omega_\upsilon^{a_\upsilon})}[t_{\upsilon+1}, x(t_{\upsilon+1})]\Big)\Bigg\}, \tag{5.3}$$

subject to (5.1).

The joint maximization problem in the time interval $[t_k, t_{k+1})$, for $k \in \{1, 2, \cdots, \upsilon - 1\}$, involving the type $\omega_k^{a_k}$ generation k player and type $\omega_{k-1}^{a_{k-1}}$ generation $k - 1$ player can be expressed as the maximization of the expected joint payoff:

$$E_{t_k}\Bigg\{\int_{t_k}^{t_{k+1}}\Bigg(g^{k-1(\omega_{k-1}^{a_{k-1}})}\Big[s, x(s), u_k^{(\omega_{k-1}^{a_{k-1}}, O)\omega_k^{a_k}}(s), u_k^{(\omega_k^{a_k}, Y)\omega_{k-1}^{a_{k-1}}}(s)\Big]$$
$$+ g^{k(\omega_k^{a_k})}\Big[s, x(s), u_k^{(\omega_{k-1}^{a_{k-1}}, O)\omega_k^{a_k}}(s), u_k^{(\omega_k^{a_k}, Y)\omega_{k-1}^{a_{k-1}}}(s)\Big]\Bigg)e^{-r(s-t_k)}ds$$
$$+ e^{-r(t_{k+1}-t_k)}\Bigg(q^{k-1(\omega_{k-1}^{a_{k+1}})}[t_{k+1}, x(t_{k+1})]$$
$$+ \sum_{\ell=1}^{\varsigma_{k+1}}\lambda_{k+1}^\ell \sum_{h=1}^{\varsigma(\omega_k^{a_k}, \omega_{k+1}^\ell)} \varpi_h^{(\omega_k^{a_k}, \omega_{k+1}^\ell)} \xi^{k(\omega_k^{a_k}, O)\omega_{k+1}^\ell[h]}(t_{k+1}, x(t_{k+1}))\Bigg)\Bigg\}, \tag{5.4}$$

subject to (5.1).

Following the analysis in Sect. 5.2 a counterpart of Theorem 2.1 characterizing an optimal solution of the problem of maximizing (5.3) and (5.4) subject to (5.1) can be obtained as Theorem 5.1 below.

Theorem 5.1 A set of controls $\Big\{\psi_{k-1}^{(\omega_{k-1}^{a_{k-1}}, O)\omega_k^{a_k}}(t, x);\ \psi_k^{(\omega_k^{a_k}, Y)\omega_{k-1}^{a_{k-1}}}(t, x)\Big\}$ constitutes an optimal solution for the control problem (5.1, 5.3 and 5.4), if there exist continuously differentiable function $W^{[t_k, t_{k+1}](\omega_{k-1}^{a_{k-1}}, \omega_k^{a_k})}(t, x) : [t_k, t_{k+1}) \times R^m \to R$ satisfying the following partial differential equations:

$$-W_t^{[t_\upsilon,t_{\upsilon+1}](\omega_{\upsilon-1}^{a_{\upsilon-1}},\omega_\upsilon^{a_\upsilon})}(t,x)-\frac{1}{2}\sum_{h,\zeta=1}^{m}\Omega^{h\zeta}(t,x)\,W_{x^h x^\zeta}^{[t_\upsilon,t_{\upsilon+1}](\omega_{\upsilon-1}^{a_{\upsilon-1}},\omega_\upsilon^{a_\upsilon})}(t,x)$$
$$=\max_{u_\upsilon^O,u_\upsilon^Y}\Big\{g^{\upsilon-1(\omega_{\upsilon-1}^{a_{\upsilon-1}})}\left[t,x,u_\upsilon^O,u_\upsilon^Y\right]e^{-r(t-t_\upsilon)}$$
$$+g^{\upsilon(\omega_\upsilon^{a_\upsilon})}\left[t,x,u_\upsilon^O,u_\upsilon^Y\right]e^{-r(t-t_\upsilon)}+W_x^{[t_\upsilon,t_{\upsilon+1}](\omega_{\upsilon-1}^{a_{\upsilon-1}},\omega_\upsilon^{a_\upsilon})}(t,x)f\left[t,x,u_\upsilon^O,u_\upsilon^Y\right]\Big\},$$
$$W^{[t_\upsilon,t_{\upsilon+1}](\omega_{\upsilon-1}^{a_{\upsilon-1}},\omega_\upsilon^{a_\upsilon})}(t_{\upsilon+1},x)=e^{-r(t_{\upsilon+1}-t_\upsilon)}\left[q^{\upsilon-1(\omega_{\upsilon-1}^{a_{\upsilon-1}})}(t_{\upsilon+1},x)+q^{\upsilon(\omega_\upsilon^{a_\upsilon}]}(t_{\upsilon+1},x)\right];\text{ and}$$
$$-W_t^{[t_k,t_{k+1}](\omega_{k-1}^{a_{k-1}},\omega_k^{a_k})}(t,x)-\frac{1}{2}\sum_{h,\zeta=1}^{m}\Omega^{h\zeta}(t,x)\,W_{x^h x^\zeta}^{[t_k,t_{k+1}](\omega_{k-1}^{a_{k-1}},\omega_k^{a_k})}(t,x)$$
$$=\max_{u_k^O,u_\upsilon^Y}\Big\{g^{k-1(\omega_{k-1}^{a_{k-1}})}\left[t,x,u_k^O,u_k^Y\right]e^{-r(t-t_k)}$$
$$+g^{k(\omega_k^{a_k})}\left[t,x,u_k^O,u_k^Y\right]e^{-r(t-t_k)}+W_x^{[t_k,t_{k+1}](\omega_{k-1}^{a_{k-1}},\omega_k^{a_k})}(t,x)f\left[t,x,u_k^O,u_k^Y\right]\Big\},$$
$$W^{[t_k,t_{k+1}](\omega_{k-1}^{a_{k-1}},\omega_k^{a_k})}(t_{k+1},x)=e^{-r(t_{k+1}-t_k)}\Big(q^{k-1(\omega_{k-1}^{a_{k-1}})}(t_{k+1},x),$$
$$+\sum_{\ell=1}^{\varsigma_{k+1}}\lambda_{k+1}^{\ell}\sum_{h=1}^{\varsigma(\omega_k^{a_k},\omega_{k+1}^{\ell})}\varpi_h^{(\omega_k^{a_k},\omega_{k+1}^{\ell})}\xi^{k(\omega_k^{a_k},O)\omega_{k+1}^{\ell}[h]}(t_{k+1},x(t_{k+1}))\Big),$$
$$\text{for } k\in\{1,2,\cdots,\upsilon-1\}. \tag{5.5}$$

Proof Follow the proof of Theorem A.3 in the Technical Appendices we obtain the conditions characterizing an optimal solution of the problem (5.1), (5.3) and (5.4) as in (5.5). ■

In particular, $W^{[t_k,t_{k+1}](\omega_{k-1}^{a_{k-1}},\omega_k^{a_k})}(t,x)$ gives the maximized expected joint payoff of the type $\omega_k^{a_k}$ generation k player and type $\omega_{k-1}^{a_{k-1}}$ generation $k-1$ player at time $t\in[t_k,t_{k+1}]$ with the state x in the stochastic control problem

$$\max_{u_k^{(\omega_{k-1}^{a_{k-1}},O)\omega_k^{a_k}},\,u_k^{(\omega_k^{a_k},Y)\omega_{k-1}^{a_{k-1}}}}$$
$$E_{t_k}\Big\{\int_t^{t_{k+1}}\Big(g^{k-1(\omega_{k-1}^{a_{k-1}})}\left[s,x(s),u_k^{(\omega_{k-1}^{a_{k-1}},O)\omega_k^{a_k}}(s),u_k^{(\omega_k^{a_k},Y)\omega_{k-1}^{a_{k-1}}}(s)\right]$$
$$+g^{k(\omega_k^{a_k})}\left[s,x(s),u_k^{(\omega_{k-1}^{a_{k-1}},O)\omega_k^{a_k}}(s),u_k^{(\omega_k^{a_k},Y)\omega_{k-1}^{a_{k-1}}}(s)\right]\Big)e^{-r(s-t_k)}ds$$
$$+e^{-r(t_{k+1}-t_k)}\Big(q^{k-1(\omega_{k-1}^{a_{k+1}})}[t_{k+1},x(t_{k+1})]$$
$$+\sum_{\ell=1}^{\varsigma_{k+1}}\lambda_{k+1}^{\ell}\sum_{h=1}^{\varsigma(\omega_k^{a_k},\omega_{k+1}^{\ell})}\varpi_h^{(\omega_k^{a_k},\omega_{k+1}^{\ell})}\xi^{k(\omega_k^{a_k},O)\omega_{k+1}^{\ell}[h]}(t_{k+1},x(t_{k+1}))\Big)\Big\},$$

subject to

$$dx(s) = f\left[s, x(s), u_{k-1}^{(\omega_{k-1},O)\omega_k}(s), u_k^{(\omega_k,Y)\omega_{k-1}}(s)\right]ds + \sigma[s, x(s)]dz(s),\ x(t) = x.$$

Substituting the set of cooperative strategies into (5.1) yields the dynamics of the cooperative state trajectory in the time interval $[t_k, t_{k+1})$

$$\begin{aligned} dx(s) = & f\left[s, x(s), \psi_{k-1}^{(\omega_{k-1}^{a_{k-1}},O)\omega_k^{a_k}}(s, x(s)), \psi_k^{(\omega_k^{a_k},Y)\omega_{k-1}^{a_{k-1}}}(s, x(s))\right]ds \\ & + \sigma[s, x(s)]dz(s) \end{aligned} \tag{5.6}$$

if type $\omega_k^{a_k}$ generation k player and type $\omega_{k-1}^{a_{k-1}}$ generation $k-1$ player coexist in $[t_k, t_{k+1})$, for $s \in [t_k, t_{k+1})$, $k \in \{1, 2, \cdots, \upsilon\}$ and $x(t_k) = x_{t_k} \in X$.

We denote the set of realizable states at time t from (5.6) under the scenarios of different players by $X_t^{\left(\omega_{k-1}^{a_{k-1}},\omega_k^{a_k}\right)*}$, for $t \in [t_k, t_{k+1})$ and $k \in \{1, 2, \cdots, \upsilon\}$. We use the term $x_t^{\left(\omega_{k-1}^{a_{k-1}},\omega_k^{a_k}\right)*}$ by $x_t^{\left(\omega_{k-1}^{a_{k-1}},\omega_k^{a_k}\right)*} \in X_t^{\left(\omega_{k-1}^{a_{k-1}},\omega_k^{a_k}\right)*}$ to denote an element in $X_t^{\left(\omega_{k-1}^{a_{k-1}},\omega_k^{a_k}\right)*}$. The term x_t^* is used to denote $x_t^{\left(\omega_{k-1}^{a_{k-1}},\omega_k^{a_k}\right)*}$ whenever there is no ambiguity. For simplicity in exposition we also use $x^{\left(\omega_{k-1}^{a_{k-1}},\omega_k^{a_k}\right)*}(t)$ and $x_t^{\left(\omega_{k-1}^{a_{k-1}},\omega_k^{a_k}\right)*}$ inter-changeably.

5.5.2 Subgame Consistent Solutions and Payoff Distribution

Now consider the case when coexisting players want to cooperate and agree to act and allocate the cooperative payoff according to a set of agreed upon optimality principles. Again in the time interval $[t_k, t_{k+1})$ the probability that the type $\omega_k^{a_k}$ generation k player and the type $\omega_{k-1}^{a_{k-1}}$ generation $k-1$ player would agree to the solution imputation

$\left[\xi^{k-1(\omega_{k-1}^{a_{k-1}},O)\omega_k^{a_k}[h]}(t,x),\ \xi^{k(\omega_k^{a_k},Y)\omega_{k-1}^{a_{k-1}}[h]}(t,x)\right]$ over the time interval $[t_k, t_{k+1})$, is $\varpi_h^{(\omega_{k-1}^{a_{k-1}},\omega_k^{a_k})}$, where $\sum_{h=1}^{\zeta(\omega_{k-1}^{a_{k-1}},\omega_k^{a_k})} \varpi_h^{(\omega_{k-1}^{a_{k-1}},\omega_k^{a_k})} = 1$. At time t_1, the agreed-upon imputation for the type ω_0^1 generation 0 player and the type ω_1^1 player are known.

Following the analysis in Sect. 5.3 a counter-part of Theorem 3.1 which derives the PDP that yields a subgame consistent solution for the cooperative game (5.1) and (5.3, 5.4) can be obtained in the theorem below.

Theorem 5.2 If the imputation vector $\left[\xi^{k-1(\omega_{k-1}^{a_{k-1}},O)\omega_k^{a_k}[h]}(t,x_t^*),\ \xi^{k(\omega_k^{a_k},O)\omega_{k-1}^{a_{k-1}}[h]}(t,x_t^*)\right]$, are functions that are continuously differentiable in t and x_t^*, a PDP with an instantaneous payment at time $t\in[t_k,t_{k+1})$:

$$B_{k-1}^{(\omega_{k-1}^{a_{k-1}},O)\omega_k^{a_k}[h]}(t) = -\xi_t^{k-1(\omega_{k-1}^{a_{k-1}},O)\omega_k^{a_k}[h]}(t,x_t^*)$$
$$-\frac{1}{2}\sum_{h,\zeta=1}^{m}\Omega^{h\zeta}(t,x_t^*)\ \xi_{x^h x^\zeta}^{k-1(\omega_{k-1}^{a_{k-1}},O)\omega_k^{a_k}}(t,x_t^*)$$
$$-\xi_x^{k-1(\omega_{k-1}^{a_{k-1}},O)\omega_k^{a_k}[h]}(t,x_t^*)f\left[t,x_t^*,\psi_{k-1}^{(\omega_{k-1}^{a_{k-1}},O)\omega_k^{a_k}}(t,x_t^*),\psi_k^{(\omega_k^{a_k},Y)\omega_{k-1}^{a_{k-1}}}(t,x_t^*)\right] \quad (5.7)$$

allocated to the type $\omega_{k-1}^{a_{k-1}}$ generation $k-1$ player;

and an instantaneous payment at time $t\in[t_k,t_{k+1})$:

$$B_k^{(\omega_k^{a_k},Y)\omega_{k-1}^{a_{k-1}}[h]}(t) = -\xi_t^{k(\omega_k^{a_k},Y)\omega_{k-1}^{a_{k-1}}[h]}(t,x_t^*) - \frac{1}{2}\sum_{h,\zeta=1}^{m}\Omega^{h\zeta}(t,x_t^*)\ \xi_{x^h x^\zeta}^{k(\omega_k^{a_k},Y)\omega_{k-1}^{a_{k-1}}}(t,x_t^*)$$
$$-\xi_x^{k(\omega_k^{a_k},Y)\omega_{k-1}^{a_{k-1}}[h]}(t,x_t^*)f\left[t,x_t^*,\psi_{k-1}^{(\omega_{k-1}^{a_{k-1}},O)\omega_k^{a_k}}(t,x_t^*),\psi_k^{(\omega_k^{a_k},Y)\omega_{k-1}^{a_{k-1}}}(t,x_t^*)\right]$$

allocated to the type $\omega_k^{a_k}$ generation k player,

yields a mechanism leading to the realization of the imputation vector

$$\left[\xi^{k-1(\omega_{k-1}^{a_{k-1}},O)\omega_k^{a_k}[h]}(t,x_t^*),\ \xi^{k(\omega_k^{a_k},Y)\omega_{k-1}^{a_{k-1}}[h]}(t,x_t^*)\right],\ \text{for } k\in\{1,2,\cdots,\upsilon\}.$$

Proof Follow the proof leading to Theorem 3.1 in Chap. 3 with the imputation vector in present value (rather than in current value). ■

5.6 Chapter Notes

This Chapter considers cooperative differential games in which players enter the game at different times and have diverse horizons. Moreover, the types of future players are not known with certainty. Subgame consistent cooperative solutions and analytically tractable payoff distribution mechanisms leading to the realization of these solutions are derived. Finally, the overlapping generations of players can be extended to more complex structures. The game horizon of the players can include more than two time intervals and be different across players. The number of players in each time interval can also be more than two and be different across intervals. Hence, the analysis can be formulated as a general class of stochastic differential games with asynchronous horizons structures. An analysis on subgame consistent cooperative solutions in stochastic differential games with asynchronous horizons and uncertain types of players can be found in Yeung (2012).

5.7 Problems

1. Consider the game in which there are 4 overlapping generations of players with generation 0 and generation 1 players in $[0,2)$, generation 1 and generation 2 players in $[2,4)$, generation 2 and generation 3 players in $[4,6]$. Players are of either type 1 or type 2. The instantaneous payoffs and terminal rewards of the type 1 generation k player and the type 2 generation k player are respectively:

$$\left[2(u_k)^{1/2} - \frac{1}{x^{1/2}}u_k\right] \text{ and } q_1 x^{1/2}; \quad \text{and } \left[(u_k)^{1/2} - \frac{2}{x^{1/2}}u_k\right] \text{ and } q_2 x^{1/2},$$

where the state variable $x(s)$ is the biomass of a renewable resource. $u_k(s)$ is the harvest of the generation k extraction firm. The type $i \in \{1,2\}$ generation k extraction firm's extraction cost is $c_i u_k(s) x(s)^{-1/2}$.

At initial time 0, it is known that the generation 0 player is of type 1 and the generation 1 player is also of type 1. It is also known that the generation 2 and generation 3 players may be of type 1 with probability $\lambda^1 = 0.4$ and of type 2 with probability $\lambda^2 = 0.6$.

The state dynamics of the game is characterized by:

$$\dot{x}(s) = 10x(s)^{1/2} - 2x(s) - u_k^{(i,O)j}(s) - u_k^{(j,Y)i}(s),$$

if the old generation $k-1$ extractor is of type i and the young generation k extractor is of type j, for $s \in [t_k, t_{k+1})$ and $k \in \{1,2,3\}$ with $t_1 = 0$, $t_2 = 2$ and $t_3 = 4$; and $x(0) = 30$,

where $u_k^{(i,O)j}(s)$ denote the harvest of the type i generation $k-1$ old extractor and $u_k^{(j,Y)i}(s)$ denote the harvest of the type j generation k young extractor. The discount rate is 0.05.

Characterize the non-cooperative feedback Nash equilibrium for the generation 0 player and generation 1 player game.
2. Construct a subgame consistent cooperative solution in which all types of players would accept the imputation which shares the excess cooperative gains (over the individual payoffs) equally among themselves.

Chapter 6
Subgame Consistent Cooperative Solution in NTU Differential Games

Subgame consistency is a fundamental element in the solution of cooperative stochastic differential games which ensures that the extension of the solution policy to a later starting time and any possible state brought about by prior optimal behavior of the players would remain optimal. In many game situations payoff (or utility) of players may not be transferable. It is well known that utility in economic study is assumed to be non-transferrable or comparable among economic agents. The Nash (1950, 1953) bargaining solution is a solution for non-transferable payoff cooperative games. Strategic interactions involving national security, social issues and political gains fall into the category of non-transferrable utility/payoff (NTU) games. In the case when payoffs are nontransferable, transfer payments cannot be made and subgame consistent solution mechanism becomes extremely complicated. In this Chapter, the issue of subgame consistency in cooperative stochastic differential games with nontransferable payoffs or utility is presented. In particular, the Chapter is an integrated exposition of the works in Yeung and Petrosyan (2005) and Yeung et al. (2007). The Chapter is organized as follows. The formulation of non-transferrable utility cooperative stochastic differential games, the corresponding Pareto optimal state trajectories and individual player's payoffs under cooperation are provided in Sect. 6.1. The notion of subgame consistency in NTU cooperative stochastic differential games under time invariant payoff weights is examined in Sect. 6.2. In Section 6.3, a class of cooperative stochastic differential games with nontransferable payoffs is developed to illustrate the derivation of subgame consistent solutions. Subgame consistent cooperative solutions of the class of NTU games developed in Sect. 6.3 are investigated in Sect. 6.4. Numerical delineations of the solutions presented in Sect. 6.4 are given in Sect. 6.5. An analysis on infinite horizon NTU cooperative stochastic differential games is provided in Sect. 6.6. A chapter appendices containing proofs are given in Sect. 6.7. Chapter notes are given Sect. 6.8 and problems in Sect. 6.9.

D.W.K. Yeung, L.A. Petrosyan, *Subgame Consistent Cooperation*,
Theory and Decision Library C 47, DOI 10.1007/978-981-10-1545-8_6

6.1 NTU Cooperative Stochastic Differential Games

Consider the two-person cooperative stochastic differential game with initial state x_0 and duration $T - t_0$. The state space of the game is $X \in R^n$, with permissible state trajectories $\{x(s),\ t_0 \leq s \leq T\}$. The state dynamics of the game is characterized by the vector-valued stochastic differential equations:

$$dx(s) = f[s, x(s), u_1(s), u_2(s)]ds + \sigma[s, x(s)]dz(s), \qquad x(t_0) = x_0, \tag{1.1}$$

where $\sigma[s, x(s)]$ is a $n \times \Theta$ matrix and $z(s)$ is a Θ-dimensional Wiener process and the initial state x_0 is given. Let $\Omega[s, x(s)] = \sigma[s, x(s)]\sigma[s, x(s)]'$ denote the covariance matrix with its element in row h and column ζ denoted by $\Omega^{h\zeta}[s, x(s)], u_i \in U_i \subset compR^{\ell}$ is the control vector of player i, for $i \in \{1, 2\}$.

At time instant $s \in [t_0, T]$, the instantaneous payoff of player i, for $i \in \{1, 2\}$, is denoted by $g^i[s, x(s), u_1(s), u_2(s)]$, and when the game terminates at time T, player i receives a terminal payment of $q^i(x(T))$. Payoffs are nontransferable across players. Given a time-varying instantaneous discount rate $r(s)$, for $s \in [t_0, T]$, values received t time after t_0 have to be discounted by the factor $\exp\left[-\int_{t_0}^{t} r(y)dy\right]$. Hence at time t_0, the expected payoff of player i, for $i \in \{1, 2\}$, is given as:

$$\begin{aligned} J^i(t_0, x_0) = E_{t_0}\Bigg\{ &\int_{t_0}^{T} g^i[s, x(s), u_1(s), u_2(s)] \exp\left[-\int_{t_0}^{s} r(y)dy\right] ds \\ &+ \exp\left[-\int_{t_0}^{T} r(y)dy\right] q^i(x(T)) \Bigg| x(t_0) = x_0 \Bigg\}, \end{aligned} \tag{1.2}$$

where E_{t_0} denotes the expectation operator performed at time t_0,

We use $\Gamma(x_0, T - t_0)$ to denote the game (1.1 and 1.2) and $\Gamma(x_\tau, T - \tau)$ to denote an alternative game with state dynamics (1.1) and payoff structure (1.2) which starts at time $\tau \in [t_0, T]$ with initial state $x_\tau \in X$. The benchmark noncooperative feedback Nash equilibrium solution can be characterized by Theorem 1.1 in Chap. 3.

6.1.1 *Pareto Optimal Trajectories*

Consider the situation when the players agree to cooperate. We use $\Gamma_c(x_0, T - t_0)$ to denote a cooperative game with dynamics (1.1) and payoffs (1.2). To achieve group optimality, the players have to consider cooperative outcomes belonging to the Pareto optimal set. Pareto optimal trajectories for $\Gamma_c(x_0, T - t_0)$ can be identified by choosing a specific weight $\alpha_1 \in (0, \infty)$ that solves the following stochastic control problem (See Leitmann (1974), Dockner and Jørgensen (1984) and Jørgensen and Zaccour (2001)):

$$\max_{u_1,u_2}\{J^1(t_0,x_0)+\alpha_1 J^2(t_0,x_0)\}\equiv$$

$$\max_{u_1,u_2} E_{t_0}\left\{\int_{t_0}^{T}\left(g^1[s,x(s),u_1(s),u_2(s)]+\alpha_1 g^2[s,x(s),u_1(s),u_2(s)]\right)\exp\left[-\int_{t_0}^{s}r(y)dy\right]ds\right.$$
$$\left.+\left[q^1(x(T))+\alpha_1 q^2(x(T))\right]\exp\left[-\int_{t_0}^{T}r(y)dy\right]\middle| x(t_0)=x_0\right\}, \tag{1.3}$$

subject to dynamics (1.1). Note that the problem $\max_{u_1,u_2}\{J^1(t_0,x_0)+\alpha_i J^2(t_0,x_0)\}$ is identical to the problem $\max_{u_1,u_2}\{J^2(t_0,x_0)+\alpha_2 J^1(t_0,x_0)\}$ when $\alpha_1=1/\alpha_2$.

Invoking the technique developed by Fleming (1969) in Theorem A.3 of the Technical Appendices, we have

Corollary 1.1 A set of controls $\left\{u_i^{\alpha_1(t_0)}(t)=\psi_i^{\alpha_1(t_0)}(t,x), \text{ for } i\in\{1,2\}\right\}$ provides an optimal solution to the stochastic control problem (1.3), if there exists twice continuously differentiable function $W^{\alpha_1(t_0)}(t,x):[t_0,T]\times R^n\to R$ satisfying the partial differential equation:

$$-W_t^{\alpha_1(t_0)}(t,x)-\frac{1}{2}\sum_{h,\zeta=1}^{n}\Omega^{h\zeta}(t,x)W_{x^h x^\zeta}^{\alpha_1(t_0)}(t,x)=$$
$$\max_{u_1,u_2}\left\{\left(g^1[t,x,u_1,u_2]+\alpha_1 g^2[t,x,u_1,u_2]\right)\exp\left[-\int_{t_0}^{t}r(y)dy\right]\right.$$
$$\left.+W_x^{\alpha_1(t_0)}(t,x)f[t,x,u^1,u^2]\right\},$$
$$W^{\alpha_1(t_0)}(T,x)=\exp[-r(T-t_0)][q^1(x)+\alpha_1 q^2(x)].$$
■

Substituting $\psi_1^{\alpha_1(t_0)}(t,x)$ and $\psi_2^{\alpha_1(t_0)}(t,x)$ into (1.1) yields the dynamics of the Pareto optimal trajectory associated with weight α_1:

$$dx(s)=f\left[s,x(s),\psi_1^{\alpha_1(t_0)}(s,x(s)),\psi_2^{\alpha_1(t_0)}(s,x(s))\right]ds+\sigma[s,x(s)]dz(s),\ x(t_0)=x_0. \tag{1.4}$$

We denote the set containing realizable values of $x^{\alpha_1 *}(t)$ by $X_t^{\alpha_1(t_0)}$, for $t\in(t_0,T]$.

The solution to (1.4) yields a Pareto optimal trajectory, which can be expressed as:

$$x(t)=x_0+\int_{t_0}^{t}f\left[s,x(s),\psi_1^{\alpha_1(t_0)}(s,x(s)),\psi_2^{\alpha_1(t_0)}(s,x(s))\right]ds+\int_{t_0}^{t}\sigma[s,x(s)]\,dz(s).$$

We denote the set containing realizable values of $x(t)$ along the optimal trajectory by $X_t^{\alpha_1(t_0)}$, for $t\in(t_0,T]$.

Now, consider the cooperative game $\Gamma_c(x_\tau,T-\tau)$ with state dynamics (1.1) and payoff structure (1.2), which starts at time $\tau\in[t_0,T]$ with initial state $x_\tau\in X_\tau^{\alpha_1(t_0)}$.

We use $\psi_i^{\alpha_1(\tau)}(t,x)$ to denote the optimal control in $\Gamma_c(x_\tau, T-\tau)$, for $\tau \in [t_0, T]$ and $t \in [\tau, T]$. Using Definition 1.1 we can characterize the solution of the control problem $\max_{u_1,u_2} \{J^1(\tau,x_\tau) + \alpha_1 J^2(\tau,x_\tau)\}$ in $\Gamma_c(x_\tau, T-\tau)$, for $\tau \in [t_0, T]$ and $t \in [\tau, T]$. In particular, we use $\left[\psi_1^{\alpha_1(\tau)}(t,x),\, \psi_2^{\alpha_1(\tau)}(t,x)\right]$ to denote the optimal control and $W^{\alpha_1(\tau)}(t,x) : [\tau, T] \times R^n \to R$ the corresponding maximized value function.

Remark 1.1 Invoking Definition 1.1, one can readily show that $\psi_i^{\alpha_1(\tau)}(t,x) = \psi_i^{\alpha_1(s)}(t,x)$ at the point (t,x), for $i \in \{1,2\}$, $t_0 \le \tau \le s \le t \le T$ and $x \in X_t^{\alpha_1(t_0)}$. ■

Remark 1.2 Invoking Definition 1.1, one can readily show that $W^{\alpha_1(\tau)}(t,x) = W^{\alpha_1(s)}(t,x)\exp[-r(\tau - s)]$, for $t_0 \le \tau \le s \le t \le T$ and $x \in X_t^{\alpha_1(t_0)}$. ■

6.1.2 *Individual Player's Payoffs Under Cooperation*

In this section, we present a methodology for the derivation of individual player's payoff under cooperation. To do this, we first substitute the optimal controls $\psi_1^{\alpha_1(t_0)}(t,x)$ and $\psi_2^{\alpha_1(t_0)}(t,x)$ into the objective functions (1.2) to derive the players' expected payoff under cooperation with α_1 being chosen as the cooperative weight.

Given that $x(t) = x \in X_t^{\alpha_1 *}$, for $t \in [\tau, T]$, we define player 1's expected cooperative payoff over the interval $[t, T]$ as:

$$\begin{aligned}
&\hat{W}^{\alpha_1(t_0)i}(t,x) = \\
&E_{t_0}\left\{ \int_t^T g^i\left[s, x(s), \psi_1^{\alpha_1(t_0)}(s,x(s)), \psi_2^{\alpha_1(t_0)}(s,x(s))\right] \exp\left[-\int_{t_0}^s r(y)dy\right] ds \right. \\
&\left. + \exp\left[-\int_{t_0}^T r(y)dy\right] q^i(x(T)) \,\middle|\, x(t) = x \right\}, \quad \text{for } i \in \{1,2\}, \qquad (1.5)
\end{aligned}$$

where

$$dx(s) = f\left[s, x(s), \psi_1^{\alpha_1(t_0)}(s,x(s)), \psi_2^{\alpha_1(t_0)}(s,x(s))\right] ds + \sigma[s, x(s)]\, dz(s), \qquad x(t) = x.$$

To facilitate the derivation individual players' cooperative payoffs a mechanism characterizing player i's cooperative payoff under payoff weights α_1 is given in the theorem below.

Theorem 1.1 If there exist continuously functions $\hat{W}^{\alpha_1(t_0)i}(t,x):[t_0,T]\times R^n\to R,\ i\in\{1,2\}$, satisfying

$$-\hat{W}_t^{\alpha_1(t_0)i}(t,x)-\frac{1}{2}\sum_{h,\zeta=1}^{n}\Omega^{h\zeta}(t,x)\hat{W}_{x^h x^\zeta}^{\alpha_1(t_0)i}(t,x)=$$
$$g^i\left[t,x,\psi_1^{\alpha_1(t_0)}(t,x),\psi_2^{\alpha_1(t_0)}(t,x)\right]\exp\left[-\int_{t_0}^{t}r(y)dy\right]$$
$$+\hat{W}_x^{\alpha_1(t_0)i}(t,x)f\left[t,x,\psi_1^{\alpha_1(t_0)}(t,x),\psi_2^{\alpha_1(t_0)}(t,x)\right]\text{ and}$$
$$\hat{W}^{\alpha_1(t_0)i}(T,x)=\exp\left[-\int_{t_0}^{T}r(y)dy\right]q^i(x)$$

then $\hat{W}^{\alpha_1(t_0)i}(t,x)$ gives player i's expected cooperative payoff over the interval $[t,T]$ with α_1 being chosen as the weight.

Proof Note that for $\Delta t\to 0$, we can express $\hat{W}^{\alpha_1(t_0)i}(t,x)$ in (1.5) as:

$$\begin{aligned}
&\hat{W}^{\alpha_1(t_0)i}(t,x)=\\
&E_{t_0}\left\{\int_t^{t+\Delta t}g^i\left[s,x(s),\psi_1^{\alpha_1(t_0)}(s,x(s)),\psi_2^{\alpha_1(t_0)}(s,x(s))\right]\exp\left[-\int_{t_0}^{s}r(y)dy\right]ds\right.\\
&\qquad\left.+\hat{W}^{\alpha_1(t_0)i}(t+\Delta t,x+\Delta x)\,\middle|\,x(t)=x\right\}\\
&=E_{t_0}\left\{g^i\left[t,x,\psi_1^{\alpha_1(t_0)}(t,x),\psi_2^{\alpha_1(t_0)}(t,x)\right]\exp\left[-\int_{t_0}^{t}r(y)dy\right]\Delta t\right.\\
&\qquad+\hat{W}^{\alpha_1(t_0)i}(t,x)+\hat{W}_t^{\alpha_1(t_0)i}(t,x)\Delta t\\
&\qquad+\hat{W}_x^{\alpha_1(t_0)i}(t,x)f\left[t,x,\psi_1^{\alpha_1(t_0)}(t,x),\psi_2^{\alpha_1(t_0)}(t,x)\right]\Delta t\\
&\qquad\left.+\hat{W}_x^{\alpha_1(t_0)i}(t,x)\sigma(t,x)\Delta z+\frac{1}{2}\sum_{h,\zeta=1}^{n}\Omega^{h\zeta}(t,x)\hat{W}_{x^h x^\zeta}^{\alpha_1(t_0)i}(t,x)+o(\Delta t)\right\}\\
&\text{for } i\in\{1,2\},
\end{aligned}\tag{1.6}$$

where

$$\Delta x=f\left[t,x,\psi_1^{\alpha_1(t_0)}(t,x),\psi_2^{\alpha_1(t_0)}(t,x)\right]\Delta t+\sigma(t,x)\Delta z+o(\Delta t),$$
$$\Delta z=z(t+\Delta t)-z(t),\ \text{and}\, E_{t_0}[o(\Delta t)]/\Delta t\to 0\text{ as }\Delta t\to 0$$

Canceling terms, performing the expectation operator, dividing throughout by Δt and taking $\Delta t\to 0$, we obtain:

$$-\hat{W}_t^{\alpha_1(t_0)i}(t,x) - \frac{1}{2}\sum_{h,\zeta=1}^{n} \Omega^{h\zeta}(t,x)\hat{W}_{x^h x^\zeta}^{\alpha_1(t_0)i}(t,x) =$$
$$g^i\left[t,x,\psi_1^{\alpha_1(t_0)}(t,x),\psi_2^{\alpha_1(t_0)}(t,x)\right]\exp\left[-\int_{t_0}^{t} r(y)dy\right]$$
$$+\hat{W}_x^{\alpha_1(t_0)i}(t,x)f\left[t,x,\psi_1^{\alpha_1(t_0)}(t,x),\psi_2^{\alpha_1(t_0)}(t,x)\right], \text{ for } i \in \{1,2\}. \quad (1.7)$$

Boundary conditions require:

$$\hat{W}^{\alpha_1(t_0)i}(T,x) = \exp\left[-\int_{t_0}^{T} r(y)dy\right]q^i(x(T)), \text{ for } i \in \{1,2\}. \quad (1.8)$$

Hence Theorem 1.1 follows. ■

6.2 Notion of Subgame Consistency

Under cooperation with nontransferable payoffs, the players negotiate to establish an agreement (optimality principle) on how to play the cooperative game and how to distribute the resulting payoff. In particular, the chosen optimality principle has to satisfy group optimality and individual rationality. Subgame consistency requires that the extension of the solution policy to a later starting time and any possible state brought about by prior optimal behavior of the players would remain optimal.

Consider the cooperative game $\Gamma_c(x_0, T-t_0)$ in which the players agree to an optimality principle. In particular, given x_0 at time t_0, according to the solution optimality principle the players will adopt

(i) a weight α_1^0 leading to a set of cooperative controls $\{\left[\psi_1^{\alpha_1^0(t_0)}(t,x),\psi_2^{\alpha_1^0(t_0)}(t,x)\right]$, for $t \in [t_0, T]\}$, and

(ii) an imputation $\left[\xi^{(t_0)1}(x_0, T-t_0; \alpha_1^0),\, \xi^{(t_0)2}(x_0, T-t_0; \alpha_1^0)\right] = \left[\hat{W}^{t_0(\alpha_1^0)1}(t_0, x_0),\right.$ $\left.\hat{W}^{t_0(\alpha_1^0)2}(t_0, x_0)\right]$.

Now consider the game $\Gamma_c(x_\tau, T-\tau)$ where $x_\tau \in X_\tau^{\alpha_1(t_0)}$ and $\tau \in [t_0, T]$, under the same solution optimality principle the players will adopt

(i) a weight α_1^τ leading to a set of cooperative controls $\{\left[\psi_1^{\alpha_1^\tau(\tau)}(t,x),\psi_2^{\alpha_1^\tau(\tau)}(t,x)\right]$, for $t \in [\tau, T]\}$, and

(ii) an imputation $\left[\xi^{(\tau)1}(\tau, T-\tau; \alpha_1^\tau),\xi^{(\tau)2}(\tau, T-\tau; \alpha_1^\tau)\right] = \left[\hat{W}^{\tau(\alpha_1^\tau)1}(\tau, x_\tau),\right.$ $\left.\hat{W}^{\tau(\alpha_1^\tau)2}(\tau, x_\tau)\right]$.

A formal definition of subgame consistency can be stated as:

Definition 2.1 An optimality principle yielding imputations $\xi^{(\tau)}\left(x_\tau, T-\tau; \alpha_1^\tau\right)$, for $\tau \in [t_0, T]$ and $x_\tau \in X_\tau^{\alpha_1^0(t_0)}$, constitutes a subgame consistent solution to the game $\Gamma_c\left(x_0, T-t_0; \alpha_1^0\right)$ if the following conditions are satisfied:

(i) $\xi^{(\tau)}\left(x_\tau, T-\tau; \alpha_1^\tau\right) = \left[\xi^{(\tau)1}\left(x_\tau, T-\tau; \alpha_1^\tau\right), \xi^{(\tau)2}\left(x_\tau, T-\tau; \alpha_1^\tau\right)\right]$, for $t_0 \le \tau \le t \le T$, is Pareto optimal;

(ii) $\xi^{(\tau)i}\left(x_\tau, T-\tau; \alpha_1^\tau\right) \ge V^{(\tau)i}(\tau, x_\tau)$, for $i \in \{1,2\}$, $\tau \in [t_0, T]$ and $x_\tau \in X_\tau^{\alpha_1^0(t_0)}$; and

(iii)
$$\xi^{(\tau)i}\left(x_t, T-t; \alpha_1^\tau\right)\exp[r(\tau - t)] = \xi^{(t)i}\left(x_t, T-t; \alpha_1^t\right),$$
for $i \in \{1,2\}$, $t_0 \le \tau \le t \le T$ and $x_t \in X_t^{\alpha_1^0(t_0)}$. ■

Part (i) of Definition 4.1 requires that according to the agreed upon optimality principle Pareto optimality is maintained at every instant of time. Hence group rationality is satisfied throughout the game interval. Part (ii) demands individual rationality to be met throughout the entire game interval. Part (iii) guarantees the consistency of the solution imputations throughout the game interval in the sense that the extension of the solution policy to a situation with a later starting time and any possible state brought about by prior optimal behavior of the players remains optimal.

6.3 A NTU Game for Illustration

Consider a two-person nonzero-sum stochastic differential game with initial state x_0 and duration $T - t_0$. The state space of the game is $X \subset R$, with permissible state trajectories $\{x(s),\, t_0 \le s \le T\}$. The state dynamics of the game is characterized by the stochastic differential equations:

$$dx(s) = [a - bx(s) - u_1(s) - u_2(s)]ds + \sigma x(s)dz(s), \quad x(t_0) = x_0 \in X, \tag{3.1}$$

where $u_i \in U_i$ is the control vector of player i, for $i \in [1,2]$, a, b and σ are positive constants, and $z(s)$ is a Wiener process. Equation (3.1) could be interpreted as the stock dynamics of a biomass of renewable resource like forest or fresh water. The state $x(s)$ represents the resource size and $u_i(s)$ the (nonnegative) amount of resource extracted by player i.

At time t_0, the expected payoff of player $i \in \{1,2\}$ is:

$$J^i(t_0, x_0) = E_{t_0}\left\{ \int_{t_0}^{T} \left[h_i u_i(s) - c_i u_i(s)^2 x(s)^{-1} + k_i x(s)\right]\exp[-r(s-t_0)]\,ds \right.$$
$$\left. + \exp[-r(T-t_0)]\, q_i x(T) \,\middle|\, x(t_0) = x_0 \right\},$$
$$\text{for } i \in \{1,2\}, \tag{3.2}$$

where h_i, c_i, k_i and q_i are positive parameters.

The term $h_i u_i(s)$ reflects player i's satisfaction level obtained from the consumption of the resource extracted, and $c_i u_i(s)^2 x(s)^{-1}$ measures the cost created in the extraction process. $k_i x(s)$ is the benefit to player i related to the existing level of the resource. Total utility of player i is the aggregate level of satisfaction. Payoffs in the form of utility are not transferable between players. There exists a time discount rate r, and utility received at time t has to be discounted by the factor $\exp[-r(t-t_0)]$. At time T, player i will receive a terminal benefit $q_i x(T)^{1/2}$, where q_i is nonnegative.

6.3.1 *Noncooperative Outcome and Pareto Optimal Trajectories*

We use $\Gamma(x_0, T-t_0)$ to denote the game (3.1 and 3.2) and $\Gamma(x_\tau, T-\tau)$ to denote an alternative game with state dynamics (3.1) and payoff structure (3.2), which starts at time $\tau \in [t_0, T]$ with initial state $x_\tau \in X$. Invoking the techniques of Isaacs (1965), Bellman (1957) and Fleming (1969) as stated in Theorem 1.1 of Chap. 3 a non-cooperative Nash equilibrium solution of the game $\Gamma(x_\tau, T-\tau)$ can be characterized as follows.

Corollary 3.1 A set of feedback strategies $\left\{u_i^{(\tau)*}(t) = \phi_i^{(\tau)*}(t,x), \text{ for } i \in \{1,2\}\right\}$ provides a Nash equilibrium solution to the game $\Gamma(x_\tau, T-\tau)$, if there exist twice continuously differentiable functions $V^{(\tau)i}(t,x) : [\tau, T] \times R \to R$, $i \in \{1,2\}$, satisfying the following partial differential equations:

$$-V_t^{(\tau)i}(t,x) - \frac{1}{2}\sigma^2 x^2 V_{xx}^{(\tau)i}(t,x)$$
$$= \max_{u_i}\left\{\left[h_i u_i - c_i u_i^2 x^{-1} + k_i x\right]\exp[-r(t-\tau)] + V_x^{(\tau)i}(t,x)\left[a - bx - u_i - u_j\right]\right\}, \text{ and}$$
$$V^{(\tau)i}(T,x) = \exp[-r(T-\tau)]\, q_i x, \quad \text{for } i \in \{1,2\}, \; j \in \{1,2\} \text{ and } j \neq i. \tag{3.3}$$

■

Performing the indicated maximization in Corollary 3.2 yields:

$$\phi_i^{(\tau)*}(t,x) = \frac{\left[h_i - V_x^{(\tau)i}\exp(r(t-\tau))\right]x}{2c_i}, \text{ for } i \in \{1,2\} \text{ and } x \in X. \tag{3.4}$$

The feedback Nash equilibrium payoffs of the players in the game $\Gamma(x_\tau, T-\tau)$ can be obtained as:

Proposition 3.1 The value function representing the feedback Nash equilibrium payoff of player i in the game $\Gamma(x_\tau, T-\tau)$ is:

$$V^{(\tau)i}(t,x) = \exp[-r(t-\tau)][A_i(t)x + B_i(t)], \quad \text{for } i \in \{1,2\} \text{ and } t \in [\tau, T], \tag{3.5}$$

where $A_i(t), B_i(t), A_j(t)$ and $B_j(t)$, for $i \in \{1,2\}$ and $j \in \{1,2\}$ and $i \neq j$, satisfy:

$$\dot{A}_i(t) = (r+b)A_i(t) - k_i - \frac{[h_i - A_i(t)]^2}{4c_i} + \frac{A_i(t)\left[h_j - A_j(t)\right]}{2c_j},$$

$$\dot{B}_i(t) = rB_i(t) - aA_i(t),$$

$$A_i(T) = q_i, \; B_i(T) = 0.$$

Proof Upon substitution of $\phi_i^{(\tau)*}(t,x)$ from (3.4) into (3.3) yields a set of partial differential equations. One can readily verify that (3.5) is a solution to this set of equations. ■

Consider the case where the players agree to cooperate in order to enhance their payoffs. Let $\Gamma_c(x_0, T-t_0)$ denote a cooperative game with payoff structure (3.1) and dynamics (3.2) starting at time t_0 with initial state x_0. If the players agree to adopt a weight $\alpha_1 > 0$, Pareto optimal trajectories for $\Gamma_c(x_0, T-t_0)$ can be identified by solving the following stochastic control problem:

$$\max_{u_1,u_2} \left\{J^1(t_0,x_0) + \alpha_1 J^2(t_0,x_0)\right\}$$
$$\equiv \max_{u_1,u_2} E_{t_0}\left\{ \int_{t_0}^{T} \left(\left[h_1 u_1(s) - c_1 u_1(s)^2 x(s)^{-1} + k_1 x(s)\right] \right.\right.$$
$$\left. + \alpha_1 \left[h_2 u_2(s) - c_2 u_2(s)^2 x(s)^{-1} + k_2 x(s)\right] \right) \exp[-r(s-t_0)]\,ds$$
$$\left. \exp[-r(T-t_0)]\left[q_1 x(T) + q_2 x(T)\right] \middle| \, x(t_0) = x_0 \right\}, \tag{3.6}$$

subject to dynamics (3.1). Note that when $\alpha_1 = 1/\alpha_2$, the problem $\max_{u_1,u_2} \left\{J^1(t_0,x_0) + \alpha_1 J^2(t_0,x_0)\right\}$ is identical to the problem $\max_{u_1,u_2} \left\{J^2(t_0,x_0) + \alpha_2 J^1(t_0,x_0)\right\}$ in the sense that $\max_{u_1,u_2} \left\{J^2(t_0,x_0) + \alpha_2 J^1(t_0,x_0)\right\} \equiv \max_{u_1,u_2} \left\{\alpha_2\left[J^1(t_0,x_0) + \alpha_1 J^2(t_0,x_0)\right]\right\}$ yields the same optimal controls as those from $\max_{u_1,u_2} \left\{J^1(t_0,x_0) + \alpha_1 J^2(t_0,x_0)\right\}$.

In $\Gamma_c(x_0, T-t_0)$, let α_1^0 be the selected weight according the agreed upon optimality principle. Invoking Corollary 1.1 in Sect. 6.1 the optimal solution of the stochastic control problem (3.1) and (3.6) can be characterized as:

Corollary 3.2 A set of controls $\{\left[\psi_1^{\alpha_1^0(t_0)}(t,x), \psi_2^{\alpha_1^0(t_0)}(t,x)\right]$, for $t \in [t_0, T]\}$ provides an optimal solution to the stochastic control problem $\max_{u_1,u_2} \left\{J^1(t_0,x_0) + \alpha_1^0 J^2(t_0,x_0)\right\}$, if there exists twice continuously differentiable function $W^{\alpha_1^0(t_0)}(t,x) : [t_0, T] \times R \to R$ satisfying the partial differential equation:

$$-W_t^{\alpha_1^0(t_0)}(t,x)-\frac{1}{2}\sigma^2x^2W_{xx}^{\alpha_1^0(t_0)}(t,x)=$$
$$\max_{u_1,u_2}\Big\{\left(\left[h_1u_1-c_1u_1^2x^{-1}+k_1x\right]+\alpha_1^0\left[h_2u_2-c_2u_2^2x^{-1}+k_2x\right]\right)\exp[-r(t-t_0)]$$
$$+W_x^{\alpha_1^0(t_0)}(t,x)\left[a-bx-u_i-u_j\right]\Big\},$$
$$W^{\alpha_1^0(t_0)}(T,x)=\exp[-r(T-t_0)]\left[q_1x(T)+\alpha_1^0q_2x(T)\right] \qquad (3.7)■$$

Performing the indicated maximization in Corollary 3.2 yields:

$$\psi_1^{\alpha_1^0(t_0)}(t,x)=\frac{\left[h_1-W_x^{\alpha_1^0(t_0)}(t,x)\exp(r(t-t_0))\right]x}{2c_1}, \text{ and}$$
$$\psi_2^{\alpha_1^0(t_0)}(t,x)=\frac{\left[\alpha_1^0h_2-W_x^{\alpha_1^0(t_0)}(t,x)\exp(r(t-t_0))\right]x}{2\alpha_1^0c_2}, \quad \text{for } t\in[t_0,T]. \qquad (3.8)$$

The maximized value function $W^{\alpha_1^0(t_0)}(t,x)$ of the control problem $\max_{u_1,u_2}\left\{J^1(t_0,x_0)+\alpha_1^0J^2(t_0,x_0)\right\}$ can be obtained as:

Proposition 3.2

$$W^{\alpha_1^0(t_0)}(t,x)=\exp[-r(t-t_0)]\left[A^{\alpha_1^0}(t)x+B^{\alpha_1^0}(t)\right], \quad \text{for } t\in[t_0,T], \qquad (3.9)$$

where $A^{\alpha_1^0}(t)$ and $B^{\alpha_1^0}(t)$ satisfy:

$$\dot{A}^{\alpha_1^0}(t)=(r+b)A^{\alpha_1^0}(t)-\frac{\left[h_1-A^{\alpha_1^0}(t)\right]^2}{4c_1}-\frac{\left[\alpha_1^0h_2-A^{\alpha_1^0}(t)\right]^2}{4\alpha_1^0c_2}-k_1-k_2,$$
$$\dot{B}^{\alpha_1^0}(t)=rB^{\alpha_1^0}(t)-A^{\alpha_1^0}(t)a,$$
$$A^{\alpha_1^0}(T)=q_1+\alpha_1^0q_2 \text{ and } B^{\alpha_1^0}(T)=0. \qquad (3.10)$$

Proof Upon substitution of $\psi_1^{\alpha_1^0(t_0)}(t,x)$ and $\psi_2^{\alpha_1^0(t_0)}(t,x)$ from (3.10) into (3.7) yields a partial differential equation. One can readily verify that (3.9) is a solution to this set of equations. ■

Substituting the partial derivatives $W_x^{\alpha_1^0(t_0)}(t,x)$ into $\psi_1^{\alpha_1^0(t_0)}(t,x)$ and $\psi_2^{\alpha_1^0(t_0)}(t,x)$ in (3.9) yields the optimal controls of the problem $\max_{u_1,u_2}\left\{J^1(t_0,x_0)+\alpha_1^0J^2(t_0,x_0)\right\}$ as:

$$\psi_1^{\alpha_1^0(t_0)}(t,x)=\frac{\left[h_1-A^{\alpha_1^0}(t)\right]x}{2c_1}, \text{ and}$$
$$\psi_2^{\alpha_1^0(t_0)}(t,x)=\frac{\left[\alpha_1^0h_2-A^{\alpha_1^0}(t)\right]x}{2\alpha_1^0c_2}, \quad \text{for } t\in[t_0,T]. \qquad (3.11)$$

Substituting these controls into (3.1) yields the dynamics of the Pareto optimal trajectory associated with a weight α_1^0. The Pareto optimal trajectory then can be solved as:

$$x^{\alpha_1^0(t_0)}(t) = \left\{ \Phi\left(\alpha_1^0; t, t_0\right) \left[x_0 + \int_{t_0}^{t} \Phi^{-1}\left(\alpha_1^0; s, t_0\right) a ds \right] \right\}^2, \tag{3.12}$$

where

$$\Phi\left(\alpha_1^0; t, t_0\right) = \exp\left[\int_{t_0}^{t} \left(-b - \frac{h_1 - A^{\alpha_1^0}(s)}{2c_1} - \frac{\alpha_1 h_2 - A^{\alpha_1^0}(s)}{2\alpha_1^0 c_2} - \frac{\sigma^2}{2} \right) ds + \int_{t_0}^{t} \sigma \, dz(s) \right].$$

We use $X_t^{\alpha_1^0(t_0)}$ to denote the set of realizable values of $x^{\alpha_1^0(t_0)}(t)$ generated by (3.12) at $t \in \left(t_0, T\right]$.

Now, consider the cooperative game $\Gamma_c(x_\tau, T - \tau)$ with state dynamics (3.1) and payoff structure (3.2), which starts at time $\tau \in \left[t_0, T\right]$ with initial state $x_\tau \in X_\tau^{\alpha_1(t_0)}$. Let α_1^τ be the selected weight according the agreed upon optimality principle.

Following previous analysis, we can obtain the maximized value function, optimal controls and optimal trajectory of the control problem $\max_{u_1, u_2} \left\{ J^1(\tau, x_\tau) + \alpha_1^\tau J^2(\tau, x_\tau) \right\}$.

Remark 3.1 One can readily show that when $\alpha_1^0 = \alpha_1^\tau = \alpha_1^*$, then $\psi_i^{\alpha_1^*(t_0)}(t, x_t) = \psi_i^{\alpha_1^*(\tau)}(t, x_t)$ at the point (t, x_t), for $i \in [1, 2]$, $t_0 \le \tau \le t \le T$ and $x_t \in X_t^{\alpha_1^*(t_0)}$. ∎

6.3.2 *Individual Player's Payoff Under Cooperation*

In order to verify individual rationality, we have to derive the players' expected payoffs in the cooperative game $\Gamma_c\left(x_\tau^*, T - \tau\right)$. Let α_1^τ be the weight dictated by the solution optimality principle. We substitute

$$\psi_1^{\alpha_1^\tau(\tau)}(t, x) = \frac{\left[h_1 - A^{\alpha_1^\tau}(t)\right] x}{2c_1} \text{ and } \psi_2^{\alpha_1^\tau(\tau)}(t, x) = \frac{\left[\alpha_1^\tau h_2 - A^{\alpha_1^\tau}(t)\right] x}{2\alpha_1^\tau c_2}$$

into the players' payoffs and define the following functions.

Definition 3.1 Given that $x(t) = x_t^{\alpha_1^\tau(\tau)} \in X_t^{\alpha_1^\tau(\tau)}$, for $t \in [\tau, T]$, player 1's expected payoff over the interval $[t, T]$ under the control problem $\max_{u_1, u_2} \left\{ J^1(\tau, x_\tau) + \alpha_1^\tau J^2(\tau, x_\tau) \right\}$ as:

$$\hat{W}^{\tau(\alpha_1^\tau)1}(t,x) =$$
$$E_\tau\left\{ \int_t^T \left[\frac{h_1\left[h_1 - A^{\alpha_1^\tau}(s)\right]x(s)}{2c_1} - \frac{\left[h_1 - A^{\alpha_1^\tau}(s)\right]^2 x(s)}{4c_1} + k_1 x(s)\right]\exp[-r(s-\tau)]\,ds\right.$$
$$\left. + \exp[-r(T-t_0)]\,q_1 x(T)\right| x(t) = x \Bigg\},$$

and the corresponding expected payoff of player 2 over the interval $[t, T]$ as:

$$\hat{W}^{\tau(\alpha_1^\tau)2}(t,x) = E_\tau\Bigg\{$$
$$\int_t^T \left[\frac{h_2\left[\alpha_1^\tau h_2 - A^{\alpha_1^\tau}(s)\right]x(s)}{2\alpha_1^\tau c_2} - \frac{\left[\alpha_1^\tau h_2 - A^{\alpha_1^\tau}(s)\right]^2 x(s)}{4\left(\alpha_1^\tau\right)^2 c_2} + k_i x(s)\right]\exp[-r(s-\tau)]\,ds$$
$$\left. + \exp[-r(T-\tau)]\,q_2 x(T)\right| x(t) = x \Bigg\},$$

where

$$dx(s) = \left[a - bx(s) - \frac{\left[h_1 - A^{\alpha_1^\tau}(s)\right]x(s)}{2c_1} - \frac{\left[\alpha_1^\tau h_2 - A^{\alpha_1^\tau}(s)\right]x(s)}{2\alpha_1^\tau c_2}\right]ds$$
$$+ \sigma x(s)dz(s), x(t) = x. \qquad \blacksquare$$

Invoking Theorem 1.1 in Sect. 6.1, player 1's expected payoff $\hat{W}^{\tau(\alpha_1^\tau)1}(t, x_\tau)$ can be characterized as:

$$-\hat{W}_t^{\tau(\alpha_1^\tau)1}(t,x_t) - \frac{1}{2}\hat{W}_{x_t x_t}^{\tau(\alpha_1^\tau)1}(t,x_t)\sigma^2 x_t^2 =$$
$$\left[\frac{h_1\left[h_1 - A^{\alpha_1^\tau}(t)\right]x_t}{2c_1} - \frac{c_1\left[h_1 - A^{\alpha_1^\tau}(t)\right]^2 x_t}{4c_1^2} + k_1 x_t\right]\exp[-r(t-\tau)]$$
$$+ \hat{W}_{x_t}^{\tau(\alpha_1^\tau)1}(t,x_t)\left[a - bx_t - \frac{\left[h_1 - A^{\alpha_1^\tau}(t)\right]x_t}{2c_1} - \frac{\left[\alpha_1^\tau h_2 - A^{\alpha_1^\tau}(t)\right]x_t}{2\alpha_1^\tau c_2}\right]. \qquad (3.13)$$

Boundary conditions require:

$$\hat{W}^{\tau(\alpha_1^\tau)1}(T,x) = \exp[-r(T-\tau)]q_1 x. \qquad (3.14)$$

If there exist continuously differentiable functions $\hat{W}^{\tau(\alpha_1^\tau)1}(t,x) : [\tau, T] \times R \to R$ satisfying (3.13) and (3.14), then player 1's expected payoff in the cooperative game $\Gamma(x_\tau, T-\tau)$ under the cooperation scheme with weight α_1^τ is indeed $\hat{W}^{\tau(\alpha_1^\tau)1}(t,x)$. The value function $\hat{W}^{\tau(\alpha_1^\tau)1}(t,x)$ indicating the expected payoff of player 1 under cooperation can be obtained as:

Proposition 3.3 The function $\hat{W}^{\tau(\alpha_1^\tau)1}(t,x) : [t_0,T] \times R \to R$ satisfying (3.13) and (3.14) can be solved as:

$$\hat{W}^{\tau(\alpha_1^\tau)1}(t,x) = \exp[-r(t-\tau)]\left[\hat{A}_1^{\alpha_1^\tau}(t)x + \hat{B}_1^{\alpha_1^\tau}(t)\right], \tag{3.15}$$

where $\hat{A}_1^{\alpha_1^\tau}(t)$ and $\hat{B}_1^{\alpha_1^\tau}(t)$ satisfy:

$$\dot{\hat{A}}_1^{\alpha_1^\tau}(t) = \left[r + b + \frac{\left[h_1 - A^{\alpha_1^\tau}(t)\right]}{2c_1} + \frac{\left[\alpha_1 h_2 - A^{\alpha_1^\tau}(t)\right]}{2\alpha_1 c_2}\right]\hat{A}_1^{\alpha_1}(t)$$
$$- \frac{\left[h_1 - A^{\alpha_1^\tau}(t)\right]\left[h_1 + A^{\alpha_1^\tau}(t)\right]}{4c_1} - k_1,$$

$$\dot{\hat{B}}_1^{\alpha_1^\tau}(t) = r\hat{B}_1^{\alpha_1^\tau}(t) - a\hat{A}_1^{\alpha_1^\tau}(t),\ \hat{A}_1^{\alpha_1^\tau}(T) = q_1 \text{ and } \hat{B}_1^{\alpha_1^\tau}(T) = 0.$$

Proof Upon calculating the derivatives $\hat{W}_t^{\tau(\alpha_1^\tau)1}(t,x)$, $\hat{W}_{xx}^{\tau(\alpha_1^\tau)1}(t,x)$, and $\hat{W}_x^{\tau(\alpha_1^\tau)1}$ (t,x) from (3.15) and then substituting them into (3.13) yield Proposition 3.3. ■

Following the above analysis, a continuously differentiable function $\hat{W}^{\tau(\alpha_1^\tau)2}$ $(t,x) : [\tau,T] \times R \to R$ giving the player 2's expected payoff under cooperation can be obtained as:

Proposition 3.4

$$\hat{W}^{\alpha_1^\tau(\tau)2}(t,x) = \exp[-r(t-\tau)]\left[\hat{A}_2^{\alpha_1^\tau}(t)x + \hat{B}_2^{\alpha_1^\tau}(t)\right], \tag{3.16}$$

where $\hat{A}_2^{\alpha_1^\tau}(t)$ and $\hat{B}_2^{\alpha_1^\tau}(t)$ has to satisfy:

$$\dot{\hat{A}}_2^{\alpha_1^\tau}(t) = \left[r + b + \frac{\left[h_1 - A^{\alpha_1^\tau}(t)\right]}{2c_1} + \frac{\left[\alpha_1 h_2 - A^{\alpha_1^\tau}(t)\right]}{2\alpha_1 c_2}\right]\hat{A}_2^{\alpha_1^\tau}(t)$$
$$- \frac{\left[\alpha_1 h_2 - A^{\alpha_1^\tau}(t)\right]\left[\alpha_1 h_2 + A^{\alpha_1^\tau}(t)\right]}{4\alpha_1^2 c_2} - k_2,$$
$$\dot{\hat{B}}_2^{\alpha_1^\tau}(t) = r\hat{B}_2^{\alpha_1^\tau}(t) - a\hat{A}_2^{\alpha_1^\tau}(t),\ \hat{A}_2^{\alpha_1^\tau}(T) = q_2 \text{ and } \hat{B}_2^{\alpha_1^\tau}(T) = 0.$$

Proof Follow the proof of Proposition 3.3. ■

6.4 Subgame Consistent Cooperative Solutions of the Game

In this section, we present subgame consistent solutions to the cooperative game $\Gamma_c(x_0, T - t_0)$. First note that group optimality will be maintained only if the solution optimality principle selects the same weight α_1 for all games

$\Gamma_c(x_\tau, T-\tau)$, $\tau \in [t_0, T]$ and $x_\tau \in X_\tau^{\alpha_1(t_0)}$. For any chosen α_1 to maintain individual rationality throughout the game interval, the following condition must be satisfied.

$$\begin{aligned}&\xi^{(\tau)i}(x_\tau, T-\tau; \alpha_1) = \hat{W}^{\tau(\alpha_1)i}(\tau, x_\tau) \geq V^{(\tau)i}(\tau, x_\tau),\\&\text{for } i \in \{1,2\},\ \tau \in [t_0, T] \text{ and } x_\tau \in X_\tau^{\alpha_1(t_0)}.\end{aligned} \tag{4.1}$$

Definition 4.1 We define the set $S_\tau^T = \bigcap_{\tau \leq t < T} S_t$, for $\tau \in [t_0, T)$. ∎

S_t represents the set of α_1 satisfying individual rationality at time $t \in [t_0, T)$ and S_τ^T represents the set of α_1 satisfying individual rationality throughout the interval $[\tau, T)$. In general $S_\tau^T \neq S_t^T$ for $\tau, t \in [t_0, T)$ where $\tau \neq t$.

6.4.1 Typical Configurations of $\mathbf{S_t}$

To find out typical configurations of the set S_t for $t \in [t_0, T)$ of the game $\Gamma_c(x_0, T-t_0)$, we perform extensive numerical simulations with a wide range of parameter specifications for $a, b, \sigma, h_1, h_2, k_1, k_2, c_1, c_2, q_1, q_2, T, r, x_0$. We calculate the time paths of $A_1(t), B_1(t), A_2(t)$ and $B_2(t)$ in Proposition 3.1 for $t \in [t_0, T]$. Then we select weights α_1 and calculate the time paths of $\hat{A}_1^{\alpha_1}(t), \hat{A}_2^{\alpha_1}(t), \hat{B}_1^{\alpha_1}(t)$ and $\hat{B}_2^{\alpha_1}(t)$ in Propositions 3.3 and 3.4, for $t \in [t_0, T]$. At each time instant $t \in [t_0, T]$, we derive the set of α_1 that yields $\hat{A}_i^{\alpha_1}(t) \geq A_i(t)$ and $\hat{B}_i^{\alpha_1}(t) \geq B_i(t)$, for $i \in [1, 2]$, to derive the set S_t, for $t \in [t_0, T)$.

We denote the locus of the values of $\underline{\alpha}_1^t$ along $t \in [t_0, T)$ as curve $\underline{\alpha}_1$ and the locus of the values of $\overline{\alpha}_1^t$ as curve $\overline{\alpha}_1$. In particular, typical patterns include:

(i) The curves $\underline{\alpha}_1$ and $\overline{\alpha}_1$ are continuous and move in the same direction over the entire game duration: either both increase monotonically or both decrease monotonically (see Fig. 6.1).
(ii) The curves $\underline{\alpha}_1$ and $\overline{\alpha}_1$ are continuous. $\underline{\alpha}_1$ declines and $\overline{\alpha}_1$ rises over the entire game duration (see Fig. 6.2).
(iii) The curves $\underline{\alpha}_1$ and $\overline{\alpha}_1$ are continuous. One of these curves would rise/fall to a peak/trough and then fall/rise (see Fig. 6.3).
(iv) The set $S_{t_0}^T$ can be nonempty or empty.

6.4.2 Examples of Subgame Consistent Solutions

In this subsection, we present some subgame consistent solutions to $\Gamma_c(x_0, T-t_0)$.

Solution 4.1 Consider the cooperative differential game $\Gamma_c(x_0, T-t_0)$ with parameters leading to a set of payoff weights as in Panel (b) of Fig. 6.1. In

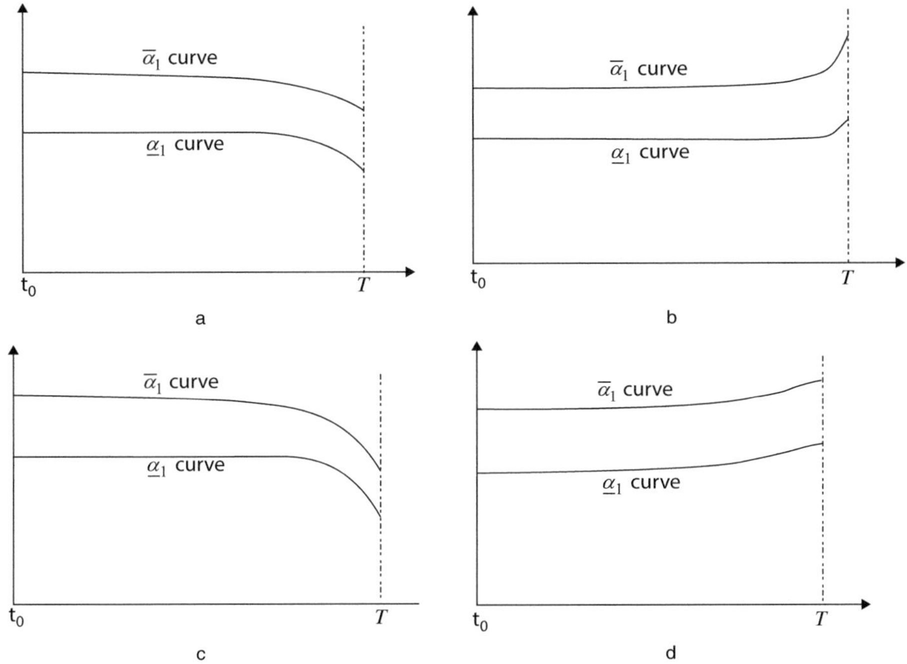

Fig. 6.1 Both upward $\underline{\alpha}_1$ and $\overline{\alpha}_1$ curves and both downward $\underline{\alpha}_1$ and $\overline{\alpha}_1$ curves

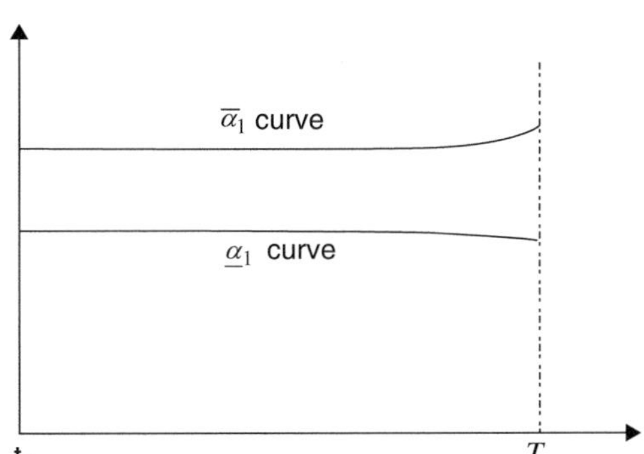

Fig. 6.2 Declining $\underline{\alpha}_1$ curve and rising $\overline{\alpha}_1$ curve

particular, there exist a set of weights $S_{t_0}^T \neq \varnothing$ under which individual rationality is satisfied throughout the game horizon $[0, T]$ and $\underline{\alpha}_1^{T-} \in S_{t_0}^T$. At initial time 0, in the cooperative $\Gamma_c(x_0, T - t_0)$, an optimality principle under which the players to choose the weight

$$\alpha_1^* = \underline{\alpha}_1^{T-}, \text{ in } \Gamma_c(x_\tau, T - \tau) \text{ for } \tau \in \left[t_0, T\right)$$

yields a subgame consistent solution to the cooperative game $\Gamma_c(x_0, T - t_0)$.

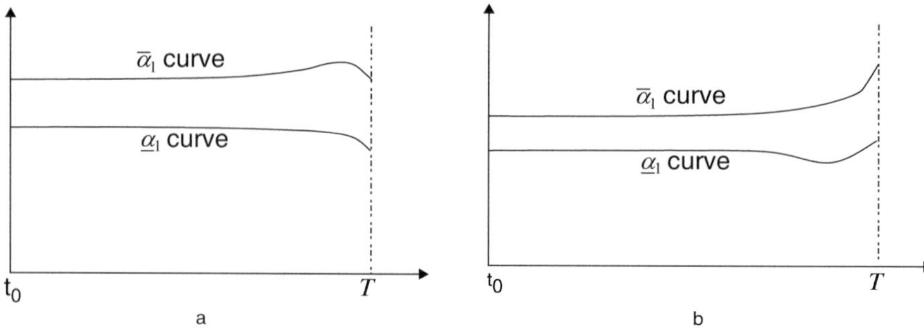

Fig. 6.3 Rising to a peak and then fall curve and falling to a trough and then rise curve

Proof According to the optimality principle in Solution 4.1, a unique $\alpha_1^* = \underline{\alpha}_1^{T-}$ will be chosen for all the subgames $\Gamma_c(x_\tau, T-\tau)$, for $t_0 \le \tau \le t < T$ and $x_\tau \in X_\tau^{\alpha_1^*(t_0)}$. The vector $\xi^{(\tau)}\left(x_\tau, T-\tau; \alpha_1^*\right) = \left[\hat{W}^{\tau(\alpha_1^*)1}(\tau, x_\tau), \hat{W}^{\tau(\alpha_1^*)2}(\tau, x_\tau)\right]$, for $\tau \in [t_0, T]$, yields a Pareto optimal pair of imputations. Hence part (i) of Definition 2.1 is proved.

One can readily verify that $\hat{W}^{\tau(\alpha_1^*)i}(t,x)\exp[r(\tau-t)] = \hat{W}^{t(\alpha_1^*)i}(t,x)$, for $i \in \{1,2\}$, $t_0 \le \tau \le t \le T$ and $x_t \in X_t^{\alpha_1^0(t_0)}$. Hence part (ii) of Definition 2.1 is satisfied.

Finally, from Definitions 4.1, one can verify that $\hat{W}^{\tau(\alpha_1^*)i}(\tau, x_\tau) = \exp[-r(t-\tau)]$ $\left[\hat{A}_i^{\alpha_1^*}(t)x^{1/2} + \hat{B}_i^{\alpha_1^*}(t)\right] \ge V^{(\tau)i}(\tau, x_\tau) = \exp[-r\ (t-\tau)]\left[A_i(t)x^{1/2} + B_i(t)\right]$, for $i \in \{1,2\}, \tau \in [t_0, T]$ and $x_\tau \in X_\tau^{\alpha_1^*(t_0)}$. Hence part (iii) of Definition 2.1 is fulfilled. ■

Solution 4.2 Consider the cooperative differential game $\Gamma_c(x_0, T-t_0)$ with parameters leading to a set of payoff weights as in Panel (a) of Fig. 6.1. In particular, there exist a set of weights $S_{t_0}^T \ne \varnothing$ under which individual rationality is satisfied throughout the game horizon $[0, T]$ and $\overline{\alpha}_1^{T-} \in S_{t_0}^T$. At initial time 0, in the cooperative $\Gamma_c(x_0, T-t_0)$, an optimality principle under which the players to choose the weight

$$\alpha_1^* = \overline{\alpha}_1^{T-}, \text{ in } \Gamma_c(x_\tau, T-\tau) \text{ for } \tau \in [t_0, T)$$

yields a subgame consistent solution to the cooperative game $\Gamma_c(x_0, T-t_0)$.

Proof Follow the proof of Solution 4.1. ■

Solution 4.3 Consider the cooperative differential game $\Gamma_c(x_0, T-t_0)$ with parameters leading to a set of payoff weights as in Fig. 6.2. In particular, there exist a set of weights $S_{t_0}^T \ne \varnothing$ under which individual rationality is satisfied throughout the game horizon $[0, T]$ and $\left(\underline{\alpha}_1^{T-}\right)^{0.5}\left(\overline{\alpha}_1^{T-}\right)^{0.5} \in S_{t_0}^T$, At initial time 0, in the cooperative $\Gamma_c(x_0, T - t_0)$, an optimality principle under which the players to choose the weight

$$\alpha_1^* = \left(\underline{\alpha}_1^{T-}\right)^{0.5}\left(\overline{\alpha}_1^{T-}\right)^{0.5} \text{ in } \Gamma_c(x_\tau, T - \tau) \text{ for } \tau \in [t_0, T)$$

yields a subgame consistent solution to the cooperative game $\Gamma_c(x_0, T - t_0)$.

Proof Follow the proof of Solution 4.1. ■

6.5 Numerical Delineation

Numerical delineations of the 4 solutions presented in Sect. 6.4 are given in the following 4 cases.

Case 5.1 Consider the cooperative game $\Gamma_c(x_0, T - t)$ with the following parameter specifications: $a = 10, b = 1, \sigma = 0.05, h_1 = 8, h_2 = 7, k_1 = 1, k_2 = 0.5, c_1 = 1, c_2 = 1.2, q_1 = 0.8, q_2 = 0.4, T = 6, r = 0.02$.

The numerical results are displayed in Fig. 6.4. The curve $\underline{\alpha}_1$ is the locus of the values of $\underline{\alpha}_1^t$ along $t \in [t_0, T)$. The curve $\overline{\alpha}_1$ is the locus of the values of $\overline{\alpha}_1^t$ along $t \in [t_0, T)$. In particular, the set $S_{t_0}^T = \bigcap_{t_0 \le t < T} S_t = \left[\underline{\alpha}_1^{t_0}, \overline{\alpha}_1^{T-}\right] = [1.182686, 1.450783]$. Note that $\underline{\alpha}_1^{T-} \in S_{t_0}^T$ and $\overline{\alpha}_1^{T-} \notin S_{t_0}^T$, for $\tau \in [t_0, T)$. According to Solution 4.1, the players would agree to the optimality principle of choosing a weight $\alpha_1^* = \underline{\alpha}_1^{T-} = 1.182686$ throughout the game interval, and a subgame consistent solution to the cooperative game $\Gamma_c(x_0, T - t_0)$ would result.

Case 5.2 Consider the cooperative game $\Gamma_c(x_0, T - t)$ with the following parameter specifications: $a = 6, b = 0.8, \sigma = 0.04, h_1 = 8, h_2 = 6, k_1 = 1, k_2 = 0.5, c_1 = 1, c_2 = 1.5, q_1 = 3, q_2 = 2, T = 3, r = 0.02$.

The numerical results are displayed in Fig. 6.5. In particular, the set $S_{t_0}^T = \bigcap_{t_0 \le t < T} S_t = \left[\underline{\alpha}_1^{t_0}, \overline{\alpha}_1^{T-}\right] = [1.246704, \ \ 1.443176]$. Note that $\overline{\alpha}_1^{T-} \in S_{t_0}^T$ and $\underline{\alpha}_1^{T-} \notin S_{t_0}^T$,

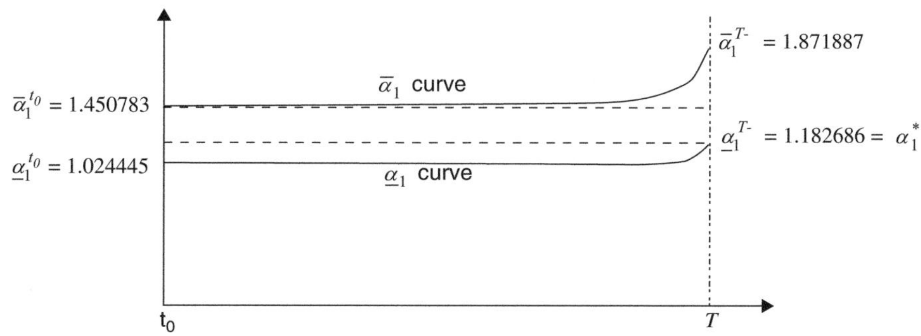

Fig. 6.4 A subgame consistent solution with optimality principle of a weight $\alpha_1{}^* = \underline{\alpha}_1{}^{T-} = 1.182686$

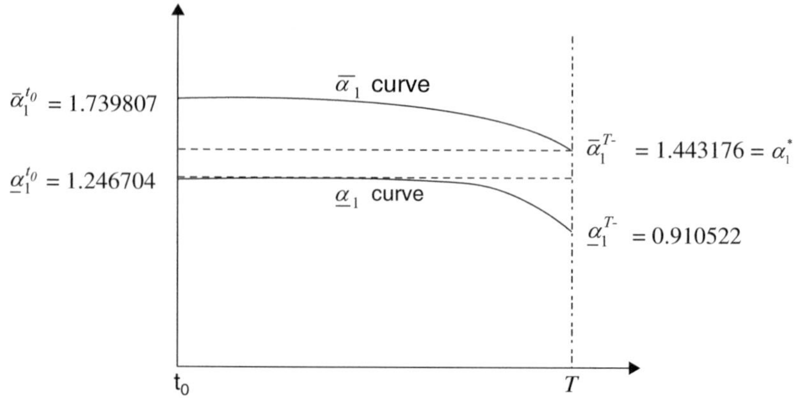

Fig. 6.5 A subgame consistent solution with optimality principle of a weight $\alpha_1{}^* = \bar{\alpha}_1{}^{T-} = 1.443176$

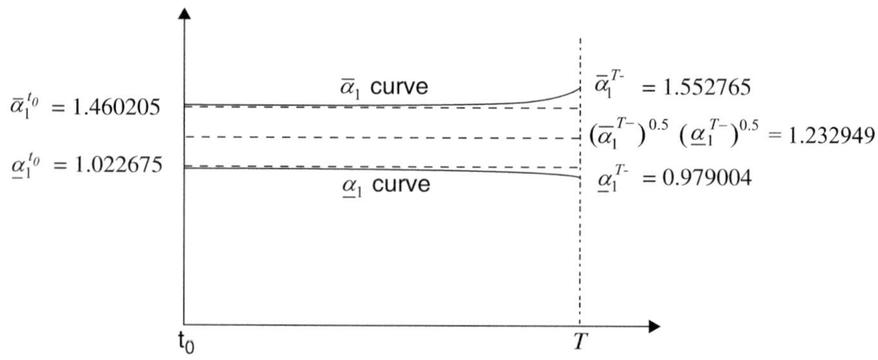

Fig. 6.6 A subgame consistent solution with optimality principle of a weight $\alpha_1{}^* = \left(\underline{\alpha}_1^{T-}\right)^{0.5} \left(\bar{\alpha}_1^{T-}\right)^{0.5} = 1.232949$

for $\tau \in [t_0, T)$. According to Solution 4.2, the players would agree to the optimality principle of choosing a weight $\alpha_1^* = \bar{\alpha}_1^{T-} = 1.443176$ throughout the game interval, and a subgame consistent solution to the cooperative game $\Gamma_c(x_0, T - t_0)$ would result.

Case 5.3 Consider the cooperative game $\Gamma_c(x_0, T - t)$ with the following parameter specifications: $a = 10, b = 1.1, \sigma = 0.04, h_1 = 8, h_2 = 7, k_1 = 1, k_2 = 0.5, c_1 = 1, c_2 = 1.2, q_1 = 3, q_2 = 2, T = 3, r = 0.02$.

The numerical results are displayed in Fig. 6.6. In particular, the set $S_{t_0}^T = \bigcap_{t_0 \le t < T} S_t = \left[\underline{\alpha}_1^{t_0}, \bar{\alpha}_1^{T-}\right] = [1.022675, \ 1.460205]$. Note that $\underline{\alpha}_1^{T-} \notin S_{t_0}^T$ and $\bar{\alpha}_1^{T-} \notin S_{t_0}^T$, for $\tau \in [t_0, T)$. According to Solution 4.3, the players would agree to the optimality principle of choosing a weight $\alpha_1^* = \left(\underline{\alpha}_1^{T-}\right)^{0.5} \left(\bar{\alpha}_1^{T-}\right)^{0.5} = 1.232949$ throughout the game interval, and a subgame consistent solution to the cooperative game $\Gamma_c(x_0, T - t_0)$ would result.

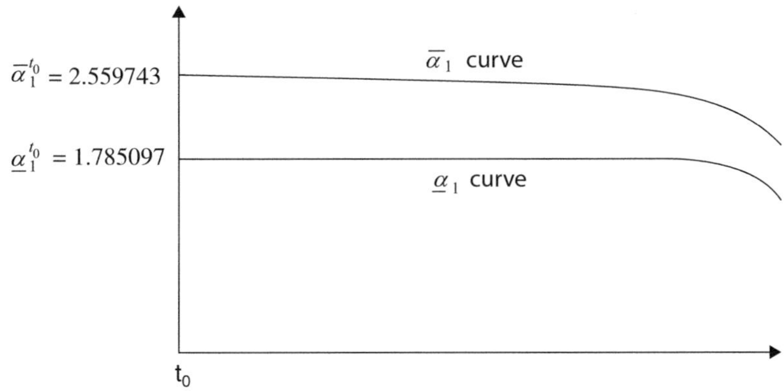

Fig. 6.7 The set $S_{t_0}^T = \bigcap_{t_0 \le t < T} S_t = \varnothing$ and no candidate for a subgame consistent solution

Case 5.4 Consider the cooperative game $\Gamma_c(x_0, T-t)$ with parameters: $a = 6, b = 1, \sigma = 0.03, h_1 = 11, h_2 = 6, k_1 = 1, k_2 = 0.5, c_1 = 1, c_2 = 1.5, q_1 = 3, q_2 = 2, T = 6, r = 0.02$.

The numerical results are displayed in Fig. 6.7. In particular, the set $S_{t_0}^T = \bigcap_{t_0 \le t < T} S_t = \varnothing$. Hence there does not exist any candidate for a subgame consistent solution for the game $\Gamma_c(x_0, T - t_0)$.

6.6 Infinite Horizon Analysis

In this Section we examine the situation when the game horizon approaches infinity. Consider an infinite-horizon cooperative stochastic differential game in which player i's payoff to be maximized is

$$J^i(x_0) = E_{t_0}\left\{ \int_{t_0}^{\infty} \left[[k_i u_i(s)]^{1/2} - \frac{c_i}{x(s)^{1/2}} u_i(s) \right] \exp[-r(s-t_0)]\, ds \,\middle|\, x(t_0) = x_0 \right\}, \quad (6.1)$$

for $i \in \{1, 2\}$.

The state dynamics of the game is characterized by the stochastic differential equations:

$$dx(s) = \left[a x(s)^{1/2} - b x(s) - u_1(s) - u_2(s) \right] ds + \sigma x(s) dz(s), \; x(t_0) = x_0 \in X \quad (6.2)$$

where $u_i \in U_i$ is the control vector of player i, for $i \in \{1, 2\}$,

a, b, and σ are positive constants, and $z(s)$ is a Wiener process. Equation (6.2) could be interpreted as the stock dynamics of a biomass of renewable resource (see Jørgensen and Yeung (1996, 1999)).

Note that the infinite-horizon autonomous problem (6.1 and 6.2) is independent of the choice of t_0 and dependent only upon the state at the starting time, that is x_0. Hence, we use $\Gamma(x, \infty)$ and $\Gamma_c(x, \infty)$ to denote respectively a noncooperative and a cooperative game with payoffs (6.1) and dynamics (6.2) with starting state x. Following the previous analysis modified for an infinite horizon problem, we can obtain the value function reflecting the expected payoff (in current value) of player $i \in \{1, 2\}$ in the noncooperative game $\Gamma(x, \infty)$ as

Proposition 6.1

$$V^i(x) = \left[\overline{A}_i x^{1/2} + \overline{B}_i\right],$$

where $\overline{A}_i$, $\overline{B}_i$, $\overline{A}_j$ and $\overline{B}_j$, for $i \in \{1, 2\}$ and $j \in \{1, 2\}$ and $i \neq j$, satisfy:

$$\left[r + \frac{1}{8}\sigma^2 + \frac{b}{2}\right]\overline{A}_i - \frac{k_i}{4\left[c_i + \overline{A}_i/2\right]} + \frac{\overline{A}_i k_j}{8\left[c_j + \overline{A}_j/2\right]^2} = 0, \text{ and } \overline{B}_i = \frac{a}{2r}\overline{A}_i.$$

Proof Applying Theorem 5.1 of Chap. 3 to the game (6.1 and 6.2) yields Proposition 6.1. ■

In the case of cooperation where α_1 is the chosen weight under the agreed optimality principle, the maximized value function reflecting the maximized expected weighted joint payoff of the stochastic control problem $\max_{u_1, u_2} \left\{J^1(x) + \alpha_1 J^2(x)\right\}$ subject to dynamics (6.1) can be obtained as:

Proposition 6.2 $W^{\alpha_1}(x) = \left[\overline{A}^{\alpha_1} x^{1/2} + \overline{B}^{\alpha_1}\right]$, where $\overline{A}^{\alpha_1}$ and $\overline{B}^{\alpha_1}$ satisfy:

$$\left[r + \frac{1}{8}\sigma^2 + \frac{b}{2}\right]\overline{A}^{\alpha_1} - \frac{k_1}{4\left[c_1 + \overline{A}^{\alpha_1}/2\right]} - \frac{\alpha_1 k_2}{4\left[c_2 + \overline{A}^{\alpha_1}/2\alpha_1\right]} = 0, \text{ and } \overline{B}^{\alpha_1} = \frac{a}{2r}\overline{A}^{\alpha_1}.$$

Proof Applying Theorem A.4 in the Technical Appendices to the stochastic control problem $\max_{u_1, u_2} \left\{J^1(x) + \alpha_1 J^2(x)\right\}$ subject to dynamics (6.1) yields Proposition 6.2. ■

The corresponding optimal controls are:

$$\psi_1^{\alpha_1(\infty)}(x) = \frac{k_1 x}{4\left[c_1 + \overline{A}^{\alpha_1}/2\right]^2} \text{ and } \psi_2^{\alpha_1(\infty)}(x) = \frac{k_2 x}{4\left[c_2 + \overline{A}^{\alpha_1}/2\alpha_1\right]^2}, \text{ for } x \in X.$$

We define player 1's expected payoff over the interval $[0, \infty)$ under the control problem $\max_{u_1, u_2} \{J^1(x) + \alpha_1 J^2(x)\}$ as:

$$\hat{W}^{\alpha_1(1)}(x) = E_0\left\{ \int_0^\infty \left[\frac{k_1 x(s)^{1/2}}{2\left[c_1 + \overline{A}^{\alpha_1}/2\right]} - \frac{c_1 k_1 x(s)^{1/2}}{4\left[c_1 + \overline{A}^{\alpha_1}/2\right]^2} \right] \exp(-rs)\, ds \right\};$$

and the corresponding expected payoff of player 2 over the interval $[0, \infty)$ as:

$$\hat{W}^{\alpha_1(2)}(x) = E_0\left\{ \int_0^\infty \left[\frac{k_2 x(s)^{1/2}}{2\left[c_2 + \overline{A}^{\alpha_1}/2\alpha_1\right]} - \frac{c_2 k_2 x(s)^{1/2}}{4\left[c_2 + \overline{A}^{\alpha_1}/2\alpha_1\right]^2} \right] \exp(-rs)\, ds \right\};$$

where

$$dx(s) = \left[a x(s)^{1/2} - \left(b + \frac{k_1}{4\left[c_1 + \overline{A}^{\alpha_1}/2\right]^2} + \frac{k_2}{4\left[c_2 + \overline{A}^{\alpha_1}/2\alpha_1\right]^2} \right) x(s) \right] ds$$
$$+ \sigma x(s) dz(s),\ x(t) = x.$$

An infinite-horizon counterpart of Theorem 1.1 characterizing player i's cooperative payoff under payoff weights α_1 is given in the theorem below.

Theorem 6.1 If there exist continuously functions $\hat{W}^{\alpha_1(i)}(x) : R^n \to R,\ i \in \{1, 2\}$, satisfying

$$r\hat{W}_t^{\alpha_1(i)}(x) - \frac{1}{2}\sum_{h,\zeta=1}^{n} \Omega^{h\zeta}(x)\hat{W}_{x^h x^\zeta}^{\alpha_1(i)}(x) = g^i\left[x, \psi_1^{\alpha_1}(x), \psi_2^{\alpha_1}(x)\right]$$
$$+ \hat{W}_x^{\alpha_1(i)}(t, x) f\left[x, \psi_1^{\alpha_1}(x), \psi_2^{\alpha_1}(x)\right],$$

then $\hat{W}^{\alpha_1(i)}(t, x)$ gives player i's expected cooperative payoff when the state is x and α_1 is chosen as the weight.

Proof Following the analysis of developing an infinite horizon counter of the stochastic control leading to Theorem A.4 in the Technical appendices one can obtain an infinite-horizon counterpart of Theorem 1.1 in Section as Theorem 6.1. ■

Using Theorem 6.1 the expected payoffs of Player 1 and Players 2 under cooperation can be obtained as follows.

Proposition 6.3 The expected payoffs of Player 1 and Player 2 (in current-value) under cooperation with bargaining weight α_1 are respectively:

$$\hat{W}^{\alpha_1(1)}(x) = \left[\hat{A}_1^{\alpha_1} x^{1/2} + \hat{B}_1^{\alpha_1}\right] \text{ and } \hat{W}^{\alpha_1(2)}(x) = \left[\hat{A}_2^{\alpha_1} x^{1/2} + \hat{B}_2^{\alpha_1}\right],$$

where

$$\left[r + \frac{1}{8}\sigma^2 + \frac{b}{2}\right]\hat{A}_1^{\alpha_1} - \frac{k_1}{2\left[c_1 + \overline{A}^{\alpha_1}/2\right]} + \frac{c_1 k_1}{4\left[c_1 + \overline{A}^{\alpha_1}/2\right]^2} + \frac{\hat{A}_1^{\alpha_1} k_1}{8\left[c_1 + \overline{A}^{\alpha_1}/2\right]^2}$$
$$+ \frac{\hat{A}_1^{\alpha_1} k_2}{8\left[c_2 + \overline{A}^{\alpha_1}/2\alpha_1\right]^2} = 0, \ \hat{B}_1^{\alpha_1}(t) = \frac{a}{2r}\hat{A}_1^{\alpha_1},$$
$$\left[r + \frac{1}{8}\sigma^2 + \frac{b}{2}\right]\hat{A}_2^{\alpha_1} - \frac{k_2}{2\left[c_2 + \overline{A}^{\alpha_1}/2\alpha_1\right]} + \frac{c_2 k_2}{4\left[c_2 + \overline{A}^{\alpha_1}/2\alpha_1\right]^2}$$
$$+ \frac{\hat{A}_2^{\alpha_1} k_1}{8\left[c_1 + \overline{A}^{\alpha_1}/2\right]^2} + \frac{\hat{A}_2^{\alpha_1} k_2}{8\left[c_2 + \overline{A}^{\alpha_1}/2\alpha_1\right]^2} = 0, \text{ and } \hat{B}_2^{\alpha_1} = \frac{a}{2r}\hat{A}_2^{\alpha_1}.$$

Proof Follow the Proof of Proposition 3.3 yields Proposition 6.3. ■

Since the solution to the control problem $\max_{u_1, u_2}\left\{J^1(x) + \alpha_1 J^2(x)\right\}$ yields a Pareto optimal outcome there exist (i) an $\underline{\alpha}_1^\infty$ such that $\hat{W}^{\underline{\alpha}_1^\infty(2)}(x) = V^2(x)$ and $\hat{W}^{\underline{\alpha}_1^\infty(1)}(x) \geq V^1(x)$, and (ii) an $\overline{\alpha}_1^\infty$ such that $\hat{W}^{\overline{\alpha}_1^\infty(1)}(x) = V^1(x)$ and $\hat{W}^{\overline{\alpha}_1^\infty(2)}(x) \geq V^2(x)$.

Comparing $\hat{W}^{\alpha_1(i)}(x)$ in Proposition 6.3 with $V^i(x)$ in Proposition 6.1 shows that $\hat{W}^{\alpha_1(i)}(x) \geq V^i(x)$ if and only if $\hat{A}_i^{\alpha_1} \geq \overline{A}_i$, for $i \in \{1, 2\}$.

A condition that would be used in subsequent analysis is:

Condition 6.1 $d\hat{A}_1^{\alpha_1}/d\alpha_1 < 0$ and $d\hat{A}_2^{\alpha_1}/d\alpha_1 > 0$.

Proof See Appendix A. ■

Therefore there exists a nonempty set S^∞ of α_1 such that $\hat{A}_i^{\alpha_1} \geq \overline{A}_i$, for $i \in \{1, 2\}$. Using Condition 6.1, we can readily show that

Corollary 6.1 $S^\infty = \left[\underline{\alpha}_1^\infty, \overline{\alpha}_1^\infty\right]$, where $\underline{\alpha}_1^\infty$ is the lowest value of α_1 in S^∞, and $\overline{\alpha}_1^\infty$ the highest. Moreover, $\hat{A}_1^{\overline{\alpha}_1^\infty} = \overline{A}_1$ and $\hat{A}_2^{\underline{\alpha}_1^\infty} = \overline{A}_2$. ■

Now consider the case where the players agree to an optimality principle which chooses the payoff weight $\alpha_1^* = \left(\underline{\alpha}_1^\infty\right)^{0.5}\left(\overline{\alpha}_1^\infty\right)^{0.5}$. We then show that such an optimality principle yields a subgame consistent solution in the following Proposition.

Proposition 6.4 An optimality principle under which the players agree to choose the weight

$$\alpha_1^* = \left(\underline{\alpha}_1^\infty\right)^{0.5}\left(\overline{\alpha}_1^\infty\right)^{0.5} \tag{6.3}$$

yields a subgame consistent solution to the cooperative game $\Gamma_c(x,\infty)$.

Proof According to the optimality principle in Proposition 6.4 a unique weight $\alpha_1^* = \left(\underline{\alpha}_1^\infty\right)^{1/2}\left(\overline{\alpha}_1^\infty\right)^{1/2}$ is chosen for any game $\Gamma_c(x,\infty)$. Since $\left(\underline{\alpha}_1^\infty\right)^{1/2}\left(\overline{\alpha}_1^\infty\right)^{1/2} \in S^\infty$, the imputation vector $\xi^{(\tau)}(x,\infty) = \left[\hat{W}^{(\alpha_1^*)1}(x), \hat{W}^{(\alpha_1^*)2}(x)\right]$ yields a Pareto optimal pair. Hence part (i) of Definition 3.2 is proved.

The present-value (at time $\tau < t$) counterpart of the current-value payoff $\hat{W}^{\alpha_1^*(i)}(x)$, $i \in \{1,2\}$, can be expressed as

$$E_\tau\left\{ \exp[-r(t-\tau)]\int_t^\infty \left[\left\{k_i\psi_i^{\alpha_1^*(\infty)}[\mathrm{x(s)}]\right\}^{1/2} - \frac{c_i}{x(s)^{1/2}}\psi_i^{\alpha_1^*(\infty)}[\mathrm{x(s)}]\right] \right.$$
$$\left. \exp[-r(s-t)]\,ds \middle| x(t) = x \right\} = \exp[-r(t-\tau)]\hat{W}^{\alpha_1^*(i)}(x).$$

Hence, part (ii) of Definition 3.2 holds.

Since $\left(\underline{\alpha}_1^\infty\right)^{1/2}\left(\overline{\alpha}_1^\infty\right)^{1/2} \in S^\infty$, $\hat{W}^{\alpha_1^*(i)}(x) \geq V^i(x)$, for $i \in \{1,2\}$ and $x \in X$. Hence, part (iii) of Definition 3.2 is satisfied. ∎

In addition, the cooperative solution in Proposition 6.4 also satisfies the axioms of symmetry in the following remark.

Remark 6.1 The Pareto optimal cooperative solution proposed in Proposition 6.4 also satisfies the axioms of symmetry. See Appendix B for proof details. ∎

6.7 Chapter Appendices

Appendix A: Proof of Condition 6.1 Note that $W^{\alpha_1}(x) = \hat{W}^{\alpha_1(1)}(x) + \alpha_1\hat{W}^{\alpha_1(2)}(x)$, therefore we have $\overline{A}^{\alpha_1} = \hat{A}_1^{\alpha_1} + \alpha_1\hat{A}_2^{\alpha_1}$. Since u_1 and u_2 are nonnegative, $\hat{W}^{\alpha_1(1)}(x) \geq 0$ and $\hat{W}^{\alpha_1(2)}(x) \geq 0$. Hence $\overline{A}^{\alpha_1}$, $\hat{A}_1^{\alpha_1}$ and $\hat{A}_2^{\alpha_1}$ are nonnegative.

Define the equation $\left[r + \frac{1}{8}\sigma^2 + \frac{b}{2}\right]\overline{A}^{\alpha_1} - \frac{k_1}{4\left[c_1 + \overline{A}^{\alpha_1}/2\right]} - \frac{\alpha_1 k_2}{4\left[c_2 + \overline{A}^{\alpha_1}/2\alpha_1\right]} = 0$ in Proposition 6.2 as $\Psi\left(\overline{A}^{\alpha_1}, \alpha_1\right) = 0$. Implicitly differentiating $\Psi\left(\overline{A}^{\alpha_1}, \alpha_1\right) = 0$ yields:

$$\frac{d\overline{A}^{\alpha_1}}{d\alpha_1} = \frac{k_2\left[c_2+\overline{A}^{\alpha_1}/\alpha_1\right]}{4\left[c_2+\overline{A}^{\alpha_1}/2\alpha_1\right]^2} \Big/ \left\{\left[r+\frac{1}{8}\sigma^2+\frac{b}{2}\right]+\frac{k_1}{8\left[c_1+\overline{A}^{\alpha_1}/2\right]^2}+\frac{k_2}{8\left[c_2+\overline{A}^{\alpha_1}/2\alpha_1\right]^2}\right\}>0 \tag{7.1}$$

Then we define the equation $\left[r+\frac{1}{8}\sigma^2+\frac{b}{2}\right]\hat{A}_1^{\alpha_1}-\frac{k_1}{2\left[c_1+\overline{A}^{\alpha_1}/2\right]}+\frac{c_1k_1}{4\left[c_1+\overline{A}^{\alpha_1}/2\right]^2}$ $+\frac{\hat{A}_1^{\alpha_1}k_1}{8\left[c_1+\overline{A}^{\alpha_1}/2\right]^2}+\frac{\hat{A}_1^{\alpha_1}k_2}{8\left[c_2+\overline{A}^{\alpha_1}/2\alpha_1\right]^2}=0$ in Proposition 6.3 as $\Psi^1\left(\hat{A}_1^{\alpha_1},\alpha_1\right)=0$.
The effect of a change in α_1 on $\hat{A}_1^{\alpha_1}$ can be obtained as:

$$\frac{d\hat{A}_1^{\alpha_1}}{d\alpha_1}=-\frac{\partial\Psi^1\left(\hat{A}_1^{\alpha_1},\alpha_1\right)/\partial\alpha_1}{\partial\Psi^1\left(\hat{A}_1^{\alpha_1},\alpha_1\right)/\partial\hat{A}_1^{\alpha_1}}, \tag{7.2}$$

Where

$$\frac{\partial\Psi^1}{\partial\hat{A}_1^{\alpha_1}}=\left[r+\frac{1}{8}\sigma^2+\frac{b}{2}\right]+\frac{k_1}{8\left[c_1+\overline{A}^{\alpha_1}/2\right]^2}+\frac{k_2}{8\left[c_2+\overline{A}^{\alpha_1}/2\alpha_1\right]^2}>0, \text{ and} \tag{7.3}$$

$$\frac{\partial\Psi^1}{\partial\alpha_1}=\frac{k_1\left[\overline{A}^{\alpha_1}-\hat{A}_1^{\alpha_1}\right]}{8\left[c_1+\overline{A}^{\alpha_1}/2\right]^3}\frac{d\overline{A}^{\alpha_1}}{d\alpha_1}+\frac{\hat{A}_1^{\alpha_1}k_2/{\alpha_1}^2}{8\left[c_2+\overline{A}^{\alpha_1}/2\alpha_1\right]^3}\left[\overline{A}^{\alpha_1}-\alpha_1\frac{d\overline{A}^{\alpha_1}}{d\alpha_1}\right] \tag{7.4}$$

From Proposition 6.3, we obtain:

$$\hat{A}_2^{\alpha_1} = \frac{k_2\left[c_2+\overline{A}^{\alpha_1}/\alpha_1\right]}{4\left[c_2+\overline{A}^{\alpha_1}/2\alpha_1\right]^2} \Big/ \left\{\left[r+\frac{1}{8}\sigma^2+\frac{b}{2}\right]+\frac{k_1}{8\left[c_1+\overline{A}^{\alpha_1}/2\right]^2}+\frac{k_2}{8\left[c_2+\overline{A}^{\alpha_1}/2\alpha_1\right]^2}\right\}. \tag{7.5}$$

Comparing (7.5) with (7.1) shows that $d\overline{A}^{\alpha_1}/d\alpha_1=\hat{A}_2^{\alpha_1}$. Upon substituting $d\overline{A}^{\alpha_1}/d\alpha_1$ by $\hat{A}_2^{\alpha_1}$ and invoking the relation $\overline{A}^{\alpha_1}=\hat{A}_1^{\alpha_1}+\alpha_1\hat{A}_2^{\alpha_1}$, we have

$$\frac{\partial\Psi^1}{\partial\alpha_1}=\frac{k_1\alpha_1\left(\hat{A}_2^{\alpha_1}\right)^2}{8\left[c_1+\overline{A}^{\alpha_1}/2\right]^3}+\frac{\left(\hat{A}_1^{\alpha_1}\right)^2k_2/{\alpha_1}^2}{8\left[c_2+\overline{A}^{\alpha_1}/2\alpha_1\right]^3}>0. \tag{7.6}$$

Therefore, $d\hat{A}_1^{\alpha_1}/d\alpha_1 < 0$. Following the above analysis, we have:

$$\frac{d\hat{A}_2^{\alpha_1}}{d\alpha_1} = \left\{ \frac{k_2\left(\hat{A}_1^{\alpha_1}\right)^2/\alpha_1{}^3}{8\left[c_2 + \overline{A}^{\alpha_1}/2\alpha_1\right]^3} + \frac{\left(\hat{A}_2^{\alpha_1}\right)^2 k_1}{8\left[c_1 + \overline{A}^{\alpha_1}/2\right]^3} \right\}$$
$$\div \left\{ \left[r + \frac{1}{8}\sigma^2 + \frac{b}{2}\right] + \frac{k_1}{8\left[c_1 + \overline{A}^{\alpha_1}/2\right]^2} + \frac{k_2}{8\left[c_2 + \overline{A}^{\alpha_1}/2\alpha_1\right]^2} \right\} > 0. \quad (7.7)$$

Hence Condition 6.1 follows. ■

Appendix B: Proof of Remark 6.1 Let $[V^{(\max)1}(x), V^2(x)]$ denote a payoff pair along the Pareto optimal trajectory. From Condition 6.1 and Corollary 6.1, in the problem $\max_{u_1,u_2}\{J^1(x) + \alpha_1 J^2(x)\}$ if $\underline{\alpha}_1^\infty$ is chosen, $\hat{W}^{\underline{\alpha}_1^\infty(2)}(x) = V^2(x)$ and $\hat{W}^{\underline{\alpha}_1^\infty(1)}(x) = V^{(\max)1}(x)$. On the other hand, in the problem $\max_{u_1,u_2}\{J^2(x) + \alpha_2 J^1(x)\}$, in order to have player 2's expected payoff being $V^2(x)$ and player 1's payoff being $V^{(\max)1}(x)$ the weight $\overline{\alpha}_2^\infty$ has to be chosen. Recall that when $\alpha_1 = 1/\alpha_2$, the problem $\max_{u_1,u_2}\{J^1(x) + \alpha_1 J^2(x)\}$ is identical to the problem $\max_{u_1,u_2}\{J^2(x) + \alpha_2 J^1(x)\}$. Since $\max_{u_1,u_2}\{J^1(x) + \underline{\alpha}_1^\infty J^2(x)\}$ and $\max_{u_1,u_2}\{J^2(x) + \overline{\alpha}_2^\infty J^1(x)\}$ both yield $V^2(x)$ and $V^{1(\max)}$ (x), it is necessary that $\underline{\alpha}_1^\infty = 1/\overline{\alpha}_2^\infty$. With similar argument, $\overline{\alpha}_1^\infty = 1/\underline{\alpha}_2^\infty$ is verified.

According to Proposition 6.4, in the problem $\max_{u_1,u_2}\{J^1(x) + \alpha_1 J^2(x)\}$ an optimality principle under which the players agree to choose the weight $\alpha_1^* = \left(\underline{\alpha}_1^\infty\right)^{0.5}\left(\overline{\alpha}_1^\infty\right)^{0.5}$ yields a subgame consistent solution to the cooperative game $\Gamma_c(x, \infty)$.

Following the same optimality principle in the problem $\max_{u_1,u_2}\{J^2(x) + \alpha_2 J^1(x)\}$ under which the players agree to choose the weight $\alpha_2^* = \left(\underline{\alpha}_2^\infty\right)^{0.5}\left(\overline{\alpha}_2^\infty\right)^{0.5}$, which is equivalent to having $1/\alpha_1^* = 1/\left[\left(\underline{\alpha}_1^\infty\right)^{0.5}\left(\overline{\alpha}_1^\infty\right)^{0.5}\right]$.

Since $\alpha_2^* = 1/\alpha_1^*$, the controls in the problems $\max_{u_1,u_2}\{J^1(x) + \alpha_1^* J^2(x)\}$ and $\max_{u_1,u_2}\{J^2(x) + \alpha_2^* J^1(x)\}$ are identical. Hence the axiom of symmetry prevails. ■

6.8 Chapter Notes

The number of studies in cooperative dynamic games with non-transferrable utility/payoff (NTU) is much less than that of cooperative dynamic games with transferrable payoffs. Leitmann (1974), Dockner and Jørgensen (1984), Hamalainen et al. (1986), Yeung and Petrosyan (2005), Yeung et al. (2007), de-Paz et al. (2013), and Marin-Solano (2014) studied continuous-time cooperative differential games with non-transferable payoffs. The stringent requirement of subgame consistency imposes additional hurdles to the derivation of solutions for cooperative stochastic differential games. In the case when players' payoffs are nontransferable, the derivation of solution candidates becomes even more complicated and intractable. In this Chapter, subgame consistent solutions of cooperative stochastic differential games with nontransferable payoffs are examined and a class of cooperative stochastic differential games with nontransferable payoffs is used to illustrate some possible solutions. Theorem 1.1 characterizing the players' expected payoff under cooperation was developed by Yeung (2004). Finally, the analysis can be applied to NTU cooperative differential games with the removal of the stochastic term $\sigma[s, x(s)]$. Finally, the notion of cooperative subgame consistency under variable payoff weights is examined in the discrete-time case in Chap. 11.

6.9 Problems

1. Consider a two-person stochastic differential game with initial state $x(0) = x_0 = 14$ and duration $[0, 4]$. The state dynamics of the game is characterized by the stochastic differential equations:

$$dx(s) = [15 - x(s) - u_1(s) - u_2(s)]ds + 0.01x(s)dz(s),$$

where $u_i \in U_i$ is the control vector of player i, for $i \in [1, 2]$, and $z(s)$ is a Wiener process. The state dynamics is the stock dynamics of a biomass of renewable resource like forest or fresh water. The state $x(s)$ represents the resource size and $u_i(s)$ the (nonnegative) amount of resource extracted by player i.

At time 0, the expected payoff of player 1 is:

$$J^1(0, x_0) = E_0\left\{ \int_0^4 \left[4u_1(s) - u_1(s)^2 x(s)^{-1} + 0.5x(s)\right] \exp[-0.05]\, ds \right.$$
$$\left. + 2\exp[-0.2]x(T) \right\}, \text{ and,}$$

the expected payoff of player 2 is:

$$J^2(0, x_0) = E_0\left\{ \int_0^4 \left[3u_1(s) - 2u_1(s)^2 x(s)^{-1} + x(s)\right] \exp[-0.05]\, ds \right.$$
$$\left. + 3\exp[-0.2]x(T) \right\}.$$

If the payoff weight $\alpha_1 = 0.4$ is chosen to maximize the expected weighted payoff

$\max_{u_1, u_2} E_0 \left\{ J^1(0, x_0) + \alpha_1 J^2(0, x_0) \right\}$, derive the individual payoffs of the players under cooperation.

2. Consider an infinite horizon stochastic differential game with initial state $x(0) = x_0 = 10$. The state dynamics of the game is characterized by the stochastic differential equations:

$$dx(s) = [9 - 2x(s) - u_1(s) - u_2(s)]ds + 0.02x(s)dz(s),$$

where $u_i \in U_i$ is the control vector of player i, for $i \in [1, 2]$, and $z(s)$ is a Wiener process.

At time 0, the expected payoff of player 1 is:

$J^1(0, x_0) = E_0 \left\{ \int_0^\infty \left[4u_1(s) - u_1(s)^2 x(s)^{-1} + 0.2x(s) \right] \exp[-0.05]\, ds \right\}$, and the expected payoff of player 2 is:

$$J^2(0, x_0) = E_0 \left\{ \int_0^\infty \left[4u_1(s) - 2u_1(s)^2 x(s)^{-1} + 1.5x(s) \right] \exp[-0.05]\, ds \right\}.$$

If the payoff weight $\alpha_1 = 0.35$ is chosen to maximize the expected weighted payoff

$\max_{u_1, u_2} E_0 \left\{ J^1(0, x_0) + \alpha_1 J^2(0, x_0) \right\}$, derive the individual payoffs of the players under cooperation.

Part II
Discrete-Time Analysis

Chapter 7
Subgame Consistent Cooperative Solution in Dynamic Games

In many game situations, the evolutionary process is in discrete time rather than in continuous time. An extension of the analysis to a discrete-time dynamic framework is provided in this chapter. In particular, it presents an analysis on subgame consistent solutions which entail group optimality and individual rationality for cooperative (deterministic and stochastic) dynamic games. It integrates the works of Yeung and Petrosyan (2010) and Chapters 12 and 13 of Yeung and Petrosyan (2012a). We first present in Sect. 7.1 a general formulation of cooperative dynamic games in discrete time with the noncooperative outcomes, and the notions of group optimality and individual rationality. Subgame consistent cooperative solutions with corresponding payoff distribution procedures are derived in Sect. 7.2. An illustration of cooperative resource extraction in discrete time is given in Sect. 7.3. A general formulation of coopeartive stochastic dynamic games in discrete time is given in Sect. 7.4. Subgame consistent cooperative solutions with corresponding payoff distribution procedures are derived in Sect. 7.5. An illustration of cooperative resource extraction under uncertainty in discrete time is given in Sect. 7.6. A heuristic approach to obtaining subgame consistent solutions for cooperative dynamic games is provided in Sect. 7.7. Section 7.8 contains Appendices of the Chapter. Chapter Notes are given in Sect. 7.9 and problems in Sect. 7.10. In addition, to make the discrete-time analysis in this Chapter fully in line with the continuous-time analyses presented in earlier chapters a terminal condition is added to each player's payoff in Yeung and Petrosyan (2010, 2012a).

7.1 Cooperative Dynamic Games

In this Section we present the basic framework of discrete-time cooperative dynamic games.

D.W.K. Yeung, L.A. Petrosyan, *Subgame Consistent Cooperation*,
Theory and Decision Library C 47, DOI 10.1007/978-981-10-1545-8_7

7.1.1 Game Formulation

Consider the general $T-$ stage $n-$ person nonzero-sum discrete-time cooperative dynamic game with initial state x^0. The state space of the game is $X \in R^m$ and the state dynamics of the game is characterized by the difference equation:

$$x_{k+1} = f_k(x_k, u_k^1, u_k^2, \cdots, u_k^n), \tag{1.1}$$

for $k \in \{1, 2, \cdots, T\} \equiv \kappa$ and $x_1 = x^0$,

where $u_k^i \in U^i \subset R^{m_i}$ is the control vector of player i at stage k, $x_k \in X \subset R^m$ is the state of the game.

The payoff of player i is

$$\sum_{\zeta=1}^{T} g_\zeta^i \left[x_\zeta, u_\zeta^1, u_\zeta^2, \cdots, u_\zeta^n\right] \left(\frac{1}{1+r}\right)^{\zeta-1} + q_{T+1}^i(x_{T+1}) \left(\frac{1}{1+r}\right)^T, \tag{1.2}$$

for $i \in \{1, 2, \cdots, n\} \equiv N$,

where r is the discount rate, and $q_{T+1}^i(x_{T+1})$ is the terminal benefit that player i received at stage $T+1$.

The payoffs of the players are transferable.

7.1.2 Noncooperative Outcome

In this subsection, we characterize the noncooperative outcome of the discrete-time economic game (1.1 and 1.2). Let $\{\phi_k^i(x), \text{ for } k \in \kappa \text{ and } i \in N\}$ denote a set of strategies that provides a feedback Nash equilibrium solution to the game (1.1 and 1.2), and

$$V^i(k, x) = \sum_{\zeta=k}^{T} g_\zeta^i \left[x_\zeta, \phi_\zeta^1(x_\zeta), \phi_\zeta^2(x_\zeta), \cdots, \phi_\zeta^n(x_\zeta)\right] \left(\tfrac{1}{1+r}\right)^{\zeta-1}$$
$$+ q_{T+1}^i(x_{T+1}) \left(\tfrac{1}{1+r}\right)^T,$$

where $x_k = x$, for $k \in K$ and $i \in N$, denote the value function indicating the game equilibrium payoff to player i over the stages from k to $T+1$. A frequently used way to characterize and derive a feedback Nash equilibrium of the game is provided in the following theorem.

Theorem 1.1 A set of strategies $\{\phi_k^i(x),\ \text{for}\, k\in\kappa\ \text{and}\ i\in N\}$ provides a feedback Nash equilibrium solution to the game (1.1 and 1.2) if there exist functions $V^i(k,x)$, for $k\in K$ and $i\in N$, such that the following recursive relations are satisfied:

$$V^i(k,x) = \max_{u_k^i} \Big\{ g_k^i\big[x,\phi_k^1(x),\phi_k^2(x),\cdots,\phi_k^{i-1}(x),u_k^i,\phi_k^{i+1}(x),\cdots,\phi_k^n(x)\big]$$

$$\left(\tfrac{1}{1+r}\right)^{k-1} + V^i\big[k+1, f_k\big[x,\phi_k^1(x),\phi_k^2(x),\cdots,\phi_k^{i-1}(x),u_k^i,\phi_k^{i+1}(x),\cdots,\phi_k^n(x)\big]\Big\}, \tag{1.3}$$

$$V^i(T+1,x) = q_{T+1}^i(x)\left(\frac{1}{1+r}\right)^T; \tag{1.4}$$

for $i\in N$ and $k\in\kappa$.

Proof Invoking the discrete-time dynamic programming technique in Theorem A.5 of the Technical Appendices, $V^i(k,x)$ is the maximized payoff of player i for given strategies $\{\phi_k^i(x),\ \text{for}\, j\in N\, \text{and}\, j\neq i\}$ of the other $n-1$ players. Hence a Nash equilibrium appears. ■

For the sake of exposition, we sidestep the issue of multiple equilibria and focus on solvable games in which a particular noncooperative Nash equilibrium is chosen by the players in the entire subgame.

7.1.3 Dynamic Cooperation

Now consider the case when the players agree to cooperate and distribute the payoff among themselves according to an optimality principle. Two essential properties that a cooperative scheme has to satisfy are group optimality and individual rationality. An agreed upon optimality principle entails group optimality and an imputation to distribute the total cooperative payoff among the players.

We first examine the group optimal solution and then the condition under which individual rationality will be maintained.

7.1.3.1 Group Optimality

Maximizing the players' joint payoff guarantees group optimality in a game where payoffs are transferable. To maximize their joint payoff the players have to solve the discrete-time dynamic programming problem of maximizing

$$\sum_{j=1}^{n}\sum_{k=1}^{T}\left[g_k^j\left[x_k, u_k^1, u_k^2, \cdots, u_k^n\right]\left(\frac{1}{1+r}\right)^{k-1}\right] + \sum_{j=1}^{n} q_{T+1}^j(x_{T+1})\left(\frac{1}{1+r}\right)^T, \quad (1.5)$$

subject to (1.1).

Invoking the discrete-time dynamic programming technique an optimal solution to the control problem (1.1) and (1.5) can be characterized by the theorem below.

Theorem 1.2 A set of strategies $\{\psi_k^i(x),\ \text{for}\, k\in\kappa \text{ and } i\in N\}$ provides an optimal solution to the problem (1.1) and (1.5) if there exist functions $W(k,x)$, for $k\in K$, such that the following recursive relations are satisfied:

$$\begin{aligned} W(k,x) = &\max_{u_k^1, u_k^2, \cdots, u_k^n}\left\{ \sum_{j=1}^{n} g_k^j\left[x_k, u_k^1, u_k^2, \cdots, u_k^n\right]\left(\frac{1}{1+r}\right)^{k-1}\right. \\ &\left. + W\left[k+1, f_k\left(x_k, u_k^1, u_k^2, \cdots, u_k^n\right)\right]\right\} \\ = &\sum_{j=1}^{n} g_k^j\left[x, \psi_k^1(x), \psi_k^2(x), \cdots, \psi_k^n(x)\right]\left(\tfrac{1}{1+r}\right)^{k-1} \\ &+ W\left[k+1, f_k\left(x, \psi_k^1(x), \psi_k^2(x), \cdots, \psi_k^n(x)\right)\right], \end{aligned} \quad (1.6)$$

$$W(T+1,x) = \sum_{j=1}^{n} q_{T+1}^j(x)\left(\frac{1}{1+r}\right)^T. \quad (1.7)$$

Proof Follow the proof of discrete-time dynamic programming technique in Theorem A.5 of the Technical Appendices. ■

Substituting the optimal control $\{\psi_k^i(x),\ \text{for}\ k\in\kappa \text{ and } i\in N\}$ into the state dynamics (1.1), one can obtain the dynamics of the cooperative trajectory as:

$$x_{k+1} = f_k\left(x_k, \psi_k^1(x_k), \psi_k^2(x_k), \cdots, \psi_k^n(x_k)\right), \quad (1.8)$$

for $k\in\kappa$ and $x_1 = x^0$.

Let $\{x_k^*\}_{k=1}^T$ denote the solution to (1.8) and hence the optimal cooperative path. The total cooperative payoff over the stages from k to $T+1$ can be expressed as:

$$\begin{aligned} W(k,x_k^*) = &\sum_{\zeta=k}^{T}\sum_{j=1}^{n} g_\zeta^j\left[x_\xi^*, \psi_\zeta^1\left(x_\zeta^*\right), \psi_\zeta^2\left(x_\zeta^*\right), \Lambda, \psi_\zeta^n\left(x_\zeta^*\right)\right]\left(\tfrac{1}{1+r}\right)^{\zeta-1} \\ &+ \sum_{j=1}^{n} q_{T+1}^j(x_{T+1})\left(\tfrac{1}{1+r}\right)^T, \qquad \text{for}\ k\in\kappa. \end{aligned} \quad (1.9)$$

We then proceed to consider individual rationality.

7.1.3.2 Individual Rationality

The players have to agree on an optimality principle in distributing the total cooperative payoff among themselves. For individual rationality to be upheld the payoffs an player receives under cooperation have to be no less than his noncooperative payoff along the cooperative state trajectory. For instance, (i) the players may share the excess of the total cooperative payoff over the sum of individual noncooperative payoffs equally, or (ii) they may share the total cooperative payoff proportionally to their noncooperative payoffs.

Let $\xi(\cdot,\cdot)$ denote the imputation vector guiding the distribution of the total cooperative payoff under the agreed-upon optimality principle along the cooperative trajectory $\{x_k^*\}_{k=1}^T$. At stage k, the imputation vector according to $\xi(\cdot,\cdot)$ is $\xi(k,x_k^*) = [\xi^1(k,x_k^*),\xi^2(k,x_k^*),\ldots,\xi^n(k,x_k^*)]$, for $k\in\kappa$.

If for example, the optimality principle specifies that the players share the excess of the total cooperative payoff over the sum of individual noncooperative payoffs equally, then the imputation to player i becomes:

$$\xi^i(k,x_k^*) = V^i(k,x_k^*) + \frac{1}{n}\left[W(k,x_k^*) - \sum_{j=1}^{n} V^j(k,x_k^*)\right], \tag{1.10}$$

for $i\in N$ and $k\in\kappa$.

If the optimality principle specifies that the players share the total cooperative proportional to their noncooperative payoffs, then the imputation to player i becomes:

$$\xi^i(k,x_k^*) = \frac{V^i(k,x_k^*)}{\sum_{j=1}^{n} V^j(k,x_k^*)} W(k,x_k^*), \tag{1.11}$$

for $i\in N$ and $k\in\kappa$.

For individual rationality to be maintained throughout all the stages $k\in\kappa$, it is required that:

$$\xi^i(k,x_k^*) \geq V^i(k,x_k^*), \text{ for } i\in N \text{ and } k\in\kappa. \tag{1.12}$$

In particular, the above condition guaranties that the payoff allocated to a player under cooperation will be no less than its noncooperative payoff.

To satisfy group optimality, the imputation vector has to satisfy

$$W(k,x_k^*) = \sum_{j=1}^{n} \xi^j(k,x_k^*), \quad \text{for } k\in\kappa. \tag{1.13}$$

This condition guarantees the highest joint payoffs for the participating players.

7.2 Subgame Consistent Solutions and Payment Mechanism

To guarantee dynamical stability in a dynamic cooperation scheme, the solution has to satisfy the property of subgame consistency. In particular, the specific agreed-upon optimality principle must remain effective at any stage of the game along the optimal state trajectory. Since at any stage of the game the players are guided by the same optimality principles and hence do not have any ground for deviation from the previously adopted optimal behavior throughout the game. Therefore for subgame consistency to be satisfied, the imputation $\xi(\cdot,\cdot)$ according to the original optimality principle has to be maintained at all the T stages along the cooperative trajectory $\{x_k^*\}_{k=1}^T$. In other words, the imputation

$$\xi(k,x_k^*) = \left[\xi^1(k,x_k^*),\xi^2(k,x_k^*),\ldots,\xi^n(k,x_k^*)\right] \text{ at stage } k, \tag{2.1}$$

for $k\in\kappa$
has to be upheld.

Crucial to the analysis is the formulation of a payment mechanism so that the imputation in (2.1) can be realized.

7.2.1 *Payoff Distribution Procedure*

Similar to the analysis of cooperative differential games, we first formulate a Payoff Distribution Procedure (PDP) so that the agreed imputations (2.1) can be realized. Let $B_k^i(x_k^*)$ denote the payment that player i will receive at stage k under the cooperative agreement along the cooperative trajectory $\{x_k^*\}_{k=1}^T$.

The payment scheme involving $B_k^i(x_k^*)$ constitutes a PDP in the sense that the imputation to player i over the stages from k to T can be expressed as:

$$\begin{aligned}\xi^i(k,x_k^*) &= B_k^i(x_k^*)\left(\tfrac{1}{1+r}\right)^{k-1}\\ &+\left\{\sum_{\zeta=k+1}^{T} B_\zeta^i\left(x_\zeta^*\right)\left(\tfrac{1}{1+r}\right)^{\zeta-1} + q_{T+1}^i(x_{T+1})\left(\tfrac{1}{1+r}\right)^{T}\right\}\end{aligned} \tag{2.2}$$

for $i\in N$ and $k\in\kappa$.

Using (2.2) one can obtain

$$\begin{aligned}\xi^i(k+1,x_{k+1}^*) &= B_{k+1}^i(x_{k+1}^*)\left(\tfrac{1}{1+r}\right)^{k}\\ &+\left\{\sum_{\zeta=k+2}^{T} B_\zeta^i\left(x_\zeta^*\right)\left(\tfrac{1}{1+r}\right)^{\zeta-1} + q_{T+1}^i(x_{T+1})\left(\tfrac{1}{1+r}\right)^{T}\right\}.\end{aligned} \tag{2.3}$$

Upon substituting (2.3) into (2.2) yields

$$\xi^i(k,x_k^*) = B_k^i(x_k^*)\left(\frac{1}{1+r}\right)^{k-1} + \xi^i\left[k+1,\ f_k(x_k^*,\psi_k(x_k^*))\right], \tag{2.4}$$

for $i\in N$ and $k\in\kappa$.

A theorem characterizing a formula for $B_k^i(x_k^*)$, for $k\in\kappa$ and $i\in N$, which yields (2.2) is provided below.

Theorem 2.1 A payment equaling

$$B_k^i(x_k^*) = (1+r)^{k-1}\left[\ \xi^i(k,x_k^*) - \xi^i\left[k+1,f_k(x_k^*,\psi_k(x_k^*))\right]\ \right], \tag{2.5}$$

for $i\in N$,
given to player i at stage $k\in\{1,2,\cdots,T\}$ along the cooperative trajectory $\{x_k^*\}_{k=1}^T$ would lead to the realization of the imputation $\{\xi(k,x_k^*)$, for $k\in\kappa\}$.

Proof From (2.4), one can readily obtain (2.5). Theorem 2.1 can also be verified alternatively by showing that from (2.2)

$$\begin{aligned}
\xi^i(k,x_k^*) &= B_k^i(x_k^*)\left(\tfrac{1}{1+r}\right)^{k-1} \\
&\quad + \left\{\sum_{\zeta=k+1}^{T} B_\zeta^i(x_\zeta^*)\left(\tfrac{1}{1+r}\right)^{\zeta-1} + q_{T+1}^i(x_{T+1})\left(\tfrac{1}{1+r}\right)^{T}\right\} \\
&= \left\{\xi^i(k,x_k^*) - \left(\xi^i\left[k+1,f_k(x_k^*,\psi_k(x_k^*))\right]\right)\right\} \\
&\quad + \sum_{\zeta=k+1}^{T}\left\{\xi^i(\zeta,x_\zeta^*) - \left(\xi^i\left[\zeta+1,f_\zeta(x_\zeta^*,\psi_\zeta(x_\zeta^*))\right]\right)\right\} \\
&= \xi^i(k,x_k^*);
\end{aligned}$$

and $\xi^i(T+1,x_{T+1}^*) = q_{T+1}^i(x_{T+1})\left(\tfrac{1}{1+r}\right)^T$.

Hence Theorem 2.1 follows. ■

The payment scheme in Theorem 2.1 gives rise to the realization of the imputation guided by the agreed-upon optimal principle and will be used to derive time (optimal-trajectory-subgame) consistent solutions in the next subsection.

7.2.2 Subgame Consistent Solution

We denote the discrete-time cooperative game with dynamics (1.1) and payoff (1.2) by $\Gamma_c(1,x_0)$. We then denote the game with dynamics (1.1) and payoff (1.2) which starts at stage υ with initial state x_υ^* by $\Gamma_c(\upsilon,x_\upsilon^*)$. Moreover, we let

$P(1, x_0) = \{u_h^i \text{ and } B_h^i \text{ for } h \in \kappa \text{ and } i \in N,\ \xi(1, x_0)\}$ denote the agreed-upon optimality principle for the cooperative game $\Gamma_c(1, x_0)$. Let $\mathrm{P}(x_\upsilon^*, \upsilon) = \{u_h^i \text{ and } B_h^i \text{ for } h \in \{\upsilon, \upsilon+1, \ldots, T\} \text{ and } i \in N,\ \xi(\upsilon, x_\upsilon^*)\}$ denote the optimality principle of the cooperative game $\Gamma_c(\upsilon, x_\upsilon^*)$ according to the original agreement.

A theorem characterizing a subgame consistent solution for the discrete-time cooperative game $\Gamma_c(1, x_0)$ is presented below.

Theorem 2.2 For the cooperative game $\Gamma_c(1, x_0)$ with optimality principle $\mathrm{P}(1, x_0) = \{u_h^i \text{ and } B_h^i \text{ for } h \in \kappa \text{ and } i \in N,\ \xi(1, x_0)\}$ in which

(i) $u_h^i = \psi_h^i(x_h^*)$, for $h \in \kappa$ and $i \in N$, is the set of group optimal strategies for the game $\Gamma_c(1, x_0)$, and
(ii) $B_h^i = B_h^i(x_h^*)$, for $h \in \kappa$ and $i \in N$, where

$$B_h^i(x_h^*) = (1+r)^{h-1}\left[\ \xi^i(h, x_h^*) - \xi^i\left[k+1,\ f_h\left(x_h^*, \psi_h(x_h^*)\right)\right]\ \right], \tag{2.6}$$

and $[\xi^1(h, x_h^*), \xi^2(h, x_h^*), \ldots, \xi^i(h, x_h^*)]$, is the imputation according to the optimality principle $P(h, x_h^*)$;

is subgame consistent.

Proof Follow the proof of the continuous-time analog in Theorem 2.2 of Chap. 3. ■

When all players are using the cooperative strategies, the payoff that player i will directly receive at stage k given that along the cooperative trajectory $\{x_k^*\}_{k=1}^T$ is

$$g_k^i\left[x_k^*, \psi_k^1(x_k^*), \psi_k^2(x_k^*), \ldots, \psi_k^n(x_k^*), x_{k+1}^*\right].$$

However, according to the agreed upon imputation, player i will receive $B_k^i(x_k^*)$ at stage k. Therefore a side-payment

$$\varpi_k^i(x_k^*) = B_k^i(x_k^*) - g_k^i\left[x_k^*, \psi_k^1(x_k^*), \psi_k^2(x_k^*), \ldots, \psi_k^n(x_k^*), x_k^*\right], \tag{2.7}$$

for $k \in \kappa$ and $i \in N$,

will be given to player i to yield the cooperative imputation $\xi^i(k, x_k^*)$.

7.3 An Illustration in Cooperative Resource Extraction

Consider an economy endowed with a renewable resource and with two resource extractors (firms). The lease for resource extraction begins at stage 1 and ends at stage 3 for these two firms. Let u_k^i denote the amount of resource extracted by firm

i at stage k, for $i \in \{1,2\}$. Let U^i be the set of admissible extraction rates, and $x_k \in X \subset R^+$ the size of the resource stock at stage k. The extraction cost for firm $i \in \{1,2\}$ depends on the quantity of resource extracted u_k^i, the resource stock size x_k, and cost parameters c_1 and c_2. The extraction cost for firm i at stage k is specified as $c_i(u_k^i)^2/x_k$. The price of the resource is P.

The profits that firm 1 and firm 2 will obtain at stage k are respectively:

$$\left[Pu_k^1 - \frac{c_1}{x_k}(u_k^1)^2\right] \text{ and } \left[Pu_k^2 - \frac{c_2}{x_k}(u_k^2)^2\right]. \tag{3.1}$$

In stage 4, the firms will receive a salvage value equaling qx_4.

The growth dynamics of the resource is governed by the difference equation:

$$x_{k+1} = x_k + a - bx_k - \sum_{j=1}^{2} u_k^j, \tag{3.2}$$

for $k \in \{1,2,3\}$ and $x_1 = x^0$.

There exists an extraction constraint that human harvesting can at most exploit Y proportion of the existing biomass, hence $u_k^1 + u_k^2 \leq Yx_k$. Moreover $b < 1 - Y$. The payoff of extractor $i \in \{1,2\}$ is to maximize the present value of the stream of future profits:

$$\sum_{k=1}^{3}\left[Pu_k^i - \frac{c_i}{x_k}(u_k^i)^2\right]\left(\frac{1}{1+r}\right)^{k-1} + \left(\frac{1}{1+r}\right)^3 qx_4, \text{ for } i \in \{1,2\}, \tag{3.3}$$

subject to (3.2).

Invoking Theorem 1.1, one can characterize the noncooperative equilibrium strategies in a feedback solution for game (3.2 and 3.3). In particular, a set of strategies $\{\phi_k^i(x), \text{ for } k \in \{1,2,3\} \text{ and } i \in \{1,2\}\}$ provides a Nash equilibrium solution to the game (3.2 and 3.3) if there exist functions $V^i(k,x)$, for $i \in \{1,2\}$ and $k \in \{1,2,3\}$, such that the following recursive relations are satisfied:

$$\begin{aligned} V^i(k,x) = \max_{u_k^i} \Bigg\{ & \left[Pu_k^i - \frac{c_i}{x}(u_k^i)^2\right]\left(\frac{1}{1+r}\right)^{k-1} \\ & + V^i\left[k+1, x + a - bx - u_k^i - \phi_k^j(x)\right] \Bigg\}, \text{ for } k \in \{1,2,3\}; \\ V^i(4,x) = {} & \left(\tfrac{1}{1+r}\right)^3 qx. \end{aligned} \tag{3.4}$$

Performing the indicated maximization in (3.4) yields:

$$\left(P - \frac{2c_i u_k^i}{x}\right)\left(\frac{1}{1+r}\right)^{k-1} - V_{x_{k+1}}^i\left[k+1, x + a - bx - u_k^i - \phi_k^j(x)\right] = 0, \tag{3.5}$$

for $i \in \{1,2\}$ and $k \in \{1,2,3\}$.

From (3.5), the game equilibrium strategies can be expressed as:

$$\phi_k^i(x) = \left(P - V_{x_{k+1}}^i\left[k+1, x+a-bx-\sum_{\ell=1}^{2}\phi_k^\ell(x)\right](1+r)^{k-1}\right)\frac{x}{2c_i}, \tag{3.6}$$

for $i\in\{1,2\}$ and $k\in\{1,2,3\}$.

The game equilibrium profits of the firms can be obtained as:

Proposition 3.1 The value function indicating the game equilibrium profit of firm i is:

$$V^i(k,x) = \left[A_k^i x + C_k^i\right], \text{ for } i\in\{1,2\} \text{ and } k\in\{1,2,3\}, \tag{3.7}$$

where A_k^i and C_k^i, for $i\in\{1,2\}$ and $k\in\{1,2,3\}$, are constants in terms of the parameters of the game (3.2 and 3.3).

Proof See Appendix A of this Chapter. ■

Substituting the relevant derivatives of the value functions in Proposition 3.1 into the game equilibrium strategies (3.6) yields a noncooperative feedback equilibrium solution of the game (3.2 and 3.3).

Now consider the case when the extractors agree to maximize their joint profit and share the excess of cooperative gains over their noncooperative payoffs equally. To maximize their joint payoff, they solve the problem of maximizing

$$\sum_{j=1}^{2}\sum_{k=1}^{3}\left[Pu_k^j - \frac{c_j}{x_k}\left(u_k^j\right)^2\right]\left(\frac{1}{1+r}\right)^{k-1} + 2\left(\frac{1}{1+r}\right)^3 qx_4 \tag{3.8}$$

subject to (3.2).

Invoking Theorem 1.2, one can characterize the optimal controls in the dynamic programming problem (3.2) and (3.8). In particular, a set of control strategies $\{\psi_k^i(x), \text{ for } k\in\{1,2,3\} \text{ and } i\in\{1,2\}\}$ provides an optimal solution to the problem (3.2) and (3.8) if there exist functions $W(k,x)$: $R \to R$, for $k\in\{1,2,3\}$, such that the following recursive relations are satisfied:

$$W(k,x) = \max_{u_k^1, u_k^2}\left\{\sum_{j=1}^{2}\left[Pu_k^j - \frac{c_j}{x}\left(u_k^j\right)^2\right]\left(\frac{1}{1+r}\right)^{k-1} + W\left[k+1, x+a-bx-\sum_{j=1}^{2}u_k^j\right]\right\}, \quad \text{for } k\in\{1,2,3\}.$$

$$W(4,x) = 2\left(\frac{1}{1+r}\right)^3 qx. \tag{3.9}$$

Performing the indicated maximization in (3.9) yields:

$$\left(P - \frac{2c_i u_k^i}{x}\right)\left(\frac{1}{1+r}\right)^{k-1} - W_{x_{k+1}}\left[k+1, x+a-bx-\sum_{j=1}^{2} u_k^j\right] = 0, \qquad (3.10)$$

for $i \in \{1,2\}$ and $k \in \{1,2,3\}$.

In particular, the optimal cooperative strategies can be obtained from (3.10) as:

$$u_k^i = \left(P - W_{x_{k+1}}\left[k+1, x+a-bx-\sum_{j=1}^{2} u_k^j\right](1+r)^{k-1}\right)\frac{x}{2c_i}, \qquad (3.11)$$

for $i \in \{1,2\}$ and $k \in \{1,2,3\}$.

The firms' joint profit under cooperation can be obtained as:

Proposition 3.2 The value function indicating the maximized joint payoff is

$$W(k,x) = [A_k x + C_k], \text{ for } k \in \{1,2,3\}, \qquad (3.12)$$

where A_k and C_k, for $k \in \{1,2,3\}$, are constants in terms of the parameters of the problem (3.8) and (3.2).

Proof See Appendix B of this Chapter. ■

Using (3.11) and Proposition 3.2, the optimal cooperative strategies of the players can be expressed as:

$$\psi_k^i(x) = \left[P - A_{k+1}(1+r)^{k-1}\right]\frac{x}{2c_i}, \quad \text{for } i \in \{1,2\} \text{ and } k \in \{1,2,3\}. \qquad (3.13)$$

Substituting $\psi_k^i(x)$ from (3.13) into (3.2) yields the optimal cooperative state trajectory:

$$x_{k+1} = x_k + a - bx_k - \sum_{j=1}^{2}\left[P - A_{k+1}(1+r)^{k-1}\right]\frac{x_k}{2c_j}, \qquad (3.14)$$

for $k \in \{1,2,3\}$ and $x_1 = x^0$.

Dynamics (3.14) is a linear difference equation readily solvable by standard techniques. Let $\{x_k^*, \text{ for } k \in \{1,2,3\}\}$ denote the solution to (3.14).

Since the extractors agree to share the excess of cooperative gains over their noncooperative payoffs equally, an imputation

$$\begin{aligned}\xi^i(k,x_k^*) &= V^i(k,x_k^*) + \frac{1}{2}\left[W(k,x_k^*) - \sum_{j=1}^{2} V^j(k,x_k^*)\right] \\ &= (A_k^i x_k^* + C_k^i) + \frac{1}{2}\left[(A_k x_k^* + C_k) - \sum_{j=1}^{2}\left(A_k^j x_k^* + C_k^j\right)\right], \end{aligned} \qquad (3.15)$$

for $k \in \{1,2,3\}$ and $i \in \{1,2\}$ has to be maintained.

Invoking Theorem 2.1, if $x_k^* \in X$ is realized at stage k a payment equaling

$$B_k^i(x_k^*) = (1+r)^{k-1}\left[\xi^i(k, x_k^*) - \xi^i[k+1, f_k(x_k^*, \psi_k(x_k^*))]\right]$$
$$= (1+r)^{k-1}\left\{(A_k^i x_k^* + C_k^i) + \frac{1}{2}\left((A_k x_k^* + C_k) - \sum_{j=1}^{2}\left(A_k^j x_k^* + C_k^j\right)\right)\right.$$
$$\left. - \left[(A_{k+1}^i x_{k+1}^* + C_{k+1}^i) + \frac{1}{2}\left((A_{k+1} x_{k+1}^* + C_{k+1}) - \sum_{j=1}^{2}\left(A_{k+1}^j x_{k+1}^* + C_{k+1}^j\right)\right)\right]\right\}, \tag{3.16}$$

for $i \in \{1,2\}$;
given to player i at stage $k \in \kappa$ would lead to the realization of the imputation(3.15).

A subgame consistent solution can be readily obtained from (3.13), (3.15) and (3.16).

7.4 Cooperative Stochastic Dynamic Games

In this Section we present the basic framework of discrete-time cooperative stochastic dynamic games.

7.4.1 Game Formulation

Consider the general $T-$ stage $n-$ person nonzero-sum discrete-time cooperative stochastic dynamic game with initial state x^0. The state space of the game is $X \in R^m$ and the state dynamics of the game is characterized by the stochastic difference equation:

$$x_{k+1} = f_k(x_k, u_k^1, u_k^2, \cdots, u_k^n) + G_k(x_k)\theta_k, \tag{4.1}$$

for $k \in \{1, 2, \cdots, T\} \equiv \kappa$ and $x_1 = x^0$,

where $u_k^i \in R^{m_i}$ is the control vector of player i at stage k, $x_k \in X$ is the state, and θ_k is a set of statistically independent random variables.

The objective of player i is

$$E_{\theta_1,\theta_2,\cdots,\theta_T}\left\{\sum_{\zeta=1}^{T} g_\zeta^i\left[x_\zeta, u_\zeta^1, u_\zeta^2, \ldots, u_\zeta^n\right]\left(\frac{1}{1+r}\right)^{\zeta-1} + q_{T+1}^i(x_{T+1})\left(\frac{1}{1+r}\right)^T\right\},$$
$$\text{for } i \in \{1, 2, \ldots, n\} \equiv N, \tag{4.2}$$

where r is the discount rate and $E_{\theta_1,\theta_2,\cdots,\theta_T}$ is the expectation operation with respect to the statistics of θ_1, θ_2, $\cdots$, θ_T.

The payoffs of the players are transferable.

We then characterize the noncooperative outcome of the discrete-time stochastic economic game (4.1 and 4.2). Let $\{\phi_k^i(x),\ \text{for}\ k\in\kappa\ \text{and}\ i\in N\}$ denote a set of strategies that provides a feedback Nash equilibrium solution (if it exists) to the game (4.1 and 4.2), and

$$V^i(k,x) = E_{\theta_k,\theta_{k+1},\cdots,\theta_T}\left\{ \sum_{\zeta=k}^{T} g_\zeta^i\left[x_\zeta, \phi_\zeta^1(x_\zeta), \phi_\zeta^2(x_\zeta), \ldots, \phi_\zeta^n(x_\zeta)\right]\left(\frac{1}{1+r}\right)^{\zeta-1} + q_{T+1}^i(x_{T+1})\left(\frac{1}{1+r}\right)^{T}\right\},$$

where $x_k = x$, for $k\in K$ and $i\in N$, denote the value function indicating the expected game equilibrium payoff to player i over the stages from k to $T+1$.

A frequently used way to characterize and derive a feedback Nash equilibrium of the game is provided in the theorem below.

Theorem 4.1 A set of strategies $\{\phi_k^i(x),\ \text{for}\ k\in\kappa\ \text{and}\ i\in N\}$ provides a feedback Nash equilibrium solution to the game (4.1 and 4.2) if there exist functions $V^i(k,x)$, for $k\in K$ and $i\in N$, such that the following recursive relations are satisfied:

$$\begin{aligned} V^i(k,x) &= \max_{u_k^i} E_{\theta_k}\left\{ g_k^i\left[x, \phi_k^1(x), \phi_k^2(x), \ldots, \phi_k^{i-1}(x), u_k^i, \phi_k^{i+1}(x), \ldots \right.\right. \\ &\quad \left.\left. \ldots, \phi_k^n(x)\right]\left(\tfrac{1}{1+r}\right)^{k-1} + V^i\left[k+1, \tilde{f}_k^i(x,u_k^i) + G_k(x)\theta_k\right]\right\} \\ &= E_{\theta_k}\left\{ g_k^i\left[x, \phi_k^1(x), \phi_k^2(x), \ldots, \phi_k^n(x)\right]\left(\tfrac{1}{1+r}\right)^{k-1} \right. \\ &\quad \left. + V^i\left[k+1, f_k(x, \phi_k^1(x), \phi_k^2(x), \ldots, \phi_k^n(x)) + G_k(x)\theta_k\right]\right\} \end{aligned} \tag{4.3}$$

$$V^i(T+1,x) = q_{T+1}^i(x)\left(\frac{1}{1+r}\right)^{T}; \tag{4.4}$$

for $i\in N$ and $k\in\kappa$,

where $\tilde{f}_k^i(x,u_k^i) = f_k\left[x, \phi_k^1(x), \phi_k^2(x), \ldots, \phi_k^{i-1}(x), u_k^i, \phi_k^{i+1}(x), \ldots, \phi_k^n(x)\right]$ and E_{θ_k} is the expectation operation with respect to the statistics of θ_k.

Proof Invoking the discrete-time stochastic dynamic programming technique in Theorem A.6 of the Technical Appendices, $V^i(k,x)$ is the maximized payoff of player i for given strategies $\{\phi_k^j(x),\ \text{for}\ j\in N\ \text{and}\ j\neq i\}$ of the other $n-1$ players. Hence a Nash equilibrium appears. ■

Again, for the sake of exposition, we sidestep the issue of multiple equilibria and focus on solvable games in which a particular noncooperative Nash equilibrium is chosen by the players in the entire subgame.

7.4.2 Dynamic Cooperation under Uncertainty

Now consider the case when the players agree to cooperate and distribute the payoff among themselves according to an optimality principle. Once again, the essential properties of group optimality and individual rationality have to be satisfied. An agreed upon optimality principle entails group optimality and an imputation to distribute the total cooperative payoff among the players.

We first examine the group optimal solution and then the condition under which individual rationality will be maintained.

7.4.2.1 Group Optimality

Maximizing the players' expected joint payoff guarantees group optimality in a game where payoffs are transferable. To maximize their expected joint payoff the players have to solve the discrete-time stochastic dynamic programming problem of maximizing

$$E_{\theta_1,\theta_2,\cdots,\theta_T}\left\{\sum_{j=1}^{n}\sum_{k=1}^{T}\left[\ g_k^j\left[x_k,u_k^1,u_k^2,\ldots,u_k^n\right]\left(\tfrac{1}{1+r}\right)^{k-1}\right]\right.$$
$$\left.+\sum_{j=1}^{n}q_{T+1}^i(x_{T+1})\left(\tfrac{1}{1+r}\right)^T\right\} \tag{4.5}$$

subject to (4.1).

Invoking the discrete-time stochastic dynamic programming technique an optimal solution to the problem (4.1) and (4.5) can be characterized in the following theorem.

Theorem 4.2 A set of strategies $\left\{\psi_k^i(x),\ \text{for } k\in\kappa \text{ and } i\in N\right\}$, provides an optimal solution to the problem (4.1) and (4.5) if there exist functions $W(k,x)$, for $k\in K$, such that the following recursive relations are satisfied:

$$W(k,x)=\max_{u_k^1,u_k^2,\ldots,u_k^n} E_{\theta_k}\left\{\sum_{j=1}^{n}g_k^j\left[x,u_k^1,u_k^2,\ldots,u_k^n\right]\left(\frac{1}{1+r}\right)^{k-1}\right.$$
$$\left.+W\left[k+1,f_k\left(x,u_k^1,u_k^2,\ldots,u_k^n\right)+G_k(x)\theta_k\right]\right\}$$
$$=E_{\theta_k}\left\{\ \sum_{j=1}^{n}g_k^j\left[x,\psi_k^1(x),\psi_k^2(x),\ldots,\psi_k^n(x)\right]\times\left(\frac{1}{1+r}\right)^{k-1}\right.$$
$$\left.+W\left[k+1,f_k\left(x,\psi_k^1(x),\psi_k^2(x),\ldots,\psi_k^n(x)\right)+G_k(x)\theta_k\right]\ \right\}, \tag{4.6}$$

$$W(T+1,x)=\sum_{j=1}^{n}q_{T+1}^j(x)\left(\frac{1}{1+r}\right)^T. \tag{4.7}$$

Proof Follow the proof of the discrete-time stochastic dynamic programming technique in Theorem A.6 of the Technical Appendices. ■

Substituting the optimal control $\{\psi_k^i(x),\ \text{for}\ k\in\kappa\ \text{and}\ i\in N\}$ into the state dynamics (4.1), one can obtain the dynamics of the cooperative trajectory as:

$$x_{k+1} = f_k\left(x_k, \psi_k^1(x_k), \psi_k^2(x_k), \ldots, \psi_k^n(x_k)\right) + G_k(x_k)\theta_k, \tag{4.8}$$

for $k\in\kappa$ and $x_1 = x^0$.

We use X_k^* to denote the set of realizable values of x_k at stage k generated by (4.8). The term $x_k^*\in X_k^*$ is used to denote an element in X_k^*.

The term $W(k, x_k^*)$ gives the expected total cooperative payoff over the stages from k to $T+1$ if $x_k^*\in X_k^*$ is realized at stage $k\in\kappa$.We then proceed to consider individual rationality.

7.4.2.2 Individual Rationality

The players have to agree to an optimality principle in distributing the total cooperative payoff among themselves. For individual rationality to be upheld the expected payoffs an player receives under cooperation have to be no less than his expected noncooperative payoff along the cooperative state trajectory. Let $\xi(\cdot,\cdot)$ denote the imputation vector guiding the distribution of the total cooperative payoff under the agreed-upon optimality principle along the cooperative trajectory $\{x_k^*\}_{k=1}^T$. At stage k, the imputation vector according to $\xi(\cdot,\cdot)$ is $\xi(k, x_k^*) = \left[\xi^1(k, x_k^*), \xi^2(k, x_k^*), \ldots, \xi^n(k, x_k^*)\right]$, for $k\in\kappa$.

For individual rationality to be maintained throughout all the stages $k\in\kappa$, it is required that:

$$\xi^i(k, x_k^*) \geq V^i(k, x_k^*),\ \text{for}\ i\in N\ \text{and}\ k\in\kappa.$$

In particular, the above condition guaranties that the expected payoff allocated to any player under cooperation will be no less than its expected noncooperative payoff.

To satisfy group optimality, the imputation vector has to satisfy

$$W(k, x_k^*) = \sum_{j=1}^n \xi^j(k, x_k^*), \quad \text{for}\ k\in\kappa.$$

This condition guarantees the highest expected joint payoffs for the participating players.

If the optimality principle specifies that the players share the excess of the expected total cooperative payoff over the sum of expected individual noncooperative payoffs equally, then the imputation to player i becomes:

$$\xi^i(k, x_k^*) = V^i(k, x_k^*) + \frac{1}{n}\left[W(k, x_k^*) - \sum_{j=1}^{n} V^j(k, x_k^*)\right],$$

for $i \in N$ and $k \in \kappa$.

If the optimality principle specifies that the players share the expected total cooperative proportional to their expected noncooperative payoffs, then the imputation to player i becomes:

$$\xi^i(k, x_k^*) = \frac{V^i(k, x_k^*)}{\sum_{j=1}^{n} V^j(k, x_k^*)} W(k, x_k^*),$$

for $i \in N$ and $k \in \kappa$.

7.5 Subgame Consistent Solutions and Payment Mechanism

Now, we proceed to consider dynamically stable solutions in cooperative stochastic dynamic games. To guarantee dynamical stability in a stochastic dynamic cooperation scheme, the solution has to satisfy the property of subgame consistency. A cooperative solution is subgame-consistent if an extension of the solution policy to a subgame starting at a later time with any realizable state brought about by prior optimal behavior would remain optimal under the agreed upon optimality principle. In particular, subgame consistency ensures that as the game proceeds players are guided by the same optimality principle at each stage of the game, and hence do not possess incentives to deviate from the previously adopted optimal behavior. Yeung and Petrosyan (2010) developed conditions leading to subgame consistent solutions in stochastic differential games.

For subgame consistency to be satisfied, the imputation $\xi(\cdot, \cdot)$ according to the original optimality principle has to be maintained at all the T stages along the cooperative trajectory $\{x_k^*\}_{k=1}^{T}$. In other words, the imputation

$$\xi(k, x_k^*) = \left[\xi^1(k, x_k^*), \xi^2(k, x_k^*), \ldots, \xi^n(k, x_k^*)\right] \text{ at stage } k, \text{ for } k \in \kappa, \tag{5.1}$$

has to be upheld.

Crucial to the analysis is the formulation of a payment mechanism so that the imputation in (5.1) can be realized.

7.5.1 Payoff Distribution Procedure

Following the analysis of Yeung and Petrosyan (2010), we formulate a discrete-time Payoff Distribution Procedure (PDP) so that the agreed imputations(5.1) can be realized. Let $B_k^i(x_k^*)$ denote the payment that player i will receive at stage k under the cooperative agreement if $x_k^* \in X_k^*$ is realized at stage $k \in \kappa$.

The payment scheme involving $B_k^i(x_k^*)$ constitutes a PDP in the sense that if $x_k^* \in X_k^*$ is realized at stage k the imputation to player i over the stages from k to T can be expressed as:

$$\xi^i(k, x_k^*) = B_k^i(x_k^*)\left(\frac{1}{1+r}\right)^{k-1} + E_{\theta_k, \theta_{k+1}, \ldots, \theta_T}\left\{ \sum_{\zeta=k+1}^{T} B_\zeta^i(x_\zeta^*)\left(\frac{1}{1+r}\right)^{\zeta-1} + q_{T+1}^i(x_{T+1})\left(\frac{1}{1+r}\right)^{T} \right\} \tag{5.2}$$

for $i \in N$ and $k \in \kappa$.

A theorem characterizing a formula for $B_k^i(x_k^*)$, for $k \in \kappa$ and $i \in N$, which yields (5.2) is provided below.

Theorem 5.1 A payment equaling

$$B_k^i(x_k^*) = (1+r)^{k-1}\left\{ \xi^i(k, x_k^*) - E_{\theta_k}\left(\xi^i\left[k+1, f_k(x_k^*, \psi_k(x_k^*)) + G_k(x_k^*)\theta_k\right] \right) \right\}, \tag{5.3}$$

for $i \in N$,
given to player i at stage $k \in \kappa$, if $x_k^* \in X_k^*$ would lead to the realization of the imputation $\{\xi(k, x_k^*),\ \text{for } k \in \kappa\}$.

Proof Using (5.2) one can obtain

$$\xi^i(k+1, x_{k+1}^*) = B_{k+1}^i(x_{k+1}^*)\left(\frac{1}{1+r}\right)^{k} + E_{\theta_{k+1}, \theta_{k+3}, \ldots, \theta_T}\left\{ \sum_{\zeta=k+2}^{T} B_\zeta^i(x_\zeta^*)\left(\frac{1}{1+r}\right)^{\zeta-1} + q_{T+1}^i(x_{T+1})\left(\frac{1}{1+r}\right)^{T} \right\}. \tag{5.4}$$

Upon substituting (5.4) into (5.2) yields

$$\xi^i(k, x_k^*) = B_k^i(x_k^*)\left(\frac{1}{1+r}\right)^{k-1} + E_{\theta_k}\left(\xi^i\left[k+1, f_k(x_k^*, \psi_k(x_k^*)) + G_k(x_k^*)\theta_k\right] \right) \tag{5.5}$$

for $i \in N$ and $k \in \kappa$.

Hence Theorem 5.1 follows. ■

The payment scheme in Theorem 5.1 gives rise to the realization of the imputation guided by the agreed-upon optimal principle and will be used to derive subgame consistent solutions in the next subsection.

7.5.2 Subgame Consistent Solution

We denote the discrete-time cooperative game with dynamics (4.1) and payoff (4.2) by $\Gamma_c(1, x_0)$. Then we denote the game with dynamics (4.1) and payoff (4.2) which starts at stage $\upsilon \geq 1$ with initial state $x_\upsilon^* \in X_\upsilon^*$ by $\Gamma_c(\upsilon, x_\upsilon^*)$. Moreover, we let $P(1, x_0) = \{u_h^i \text{ and } B_h^i \text{ for } h \in \kappa \text{ and } i \in N, \xi(1, x_0)\}$ denote the agreed-upon optimality principle for the cooperative game $\Gamma_c(1, x_0)$. Let $\mathrm{P}(x_\upsilon^*, \upsilon) = \{u_h^i \text{ and } B_h^i \text{ for } h \in \{\upsilon, \upsilon+1, \ldots, T\} \text{ and } i \in N, \xi(\upsilon, x_\upsilon^*)\}$ denote the optimality principle of the cooperative game $\Gamma_c(\upsilon, x_\upsilon^*)$ according to the original agreement.

A theorem characterizing a subgame consistent solution for the discrete-time cooperative game $\Gamma_c(1, x_0)$ is presented below.

Theorem 5.2 For the cooperative game $\Gamma_c(1, x_0)$ with optimality principle $P(1, x_0) = \{u_h^i(x_h^*) \text{ and } B_h^i(x_h^*) \text{ for } h \in \kappa \text{ and } i \in N \text{ and } x_h^* \in X_h^*, \xi(1, x_0)\}$ in which

(i) $u_h^i(x_h^*) = \psi_h^i(x_h^*)$, for $h \in \kappa$ and $i \in N$ and $x_h^* \in X_h^*$, is the set of group optimal strategies for the game $\Gamma_c(1, x_0)$, and
(ii) $B_h^i(x_h^*) = B_h^i(x_h^*)$, for $h \in \kappa$ and $i \in N$ and $x_h^* \in X_h^*$, where

$$B_h^i(x_h^*) = (1+r)^{h-1}\left\{\xi^i(h, x_h^*) - E_{\theta_h}\left(\xi^i\left[h+1, f_h(x_h^*, \psi_h(x_h^*)) + G_h(x_h^*)\theta_h\right]\right)\right\}, \tag{5.6}$$

and $[\xi^1(h, x_h^*), \xi^2(h, x_h^*), \ldots, \xi^i(h, x_h^*)] \in P(h, x_h^*)$ is the imputation according to optimality principle $P(h, x_h^*)$;
is subgame consistent.

Proof Follow the proof of the continuous-time analog in Theorem 5.2 of Chap. 7. ■

When all players are using the cooperative strategies, the payoff that player i will directly receive at stage k given that $x_k^* \in X_k^*$ is

$$g_k^i\left[x_k^*, \psi_k^1(x_k^*), \psi_k^2(x_k^*), \cdots, \psi_k^n(x_k^*)\right].$$

However, according to the agreed upon imputation, player i will receive $B_k^i(x_k^*)$ at stage k. Therefore a side-payment

$$\varpi_k^i(x_k^*) = B_k^i(x_k^*) - g_k^i[x_k^*, \psi_k^1(x_k^*), \psi_k^2(x_k^*), \ldots, \psi_k^n(x_k^*)], \tag{5.7}$$

for $k \in \kappa$ and $i \in N$,

will be given to player i to yield the cooperative imputation $\xi^i(k, x_k^*)$.

7.6 Cooperative Resource Extraction under Uncertainty

Consider an economy endowed with a renewable resource and with two resource extractors (firms). The lease for resource extraction begins at stage 1 and ends at stage 3 for these two firms. Let u_k^i denote the rate of resource extraction of firm i at stage k, for $i \in \{1, 2\}$. Let U^i be the set of admissible extraction rates, and $x_k \in X \subset R^+$ the size of the resource stock at stage k. The extraction cost for firm $i \in \{1, 2\}$ depends on the quantity of resource extracted u_k^i, the resource stock size x_k, and cost parameters c_1 and c_2. In particular, extraction cost for firm i at stage k is specified as $c_i(u_k^i)^2/x_k$. The price of the resource is P.

The profits that firm 1 and firm 2 will obtain at stage k are respectively:

$$\left[Pu_k^1 - \frac{c_1}{x_k}(u_k^1)^2\right] \text{ and } \left[Pu_k^2 - \frac{c_2}{x_k}(u_k^2)^2\right]. \tag{6.1}$$

In stage 4, the firms will receive a salvage value equaling qx_4. The growth dynamics of the resource is governed by the stochastic difference equation:

$$x_{k+1} = x_k + a - \theta_k x_k - \sum_{j=1}^{2} u_k^j, \tag{6.2}$$

for $k \in \{1, 2, 3\}$ and $x_1 = x^0$,

where θ_k is a random variable with non-negative range $\{\theta_k^1, \theta_k^2, \theta_k^3\}$ and corresponding probabilities $\{\lambda_k^1, \lambda_k^2, \lambda_k^3\}$.

With no human harvesting, the natural growth of the resource stock is $x_{k+1} - x_k = a - \theta_k x_k$. The natural growth of the resource is while the death rate exhibits stochasticity. There exists an extraction constraint that human harvesting can at most exploit b proportion of the existing biomass, hence $u_k^1 + u_k^2 \leq bx_k$. In addition, the highest value of $\theta_k^y < (1 - b)$ for $k \in \{1, 2, 3\}$ and $y \in \{1, 2, 3\}$.

The objective of extractor $i\in\{1,2\}$ is to maximize the present value of the expected stream of future profits:

$$E_{\theta_1\theta_2\theta_3}\left\{\sum_{k=1}^{3}\left[Pu_k^i-\frac{c_i}{x_k}\left(u_k^i\right)^2\right]\left(\frac{1}{1+r}\right)^{k-1}+\left(\frac{1}{1+r}\right)^3 qx_4\right\},\ \text{for } i\in\{1,2\}, \quad (6.3)$$

subject to (6.2).

Invoking Theorem 4.2, one can characterize the noncooperative equilibrium strategies in a feedback solution for game (6.2 and 6.3). In particular, a set of strategies $\{\phi_k^i(x),\ \text{for } k\in\{1,2,3\} \text{ and } i\in\{1,2\}\}$ provides a Nash equilibrium solution to the game (6.2 and 6.3) if there exist functions $V^i(k,x)$, for $i\in\{1,2\}$ and $k\in\{1,2,3\}$, such that the following recursive relations are satisfied:

$$\begin{aligned}V^i(k,x) &= \max_{u_k^i} E_{\theta_k}\left\{\left[Pu_k^i-\frac{c_i}{x}\left(u_k^i\right)^2\right]\left(\frac{1}{1+r}\right)^{k-1}\right.\\ &\quad \left.+V^i\left[k+1,x+a-\theta_k x-u_k^i-\phi_k^j(x)\right]\right\}\\ &= \max_{u_k^i}\left\{\left[Pu_k^i-\frac{c_i}{x}\left(u_k^i\right)^2\right]\left(\frac{1}{1+r}\right)^{k-1}\right.\\ &\quad \left.+\sum_{y=1}^{3}\lambda_k^y V^i\left[k+1,x+a-\theta_k^y x-u_k^i-\phi_k^j(x)\right]\right\};\\ V^i(T+1,x) &= \left(\frac{1}{1+r}\right)^3 qx_4. \end{aligned} \quad (6.4)$$

Performing the indicated maximization in (6.4) yields:

$$\left(P-\frac{2c_i u_k^i}{x}\right)\left(\frac{1}{1+r}\right)^{k-1}-\sum_{y=1}^{3}\lambda_k^y V_{x_{k+1}}^i\left[k+1,x+a-\theta_k^y x-u_k^i-\phi_k^j(x)\right]=0, \quad (6.5)$$

for $i\in\{1,2\}$ and $k\in\{1,2,3\}$.

From (6.5), the game equilibrium strategies can be expressed as:

$$\phi_k^i(x)=\left(P-\sum_{y=1}^{3}\lambda_k^y V_{x_{k+1}}^i\left[k+1,x+a-\theta_k^y x-\sum_{\ell=1}^{2}\phi_k^\ell(x)\right](1+r)^{k-1}\right)\frac{x}{2c_i}, \quad (6.6)$$

for $i\in\{1,2\}$ and $k\in\{1,2,3\}$.

The expected game equilibrium profits of the firms can be obtained as:

Proposition 6.1 The value function indicating the expected game equilibrium profit of firm i is

$$V^i(k,x)=\left[A_k^i x+C_k^i\right],\ \text{for } i\in\{1,2\} \text{ and } k\in\{1,2,3\}, \quad (6.7)$$

where A_k^i and C_k^i, for $i \in \{1,2\}$ and $k \in \{1,2,3\}$, are constants in terms of the parameters of the game (6.2 and 6.3).

Proof See Appendix C of this Chapter. ■

Substituting the relevant derivatives of the value functions in Proposition 6.1 into the game equilibrium strategies (6.6) yields a noncooperative feedback equilibrium solution of the game (6.2 and 6.3).

Now consider the case when the extractors agree to maximize their expected joint profit and share the excess of cooperative gains over their expected noncooperative payoffs equally. To maximize their expected joint payoff, they solve the problem of maximizing

$$E_{\theta_1\theta_2\theta_3}\left\{\sum_{j=1}^{2}\sum_{k=1}^{3}\left[Pu_k^j - \frac{c_j}{x_k}\left(u_k^j\right)^2\right]\left(\frac{1}{1+r}\right)^{k-1} + 2\left(\frac{1}{1+r}\right)^3 qx_4\right\} \tag{6.8}$$

subject to (6.2).

Invoking Theorem 4.2, one can characterize the optimal controls in the stochastic dynamic programming problem (6.2) and (6.8). In particular, a set of control strategies $\{\psi_k^i(x)$, for $k \in \{1,2,3\}$ and $i \in \{1,2\}\}$ provides an optimal solution to the problem (6.2) and (6.8) if there exist functions $W(k,x): R \to R$, for $k \in \{1,2,3\}$, such that the following recursive relations are satisfied:

$$\begin{aligned}
W(k,x) &= \max_{u_k^1,\, u_k^2} E_{\theta_{k+1}}\left\{\sum_{j=1}^{2}\left[Pu_k^j - \frac{c_j}{x}\left(u_k^j\right)^2\right]\left(\frac{1}{1+r}\right)^{k-1}\right. \\
&\quad \left. + W\left[k+1, x+a-\theta_k x - \sum_{j=1}^{2} u_k^j\right]\right\} \\
&= \max_{u_k^1,\, u_k^2}\left\{\sum_{j=1}^{2}\left[Pu_k^j - \frac{c_j}{x}\left(u_k^j\right)^2\right]\left(\frac{1}{1+r}\right)^{k-1}\right. \\
&\quad \left. + \sum_{y=1}^{3}\lambda_k^y\, W\left[k+1, x+a-\theta_k^y x - \sum_{j=1}^{2} u_k^j\right]\right\}, \quad \text{for } k \in \{1,2,3\}. \\
W(T+1,x) &= 2\left(\tfrac{1}{1+r}\right)^3 qx_4.
\end{aligned} \tag{6.9}$$

Performing the indicated maximization in (6.9) yields:

$$\left(P - \frac{2c_i u_k^i}{x}\right)\left(\frac{1}{1+r}\right)^{k-1} - \sum_{y=1}^{3}\lambda_k^y\, W_{x_{k+1}}\left[k+1, x+a-\theta_k^y x - \sum_{j=1}^{2} u_k^j\right] = 0, \tag{6.10}$$

for $i \in \{1,2\}$ and $k \in \{1,2,3\}$.

In particular, the optimal cooperative strategies can be obtained from (6.10) as:

$$u_k^i\left(P - \sum_{y=1}^{3} \lambda_k^y \ W_{x_{k+1}} \left[k+1, x+a-\theta_k^y x - \sum_{j=1}^{2} u_k^j \right] (1+r)^{k-1} \right) \frac{x}{2c_i}, \tag{6.11}$$

for $i \in \{1,2\}$ and $k \in \{1,2,3\}$.

The expected joint profit under cooperation is given below.

Proposition 6.2 The value function indicating the maximized expected joint payoff is

$$W(k,x) = [A_k x + C_k], \text{ for } k \in \{1,2,3\}, \tag{6.12}$$

where A_k and C_k, for $k \in \{1,2,3\}$, are constants in terms of the parameters of the problem (6.8) and (6.2).

Proof See Appendix D of this Chapter. ■

Using (6.11) and Proposition 6.2, the optimal cooperative strategies of the extracting firms can be expressed as:

$$\psi_k^i(x) = \left[P - A_{k+1}(1+r)^{k-1} \right] \frac{x}{2c_i}, \quad \text{for } i \in \{1,2\} \text{ and } k \in \{1,2,3\}. \tag{6.13}$$

Substituting $\psi_k^i(x)$ from (6.13) into (6.2) yields the optimal cooperative state trajectory:

$$x_{k+1} = x_k + a - \theta_k x_k - \sum_{j=1}^{2} \left[P - A_{k+1}(1+r)^{k-1} \right] \frac{x_k}{2c_j}, \tag{6.14}$$

for $k \in \{1,2,3\}$ and $x_1 = x^0$.

Dynamics (6.14) is a linear stochastic difference equation readily solvable by standard techniques. Let $\{x_k^*, \text{ for } k \in \{1,2,3\}\}$ denote the solution to (6.14).

Since the extractors agree to share the excess of cooperative gains over their expected noncooperative payoffs equally, an imputation

$$\begin{aligned} \xi^i(k, x_k^*) &= V^i(k, x_k^*) + \frac{1}{2}\left[W(k, x_k^*) - \sum_{j=1}^{2} V^j(k, x_k^*) \right] \\ &= (A_k^i x_k^* + C_k^i) + \frac{1}{2}\left[(A_k x_k^* + C_k) - \sum_{j=1}^{2} \left(A_k^j x_k^* + C_k^j \right) \right], \end{aligned} \tag{6.15}$$

for $k \in \{1,2,3\}$ and $i \in \{1,2\}$ has to be maintained.

Invoking Theorem 4.1, if $x_k^* \in X$ is realized at stage k a payment equaling

$$B_k^i(x_k^*) = (1+r)^{k-1}\left[\xi^i(k, x_k^*) - E_{\theta_k}\left(\xi^i\left[k+1, x_{k+1}^{(\theta_y)*}\right]\right)\right]$$
$$= (1+r)^{k-1}\left\{(A_k^i x_k^* + C_k^i) + \frac{1}{2}\left((A_k x_k^* + C_k) - \sum_{j=1}^{2}\left(A_k^j x_k^* + C_k^j\right)\right)\right.$$
$$- \sum_{y=1}^{3}\lambda_k^y\left[\left(A_{k+1}^i x_{k+1}^{*(\theta_k^y)} + C_{k+1}^i\right)\right.$$
$$\left.\left.+\frac{1}{2}\left(\left(A_{k+1} x_{k+1}^{*(\theta_k^y)} + C_{k+1}\right) - \sum_{j=1}^{2}\left(A_{k+1}^j x_{k+1}^{*(\theta_k^y)} + C_{k+1}^j\right)\right)\right]\right\},$$

for $i \in \{1, 2\}$; (6.16)

where $x_{k+1}^{*(\theta_k^y)} = x_k^* + a - \theta_k^y x_k^* - \sum_{j=1}^{2}\left[P - A_{k+1}(1+r)^{k-1}\right]\frac{x_k^*}{2c_j}$, for $y \in \{1, 2, 3\}$,

given to firm i at stage $k \in \kappa$ would lead to the realization of the imputation (6.15).

A subgame consistent solution can be readily obtained from (6.13), (6.15) and (6.16).

7.7 A Heuristic Approach

In some game situations it may not be possible or practical to obtain all the information needed in this Chapter. Therefore a heuristic method may have to be considered to resolve the problem. To solve the problem in concern a heuristic method employs a practical methodology not guaranteed to be optimal or perfect, but sufficient for the immediate goals. Where finding an optimal solution is impossible or impractical, heuristic methods often prove to be able to speed up the process of finding a satisfactory solution. In particular, heuristic methods use strategies and information that are readily accessible (though not a 100% exact and accurate) to obtain a solution.

Consider the case of a heuristic approach to solving a subgame consistent solution in a situation where the differentiable functions

$$f_k(x_k, u_k^1, u_k^2, \cdots, u_k^n),$$
$$G_k(x_k)\theta_k, \text{ and}$$

$g_k^i[x_k, u_k^1, u_k^2, \cdots, u_k^n]$, for $i \in \{1, 2, \cdots, n\} \equiv N$ and $k \in \{1, 2, \cdots, T\} \equiv \kappa$, in (4.1 and 4.2)

are not available.

However, the players concur with the adoption of a set of cooperative strategies $\{\hat{\psi}_k^i(x_k)$, for $k\in\kappa$ and $i\in N\}$. Though the cooperative strategies may not be the set of theoretically optimal controls they are perceived to be certainly beneficial to the joint well-being of all players.

In addition, with expert knowledge and statistical techniques the expected value of cooperative payment $\sum_{j=1}^{n}\hat{g}_\tau^j[\hat{x}_\tau,\hat{\psi}_\tau^1(\hat{x}_\tau),\hat{\psi}_\tau^2(\hat{x}_\tau),\cdots,\hat{\psi}_\tau^n(\hat{x}_\tau)]$ received in each stage $\tau\in\{k,k+1,k+2,\cdots,T\}$ can be estimated with acceptable degrees of accuracy. The value $\hat{W}(k,\hat{x}_k)$ can be obtained by summing the cooperative payments $\sum_{j=1}^{n}\hat{g}_\tau^j[\hat{x}_\tau,\hat{\psi}_\tau^1(\hat{x}_\tau),\hat{\psi}_\tau^2(\hat{x}_\tau),\cdots,\hat{\psi}_\tau^n(\hat{x}_\tau)]$ expected to be received in each stage from stage k to stage T for $k\in\kappa$ along the cooperation path $\{\hat{x}_\tau\}_{\tau=k}^{T}$, that is:

$$\hat{W}(k,\hat{x}_k)=\sum_{\tau=k}^{T}\sum_{j=1}^{n}\hat{g}_\tau^j[\hat{x}_\tau,\hat{\psi}_\tau^1(\hat{x}_\tau),\hat{\psi}_\tau^2(\hat{x}_\tau),\ldots,\hat{\psi}_\tau^n(\hat{x}_\tau)]$$
$$+\sum_{j=1}^{n}q_{T+1}^j(\hat{x}_{T+1})\left(\tfrac{1}{1+r}\right)^T,\ \text{for } k\in\kappa. \tag{7.1}$$

Again, with expert knowledge and statistical techniques the expected value of non-cooperative payment $\overline{g}_\tau^i\left[\overline{x}_\tau,\overline{\phi}_\tau^1(\overline{x}_\tau),\overline{\phi}_\tau^1(\overline{x}_\tau),\cdots,\overline{\phi}_\tau^1(\overline{x}_\tau)\right]$ of player $i\in N$ received in each stage $\tau\in\{k,k+1,k+2,\cdots,T\}$ if the players revert to non-cooperation from stage k to stage T for $k\in\kappa$ can be estimated with acceptable degrees of accuracy. The value $\overline{V}^i(k,\hat{x}_k)$ can be obtained by summing of the expected payments to be received by player i in each stage from stage k to stage T for $k\in\kappa$ along the non-cooperation path $\{\overline{x}_\tau\}_{\tau=k}^{T}$ where $\overline{x}_k=\hat{x}_k$, that is

$$\overline{V}^i(k,\hat{x}_k)=\sum_{\tau=k}^{T}\overline{g}_\tau^i\left[\overline{x}_\tau,\overline{\phi}_\tau^1(\overline{x}_\tau),\overline{\phi}_\tau^1(\overline{x}_\tau),\ldots,\overline{\phi}_\tau^1(\overline{x}_\tau)\right]+q_{T+1}^i(\overline{x}_{T+1})\left(\tfrac{1}{1+r}\right)^T$$
$$\text{for } i\in N. \tag{7.2}$$

If the agreed upon optimality principle specifies that the players share the expected total cooperative proportional to their expected noncooperative payoffs, then the imputation to player i becomes:

$$\hat{\xi}^i(k,\hat{x}_k)=\frac{\overline{V}^i(k,\hat{x}_k)}{\sum_{j=1}^{n}\overline{V}^j(k,\hat{x}_k)}\hat{W}(k,\hat{x}_k), \tag{7.3}$$

for $i\in N$ and $k\in\kappa$.

Invoking Theorem 5.1 a theoretically subgame consistent payment distribution procedure can be obtained with:

$$B_k^i(x_k^*) = (1+r)^{k-1}\left\{\xi^i(k,x_k^*) - E_{\theta_k}\left(\xi^i\left[k+1, f_k(x_k^*, \psi_k(x_k^*)) + G_k(x_k^*)\theta_k\right]\right)\right\}, \tag{7.4}$$

for $i \in N$,

given to player i at stage $k \in \kappa$, if $x_k^* \in X_k^*$.

Using (7.1, 7.2, 7.3 and 7.4) a subgame consistent PDP under a heuristic scheme can be obtained with:

$$B_k^i(\hat{x}_k) = (1+r)^{k-1}\left\{ \frac{\overline{V}^i(k,\hat{x}_k)}{\sum_{j=1}^{n}\overline{V}^j(k,\hat{x}_k)}\hat{W}(k,\hat{x}_k) - \frac{\overline{V}^i(k+1,\hat{x}_{k+1})}{\sum_{j=1}^{n}\overline{V}^j(k+1,\hat{x}_{k+1})}\hat{W}(k+1,\hat{x}_{k+1}) \right\} \tag{7.5}$$

given to player $i \in N$ at stage $k \in \kappa$, along the cooperation path $\{\hat{x}_k\}_{k=1}^{T}$.

The heuristic approach allows the application of subgame consistent solution in dynamic game situations if estimates of the expected cooperative payoffs and individual non-cooperative payoffs with acceptable degrees of accuracy are available. This approach would be helpful to resolving the unstable elements in cooperative schemes for a wide range of game theoretic real-world problems.

7.8 Chapter Appendices

Appendix A. Proof of Proposition 3.1

Consider first the last stage, that is stage 3. Invoking that $V^i(3,x) = \left[A_3^i x + C_3^i\right]$ from Proposition 3.1 and $V^i(4,x) = \left(\frac{1}{1+r}\right)^3 qx$, the conditions in Eq. (3.4) become

$$V^i(3,x) = \left[A_3^i x + C_3^i\right] = \max_{u_3^i}\left\{\left[Pu_3^i - \frac{c_i}{x}\left(u_3^i\right)^2\right]\left(\frac{1}{1+r}\right)^2 + \left(\frac{1}{1+r}\right)^3 q\left[x + a - bx - u_3^i - \phi_3^j(x)\right]\right\}, \quad \text{for } i \in \{1,2\}. \tag{8.1}$$

Performing the indicated maximization in (8.1) yields the game equilibrium strategies in stage 3 as:

$$\phi_3^i(x) = \frac{\left[P - (1+r)^{-1}q\right]x}{2c_i}, \quad \text{for } i \in \{1,2\}. \tag{8.2}$$

Substituting (8.2) into (8.1) yields:

$$\begin{aligned} V^i(3,x) = \left[A_3^i x + C_3^i\right] &= \left(\tfrac{1}{1+r}\right)^2 \Bigg(\left[P - (1+r)^{-1}q\right]\frac{P}{2c_i}x \\ &- \left[P - (1+r)^{-1}q\right]^2 \frac{1}{4c_i}x \Bigg) \\ &+ q\Bigg(x + a - bx - \left[P - (1+r)^{-1}q\right]\frac{1}{2c_i}x \\ &- \left[P - (1+r)^{-1}q\right]\frac{1}{2c_j}x \Bigg)\left(\frac{1}{1+r}\right)^3 \text{ for } i,j \in \{1,2\} \text{ and } i \neq j. \end{aligned} \tag{8.3}$$

Using (8.3), we can obtain A_3^i and C_3^i, for $i \in \{1,2\}$.

Now we proceed to stage 2, the conditions in Eq. (3.4) become

$$\begin{aligned} V^i(2,x) = \left[A_2^i x + C_2^i\right] &= \max_{u_2^i} \Bigg\{ \left[P u_2^i - \frac{c_i}{x}(u_2^i)^2\right]\left(\frac{1}{1+r}\right) \\ &+ A_3^i\left[x + a - bx - u_2^i - \phi_2^j(x)\right]\Bigg\}, \text{ for } i,j \in \{1,2\} \text{ and } i \neq j. \end{aligned} \tag{8.4}$$

Performing the indicated maximization in (8.4) yields the game equilibrium strategies in stage 2 as:

$$\phi_2^i(x) = \left[P - (1+r)A_3^i\right]\frac{x}{2c_i}, \quad \text{for } i \in \{1,2\}. \tag{8.5}$$

Substituting (8.5) into (8.4) yields

$$\begin{aligned} V^i(2,x) = \left[A_2^i x + C_2^i\right] &= \Bigg\{ \left(\frac{1}{1+r}\right)\left[P - (1+r)A_3^i\right]\frac{P + (1+r)A_3^i}{4c_i} \\ &+ A_3^i(1-b) - \left[P - (1+r)A_3^i\right]\frac{A_3^i}{2c_i} - \left[P - (1+r)A_3^j\right]\frac{A_3^i}{2c_j}\Bigg\}x + aA_3^i, \\ &\text{for } i,j \in \{1,2\} \text{ and } i \neq j. \end{aligned} \tag{8.6}$$

Substituting A_3^i for $i \in \{1,2\}$ into (8.6), A_2^i and C_2^i for $i \in \{1,2\}$ are obtained in explicit terms.

Finally, we proceed to the first stage, the conditions in Eq. (3.4) become

$$\begin{aligned} V^i(1,x) = \left[A_1^i x + C_1^i\right] &= \max_{u_1^i} \Bigg\{ \left[P u_1^i - \frac{c_i}{x}(u_1^i)^2\right] \\ &+ \left(A_2^i\left[x + a - bx - u_1^i - \phi_1^j(x)\right] + C_2^i \right) \Bigg\}, \text{ for } i,j \in \{1,2\} \text{ and } i \neq j. \end{aligned} \tag{8.7}$$

Performing the indicated maximization in (8.7) yields the game equilibrium strategies in stage 1 as:

$$\phi_1^i(x) = \left[P - A_2^i\right]\frac{x}{2c_i}, \quad \text{for } i \in \{1, 2\}. \tag{8.8}$$

Substituting (8.8) into (8.7) yields:

$$\begin{aligned} &V^i(3,x) = \left[A_1^i x + C_1^i\right] = \\ &\left[\left(P - A_2^i\right)\frac{P + A_2^i}{4c_i} + A_2^i(1-b) - \left(P - A_2^i\right)\frac{A_2^i}{2c_i} - \left(P - A_2^j\right)\frac{A_2^i}{2c_j}\right]x \\ &+ aA_2^i + C_2^i, \quad \text{for } i,j \in \{1,2\} \text{ and } i \neq j. \end{aligned} \tag{8.9}$$

Substituting the explicit terms for A_2^i, A_2^j, C_2^i and C_2^j from (8.6) into (8.9), A_1^i and C_1^i for $i \in \{1, 2\}$ are obtained in explicit terms. ■

Appendix B. Proof of Proposition 3.2

Consider first the last stage, that is stage 3. Invoking that $W(3, x) = [A_3 x + C_3]$ from Proposition 3.2 and $W(4, x) = 2\left(\frac{1}{1+r}\right)^3 qx$, the conditions in Eq. (3.9) become

$$\begin{aligned} W(3,x) = [A_3 x + C_3] = \max_{u_3^1, u_3^2} \Bigg\{ &\sum_{j=1}^{2}\left[Pu_3^j - \frac{c_j}{x}\left(u_3^j\right)^2\right]\left(\frac{1}{1+r}\right)^2 \\ &+ 2\left(\frac{1}{1+r}\right)^3 q\left[x + a - bx - u_3^1 - u_3^2\right] \Bigg\}. \end{aligned} \tag{8.10}$$

Performing the indicated maximization in (8.10) yields the optimal cooperative strategies in stage 3 as:

$$\psi_3^i(x) = \frac{\left[P - (1+r)^{-1}2q\right]x}{2c_i}, \quad \text{for } i \in \{1, 2\}. \tag{8.11}$$

Substituting (8.11) into (8.10) yields:

$$\begin{aligned} W(3,x) = [A_3 x + C_3] = &\left(\frac{1}{1+r}\right)^2 \sum_{j=1}^{2}\Big\{\left[P - (1+r)^{-1}q\right]\frac{P}{2c_j}x \\ &- \left[P - (1+r)^{-1}q\right]^2\frac{1}{4c_j}x\Big\} + 2q\Big(x + a - bx - \left[P - (1+r)^{-1}q\right]\frac{1}{2c_i}x \\ &- \left[P - (1+r)^{-1}q\right]\frac{1}{2c_j}x\Big)\left(\frac{1}{1+r}\right)^3. \end{aligned} \tag{8.12}$$

Using (8.12), we obtain A_3 and C_3.

Now we proceed to stage 2, the conditions in Eq. (3.9) become

$$W(2,x) = [A_2x + C_2] = \max_{u_2^1, u_2^2} \left\{ \sum_{j=1}^{2} \left[Pu_2^j - \frac{c_j}{x}\left(u_2^j\right)^2 \right] \left(\frac{1}{1+r}\right) + A_3\left[x + a - bx - \sum_{j=1}^{2} u_2^j\right] \right\} \tag{8.13}$$

Performing the indicated maximization in (8.13) yields the optimal cooperative strategies in stage 2 as:

$$\psi_2^i(x) = [P - (1+r)A_3]\frac{x}{2c_i}, \quad \text{for } i \in \{1,2\}. \tag{8.14}$$

Substituting (8.14) into (8.13) yields:

$$W(2,x) = [A_2x + C_2] = \left[\left(\frac{1}{1+r}\right)\sum_{j=1}^{2}[P - (1+r)A_3]\frac{P + (1+r)A_3}{4c_j} + A_3(1-b) - [P - (1+r)A_3]\frac{A_3}{2c_1} - [P - (1+r)A_3]\frac{A_3^i}{2c_2} \right]x + aA_3 \Bigg). \tag{8.15}$$

Substituting A_3 into (8.15), A_2 and C_2 are obtained in explicit terms.

Finally, we proceed to the first stage, the conditions in Eq. (3.9) become

$$W(1,x) = [A_1x + C_1] = \max_{u_1^1, u_1^2} \left\{ \sum_{j=1}^{2} \left[Pu_1^j - \frac{c_j}{x}\left(u_1^j\right)^2 \right] + \left(A_2\left[x + a - bx - \sum_{j=1}^{2} u_1^j\right] + C_2 \right) \right\}. \tag{8.16}$$

Performing the indicated maximization in (8.16) yields the optimal cooperative strategies in stage 1 as:

$$\psi_1^i(x) = (P - A_2)\frac{x}{2c_i}, \quad \text{for } i \in \{1,2\}. \tag{8.17}$$

Substituting (8.17) into (8.16) yields:

$$W(1,x) = [A_1x + C_1] = \left[\sum_{j=1}^{2}(P - A_2)\frac{P + A_2}{4c_j} + A_2(1-b) - (P - A_2)\frac{A_2}{2c_1} - (P - A_2)\frac{A_2}{2c_2} \right]x + aA_2 + C_2. \tag{8.18}$$

Substituting the explicit terms for A_2 and C_2 from (8.15) into (8.18), A_1 and C_1 are obtained in explicit terms. ■

Appendix C. Proof of Proposition 6.1

Consider first the last operating stage, that is stage 3. Invoking that $V^i(3,x) = [A_3^i x + C_3^i]$ from Proposition 6.1 and $V^i(4,x) = \left(\frac{1}{1+r}\right)^3 q x_4$, the conditions in Eq. (6.4) become

$$V^i(3,x) = [A_3^i x + C_3^i] = \max_{u_3^i} \left\{ \left[P u_3^i - \frac{c_i}{x}(u_3^i)^2\right] \left(\frac{1}{1+r}\right)^2 + \sum_{y=1}^{3} \lambda_3^y \left(\frac{1}{1+r}\right)^3 q \left[x + a - \theta_3^y x - u_3^i - \phi_3^j(x)\right] \right\}, \quad \text{for } i \in \{1,2\}. \tag{8.19}$$

Performing the indicated maximization in (8.19) yields the game equilibrium strategies in stage 3 as:

$$\phi_3^i(x) = \frac{\left[P - (1+r)^{-1} q\right] x}{2c_i}, \quad \text{for } i \in \{1,2\}. \tag{8.20}$$

Substituting (8.20) into (8.19) yields:

$$V^i(3,x) = [A_3^i x + C_3^i] = \left(\frac{1}{1+r}\right)^2 \left(\left[P - (1+r)^{-1} q\right] \frac{P}{2c_i} x - \left[P - (1+r)^{-1} q\right]^2 \frac{1}{4c_i} x \right) + q \left(x + a - \sum_{y=1}^{3} \lambda_3^y \theta_3^y x - \left[P - (1+r)^{-1} q\right] \frac{1}{2c_i} x - \left[P - (1+r)^{-1} q\right] \frac{1}{2c_j} x \right) \left(\frac{1}{1+r}\right)^3$$
$$\text{for } i,j \in \{1,2\} \text{ and } i \neq j. \tag{8.21}$$

Using (8.21), we can obtain A_3^i and C_3^i, for $i \in \{1,2\}$.

Now we proceed to stage 2, the conditions in Eq. (6.4) become

$$V^i(2,x) = [A_2^i x + C_2^i] = \max_{u_2^i} \left\{ \left[P u_2^i - \frac{c_i}{x}(u_2^i)^2\right] \left(\frac{1}{1+r}\right) + \sum_{y=1}^{3} \lambda_2^y A_3^i \left[x + a - \theta_2^y x - u_2^i - \phi_2^j(x)\right] \right\},$$
$$\text{for } i,j \in \{1,2\} \text{ and } i \neq j. \tag{8.22}$$

Performing the indicated maximization in (8.22) yields the game equilibrium strategies in stage 2 as:

$$\phi_2^i(x) = \left[P - (1+r) A_3^i\right] \frac{x}{2c_i}, \quad \text{for } i \in \{1,2\}. \tag{8.23}$$

Substituting (8.23) into (8.22) yields:

$$V^i(2,x) = \left[A_2^i x + C_2^i\right] = \left\{ \left(\frac{1}{1+r}\right)\left[P-(1+r)A_3^i\right]\frac{P+(1+r)A_3^i}{4c_i} \right.$$
$$\left. +A_3^i\left(1-\sum_{y=1}^{3}\lambda_2^y\theta_2^y\right) - \left[P-(1+r)A_3^i\right]\frac{A_3^i}{2c_i} - \left[P-(1+r)A_3^j\right]\frac{A_3^i}{2c_j} \right\}x + aA_3^i,$$
for $i,j\in\{1,2\}$ and $i\neq j$. (8.24)

Substituting A_3^i for $i\in\{1,2\}$ into (8.24), A_2^i and C_2^i for $i\in\{1,2\}$ are obtained in explicit terms.

Finally, we proceed to the first stage, the conditions in Eq. (6.4) become

$$V^i(1,x) = \left[A_1^i x + C_1^i\right] = \max_{u_1^i}\left\{ \left[P u_1^i - \frac{c_i}{x}\left(u_1^i\right)^2\right] \right.$$
$$\left. +\sum_{y=1}^{3}\lambda_1^y\left(A_2^i\left[x+a-\theta_1^y x - u_1^i - \phi_1^j(x)\right] + C_2^i \right) \right\},$$
for $i,j\in\{1,2\}$ and $i\neq j$. (8.25)

Performing the indicated maximization in (8.25) yields the game equilibrium strategies in stage 1 as:

$$\phi_1^i(x) = \left[P-A_2^i\right]\frac{x}{2c_i}, \quad \text{for } i\in\{1,2\}. \tag{8.26}$$

Substituting (8.26) into (8.25) yields:

$$V^i(3,x) = \left[A_1^i x + C_1^i\right] =$$
$$\left[\left(P-A_2^i\right)\frac{P+A_2^i}{4c_i} + A_2^i\left(1-\sum_{y=1}^{3}\lambda_1^y\theta_1^y\right) - \left(P-A_2^i\right)\frac{A_2^i}{2c_i} - \left(P-A_2^j\right)\frac{A_2^i}{2c_j} \right]x$$
$$+aA_2^i + C_2^i, \quad \text{for } i,j\in\{1,2\} \text{ and } i\neq j. \tag{8.27}$$

Substituting the explicit terms for A_2^i, A_2^j, C_2^i and C_2^j from (8.24) into (8.27), A_1^i and C_1^i for $i\in\{1,2\}$ are obtained in explicit terms.

Appendix D. Proof of Proposition 6.2

Consider first the last stage, that is stage 3. Invoking that $W(3,x) = [A_3 x + C_3]$ from Proposition 6.2 and $W(4,x) = 2\left(\frac{1}{1+r}\right)^3 q x_4$, the conditions in Eq. (6.9) become

$$W(3,x) = [A_3x + C_3] = \max_{u_3^1, u_3^2} \left\{ \sum_{j=1}^{2} \left[Pu_3^j - \frac{c_j}{x}\left(u_3^j\right)^2 \right] \left(\frac{1}{1+r}\right)^2 \right.$$

$$\left. + \sum_{y=1}^{3} \lambda_3^y \, 2\left(\tfrac{1}{1+r}\right)^3 q \left[x + a - \theta_3^y x - \sum_{j=1}^{2} u_3^j \right] \right\}. \tag{8.28}$$

Performing the indicated maximization in (8.28) yields the optimal cooperative strategies in stage 3 as:

$$\psi_3^i(x) = \frac{\left[P - (1+r)^{-1} 2q\right] x}{2c_i}, \quad \text{for } i \in \{1,2\}. \tag{8.29}$$

Substituting (8.29) into (8.28) yields:

$$W(3,x) = [A_3x + C_3] = \left(\tfrac{1}{1+r}\right)^2 \sum_{j=1}^{2} \left\{ \left[P - (1+r)^{-1} q\right] \frac{P}{2c_j} x \right.$$

$$\left. - \left[P - (1+r)^{-1} q\right]^2 \frac{1}{4c_j} x \right\}$$

$$+ 2q \left(x + a - \sum_{y=1}^{3} \lambda_3^y \theta_3^y x - \left[P - (1+r)^{-1} q\right] \frac{1}{2c_i} x \right.$$

$$\left. - \left[P - (1+r)^{-1} q\right] \frac{1}{2c_j} x \right) \left(\frac{1}{1+r}\right)^3. \tag{8.30}$$

Using (8.30), we obtain A_3 and C_3.

Now we proceed to stage 2, the conditions in Eq. (6.9) become

$$W(2,x) = [A_2x + C_2] = \max_{u_2^1, u_2^2} \left\{ \sum_{j=1}^{2} \left[Pu_2^j - \frac{c_j}{x}\left(u_2^j\right)^2 \right] \left(\frac{1}{1+r}\right) \right.$$

$$\left. + \sum_{y=1}^{3} \lambda_2^y A_3 \left[x + a - \theta_2^y x - \sum_{j=1}^{2} u_2^j \right] \right\}. \tag{8.31}$$

Performing the indicated maximization in (8.31) yields the optimal cooperative strategies in stage 2 as:

$$\psi_2^i(x) = [P - (1+r)A_3] \frac{x}{2c_i}, \quad \text{for } i \in \{1,2\}. \tag{8.32}$$

Substituting (8.32) into (8.31) yields:

$$W(2,x) = [A_2x + C_2] = \left[\left(\frac{1}{1+r}\right) \sum_{j=1}^{2} [P - (1+r)A_3] \frac{P + (1+r)A_3}{4c_j} \right.$$
$$\left. + A_3\left(1 - \sum_{y=1}^{3} \lambda_2^y \theta_2^y\right) - [P - (1+r)A_3]\frac{A_3}{2c_1} - [P - (1+r)A_3]\frac{A_3^i}{2c_2} \right] x$$
$$+ aA_3. \tag{8.33}$$

Substituting A_3 into (8.33), A_2 and C_2 are obtained in explicit terms.

Finally, we proceed to the first stage, the conditions in Eq. (6.9) become

$$W(1,x) = [A_1x + C_1] = \max_{u_1^1, u_1^2} \left\{ \sum_{j=1}^{2} \left[Pu_1^j - \frac{c_j}{x}\left(u_1^j\right)^2 \right] \right.$$
$$\left. + \sum_{y=1}^{3} \lambda_1^y \left(A_2 \left[x + a - \theta_1^y x - \sum_{j=1}^{2} u_1^j \right] + C_2 \right) \right\}. \tag{8.34}$$

Performing the indicated maximization in (8.34) yields the optimal cooperative strategies in stage 1 as:

$$\psi_1^i(x) = (P - A_2)\frac{x}{2c_i}, \quad \text{for } i \in \{1, 2\}. \tag{8.35}$$

Substituting (8.35) into (8.34) yields:

$$W(1,x) = [A_1x + C_1] = \left[\sum_{j=1}^{2} (P - A_2)\frac{P + A_2}{4c_j} + A_2\left(1 - \sum_{y=1}^{3} \lambda_1^y \theta_1^y\right) \right.$$
$$\left. - (P - A_2)\frac{A_2}{2c_1} - (P - A_2)\frac{A_2}{2c_2} \right] x + aA_2 + C_2. \tag{8.36}$$

Substituting the explicit terms for A_2 and C_2 from (8.33) into (8.36), A_1 and C_1 are obtained in explicit terms.

7.9 Chapter Notes

Discrete-time dynamic games often are more suitable for real-life applications and operations research analyses. Properties of Nash equilibria in dynamic games are examined in Basar (1974, 1976). Solution algorithm for solving dynamic games can be found in Basar (1977a, b). Petrosyan and Zenkevich (1996) presented an analysis on cooperative dynamic games in discrete time framework. The SIAM Classics on Dynamic Noncoperative Game Theory by Basar and Olsder (1995) gave a comprehensive treatment of discrete-time noncooperative dynamic games. Bylka et al. (2000) analyzed oligopolistic price competition in a dynamic game model. Wie and Choi (2000) examined discrete-time traffic network. Beard and McDonald (2007) investigated water sharing agreements, and Amir and Nannerup (2006) considered resource extraction problems in a discrete-time dynamic framework. Krawczyk and Tidball (2006) considered a dynamic game of water allocation. Nie et al. (2006) considered dynamic programmingapproach to discrete time dynamic Stackelberg games. Dockner and Nishimura (1999) and Rubio and Ulph (2007) presented discrete-time dynamic game for pollution management. Dutta and Radner (2006) presented a discrete-time dynamic game to study global warming. Ehtamo and Hamalainen (1993) examined cooperative incentive equilibrium for a dynamic resource game. Yeung (2014) examined dynamically consistent collaborative environmental management with technology selection in a discrete-time dynamic game framework. Lehrer and Scarsini (2013) considered the core of dynamic cooperative games.

Discrete-time stochastic differential game analyses are less frequent than its continuous-time counterpart. Basar and Ho (1974) examined informational properties of the Nash solutions of stochastic nonzero-sum games. Elimination of informational nonuiqueness in Nash equilibrium through a stochastic formulation was first discussed in Basar (1976) and further examined in Basar (1975, 1979, 1989). Basar and Mintz (1972, 1973) and Basar (1978) developed equilibrium solution of linear-quadratic stochastic dynamic games with noisy observation. Bauso and Timmer (2009) considered robust dynamic cooperative games where at each point in time the coalitional values are unknown but bounded by a polyhedron. Smith and Zenou (2003) considered a discrete-time stochastic job searching model. Esteban-Bravo and Nogales (2008) analyzed mathematical programming for stochastic discrete-time dynamics arising in economic systems including examples in a stochastic national growth model and international growth model with uncertainty. Basar and Olsder (1995) gave a comprehensive treatment of noncooperative stochastic dynamic games. Yeung and Petrosyan (2010) provided the techniques in characterizing subgame consistent solutions for stochastic dynamic gamessubgame consistent solutions for stochastic dynamic games. Finally, a heuristic approach of obtaining subgame consistent solutions is provided in Sect. 7.7 to widen the application to a wide range of cooperative game problems in which only estimates of the expected cooperative payoffs and individual non-cooperative payoffs with acceptable degrees of accuracy are available.

7.10 Problems

(1) Consider an economy endowed with a renewable resource and with 2 resource extractors (firms). The lease for resource extraction begins at stage 1 and ends at stage 3 for these two firms. Let u_k^i denote the rate of resource extraction of firm i at stage k, for $i \in \{1, 2\}$. Let U^i be the set of admissible extraction rates, and

$x_k \in X \subset R^+$ the size of the resource stock at stage k. In particular, we have $U^i \in$ R$^+$and $u_k^1 + u_k^2 \leq x_k$. The extraction cost for firms 1 and 2 are respectively $(u_k^1)^2/x_k$ and $1.5(u_k^i)^2/x_k$.

The profits that firm 1 and firm 2 will obtain at stage k are respectively:

$$\left[10u_k^1 - \frac{4}{x_k}\left(u_k^1\right)^2\right] \text{ and } \left[4u_k^2 - \frac{2}{x_k}\left(u_k^2\right)^2\right].$$

A terminal payment of $4x_4$ will be given to each firm after stage 3.

The growth dynamics of the resource is governed by the difference equation:

$$x_{k+1} = x_k + 20 - 0.1x_k - \sum_{j=1}^{2} u_k^j, \quad \text{for } k \in \{1, 2, 3\} \text{ and } x_1 = 24.$$

Characterize the feedback Nash equilibriumsolution for the above resource economy.

(2) If the extractors agree to cooperate and maximize their joint payoff, derive the optimal cooperative strategies and the optimal resource trajectory.
(3) Consider the case when the extractors agree to share the excess of cooperative gains over their noncooperative payoffs equally. Derive a subgame consistent solution.
(4) Consider an economy endowed with a renewable resource and with two resource extractors (firms). The lease for resource extraction begins at stage 1 and ends at stage 4 for these two firms. Let u_k^i denote the rate of resource extraction of firm i at stage k, for $i \in \{1, 2\}$. Let U^i be the set of admissible extraction rates, and $x_k \in X \subset R^+$ the size of the resource stock at stage k. In particular, we have $U^i \in$ R+ and $u_k^1 + u_k^2 \leq x_k$.

The profits that firm 1 and firm 2 will obtain at stage k are respectively:

$$\left[5u_k^1 - \frac{2}{x_k}\left(u_k^1\right)^2\right] \text{ and } \left[3u_k^2 - \frac{1}{x_k}\left(u_k^2\right)^2\right].$$

A terminal payment of $3x_4$ will be given to each firm after stage 4.

The growth dynamics of the resource is governed by the stochastic difference equation:

$$x_{k+1} = x_k + 15 - 0.1x_k - \sum_{j=1}^{2} u_k^j + \theta_k x_k,$$

for $k \in \{1, 2, 3, 4\}$ and $x_1 = 55$,

where θ_k is a random variable with range $\{0, 0.1, 0.2\}$ and corresponding probabilities $\{0.3, 0.5, 0.2\}$

Characterize a Nash equilibrium solution for the above discrete-time stochastic market game.

(5) If the extractors agree to cooperate and maximize their expected joint payoff, derive the group optimal cooperative strategies.

(6) Consider the case when the extractors agree to share the excess of expected cooperative gains proportional to their expected noncooperative payoffs. Derive a subgame consistent solution.

Chapter 8
Subgame Consistent Cooperative Solution in Random Horizon Dynamic Games

In many game situations, the terminal time of the game is not known with certainty. Examples of this kind of problems include uncertainty in the renewal of lease, the terms of offices of elected authorities, contract renewal and continuation of agreements subjected to periodic negotiations. This Chapter presents subgame consistent solutions in discrete-time cooperative dynamic cooperative games with random horizon. The analysis is based on the work in Yeung and Petrosyan (2011). In Sect. 8.1, a discrete-time dynamic games with random duration is formulated and a dynamic programming technique for solving inter-temporal problems with random horizon is developed to serve as the foundation of solving the game problem. In Sect. 8.2, the noncooperative equilibrium is characterized with a set of random duration discrete-time Isaacs-Bellman equations. Dynamic cooperation under random horizon, group optimality and individual rationality are analyzed in Sect. 8.3. Subgame consistent solutions and their corresponding payment mechanism are presented in Sect. 8.4. An illustration in a resource extraction game with random duration lease is provided. The chapter appendices are given in Sect. 8.6. Chapter notes are provided in Sect. 8.7 and problems in Sect. 8.8.

8.1 Random Horizon Dynamic Games

In this section, we first formulate a class of dynamic games with random duration. Then we develop a dynamic programming technique for solving inter-temporal problems with random horizon which will serve as the foundation of solving the game problem.

8.1.1 Game Formulation

Consider the $n-$person dynamic game with $\hat{T}$ stages where $\hat{T}$ is a random variable with range $\{1, 2, \cdots, T\}$ and corresponding probabilities $\{\theta_1, \theta_2, \cdots, \theta_T\}$.

D.W.K. Yeung, L.A. Petrosyan, *Subgame Consistent Cooperation*,
Theory and Decision Library C 47, DOI 10.1007/978-981-10-1545-8_8

Conditional upon the reaching of stage τ, the probability of the game would last up to stages $\tau, \tau+1, \cdots, T$ becomes respectively

$$\frac{\theta_\tau}{\sum_{\zeta=\tau}^{T}\theta_\zeta}, \frac{\theta_{\tau+1}}{\sum_{\zeta=\tau}^{T}\theta_\zeta}, \cdots, \frac{\theta_T}{\sum_{\zeta=\tau}^{T}\theta_\zeta}.$$

The payoff of player i at stage $k \in \{1, 2, \cdots, T\}$ is $g_k^i\left[x_k, u_k^1, u_k^2, \cdots, u_k^n\right]$. When the game ends after stage $\hat{T}$, player i will receive a terminal payment $q_{\hat{T}+1}^i\left(x_{\hat{T}+1}\right)$ in stage $\hat{T}+1$.

The state space of the game is $X \in R^m$ and the state dynamics of the game is characterized by the difference equation:

$$x_{k+1} = f_k\left(x_k, u_k^1, u_k^2, \cdots, u_k^n\right), \tag{1.1}$$

for $k \in \{1, 2, \cdots, T\} \equiv \mathrm{T}$ and $x_1 = x^0$,

where $u_k^i \in R^{m_i}$ is the control vector of player i at stage k and $x_k \in X$ is the state.

The objective of player i is

$$E\left\{\sum_{k=1}^{\hat{T}} g_k^i\left[x_k, u_k^1, u_k^2, \cdots, u_k^n\right] + q_{\hat{T}+1}^i\left(x_{\hat{T}+1}\right)\right\}$$

$$= \sum_{\hat{T}=1}^{T}\theta_{\hat{T}}\left\{\sum_{k=1}^{\hat{T}} g_k^i\left[x_k, u_k^1, u_k^2, \cdots, u_k^n\right] + q_{\hat{T}+1}^i\left(x_{\hat{T}+1}\right)\right\},$$
$$\text{for } i \in \{1, 2, \cdots, n\} \equiv N. \tag{1.2}$$

To solve the game (1.1) and (1.2), we first develop a dynamic programming technique for solving a random horizon problem.

8.1.2 Dynamic Programming for Random Horizon Problem

Consider the case when $n = 1$ in the system (1.1) and (1.2). The payoff at stage $k \in \{1, 2, \cdots, T\}$ is $g_k[x_k, u_k]$. If the game ends after stage $\hat{T}$, the decision maker will receive a terminal payment $q_{\hat{T}+1}\left(x_{\hat{T}+1}\right)$ in stage $\hat{T}+1$.

The problem can be formulized as the maximization of the expected payoff:

$$E\left\{\sum_{k=1}^{\hat{T}} g_k[x_k, u_k] + q_{\hat{T}+1}\left(x_{\hat{T}+1}\right)\right\} = \sum_{\hat{T}=1}^{T}\theta_{\hat{T}}\left\{\sum_{k=1\hat{T}} g_k[x_k, u_k] + q_{\hat{T}+1}\left(x_{\hat{T}+1}\right)\right\}, \tag{1.3}$$

subject to the dynamics

$$x_{k+1} = f_k(x_k, u_k), x_1 = \overline{x}_0. \tag{1.4}$$

Now consider the case when stage τ has arrived and the state is $\overline{x}_\tau$. The problem can formulized as the maximization of the payoff:

$$E\left\{\sum_{k=\tau}^{\hat{T}} g_k[x_k, u_k] + q_{\hat{T}+1}\left(x_{\hat{T}+1}\right)\right\}$$
$$= \sum_{\hat{T}=\tau}^{T} \frac{\theta_{\hat{T}}}{\sum_{\zeta=\tau}^{T}\theta_\zeta}\left\{\sum_{k=\tau}^{\hat{T}} g_k[x_k, u_k] + q_{\hat{T}+1}\left(x_{\hat{T}+1}\right)\right\}, \tag{1.5}$$

subject to the dynamics

$$x_{k+1} = f_k(x_k, u_k), x_\tau = \overline{x}_\tau. \tag{1.6}$$

We define the value function $V(\tau, x)$ and the set of strategies $\{u_k = \psi_k(x)$, for $k \in \{\tau, \tau+1, \cdots, \mathrm{T}\}\}$ which provides an optimal solution to (1.5) and (1.6) as follows:

$$V(\tau, x) = \max_{u_\tau, u_{\tau+1}, \cdots, u_{\hat{T}}} E\left\{\sum_{k=\tau}^{\hat{T}} g_k[x_k, u_k] + q_{\hat{T}+1}\left(x_{\hat{T}+1}\right) | x_\tau = x\right\}$$
$$= \max_{u_\tau, u_{\tau+1}, \cdots, u_{\hat{T}}} \sum_{\hat{T}=\tau}^{T} \frac{\theta_{\hat{T}}}{\sum_{\zeta=\tau}^{T}\theta_\zeta}\left\{\sum_{k=\tau}^{\hat{T}} g_k[x_k, u_k] + q_{\hat{T}+1}\left(x_{\hat{T}+1}\right) | x_\tau = x\right\}$$
$$= \sum_{\hat{T}=\tau}^{T} \frac{\theta_{\hat{T}}}{\sum_{\zeta=\tau}^{T}\theta_\zeta}\left\{\sum_{k=\tau}^{\hat{T}} g_k\left[x_k^*, \psi_k(x_k)\right] + q_{\hat{T}+1}\left(x_{\hat{T}+1}^*\right) | x_\tau^* = x\right\}, \tag{1.7}$$

for $\tau \in \mathrm{T}$, where $x_{k+1}^* = f_k\left[x_k^*, \psi_k\left(x_k^*\right)\right], x_1^* = \overline{x}_0$.

A theorem characterizing an optimal solution to the random-horizon problem (1.3) and (1.4) is provided below.

Theorem 1.1 A set of strategies $\left\{u_k = \psi_k(x), \text{ for } k \in \mathrm{T}\right\}$ provides an optimal solution to the problem (1.3) and (1.4) if there exist functions $V(k, x)$, for $k \in \mathrm{T}$, such that the following recursive relations are satisfied:

$$V(T+1,x) = q_{T+1}(x),$$
$$V(T,x) = \max_{u_T}\left\{g_T[x,u_T] + V[T+1,f_T(x,u_T)]\right\},$$
$$V(\tau,x) = \max_{u_\tau}\left\{g_\tau[x,u_\tau] + \frac{\theta_\tau}{\sum_{\zeta=\tau}^{T}\theta_\zeta} q_{\tau+1}[f_\tau(x,u_\tau)]\right.$$
$$\left. + \frac{\sum_{\zeta=\tau+1}^{T}\theta_\zeta}{\sum_{\zeta=\tau}^{T}\theta_\zeta} V[\tau+1,f_\tau(x,u_\tau)]\right\}, \text{ for } \tau\in\{1,2,\cdots,T-1\}. \tag{1.8}$$

Proof By definition, the value function at stage $T+1$ is

$$V(T+1,x) = q_{T+1}(x).$$

We first consider the case when the last stage T has arrived.

The problem then becomes

$$\max_{u_T}\left\{g_T[x_T,u_T] + q_{T+1}(x_{T+1})\right\}$$
$$\text{subject to } x_{T+1} = f_T(x_T,u_T), x_T = \bar{x}_T. \tag{1.9}$$

Using $V(T+1,x) = q_{T+1}(x)$, the problem in (1.9) can be formulated as a single stage problem

$$\max_{u_T}\left\{g_T[x_T,u_T] + V(T+1,f_T(x,u_T)\right\},$$

with $x_T = x$.

Hence we have $V(T,x) = \max_{u_T}\left\{g_T[x,u_T] + V[T+1,f_T(x,u_T)]\right\}$.

Now consider the problem in stage $\tau\in\{1,2,\cdots,T-1\}$ in which one have to maximize

$$\sum_{\hat{T}=\tau}^{T}\frac{\theta_{\hat{T}}}{\sum_{\zeta=\tau}^{T}\theta_\zeta}\left\{\sum_{k=\tau}^{\hat{T}} g_k[x_k,u_k] + q_{\hat{T}+1}(x_{\hat{T}+1})\right\}$$
$$= g_\tau[x_\tau,u_\tau] + \frac{\theta_\tau}{\sum_{\zeta=\tau}^{T}\theta_\zeta} q_{\tau+1}(x_{\tau+1})$$

$$+\frac{\sum_{\hat{T}=\tau+1}^{T}\theta_{\hat{T}}}{\sum_{\zeta=\tau}^{T}\theta_{\zeta}}\left\{\sum_{k=\tau+1}^{\hat{T}} g_k[x_k,u_k]+q_{\hat{T}+1}(x_{\hat{T}+1})\right\}$$

$$=g_\tau[x_\tau,u_\tau]+\frac{\theta_\tau}{\sum_{\zeta=\tau}^{T}\theta_\zeta}q_{\tau+1}(x_{\tau+1})$$

$$+\frac{\sum_{\zeta=\tau+1}^{T}\theta_\zeta}{\sum_{\zeta=\tau}^{T}\theta_\zeta}\frac{\sum_{\hat{T}=\tau+1}^{T}\theta_{\hat{T}}}{\sum_{\zeta=\tau+1}^{T}\theta_\zeta}\left\{\sum_{k=\tau+1}^{\hat{T}} g_k[x_k,u_k]+q_{\hat{T}+1}(x_{\hat{T}+1})\right\}. \qquad (1.10)$$

Characterizing $V(\tau+1,x)$ according to (1.7), the problem (1.10) can be expressed as a single stage problem

$$\max_{u_\tau}\left\{g_\tau[x,u_\tau]+\frac{\theta_\tau}{\sum_{\zeta=\tau}^{T}\theta_\zeta}q_{\tau+1}[f_\tau(x,u_\tau)]+\frac{\sum_{\zeta=\tau+1}^{T}\theta_\zeta}{\sum_{\zeta=\tau}^{T}\theta_\zeta}V[\tau+1,f_\tau(x,u_\tau)]\right\}, \qquad (1.11)$$

with $x_\tau=x$.

Hence we have

$$V(\tau,x)=\max_{u_\tau}\left\{g_\tau[x,u_\tau]+\frac{\theta_\tau}{\sum_{\zeta=\tau}^{T}\theta_\zeta}q_{\tau+1}[f_\tau(x,u_\tau)]\right.$$

$$\left.+\frac{\sum_{\zeta=\tau+1}^{T}\theta_\zeta}{\sum_{\zeta=\tau}^{T}\theta_\zeta}V[\tau+1,f_\tau(x,u_\tau)]\right\}, \text{ for } \tau\in\{1,2,\cdots,T-2\}. \qquad (1.12)$$

and Theorem 1.1 follows. ■

Theorem 1.1 yields a set of Bellman equations for random horizon problems (1.3) and (1.4).

8.2 Random Horizon Feedback Nash Equilibrium

In this subsection, we investigate the noncooperative outcome of the random horizon discrete-time game (1.1) and (1.2). In particular, a feedback Nash equilibrium of the game can be characterized by the following theorem.

Theorem 2.1 A set of strategies $\{\phi_k^i(x), \text{ for } k \in T \text{ and } i \in N\}$ provides a feedback Nash equilibrium solution to the game (1.1) and (1.2) if there exist functions $V^i(k, x)$, for $k \in T$ and $i \in N$, such that the following recursive relations are satisfied:

$$V^i(T,x) = \max_{u_T^i}\left\{g_T^i\left[x, \phi_T^1(x), \phi_T^2(x), \cdots, \phi_T^{i-1}(x), u_T^i, \phi_T^{i+1}(x), \cdots, \phi_T^n(x)\right]\right.$$
$$\left. + q_{T+1}^i\left[\tilde{f}_T^i(x, u_T^i)\right]\right\},$$
$$V^i(\tau,x) = \max_{u_\tau^i}\left\{g_\tau^i\left[x, \phi_\tau^1(x), \phi_\tau^2(x), \cdots, \phi_\tau^{i-1}(x), u_\tau^i, \phi_\tau^{i+1}(x), \cdots, \phi_\tau^n(x)\right]\right.$$
$$+\frac{\theta_\tau}{\sum_{\zeta=\tau}^{T}\theta_\zeta} q_{\tau+1}^i\left[\tilde{f}_\tau^i(x, u_\tau^i)\right]$$
$$\left.+\frac{\sum_{\zeta=\tau+1}^{T}\theta_\zeta}{\sum_{\zeta=\tau}^{T}\theta_\zeta} V^i\left[\tau+1, f_k\left(x, \phi_k^1(x), \phi_k^2(x), \cdots, \phi_k^{i-1}(x), u_k^i, \phi_k^{i+1}(x), \cdots, \phi_k^n(x)\right)\right]\right\},$$
$$\text{for } \tau \in \{1, 2, \cdots, T-1\}. \tag{2.1}$$

Proof The conditions in (2.1) shows that the random horizon dynamic programming result in Theorem 1.1 holds for each player given other players' equilibrium strategies. Hence the conditions of a Nash (1951) equilibrium are satisfied and Theorem 2.1 follows. ■

The set of equations in (2.1) represents the discrete analogue of the Isaacs-Bellman equations under random horizon.

Substituting the set of feedback Nash equilibrium strategies $\{\phi_k^i(x), \text{ for } k \in \mathrm{T} \text{ and } i \in N\}$ into the players' payoff yields

$$V^i(\tau,x) = E\left\{\sum_{k=\tau}^{\hat{T}} g_k^i\left[x_k, \phi_k^1(x_k), \phi_k^2(x_k), \cdots, \phi_k^n(x_k)\right] + q_{\hat{T}+1}^i\left(x_{\hat{T}+1}\right)\right\}$$
$$= \sum_{\hat{T}=\tau}^{T} \frac{\theta_{\hat{T}}}{\sum_{\zeta=\tau}^{T}\theta_\zeta}\left\{\sum_{k=\tau}^{\hat{T}} g_k^i\left[x_k, \phi_k^1(x_k), \phi_k^2(x_k), \cdots, \phi_k^n(x_k)\right] + q_{\hat{T}+1}^i\left(x_{\hat{T}+1}\right)\right\}, \; i \in N,$$
$$\tag{2.2}$$

where $x_\tau = x$. The $V^i(\tau, x)$ value function gives the expected game equilibrium payoff to player i from stage τ to the end of the game.

8.3 Dynamic Cooperation under Random Horizon

Now consider the case when the players agree to cooperate and distribute the payoff among themselves according to an optimality principle. Two essential properties that a cooperative scheme has to satisfy are group optimality and individual rationality.

8.3.1 Group Optimality

Maximizing the players' expected joint payoff guarantees group optimality in a game where payoffs are transferable. To maximize their expected joint payoff the players have to solve the discrete-time dynamic programming problem of maximizing

$$E\left\{\sum_{j=1}^{n}\left[\sum_{k=1}^{\hat{T}} g_k^j\left[x_k, u_k^1, u_k^2, \cdots, u_k^n\right] + q_{\hat{T}+1}^j\left(x_{\hat{T}+1}\right)\right]\right\}$$

$$= \sum_{j=1}^{n}\sum_{\hat{T}=1}^{T}\theta_{\hat{T}}\left\{\sum_{k=1}^{\hat{T}} g_k^j\left[x_k, u_k^1, u_k^2, \cdots, u_k^n\right] + q_{\hat{T}+1}^j\left(x_{\hat{T}+1}\right)\right\} \tag{3.1}$$

subject to (1.1).

Invoking the random horizon dynamic programming method in Theorem 1.1 we can characterize an optimal solution to the problem (3.1) to (1.1) as

Corollary 3.1 A set of strategies $\left\{u_k^{i*} = \psi_k^i(x), \text{ for } k \in \kappa \text{ and } i \in N\right\}$ provides an optimal solution to the problem (3.1) to (1.1) if there exist functions $W(k, x)$, for $k \in \mathrm{T}$, such that the following recursive relations are satisfied:

$$V(T+1, x) = \sum_{j=1}^{n} q_{T+1}^j(x),$$

$$W(T, x) = \max_{u_T^1, u_T^2, \cdots, u_T^n}\left\{\sum_{j=1}^{n} g_T^j\left[x, u_T^1, u_T^2, \cdots, u_T^n\right]\right.$$

$$\left. + q_{T+1}\left[f_T\left(x, u_T, u_T^1, u_T^2, \cdots, u_T^n\right)\right]\right\},$$

$$W(\tau,x) = \max_{u_\tau^1, u_\tau^2, \cdots, u_\tau^n} \left\{ \sum_{j=1}^{n} \left[g_\tau^j\left[x, u_\tau^1, u_\tau^2, \cdots, u_\tau^n\right] + \frac{\theta_\tau}{\sum_{\zeta=\tau}^{T}\theta_\zeta} q_{\tau+1}^j\left[f_\tau\left(x, u_\tau^1, u_\tau^2, \cdots, u_\tau^n\right)\right] \right] + \frac{\sum_{\zeta=\tau+1}^{T}\theta_\zeta}{\sum_{\zeta=\tau}^{T}\theta_\zeta} W\left[\tau+1, f_\tau\left(x, u_\tau^1, u_\tau^2, \cdots, u_\tau^n\right)\right] \right\}, \text{ for } \tau \in \{1, 2, \cdots, T-1\}. \quad (3.2) \blacksquare$$

Substituting the optimal control $\{\psi_k^i(x)$, for $k \in \mathrm{T}$ and $i \in N\}$ into the state dynamics (1.1), one can obtain the dynamics of the cooperative trajectory as:

$$x_{k+1} = f_k\left[x_k, \psi_k^1(x_k), \psi_k^2(x_k), \cdots, \psi_k^n(x_k)\right], \quad (3.3)$$

for $k \in \mathrm{T}$ and $x_1 = x^0$.

We use x_k^* to denote the solution generated by (3.3).

Using the set of optimal strategies $\{\psi_k^i(x_k^*)$, for $k \in \mathrm{T}$ and $i \in N\}$ one can obtain the expected cooperative payoff as

$$W(\tau,x) = E\left\{ \sum_{j=1}^{n} \left[\sum_{k=\tau}^{\hat{T}} g_k^j\left[x_k^*, \psi_k^1(x_k^*), \psi_k^2(x_k^*), \cdots, \psi_k^n(x_k^*)\right] + q_{\hat{T}+1}^j\left(x_{\hat{T}+1}^*\right) \right] \right\}$$
$$= \sum_{j=1}^{n} \sum_{\hat{T}=\tau}^{T} \frac{\theta_{\hat{T}}}{\sum_{\zeta=\tau}^{T}\theta_\zeta} \left\{ \sum_{k=\tau}^{\hat{T}} g_k^j\left[x_k^*, \psi_k^1(x_k^*), \psi_k^2(x_k^*), \cdots, \psi_k^n(x_k^*)\right] + q_{\hat{T}+1}^j\left(x_{\hat{T}+1}^*\right) \right\}, \quad (3.4)$$

for $\tau \in \{1, 2, \cdots, t\}$.

8.3.2 *Individual Rationality*

The players then have to agree to an optimality principle in distributing the total cooperative payoff among themselves. For individual rationality to be upheld the imputation (see von Neumann and Morgenstern 1944) a player receives under cooperation have to be no less than his expected noncooperative payoff along the cooperative state trajectory.

Let $\xi(\cdot,\cdot)$ denote the imputation vector guiding the distribution of the total cooperative payoff under the agreed-upon optimality principle along the cooperative trajectory $\{x_k^*\}_{k=1}^T$. At stage τ, the imputation vector according to $\xi(\cdot,\cdot)$ is

$\xi(\tau,x_\tau^*) = [\xi^1(\tau,x_\tau^*), \xi^2(\tau,x_\tau^*), \cdots, \xi^n(\tau,x_\tau^*)]$, for $\tau \in \mathrm{T}$.

There is a variety of imputations that the players can agree upon. For examples, (i) the players may share the excess of the total cooperative payoff over the sum of individual noncooperative payoffs equally, or (ii) they may share the total cooperative payoff proportional to their noncooperative payoffs or a linear combination of (i) and (ii).

For individual rationality to be maintained throughout all the stages $\tau \in \mathrm{T}$, it is required that:

$\xi^i(\tau,x_\tau^*) \geq V^i(\tau,x_\tau^*)$, for $i \in N$ and $\tau \in \mathrm{T}$.

To satisfy group optimality, the imputation vector has to satisfy

$$W(\tau,x_\tau^*) = \sum_{j=1}^{n} \xi^j(\tau,x_\tau^*), \text{ for } \tau \in \mathrm{T}.$$

8.4 Subgame Consistent Solutions and Payment Mechanism

To guarantee dynamical stability in a dynamic cooperation scheme, the solution has to satisfy the property of subgame consistency. A cooperative solution is subgame-consistent if an extension of the solution policy to a subgame starting at a later time with a state along the optimal cooperative trajectory would remain optimal. In particular, subgame consistency ensures that as the game proceeds players are guided by the same optimality principle at each stage of the game, and hence do not possess incentives to deviate from the previously adopted optimal behavior. Therefore for subgame consistency to be satisfied, the imputation $\xi(\cdot,\cdot)$ according to the original optimality principle has to be maintained along the cooperative trajectory $\{x_k^*\}_{k=1}^T$. Let the imputation governed by the agreed upon optimality principle be

$$\xi(k,x_k^*) = [\xi^1(k,x_k^*), \xi^2(k,x_k^*), \cdots, \xi^n(k,x_k^*)] \quad \text{at stage} \quad k, \text{ for } \quad k \in \mathrm{T}. \tag{4.1}$$

Crucial to the analysis is the formulation of a payment mechanism so that the imputation in (4.1) can be realized as the game proceeds.

Following the analysis of Yeung and Petrosyan (2010), we formulate a discrete-time random-horizon Payoff Distribution Procedure (PDP) so that the agreed-upon imputations (4.1) can be realized. Let $B_k^i(x_k^*)$ denote the payment that player i will received at stage k under the cooperative agreement.

The payment scheme involving $B_k^i(x_k^*)$ constitutes a PDP in the sense that along the cooperative trajectory $\{x_k^*\}_{k=1}^{T}$ the imputation to player i over the stages from k to T can be expressed as:

$$\begin{aligned}\xi^i(\tau, x_\tau^*) &= E\left\{\sum_{k=\tau}^{\hat{T}} B_k^i(x_k^*) + q_{\hat{T}+1}^i\left(x_{\hat{T}+1}^*\right)\right\}\\ &= \sum_{\hat{T}=\tau}^{T} \frac{\theta_{\hat{T}}}{\sum_{\zeta=\tau}^{T}\theta_\zeta}\left\{\sum_{k=\tau}^{\hat{T}} B_k^i(x_k^*) + q_{\hat{T}+1}^i\left(x_{\hat{T}+1}^*\right)\right\}, i \in N \text{ and } k \in \mathrm{T}.\end{aligned} \tag{4.2}$$

If the game lasts up to stage T, then at stage $T+1$, player i will receive a terminal payment $q_{T+1}^i(x_{T+1}^*)$ and $B_{T+1}^i(x_{T+1}^*) = 0$. Hence the imputation $\xi^i(T+1, x_{T+1}^*)$ equals $q_{T+1}^i(x_{T+1}^*)$.

A theorem characterizing a formula for $B_k^i(x_k^*)$, for $k \in \mathrm{T}$ and $i \in N$, which yields (4.2) is provided below.

Theorem 4.1 A payment equaling

$$\begin{aligned}B_\tau^i(x_\tau^*) = \xi^i(\tau, x_\tau^*) &- \frac{\sum_{\zeta=\tau+1}^{T}\theta_\zeta}{\sum_{\zeta=\tau}^{T}\theta_\zeta}\xi^i\left(\tau+1, f_\tau\left[x_\tau, \psi_\tau^1(x_\tau), \psi_\tau^2(x_\tau), \cdots, \psi_\tau^n(x_\tau)\right]\right)\\ &- \frac{\theta_\tau}{\sum_{\zeta=\tau}^{T}\theta_\zeta} q_{\tau+1}^i\left(f_\tau\left[x_\tau, \psi_\tau^1(x_\tau), \psi_\tau^2(x_\tau), \cdots, \psi_\tau^n(x_\tau)\right]\right), \text{ for } i \in N,\end{aligned} \tag{4.3}$$

given to player i at stage $\tau \in \mathrm{T}$ would lead to the realization of the imputation $\xi(\tau, x_\tau^*)$ in (4.1).

Proof Using (4.2) we obtain

$$\begin{aligned}\xi^i(\tau, x_\tau^*) &= B_\tau^i(x_\tau^*) + \frac{\theta_\tau}{\sum_{\zeta=\tau}^{T}\theta_\zeta} q_{\tau+1}^i(x_{\tau+1}^*) + \frac{\sum_{\zeta=\tau+1}^{T}\theta_{\hat{T}}}{\sum_{\zeta=\tau}^{T}\theta_\zeta}\left\{\sum_{k=\tau+1}^{\hat{T}} B_k^i(x_k^*) + q_{\hat{T}+1}^i\left(x_{\hat{T}+1}^*\right)\right\}\\ &= B_\tau^i(x_\tau^*) + \frac{\theta_\tau}{\sum_{\zeta=\tau}^{T}\theta_\zeta} q_{\tau+1}^i(x_{\tau+1}^*) + \frac{\sum_{\zeta=\tau+1}^{T}\theta_\zeta}{\sum_{\zeta=\tau}^{T}\theta_\zeta}\frac{\sum_{\zeta=\tau+1}^{T}\theta_{\hat{T}}}{\sum_{\zeta=\tau+1}^{T}\theta_\zeta}\left\{\sum_{k=\tau+1}^{\hat{T}} B_k^i(x_k^*) + q_{\hat{T}+1}^i\left(x_{\hat{T}+1}^*\right)\right\}.\end{aligned} \tag{4.4}$$

Invoking $x_{\tau+1}^{*} = f_{\tau}\left[x_{\tau}, \psi_{\tau}^{1}(x_{\tau}), \psi_{\tau}^{2}(x_{\tau}), \cdots, \psi_{\tau}^{n}(x_{\tau})\right]$ and the definition of $\xi^{i}(\tau, x_{\tau}^{*})$ in (4.2), we can express (4.4) as

$$\xi^{i}\left(\tau, x_{\tau}^{*}\right) = B_{\tau}^{i}\left(x_{\tau}^{*}\right) + \frac{\theta_{\tau}}{\sum_{\zeta=\tau}^{T}\theta_{\zeta}} q_{\tau+1}^{i}\left(f_{\tau}\left[x_{\tau}, \psi_{\tau}^{1}(x_{\tau}), \psi_{\tau}^{2}(x_{\tau}), \cdots, \psi_{\tau}^{n}(x_{\tau})\right]\right)$$
$$+ \frac{\sum_{\zeta=\tau+1}^{T}\theta_{\zeta}}{\sum_{\zeta=\tau}^{T}\theta_{\zeta}} \xi^{i}\left(\tau+1, f_{\tau}\left[x_{\tau}, \psi_{\tau}^{1}(x_{\tau}), \psi_{\tau}^{2}(x_{\tau}), \cdots, \psi_{\tau}^{n}(x_{\tau})\right]\right). \tag{4.5}$$

Hence Theorem 4.1 follows. ■

Note that the payoff distribution procedure $B_{\tau}^{i}(x_{\tau}^{*})$ in (4.3) would give rise to the agreed-upon imputation

$\xi(k, x_{k}^{*}) = \left[\xi^{1}\left(k, x_{k}^{*}\right), \xi^{2}\left(k, x_{k}^{*}\right), \cdots, \xi^{n}\left(k, x_{k}^{*}\right)\right]$ at stage k, for $k \in \mathrm{T}$.

Therefore subgame consistency is satisfied,

When all players are using the cooperative strategies, the payoff that player i will directly receive at stage $k \in \mathrm{T}$ is

$$g_{k}^{i}\left[x_{k}^{*}, \psi_{k}^{1}\left(x_{k}^{*}\right), \psi_{k}^{2}\left(x_{k}^{*}\right), \cdots, \psi_{k}^{n}\left(x_{k}^{*}\right)\right].$$

However, according to the agreed upon imputation, player i is to receive $B_{k}^{i}(x_{k}^{*})$ at stage k. Therefore a side-payment

$$\varpi_{k}^{i}\left(x_{k}^{*}\right) = B_{k}^{i}\left(x_{k}^{*}\right) - g_{k}^{i}\left[x_{k}^{*}, \psi_{k}^{1}\left(x_{k}^{*}\right), \psi_{k}^{2}\left(x_{k}^{*}\right), \cdots, \psi_{k}^{n}\left(x_{k}^{*}\right)\right], \ for \ k \in \mathrm{T} \text{ and } i \in N, \tag{4.6}$$

will be given to player i.

8.5 An Illustration in Random Duration Lease

Consider the case when two firms are given the lease to extract a renewable resource from a resource pool. The lease for resource extraction has to be renewed after each stage (year) for up to a maximum of 4 stages. At stage 1, it is known that the probabilities that the lease will be 1,2,3 or 4 years long are respectively $\theta_1, \theta_2, \theta_3$ and θ_4. Conditional upon the of reaching stage $\tau > 1$, the probability of the game would last up to stages $\tau, \tau+1$, to 4 are

$$\frac{\theta_{\tau}}{\sum_{\zeta=\tau}^{4}\theta_{\zeta}}, \frac{\theta_{\tau+1}}{\sum_{\zeta=\tau}^{4}\theta_{\zeta}}, \frac{\theta_{4}}{\sum_{\zeta=\tau}^{4}\theta_{\zeta}}.$$

Let u_k^i denote the amount of resource extraction of firm i at stage k, for $i \in \{1,2\}$ and $x_k \in X \subset R^+$ the size of the resource stock at stage k. The extraction cost for firm $i \in \{1,2\}$ depends on the quantity of resource extracted u_k^i, the resource stock size x_k, and cost parameters c_i. In particular, extraction cost for firm i at stage k is $c_i(u_k^i)^2/x_k$. The price of the resource is P.

The profits that firm 1 and firm 2 will obtain at stage k are respectively:

$$\left[Pu_k^1 - \frac{c_1}{x_k}\left(u_k^1\right)^2\right] \text{ and } \left[Pu_k^2 - \frac{c_2}{x_k}\left(u_k^2\right)^2\right]. \tag{5.1}$$

If the game ends after stage τ, a terminal payment $q^i x_{\tau+1}$ will be received by firm i in stage $\tau+1$. The growth dynamics of the resource is governed by the difference equation:

$$x_{k+1} = x_k + a - bx_k - \sum_{j=1}^{2} u_k^j, x_1 = x^0 \tag{5.2}$$

for $k \in \{1,2,3,4\}$.

In particular, there is an extraction constraint $u_k^1 + u_k^2 \le (1-b)x_k + a$.The discount rate is r. The objective of extractor $i \in \{1,2\}$ is to maximize the present value of the expected stream of future profits:

$$\begin{aligned}
&E\left\{\sum_{k=1}^{\hat{T}}\left[Pu_k^i - \frac{c_i}{x_k}\left(u_k^i\right)^2\right]\left(\frac{1}{1+r}\right)^{k-1} + q^i x_{\hat{T}+1}\left(\frac{1}{1+r}\right)^{\hat{T}}\right\} \\
&= \sum_{\hat{T}=1}^{4} \theta_{\hat{T}} \left\{\sum_{k=1}^{\hat{T}}\left[Pu_k^i - \frac{c_i}{x_k}\left(u_k^i\right)^2\right]\left(\frac{1}{1+r}\right)^{k-1} + q^i x_{\hat{T}+1}\left(\frac{1}{1+r}\right)^{\hat{T}}\right\},
\end{aligned} \tag{5.3}$$

subject to (5.2).

Invoking Theorem 2.1, one can characterize the noncooperative equilibrium strategies in a feedback solution for game (5.2) and (5.3). In particular, a set of strategies $\{\phi_k^i(x)$, for $k \in \{1,2,3,4\}$ and $i \in \{1,2\}\}$ provides a Nash equilibrium solution to the game (5.2) and (5.3) if there exist functions $V^i(k,x)$, for $i \in \{1,2\}$ and $k \in \{1,2,3,4,5\}$, such that the following recursive relations are satisfied:

$$\begin{aligned}
V^i(5,x) &= q^i x\left(\frac{1}{1+r}\right)^4, \\
V^i(4,x) &= \max_{u_4^i}\left\{\left[Pu_4^i - \frac{c_i}{x}\left(u_4^i\right)^2\right]\left(\frac{1}{1+r}\right)^3\right. \\
&\quad \left. + q^i\left[x + a - bx - u_4^i - \phi_4^j(x)\right]\left(\frac{1}{1+r}\right)^4\right\},
\end{aligned}$$

$$V^i(\tau, x) = \max_{u_\tau^i} \left\{ \left[P u_\tau^i - \frac{c_i}{x} (u_\tau^i)^2 \right] \left(\frac{1}{1+r} \right)^{\tau-1} + \frac{\theta_\tau}{\sum_{\zeta=\tau}^{T} \theta_\zeta} q^i \left[x + a - bx - u_\tau^i - \phi_\tau^j(x) \right] \left(\frac{1}{1+r} \right)^{\tau} + \frac{\sum_{\zeta=\tau+1}^{T} \theta_\zeta}{\sum_{\zeta=\tau}^{T} \theta_\zeta} V^i \left[\tau + 1, x + a - bx - u_\tau^i - \phi_\tau^j(x) \right] \right\}, \text{ for } \tau \in \{1, 2, 3\}. \tag{5.4}$$

Performing the indicated maximization in (5.4) yields:

$$\begin{aligned} &\left(P - \frac{2c_i \phi_4^i(x)}{x} \right) - q^i \left(\frac{1}{1+r} \right) = 0, \\ &\left(P - \frac{2c_i \phi_\tau^i(x)}{x} \right) \left(\frac{1}{1+r} \right)^{\tau-1} - \frac{\theta_\tau}{\sum_{\zeta=\tau}^{T} \theta_\zeta} q^i \left(\frac{1}{1+r} \right)^{\tau} \\ &- \frac{\sum_{\zeta=\tau+1}^{T} \theta_\zeta}{\sum_{\zeta=\tau}^{T} \theta_\zeta} V_{x_{\tau+1}}^i \left[\tau + 1, x + a - bx - \phi_\tau^i(x) - \phi_\tau^j(x) \right] = 0, \quad \text{for} \quad \tau \in \{1, 2, 3\}. \end{aligned} \tag{5.5}$$

From (5.5), the game equilibrium strategies can be expressed as:

$$\begin{aligned} \phi_4^i(x) &= \left[P - q^i \left(\frac{1}{1+r} \right) \right] \frac{x}{2c_i}, \text{ for } i \in \{1, 2\}. \\ \phi_\tau^i(x) &= \left[P - \frac{\theta_\tau}{\sum_{\zeta=\tau}^{T} \theta_\zeta} q^i \left(\frac{1}{1+r} \right) \right. \\ &\left. - \frac{\sum_{\zeta=\tau+1}^{T} \theta_\zeta}{\sum_{\zeta=\tau}^{T} \theta_\zeta} V_{x_{\tau+1}}^i \left[\tau + 1, x + a - bx - \phi_\tau^i(x) - \phi_\tau^j(x) \right] (1+r)^{\tau-1} \right] \frac{x}{2c_i}, \end{aligned} \tag{5.6}$$

for $i \in \{1, 2\}$ and $\tau \in \{1, 2, 3\}$.

The value function $V^i(\tau, x)$ indicating the game equilibrium payoff of firm $i \in \{1, 2\}$ and $\tau \in \{1, 2, 3, 4\}$ can be obtained as:

Proposition 5.1

$$V^i(\tau,x) = \left[A^i_\tau x + C^i_\tau\right],\ \text{for}\, i\in\{1,2\}\,\text{and}\,\tau\in\{1,2,3,4\}, \tag{5.7}$$

where A^i_τ and C^i_τ, for $i\in\{1,2\}$ and $\tau\in\{1,2,3,4\}$, are constants in terms of the parameters of the game (5.2) and (5.3).

Proof See Appendix A. ■

Now consider the case when the extractors agree to maximize their expected joint profit and share the excess of cooperative gains over their expected noncooperative payoffs equally. To maximize their expected joint payoff, they solve the problem of maximizing

$$E\left\{\sum_{j=1}^{2}\left[\sum_{k=1}^{\hat{T}}\left[Pu^j_k - \frac{c_j}{x_k}\left(u^j_k\right)^2\right]\left(\tfrac{1}{1+r}\right)^{k-1} + q^j x_{\hat{T}+1}\left(\tfrac{1}{1+r}\right)^{\hat{T}}\right]\right\}$$

$$= \sum_{j=1}^{2}\sum_{\hat{T}=1}^{4}\theta_{\hat{T}}\left\{\sum_{k=1}^{\hat{T}}\left[Pu^j_k - \frac{c_j}{x_k}\left(u^j_k\right)^2\right]\left(\frac{1}{1+r}\right)^{k-1} + q^j x_{\hat{T}+1}\left(\frac{1}{1+r}\right)^{\hat{T}}\right\} \tag{5.8}$$

subject to (5.2).

Invoking Theorem 1.1, one can characterize the optimal controls in the dynamic programming problem (5.2) and (5.8). In particular, a set of control strategies $\{\psi^i_k(x),\ \text{for}\, k\in\{1,2,3,4\}\,\text{and}\, i\in\{1,2\}\}$ provides an optimal solution to the control problem (5.2) and (5.8) if there exist functions $W(k,x): R \to R$, for $k\in\{1,2,3,4\}$, such that the following recursive relations are satisfied:

$$W(5,x) = \sum_{j=1}^{n} q^j_{T+1}(x),$$

$$W(4,x) = \max_{u^1_4,u^2_4}\left\{\sum_{j=1}^{n}\left[\left[Pu^j_4 - \frac{c_j}{x}\left(u^j_4\right)^2\right]\left(\frac{1}{1+r}\right)^3 + q^j\left[x+a-bx-u^1_4-u^2_4\right]\left(\tfrac{1}{1+r}\right)^4\right]\right\},$$

$$W(\tau,x) = \max_{u^1_\tau,u^2_\tau}\left\{\sum_{j=1}^{n}\left[\left[Pu^j_\tau - \frac{c_j}{x}\left(u^j_\tau\right)^2\right]\left(\frac{1}{1+r}\right)^{\tau-1} + \frac{\theta_\tau}{\sum_{\zeta=\tau}^{T}\theta_\zeta} q^j\left[x+a-bx-u^1_\tau-u^2_\tau\right]\left(\frac{1}{1+r}\right)^{\tau}\right] + \frac{\sum_{\zeta=\tau+1}^{T}\theta_\zeta}{\sum_{\zeta=\tau}^{T}\theta_\zeta} W\left[\tau+1, x+a-bx-u^1_\tau-u^2_\tau\right]\right\},\ \text{for}\ \tau\in\{1,2,3\}. \tag{5.9}$$

Performing the indicated maximization in (5.9) yields:

$$\left(P - \frac{2c_i\psi_4^i(x)}{x}\right) - (q^1 + q^2)\left(\frac{1}{1+r}\right) = 0, \text{ for } i \in \{1,2\}$$

$$\left(P - \frac{2c_i\psi_\tau^i(x)}{x}\right)\left(\frac{1}{1+r}\right)^{\tau-1} - \frac{\theta_\tau}{\sum_{\zeta=\tau}^{T}\theta_\zeta}(q^1 + q^2)\left(\frac{1}{1+r}\right)^{\tau}$$

$$-\frac{\sum_{\zeta=\tau+1}^{T}\theta_\zeta}{\sum_{\zeta=\tau}^{T}\theta_\zeta} W_{x_{\tau+1}}\left[\tau+1, x+a-bx-\psi_\tau^1(x)-\psi_\tau^2(x)\right] = 0, \tag{5.10}$$

for $i \in \{1,2\}$ and $\tau \in \{1,2,3\}$.

In particular, the optimal cooperative strategies can be obtained from (5.10) as:

$$\psi_4^i(x) = \left[P - (q^1 + q^2)\left(\frac{1}{1+r}\right)\right]\frac{x}{2c_i}, \text{ for } i \in \{1,2\}.$$

$$\psi_\tau^i(x) = \left[P - \frac{\theta_\tau}{\sum_{\zeta=\tau}^{T}\theta_\zeta}(q^1 + q^2)\left(\frac{1}{1+r}\right)\right.$$

$$\left. -\frac{\sum_{\zeta=\tau+1}^{T}\theta_\zeta}{\sum_{\zeta=\tau}^{T}\theta_\zeta} W_{x_{\tau+1}}\left[\tau+1, x+a-bx-\psi_\tau^1(x)-\psi_\tau^2(x)\right](1+r)^{\tau-1}\right]\frac{x}{2c_i}, \tag{5.11}$$

for $i \in \{1,2\}$ and $\tau \in \{1,2,3\}$.

The joint payoff of the firms under cooperation can be obtained as:

Proposition 5.2 The value function indicating the maximized joint payoff is

$$W(\tau, x) = [A_\tau x + C_\tau], \text{ for } \tau \in \{1,2,3,4\}, \tag{5.12}$$

where A_τ and C_τ, for $\tau \in \{1,2,3,4\}$, are constants in terms of the parameters of the problem (5.8) and (5.2).

Proof See Appendix B. ∎

Using (5.11) and Proposition 5.2, the optimal cooperative strategies of the players can be expressed as:

$$\psi_4^i(x) = \left[P - (q^1 + q^2)\left(\frac{1}{1+r}\right)\right]\frac{x}{2c_i}, \text{ for } i \in \{1,2\}, \text{ and}$$

$$\psi_\tau^i(x) = \left[P - \frac{\theta_\tau}{\sum_{\zeta=\tau}^{T}\theta_\zeta}(q^1 + q^2)\left(\frac{1}{1+r}\right) - \frac{\sum_{\zeta=\tau+1}^{T}\theta_\zeta}{\sum_{\zeta=\tau}^{T}\theta_\zeta}A_{x_{\tau+1}}(1+r)^{\tau-1}\right]\frac{x}{2c_i}, \quad (5.13)$$

for $i \in \{1,2\}$ and $\tau \in \{1,2,3\}$.

Substituting $\psi_k^i(x)$ from (5.13) into (5.2) yields the optimal cooperative state trajectory:

$$x_5 = x_4 + a - bx_4 - \sum_{j=1}^{2}\left[P - (q^1 + q^2)\left(\frac{1}{1+r}\right)\right]\frac{x_4}{2c_j},$$

$$x_{\tau+1} = x_\tau + a - bx_\tau - \sum_{j=1}^{2}\left[P - \frac{\theta_\tau}{\sum_{\zeta=\tau}^{T}\theta_\zeta}(q^1 + q^2)\left(\frac{1}{1+r}\right) - \frac{\sum_{\zeta=\tau+1}^{T}\theta_\zeta}{\sum_{\zeta=\tau}^{T}\theta_\zeta}A_{x_{\tau+1}}\right.$$
$$\left.(1+r)^{\tau-1}\right]\frac{x_\tau}{2c_j}, \text{ for } \tau \in \{1,2,3\} \text{ and } x_1 = x^0. \quad (5.14)$$

Dynamics (5.14) is a linear first-order difference equation which is readily solvable by standard techniques. Let $\{x_k^*, \text{ for } k \in \{1,2,\cdots,5\}\}$ denote the solution to (5.14).

Since the extractors agree to share the excess of cooperative gains over their expected noncooperative payoffs equally, an imputation

$$\begin{aligned}\xi^i(\tau, x_\tau^*) &= V^i(\tau, x_k^*) + \frac{1}{2}\left[W(\tau, x_\tau^*) - \sum_{j=1}^{2}V^j(\tau, x_\tau^*)\right]\\ &= (A_\tau^i x_\tau^* + C_\tau^i) + \frac{1}{2}\left[(A_\tau x_\tau^* + C_\tau) - \sum_{j=1}^{2}(A_\tau^j x_\tau^* + C_\tau^j)\right],\end{aligned} \quad (5.15)$$

for $\tau \in \{1,2,3,4\}$ and $i \in \{1,2\}$ has to be maintained.

Invoking Theorem 4.1, a payment equaling

$$B_\tau^i(x_\tau^*) = \xi^i(\tau, x_\tau^*) - \frac{\sum_{\zeta=\tau+1}^{T}\theta_\zeta}{\sum_{\zeta=\tau}^{T}\theta_\zeta}\xi^i(\tau+1, x_{\tau+1}^*) - \frac{\theta_\tau}{\sum_{\zeta=\tau}^{T}\theta_\zeta}q_{\tau+1}^i(x_{\tau+1}^*)$$

$$= \left(A_\tau^i x_\tau^* + C_\tau^i\right) + \frac{1}{2}\left[\left(A_\tau x_\tau^* + C_\tau\right) - \sum_{j=1}^{2}\left(A_\tau^j x_\tau^* + C_\tau^j\right)\right]$$

$$- \frac{\sum_{\zeta=\tau+1}^{T} \theta_\zeta}{\sum_{\zeta=\tau}^{T} \theta_\zeta} \left\{ \left(A_{\tau+1}^i x_{\tau+1}^* + C_{\tau+1}^i\right) \right.$$

$$\left. + \frac{1}{2}\left[\left(A_{\tau+1} x_{\tau+1}^* + C_{\tau+1}\right) - \sum_{j=1}^{2}\left(A_{\tau+1}^j x_{\tau+1}^* + C_{\tau+1}^j\right)\right] \right\} - \frac{\theta_\tau}{\sum_{\zeta=\tau}^{T} \theta_\zeta} q^i \left(x_{\tau+1}^*\right) \left(\frac{1}{1+r}\right)^\tau, \tag{5.16}$$

given to player $i \in \{1, 2\}$ at stage $\tau \in \{1, 2, 3, 4\}$ would lead to the realization of the imputation in (5.15).

8.6 Chapter Appendices

Appendix A: Proof of Proposition 5.1 Consider first the last stage, that is stage 4. Invoking that $V^i(4, x) = \left[A_4^i x + C_4^i\right]$ from Proposition 5.1 and $V^i(5, x) = q^i x \left(\frac{1}{1+r}\right)^4$ and (5.6), the second equation in (5.4) become

$$\begin{aligned}
\left[A_4^i x + C_4^i\right] = &\left\{ P\left[P - q^i \left(\frac{1}{1+r}\right) \right] \frac{x}{2c_i} \right. \\
&\left. - \left[P - q^i \left(\frac{1}{1+r}\right) \right]^2 \frac{x}{4c_i} \right\} \left(\frac{1}{1+r}\right)^3 \\
&+ q^i \left\{ x + a - bx - \left[P - q^i \left(\frac{1}{1+r}\right) \right] \frac{x}{2c_i} \right. \\
&\left. - \left[P - q^j \left(\frac{1}{1+r}\right) \right] \frac{x}{2c_j} \right\} \left(\frac{1}{1+r}\right)^4 .
\end{aligned} \tag{6.1}$$

Note that the expressions on both the left-hand-side and right-hand-side of equation (6.1) are linear in x, one can readily obtain A_4^i and C_4^i explicitly as:

$$A_4^i = \left\{P\left[P - q^i\left(\frac{1}{1+r}\right)\right]\frac{1}{2c_i} - \left[P - q^i\left(\frac{1}{1+r}\right)\right]^2\frac{1}{4c_i}\right\}\left(\frac{1}{1+r}\right)^3$$
$$+ q^i\left\{1 - b - \left[P - q^i\left(\frac{1}{1+r}\right)\right]\frac{1}{2c_i} - \left[P - q^j\left(\frac{1}{1+r}\right)\right]\frac{1}{2c_j}\right\}\left(\frac{1}{1+r}\right)^4,$$
and
$$C_4^i = q^i a. \tag{6.2}$$

Using (5.6) and Proposition 5.1, the game equilibrium strategy of player i in stage $\tau \in \{1, 2, 3\}$ can be obtained as:

$$\phi_\tau^i(x) = \left[P - \frac{\theta_\tau}{\sum_{\zeta=\tau}^{T}\theta_\zeta} q^i\left(\frac{1}{1+r}\right) - \frac{\sum_{\zeta=\tau+1}^{T}\theta_\zeta}{\sum_{\zeta=\tau}^{T}\theta_\zeta} A_{x_{\tau+1}}^i (1+r)^{\tau-1}\right]\frac{x}{2c_i}, i \in \{1, 2\}. \tag{6.3}$$

Once again invoking Proposition 5.1 the third set of equations in (5.4) can be expressed as:

$$[A_\tau^i x + C_\tau^i] = \left\{P\left[P - \frac{\theta_\tau}{\sum_{\zeta=\tau}^{T}\theta_\zeta} q^i\left(\frac{1}{1+r}\right) - \frac{\sum_{\zeta=\tau+1}^{T}\theta_\zeta}{\sum_{\zeta=\tau}^{T}\theta_\zeta} A_{x_{\tau+1}}^i (1+r)^{\tau-1}\right]\frac{x}{2c_i}\right.$$
$$\left. - \left[P - \frac{\theta_\tau}{\sum_{\zeta=\tau}^{T}\theta_\zeta} q^i\left(\frac{1}{1+r}\right) - \frac{\sum_{\zeta=\tau+1}^{T}\theta_\zeta}{\sum_{\zeta=\tau}^{T}\theta_\zeta} A_{x_{\tau+1}}^i (1+r)^{\tau-1}\right]^2\frac{x}{4c_i}\right\}\left(\frac{1}{1+r}\right)^{\tau-1}$$
$$+ \frac{\theta_\tau}{\sum_{\zeta=\tau}^{T}\theta_\zeta} q^i\left\{x + a - bx - \left[P - \frac{\theta_\tau}{\sum_{\zeta=\tau}^{T}\theta_\zeta} q^i\left(\frac{1}{1+r}\right)\right.\right.$$
$$\left. - \frac{\sum_{\zeta=\tau+1}^{T}\theta_\zeta}{\sum_{\zeta=\tau}^{T}\theta_\zeta} A_{x_{\tau+1}}^i (1+r)^{\tau-1}\right]\frac{x}{2c_i}$$
$$\left. - \left[P - \frac{\theta_\tau}{\sum_{\zeta=\tau}^{T}\theta_\zeta} q^j\left(\frac{1}{1+r}\right) - \frac{\sum_{\zeta=\tau+1}^{T}\theta_\zeta}{\sum_{\zeta=\tau}^{T}\theta_\zeta} A_{x_{\tau+1}}^j (1+r)^{\tau-1}\right]\frac{x}{2c_j}\right\}\left(\frac{1}{1+r}\right)^{\tau}$$

$$+\frac{\sum_{\zeta=\tau+1}^{T}\theta_\zeta}{\sum_{\zeta=\tau}^{T}\theta_\zeta}\left(A^i_{\tau+1}\left\{x+a-bx-\left[P-\frac{\theta_\tau}{\sum_{\zeta=\tau}^{T}\theta_\zeta}q^i\left(\frac{1}{1+r}\right)\right.\right.\right.$$

$$\left.-\frac{\sum_{\zeta=\tau+1}^{T}\theta_\zeta}{\sum_{\zeta=\tau}^{T}\theta_\zeta}A^i_{x_{\tau+1}}(1+r)^{\tau-1}\right]\frac{x}{2c_i}$$

$$\left.\left.-\left[P-\frac{\theta_\tau}{\sum_{\zeta=\tau}^{T}\theta_\zeta}q^j\left(\frac{1}{1+r}\right)-\frac{\sum_{\zeta=\tau+1}^{T}\theta_\zeta}{\sum_{\zeta=\tau}^{T}\theta_\zeta}A^j_{x_{\tau+1}}(1+r)^{\tau-1}\right]\frac{x}{2c_j}\right\}+C^i_{\tau+1}\right). \tag{6.4}$$

Once again, both the left-hand-side and right-hand-side of equation (6.4) are linear in x, one can readily obtain A^i_τ and C^i_τ in terms of the model parameters and $A^i_{\tau+1}$ and $C^i_{\tau+1}$.

Using A^i_4 and C^i_4 in (6.2), one can obtain A^i_3 and C^i_3 explicitly. Using A^i_3 and C^i_3, one can obtain A^i_2 and C^i_2 explicitly. Using A^i_2 and C^i_2, one can obtain A^i_1 and C^i_1 explicitly. Hence Proposition 5.1 follows. Q.E.D.

Appendix B: Proof of Proposition 5.2 Consider first the last stage, that is stage 4. Invoking that $W(4,x)=[A_4x+C_4]$ from Proposition 5.2 and $W(5,x)=\sum_{j=1}^{2}q^jx\left(\frac{1}{1+r}\right)^4$ and (5.11), the second equation in (5.9) become

$$\begin{aligned}[A_4x+C_4]=\sum_{j=1}^{2}&\left\{P\left[P-(q^1+q^2)\left(\frac{1}{1+r}\right)\right]\frac{x}{2c_j}\right.\\
&\left.-\left[P-(q^1+q^2)\left(\frac{1}{1+r}\right)\right]^2\frac{x}{4c_j}\right\}\left(\frac{1}{1+r}\right)^3\\
&+\sum_{j=1}^{2}q^j\left\{x+a-bx-\left[P-(q^1+q^2)\left(\frac{1}{1+r}\right)\right]\frac{x}{2c_1}\right.\\
&\left.-\left[P-(q^1+q^2)\left(\frac{1}{1+r}\right)\right]\frac{x}{2c_2}\right\}\left(\frac{1}{1+r}\right)^4.\end{aligned} \tag{6.5}$$

Note that the expressions on both the left-hand-side and right-hand-side of equation (6.5) are linear in x, one can readily obtain A_4 and C_4 explicitly as:

$$A_4 = \sum_{j=1}^{2}\left\{P\left[P-(q^1+q^2)\left(\frac{1}{1+r}\right)\right]\frac{1}{2c_j}\right.$$
$$\left.-\left[P-(q^1+q^2)\left(\frac{1}{1+r}\right)\right]^2\frac{1}{4c_j}\right\}\left(\frac{1}{1+r}\right)^3$$
$$+\sum_{j=1}^{2}q^j\left\{1-b-\left[P-(q^1+q^2)\left(\frac{1}{1+r}\right)\right]\frac{1}{2c_1}\right.$$
$$\left.-\left[P-(q^1+q^2)\left(\frac{1}{1+r}\right)\right]\frac{1}{2c_2}\right\}\left(\frac{1}{1+r}\right)^4, \text{ and}$$
$$C_4 = (q^1+q^2)a. \tag{6.6}$$

Using (5.11) and Proposition 5.2, the cooperative strategy of player i in stage $\tau\in\{1,2,3\}$ can be obtained as:

$$\psi_\tau^i(x) = \left[P - \frac{\theta_\tau}{\sum_{\zeta=\tau}^{T}\theta_\zeta}(q^1+q^2)\left(\frac{1}{1+r}\right) - \frac{\sum_{\zeta=\tau+1}^{T}\theta_\zeta}{\sum_{\zeta=\tau}^{T}\theta_\zeta}A_{x_{\tau+1}}(1+r)^{\tau-1}\right]\frac{x}{2c_i}, \tag{6.7}$$

for $i\in\{1,2\}$ and $\tau\in\{1,2,3\}$.

Invoking Proposition 5.2 the third set of equations in (5.9) can be expressed as:

$$[A_\iota x + C_\iota] - \sum_{j=1}^{n}\left\{P\left[P - \frac{\theta_\tau}{\sum_{\zeta=\tau}^{T}\theta_\zeta}(q^1+q^2)\left(\frac{1}{1+r}\right)\right.\right.$$
$$\left. - \frac{\sum_{\zeta=\tau+1}^{T}\theta_\zeta}{\sum_{\zeta=\tau}^{T}\theta_\zeta}A_{x_{\tau+1}}(1+r)^{\tau-1}\right]\frac{x}{2c_j}$$
$$\left. - \left[P - \frac{\theta_\tau}{\sum_{\zeta=\tau}^{T}\theta_\zeta}(q^1+q^2)\left(\frac{1}{1+r}\right) - \frac{\sum_{\zeta=\tau+1}^{T}\theta_\zeta}{\sum_{\zeta=\tau}^{T}\theta_\zeta}A_{x_{\tau+1}}(1+r)^{\tau-1}\right]^2\frac{x}{4c_j}\right\}\left(\frac{1}{1+r}\right)^{\tau-1}$$
$$+\sum_{j=1}^{n}\frac{\theta_\tau}{\sum_{\zeta=\tau}^{T}\theta_\zeta}q^j\left\{x+a-bx\right.$$

$$
-\left[P-\frac{\theta_\tau}{\sum_{\zeta=\tau}^{T}\theta_\zeta}(q^1+q^2)\left(\frac{1}{1+r}\right)-\frac{\sum_{\zeta=\tau+1}^{T}\theta_\zeta}{\sum_{\zeta=\tau}^{T}\theta_\zeta}A_{x_{\tau+1}}(1+r)^{\tau-1}\right]\frac{x}{2c_1}
$$

$$
-\left[P-\frac{\theta_\tau}{\sum_{\zeta=\tau}^{T}\theta_\zeta}(q^1+q^2)\left(\frac{1}{1+r}\right)-\frac{\sum_{\zeta=\tau+1}^{T}\theta_\zeta}{\sum_{\zeta=\tau}^{T}\theta_\zeta}A_{x_{\tau+1}}(1+r)^{\tau-1}\right]\frac{x}{2c_2}\Bigg\}\left(\frac{1}{1+r}\right)^{\tau}
$$

$$
+\frac{\sum_{\zeta=\tau+1}^{T}\theta_\zeta}{\sum_{\zeta=\tau}^{T}\theta_\zeta}\Bigg(A_{\tau+1}\Bigg\{x+a-bx
$$

$$
-\left[P-\frac{\theta_\tau}{\sum_{\zeta=\tau}^{T}\theta_\zeta}(q^1+q^2)\left(\frac{1}{1+r}\right)-\frac{\sum_{\zeta=\tau+1}^{T}\theta_\zeta}{\sum_{\zeta=\tau}^{T}\theta_\zeta}A_{x_{\tau+1}}(1+r)^{\tau-1}\right]\frac{x}{2c_1}
$$

$$
-\left[P-\frac{\theta_\tau}{\sum_{\zeta=\tau}^{T}\theta_\zeta}(q^1+q^2)\left(\frac{1}{1+r}\right)-\frac{\sum_{\zeta=\tau+1}^{T}\theta_\zeta}{\sum_{\zeta=\tau}^{T}\theta_\zeta}A_{x_{\tau+1}}(1+r)^{\tau-1}\right]\frac{x}{2c_2}\Bigg\}+C_{\tau+1}\Bigg), \tag{6.8}
$$

for $\tau\in\{1,2,3\}$.

Both the left-hand-side and right-hand-side of equation (6.8) are linear in x, one can readily obtain A_τ and C_τ in terms of the model parameters and $A_{\tau+1}$ and $C_{\tau+1}$.

Using A_4 and C_4 in (6.6), one can obtain A_3 and C_3 explicitly. Using A_3 and C_3, one can obtain A_2 and C_2 explicitly. Using A_2 and C_2, one can obtain A_1 and C_1 explicitly. Hence Proposition 5.2 follows. Q.E.D.

8.7 Chapter Notes

Petrosyan and Murzov (1966) first developed the Bellman Isaacs equations under random horizon for zero-sum differential games. Petrosyan and Shevkoplyas (2003) gave the first analysis of dynamically consistent solutions for cooperative differential with random duration. Shevkoplyas (2011) considered Shapley value in cooperative differential games with random horizon in an infinite horizon framework. In this Chapter, we extend subgame consistent solutions to dynamic cooperative games with random horizon. Random horizon Bellman equation and the Isaacs-Bellman equations are derived. Subgame consistent cooperative solutions are derived for dynamic games with random horizon. Analytically tractable payoff distribution mechanisms which lead to the realization of these solutions are derived. The analysis widens the application of cooperative dynamic game theory to problems where the game horizon is random.

8.8 Problems

1. Consider the case when two firms are given the lease to extract a renewable resource from a resource pool. The lease for resource extraction has to be renewed after each stage (year) for up to a maximum of 4 stages. At stage 1, it is known that the probabilities that the lease will be 1,2,3 or 4 years long are respectively 0.1, 0.3, 0.5 and 0.2.

   Let u_k^i denote the amount of resource extraction of firm i at stage k, for $i \in \{1,2\}$ and $x_k \in X \subset R^+$ the size of the resource stock at stage k.

   The profits that firm 1 and firm 2 will obtain at stage k are respectively:

$$\left[2u_k^1 - \frac{1}{x_k}\left(u_k^1\right)^2\right] \text{ and } \left[u_k^2 - \frac{1}{x_k}\left(u_k^2\right)^2\right].$$

   If the game ends after stage $\tau \in \{1,2,3,4\}$, a terminal payment $1.5x_{\tau+1}$ will be received by firm 1 and a terminal payment $x_{\tau+1}$ will be received by firm 2 in stage $\tau+1$. The growth dynamics of the resource is governed by the difference equation:

$$x_{k+1} = x_k + 10 - 0.1x_k - \sum_{j=1}^{2} u_k^j, x_1 = x^0$$

   for $k \in \{1,2,3,4\}$.

   In particular, there is an extraction constraint $u_k^1 + u_k^2 \le 0.9x_k + 10$. The discount rate is 0.05.

   Characterize the feedback Nash equilibrium.
2. Obtain a group optimal solution that maximizes the joint expected payoff.
3. Consider the case when the extractors agree to share the excess of cooperative gains over their noncooperative payoffs equally. Derive a subgame consistent solution.

Chapter 9
Subgame Consistency in Randomly-Furcating Cooperative Stochastic Dynamic Games

This Chapter considers subgame consistent cooperative solutions in randomly furcating stochastic discrete-time dynamic games. In particular, in this type of games the evolution of the state is stochastic and future payoff structures are not known with certainty. The presence of random elements in future payoff structures and stock dynamics are prevalent in many practical game situations like regional economic cooperation, corporate joint ventures and transboundary environmental management. The analysis is based on Yeung and Petrosyan (2013a). It first develops a class of randomly furcating stochastic dynamic games in which future payoff structures of the game furcates or branches out randomly and the discrete-time game dynamics evolves stochastically. Nash equilibria of this class of games are characterized for non-cooperative outcomes and subgame-consistent solutions are derived for cooperative paradigms. A discrete-time analytically tractable payoff distribution procedure contingent upon specific random realizations of the state and payoff structure is derived. Worth mentioning is that in computer modeling and operations research discrete-time analysis often proved to be more applicable and compatible with actual data than continuous-time analysis. The Chapter is organized as follows. The game formulation and non-cooperative equilibria are given in Sect. 9.1. Group optimality and individual rationality under dynamic cooperation are discussed in Sect. 9.2. Subgame consistent solutions and payment mechanism leading to the realization of these solutions are obtained in Sect. 9.3. Section 9.4 presents an illustration in cooperative resource extraction. Extensions of the model are provided in Sect. 9.5. Chapter appendices, chapter notes and problems are presented in Sect. 9.6, Sect. 9.7, and Sect. 9.8 respectively.

9.1 Game Formulation and Non-cooperative Outcome

In this Section, we first consider the formulation of a general class of randomly-furcating stochastic dynamic games and then derive the non-cooperative outcome.

D.W.K. Yeung, L.A. Petrosyan, *Subgame Consistent Cooperation*,
Theory and Decision Library C 47, DOI 10.1007/978-981-10-1545-8_9

9.1.1 Randomly-Furcating Stochastic Dynamic Games

Consider the $T-$ stage $n-$ person nonzero-sum dynamic game with initial state x^0. The state space of the game is $X \in R^m$ and the state dynamics of the game is characterized by the stochastic difference equation:

$$x_{k+1} = f_k(x_k, u_k^1, u_k^2, \cdots, u_k^n) + \vartheta_k, \tag{1.1}$$

for $k \in \{1, 2, \cdots, T\}$ and $x_1 = x^0$,

where $u_k^i \in U^i \subset R^{m_i}$ is the control vector of player i at stage k, $x_k \in X$ is the state, and ϑ_k is a sequence of statistically independent random variables.

The payoff of player i at stage k is $g_k^i(x_k, u_k^1, u_k^2, \cdots, u_k^n; \theta_k)$ which is affected by a random variable θ_k. In particular, θ_k for $k \in \{1, 2, \cdots, T\}$ are independent discrete random variables with range $\{\theta_k^1, \theta_k^2, \cdots, \theta_k^{\eta_k}\}$ and corresponding probabilities $\{\lambda_k^1, \lambda_k^2, \cdots, \lambda_k^{\eta_k}\}$, where η_k is a positive integer for $k \in \{1, 2, \cdots, T\}$. In stage 1, it is known that θ_1 equals θ_1^1 with probability $\lambda_1^1 = 1$.

The objective that player i seeks to maximize is

$$E_{\theta_1, \theta_2, \cdots, \theta_T; \vartheta_1, \vartheta_2, \cdots, \vartheta_T}\left\{\sum_{k=1}^{T} g_k^i(x_k, u_k^1, u_k^2, \cdots, u_k^n; \theta_k) + q^i(x_{T+1})\right\},$$
$$\text{for } i \in \{1, 2, \cdots, n\} \equiv N, \tag{1.2}$$

where $E_{\theta_1, \theta_2, \cdots, \theta_T; \vartheta_1, \vartheta_2, \cdots, \vartheta_T}$ is the expectation operation with respect to the random variables $\theta_1, \theta_2, \cdots, \theta_T$ and $\vartheta_1, \vartheta_2, \cdots, \vartheta_T$, and $q^i(x_{T+1})$ is a terminal payment given at stage $T+1$. The payoffs of the players are transferable.

9.1.2 Noncooperative Equilibria

Let $u_t^{(\sigma_t)i}$ denote the strategy of player i at stage t given that the realized random variable affecting the players' payoffs is $\theta_t^{\sigma_t}$. In a stochastic dynamic game framework, a strategy space with state-dependent property has to be considered. In particular, a pre-specified class Γ^i of mapping $\phi_t^{(\sigma_t)i}(\cdot) : X \to U^i$ with the property $u_t^{(\sigma_t)i} = \phi_t^{(\sigma_t)i}(x) \in \Gamma^i$ is the strategy space of player i and each of its elements is a permissible strategy.

To solve the game, we invoke the principle of optimality in Bellman's (1957) technique of dynamic programming and begin with the subgame starting at the last operating stage, that is stage T. If $\theta_T^{\sigma_T} \in \{\theta_T^1, \theta_T^2, \cdots, \theta_T^{\eta_T}\}$ has occurred at stage T and the state $x_T = x$, the subgame becomes:

$$\max_{u_T^i} E_{\vartheta_T}\left\{g_T^i\left(x, u_T^1, u_T^2, \cdots, u_T^n; \theta_T^{\sigma_T}\right) + q^i(x_{T+1})\right\}, \text{ for } i \in N,$$

subject to

$$x_{T+1} = f_T\left(x, u_T^1, u_T^2, \cdots, u_T^n\right) + \vartheta_T. \tag{1.3}$$

A set of state-dependent strategies $\phi_T^{(\sigma_T)*}(x) = \left\{\phi_T^{(\sigma_T)1*}(x), \phi_T^{(\sigma_T)2*}(x), \cdots, \phi_T^{(\sigma_T)n*}(x)\right\}$ constitutes a Nash equilibrium solution to the subgame (1.3) if the following conditions are satisfied:

$$\begin{aligned} V^{(\sigma_T)i}(T, x) &= E_{\vartheta_T}\left\{g_T^i\left[x, \phi_T^{(\sigma_T)*}(x); \theta_T^{\sigma_T}\right] + q^i(x_{T+1})\right\} \\ &\geq E_{\vartheta_T}\left\{g_T^i\left[x, \phi_T^{(\sigma_T)\neq i*}(x); \theta_T^{\sigma_T}\right] + q^i(\tilde{x}_{T+1})\right\}, \text{ for } i \in N, \end{aligned}$$

where $x_{T+1} = f_T\left[x, \phi_T^{(\sigma_T)*}(x)\right] + \vartheta_T$,

$$\begin{aligned} &\phi_T^{(\sigma_T)\neq i*}(x) \\ &= \left[\phi_T^{(\sigma_T)1*}(x), \phi_T^{(\sigma_T)2*}(x), \cdots, \phi_T^{(\sigma_T)i-1*}(x), u_T^{(\sigma_T)i}, \phi_T^{(\sigma_T)i+1*}(x), \cdots, \phi_T^{(\sigma_T)n*}(x)\right], \end{aligned}$$

and $\tilde{x}_{T+1} = f_T\left[x, \phi_T^{(\sigma_T)\neq i*}(x)\right] + \vartheta_T$.

A characterization of the Nash equilibrium of the subgame (1.3) is provided in the following lemma.

Lemma 1.1 A set of strategies $\phi_T^{(\sigma_T)*}(x) = \left\{\phi_T^{(\sigma_T)1*}(x), \phi_T^{(\sigma_T)2*}(x), \cdots, \phi_T^{(\sigma_T)n*}(x)\right\}$ provides a Nash equilibrium solution to the subgame (1.3) if there exist functions $V^{(\sigma_T)i}(T, x)$, for $i \in N$, such that the following conditions are satisfied:

$$\begin{aligned} V^{(\sigma_T)i}(T, x) &= \max_{u_T^{(\sigma_T)i}} E_{\vartheta_T}\left\{g_T^i\left[x, \phi_T^{(\sigma_T)\neq i*}(x); \theta_T^{\sigma_T}\right]\right. \\ &\quad \left. + V^{(\sigma_{T+1})i}\left[T+1, f_T\left(x, \phi_T^{(\sigma_T)\neq i*}(x)\right) + \vartheta_T\right]\right\}, \\ V^{(\sigma_T)i}(T+1, x) &= q^i(x); \quad \text{for } i \in N. \end{aligned} \tag{1.4}$$

Proof The system of equations in (1.4) satisfies the standard stochastic dynamic programming property and the Nash property for each player $i \in N$. Hence a Nash

(1951) equilibrium of the subgame (1.3) is characterized. Details of the proof of the results can be found in Theorem 6.10 in Basar and Olsder (1999). ■

For the sake of exposition, we sidestep the issue of multiple equilibria and focus on games in which there is a unique noncooperative Nash equilibrium in each subgame. Using Lemma 1.1, one can characterize the value functions $V^{(\sigma_T)i}(T,x)$ for all $\sigma_T \in \{1,2,\cdots,\eta_T\}$ if they exist. In particular, $V^{(\sigma_T)i}(T,x)$ yields player i's expected game equilibrium payoff in the subgame starting at stage T given that $\theta_T^{\sigma_T}$ occurs and $x_T = x$.

Then we proceed to the subgame starting at stage $T-1$ when $\theta_{T-1}^{\sigma_{T-1}} \in \{\theta_{T-1}^1, \theta_{T-1}^2, \cdots, \theta_{T-1}^{\eta_{T-1}}\}$ occurs and $x_{T-1} = x$. In this subgame player $i \in N$ seeks to maximize his expected payoff

$$E_{\theta_T;\vartheta_{T-1},\vartheta_T}\left\{g_{T-1}^i\left(x, u_{T-1}^1, u_{T-1}^2, \cdots, u_{T-1}^n; \theta_{T-1}^{\sigma_{T-1}}\right)\right.$$
$$\left. + g_T^i\left(x_T, u_T^1, u_T^2, \cdots, u_T^n; \theta_T\right) + q^i(x_{T+1})\right\}$$
$$= E_{\vartheta_{T-1},\vartheta_T}\left\{g_{T-1}^i\left(x, u_{T-1}^1, u_{T-1}^2, \cdots, u_{T-1}^n; \theta_{T-1}^{\sigma_{T-1}}\right)\right.$$
$$\left. + \sum_{\sigma_T=1}^{\eta_T} \lambda_T^{\sigma_T} g_T^i\left(x_T, u_T^1, u_T^2, \cdots, u_T^n; \theta_T^{\sigma_T}\right) + q^i(x_{T+1})\right\}, \text{ for } i \in N, \tag{1.5}$$

subject to

$$x_{k+1} = f_k\left(x_k, u_k^1, u_k^2, \cdots, u_k^n\right) + \vartheta_k, \text{ for } k \in \{T-1, T\} \text{ and } x_{T-1} = x. \tag{1.6}$$

If the functions $V^{(\sigma_T)i}(T,x)$ for all $\sigma_T \in \{1,2,\cdots,\eta_T\}$ characterized in Lemma 1.1 exist, the subgame (1.5 and 1.6) can be expressed as a game in which player i seeks to maximize the expected payoff

$$E_{\vartheta_{T-1}}\left\{g_{T-1}^i\left(x, u_{T-1}^1, u_{T-1}^2, \cdots, u_{T-1}^n; \theta_{T-1}^{\sigma_{T-1}}\right)\right.$$
$$\left. + \sum_{\sigma_T=1}^{\eta_T} \lambda_T^{\sigma_T} V^{(\sigma_T)i}\left[T, f_{T-1}\left(x, u_{T-1}^1, u_{T-1}^2, \cdots, u_{T-1}^n\right) + \vartheta_{T-1}\right]\right\}, \text{ for } i \in N, \tag{1.7}$$

using his control u_{T-1}^i.

A set of strategies $\phi_{T-1}^{(\sigma_{T-1})*}(x) = \left\{\phi_{T-1}^{(\sigma_{T-1})1*}(x), \phi_{T-1}^{(\sigma_{T-1})2*}(x), \cdots, \phi_{T-1}^{(\sigma_{T-1})n*}(x)\right\}$ constitutes a Nash equilibrium solution to the subgame (1.7) if the following conditions are satisfied:

$$V^{(\sigma_{T-1})i}(T-1,x) = E_{\vartheta_{T-1}}\left\{g_{T-1}^{i}\left[x,\phi_{T-1}^{(\sigma_{T-1})*}(x);\theta_{T-1}^{\sigma_{T-1}}\right]\right.$$
$$\left.+\sum_{\sigma_T=1}^{\eta_T}\lambda_T^{\sigma_T}V^{(\sigma_T)i}\left[T,f_{T-1}\left[x,\phi_{T-1}^{(\sigma_{T-1})*}(x)\right]+\vartheta_{T-1}\right]\right\}$$
$$\geq E_{\vartheta_{T-1}}\left\{g_{T-1}^{i}\left[x,\phi_{T-1}^{(\sigma_{T-1})\neq i*}(x);\theta_{T-1}^{\sigma_{T-1}}\right]\right.$$
$$\left.+\sum_{\sigma_T=1}^{\eta_T}\lambda_T^{\sigma_T}V^{(\sigma_T)i}\left[T,f_{T-1}\left(x,\phi_{T-1}^{(\sigma_{T-1})\neq i*}(x)\right)+\vartheta_{T-1}\right]\right\},\ \text{for } i\in N, \tag{1.8}$$

where

$$\phi_{T-1}^{(\sigma_{T-1})\neq i*}(x) =$$
$$\left[\phi_{T-1}^{(\sigma_{T-1})1*}(x),\phi_{T-1}^{(\sigma_{T-1})2*}(x),\cdots,\phi_{T-1}^{(\sigma_{T-1})i-1*}(x),u_{T-1}^{(\sigma_{T-1})i},\phi_{T-1}^{(\sigma_{T-1})i+1*}(x),\cdots,\phi_{T-1}^{(\sigma_{T-1})n*}(x)\right].$$

A characterization of the Nash equilibrium of the subgame (1.7) is provided in the following lemma.

Lemma 1.2 A set of strategies $\phi_{T-1}^{(\sigma_{T-1})*}(x) = \left\{\phi_{T-1}^{(\sigma_{T-1})1*}(x),\phi_{T-1}^{(\sigma_{T-1})2*}(x),\cdots,\right.$ $\left.\phi_{T-1}^{(\sigma_{T-1})n*}(x)\right\}$ provides a Nash equilibrium solution to the subgame (1.7) if there exist functions $V^{(\sigma_T)i}(T,x_T)$ for $i\in N$ and $\sigma_T=\{1,2,\cdots,\eta_T\}$ characterized in Lemma 1.1, and functions $V^{(\sigma_{T-1})i}(T-1,x)$, for $i\in N$, such that the following conditions are satisfied:

$$V^{(\sigma_{T-1})i}(T-1,x) = \max_{u_{T-1}^{(\sigma_{T-1})i}} E_{\vartheta_{T-1}}\left\{g_{T-1}^{i}\left[x,\phi_{T-1}^{(\sigma_{T-1})\neq i*}(x);\theta_{T-1}^{\sigma_{T-1}}\right]\right.$$
$$\left.+\sum_{\sigma_T=1}^{\eta_T}\lambda_T^{\sigma_T}V^{(\sigma_T)i}\left[T,f_{T-1}\left(x,\phi_{T-1}^{(\sigma_{T-1})\neq i*}\right)+\vartheta_{T-1}\right]\right\},\ \text{for } i\in N. \tag{1.9}$$

Proof The conditions in Lemma 1.1 and the system of equations in (1.9) satisfies the standard discrete-time stochastic dynamic programming property and the Nash property for each player $i\in N$. Hence a Nash equilibrium of the subgame (1.7) is characterized. ■

In particular, $V^{(\sigma_{T-1})i}(T-1,x)$, if it exists, yields player i's expected game equilibrium payoff in the subgame starting at stage $T-1$ given that $\theta_{T-1}^{\sigma_{T-1}}$ occurs and $x_{T-1}=x$.

Consider the subgame starting at stage $t\in\{T-2,T-3,\cdots,1\}$ when $\theta_t^{\sigma_t}\in\{\theta_t^1,\theta_t^2,\cdots,\theta_t^{\eta_t}\}$ occurs and $x_t=x$, in which player $i\in N$ maximizes his expected payoff

$$E_{\theta_{t+1};\vartheta_t,\vartheta_{t+1},\cdots,\vartheta_T}\left\{g_t^i\left(x,u_t^1,u_t^2,\cdots,u_t^n;\theta_t^{\sigma_t}\right)\right.$$
$$\left.+\sum_{\zeta=t+1}^{T}g_\zeta^i\left(x_\zeta,u_\zeta^1,u_\zeta^2,\cdots,u_\zeta^n;\theta_\zeta\right)+q^i(x_{T+1})\right\},\ \text{for}\ i\in N, \tag{1.10}$$

subject to

$$x_{k+1}=f_k\left(x_k,u_k^1,u_k^2,\cdots,u_k^n\right)+\vartheta_k,\ \text{for}\ k\in\{t,t+1,\cdots,T\}\ \text{and}\ x_t=x. \tag{1.11}$$

Following the above analysis, the subgame (1.10 and 1.11) can be expressed as a game in which player $i\in N$ maximizes his expected payoff

$$E_{\vartheta_t}\left\{g_t^i\left(x,u_t^1,u_t^2,\cdots,u_t^n;\theta_t^{\sigma_t}\right)\right.$$
$$\left.+\sum_{\sigma_{t+1}=1}^{\eta_{t+1}}\lambda_{t+1}^{\sigma_{t+1}}V^{(\sigma_{t+1})i}\left[t+1,f_t\left(x,u_t^1,u_t^2,\cdots,u_t^n\right)+\vartheta_t\right]\right\},\ \text{for}\ i\in N, \tag{1.12}$$

with his control u_t^i,

where $V^{(\sigma_{t+1})i}\left[t+1,f_t\left(x,u_t^1,u_t^2,\cdots,u_t^n\right)+\vartheta_t\right]$ is player i's expected game equilibrium payoff in the subgame starting at stage $t+1$ given that $\theta_{t+1}^{\sigma_{t+1}}$ occurs and $x_{t+1}=f_t\left(x,u_t^1,u_t^2,\cdots,u_t^n\right)+\vartheta_t$.

A set of strategies $\phi_t^{(\sigma_t)*}(x)=\left\{\phi_t^{(\sigma_t)1*}(x),\phi_t^{(\sigma_t)2*}(x),\cdots,\phi_t^{(\sigma_t)n*}(x)\right\}$, constitutes a Nash equilibrium solution to the subgame (1.12) if the following conditions are satisfied:

$$V^{(\sigma_t)i}(t,x)=E_{\vartheta_t}\left\{g_t^i\left[x,\phi_t^{(\sigma_t)*}(x);\theta_t^{\sigma_t}\right]\right.$$
$$\left.+\sum_{\sigma_{t+1}=1}^{\eta_{t+1}}\lambda_{t+1}^{\sigma_{t+1}}V^{(\sigma_{t+1})i}\left[t+1,f_t\left[x,\phi_t^{(\sigma_t)*}(x)\right]+\vartheta_t\right]\right\}$$
$$\geq E_{\vartheta_t}\left\{g_t^i\left[x,\phi_t^{(\sigma_t)\neq i*}(x);\theta_t^{\sigma_t}\right]+\sum_{\sigma_{t+1}=1}^{\eta_{t+1}}\lambda_{t+1}^{\sigma_{t+1}}V^{(\sigma_{t+1})i}\left[t+1,f_t\left(x,\phi_t^{(\sigma_t)\neq i*}(x)\right)+\vartheta_t\right]\right\}$$

where

$$\phi_t^{(\sigma_t)\neq i*}(x)=\left\{\phi_t^{(\sigma_t)1*}(x),\phi_t^{(\sigma_t)2*}(x),\cdots,\phi_t^{(\sigma_t)i-1*}(x),u_t^{(\sigma_t)i},\phi_t^{(\sigma_t)i+1*}(x),\cdots,\phi_t^{(\sigma_t)n*}(x)\right\}.$$

A Nash equilibrium solution for the game (1.1 and 1.2) can be characterized by the following theorem.

Theorem 1.1 A set of strategies $\phi_t^{(\sigma_t)*}(x) = \left\{\phi_t^{(\sigma_t)1*}(x), \phi_t^{(\sigma_t)2*}(x), \cdots, \phi_t^{(\sigma_t)n*}(x)\right\}$, for $\sigma_t \in \{1, 2, \cdots, \eta_t\}$ and $t \in \{1, 2, \cdots, T\}$, constitutes a Nash equilibrium solution to the game (1.1 and 1.2) if there exist functions $V^{(\sigma_t)i}(t, x)$, for $\sigma_t \in \{1, 2, \cdots, \eta_t\}$, $t \in \{1, 2, \cdots, T\}$ and $i \in N$, such that the following recursive relations are satisfied:

$$
\begin{aligned}
V^{(\sigma_T)i}(T+1, x) &= q^i(x), \\
V^{(\sigma_T)i}(T, x) &= \max_{u_T^{(\sigma_T)i}} E_{\vartheta_T}\left\{ g_T^i\left[x, \phi_T^{(\sigma_T)\neq i*}(x); \theta_T^{\sigma_T}\right]\right. \\
&\quad \left. + V^{(\sigma_{T+1})i}\left[T+1, f_T\left(x, \phi_T^{(\sigma_T)\neq i*}(x)\right) + \vartheta_T\right]\right\}, \\
V^{(\sigma_t)i}(t, x) &= \max_{u_t^{(\sigma_t)i}} E_{\vartheta_t}\left\{ g_t^i\left[x, \phi_t^{(\sigma_t)\neq i*}(x); \theta_t^{\sigma_t}\right]\right. \\
&\quad \left. + \sum_{\sigma_{t+1}=1}^{\eta_{t+1}} \lambda_{t+1}^{\sigma_{t+1}} V^{(\sigma_{t+1})i}\left[t+1, f_t\left(x, \phi_t^{(\sigma_t)\neq i*}(x)\right) + \vartheta_t\right]\right\}; \\
&\text{for } \sigma_t \in \{1, 2, \cdots, \eta_t\},\ t \in \{1, 2, \cdots, T-1\} \text{ and } i \in N.
\end{aligned}
\tag{1.13}
$$

Proof The results in (1.13) characterizing the game equilibrium in stage T and stage $T-1$ are proved in Lemma 1.1 and 1.2. Invoking the subgame in stage $t \in \{1, 2, \cdots, T-2\}$ as expressed in (1.12), the results in (1.13) satisfy the optimality conditions in stochastic dynamic programming and the Nash equilibrium property for each player in each of these subgames. Therefore, a feedback Nash equilibrium of the game (1.1 and 1.2) is characterized. ■

Theorem 1.1 is the discrete-time analog of the Nash equilibrium in the continuous-time randomly furcating stochastic differential games in Chap. 4.

9.2 Dynamic Cooperation

Now consider the case when the players agree to cooperate and distribute the joint payoff among themselves according to an optimality principle. As pointed out before two essential properties that a cooperative scheme has to satisfy are group optimality and individual rationality.

9.2.1 Group Optimality

In this subsection, we consider the issue of ensuring group optimality in a cooperative scheme. To achieve group optimality by maximizing their expected joint payoff the players have to solve the discrete-time stochastic dynamic programming problem of maximizing

$$E_{\theta_1,\theta_2,\cdots,\theta_T;\vartheta_1,\vartheta_2,\cdots,\vartheta_T}\left\{\sum_{j=1}^{n}\sum_{k=1}^{T}\left[g_k^j\left(x_k,u_k^1,u_k^2,\cdots,u_k^n;\theta_k\right)\right]+\sum_{j=1}^{n}q^j(x_{T+1})\right\} \tag{2.1}$$

subject to (1.1).

The stochastic dynamic programming problem (1.1, 1.2, 1.3, 1.4, 1.5, 1.6, 1.7, 1.8, 1.9, 1.10, 1.11, 1.12, 1.13 and 2.1) can be regarded as a single-player case of the game problem (1.1 and 1.2) with $n = 1$ and the payoff being the sum of the all the players' payoffs. In a stochastic dynamic framework, again strategy space with state-dependent property has to be considered. In particular, a pre-specified class $\hat{\Gamma}^i$ of mapping $\psi_i^{(\sigma_t)^i}(\cdot):X\to U^i$ with the property $u_t^{(\sigma_t)i}=\psi_t^{(\sigma_t)i}(x)\in\hat{\Gamma}^i$, for $\sigma_t\in\{1,2,\cdots,\eta_t\}$ and $t\in\{1,2,\cdots,T\}$, is the strategy space of player i and each of its elements is a permissible strategy.

To solve the dynamic programming problem (1.1) and (2.1), we first consider the problem starting at stage T. If $\theta_T^{\sigma_T}\in\{\theta_T^1,\theta_T^2,\cdots,\theta_T^{\eta_T}\}$ has occurred at stage T and the state $x_T = x$, the problem becomes:

$$\max_{u_T^1,u_T^2,\cdots,u_T^n} E_{\vartheta_T}\left\{\sum_{j=1}^{n} g_T^j\left(x,u_T^1,u_T^2,\cdots,u_T^n;\theta_T^{\sigma_T}\right)+\sum_{j=1}^{n}q^j(x_{T+1})\right\} \tag{2.2}$$

subject to

$$x_{T+1}=f_T\left(x,u_T^1,u_T^2,\cdots,u_T^n\right)+\vartheta_T. \tag{2.3}$$

An optimal solution to the stochastic control problem (2.2 and 2.3) is characterized by the following lemma.

Lemma 2.1 A set of controls $u_T^{(\sigma_T)*}=\psi_T^{(\sigma_T)*}(x)=\left\{\psi_T^{(\sigma_T)1*}(x),\psi_T^{(\sigma_T)2*}(x),\cdots\cdots,\psi_T^{(\sigma_T)n*}(x)\right\}$ provides an optimal solution to the stochastic control problem (2.2 and 2.3) if there exist functions $W^{(\sigma_{T+1})}(T,x)$, for $i\in N$, such that the following conditions are satisfied:

$$\begin{aligned}W^{(\sigma_T)}(T,x)=&\max_{u_T^{(\sigma_T)1},u_T^{(\sigma_T)2},\cdots,u_T^{(\sigma_T)n}} E_{\vartheta_T}\left\{\sum_{j=1}^{n}g_T^j\left[x,u_T^{(\sigma_T)1},u_T^{(\sigma_T)2},\cdots,u_T^{(\sigma_T)n};\theta_T^{\sigma_T}\right]\right.\\&\left.+W^{(\sigma_{T+1})}\left[T+1,f_T\left(x,u_T^{(\sigma_T)1},u_T^{(\sigma_T)2},\cdots,u_T^{(\sigma_T)n}\right)+\vartheta_T\right]\right\},\\W^{(\sigma_T)}(T+1,x)+&\sum_{j=1}^{n}q^j(x).\end{aligned} \tag{2.4}$$

Proof The system of equations in (2.4) satisfies the standard discrete-time stochastic dynamic programming property. Details of the proof of the results can be found in Basar and Olsder (1999). ■

Note that $W^{(\sigma_T)}(T,x)$ yields the expected cooperative payoff starting at stage T given that $\theta_T^{\sigma_T}$ occurs and $x_T = x$ according to the dynamic programming problem (2.2 and 2.3) if $\theta_T^{\sigma_T}$. Using Lemma 2.1, one can characterize the functions $W^{(\sigma_T)}(T,x)$ for all $\theta_T^{\sigma_T} \in \{\theta_T^1, \theta_T^2, \cdots, \theta_T^{\eta_T}\}$, if they exist. Following the analysis in Sect. 9.2, the control problem starting at stage t when $\theta_t^{\sigma_t} \in \{\theta_t^1, \theta_t^2, \cdots, \theta_t^{\eta_t}\}$ occurs and $x_t = x$ can be expressed as:

$$\max_{u_t} E_{\vartheta_t}\left\{\sum_{j=1}^{n} g_t^j\left(x, u_t^1, u_t^2, \cdots, u_t^n; \theta_t^{\sigma_t}\right)\right.$$
$$\left.+\sum_{\sigma_{t+1}=1}^{\eta_{t+1}} \lambda_{t+1}^{\sigma_{t+1}} W^{(\sigma_{t+1})}\left[t+1, f_t\left(x, u_t^1, u_t^2, \cdots, u_t^n\right)+\vartheta_t\right]\right\}, \tag{2.5}$$

where $W^{(\sigma_{t+1})}\left[t+1, f_t\left(x, u_t^1, u_t^2, \cdots, u_t^n\right)+\vartheta_t\right]$ is the expected optimal cooperative payoff in the control problem starting at stage $t+1$ when $\theta_{t+1}^{\sigma_{t+1}} \in \{\theta_{t+1}^1, \theta_{t+1}^2, \cdots, \theta_{t+1}^{\eta_{t+1}}\}$ occurs and $x_{t+1} = f_t\left(x, u_t^1, u_t^2, \cdots, u_t^n\right)+\vartheta_t$.

An optimal solution for the stochastic control problem (1.1) and (2.1) can be characterized by the following theorem.

Theorem 2.1 A set of controls $u_t^{(\sigma_t)i*} = \psi_t^{(\sigma_t)*}(x) = \{\psi_t^{(\sigma_t)1*}(x), \psi_t^{(\sigma_t)2*}(x), \cdots\cdots, \psi_t^{(\sigma_t)n*}(x)\}$, for $\sigma_t \in \{1, 2, \cdots, \eta_t\}$ and $t \in \{1, 2, \cdots, T\}$ provides an optimal solution to the stochastic control problem (1.1) and (2.1) if there exist functions $W^{(\sigma_t)}(t,x)$, for $\sigma_t \in \{1, 2, \cdots, \eta_t\}$ and $t \in \{1, 2, \cdots, T\}$, such that the following recursive relations are satisfied:

$$W^{(\sigma_T)}(T+1,x) = \sum_{j=1}^{n} q^j(x),$$

$$W^{(\sigma_T)}(T,x) = \max_{u_T^{(\sigma_T)1}, u_T^{(\sigma_T)2}, \cdots, u_T^{(\sigma_T)n}} E_{\vartheta_T}\left\{\sum_{j=1}^{n} g_T^j\left[x, u_T^{(\sigma_T)1}, u_T^{(\sigma_T)2}, \cdots, u_T^{(\sigma_T)n}; \theta_T^{\sigma_T}\right]\right.$$
$$\left.+W^{(\sigma_{T+1})}\left[T+1, f_T\left(x, u_T^{(\sigma_T)1}, u_T^{(\sigma_T)2}, \cdots, u_T^{(\sigma_T)n}\right)+\vartheta_T\right]\right\},$$

$$W^{(\sigma_t)}(t,x) = \max_{u_t^{(\sigma_t)1}, u_t^{(\sigma_t)2}, \cdots, u_t^{(\sigma_t)n}} E_{\vartheta_t}\left\{\sum_{j=1}^{n} g_t^j\left[x, u_t^{(\sigma_t)1}, u_t^{(\sigma_t)2}, \cdots, u_t^{(\sigma_t)n}; \theta_t^{\sigma_t}\right]\right.$$
$$\left.+\sum_{\sigma_{t+1}=1}^{\eta_{t+1}} \lambda_{t+1}^{\sigma_{t+1}} W^{(\sigma_{t+1})}\left[t+1, f_t\left(x, u_t^{(\sigma_t)1}, u_t^{(\sigma_t)2}, \cdots, u_t^{(\sigma_t)n}\right)+\vartheta_t\right]\right\},$$

for $\sigma_t \in \{1, 2, \cdots, \eta_t\}$ and $t \in \{1, 2, \cdots, T-1\}$. (2.6)

Proof The results in (2.6) characterizing the optimal solution in stage T is proved in Lemma 2.1. Invoking the specification of the control problem starting in stage $t \in \{1, 2, \cdots, T-1\}$ as expressed in (2.5), the results in (2.6) satisfy the optimality conditions in stochastic dynamic programming. Therefore, an optimal solution of the stochastic control problem (1.1) and (2.1) is characterized. ■

Theorem 2.1 is the discrete-time analog of the optimal cooperative scheme in randomly furcating stochastic differential games in Petrosyan and Yeung (2007).

Substituting the optimal control $\{\psi_k^{(\sigma_k)i*}(x)$, for $k \in \{1, 2, \cdots T\}$ and $i \in N\}$ into the state dynamics (1.1), one can obtain the dynamics of the cooperative trajectory as:

$$x_{k+1} = f_k\left(x_k, \psi_k^{(\sigma_k)1*}(x_k), \psi_k^{(\sigma_k)2*}(x_k), \cdots, \psi_k^{(\sigma_k)n*}(x_k)\right) + \vartheta_k, \text{ if } \theta_k^{\sigma_k} \text{ occurs}, \quad (2.7)$$

for $k \in \{1, 2, \cdots, T\}, \sigma_k \in \{1, 2, \cdots, \eta_k\}$ and $x_1 = x^0$.

We use X_k^* to denote the set of realizable values of x_k^* at stage k generated by (2.7). The term $x_k^* \in X_k^*$ is used to denote an element in X_k^*.

The term $W^{(\sigma_k)}(k, x_k^*)$ gives the expected total cooperative payoff over the stages from k to T if $\theta_k^{\sigma_k}$ occurs and $x_k^* \in X_k^*$ is realized at stage k.

9.2.2 *Individual Rationality*

The players then have to agree to an optimality principle in distributing the total cooperative payoff among themselves. For individual rationality to be upheld the expected payoffs a player receives under cooperation have to be no less than his expected noncooperative payoff along the cooperative state trajectory $\{x_k^*\}_{k=1}^{T+1}$. The players may (i) share the excess of the total expected cooperative payoff over the expected sum of individual noncooperative payoffs equally, or (ii) share the total expected cooperative payoff proportional to their expected noncooperative payoffs.

Let $\xi^{(\sigma_k)}(k, x_k^*) = [\xi^{(\sigma_k)1}(k, x_k^*), \xi^{(\sigma_k)2}(k, x_k^*), \cdots, \xi^{(\sigma_k)n}(k, x_k^*)]$ denote the imputation vector guiding the distribution of the total expected cooperative payoff under the agreed-upon optimality principle along the cooperative trajectory given that $\theta_k^{\sigma_k}$ has occurred in stage k, for $\sigma_k \in \{1, 2, \cdots, \eta_k\}$ and $k \in \{1, 2, \cdots, T\}$. In particular, the imputation $\xi^{(\sigma_k)i}(k, x_k^*)$ gives the expected cumulative payments that player i will receive from stage k to stage $T+1$ under cooperation.

If for example, the optimality principle specifies that the players share the excess of the total cooperative payoff over the sum of individual noncooperative payoffs equally, then the imputation to player i becomes:

$$\xi^{(\sigma_k)i}(k, x_k^*) = V^{(\sigma_k)i}(k, x_k^*) + \frac{1}{n}\left[W^{(\sigma_k)}(k, x_k^*) - \sum_{j=1}^{n} V^{(\sigma_k)j}(k, x_k^*)\right], \quad (2.8)$$

for $i \in N$ and $k \in \{1, 2, \cdots, T\}$.

For individual rationality to be maintained throughout all the stages $k \in \{1,2,\cdots,T\}$, it is required that the imputation satisfies:

$$\begin{aligned}&\xi^{(\sigma_k)i}\left(k,x_k^*\right) \geq V^{(\sigma_k)i}\left(k,x_k^*\right),\\&\text{for } i\in N,\ \sigma_k \in \{1,2,\cdots,\eta_k\} \text{ and } k \in \{1,2,\cdots,T\}.\end{aligned} \tag{2.9}$$

To guarantee group optimality, the imputation vector has to satisfy

$$\begin{aligned}&W^{(\sigma_k)}\left(k,x_k^*\right) = \sum_{j=1}^{n}\xi^{(\sigma_k)j}\left(k,x_k^*\right),\\&\text{for } \sigma_k \in \{1,2,\cdots,\eta_k\} \text{ and } k \in \{1,2,\cdots,T\}.\end{aligned} \tag{2.10}$$

Hence, a valid imputation $\xi^{(\sigma_k)i}\left(k,x_k^*\right)$, for $i\in N,\ \sigma_k \in \{1,2,\cdots,\eta_k\}$ and $k \in \{1,2,\cdots,T\}$, has to satisfy conditions (2.9) and (2.10).

9.3 Subgame Consistent Solutions and Payment Mechanism

As demonstrated in Chap. 7, to guarantee dynamical stability in a stochastic dynamic cooperation scheme, the solution has to satisfy the property of subgame consistency in addition to group optimality and individual rationality. In particular, an extension of a subgame-consistent cooperative solution policy to a subgame starting at a later time with a feasible state brought about by prior optimal behavior would remain optimal. Thus subgame consistency ensures that as the game proceeds players are guided by the same optimality principle at each stage of the game, and hence do not possess incentives to deviate from the previously adopted optimal behavior. For subgame consistency to be satisfied, the imputation according to the original optimality principle has to be maintained at all the T stages along the cooperative trajectory $\left\{x_k^*\right\}_{k=1}^{T}$. In other words, the imputation

$$\xi^{(\sigma_k)}\left(k,x_k^*\right) = \left[\xi^{(\sigma_k)1}\left(k,x_k^*\right),\xi^{(\sigma_k)2}\left(k,x_k^*\right),\cdots,\xi^{(\sigma_k)n}\left(k,x_k^*\right)\right], \tag{3.1}$$

for $\sigma_k \in \{1,2,\cdots,\eta_k\}$, $x_k^* \in X_k^*$ and $k \in \{1,2,\cdots,T\}$, has to be upheld.

9.3.1 Payoff Distribution Procedure

Following the analysis of Yeung and Petrosyan (2010 and 2011), we formulate a Payoff Distribution Procedure (PDP) so that the agreed-upon imputation (3.1) can be realized. Let $B_k^{(\sigma_k)i}\left(x_k^*\right)$ denote the payment that player i will received at stage

k under the cooperative agreement if $\theta_k^{\sigma_k} \in \{\theta_k^1, \theta_k^2, \cdots, \theta_k^{\eta_k}\}$ occurs and $x_k^* \in X_k^*$ is realized at stage $k \in \{1, 2, \cdots, T\}$. The payment scheme $\{B_k^{(\sigma_k)i}(x_k^*)$ contingent upon the event $\theta_k^{\sigma_k}$ and state x_k^*, for $k \in \{1, 2, \cdots, T\}\}$ constitutes a PDP in the sense that the imputation to player i over the stages 1 to $T+1$ can be expressed as:

$$\xi^{(\sigma_1)i}(1, x_{1(0)}) = B_1^{(\sigma_1)i}(x_{1(0)}) + E_{\theta_2, \cdots, \theta_T; \vartheta_1, \vartheta_2, \cdots, \vartheta_T}\left(\sum_{\zeta=2}^{T} B_\zeta^{(\sigma_\zeta)i}(x_\zeta^*) + q^i(x_{T+1}^*)\right), \tag{3.2}$$

for $i \in N$.

Moreover, according to the agreed-upon optimality principle in (3.1), if $\theta_k^{\sigma_k}$ occurs and $x_k^* \in X_k^*$ is realized at stage k the imputation to player i is $\xi^{(\sigma_k)i}(k, x_k^*)$. For subgame consistency to be satisfied, the imputation according to the agreed-upon optimality principle has to be maintained at all the T stages along the cooperative trajectory $\{x_k^*\}_{k=1}^{T}$. Therefore to guarantee subgame consistency, the payment scheme $\left\{B_k^{(\sigma_k)i}(x_k^*)\right\}$ has to satisfy the conditions

$$\xi^{(\sigma_k)i}(k, x_k^*) = B_k^{(\sigma_k)i}(x_k^*) + E_{\theta_{k+1}, \theta_{k+2}, \cdots, \theta_T; \vartheta_k, \vartheta_{k+1}, \cdots, \vartheta_T}\left(\sum_{\zeta=k+1}^{T} B_\zeta^{(\sigma_\zeta)i}(x_\zeta^*) + q^i(x_{T+1}^*)\right) \tag{3.3}$$

for $i \in N$ and $k \in \{1, 2, \cdots, T\}$.

Using (3.3) one can readily obtain $\xi^{(\sigma_{T+1})i}(T+1, x_{T+1}^*)$ equals $q^i(x_{T+1}^*)$ with probability 1. Crucial to the formulation of a subgame consistent solution is the derivation of a payment scheme $\{B_k^{(\sigma_k)i}(x_k^*)$, for $i \in N$, $\sigma_k \in \{1, 2, \cdots, \eta_k\}$, $x_k^* \in X_k^*$ and $k \in \{1, 2, \cdots, T\}$ so that the imputation in (3.3) can be realized. This will be done in the sequel.

A theorem for the derivation of a subgame consistent PDP can be established as follows.

Theorem 3.1 A payment equaling

$$B_k^{(\sigma_k)i}(x_k^*) = \xi^{(\sigma_k)i}(k, x_k^*) - E_{\vartheta_k}\left[\sum_{\sigma_{k+1}=1}^{\eta_{k+1}} \lambda_{k+1}^{\sigma_{k+1}}\left(\xi^{(\sigma_{k+1})i}\left[k+1, f_k\left(x_k^*, \psi_k^{(\sigma_k)*}(x_k^*)\right) + \vartheta_k\right]\right)\right], \tag{3.4}$$

for $i \in N$,

given to player i at stage $k\in\{1,2,\cdots,T\}$, if $\theta_k^{\sigma_k}$ occurs and $x_k^*\in X_k^*$, leads to the realization of the imputation in (3.3).

Proof To construct the proof of Theorem 3.1, we first consider the imputation

$$E_{\theta_{k+1},\theta_{k+2},\cdots,\theta_T;\vartheta_k,\vartheta_{k+1},\cdots,\vartheta_T}\left(\sum_{\zeta=k+1}^{T} B_\zeta^{(\sigma_\zeta)i}\left(x_\zeta^*\right)+q^i\left(x_{T+1}^*\right)\right)$$
$$=E_{\vartheta_k}\left\{\sum_{\sigma_{k+1}=1}^{\eta_{k+1}}\lambda_{k+1}^{\sigma_{k+1}}\left[B_{k+1}^{(\sigma_{k+1})i}\left(x_{k+1}^*\right)\right.\right.$$
$$\left.\left.+E_{\theta_{k+2},\theta_{k+3},\cdots,\theta_T;\vartheta_{k+2},\vartheta_{k+3},\cdots,\vartheta_T}\left(\sum_{\zeta=k+2}^{T} B_\zeta^{(\sigma_\zeta)i}\left(x_\zeta^*\right)+q^i\left(x_{T+1}^*\right)\right)\right]\right\}. \tag{3.5}$$

Then, using (3.3) we can derive the term $\xi^{(\sigma_{k+1})i}\left(k+1,x_{k+1}^*\right)$ as

$$\xi^{(\sigma_{k+1})i}\left(k+1,x_{k+1}^*\right)=B_{k+1}^{(\sigma_{k+1})i}\left(x_{k+1}^*\right)$$
$$+E_{\theta_{k+2},\theta_{k+3},\cdots,\theta_T;\vartheta_{k+2},\vartheta_{k+3},\cdots,\vartheta_T}\left(\sum_{\zeta=k+2}^{T} B_\zeta^{(\sigma_\zeta)i}\left(x_\zeta^*\right)+q^i\left(x_{T+1}^*\right)\right) \tag{3.6}$$

The expression on the right-hand-side of equation (3.6) is the same as the expression inside the square brackets of (3.5). Invoking equation (3.6) we can replace the expression inside the square brackets of (3.5) by $\xi^{(\sigma_{k+1})i}\left(k+1,x_{k+1}^*\right)$ and obtain:

$$E_{\theta_{k+1},\theta_{k+2},\cdots,\theta_T;\vartheta_k,\vartheta_{k+1},\cdots,\vartheta_T}\left(\sum_{\zeta=k+1}^{T} B_\zeta^{(\sigma_\zeta)i}\left(x_\zeta^*\right)+q^i\left(x_{T+1}^*\right)\right)$$
$$=E_{\vartheta_k}\left[\sum_{\sigma_{k+1}=1}^{\eta_{k+1}}\lambda_{k+1}^{\sigma_{k+1}}\left(\xi^{(\sigma_{k+1})i}\left[k+1,f_k\left(x_k^*,\psi_k^{(\sigma_k)*}\left(x_k^*\right)\right)+\vartheta_k\right]\right)\right].$$

Substituting the term $E_{\theta_{k+1},\theta_{k+2},\cdots,\theta_T;\vartheta_k,\vartheta_{k+1},\cdots,\vartheta_T}\left(\sum_{\zeta=k+1}^{T} B_\zeta^{(\sigma_\zeta)i}\left(x_\zeta^*\right)+q^i\left(x_{T+1}^*\right)\right)$ by E_{ϑ_k} $\left[\sum_{\sigma_{k+1}=1}^{\eta_{k+1}}\lambda_{k+1}^{\sigma_{k+1}}\left(\xi^{(\sigma_{k+1})i}\left[k+1,f_k\left(x_k^*,\psi_k^{(\sigma_k)*}\left(x_k^*\right)\right)+\vartheta_k\right]\right)\right]$ in (3.3) we can express (3.3) as:

$$\xi^{(\sigma_k)i}\left(k,x_k^*\right)=B_k^{(\sigma_k)i}\left(x_k^*\right)$$
$$+E_{\vartheta_k}\left[\sum_{\sigma_{k+1}=1}^{\eta_{k+1}}\lambda_{k+1}^{\sigma_{k+1}}\left(\xi^{(\sigma_{k+1})i}\left[k+1,f_k\left(x_k^*,\psi_k^{(\sigma_k)*}\left(x_k^*\right)\right)+\vartheta_k\right]\right)\right]. \tag{3.7}$$

For condition (3.7), which is an alternative form of (3.3), to hold it is required that:

$$B_k^{(\sigma_k)i}(x_k^*) = \xi^{(\sigma_k)i}(k, x_k^*) - E_{\vartheta_k}\left[\sum_{\sigma_{k+1}=1}^{\eta_{k+1}} \lambda_{k+1}^{\sigma_{k+1}} \left(\xi^{(\sigma_{k+1})i}\left[k+1, f_k\left(x_k^*, \psi_k^{(\sigma_k)*}(x_k^*)\right) + \vartheta_k\right]\right)\right], \quad (3.8)$$

for $i \in N$ and $k \in \{1, 2, \cdots, T\}$.

Therefore by paying $B_k^{(\sigma_k)i}(x_k^*)$ to player $i \in N$ at stage $k \in \{1, 2, \cdots, T\}$, if $\theta_k^{\sigma_k}$ occurs and $x_k^* \in X_k^*$, leads to the realization of the imputation in (3.3). Hence Theorem 3.1 follows. ■

For a given imputation vector

$$\xi^{(\sigma_k)}(k, x_k^*) = \left[\xi^{(\sigma_k)1}(k, x_k^*), \xi^{(\sigma_k)2}(k, x_k^*), \cdots, \xi^{(\sigma_k)n}(k, x_k^*)\right],$$

for $\sigma_k \in \{1, 2, \cdots, \eta_k\}$ and $k \in \{1, 2, \cdots, T\}$,

Theorem 3.1 can be used to derive the PDP that leads to the realization this vector.

9.3.2 Transfer Payments

When all players are using the cooperative strategies, the payoff that player i will directly received at stage k given that $x_k^* \in X_k^*$ and $\theta_k^{\sigma_k}$ occurs becomes

$$g_k^i\left[x_k^*, \psi_k^{(\sigma_k)1*}(x_k^*), \psi_k^{(\sigma_k)2*}(x_k^*), \cdots, \psi_k^{(\sigma_k)n*}(x_k^*); \theta_k^{\sigma_k}\right].$$

However, according to the agreed upon imputation, player i is supposed to received $B_k^{(\sigma_k)i}(x_k^*)$ at stage k as given in Theorem 3.1. Therefore a transfer payment (which can be positive or negative)

$$\varpi_k^{(\sigma_k)i}(x_k^*) = B_k^{(\sigma_k)i}(x_k^*) - g_k^i\left[x_k^*, \psi_k^{(\sigma_k)1*}(x_k^*), \psi_k^{(\sigma_k)2*}(x_k^*), \cdots, \psi_k^{(\sigma_k)n*}(x_k^*); \theta_k^{\sigma_k}\right], \quad (3.9)$$

for $k \in \{1, 2, \cdots, T\}$ and $i \in N$,

will be assigned to player i to yield the cooperative imputation $\xi^i(k, x_k^*)$.

The transfer payments system in (3.9) constitutes an instrument to guide the execution of the agreed-upon payoff sharing mechanism. Coordination of payments is jointly performed by the participating players.

9.4 An Illustration in Cooperative Resource Extraction Under Uncertainty

Consider an economy endowed with a renewable resource and with 2 resource extractors (firms). The lease for resource extraction begins at stage 1 and ends at stage 3 for these two firms. Let u_k^i denote the resource extracted by firm i at stage k, for $i \in \{1, 2\}$. Let U^i be the set of admissible amount of resource extracted by firm i, and $x_k \in X \subset R^+$ be the size of the resource stock at stage k.

It is known at each stage there is a random element, θ_k for $k \in \{1, 2, 3\}$, affecting the prices of the outputs produced by these firms and their costs of extraction. If $\theta_k^{\sigma_k} \in \{\theta_k^1, \theta_k^2\}$ happens at stage $k \in \{2, 3\}$ the profits (in present-value) that firm 1 and firm 2 will obtain at stage k are respectively:

$$\left[P_k^{(\sigma_k)1} u_k^1 - \frac{c_k^{(\sigma_k)1}}{x_k}\left(u_k^1\right)^2\right]\left(\frac{1}{1+r}\right)^{k-1}$$
$$\text{and}\left[P_k^{(\sigma_k)2} u_k^2 - \frac{c_k^{(\sigma_k)2}}{x_k}\left(u_k^2\right)^2\right]\left(\frac{1}{1+r}\right)^{k-1}, \tag{4.1}$$

where $P_k^{(\sigma_k)i}$ is the price of the resource extracted and processed by firm i, and $c_k^{(\sigma_k)i}$ $\left(u_k^i\right)^2/x_k$ is the production cost of firm i in stage k if $\theta_k^{\sigma_k}$ occurs.

It is known in stage 1 that θ_1 is θ_1^1 with probability $\lambda_1^1 = 1$. The probability that $\theta_k^{\sigma_k} \in \{\theta_k^1, \theta_k^2\}$ will occur at stage $k \in \{2, 3\}$ is $\lambda_k^{\sigma_k}$. In stage 4, a terminal payment (again in present-value) contingent upon the resource size equaling $q^i x_4 \left(\frac{1}{1+r}\right)^3$ will be paid to firm i.

The growth dynamics of the resource is governed by the stochastic difference equation:

$$x_{k+1} = x_k + a - bx_k - \sum_{j=1}^{2} u_k^j + \vartheta_k, \tag{4.2}$$

for $k \in \{1, 2, 3\}$ and $x_1 = x^0$,

where ϑ_k is a random variable with non-negative range $\{\vartheta_k^1, \vartheta_k^2, \vartheta_k^3\}$ and corresponding probabilities $\{\gamma_k^1, \gamma_k^2, \gamma_k^3\}$; moreover $\vartheta_1, \vartheta_2, \vartheta_3$ are independent. Moreover, we have the constraint $u_k^1 + u_k^2 \leq (1 - b)x_k + a$.

The objective of extractor $i \in \{1, 2\}$ is to maximize the present value of the expected stream of future profits:

$$E_{\theta_1,\theta_2,\theta_3;\vartheta_1,\vartheta_2,\vartheta_3}\left\{\sum_{k=1}^{3}\left[P_k^{(\sigma_k)i}u_k^i-\frac{c_k^{(\sigma_k)i}}{x_k}\left(u_k^i\right)^2\right]\left(\frac{1}{1+r}\right)^{k-1}+q^i x_4\left(\frac{1}{1+r}\right)^3\right\} \quad (4.3)$$

subject to the stochastic dynamics (4.2).

Invoking Lemma 1.2, one can characterize the noncooperative Nash equilibrium strategies for the game (4.2 and 4.3) as follows. In particular, a set of strategies $\{u_k^{(\sigma_k)i*}=\phi_k^{(\sigma_k)i*}(x)\in\Gamma^i$, for $\sigma_1\in\{1\},\sigma_2,\sigma_3\in\{1,2\}$, $k\in\{1,2,3\}$ and $i\in\{1,2\}\}$ provides a Nash equilibrium solution to the game (4.2 and 4.3) if there exist functions $V^{(\sigma_k)i}(k,x)$, for $i\in\{1,2\}$ and $k\in\{1,2,3\}$, such that the following recursive relations are satisfied:

$$\begin{aligned}
V^{(\sigma_k)i}(k,x) &= \max_{u_k^{(\sigma_k)i}} E_{\vartheta_k}\left\{\left[P_k^{(\sigma_k)i}u_k^{(\sigma_k)i}-\frac{c_k^{(\sigma_k)i}}{x}\left(u_k^{(\sigma_k)i}\right)^2\right]\left(\frac{1}{1+r}\right)^{k-1}\right.\\
&\quad\left.+\sum_{\sigma_{k+1}=1}^{2}\lambda_{k+1}^{\sigma_{k+1}}V^{(\sigma_{k+1})i}\left[k+1,x+a-bx-u_k^{(\sigma_k)i}-\phi_k^{(\sigma_k)j*}(x)+\vartheta_k\right]\right\}\\
&= \max_{u_k^{(\sigma_k)i}}\left\{\left[P_k^{(\sigma_k)i}u_k^{(\sigma_k)i}-\frac{c_k^{(\sigma_k)i}}{x}\left(u_k^{(\sigma_k)i}\right)^2\right]\left(\frac{1}{1+r}\right)^{k-1}\right.\\
&\quad\left.+\sum_{y=1}^{3}\gamma_k^y\sum_{\sigma_{k+1}=1}^{2}\lambda_{k+1}^{\sigma_{k+1}}V^{(\sigma_{k+1})i}\left[k+1,x+a-bx-u_k^{(\sigma_k)i}-\phi_k^{(\sigma_k)j*}(x)+\vartheta_k^y\right]\right\};\\
V^{(\sigma_3)i}(4,x) &= q^i x\left(\frac{1}{1+r}\right)^3.
\end{aligned} \quad (4.4)$$

Performing the indicated maximization in (4.4) yields:

$$\begin{aligned}
&\left[P_k^{(\sigma_k)i}-\frac{2c_k^{(\sigma_k)i}u_k^{(\sigma_k)i}}{x}\right]\left(\frac{1}{1+r}\right)^{k-1}\\
&-\sum_{y=1}^{3}\gamma_k^y\sum_{\sigma_{k+1}=1}^{2}\lambda_{k+1}^{\sigma_{k+1}}V_{x_{k+1}}^{(\sigma_{k+1})i}\left[k+1,x+a-bx-u_k^{(\sigma_k)i}-\phi_k^{(\sigma_k)j*}(x)+\vartheta_k^y\right]=0;
\end{aligned} \quad (4.5)$$

for $i\in\{1,2\}$ and $k\in\{1,2,3\}$.

From (4.5), the game equilibrium strategies can be expressed as:

$$\begin{aligned}
\phi_k^{(\sigma_k)i*}(x) &= \frac{x}{2c_k^{(\sigma_k)i}}\left(P_k^{(\sigma_k)i}-(1+r)^{k-1}\sum_{y=1}^{3}\gamma_k^y\right.\\
&\left.\sum_{\sigma_{k+1}=1}^{2}\lambda_{k+1}^{\sigma_{k+1}}V_{x_{k+1}}^{(\sigma_{k+1})i}\left[k+1,x+a-bx-\phi_k^{(\sigma_k)1*}(x)-\phi_k^{(\sigma_k)2*}(x)+\vartheta_k^y\right]\right),
\end{aligned} \quad (4.6)$$

for $i\in\{1,2\}$ and $k\in\{1,2,3\}$.

The expected game equilibrium payoffs of the extractors can be obtained as:

Proposition 4.1 The value function indicating the expected game equilibrium payoff of player i is

$$V^{(\sigma_k)i}(k,x) = \left[A_k^{(\sigma_k)i}x + C_k^{(\sigma_k)i}\right], \text{ for } i\in\{1,2\} \text{ and } k\in\{1,2,3\}, \tag{4.7}$$

where $A_k^{(\sigma_k)i}$ and $C_k^{(\sigma_k)i}$, for $i\in\{1,2\}$ and $k\in\{1,2,3\}$, are constants in terms of the parameters of the game (4.2 and 4.3).

Proof See Appendix A. ■

Substituting the relevant derivatives of the value functions in Proposition 4.1 into the game equilibrium strategies (4.6) yields a noncooperative Nash equilibrium solution of the game (4.2 and 4.3).

Now consider the case when the extractors agree to maximize their expected joint profit and share the excess of cooperative gains over their expected noncooperative payoffs equally. To maximize their expected joint payoff, they solve the problem of maximizing

$$E_{\theta_1,\theta_2,\theta_3;\vartheta_1,\vartheta_2,\vartheta_3}\left\{\sum_{j=1}^{2}\left[\sum_{k=1}^{3}\left(P_k^{(\sigma_k)j}u_k^j - \frac{c_k^{(\sigma_k)j}}{x_k}\left(u_k^j\right)^2\right)\left(\tfrac{1}{1+r}\right)^{k-1} + q^j x_4\left(\tfrac{1}{1+r}\right)^3\right]\right\} \tag{4.8}$$

subject to (4.2).

Invoking Theorem 2.1, one can characterize the optimal controls in the stochastic dynamic programming problem (4.2) and (4.8). In particular, a set of control strategies $\left\{u_k^{(\sigma_k)i*} = \psi_k^{(\sigma_k)i*}(x)\in\hat{\Gamma}^i, \text{ for } \sigma_k\in\{1,2\}, k\in\{1,2,3\} \text{ and } i\in\{1,2\}\right\}$ provides an optimal solution to the problem (4.2) and (4.8) if there exist functions $W^{(\sigma_k)}(k,x)$, for $k\in\{1,2,3\}$, such that the following recursive relations are satisfied:

$$\begin{aligned}
W^{(\sigma_4)}(4,x) &= \sum_{j=1}^{2} q^j x\left(\tfrac{1}{1+r}\right)^3,\\
W^{(\sigma_k)}(k,x) &= \max_{u_k^1,u_k^2} E_{\vartheta_k}\left\{\sum_{j=1}^{2}\left(P_k^{(\sigma_k)j}u_k^j - \frac{c_k^{(\sigma_k)j}}{x}\left(u_k^j\right)^2\right)\left(\frac{1}{1+r}\right)^{k-1}\right.\\
&\quad \left. + \sum_{\sigma_{k+1}=1}^{2}\lambda_{k+1}^{\sigma_{k+1}}W^{(\sigma_{k+1})}\left[k+1, x+a-bx-u_k^1-u_k^2+\vartheta_k\right]\right\}\\
&= \max_{u_k^1,u_k^2}\left\{\sum_{j=1}^{2}\left(P_k^{(\sigma_k)j}u_k^j - \frac{c_k^{(\sigma_k)j}}{x}\left(u_k^j\right)^2\right)\left(\frac{1}{1+r}\right)^{k-1}\right.\\
&\quad \left. + \sum_{y=1}^{3}\gamma_k^y\sum_{\sigma_{k+1}=1}^{2}\lambda_{k+1}^{\sigma_{k+1}}W^{(\sigma_{k+1})}\left[k+1, x+a-bx-u_k^1-u_k^2+\vartheta_k^y\right]\right\},
\end{aligned}$$

for $k\in\{1,2,3\}$ and $\sigma_k\in\{1,2\}$. (4.9)

Performing the indicated maximization in (4.9) yields:

$$\left(P_k^{(\sigma_k)j} - \frac{2c_k^{(\sigma_k)j}u_k^j}{x_k}\right)\left(\frac{1}{1+r}\right)^{k-1}$$
$$-\sum_{y=1}^{3}\gamma_k^y\sum_{\sigma_{k+1}=1}^{2}\lambda_{k+1}^{\sigma_{k+1}}W_{x_{k+1}}^{(\sigma_{k+1})}\left[k+1, x+a-bx-u_k^1-u_k^2+\vartheta_k^y\right]=0, \qquad (4.10)$$

for $k\in\{1,2,3\}$ and $\sigma_k\in\{1,2\}$.

In particular, the optimal cooperative strategies can be obtained from (4.10) as:

$$\psi_k^{(\sigma_k)i*}(x) = \frac{x}{2c_k^{(\sigma_k)i}}\left(P_k^{(\sigma_k)i} - \sum_{y=1}^{3}\gamma_k^y\sum_{\sigma_{k+1}=1}^{2}\lambda_{k+1}^{\sigma_{k+1}}W_{x_{k+1}}^{(\sigma_{k+1})}\left[k+1, x+a-bx\right.\right.$$
$$\left.\left.-\psi_k^{(\sigma_k)1*}(x)-\psi_k^{(\sigma_k)2*}(x)+\vartheta_k^y\right](1+r)^{k-1}\right), \qquad (4.11)$$

for $k\in\{1,2,3\}$ and $\sigma_k\in\{1,2\}$.

The expected joint payoff under cooperation can be obtained as:

Proposition 4.2 The value function indicating the maximized expected joint payoff is

$$W^{(\sigma_k)}(k,x) = \left[\widetilde{A}_k^{(\sigma_k)}x + \widetilde{C}_k^{(\sigma_k)}\right], \text{ for } k\in\{1,2,3\} \text{ and } \sigma_k\in\{1,2\}, \qquad (4.12)$$

where $\widetilde{A}_k^{(\sigma_k)}$ and $\widetilde{C}_k^{(\sigma_k)}$, for $k\in\{1,2,3\}$ and $\sigma_k\in\{1,2\}$, are constants in terms of the parameters of the problem (4.8) and (4.2).

Proof See Appendix B. ■

Using (4.11) and Proposition 4.2, the optimal cooperative strategies of the agents can be expressed as:

$$\psi_k^{(\sigma_k)i*}(x) = \frac{x}{2c_k^{(\sigma_k)i}}\left(P_k^{(\sigma_k)i} - \sum_{\sigma_{k+1}=1}^{2}\lambda_{k+1}^{\sigma_{k+1}}A_{k+1}^{(\sigma_{k+1})}(1+r)^{k-1}\right), \qquad (4.13)$$

for $i\in\{1,2\}$, $k\in\{1,2,3\}$ and $\sigma_k\in\{1,2\}$.

Substituting $\psi_k^{(\sigma_k)i}(x)$ from (4.13) into (4.2) yields the optimal cooperative state trajectory:

$$x_{k+1} = x_k + a - bx_k$$
$$-\sum_{j=1}^{2}\frac{x}{2c_k^{(\sigma_k)j}}\left(P_k^{(\sigma_k)j} - \sum_{\sigma_{k+1}=1}^{2}\lambda_{k+1}^{\sigma_{k+1}}\widetilde{A}_{k+1}^{(\sigma_{k+1})}(1+r)^{k-1}\right) + \vartheta_k, \qquad (4.14)$$

if $\theta_k^{\sigma_k}$ occurs at stage k for $k \in \{1,2,3\}$ and $x_1 = x^0$.

Dynamics (4.14) is a linear stochastic difference equation readily solvable by standard techniques. Let $\{x_k^*, \text{ for } k \in \{1,2,3\}\}$ denote the solution to (4.14).

Since the extractors agree to share the excess of cooperative gains over their expected noncooperative payoffs equally, an imputation

$$\xi^{(\sigma_k)i}(k, x_k^*) = V^{(\sigma_k)i}(k, x_k^*) + \frac{1}{2}\left[W^{(\sigma_k)}(k, x_k^*) - \sum_{j=1}^{2} V^{(\sigma_k)j}(k, x_k^*)\right]$$
$$= \left(A_k^{(\sigma_k)i} x_k^* + C_k^{(\sigma_k)i}\right) + \frac{1}{2}\left[\left(\widetilde{A}_k^{(\sigma_k)} x_k^* + \widetilde{C}_k^{(\sigma_k)}\right) - \sum_{j=1}^{2}\left(A_k^{(\sigma_k)j} x_k^* + C_k^{(\sigma_k)j}\right)\right], \quad (4.15)$$

if $\theta_k^{\sigma_k}$ occurs at stage k for $k \in \{1,2,3\}$, $\sigma_k \in \{1,2\}$ and $i \in \{1,2\}$ has to be maintained.

Invoking Theorem 3.1, if $\theta_k^{\sigma_k}$ occurs and $x_k^* \in X$ is realized at stage k a payment equaling

$$B_k^{(\sigma_k)i}(x_k^*) = (1+r)^{k-1}\Big[\xi^i(k, x_k^*)$$
$$- E_{\theta_{k+1}}\left(\xi^i\left[k+1, f_k\left(x_k^*, \psi_k^{(\sigma_k)*}(x_k^*)\right) + \theta_k\right]\right)\Big]$$
$$= (1+r)^{k-1}\Bigg\{\left(A_k^{(\sigma_k)i} x_k^* + C_k^{(\sigma_k)i}\right)$$
$$+\frac{1}{2}\left(\left(\widetilde{A}_k^{(\sigma_k)} x_k^* + \widetilde{C}_k^{(\sigma_k)}\right) - \sum_{j=1}^{2}\left(A_k^{(\sigma_k)j} x_k^* + C_k^{(\sigma_k)j}\right)\right)$$
$$-\sum_{y=1}^{3}\gamma_k^y \sum_{\sigma_{k+1}=1}^{\eta_{k+1}} \lambda_{k+1}^{\sigma_{k+1}}\Bigg[\left(A_{k+1}^{(\sigma_{k+1})i} x_{k+1}^{*(\vartheta_k^y)} + C_{k+1}^{(\sigma_{k=1})i}\right)$$
$$+\frac{1}{2}\left(\left(\widetilde{A}_{k+1}^{(\sigma_{k+1})} x_{k+1}^{*(\vartheta_k^y)} + \widetilde{C}_{k+1}^{(\sigma_{k+1})}\right) - \sum_{j=1}^{2}\left(A_{k+1}^{(\sigma_{k+1})j} x_{k+1}^{*(\vartheta_k^y)} + C_{k+1}^{(\sigma_{k+1})j}\right)\right)\Bigg]\Bigg\}, \quad (4.16)$$

where

$$x_{k+1}^{*(\vartheta_k^y)} = x_k^* + a - bx_k^* - \sum_{j=1}^{2}\frac{x_k^*}{2c_k^{(\sigma_k)j}}\left(P_k^{(\sigma_k)j} - \sum_{\sigma_{k+1}=1}^{2}\lambda_{k+1}^{\sigma_{k+1}}\widetilde{A}_{k+1}^{(\sigma_{k+1})}(1+r)^{k-1}\right) + \vartheta_k^*$$

for $y \in \{1,2,3\}$,

given to firm i at stage $k \in \{1,2,3\}$ would lead to the realization of the imputation (4.15).

A subgame consistent solution can be readily obtained using (4.13), (4.15) and (4.16).

9.5 Extensions

The analysis can be expanded in a few directions.

Case 1: Random Changes in the State Dynamics Structures
Following Yeung (2011) one allow the structure of the state dynamics in (1.1) be affected by the random variable θ_k for $k \in \{1, 2, \cdots, T\}$. In particular the state dynamics become:

$$x_{k+1} = f_k\left(x_k, u_k^1, u_k^2, \cdots, u_k^n; \theta_k\right) + \vartheta_k, \tag{5.1}$$

for $k \in \{1, 2, \cdots, T\}$ and $x_1 = x^0$,

where $u_k^i \in U^i \subset R^{m_i}$ is the control vector of player i at stage k, $x_k \in X$ is the state, ϑ_k is a sequence of statistically independent random variables, and θ_k is an independent discrete random variables with range $\left\{\theta_k^1, \theta_k^2, \cdots, \theta_k^{\eta_k}\right\}$ and corresponding probabilities $\left\{\lambda_k^1, \lambda_k^2, \cdots, \lambda_k^{\eta_k}\right\}$.

Following the analyses in Sects. 9.1, 9.2 and 9.3, a theorem deriving a subgame consistent PDP can be established as follows.

Theorem 5.1 A payment equaling

$$B_k^{(\sigma_k)i}\left(x_k^*\right) = \xi^{(\sigma_k)i}\left(k, x_k^*\right)$$
$$- E_{\vartheta_k}\left[\sum_{\sigma_{k+1}-1}^{\eta_{k+1}} \lambda_{k+1}^{\sigma_{k+1}}\left(\xi^{(\sigma_{k+1})i}\left[k+1, f_k\left(x_k^*, \psi_k^{(\sigma_k)*}\left(x_k^*\right); \theta_k^{\sigma_k}\right) + \vartheta_k\right]\right)\right], \tag{5.2}$$

for $i \in N$,

given to player i at stage $k \in \{1, 2, \cdots, T\}$, if $\theta_k^{\sigma_k}$ occurs and $x_k^* \in X_k^*$, leads to the realization of the imputation according to the agreed upon optimality principle. ■

Case 2: More Complex Branching Processes
The random event θ_k affecting the payoff structures of the players in stage k may be more complex branching processes. For instance, the random variables may not be independent and may stem from a branching process in which the random variable θ_k for $k \in \{1, 2, \cdots, T\}$ is conditional upon the realization of the random variables in its preceding stages. An example of this type of processes is the one adopted in Yeung (2003) as a random variable stemming from the branching process as described below.

$\theta^1 = \left\{\theta_1^1, \theta_2^1, \ \ldots, \theta_{\eta_1}^1\right\}$ with corresponding probabilities $\left\{\lambda_1^1, \lambda_2^1, \ \ldots, \lambda_{\eta_1}^1\right\}$.

Given that $\theta_{a_1}^1$ is realized in time interval $[t_1, t_2)$, for $a_1 = 1, 2, \ldots, \eta_1$,

$\theta^2 = \left\{\theta_1^{2[(1,a_1)]}, \theta_2^{2[(1,a_1)]}, \ldots, \theta_{\eta_{2[(1,a_1)]}}^{2[(1,a_1)]}\right\}$ would be realized with the corresponding probabilities $\left\{\lambda_1^{2[(1,a_1)]}, \lambda_2^{2[(1,a_1)]}, \ldots, \lambda_{\eta_{2[(1,a_1)]}}^{2[(1,a_1)]}\right\}$.

Given that $\theta_{a_1}^1$ is realized in time interval $[t_1, t_2)$ and $\theta_{a_2}^{2[(1,a_1)]}$ is realized in time interval $[t_2, t_3)$, for $a_1 = 1, 2, \ldots, \eta_1$ and $a_2 = 1, 2, \ldots, \eta_{2[(1,a_1)]}$,

$\theta^3 = \left\{\theta_1^{3[(1,a_1)(2,a_2)]}, \theta_2^{3[(1,a_1)(2,a_2)]}, \ldots, \theta_{\eta_{3[(1,a_1)(2,a_2)]}}^{3[(1,a_1)(2,a_2)]}\right\}$ would be realized with the corresponding probabilities

$\left\{\lambda_1^{3[(1,a_1)(2,a_2)]}, \lambda_2^{3[(1,a_1)(2,a_2)]}, \ldots, \lambda_{\eta_{3[(1,a_1)(2,a_2)]}}^{3[(1,a_1)(2,a_2)]}\right\}$.

In general, given that $\theta_{a_1}^1$ is realized in time interval $[t_1, t_2)$, $\theta_{a_2}^{2[(1,a_1)]}$ is realized in time interval $[t_2, t_3)$, ..., and $\theta_{a_{k-1}}^{k-1[(1,a_1)(2,a_2)\ldots(k-2,a_{k-2})]}$ is realized in time interval $[t_{k-1}, t_k)$, for $a_1 = 1, 2, \ldots, \eta_1$, $a_2 = 1, 2, \ldots, \eta_{2[(1,a_1)]}$, $\ldots$, $a_{k-1} = 1, 2, \ldots, \eta_{k-1[(1,a_1)(2,a_2)\ldots(k-1,a_{k-1})]}$,

$$\theta^k = \left\{\theta_1^{k[(1,a_1)(2,a_2)\ldots(k-1,a_{k-1})]}, \theta_2^{k[(1,a_1)(2,a_2)\ldots(k-1,a_{k-1})]}, \ldots, \theta_{\eta_{k[(1,a_1)(2,a_2)\ldots(k-1,a_{k-1})]}}^{k[(1,a_1)(2,a_2)\ldots(k-1,a_{k-1})]}\right\}$$

would be realized with the corresponding probabilities

$$\left\{\lambda_1^{k[(1,a_1)(2,a_2)\ldots(k-1,a_{k-1})]}, \lambda_2^{k[(1,a_1)(2,a_2)\ldots(k-1,a_{k-1})]}, \ldots, \lambda_{\eta_{k[(1,a_1)(2,a_2)\ldots(k-1,a_{k-1})]}}^{k[(1,a_1)(2,a_2)\ldots(k-1,a_{k-1})]}\right\}$$

for $k = 1, 2, \ldots, \tau$.

Applying the techniques derived in the analysis in this paper, subgame consistent solutions can be derived accordingly.

Case 3: Games with Deterministic Dynamics

The analysis can be readily applied to derive subgame consistent solutions in randomly-furcating dynamic games in which the random variables ϑ_k in the stock dynamics are not present. In particular, the objective that player i seeks to maximize becomes

$$E_{\theta_1,\theta_2,\cdots,\theta_T}\left\{\sum_{k=1}^{T} g_k^i\left[x_k, u_k^1, u_k^2, \cdots, u_k^n; \theta_k\right] + q^i(x_{T+1})\right\}, \text{ for } i \in N \tag{5.3}$$

subject to the deterministic dynamics:

$$x_{k+1} = f_k\left(x_k, u_k^1, u_k^2, \cdots, u_k^n\right). \tag{5.4}$$

Following the analysis in Sects 9.3 and 9.4 and the proof of Theorem 3.1, a theorem deriving a subgame consistent PDP for the randomly-furcating dynamic game (5.3 and 5.4) can be established as follows.

Theorem 5.2 A payment equaling

$$B_k^{(\sigma_k)i}\left(x_k^*\right) = \xi^{(\sigma_k)i}\left(k, x_k^*\right) - \left[\sum_{\sigma_{k+1}=1}^{\eta_{k+1}} \lambda_{k+1}^{\sigma_{k+1}}\left(\xi^{(\sigma_{k+1})i}\left[k+1, f_k\left(x_k^*, \psi_k^{(\sigma_k)*}\left(x_k^*\right)\right)\right]\right)\right], \tag{5.5}$$

for $i \in N$,

given to player i at stage $k \in \{1, 2, \cdots, T\}$, if $\theta_k^{\sigma_k}$ occurs and $\in X_k^*$, would yield the PDP leading to a subgame consistent solution of the game (5.3 and 5.4). ■

9.6 Chapter Appendices

Appendix A. Proof of Proposition 4.1 Consider first the last stage, that is stage 3, when $\theta_3^{\sigma_3}$ occurs. Invoking that $V^{(\sigma_3)i}(3, x) = \left[A_3^{(\sigma_3)i} x + C_3^{(\sigma_3)i}\right]$ from Proposition 4.1 and $V^{(\sigma_3)i}(4, x) = q^i x \left(\frac{1}{1+r}\right)^3$, the conditions in equation (4.4) become

$$\left[A_3^{(\sigma_3)i} x + C_3^{(\sigma_3)i}\right] = \max_{u_3^{(\sigma_3)i}} \left\{\left[P_3^{(\sigma_3)i} u_3^{(\sigma_3)i} - \frac{c_3^{(\sigma_3)i}}{x}\left(u_3^{(\sigma_3)i}\right)^2\right]\left(\frac{1}{1+r}\right)^{k-1}\right.$$
$$\left. + \sum_{y=1}^{3} \gamma_3^y q^i \left[x + a - bx - u_3^{(\sigma_3)i} - \phi_3^{(\sigma_3)j*}(x) + \vartheta_3^y\right]\right\}, \text{ for } i \in \{1, 2\}. \tag{A.1}$$

Performing the indicated maximization in (A.1) yields:

$$\left[P_3^{(\sigma_3)i} - \frac{2c_3^{(\sigma_3)i} u_3^{(\sigma_3)i}}{x}\right]\left(\frac{1}{1+r}\right)^{k-1} - \sum_{y=1}^{3} \gamma_3^y q^i = 0, \text{ for } i \in \{1, 2\}. \tag{A.2}$$

The game equilibrium strategies in stage 3 can then be expressed as:

$$\phi_3^{(\sigma_3)i*}(x) = \left[P_3^{(\sigma_3)i} - (1+r)^2 q^i\right]\frac{x}{2c_3^{(\sigma_3)i}}, \text{ for } i \in \{1, 2\}. \tag{A.3}$$

Substituting (A.3) into (A.1) yields:

$$\left[A_3^{(\sigma_3)i} x + C_3^{(\sigma_3)i}\right] = \left[P_3^{(\sigma_3)i}\left[P_3^{(\sigma_3)i} - (1+r)^2 q^i\right]\frac{x}{2c_3^{(\sigma_3)i}}\right.$$
$$\left. - \frac{1}{4c_3^{(\sigma_3)i}}\left[P_3^{(\sigma_3)i} - (1+r)^2 q^i\right]^2 x\right]\left(\frac{1}{1+r}\right)^{k-1}$$
$$+ \sum_{y=1}^{3} \gamma_3^y q^i \left(x + a - bx - \sum_{j=1}^{2}\left[P_3^{(\sigma_3)j} - (1+r)^2 q^j\right]\frac{x}{2c_3^{(\sigma_3)j}} + \vartheta_3^y\right), \tag{A.4}$$

for $i \in \{1, 2\}$.

Note that both sides of equation (A.4) are linear expression of x, the terms $A_3^{(\sigma_3)i}$ and $C_3^{(\sigma_3)i}$, for $i \in \{1,2\}$ and $\sigma_3 \in \{1,2\}$, are explicitly given in (A.4).

Now we proceed to stage 2, the conditions in equation (4.4) become

$$\left[A_2^{(\sigma_2)i}x + C_2^{(\sigma_2)i}\right] = \max_{u_2^{(\sigma_2)i}} \left\{ \left[P_2^{(\sigma_2)i}u_2^{(\sigma_2)i} - \frac{c_2^{(\sigma_2)i}}{x}\left(u_2^{(\sigma_2)i}\right)^2\right]\left(\frac{1}{1+r}\right) \right.$$
$$\left. + \sum_{y=1}^{3}\gamma_2^y \sum_{\sigma_3=1}^{2} \lambda_3^{\sigma_3}\left[A_3^{(\sigma_3)i}\left[x + a - bx - u_2^{(\sigma_2)i} - \phi_2^{(\sigma_2)j*}(x) + \vartheta_2^y\right] + C_3^{(\sigma_3)i}\right]\right\}, \quad \text{(A.5)}$$

for $i \in \{1,2\}$.

Performing the indicated maximization in (A.5) yields:

$$\left[P_2^{(\sigma_2)i} - \frac{2c_2^{(\sigma_2)i}u_2^{(\sigma_2)i}}{x}\right]\left(\frac{1}{1+r}\right) - \sum_{y=1}^{3}\gamma_2^y \sum_{\sigma_3=1}^{2}\lambda_3^{\sigma_3}A_3^{(\sigma_3)i} = 0, \text{ for } i \in \{1,2\}. \quad \text{(A.6)}$$

The game equilibrium strategies in stage 2 can then be expressed as:

$$\phi_2^{(\sigma_2)i*}(x) = \left[P_2^{(\sigma_2)i} - (1+r)\sum_{\sigma_3=1}^{2}\lambda_3^{\sigma_3}A_3^{(\sigma_3)i}\right]\frac{x}{2c_2^{(\sigma_2)i}}, \text{ for } i \in \{1,2\}. \quad \text{(A.7)}$$

Substituting (A.7) into (A.5) yields:

$$\left[A_2^{(\sigma_2)i}x + C_2^{(\sigma_2)i}\right] = \left[P_2^{(\sigma_2)i}\left[P_2^{(\sigma_2)i} - (1+r)\sum_{\sigma_3=1}^{2}\lambda_3^{\sigma_3}A_3^{(\sigma_3)i}\right]\frac{x}{2c_2^{(\sigma_2)i}}\right.$$
$$\left. - \frac{1}{4c_2^{(\sigma_2)i}}\left[P_2^{(\sigma_2)i} - (1+r)\sum_{\sigma_3=1}^{2}\lambda_3^{\sigma_3}A_3^{(\sigma_3)i}\right]^2 x\right]\left(\frac{1}{1+r}\right)$$
$$+ \sum_{y=1}^{3}\gamma_2^y\sum_{\sigma_3=1}^{2}\lambda_3^{\sigma_3}\left[A_3^{(\sigma_3)i}\left(x + a - bx\right.\right.$$
$$\left.\left. - \sum_{j=1}^{2}\left[P_2^{(\sigma_2)j} - (1+r)\sum_{\sigma_3=1}^{2}\lambda_3^{\sigma_3}A_3^{(\sigma_3)i}\right]\frac{x}{2c_2^{(\sigma_2)j}} + \vartheta_2^y\right) + C_3^{(\sigma_3)i}\right],$$
$$\text{(A.8)}$$

for $i \in \{1,2\}$.

Once again, both sides of equation (A.8) are linear expression of x, the terms $A_2^{(\sigma_2)i}$ and $C_2^{(\sigma_2)i}$, for $i \in \{1,2\}$ and $\sigma_2 \in \{1,2\}$, can be obtained explicitly using (A.8).

Finally, we proceed to the first stage, the conditions in equation (4.4) become

$$\left[A_1^{(\sigma_1)i}x + C_1^{(\sigma_1)i}\right] = \max_{u_1^{(\sigma_1)i}}\left\{\left[P_1^{(\sigma_1)i}u_1^{(\sigma_1)i} - \frac{c_1^{(\sigma_1)i}}{x}\left(u_1^{(\sigma_1)i}\right)^2\right]\right.$$
$$\left.+\sum_{y=1}^{3}\gamma_1^y\sum_{\sigma_2=1}^{2}\lambda_2^{\sigma_2}\left[A_2^{(\sigma_2)i}\left[x+a-bx-u_1^{(\sigma_1)i}-\phi_1^{(\sigma_1)j}(x)+\vartheta_1^y\right]+C_2^{(\sigma_2)i}\right]\right\} \quad \text{(A.9)}$$

for $i\in\{1,2\}$.

Following the analysis in (A.6 and A.7), the game equilibrium strategies in stage 1 can then be expressed as:

$$\phi_1^{(\sigma_1)i*}(x) = \left[P_1^{(\sigma_1)i} - \sum_{\sigma_2=1}^{2}\lambda_2^{\sigma_2}A_2^{(\sigma_2)i}\right]\frac{x}{2c_1^{(\sigma_1)i}}, \quad \text{for } i\in\{1,2\}. \quad \text{(A.10)}$$

Substituting (A.10) into (A.9) yields:

$$\left[A_1^{(\sigma_1)i}x + C_1^{(\sigma_1)i}\right] = \left[P_1^{(\sigma_1)i}\left[P_1^{(\sigma_1)i} - \sum_{\sigma_2=1}^{2}\lambda_2^{\sigma_2}A_2^{(\sigma_2)i}\right]\frac{x}{2c_1^{(\sigma_1)i}}\right.$$
$$\left.-\frac{1}{4c_1^{(\sigma_1)i}}\left[P_1^{(\sigma_1)i} - \sum_{\sigma_2=1}^{2}\lambda_2^{\sigma_2}A_2^{(\sigma_2)i}\right]^2 x\right] + \sum_{y=1}^{3}\gamma_1^y\sum_{\sigma_3=1}^{2}\lambda_2^{\sigma_2}$$
$$\left[A_2^{(\sigma_2)i}\left(x+a-bx\right.\right.$$
$$\left.\left.-\sum_{j=1}^{2}\left[P_1^{(\sigma_1)j} - (1+r)\sum_{\sigma_2=1}^{2}\lambda_2^{\sigma_2}A_2^{(\sigma_2)i}\right]\frac{x}{2c_1^{(\upsilon_1)j}} + \vartheta_1^y\right) + C_2^{(\sigma_2)i}\right], \quad \text{(A.11)}$$

for $i\in\{1,2\}$.

Once again, both sides of equation (A.11) are linear expression of x, the terms $A_1^{(\sigma_1)i}$ and $C_1^{(\sigma_1)i}$, for $i\in\{1,2\}$ and $\sigma_1 = 1$, can be obtained explicitly using (A.11).

Appendix B. Proof of Proposition 4.2 Consider first the last stage, that is stage 3, when $\theta_3^{\sigma_3}$ occurs. Invoking that $W^{(\sigma_3)}(3,x) = \left[\widetilde{A}_3^{(\sigma_3)}x + \widetilde{C}_3^{(\sigma_3)}\right]$ from Proposition 4.2 and $W^{(\sigma_3)}(4,x) = \sum_{j=1}^{2}q^j x\left(\frac{1}{1+r}\right)^3$, the condition in equation (4.9) becomes

$$\left[\widetilde{A}_3^{(\sigma_3)}x + \widetilde{C}_3^{(\sigma_3)}\right] = \max_{u_3^{(\sigma_3)1},\,u_3^{(\sigma_3)2}}\left\{\sum_{j=1}^{2}\left[P_3^{(\sigma_3)j}u_3^{(\sigma_3)j} - \frac{c_3^{(\sigma_3)j}}{x}\left(u_3^{(\sigma_3)j}\right)^2\right]\left(\frac{1}{1+r}\right)^{k-1}\right.$$
$$\left.+\sum_{y=1}^{3}\gamma_3^y\sum_{j=1}^{2}q^j\left[x+a-bx-\sum_{\ell=1}^{2}u_3^{(\sigma_3)\ell}+\vartheta_3^y\right]\right\}. \quad \text{(B.1)}$$

Performing the indicated maximization in (B.1) yields:

$$\left[P_3^{(\sigma_3)i} - \frac{2c_3^{(\sigma_3)i}u_3^{(\sigma_3)i}}{x}\right]\left(\frac{1}{1+r}\right)^{k-1} - \sum_{y=1}^{3}\gamma_3^y\sum_{j=1}^{2}q^j = 0, \text{ for } i \in \{1,2\}. \quad \text{(B.2)}$$

The optimal cooperative strategies in stage 3 can then be expressed as:

$$\psi_3^{(\sigma_3)i*}(x) = \left[P_3^{(\sigma_3)i} - (1+r)^2\sum_{j=1}^{2}q^j\right]\frac{x}{2c_3^{(\sigma_3)i}}, \text{ for } i \in \{1,2\}. \quad \text{(B.3)}$$

Substituting (B.3) into (B.1) yields:

$$\begin{aligned}\left[\widetilde{A}_3^{(\sigma_3)}x + \widetilde{C}_3^{(\sigma_3)}\right] &= \sum_{j=1}^{2}\left[P_3^{(\sigma_3)j}\left[P_3^{(\sigma_3)j} - (1+r)^2\sum_{\ell=1}^{2}q^\ell\right]\frac{x}{2c_3^{(\sigma_3)j}}\right.\\ &\quad \left.-\frac{1}{4c_3^{(\sigma_3)j}}\left[P_3^{(\sigma_3)j} - (1+r)^2\sum_{\ell=1}^{2}q^\ell\right]^2 x\right]\left(\frac{1}{1+r}\right)^{k-1}\\ &\quad +\sum_{y=1}^{3}\gamma_3^y\sum_{j=1}^{2}q^j\left(x + a - bx\right.\\ &\quad \left.-\sum_{\ell=1}^{2}\left[P_3^{(\sigma_3)\ell} - (1+r)^2\sum_{\zeta=1}^{2}q^\zeta\right]\frac{x}{2c_3^{(\sigma_3)j}} + \vartheta_3^y\right),\end{aligned} \quad \text{(B.4)}$$

for $i \in \{1,2\}$.

Note that both sides of equation (B.4) are linear expression of x, the terms $\widetilde{A}_3^{(\sigma_3)}$ and $\widetilde{C}_3^{(\sigma_3)}$, for $\sigma_3 \in \{1,2\}$, are explicitly given in (B.4).

Now we proceed to stage 2, the condition in equation (4.9) becomes

$$\begin{aligned}\left[\widetilde{A}_2^{(\sigma_2)}x + \widetilde{C}_2^{(\sigma_2)}\right] &= \max_{u_2^{(\sigma_2)1}, u_2^{(\sigma_2)2}}\left\{\sum_{j=1}^{2}\left[P_2^{(\sigma_2)j}u_2^{(\sigma_2)j} - \frac{c_2^{(\sigma_2)j}}{x}\left(u_2^{(\sigma_2)j}\right)^2\right]\left(\frac{1}{1+r}\right)\right.\\ &\left.+\sum_{y=1}^{3}\gamma_2^y\sum_{\sigma_3=1}^{2}\lambda_3^{\sigma_3}\left[\widetilde{A}_3^{(\sigma_3)}\left[x + a - bx - \sum_{j=1}^{2}u_2^{(\sigma_2)j} + \vartheta_2^y\right] + \widetilde{C}_3^{(\sigma_3)}\right]\right\}.\end{aligned} \quad \text{(B.5)}$$

Performing the indicated maximization in (B.5) yields:

$$\left[P_2^{(\sigma_2)i} - \frac{2c_2^{(\sigma_2)i}u_2^{(\sigma_2)i}}{x}\right]\left(\frac{1}{1+r}\right) - \sum_{y=1}^{3}\gamma_2^y\sum_{\sigma_3=1}^{2}\lambda_3^{\sigma_3}\widetilde{A}_3^{(\sigma_3)} = 0, \text{ for } i \in \{1,2\}. \quad \text{(B.6)}$$

The optimal cooperative strategies in stage 2 can then be expressed as:

$$\psi_2^{(\sigma_2)i*}(x) = \left[P_2^{(\sigma_2)i} - (1+r)\sum_{\sigma_3=1}^{2}\lambda_3^{\sigma_3}\widetilde{A}_3^{(\sigma_3)}\right]\frac{x}{2c_2^{(\sigma_2)i}}, \text{ for } i\in\{1,2\}. \tag{B.7}$$

Substituting (B.7) into (B.5) yields:

$$\begin{aligned}\left[\widetilde{A}_2^{(\sigma_2)}x + \widetilde{C}_2^{(\sigma_2)}\right] &= \sum_{j=1}^{2}\left[P_2^{(\sigma_2)j}\left[P_2^{(\sigma_2)j} - (1+r)\sum_{\sigma_3=1}^{2}\lambda_3^{\sigma_3}\widetilde{A}_3^{(\sigma_3)}\right]\frac{x}{2c_2^{(\sigma_2)j}}\right.\\
&\quad -\frac{1}{4c_2^{(\sigma_2)j}}\left[P_2^{(\sigma_2)j} - (1+r)\sum_{\sigma_3=1}^{2}\lambda_3^{\sigma_3}\widetilde{A}_3^{(\sigma_3)}\right]^2 x\Bigg]\left(\frac{1}{1+r}\right)\\
&\quad +\sum_{y=1}^{3}\gamma_2^y\sum_{\sigma_3=1}^{2}\lambda_3^{\sigma_3}\left[\widetilde{A}_3^{(\sigma_3)}\left(x + a - bx\right.\right.\\
&\quad \left.\left.-\sum_{j=1}^{2}\left[P_2^{(\sigma_2)j} - (1+r)\sum_{\sigma_3=1}^{2}\lambda_3^{\sigma_3}\widetilde{A}_3^{(\sigma_3)}\right]\frac{x}{2c_2^{(\sigma_2)j}} + \vartheta_2^y\right) + \widetilde{C}_3^{(\sigma_3)}\right].\end{aligned} \tag{B.8}$$

Once again, both sides of equation (B.8) are linear expression of x, the terms $\widetilde{A}_2^{(\sigma_2)}$ and $\widetilde{C}_2^{(\sigma_2)}$, for $\sigma_2\in\{1,2\}$, can be obtained explicitly using (B.8).

Finally, we proceed to the first stage, the conditions in equation (4.9) become

$$\begin{aligned}\left[\widetilde{A}_1^{(\sigma_1)}x + \widetilde{C}_1^{(\sigma_1)}\right] &= \max_{u_1^{(\sigma_1)1},\,u_1^{(\sigma_1)2}}\left\{\sum_{j=1}^{2}\left[P_1^{(\sigma_1)j}u_1^{(\sigma_1)j} - \frac{c_1^{(\sigma_1)j}}{x}\left(u_1^{(\sigma_1)j}\right)^2\right]\right.\\
&\left. +\sum_{y=1}^{3}\gamma_1^y\sum_{\sigma_2=1}^{2}\lambda_2^{\sigma_2}\left[\widetilde{A}_2^{(\sigma_2)}\left[x + a - bx - \sum_{j=1}^{2}u_1^{(\sigma_1)j} + \vartheta_1^y\right] + \widetilde{C}_2^{(\sigma_2)}\right]\right\}.\end{aligned} \tag{B.9}$$

Following the analysis in (B.6 and B.7), the optimal cooperative strategies in stage 1 can then be expressed as:

$$\psi_1^{(\sigma_1)i*}(x) = \left[P_1^{(\sigma_1)i} - \sum_{\sigma_1=1}^{2}\lambda_2^{\sigma_2}\widetilde{A}_2^{(\sigma_2)}\right]\frac{x}{2c_1^{(\sigma_1)i}}, \text{ for } i\in\{1,2\}. \tag{B.10}$$

Substituting (B.10) into (B.9) yields:

$$\begin{aligned}\left[\widetilde{A}_1^{(\sigma_1)}x + \widetilde{C}_1^{(\sigma_1)}\right] &= \sum_{j=1}^{2}\left[P_1^{(\sigma_1)j}\left[P_1^{(\sigma_1)j} - \sum_{\sigma_2=1}^{2}\lambda_2^{\sigma_2}\widetilde{A}_2^{(\sigma_2)}\right]\frac{x}{2c_1^{(\sigma_1)j}}\right.\\
&\quad \left.-\frac{1}{4c_1^{(\sigma_1)j}}\left[P_1^{(\sigma_1)j} - \sum_{\sigma_2=1}^{2}\lambda_2^{\sigma_2}\widetilde{A}_2^{(\sigma_2)}\right]^2 x\right]\\
&\quad +\sum_{y=1}^{3}\gamma_1^y\sum_{\sigma_2=1}^{2}\lambda_2^{\sigma_2}\left[\widetilde{A}_2^{(\sigma_2)}\left(x + a - bx\right.\right.\\
&\quad \left.\left.-\sum_{j=1}^{2}\left[P_1^{(\sigma_1)j} - \sum_{\sigma_2=1}^{2}\lambda_2^{\sigma_2}\widetilde{A}_2^{(\sigma_2)}\right]\frac{x}{2c_1^{(\sigma_1)j}} + \vartheta_1^y\right) + \widetilde{C}_2^{(\sigma_2)}\right].\end{aligned} \tag{B.11}$$

Once again, both sides of equation (B.11) are linear expression of x, the terms $\widetilde{A}_1^{(\sigma_1)}$ and $\widetilde{C}_1^{(\sigma_1)}$, for $\sigma_1 = 1$, can be obtained explicitly using (B.11).

9.7 Chapter Notes

This Chapter considers subgame-consistent cooperative solutions in randomly furcating stochastic dynamic games developed by Yeung and Petrosyan (2013a). The extension of continuous-time randomly furcating stochastic differential games to an analysis in discrete time is not just of theoretical interest but also for practical reasons in applications in operations research. In the process of obtaining the main results for subgame consistent solution, Nash equilibrium for randomly furcating stochastic dynamic games and optimal control for randomly furcating stochastic control problems are also derived. Yeung and Petrosyan (2014b) considered subgame consistent cooperative provision of public goods under accumulation and payoff uncertainties. Yeung and Petrosyan (2014a) examined subgame consistent solution for a dynamic game of pollution management in which future environmental costs are not known with certainty.

9.8 Problems

1. Consider an economy endowed with a renewable resource and with 2 resource extractors (firms). The lease for resource extraction begins at stage 1 and ends at stage 3 for these two firms. Let u_k^i denote the resource extracted by firm i at stage k, for $i \in \{1, 2\}$. Let U^i be the set of admissible amount of resource extracted by firm i, and $x_k \in X \subset R^+$ be the size of the resource stock at stage k.

   It is known at each stage there is a random element, θ_k for $k \in \{1, 2, 3\}$, affecting the revenues of the outputs produced by these firms and their costs of extraction. If θ_k^1 happens at stage $k \in \{2, 3\}$ the profits (in present-value) that firm 1 and firm 2 will obtain at stage k are respectively:

$$\left[4u_k^1 - \frac{2}{x_k}(u_k^1)^2\right]\left(\frac{1}{1+r}\right)^{k-1} \text{ and } \left[2u_k^2 - \frac{1}{x_k}(u_k^2)^2\right]\left(\frac{1}{1+r}\right)^{k-1},$$

   where $r = 0.05$ is the discount rate.

   If θ_k^2 happens at stage $k \in \{2, 3\}$ the profits (in present-value) that firm 1 and firm 2 will obtain at stage k are respectively:

$$\left[2u_k^1 - \frac{2}{x_k}(u_k^1)^2\right]\left(\frac{1}{1+r}\right)^{k-1} \text{ and } \left[3u_k^2 - \frac{2}{x_k}(u_k^2)^2\right]\left(\frac{1}{1+r}\right)^{k-1}.$$

It is known in stage 1 that θ_1^1 has occurred. The probability that θ_k^1 will occur at stage $k \in \{2,3\}$ is 0.6 and the probability that θ_k^2 will occur at stage $k \in \{2,3\}$ is 0.4. In stage 4, a terminal payment (again in present-value) equaling $x_4\left(\frac{1}{1+r}\right)^3$ will be paid to firm 1 and a terminal payment (again in present-value) equaling $0.5x_4\left(\frac{1}{1+r}\right)^3$ will be paid to firm 2.

The growth dynamics of the resource is governed by the stochastic difference equation:

$$x_{k+1} = x_k + 15 - 0.1x_k - \sum_{j=1}^{2} u_k^j + \vartheta_k,$$

for $k \in \{1,2,3\}$ and $x_1 = 12$,

where ϑ_k is a random variable with non-negative range $\{0,1,2\}$ and corresponding probabilities $\{0.1, 0.7, 0.2\}$; moreover $\vartheta_1, \vartheta_2, \vartheta_3$ are independent. Moreover, we have the constraint $u_k^1 + u_k^2 \leq 0.9x_k + 15$.

The objective of extractor $i \in \{1,2\}$ is to maximize the present value of the expected stream of future profits:

Characterize the feedback Nash equilibrium.

2. Obtain a group optimal solution that maximizes the joint expected profit.
3. Consider the case when the extractors agree to share the excess of cooperative gains over their expected noncooperative profits equally. Derive a subgame consistent solution.

Chapter 10
Subgame Consistency Under Furcating Payoffs, Stochastic Dynamics and Random Horizon

This Chapter investigates the class of randomly furcating stochastic dynamic games with uncertain game horizon. In particular, in this class of games, there exist uncertainties in the state dynamics, future payoff structures and game horizon. The non-cooperative Nash equilibrium is characterized and subgame-consistent cooperative solutions is derived. A discrete-time analytically tractable payoff distribution procedures contingent upon specific random realizations of the state and payoff structure are derived. This approach widens the application of cooperative dynamic game theory to discrete-time random horizon problems where the evolution of the state and future environments are not known with certainty. In addition, a corresponding form of Bellman equations for solving inter-temporal problems with randomly furcating payoffs and random horizon is developed to serve as the foundation of solving the game problem. To characterize a noncooperative game equilibrium, a set of random duration discrete-time Hamilton-Jacobi-Bellman equations is presented. Subgame consistent solution and corresponding Payoff Distribution Procedures are provided. The analysis is developed along the work of Yeung and Petrosyan (2014c).

The Chapter is organized as follows. The game formulation and the development of a Bellman equation to characterize the stochastic control problem are provided in Sect. 10.1. The non-cooperative game outcome is derived in Sect. 10.2. The issues of group optimality and individual rationality in dynamic cooperation are discussed in Sect. 10.3. Subgame consistent solutions and payment mechanism leading to the realization of these solutions are analyzed in Sect. 10.4. Section 10.5 presents an illustration in cooperative resource extraction under uncertainty. Chapter appendices are provided in Sect. 10.6. Chapter notes are given in Sect. 10.7 and problems in Sect. 10.8.

D.W.K. Yeung, L.A. Petrosyan, *Subgame Consistent Cooperation*,
Theory and Decision Library C 47, DOI 10.1007/978-981-10-1545-8_10

10.1 Game Formulation and Control Techniques

In this section, we first present the formulation of a stochastic dynamic game with randomly furcating future payoffs and random horizon. Then we develop a stochastic control technique for solving intertemporal problem with randomly furcating payoffs and uncertain horizon to serve as the foundation of solving the game problem.

10.1.1 Game Formulation

Consider the discrete time $\hat{T}$ – stage dynamic optimization problem where $\hat{T}$ is a random variable with range $\{1, 2, \cdots, T\}$ and corresponding probabilities $\{\varpi_1, \varpi_2, \cdots, \varpi_T\}$. Conditional upon the reaching of stage τ, the probability of the game would last up to stages $\tau, \tau + 1, \cdots, T$ becomes respectively

$$\frac{\varpi_\tau}{\sum_{\zeta=\tau}^{T} \varpi_\zeta}, \frac{\varpi_{\tau+1}}{\sum_{\zeta=\tau}^{T} \varpi_\zeta}, \cdots, \frac{\varpi_T}{\sum_{\zeta=\tau}^{T} \varpi_\zeta}. \tag{1.1}$$

The state space of the game is $X \in R^m$ and the state dynamics of the game is characterized by the stochastic difference equation:

$$x_{k+1} = f_k(x_k, u_k^1, u_k^2, \cdots, u_k^n) + \vartheta_k, \tag{1.2}$$

for $k \in \{1, 2, \cdots, T\}$ and $x_1 = x^0$,

where $u_k^i \in U^i \subset R^{m_i}$ is the control vector of player i at stage k, $x_k \in X$ is the state, and ϑ_k is a sequence of statistically independent random variables.

The payoff of player i at stage k is $g_k^i[x_k, u_k^1, u_k^2, \cdots, u_k^n; \theta_k]$ which is affected by a random variable θ_k. In particular, θ_k for $k \in \{1, 2, \cdots, T\}$ are independent random variables with range $\{\theta_k^1, \theta_k^2, \cdots, \theta_k^{\eta_k}\}$ and corresponding probabilities $\{\lambda_k^1, \lambda_k^2, \cdots, \lambda_k^{\eta_k}\}$. In stage 1, it is known that θ_1 equals θ_1^1 with probability $\lambda_1^1 = 1$. When the game ends after stage $\hat{T}$, a terminal payment $q_{\hat{T}+1}^i(x_{\hat{T}+1})$ will be given to player i in stage $\hat{T} + 1$.

The objective that player i seeks to maximize is

$$E_{\theta_1, \theta_2, \cdots, \theta_T; \vartheta_1, \vartheta_2, \cdots, \vartheta_T}\left\{\sum_{\hat{T}=1}^{T} \varpi_{\hat{T}} \left[\sum_{k=1}^{\hat{T}} g_k^i[x_k, u_k^1, u_k^2, \cdots, u_k^n; \theta_k] + q^i(x_{\hat{T}+1})\right]\right\},$$

for $i \in \{1, 2, \cdots, n\} \equiv N$,

where $E_{\theta_1,\theta_2,\cdots,\theta_T;\vartheta_1,\vartheta_2,\cdots,\vartheta_T}$ is the expectation operation with respect to the random variables $\theta_1,\theta_2,\cdots,\theta_T$ and $\vartheta_1,\vartheta_2,\cdots,\vartheta_T$. Since there is no uncertainty in the payoff structure in stage $T+1$, we denote $\sigma_{T+1}=1, \theta_{T+1}^{\sigma_{T+1}}=\theta_{T+1}^1$ with probability $\lambda_{T+1}^{\sigma_{T+1}}=\lambda_{T+1}^1=1$ for notational convenience. The payoffs of the players are transferable.

The objective function of player i can be expressed as

$$E_{\vartheta_1,\vartheta_2,\cdots,\vartheta_T}\left\{\sum_{\hat{T}=1}^{T}\varpi_{\hat{T}}\left[g_1^i\left[x_1,u_1^1,u_1^2,\cdots,u_1^n;\theta_1^1\right]\right.\right.$$

$$\left.\left.+\sum_{k=2}^{\hat{T}}\sum_{\sigma_k=1}^{\eta_k}\lambda_k^{\sigma_k}g_k^i\left[x_k,u_k^1,u_k^2,\cdots,u_k^n;\theta_k^{\sigma_k}\right]+q^i\left(x_{\hat{T}+1}\right)\right]\right\}$$

$$=E_{\vartheta_1,\vartheta_2,\cdots,\vartheta_T}\left\{g_1^i\left[x_1,u_1^1,u_1^2,\cdots,u_1^n;\theta_1^1\right]+\frac{\varpi_1}{\sum_{\zeta=1}^{T}\varpi_\zeta}q_2^i(x_{\tau+1})\right.$$

$$\left.+\frac{\sum_{\hat{T}=2}^{T}\varpi_{\hat{T}}}{\sum_{\zeta=1}^{T}\varpi_\zeta}\left[\sum_{k=2}^{\hat{T}}\sum_{\sigma_k=1}^{\eta_k}\lambda_k^{\sigma_k}g_k^i\left[x_k,u_k^1,u_k^2,\cdots,u_k^n;\theta_k^{\sigma_k}\right]+q^i\left(x_{\hat{T}+1}\right)\right]\right\}, \quad (1.3)$$

where the notation of the sum $\sum_{k=\tau+1}^{\tau}$ being an empty sum is adopted.

The game (1.2 and 1.3) is a randomly furcating stochastic dynamic game with random horizon. To solve the game (1.2 and 1.3), we first have to derive a stochastic dynamic programming technique for solving a stochastic dynamic programming problem with random horizon and randomly furcating payoffs.

10.1.2 Random Horizon Stochastic Dynamic Programming with Uncertain Payoffs

Consider the case when $n=1$ in the system (1.2 and 1.3). The problem can be formulized as the maximization of the expected payoff:

$$E_{\vartheta_1,\vartheta_2,\cdots,\vartheta_T}\left\{g_1\left[x_1,u_1;\theta_1^1\right]+\frac{\varpi_1}{\sum_{\zeta=1}^{T}\varpi_\zeta}q_2(x_2)\right.$$

$$\left.+\frac{\sum_{\hat{T}=2}^{T}\varpi_{\hat{T}}}{\sum_{\zeta=1}^{T}\varpi_\zeta}\left[\sum_{k=2}^{\hat{T}}\sum_{\sigma_k=1}^{\eta_k}\lambda_k^{\sigma_k}g_k\left[x_k,u_k;\theta_k^{\sigma_k}\right]+q\left(x_{\hat{T}+1}\right)\right]\right\} \tag{1.4}$$

subject to

$$x_{k+1}=f_k(x_k,u_k)+\vartheta_k, \tag{1.5}$$

for $k\in\{1,2,\cdots,T\}$ and $x_1=x^0$.

Now consider the case when stage τ has arrived, if $\theta_\tau^{\sigma_\tau}\in\left\{\theta_\tau^1,\theta_\tau^2,\cdots,\theta_\tau^{\eta_\tau}\right\}$ occurs and the state $x_\tau=x$, the problem can be formulized as the maximization of the expected payoff:

$$E_{\vartheta_\tau,\vartheta_{\tau+1},\cdots,\vartheta_T}\left\{\sum_{\hat{T}=\tau}^{T}\frac{\varpi_{\hat{T}}}{\sum_{\zeta=\tau}^{T}\varpi_\zeta}\left[g_\tau\left(x,u_\tau;\theta_\tau^{\sigma_\tau}\right)\right.\right.$$

$$\left.\left.+\sum_{k=\tau+1}^{\hat{T}}\sum_{\sigma_k=1}^{\eta_k}\lambda_k^{\sigma_k}g_k(x_k,u_k;\theta_k)+q_{\hat{T}+1}\left(x_{\hat{T}+1}\right)\right]\right\}$$

$$=E_{\vartheta_\tau,\vartheta_{\tau+1},\cdots,\vartheta_T}\left\{g_\tau\left[x_\tau,u_\tau;\theta_\tau^{\sigma_\tau}\right]+\frac{\varpi_\tau}{\sum_{\zeta=\tau}^{T}\varpi_\zeta}q_{\tau+1}(x_{\tau+1})\right.$$

$$\left.+\frac{\sum_{\hat{T}=\tau+1}^{T}\varpi_{\hat{T}}}{\sum_{\zeta=\tau}^{T}\varpi_\zeta}\left[\sum_{k=\tau+1}^{\hat{T}}\sum_{\sigma_k=1}^{\eta_k}\lambda_k^{\sigma_k}g_k\left[x_k,u_k;\theta_k^{\sigma_k}\right]+q\left(x_{\hat{T}+1}\right)\right]\right\} \tag{1.6}$$

subject to the dynamics

$$x_{k+1}=f_k(x_k,u_k)+\vartheta_k,\quad x_\tau=x,\ \text{for}\,k\in\{\tau,\tau+1,\cdots,T\}. \tag{1.7}$$

We use $V^{(\sigma_t)}(\tau,x)$ to denote the value function (if it exist)

$$\max_{u_\tau, u_{\tau+1}, \cdots, u_T} E_{\vartheta_\tau, \vartheta_{\tau+1}, \cdots, \vartheta_T} \left\{ g_\tau\left[x_\tau, u_\tau; \theta_\tau^{\sigma_\tau}\right] + \frac{\varpi_\tau}{\sum_{\zeta=\tau}^{T} \varpi_\zeta} q_{\tau+1}(x_{\tau+1}) \right.$$

$$\left. + \frac{\sum_{\hat{T}=\tau+1}^{T} \varpi_{\hat{T}}}{\sum_{\zeta=\tau}^{T} \varpi_\zeta} \left[\sum_{k=\tau+1}^{\hat{T}} \sum_{\sigma_k=1}^{\eta_k} \lambda_k^{\sigma_k} g_k\left[x_k, u_k; \theta_k^{\sigma_k}\right] + q\left(x_{\hat{T}+1}\right)\right]\right\} \tag{1.8}$$

A theorem characterizing a solution to the stochastic dynamic programming problem with uncertain future payoffs and random horizon in (1.4 and 1.5) is provided as follows.

Theorem 1.1 A set of strategies $\left\{u_k^{(\sigma_k)*} = \phi_k^{(\sigma_k)*}(x) \in \Gamma, \text{ for } \sigma_k \in \{1, 2, \cdots, \eta_k\} \text{ and } k \in \{1, 2, \cdots, T\}\right\}$ provides an optimal solution to the problem (1.4 and 1.5) if there exist functions $V^{(\sigma_k)}(k, x)$, for $k \in \{1, 2, \cdots, T\}$, such that the following recursive relations are satisfied:

$$V^{(\sigma_{T+1})}(T+1, x) = q_{T+1}(x),$$

$$V^{(\sigma_T)}(T, x) = \max_{u_T} E_{\vartheta_T}\left\{ g_T(x, u_T; \theta_T^{\sigma_T}) + V^{(\sigma_{T+1})}[T+1, f_T(x, u_T) + \vartheta_T]\right\}$$

$$= E_{\vartheta_T}\left\{ g_T\left(x, \phi_T^{(\sigma_T)*}(x); \theta_T^{\sigma_T}\right) + V^{(\sigma_{T+1})}\left[T+1, f_T\left(x, \phi_T^{(\sigma_T)*}(x)\right) + \vartheta_T\right]\right\},$$

$$V^{(\sigma_\tau)}(\tau, x) = \max_{u_\tau} E_{\vartheta_\tau}\left\{ g_\tau(x, u_\tau; \theta_\tau^{\sigma_\tau}) + \frac{\varpi_\tau}{\sum_{\zeta=\tau}^{T} \varpi_\zeta} q_{\tau+1}[f_\tau(x, u_\tau) + \vartheta_\tau] \right.$$

$$\left. + \frac{\sum_{\zeta=\tau+1}^{T} \varpi_\zeta}{\sum_{\zeta=\tau}^{T} \varpi_\zeta} \sum_{\sigma_{\tau+1}=1}^{\eta_{\tau+1}} \lambda_{\tau+1}^{\sigma_{\tau+1}} V^{(\sigma_{\tau+1})}[\tau+1, f_\tau(x, u_\tau) + \vartheta_\tau]\right\}$$

$$= E_{\vartheta_\tau}\left\{ g_\tau(x, \phi_\tau^{(\sigma_\tau)*}(x); \theta_\tau^{\sigma_\tau}) + \frac{\varpi_\tau}{\sum_{\zeta=\tau}^{T} \varpi_\zeta} q_{\tau+1}\left[f_\tau\left(x, \phi_\tau^{(\sigma_\tau)*}(x)\right) + \vartheta_\tau\right] \right.$$

$$\left. + \frac{\sum_{\zeta=\tau+1}^{T} \varpi_\zeta}{\sum_{\zeta=\tau}^{T} \varpi_\zeta} \sum_{\sigma_{\tau+1}=1}^{\eta_{\tau+1}} \lambda_{\tau+1}^{\sigma_{\tau+1}} V^{(\sigma_{\tau+1})}\left[\tau+1, f_\tau(x, \phi_\tau^{(\sigma_\tau)*}(x)) + \vartheta_\tau\right]\right\}, \text{ for } \tau \in \{1, 2, \cdots, T-1\}.$$

$$\tag{1.9}$$

Proof See Appendix A. ■

Theorem 1.1 yields a set of optimality equations for discrete-time random horizon stochastic dynamic programming problem with uncertain future payoffs.

10.2 Noncooperative Outcome

To solve the noncoperative outcome of the game (1.2 and 1.3), we invoke the technique of backward induction and begin with the subgame starting at last operating stages, that is stage T. If $\theta_T^{\sigma_T} \in \{\theta_T^1, \theta_T^2, \cdots, \theta_T^{\eta_T}\}$ has occurred at stage T and the state $x_T = x$, the subgame becomes:

$$\max_{u_T^i} E_{\vartheta_T}\left\{g_T^i\left[x, u_T^1, u_T^2, \cdots, u_T^n; \theta_T^{\sigma_T}\right] + q^i(x_{T+1})\right\}, \text{ for } i \in N,$$

subject to

$$x_{T+1} = f_T(x, u_T^1, u_T^2, \cdots, u_T^n) + \vartheta_T. \tag{2.1}$$

A set of state-dependent strategies $\{\phi_T^{(\sigma_T)i*}(x) \in \Gamma^i, \text{ for } i \in N\}$ constitutes a Nash equilibrium solution to the subgame (2.1) if the following conditions are satisfied:

$$\begin{aligned}
&V^{(\sigma_T)i}(T,x) = \\
&E_{\vartheta_T}\left\{g_T^i\left[x, \phi_T^{(\sigma_T)1*}(x), \phi_T^{(\sigma_T)2*}(x), \cdots, \phi_T^{(\sigma_T)n*}(x); \theta_T^{\sigma_T}\right] + q^i(x_{T+1})\right\} \\
&\geq E_{\vartheta_T}\left\{g_T^i\left[x, \phi_T^{(\sigma_T)1*}(x), \phi_T^{(\sigma_T)2*}(x), \cdots, \phi_T^{(\sigma_T)i-1*}(x), \phi_T^{(\sigma_T)i*}(x), \phi_T^{(\sigma_T)i+1*}(x), \cdots\right.\right. \\
&\left.\left. \cdots, \phi_T^{(\sigma_T)n*}(x); \theta_T^{\sigma_T}\right] + q^i(\tilde{x}_{T+1})\right\},
\end{aligned}$$

for $i \in N$,

$$\text{where } x_{T+1} = f_T\left[x, \phi_T^{(\sigma_T)1*}(x), \phi_T^{(\sigma_T)2*}(x), \cdots, \phi_T^{(\sigma_T)n*}(x)\right] + \vartheta_T$$

$$\begin{aligned}
&\tilde{x}_{T+1} = f_T\left[x, \phi_T^{(\sigma_T)1*}(x), \phi_T^{(\sigma_T)2*}(x), \cdots, \phi_T^{(\sigma_T)i-1*}(x), \phi_T^{(\sigma_T)i}(x), \phi_T^{(\sigma_T)i+1*}(x), \cdots\right. \\
&\left. \cdots, \phi_T^{(\sigma_T)n*}(x)\right] + \vartheta_T.
\end{aligned}$$

A characterization of the Nash equilibrium of the subgame (2.1) is provided in the following lemma.

Lemma 2.1 A set of strategies $\{u_T^{i*} = \phi_T^{(\sigma_T)i*}(x) \in \Gamma^i, \text{ for } i \in N\}$ provides a Nash equilibrium solution to the subgame (2.1) if there exist functions $V^{(\sigma_T)i}(T,x)$, for $i \in N$, such that the following conditions are satisfied:

$$V^{(\sigma_T)i}(T,x) = \max_{u_T^i} E_{\vartheta_T}\Big\{$$
$$g_T^i\big[x, \phi_T^{(\sigma_T)1*}(x), \phi_T^{(\sigma_T)2*}(x), \cdots, \phi_T^{(\sigma_T)i-1*}(x), u_T^i, \phi_T^{(\sigma_T)i+1*}(x), \cdots$$
$$\cdots, \phi_T^{(\sigma_T)n*}(x); \theta_T^{\sigma_T}\big] + V^{(\sigma_{T+1})i}\Big[T+1, f_T\Big(x, \underline{\phi}_T^{(\sigma_T)*\neq i}(x)\Big) + \vartheta_T\Big]\Big\},$$
$$V^{(\sigma_T)i}(T+1,x) = q^i(x); \qquad \text{for } i \in N; \tag{2.2}$$

where $\underline{\phi}_T^{(\sigma_T)*\neq i}(x)$

$$= \Big[\phi_T^{(\sigma_T)1*}(x), \phi_T^{(\sigma_T)2*}(x), \cdots, \phi_T^{(\sigma_T)i-1*}(x), u_T^i(x), \phi_T^{(\sigma_T)i+1*}(x), \cdots, \phi_T^{(\sigma_T)n}(x)\Big].$$

Proof The system of equations in (2.2) satisfies the standard stochastic dynamic programming property in Theorem A.6 of the technical Appendices and the Nash equilibrium (1951) property for each player $i \in N$. Hence a Nash equilibrium of the subgame (2.1) is characterized. ■

For the sake of exposition, we sidestep the issue of multiple equilibria and focus on playable games in which a particular noncooperative Nash equilibrium is chosen by the players in the subgame. Using Lemma 2.1, one can characterize the value functions $V^{(\sigma_T)i}(T,x)$ for all $\sigma_T \in \{1,2,\cdots,\eta_T\}$ if they exist. In particular, $V^{(\sigma_T)i}(T,x)$ yields player i's expected game equilibrium payoff in the subgame starting at stage T given that $\theta_T^{\sigma_T}$ occurs and $x_T = x$.

Then we proceed to the subgame starting at stage $T-1$ when $\theta_{T-1}^{\sigma_{T-1}} \in \{\theta_{T-1}^1, \theta_{T-1}^2, \cdots, \theta_{T-1}^{\eta_{T-1}}\}$ occurs and $x_{T-1} = x$. In this subgame player $i \in N$ seeks to maximize his expected payoff

$$E_{\vartheta_{T-1},\vartheta_T}\Bigg\{g_{T-1}^i\big(x_{T-1}, u_{T-1}^1, u_{T-1}^2, \cdots, u_{T-1}^n; \theta_{T-1}^{\sigma_{T-1}}\big) + \frac{\varpi_{T-1}}{\sum_{\zeta=T-1}^{T}\varpi_\zeta} q_T^i(x_T)$$
$$+ \frac{\varpi_T}{\sum_{\zeta=T-1}^{T}\varpi_\zeta}\sum_{\sigma_T=1}^{\eta_T}\lambda_T^{\sigma_T}\Big[g_T^i\big(x_T, u_T^1, u_T^2, \cdots, u_T^n; \theta_T^{\sigma_T}\big) + q_{T+1}^i(x_{T+1})\Big]\Bigg\} \tag{2.3}$$

subject to

$$x_{k+1} = f_k\big(x_k, u_k^1, u_k^2, \cdots, u_k^n\big) + \vartheta_k, \text{for } k \in \{T-1, T\} \text{ and } x_{T-1} = x. \tag{2.4}$$

If the functions $V^{(\sigma_T)i}(T,x)$ for all $\sigma_T \in \{1,2,\cdots,\eta_T\}$ characterized in Lemma 2.1 exist, the subgame (2.3 and 2.4) can be expressed as a game in which player $i \in N$ seeks to maximize the expected payoff

$$E_{\vartheta_{T-1}}\left\{g^i_{T-1}\left(x_{T-1}, u^1_{T-1}, u^2_{T-1}, \cdots, u^n_{T-1}; \theta^{\sigma_{T-1}}_{T-1}\right) + \frac{\varpi_{T-1}}{\sum_{\zeta=T-1}^{T}\varpi_\zeta} q^i_T(x_T)\right.$$
$$\left. + \frac{\varpi_T}{\sum_{\zeta=T-1}^{T}\varpi_\zeta}\sum_{\sigma_T=1}^{\eta_T}\lambda^{\sigma_T}_T V^{(\sigma_T)i}\left[T, f_{T-1}\left(x, u^1_{T-1}, u^2_{T-1}, \cdots, u^n_{T-1}\right) + \vartheta_{T-1}\right]\right\} \tag{2.5}$$

using his control u^i_{T-1}.

A characterization of the Nash equilibrium of the subgame (2.5) is provided in the following lemma.

Lemma 2.2 A set of strategies $\left\{u^{i*}_{T-1} = \phi^{(\sigma_{T-1})i*}_{T-1}(x) \in \Gamma^i, \text{ for } i \in N\right\}$ provides a Nash equilibrium solution to the subgame (2.5) if there exist functions $V^{(\sigma_T)i}(T, x_T)$ for $i \in N$ and $\sigma_T = \{1, 2, \cdots, \eta_T\}$ characterized in Lemma 2.1, and functions $V^{(\sigma_{T-1})i}(T-1, x)$, for $i \in N$, such that the following conditions are satisfied:

$$V^{(\sigma_{T-1})i}(T-1, x) = \max_{u^i_{T-1}} E_{\vartheta_{T-1}}\left\{g^i_{T-1}\left[x, \phi^{(\sigma_{T-1})1*}_{T-1}(x), \phi^{(\sigma_{T-1})2*}_{T-1}(x), \cdots\right.\right.$$
$$\left.\cdots, \phi^{(\sigma_{T-1})i-1*}_{T-1}(x), u^i_{T-1}, \phi^{(\sigma_{T-1})i+1*}_{T-1}(x), \cdots, \phi^{(\sigma_{T-1})n*}_{T-1}(x); \theta^{\sigma_{T-1}}_{T-1}\right]$$
$$+ \frac{\varpi_{T-1}}{\sum_{\zeta=T-1}^{T}\varpi_\zeta} q^i_T\left[T, f_{T-1}\left(x, \underline{\phi}^{(\sigma_{T-1})*\neq i}_{T-1}\right) + \vartheta_{T-1}\right]$$
$$\left. + \frac{\varpi_T}{\sum_{\zeta=T-1}^{T}\varpi_\zeta}\sum_{\sigma_T=1}^{\eta_T}\lambda^{\sigma_T}_T V^{(\sigma_T)i}\left[T, f_{T-1}\left(x, \underline{\phi}^{(\sigma_{T-1})*\neq i}_{T-1}\right) + \vartheta_{T-1}\right]\right\}, \tag{2.6}$$

for $i \in N$,

where $\underline{\phi}^{(\sigma_{T-1})*\neq i}_{T-1}$

$$= \left[\phi^{(\sigma_{T-1})1*}_{T-1}(x), \phi^{(\sigma_{T-1})2*}_{T-1}(x), \cdots, \phi^{(\sigma_{T-1})i-1*}_{T-1}(x), u^i_{T-1}, \phi^{(\sigma_{T-1})i+1*}_{T-1}(x), \cdots, \phi^{(\sigma_{T-1})n*}_{T-1}(x)\right].$$

Proof The conditions in Lemma 2.1 and the system of equations in (2.6) satisfies the random horizon dynamic programming property in Theorem 1.1 of Chap. 7 and the discrete-time stochastic dynamic programming property in Theorem A.6 in the Technical Appendices and the Nash equilibrium property for each player $i \in N$. Hence a Nash equilibrium of the subgame (2.5) is characterized. ■

In particular, $V^{(\sigma_{T-1})i}(T-1, x)$ yields player i's expected game equilibrium payoff in the subgame starting at stage $T-1$ given that $\theta^{\sigma_{T-1}}_{T-1}$ occurs and $x_{T-1} = x$.

Consider the subgame starting at stage $\tau \in \{T-2, T-3, \cdots, 1\}$ when $\theta_\tau^{\sigma_\tau} \in \{\theta_\tau^1, \theta_\tau^2, \cdots, \theta_\tau^{\eta_\tau}\}$ occurs and $x_\tau = x$, in which player $i \in N$ maximizes his expected payoff

$$E_{\vartheta_\tau, \vartheta_{\tau+1}, \cdots, \vartheta_T}\left\{ g_\tau^i\left(x, u_\tau^1, u_\tau^2, \cdots, u_\tau^n; \theta_\tau^{\sigma_\tau}\right) + \frac{\varpi_\tau}{\sum_{\zeta=\tau}^{T} \varpi_\zeta} q_{\tau+1}^i(x_{\tau+1}) \right.$$
$$\left. + \frac{\sum_{\hat{T}=\tau+1}^{T} \varpi_{\hat{T}}}{\sum_{\zeta=\tau}^{T} \varpi_\zeta} \left[\sum_{k=\tau+1}^{\hat{T}} \sum_{\sigma_k=1}^{\eta_k} \lambda_k^{\sigma_k} g_k^i\left(x_k, u_k^1, u_k^2, \cdots, u_k^n, \theta_k^{\sigma_k}\right) + q_{\hat{T}+1}^i\left(x_{\hat{T}+1}\right) \right] \right\} \tag{2.7}$$

subject to

$$x_{k+1} = f_k\left(x_k, u_k^1, u_k^2, \cdots, u_k^n\right) + \vartheta_k, \quad \text{for } k \in \{\tau, \tau+1, \cdots, T\} \text{ and } x_\tau = x \tag{2.8}$$

Following the above analysis, the subgame (2.7 and 2.8) can be expressed as a game in which player $i \in N$ maximizes his expected payoff

$$E_{\vartheta_\tau}\left\{ g_\tau^i\left(x, u_\tau^1, u_\tau^2, \cdots, u_\tau^n; \theta_\tau^{\sigma_\tau}\right) + \frac{\varpi_\tau}{\sum_{\zeta=\tau}^{T} \varpi_\zeta} q_{\tau+1}^i\left[f_\tau\left(x, u_\tau^1, u_\tau^2, \cdots, u_\tau^n\right) + \vartheta_\tau\right] \right.$$
$$\left. + \frac{\sum_{\hat{T}=\tau+1}^{T} \varpi_{\hat{T}}}{\sum_{\zeta=\tau}^{T} \varpi_\zeta} \sum_{\sigma_{\tau+1}=1}^{\eta_{\tau+1}} \lambda_{\tau+1}^{\sigma_{\tau+1}} V^{(\sigma_{\tau+1})i}\left[\tau+1, f_\tau\left(x, u_\tau^1, u_\tau^2, \cdots, u_\tau^n\right) + \vartheta_\tau\right] \right\} \tag{2.9}$$

with his control u_t^i.

A Nash equilibrium solution for the randomly furcating stochastic dynamic games with random horizon (1.2 and 1.3) can be characterized by the following theorem.

Theorem 2.1 A set of strategies $\left\{u_\tau^{i*} = \phi_t^{(\sigma_t)i*}(x) \in \Gamma^i, \text{ for } \sigma_t \in \{1, 2, \cdots, \eta_t\},\ t \in \{1, 2, \cdots, T\} \text{ and } i \in N\right\}$ constitutes a Nash equilibrium solution to the game (1.2 and 1.3) if there exist functions $V^{(\sigma_\tau)i}(\tau, x)$, for $\sigma_t \in \{1, 2, \cdots, \eta_t\}$, $t \in \{1, 2, \cdots, T\}$ and $i \in N$, such that the following recursive relations are satisfied:

$$V^{(\sigma_{T+1})i}(T+1,x) = q_{T+1}(x),$$
$$V^{(\sigma_T)i}(T,x) =$$
$$\max_{u_T^i} E_{\vartheta_T}\Big\{ g_T^i\big[x, \phi_T^{(\sigma_T)1*}(x), \phi_T^{(\sigma_T)2*}(x), \cdots, \phi_T^{(\sigma_T)i-1*}(x), u_T^{(\sigma_T)i}, \phi_T^{(\sigma_T)i+1*}(x), \cdots$$
$$\cdots, \phi_T^{(\sigma_T)n*}(x); \theta_T^{\sigma_T}\big] + V^{(\sigma_{T+1})i}\Big[T+1, f_T\Big(x, \underline{\phi}_T^{(\sigma_T)*\neq i}\Big) + \vartheta_T\Big]\Big\},$$

$$V^{(\sigma_\tau)i}(\tau,x) = \max_{u_\tau^i} E_{\vartheta_\tau}\Big\{ g_\tau^i\big[x, \phi_\tau^{(\sigma_\tau)1*}(x), \phi_\tau^{(\sigma_\tau)2*}(x), \cdots, \phi_\tau^{(\sigma_\tau)i-1*}(x), u_\tau^{(\sigma_\tau)i},$$
$$\phi_\tau^{(\sigma_\tau)i+1*}(x), \cdots \quad \cdots, \phi_\tau^{(\sigma_\tau)n*}(x); \theta_\tau^{\sigma_\tau}\big]$$
$$+ \frac{\varpi_\tau}{\sum_{\zeta=\tau}^{T} \varpi_\zeta} q_{\tau+1}\Big[f_\tau\Big(x, \underline{\phi}_\tau^{(\sigma_\tau)*\neq i}(x)\Big) + \vartheta_\tau\Big]$$

$$+ \frac{\sum_{\zeta=\tau+1}^{T} \varpi_\zeta}{\sum_{\zeta=\tau}^{T} \varpi_\zeta} \sum_{\sigma_{\tau+1}=1}^{\eta_{\tau+1}} \lambda_{\tau+1}^{\sigma_{\tau+1}} V^{(\sigma_{\tau+1})}\Big[\tau+1, f_\tau\Big(x, \underline{\phi}_\tau^{(\sigma_\tau)*\neq i}(x)\Big) + \vartheta_\tau\Big]\Big\},$$

$$\tau \in \{1,2,\cdots,T-1\}; \quad \text{for}\, \sigma_t \in \{1,2,\cdots,\eta_t\}, t \in \{1,2,\cdots,T\} \,\text{and}\, i \in N; \qquad (2.10)$$

where $\underline{\phi}_t^{(\sigma_t)*\neq i}(x)$

$$= \Big[\phi_t^{(\sigma_t)1*}(x), \phi_t^{(\sigma_t)2*}(x), \cdots, \phi_t^{(\sigma_t)i-1*}(x), u_t^{(\sigma_t)i}, \phi_t^{(\sigma_t)i+1*}(x), \cdots, \phi_t^{(\sigma_t)n*}(x)\Big];$$

for $t \in \{1,2,\cdots,T\}$.

Proof The results in (2.10) characterizing the game equilibrium in stage T and stage $T-1$ are proved in Lemma 2.1 and Lemma 2.2. Invoking the subgame in stage $\tau \in \{1,2,\cdots,T-2\}$ as expressed in (2.5), the results in (2.10) satisfy the optimality conditions in stochastic dynamic programming and the property of random horizon dynamic programming and the Nash (1951) equilibrium property for each player in each of these subgames. Therefore, a feedback Nash equilibrium of the game (1.2 and 1.3) is characterized. ∎

10.3 Dynamic Cooperation

Now consider the case where the players agree to cooperate and distribute the joint payoff among themselves according to an optimality principle. As pointed out in earlier Chapters two essential properties that a cooperative scheme has to satisfy are group optimality and individual rationality.

10.3.1 Group Optimality

To achieve group optimality the players have to solve the random horizon stochastic dynamic programming problem with uncertain future payoffs and horizon in which they jointly maximizes the expected payoff

$$E_{\vartheta_1,\vartheta_2,\cdots,\vartheta_T}\left\{\sum_{j=1}^{n} g_1^j\left[x_1,u_1^1,u_1^2,\cdots,u_1^n;\theta_1^1\right]+\frac{\varpi_1}{\sum_{\zeta=1}^{T}\varpi_\zeta}\sum_{j=1}^{n} q_2^j(x_{\tau+1})\right.$$

$$\left.+\frac{\sum_{\hat{T}=2}^{T}\varpi_{\hat{T}}}{\sum_{\zeta=1}^{T}\varpi_\zeta}\sum_{j=1}^{n}\left[\sum_{k=2}^{\hat{T}}\sum_{\sigma_k=1}^{\eta_k}\lambda_k^{\sigma_k} g_k^j\left[x_k,u_k^1,u_k^2,\cdots,u_k^n;\theta_k^{\sigma_k}\right]+q^j\left(x_{\hat{T}+1}\right)\right]\right\} \tag{3.1}$$

subject to the stochastic dynamics (1.2).

Again, in a stochastic dynamic framework, strategy space with state-dependent property has to be considered. In particular, a pre-specified class $\hat{\Gamma}^i$ of mapping $\psi_t^{(\sigma_t)i}(\cdot): X \to U^i$ with the property $u_t^{(\sigma_t)i} = \psi_t^{(\sigma_t)i}(x) \in \hat{\Gamma}^i$, for $\sigma_t \in \{1,2,\cdots,\eta_t\}$ and $t \in \{1,2,\cdots,T\}$, is the strategy space of player i and each of its elements is a permissible strategy.

Invoking Theorem 1.1 which characterizes the solution to stochastic dynamic programming problem with uncertain future payoffs and random horizon we have the following Corollary:

Corollary 3.1 A set of controls $u_t^* = \psi_t^{(\sigma_t)*}(x) = \{\psi_t^{(\sigma_t)1*}(x),\ \psi_t^{(\sigma_t)2*}(x),\cdots$ $\cdots,\psi_t^{(\sigma_t)n*}(x)\} \in \hat{\Gamma}^i$, for $\sigma_t \in \{1,2,\cdots,\eta_t\}$ and $t \in \{1,2,\cdots,T\}$ provides an optimal solution to the stochastic control problem (1.2) and (3.1) if there exist functions $W^{(\sigma_t)}(t,x)$, for $\sigma_t \in \{1,2,\cdots,\eta_t\}$ and $t \in \{1,2,\cdots,T\}$, such that the following recursive relations are satisfied:

$$W^{(\sigma_{T+1})}(T+1,x)=\sum_{j=1}^{n}q_{T+1}^{j}(x),$$

$$W^{(\sigma_T)}(T,x)=\max_{u_T^1,u_T^2,\cdots,u_T^n}E_{\vartheta_T}\left\{\sum_{j=1}^{n}g_T^j\left(x,u_T^1,u_T^2,\cdots,u_T^n;\theta_T^{\sigma_T}\right)\right.$$
$$\left.+W^{(\sigma_{T+1})}\left[T+1,f_T\left(x,u_T^1,u_T^2,\cdots,u_T^n\right)+\vartheta_T\right]\right\},$$

$$W^{(\sigma_\tau)}(\tau,x)=\max_{u_\tau^1,u_\tau^2,\cdots,u_\tau^n}E_{\vartheta_\tau}\left\{\sum_{j=1}^{n}g_\tau^j\left(x,u_\tau^1,u_\tau^2,\cdots,u_\tau^n;\theta_\tau^{\sigma_\tau}\right)\right.$$
$$+\sum_{j=1}^{n}\frac{\varpi_\tau}{\sum_{\zeta=\tau}^{T}\varpi_\zeta}q_{\tau+1}^{j}\left[f_\tau\left(x,u_\tau^1,u_\tau^2,\cdots,u_\tau^n\right)+\vartheta_\tau\right]$$

$$\left.+\frac{\sum_{\zeta=\tau+1}^{T}\varpi_\zeta}{\sum_{\zeta=\tau}^{T}\varpi_\zeta}\sum_{\sigma_{\tau+1}=1}^{\eta_{\tau+1}}\lambda_{\tau+1}^{\sigma_{\tau+1}}W^{(\sigma_{\tau+1})}\left[\tau+1,f_\tau\left(x,u_\tau^1,u_\tau^2,\cdots,u_\tau^n\right)+\vartheta_\tau\right]\right\} \quad (3.2)$$

for $\tau\in\{1,2,\cdots,T-1\}$. ■

Substituting the optimal control

$$\psi_k^{(\sigma_k)*}(x)=\left[\psi_k^{(\sigma_k)1*}(x_k),\psi_k^{(\sigma_k)2*}(x_k),\cdots,\psi_k^{(\sigma_k)n*}(x_k)\right]$$

into the state dynamics (1.2), one can obtain the dynamics of the cooperative trajectory as:

$$x_{k+1}=f_k\left(x_k,\psi_k^{(\sigma_k)1*}(x_k),\psi_k^{(\sigma_k)2*}(x_k),\cdots,\psi_k^{(\sigma_k)n*}(x_k)\right)+\vartheta_k \quad \text{if } \theta_k^{\sigma_k}\text{ occurs}, \quad (3.3)$$

for $k\in\{1,2,\cdots,T\}$, $\sigma_k\in\{1,2,\cdots,\eta_k\}$ and $x_1=x^0$.

We use X_k^* to denote the set of realizable values of x_k^* at stage k generated by (3.3). The term $x_k^*\in X_k^*$ is used to denote an element in X_k^*.

The term $W^{(\sigma_k)}(k,x_k^*)$ gives the expected total cooperative payoff over the stages from k to T if $\theta_k^{\sigma_k}$ occurs and $x_k^*\in X_k^*$ is realized at stage k.

10.3.2 Individual Rationality

The players then have to agree to an optimality principle in distributing the total cooperative payoff among themselves. For individual rationality to be upheld the expected payoffs a player receives under cooperation have to be no less than his expected noncooperative payoff along the cooperative state trajectory $\left\{x_k^*\right\}_{k=1}^{T+1}$. For instance, the players may (i) share the excess of the total expected cooperative payoff over the expected sum of individual noncooperative payoffs equally, or (ii) share the total expected cooperative payoff proportional to their expected noncooperative payoffs.

Let $\xi^{(\sigma_k)}(k, x_k^*) = \left[\xi^{(\sigma_k)1}(k, x_k^*), \xi^{(\sigma_k)2}(k, x_k^*), \cdots, \xi^{(\sigma_k)n}(k, x_k^*)\right]$ denote the imputation vector guiding the distribution of the total expected cooperative payoff under the agreed-upon optimality principle along the cooperative trajectory given that $\theta_k^{\sigma_k}$ has occurred in stage k, for $\sigma_k \in \{1, 2, \cdots, \eta_k\}$ and $k \in \{1, 2, \cdots, T\}$. In particular, the imputation $\xi^{(\sigma_k)i}(k, x_k^*)$ gives the expected cumulative payments that player i will receive from stage k to stage $T+1$ under cooperation.

If for example, the optimality principle specifies that the players share the excess of the total cooperative payoff over the sum of individual noncooperative payoffs equally, then the imputation to player i becomes:

$$\xi^{(\sigma_k)i}(k, x_k^*) = V^{(\sigma_k)i}(k, x_k^*) + \frac{1}{n}\left[W^{(\sigma_k)}(k, x_k^*) - \sum_{j=1}^{n} V^{(\sigma_k)j}(k, x_k^*)\right], \tag{3.4}$$

for $i \in N$ and $k \in \{1, 2, \cdots, T\}$.

For individual rationality to be maintained throughout all the stages $k \in \{1, 2, \cdots, T\}$, it is required that the imputation satisfies:

$$\begin{aligned}&\xi^{(\sigma_k)i}(k, x_k^*) \geq V^{(\sigma_k)i}(k, x_k^*),\\ &\text{for } i \in N, \sigma_k \in \{1, 2, \cdots, \eta_k\} \text{ and } k \in \{1, 2, \cdots, T\}.\end{aligned} \tag{3.5}$$

To guarantee group optimality, the imputation vector has to satisfy

$$\begin{aligned}&W^{(\sigma_k)}(k, x_k^*) = \sum_{j=1}^{n} \xi^{(\sigma_k)j}(k, x_k^*),\\ &\text{for } \sigma_k \in \{1, 2, \cdots, \eta_k\} \text{ and } k \in \{1, 2, \cdots, T\}.\end{aligned} \tag{3.6}$$

Hence, a valid imputation $\xi^{(\sigma_k)i}(k, x_k^*)$, for $i \in N$, $\sigma_k \in \{1, 2, \cdots, \eta_k\}$ and $k \in \{1, 2, \cdots, T\}$, has to satisfy conditions (3.5) and (3.6).

10.4 Subgame Consistent Solutions and Payment Mechanism

As demonstrated in Chap. 7, to guarantee dynamical stability in a stochastic dynamic cooperation scheme, the solution has to satisfy the property of subgame consistency in addition to group optimality and individual rationality. In particular, an extension of a subgame-consistent cooperative solution policy to a subgame starting at a later time with a feasible state brought about by prior optimal behavior would remain optimal. For subgame consistency to be satisfied, the imputation according to the original optimality principle has to be maintained at all the T stages along the cooperative trajectory $\{x_k^*\}_{k=1}^T$. In other words, the imputation

$$\xi^{(\sigma_k)}(k, x_k^*) = \left[\xi^{(\sigma_k)1}(k, x_k^*), \xi^{(\sigma_k)2}(k, x_k^*), \cdots, \xi^{(\sigma_k)n}(k, x_k^*)\right], \tag{4.1}$$

for $\sigma_k \in \{1, 2, \cdots, \eta_k\}$, $x_k^* \in X_k^*$ and $k \in \{1, 2, \cdots, T\}$, has to be upheld.

10.4.1 Payoff Distribution Procedure

Following the analyses in Chaps. 8 and 9, we formulate a Payoff Distribution Procedure (PDP) so that the agreed-upon imputation (4.1) can be realized. Let $B_k^{(\sigma_k)i}(x_k^*)$ denote the payment that player i will received at stage k under the cooperative agreement if $\theta_k^{\sigma_k} \in \{\theta_k^1, \theta_k^2, \cdots, \theta_k^{\eta_k}\}$ occurs and $x_k^* \in X_k^*$ is realized at stage $k \in \{1, 2, \cdots, T\}$. The payment scheme $\left\{B_k^{(\sigma_k)i}(x_k^*)\right\}_{k=1}^T$ contingent upon the event $\theta_k^{\sigma_k}$ and state x_k^*, for $k \in \{1, 2, \cdots, T\}\}$ constitutes a PDP in the sense that the imputation to player i can be expressed as:

$$\xi^{(\sigma_k)i}(k, x_k^*) = B_k^{(\sigma_k)i}(x_k^*) + E_{\vartheta_k, \vartheta_{k+1}, \cdots, \vartheta_T}\left\{\frac{\varpi_k}{\sum_{\zeta=k}^T \varpi_\zeta} q_{k+1}^i(x_{k+1}^*)\right.$$
$$\left. + \frac{\sum_{\hat{T}=k+1}^T \varpi_{\hat{T}}}{\sum_{\zeta=k}^T \varpi_\zeta}\left[\sum_{\tau=k+1}^{\hat{T}} \sum_{\sigma_\tau=1}^{\eta_\tau} \lambda_\tau^{\sigma_\tau} B_\tau^{(\sigma_\tau)i}(x_\tau^*) + q_{\hat{T}+1}^i\left(x_{\hat{T}+1}^*\right)\right]\right\},$$
$$\text{for } i \in N. \tag{4.2}$$

For subgame consistency to be satisfied, $\xi^{(\sigma_k)i}(k, x_k^*)$ in (4.2) must be the same as the imputation in (4.1). Crucial to the formulation of a subgame consistent solution is the derivation of a payment scheme $\left\{B_k^{(\sigma_k)i}(x_k^*)\right\}$, for $i \in N$, $\sigma_k \in \{1, 2, \cdots, \eta_k\}$, x_k^*

$\in X_k^*$ and $k \in \{1, 2, \cdots, T\}\}$ so that the imputation in (4.1) can be realized. This is derived in the following Theorem.

Theorem 4.1 A payment equaling

$$B_k^{(\sigma_k)i}(x_k^*) = \xi^{(\sigma_k)i}(k, x_k^*) - E_{\vartheta_k}\left\{\frac{\varpi_k}{\sum_{\zeta=k}^{T}\varpi_\zeta} q_{k+1}^i\left[f_k\left(x_k^*, \psi_k^{(\sigma_k)*}(x_k^*)\right) + \vartheta_k\right]\right.$$
$$\left. + \frac{\sum_{\mu=k+1}^{T}\varpi_\mu}{\sum_{\zeta=k}^{T}\varpi_\zeta} \sum_{\sigma_{k+1}=1}^{\eta_k} \lambda_{k+1}^{\sigma_{k+1}} \xi^{(\sigma_{k+1})i}\left[k+1, f_k\left(x_k^*, \psi_k^{(\sigma_k)*}(x_k^*)\right) + \vartheta_k\right]\right\}, \tag{4.3}$$

given to player $i \in N$ at stage $k \in \{1, 2, \cdots, T\}$ if $\theta_k^{\sigma_k} \in \{\theta_k^1, \theta_k^2, \cdots, \theta_k^{\eta_k}\}$ occurs would lead to the realization of the imputation in (4.1).

Proof To guarantee subgame consistency, the payment scheme $\left\{B_k^{(\sigma_k)i}(x_k^*)\right\}$ defined in (4.2) has also to satisfy the conditions

$$\xi^{(\sigma_k)i}(k, x_k^*) = B_k^{(\sigma_k)i}(x_k^*) + E_{\vartheta_k, \vartheta_{k+1}, \cdots, \vartheta_T}\left\{\frac{\varpi_k}{\sum_{\zeta=k}^{T}\varpi_\zeta} q_{k+1}^i(x_{k+1}^*)\right.$$
$$\left. + \frac{\sum_{\hat{T}=k+1}^{T}\varpi_{\hat{T}}}{\sum_{\zeta=k}^{T}\varpi_\zeta}\left[\sum_{\tau=k+1}^{\hat{T}} \sum_{\sigma_\tau=1}^{\eta_\tau} \lambda_\tau^{\sigma_\tau} B_\tau^{(\sigma_\tau)i}(x_\tau) + q_{\hat{T}+1}^i(x_{\hat{T}+1})\right]\right\},$$
$$\text{for } i \in N \tag{4.4}$$

for $i \in N$, $x_k^* \in X_k^*$, $\sigma_k \in \{1, 2, \cdots, \eta_k\}$ and $k \in \{1, 2, \cdots, T\}$.

First we consider the case when the game lasts up to the final operation stage T, then at stage $T+1$, player i will receive a terminal payment $q_{T+1}^i(x_{T+1}^*)$ with probability one if the state is $x_{T+1}^* \in X_{T+1}^*$. Hence we would have $\xi^{(\sigma_{T+1})i}(T+1, x_{T+1}^*)$ equals $q^i(x_{T+1}^*)$ with probability one.

Next, note that using (4.4) we can express $\xi^{(\sigma_{k+1})i}(k+1, x_{k+1}^*)$ as

$$\xi^{(\sigma_{k+1})i}\left(k+1,x_{k+1}^{*}\right)=B_{k+1}^{(\sigma_{k+1})i}\left(x_{k+1}^{*}\right)+E_{\vartheta_{k+1},\vartheta_{k+2},\cdots,\vartheta_{T}}\left\{\frac{\varpi_{k+1}}{\sum_{\zeta=k+1}^{T}\varpi_{\zeta}^{k+1}}q_{k+1}^{i}\left(x_{k+1}^{*}\right)\right.$$

$$\left.+\frac{\sum_{\hat{T}=k+2}^{T}\varpi_{\hat{T}}}{\sum_{\zeta=k+1}^{T}\varpi_{\zeta}}\left[\sum_{\tau=k+2}^{\hat{T}}\sum_{\sigma_\tau=1}^{\eta_\tau}\lambda_{\tau}^{\sigma_\tau}B_{\tau}^{(\sigma_\tau)i}(x_\tau)+q_{\hat{T}+1}^{i}\left(x_{\hat{T}+1}\right)\right]\right\}, \tag{4.5}$$

for $\sigma_{k+1}\in\left\{1,2,\cdots,\eta_{k+1}\right\}$ and $i\in N$.

Using (4.5), we can obtain:

$$\sum_{\sigma_{k+1}=1}^{\eta_{k+1}}\lambda_{k+1}^{\sigma_{k+1}}\xi^{(\sigma_{k+1})i}\left(k+1,x_{k+1}^{*}\right)=\sum_{\tau=k+1}^{\hat{T}}\sum_{\sigma_\tau=1}^{\eta_\tau}\lambda_{\tau}^{\sigma_\tau}B_{\tau}^{(\sigma_\tau)i}(x_\tau)+q_{\hat{T}+1}^{i}\left(x_{\hat{T}+1}\right). \tag{4.6}$$

The expression on the right-hand-side of (4.6) is the same as the term in square-brackets in (4.4). Substituting the term in square-brackets in (4.4) by the expression on the left-hand-side of (4.6) yields:

$$\xi^{(\sigma_k)i}\left(k,x_k^{*}\right)=B_k^{(\sigma_k)i}\left(x_k^{*}\right)+E_{\vartheta_k}\left\{\frac{\varpi_k}{\sum_{\zeta=k}^{T}\varpi_{\zeta}^{k}}q_{k+1}^{i}\left(x_{k+1}^{*}\right)\right.$$

$$\left.+\frac{\sum_{\hat{T}=k+1}^{T}\varpi_{\hat{T}}}{\sum_{\zeta=k}^{T}\varpi_{\zeta}}\sum_{\sigma_{k+1}=1}^{\eta_{k+1}}\lambda_{k+1}^{\sigma_{k+1}}\xi^{(\sigma_{k+1})i}\left(k+1,x_{k+1}^{*}\right)\right\},\ \text{for}\, i\in N. \tag{4.7}$$

Replacing x_{k+1}^{*} in (4.7) by $f_k\left(x_k^{*},\psi_k^{(\sigma_k)*}\left(x_k^{*}\right)\right)+\vartheta_k$ yields (4.3). Hence Theorem 4.1 follows. ■

For a given imputation vector

$$\xi^{(\sigma_k)}\left(k,x_k^{*}\right)=\left[\xi^{(\sigma_k)1}\left(k,x_k^{*}\right),\xi^{(\sigma_k)2}\left(k,x_k^{*}\right),\cdots,\xi^{(\sigma_k)n}\left(k,x_k^{*}\right)\right],$$

for $\sigma_k\in\{1,2,\cdots,\eta_k\}$ and $k\in\{1,2,\cdots,T\}$,

Theorem 4.1 can be used to derive the corresponding PDP leading to the realization of the imputation vector.

10.4.2 Transfer Payments

When all players are using the cooperative strategies, the payoff that player i will directly receive at stage k given that $x_k^* \in X_k^*$ and $\theta_k^{\sigma_k}$ occurs becomes

$$g_k^i\left[x_k^*, \psi_k^{(\sigma_k)1*}(x_k^*), \psi_k^{(\sigma_k)2*}(x_k^*), \cdots, \psi_k^{(\sigma_k)n*}(x_k^*); \theta_k^{\sigma_k}\right].$$

However, according to the agreed upon imputation, player i is supposed to receive $B_k^{(\sigma_k)i}(x_k^*)$ at stage k as given in Theorem 4.1. Therefore a transfer payment (which can be positive or negative)

$$\varpi_k^{(\sigma_k)i}(x_k^*) = B_k^{(\sigma_k)i}(x_k^*) - g_k^i\left[x_k^*, \psi_k^{(\sigma_k)1*}(x_k^*), \psi_k^{(\sigma_k)2*}(x_k^*), \cdots, \psi_k^{(\sigma_k)n*}(x_k^*); \theta_k^{\sigma_k}\right], \quad (4.8)$$

for $k \in \{1, 2, \cdots, T\}$ and $i \in N$,

will be assigned to player i to yield the cooperative imputation $\xi^i(k, x_k^*)$.

The transfer payments system in (4.8) constitutes an instrument to guide the execution of the agreed-upon payoff sharing mechanism.

10.5 Random Lease Cooperative Resource Extraction under Uncertainty

Consider an economy endowed with a renewable resource and there are two resource extractors (firms). These firms are given the lease to extract the resource. The lease for resource extraction has to be renewed after each stage (year) for up to a maximum of four stages. At stage 1, it is known that the probabilities that the lease will last up to 1, 2, 3 or 4 years long are respectively $\varpi_1, \varpi_2, \varpi_3$ and ϖ_4. Conditional upon the of reaching stage $\tau > 1$, the probability of the game would last up to stages $\tau, \tau + 1$, to four are

$$\frac{\varpi_\tau}{\sum_{\zeta=\tau}^{4} \varpi_\zeta}, \frac{\varpi_{\tau+1}}{\sum_{\zeta=\tau}^{4} \varpi_\zeta}, \frac{\varpi_4}{\sum_{\zeta=\tau}^{4} \varpi_\zeta}.$$

Let u_k^i denote the resource extracted by firm i at stage k, for $i \in \{1, 2\}$. Let U^i be the set of admissible amount of resource extracted by firm i, and $x_k \in X \subset R^+$ be the size of the resource stock at stage k.

It is known at each stage there is a random element, θ_k for $k \in \{1, 2, 3, 4\}$, affecting the prices of the outputs produced by these firms and their costs of extraction. If $\theta_k^{\sigma_k} \in \{\theta_k^1, \theta_k^2\}$ happens at stage $k \in \{2, 3, 4\}$ the profits that firm 1 and firm 2 will obtain at stage k are respectively:

$$\left[P_k^{(\sigma_k)1}u_k^1-\frac{c_k^{(\sigma_k)1}}{x_k}\left(u_k^1\right)^2\right] \text{ and } \left[P_k^{(\sigma_k)2}u_k^2-\frac{c_k^{(\sigma_k)2}}{x_k}\left(u_k^2\right)^2\right], \tag{5.1}$$

where $P_k^{(\sigma_k)i}$ is the price of the resource extracted and processed by firm i, and $c_k^{(\sigma_k)i}$ $\left(u_k^i\right)^2/x_k$ is the production cost of firm i in stage k if $\theta_k^{\sigma_k}$ occurs.

It is known in stage 1 that θ_1 is θ_1^1 with probability $\lambda_1^1=1$. The probability that $\theta_k^{\sigma_k}\in\left\{\theta_k^1,\theta_k^2\right\}$ will occur at stage $k\in\{2,3,4\}$ is $\lambda_k^{\sigma_k}$. A terminal payment contingent upon the resource size equaling $q^i x_{k+1}$ will be paid to firm i in stage $k+1$ if the lease ends at stage k.

The growth dynamics of the resource is governed by the stochastic difference equation:

$$x_{k+1}=x_k+a-bx_k-\sum_{j=1}^{2}u_k^j+\vartheta_k, \tag{5.2}$$

for $k\in\{1,2,3,4\}$ and $x_1=x^0$,

where ϑ_k is a random variable with non-negative range $\{\vartheta_k^1,\vartheta_k^2,\vartheta_k^3\}$ and corresponding probabilities $\{\gamma_k^1,\gamma_k^2,\gamma_k^3\}$; moreover ϑ_1, ϑ_2, ϑ_3 are independent.

There is an extraction constraint $u_k^1+u_k^2\le(1-b)x_k$. Moreover, $\left|\vartheta_k^w\right|\le a$ for $k\in\{1,2,3,4\}$ and $w\in\{1,2,3\}$, and $\sum_{w=1}^{3}\gamma_k^w\vartheta_k^w=\overline{\vartheta}_k$. The discount rate is r. The objective of the firm is to maximize the present value of the expected stream of future profits:

$$E_{\vartheta_1,\vartheta_2,\vartheta_3,\vartheta_4}\left\{\left[P_1^{(1)i}u_1^i-\frac{c_1^{(\sigma_1)i}}{x_1}\left(u_1^i\right)^2\right]+\frac{\varpi_1}{\sum_{\zeta=1}^{4}\varpi_\zeta}q_2^i(x_2)\right.$$

$$\left.+\frac{\sum_{\hat{T}=2}^{4}\varpi_{\hat{T}}}{\sum_{\zeta=1}^{4}\varpi_\zeta}\left[\sum_{k=2}^{\hat{T}}\sum_{\sigma_k=1}^{\eta_k}\lambda_k^{\sigma_k}\left(P_k^{(\sigma_k)i}u_k^i-\frac{c_k^{(\sigma_k)i}}{x_k}\left(u_k^i\right)^2\right)\left(\frac{1}{1+r}\right)^{k-1}+q_{\hat{T}+1}^i\left(x_{\hat{T}+1}\right)\right]\right\} \tag{5.3}$$

subject to the stochastic dynamics (5.2).

10.5.1 Noncooperative Market Outcome

Invoking Theorem 2.1, one can characterize the noncooperative Nash equilibrium strategies for the game (5.2 and 5.3) as follows. In particular, a set of strategies $\{u_k^{(\sigma_k)i*} = \phi_k^{(\sigma_k)i*}(x) \in \Gamma^i$, for $\sigma_1, \sigma_5 \in \{1\}$, $\sigma_2, \sigma_3, \sigma_4 \in \{1,2\}$, $k \in \{1,2,3,4\}$ and $i \in \{1,2\}$ provides a Nash equilibrium solution to the game (5.2 and 5.3) if there exist functions $V^{(\sigma_k)i}(k,x)$, for $i \in \{1,2\}$, $\sigma_1, \sigma_5 \in \{1\}$, $\sigma_2, \sigma_3, \sigma_4 \in \{1,2\}$, and $k \in \{1,2,3,4\}$, such that the following recursive relations are satisfied:

$$V^{(\sigma_5)i}(5,x) = q^i x\left(\frac{1}{1+r}\right)^4,$$

$$V^{(\sigma_k)i}(k,x) = \max_{u_k^{(\sigma_k)i}}\left\{\left[P_k^{(\sigma_k)i}u_k^i - \frac{c_k^{(\sigma_k)i}}{x_k}\left(u_k^i\right)^2\right]\left(\frac{1}{1+r}\right)^{k-1}\right.$$
$$+\frac{\varpi_k}{\sum_{\zeta=k}^{4}\varpi_\zeta}\sum_{y=1}^{3}\gamma_k^y q^i\left[x+a-bx-u_k^{(\sigma_k)i}-\phi_k^{(\sigma_k)j*}(x)+\vartheta_k^y\right]\left(\frac{1}{1+r}\right)^k$$

$$\left.+\frac{\sum_{\mu=k+1}^{4}\varpi_\mu}{\sum_{\zeta=k}^{4}\varpi_\zeta}\sum_{y=1}^{3}\gamma_k^y\sum_{\sigma_{k+1}=1}^{2}\lambda_{k+1}^{\sigma_{k+1}}V^{(\sigma_{k+1})i}\left[k+1, x+a-bx-u_k^{(\sigma_k)i}-\phi_k^{(\sigma_k)j*}(x)+\vartheta_k^y\right]\right\}. \tag{5.4}$$

Performing the indicated maximization in (5.4) yields:

$$\left[P_k^{(\sigma_k)i} - \frac{2c_k^{(\sigma_k)i}u_k^{(\sigma_k)i}}{x}\right]\left(\frac{1}{1+r}\right)^{k-1} - \frac{\varpi_k}{\sum_{\zeta=k}^{4}\varpi_\zeta}q^i\left(\frac{1}{1+r}\right)^k$$

$$-\frac{\sum_{\mu=k+1}^{4}\varpi_\mu}{\sum_{\zeta=k}^{4}\varpi_\zeta}\sum_{y=1}^{3}\gamma_k^y\sum_{\sigma_{k+1}=1}^{2}\lambda_{k+1}^{\sigma_{k+1}}V_{x_{k+1}}^{(\sigma_{k+1})i}\left[k+1, x+a-bx-u_k^{(\sigma_k)i}-\phi_k^{(\sigma_k)j*}(x)+\vartheta_k^y\right]=0; \tag{5.5}$$

for $i \in \{1,2\}$, $\sigma_1, \sigma_5 \in \{1\}$, $\sigma_2, \sigma_3, \sigma_4 \in \{1,2\}$, and $k \in \{1,2,3,4\}$.

From (5.5), the game equilibrium strategies can be expressed as:

$$\phi_k^{(\sigma_k)i*}(x) = \frac{x}{2c_k^{(\sigma_k)i}}\Bigg(P_k^{(\sigma_k)i} - \frac{\varpi_k}{\sum_{\zeta=k}^{4}\varpi_\zeta} q^i(1+r)^{-1} - (1+r)^{k-1}\frac{\sum_{\mu=k+1}^{4}\varpi_\mu}{\sum_{\zeta=k}^{4}\varpi_\zeta}\sum_{y=1}^{3}\gamma_k^y$$

$$\times \sum_{\sigma_{k+1}=1}^{2}\lambda_{k+1}^{\sigma_{k+1}} V_{x_{k+1}}^{(\sigma_{k+1})i}\left[k+1, x+a-bx-\phi_k^{(\sigma_k)1*}(x)-\phi_k^{(\sigma_k)2*}(x)+\vartheta_k^y\right]\Bigg), \tag{5.6}$$

for $i \in \{1,2\}$, $\sigma_1, \sigma_5 \in \{1\}$, $\sigma_2, \sigma_3, \sigma_4 \in \{1,2\}$, and $k \in \{1,2,3,4\}$.

The expected game equilibrium payoffs can be obtained as:

Proposition 5.1 The value function indicating the expected game equilibrium payoff of player i is

$$V^{(\sigma_k)i}(k,x) = \left[A_k^{(\sigma_k)i}x + C_k^{(\sigma_k)i}\right]\left(\frac{1}{1+r}\right)^{k-1}, \text{ for } i \in \{1,2\} \text{ and}$$
$$k \in \{1,2,3,4\}, \tag{5.7}$$

where $A_k^{(\sigma_k)i}$ and $C_k^{(\sigma_k)i}$, for $i \in \{1,2\}$ and $k \in \{1,2,3,4\}$, are constants in terms of the parameters of the game given in Appendix B.

Proof See Appendix B. ■

Substituting the relevant derivatives of the value functions in Proposition 5.1 into the game equilibrium strategies (5.6) yields a noncooperative Nash equilibrium solution of the game (5.2 and 5.3).

10.5.2 Subgame Consistent Cooperative Extraction

Now consider the case where the extractors collaborate to maximize their expected joint profit and share the excess of cooperative gains over their expected noncooperative payoffs equally. To maximize their expected joint payoff, they solve the problem of maximizing

$$E_{\vartheta_1,\vartheta_2,\vartheta_3,\vartheta_4}\left\{\sum_{j=1}^{2}\left[P_1^{(1)j}u_1^j - \frac{c_1^{(\sigma_1)i}}{x_1}\left(u_1^j\right)^2\right] + \frac{\varpi_1}{\sum_{\zeta=1}^{4}\varpi_\zeta}\sum_{j=1}^{2}q_2^j(x_2)\right.$$

$$+\frac{\sum_{\hat{T}=2}^{4}\varpi_{\hat{T}}}{\sum_{\zeta=1}^{4}\varpi_\zeta}\left[\sum_{j=1}^{2}\sum_{k=2}^{\hat{T}}\sum_{\sigma_k=1}^{\eta_k}\lambda_k^{\sigma_k}\left(P_k^{(\sigma_k)j}u_k^j - \frac{c_k^{(\sigma_k)j}}{x_k}\left(u_k^j\right)^2\right)\left(\frac{1}{1+r}\right)^{k-1}\right.$$

$$\left.\left.+\sum_{j=1}^{2}q_{\hat{T}+1}^j\left(x_{\hat{T}+1}\right)\right]\right\}, \tag{5.8}$$

subject to (5.2).

Invoking Theorem 1.1, one can characterize the optimal controls in the random horizon stochastic dynamic programming problem (5.2) and (5.8) as follows. In particular, a set of control strategies $\{u_k^{(\sigma_k)i*} = \psi_k^{(\sigma_k)i*}(x) \in \hat{\Gamma}^i$, for $\sigma_1, \sigma_5 \in \{1\}$, $\sigma_2, \sigma_3, \sigma_4 \in \{1,2\}, k \in \{1,2,3,4\}$ and $i \in \{1,2\}$ provides an optimal solution to the problem (5.2) and (5.8) if there exist functions $W^{(\sigma_k)}(k,x)$, for $k \in \{1,2,3,4\}$, such that the following recursive relations are satisfied:

$$W^{(\sigma_5)}(5,x) = \sum_{j=1}^{2}q^j x\left(\frac{1}{1+r}\right)^4,$$

$$W^{(\sigma_k)}(k,x) = \max_{u_k^1,u_k^1}\left\{\sum_{j=1}^{2}\left[P_k^{(\sigma_k)j}u_k^j - \frac{c_k^{(\sigma_k)j}}{x_k}\left(u_k^j\right)^2\right]\left(\frac{1}{1+r}\right)^{k-1}\right.$$

$$+\frac{\varpi_k}{\sum_{\zeta=k}^{4}\varpi_\zeta}\sum_{y=1}^{3}\gamma_k^y\sum_{j=1}^{2}q^j\left[x + a - bx - u_k^1 - u_k^2 + \vartheta_k^y\right]\left(\frac{1}{1+r}\right)^k$$

$$\left.+\frac{\sum_{\mu=k+1}^{4}\varpi_\mu}{\sum_{\zeta=k}^{4}\varpi_\zeta}\sum_{y=1}^{3}\gamma_k^y\sum_{\sigma_{k+1}=1}^{2}\lambda_{k+1}^{\sigma_{k+1}}W^{(\sigma_{k+1})}\left[k+1, x + a - bx - u_k^1 - u_k^2 + \vartheta_k^y\right]\right\}. \tag{5.9}$$

Performing the indicated maximization in (5.9) yields:

$$\left[P_k^{(\sigma_k)i} - \frac{2c_k^{(\sigma_k)i}u_k^i}{x}\right]\left(\frac{1}{1+r}\right)^{k-1} - \frac{\varpi_k}{\sum_{\zeta=k}^{4}\varpi_\zeta}\sum_{j=1}^{2}q^j\left(\frac{1}{1+r}\right)^k$$

$$-\frac{\sum_{\mu=k+1}^{4}\varpi_\mu}{\sum_{\zeta=k}^{4}\varpi_\zeta}\sum_{y=1}^{3}\gamma_k^y\sum_{\sigma_{k+1}=1}^{2}\lambda_{k+1}^{\sigma_{k+1}}W_{x_{k+1}}^{(\sigma_{k+1})}\left[k+1, x+a-bx-u_k^1-u_k^2(x)+\vartheta_k^y\right]=0;$$
(5.10)

for $i \in \{1,2\}$, $\sigma_1, \sigma_5 \in \{1\}$, $\sigma_2, \sigma_3, \sigma_4 \in \{1,2\}$, and $k \in \{1,2,3,4\}$.

From (5.10), the game equilibrium strategies can be expressed as:

$$\psi_k^{(\sigma_k)i*}(x) = \frac{x}{2c_k^{(\sigma_k)i}}\left(P_k^{(\sigma_k)i} - \frac{\varpi_k}{\sum_{\zeta=k}^{4}\varpi_\zeta}\sum_{j=1}^{2}q^j(1+r)^{-1} - (1+r)^{k-1}\frac{\sum_{\mu=k+1}^{4}\varpi_\mu}{\sum_{\zeta=k}^{4}\varpi_\zeta}\sum_{y=1}^{3}\gamma_k^y\right.$$

$$\left.\times\sum_{\sigma_{k+1}=1}^{2}\lambda_{k+1}^{\sigma_{k+1}}W_{x_{k+1}}^{(\sigma_{k+1})}\left[k+1, x+a-bx-\psi_k^{(\sigma_k)1*}(x)-\psi_k^{(\sigma_k)2*}(x)+\vartheta_k^y\right]\right),$$
(5.11)

for $i \in \{1,2\}$, $\sigma_1, \sigma_5 \in \{1\}$, $\sigma_2, \sigma_3, \sigma_4 \in \{1,2\}$, and $k \in \{1,2,3,4\}$.

The value function $W^{(\sigma_k)}(k,x)$ representing the maximized expected joint profit can be obtained as:

Proposition 5.2 The value function

$$W^{(\sigma_k)}(k,x) = \left[A_k^{(\sigma_k)}x + C_k^{(\sigma_k)}\right]\left(\frac{1}{1+r}\right)^{k-1}, \tag{5.12}$$

for $\sigma_1 \in \{1\}$, $\sigma_2, \sigma_3, \sigma_4 \in \{1,2\}$, and $k \in \{1,2,3,4\}$,

where $A_k^{(\sigma_k)}$ and $C_k^{(\sigma_k)}$, for $\sigma_1 \in \{1\}$, $\sigma_2, \sigma_3, \sigma_4 \in \{1,2\}$, and $k \in \{1,2,3,4\}$, are constants in terms of the parameters of the problem (5.2) and (5.8).

Proof Follow the proof of Proposition 5.1 in Appendix B. ■

Using (5.11) and Proposition 5.2, the optimal cooperative strategies of the agents can be expressed as:

$$\psi_k^{(\sigma_k)i*}(x) = \frac{x}{2c_k^{(\sigma_k)i}}\left(P_k^{(\sigma_k)i} - \frac{\varpi_k}{\sum_{\zeta=k}^{4}\varpi_\zeta}\sum_{j=1}^{2} q^j(1+r)^{-1}\right.$$

$$\left. -(1+r)^{-1}\frac{\sum_{\mu=k+1}^{4}\varpi_\mu}{\sum_{\zeta=k}^{4}\varpi_\zeta}\sum_{y=1}^{3}\gamma_k^y\sum_{\sigma_{k+1}=1}^{2}\lambda_{k+1}^{\sigma_{k+1}}A_{k+1}^{(\sigma_{k+1})}\right), \tag{5.13}$$

for $i \in \{1,2\}$, $\sigma_1, \sigma_5 \in \{1\}$, $\sigma_2, \sigma_3, \sigma_4 \in \{1,2\}$, and $k \in \{1,2,3,4\}$.

Substituting $\psi_k^{(\sigma_k)i}(x)$ from (5.13) into (5.2) yields the optimal cooperative state trajectory:

$$x_{k+1} = x_k + a - bx_k - \sum_{\ell=1}^{2}\frac{x_k}{2c_k^{(\sigma_k)\ell}}\left(P_k^{(\sigma_k)\ell} - \frac{\varpi_k}{\sum_{\zeta=k}^{4}\varpi_\zeta}\sum_{j=1}^{2} q^j(1+r)^{-1}\right.$$

$$\left. -(1+r)^{-1}\frac{\sum_{\mu=k+1}^{4}\varpi_\mu}{\sum_{\zeta=k}^{4}\varpi_\zeta}\sum_{y=1}^{3}\gamma_k^y\sum_{\sigma_{k+1}=1}^{2}\lambda_{k+1}^{\sigma_{k+1}}A_{k+1}^{(\sigma_{k+1})}\right) + \vartheta_k. \tag{5.14}$$

if $\theta_k^{\sigma_k}$ occurs at stage k for $k \in \{1,2,3,4\}$ and $x_1 = x^0$.

Dynamics (5.14) is a linear stochastic difference equation readily solvable by standard techniques. We use X_k^* to denote the set of realizable values of x_k^* at stage k generated by (5.14). The term $x_k^* \in X_k^*$ is used to denote an element in X_k^*.

Since the extractors agree to share the excess of cooperative gains over their expected noncooperative payoffs equally, an imputation

$$\xi^{(\sigma_k)i}(k, x_k^*) = V^{(\sigma_k)i}(k, x_k^*) + \frac{1}{2}\left[W^{(\sigma_k)}(k, x_k^*) - \sum_{j=1}^{2} V^{(\sigma_k)j}(k, x_k^*)\right]$$

$$= \left(A_k^{(\sigma_k)i}x_k^* + C_k^{(\sigma_k)i}\right)\left(\tfrac{1}{1+r}\right)^{k-1}$$

$$+\frac{1}{2}\left[\left(A_k^{(\sigma_k)}x_k^* + C_k^{(\sigma_k)}\right) - \sum_{j=1}^{2}\left(A_k^{(\sigma_k)j}x_k^* + C_k^{(\sigma_k)j}\right)\right]\left(\frac{1}{1+r}\right)^{k-1}, \tag{5.15}$$

if $\theta_k^{\sigma_k}$ occurs at stage k for $k \in \{1,2,3,4\}$, $\sigma_1 \in \{1\}$, $\sigma_2, \sigma_3, \sigma_4 \in \{1,2\}$ and $i \in \{1,2\}$ has to be maintained.

Invoking Theorem 4.1, if $\theta_k^{\sigma_k}$ occurs and $x_k^* \in X$ is realized at stage k the payment scheme

$$B_k^{(\sigma_k)i}(x_k^*) = \left(A_k^{(\sigma_k)i}x_k^* + C_k^{(\sigma_k)i}\right) + \frac{1}{2}\left(\left(A_k^{(\sigma_k)}x_k^* + C_k^{(\sigma_k)}\right) - \sum_{j=1}^{2}\left(A_k^{(\sigma_k)j}x_k^* + C_k^{(\sigma_k)j}\right)\right)$$

$$-\left\{\frac{\varpi_k}{\sum_{\zeta=k}^{T}\varpi_\zeta}\sum_{y=1}^{3}\gamma_k^y q^i\left(x_{k+1}^{*(\vartheta_k^y)}\right) + \frac{\sum_{\mu=k+1}^{T}\varpi_\mu}{\sum_{\zeta=k}^{T}\varpi_\zeta}\sum_{\sigma_{k+1}=1}^{\eta_k}\lambda_{k+1}^{\sigma_{k+1}}\sum_{y=1}^{3}\gamma_k^y\left[\left(A_{k+1}^{(\sigma_{k+1})i}x_{k+1}^{*(\vartheta_k^y)} + C_{k+1}^{(\sigma_{k+1})i}\right)\right.\right.$$

$$\left.\left.+\frac{1}{2}\left(\left(A_{k+1}^{(\sigma_{k+1})}x_{k+1}^{*(\vartheta_k^y)} + C_{k+1}^{(\sigma_{k+1})}\right) - \sum_{j=1}^{2}\left(A_{k+1}^{(\sigma_{k+1})j}x_{k+1}^{*(\vartheta_k^y)} + C_{k+1}^{(\sigma_{k+1})j}\right)\right)\right]\left(\frac{1}{1+r}\right)\right\} \tag{5.16}$$

where

$$x_{k+1}^{*(\vartheta_k^y)} = x_k^* + a - bx_k^* - \sum_{\ell=1}^{2}\frac{x_k^*}{2c_k^{(\sigma_k)\ell}}\left(P_k^{(\sigma_k)\ell} - \frac{\varpi_k}{\sum_{\zeta=k}^{4}\varpi_\zeta}\sum_{j=1}^{2}q^j(1+r)^{-1}\right.$$

$$\left.-(1+r)^{-1}\frac{\sum_{\mu=k+1}^{4}\varpi_\mu}{\sum_{\zeta=k}^{4}\varpi_\zeta}\sum_{y=1}^{3}\gamma_k^y\sum_{\sigma_{k+1}=1}^{2}\lambda_{k+1}^{\sigma_{k+1}}A_{k+1}^{(\sigma_{k+1})}\right) + \vartheta_k^y,$$

for $k \in \{1,2,3,4\}$, $\sigma_1 \in \{1\}$, $\sigma_2, \sigma_3, \sigma_4 \in \{1,2\}$ and $i \in \{1,2\}$.

A subgame consistent solution is then obtained.

10.6 Chapter Appendices

Appendix A: Proof of Theorem 1.1

The proof follows the standard analysis in stochastic dynamic programming (see Bertsekas and Shreve (1996), Puterman (1994) and Fleming and Rishel (1975)). By definition, the value function at stage $T+1$ is

$$V^{(\sigma_{T+1})}(T+1,x) = q_{T+1}(x).$$

We first consider the case when the last operation stage T has arrived and $\theta_T^{\sigma_T} \in \{\theta_T^1, \theta_T^2, \cdots, \theta_T^{\eta_T}\}$ occurs and the state is $x_T = x$. The problem then becomes

$$\max_{u_T} E_{\vartheta_T}\left[g_T(x,u_T;\theta_T^{\sigma_T}) + q_{T+1}(x_{T+1})\right] \tag{6.1}$$

subject to

$$x_{T+1} = f_T(x,u_T) + \vartheta_T, \qquad x_T = x. \tag{6.2}$$

Using $V^{(\sigma_{T+1})}(T+1,x) = q_{T+1}(x_{T+1})$, the problem in (6.1 and 6.2) can be formulated as a single stage problem

$$\max_{u_T} E_{\vartheta_T}\left[g_T(x,u_T,\theta_T^{\sigma_T}) + V^{(\sigma_{T+1})}[T+1,f_T(x,u_T)+\vartheta_T]\right]. \tag{6.3}$$

If the value function $V^{(\sigma_T)}(T,x)$, for $\theta_T^{\sigma_T} \in \{\theta_T^1,\theta_T^2,\cdots,\theta_T^{\eta_T}\}$, characterizing the solution to the problem in (6.1 and 6.2) exists, we have

$$V^{(\sigma_T)}(T,x) = \max_{u_T} E_{\vartheta_T}\left[g_T(x,u_T,\theta_T^{\sigma_T}) + V^{(\sigma_{T+1})}[T+1,f_T(x,u_T)+\vartheta_T]\right], \tag{6.4}$$

for $\theta_T^{\sigma_T} \in \{\theta_T^1,\theta_T^2,\cdots,\theta_T^{\eta_T}\}$.

Equation (6.4) yields the optimality equation in a standard stochastic optimal control problem.

Now consider the problem in stage $T-1$. Invoking the probabilities that the game would last up to stages $T-1$ and T conditional upon the reaching of stage $T-1$ and $\theta_{T-1}^{\sigma_{T-1}} \in \{\theta_{T-1}^1,\theta_{T-1}^2,\cdots,\theta_{T-1}^{\eta_{T-1}}\}$ occurs and $x_{T-1} = x$, the problem in stage $T-1$ can be expressed as maximizing

$$E_{\vartheta_{T-1},\vartheta_T}\Bigg\{g_{T-1}\left(x,u_{T-1};\theta_{T-1}^{\sigma_{T-1}}\right) + \frac{\varpi_{T-1}}{\sum_{\zeta=T-1}^{T}\varpi_\zeta}q_T(x_T)$$
$$+\frac{\varpi_T}{\sum_{\zeta=T-1}^{T}\varpi_\zeta}\sum_{\sigma_T=1}^{\eta_T}\lambda_T^{\sigma_T}\left[g_T(x_T,u_T;\theta_T^{\sigma_T}) + q_{T+1}(x_{T+1})\right]\Bigg\} \tag{6.5}$$

subject to

$$x_{k+1} = f_k(x_k,u_k) + \vartheta_k, \qquad \text{for } k \in \{T-1,T\} \text{ and } x_{T-1} = x. \tag{6.6}$$

Using (6.4), the problem (6.5 and 6.6) can be expressed as a single stage problem of maximizing the expected payoff

$$E_{\vartheta_{T-1}}\left\{g_{T-1}\left(x,u_{T-1};\theta_{T-1}^{\sigma_{T-1}}\right)+\frac{\varpi_{T-1}}{\sum_{\zeta=T-1}^{T}\varpi_\zeta}q_T\left[f_{T-1}(x,u_{T-1})+\vartheta_{T-1}\right]\right.$$
$$\left.+\frac{\varpi_T}{\sum_{\zeta=T-1}^{T}\varpi_\zeta}\sum_{\sigma_T=1}^{\eta_T}\lambda_T^{\sigma_T}V^{(\sigma_T)}\left[T,f_{T-1}(x,u_{T-1})+\vartheta_{T-1}\right]\right\}. \tag{6.7}$$

If the value function $V^{(\sigma_{T-1})}(T-1,x)$, for $\theta_{T-1}^{\sigma_{T-1}}\in\left\{\theta_{T-1}^1,\theta_{T-1}^2,\cdots,\theta_{T-1}^{\eta_{T-1}}\right\}$ characterizing the solution to problem (6.5 and 6.6) exists, we have

$$V^{(\sigma_{T-1})}(T-1,x)=\max_{u_{T-1}}E_{\vartheta_{T-1}}\left\{g_{T-1}\left(x,u_{T-1};\theta_{T-1}^{\sigma_{T-1}}\right)\right.$$
$$+\frac{\varpi_{T-1}}{\sum_{\zeta=T-1}^{T}\varpi_\zeta}q_T\left[f_{T-1}(x,u_{T-1})+\vartheta_{T-1}\right]$$
$$\left.+\frac{\varpi_T}{\sum_{\zeta=T-1}^{T}\varpi_\zeta}\sum_{\sigma_T=1}^{\eta_T}\lambda_T^{\sigma_T}V^{(\sigma_T)}\left[T,f_{T-1}(x,u_{T-1})+\vartheta_{T-1}\right]\right\}. \tag{6.8}$$

Now consider the problem in stage $\tau\in\{1,2,\cdots,T-2\}$. Following the analysis above, given that $\theta_\tau^{\sigma_\tau}\in\left\{\theta_\tau^1,\theta_\tau^2,\cdots,\theta_\tau^{\eta_\tau}\right\}$ occurs, the problem in stage τ becomes the maximization of the expected payoff

$$E_{\vartheta_\tau,\vartheta_{\tau+1},\cdots,\vartheta_T}\left\{g_\tau\left(x,u_\tau;\theta_\tau^{\sigma_\tau}\right)+\frac{\varpi_\tau}{\sum_{\zeta=\tau}^{T}\varpi_\zeta}q_{\tau+1}(x_{\tau+1})\right.$$
$$\left.+\frac{\sum_{\hat{T}=\tau+1}^{T}\varpi_{\hat{T}}}{\sum_{\zeta=\tau}^{T}\varpi_\zeta}\left[\sum_{k=\tau+1}^{\hat{T}}\sum_{\sigma_k=1}^{\eta_k}\lambda_k^{\sigma_k}g_k\left(x_k,u_k,\theta_k^{\sigma_k}\right)+q_{\hat{T}+1}\left(x_{\hat{T}+1}\right)\right]\right\} \tag{6.9}$$

subject to

$$x_{k+1}=f_k(x_k,u_k)+\vartheta_k,\quad \text{for}\, k\in\{\tau,\tau+1,\cdots,T\}\,\text{and}\, x_\tau=x. \tag{6.10}$$

Note that the maximized value function representing the term

$$E_{\vartheta_{\tau+1},\vartheta_{\tau+2},\cdots,\vartheta_T}\left\{\frac{\sum_{\hat{T}=\tau+1}^{T}\varpi_{\hat{T}}}{\sum_{\zeta=\tau+1}^{T}\varpi_\zeta}\left[\sum_{k=\tau+1}^{\hat{T}}\sum_{\sigma_k=1}^{\eta_k}\lambda_k^{\sigma_k}g_k\left(x_k,u_k;\theta_k^{\sigma_k}\right)+q_{\hat{T}+1}\left(x_{\hat{T}+1}\right)\right]\right\}$$

can be expressed as

$$\frac{\sum_{\hat{T}=\tau+1}^{T}\varpi_{\hat{T}}}{\sum_{\zeta=\tau}^{T}\varpi_\zeta}\sum_{\sigma_{\tau+1}=1}^{\eta_{\tau+1}}\lambda_{\tau+1}^{\sigma_{\tau+1}}V^{(\sigma_{t+1})}(\tau+1,x). \tag{6.11}$$

If the value functions $V^{(\sigma_{t+1})}(\tau+1,x)$, for $\theta_{\tau+1}^{\sigma_{\tau+1}}\in\left\{\theta_{\tau+1}^1,\theta_{\tau+1}^2,\cdots,\theta_{\tau+1}^{\eta_{\tau+1}}\right\}$, exist, we can expressed the expected payoff to be maximized in (6.9) as

$$E_{\vartheta_\tau}\left\{g_\tau\left(x,u_\tau;\theta_\tau^{\sigma_\tau}\right)+\frac{\varpi_\tau}{\sum_{\zeta=\tau}^{T}\varpi_\zeta}q_{\tau+1}[f_\tau(x,u_\tau)+\vartheta_\tau]\right.$$
$$\left.+\frac{\sum_{\hat{T}=\tau+1}^{T}\varpi_{\hat{T}}}{\sum_{\zeta=\tau}^{T}\varpi_\zeta}\sum_{\sigma_{\tau+1}=1}^{\eta_{\tau+1}}\lambda_{\tau+1}^{\sigma_{\tau+1}}V^{(\sigma_{\tau+1})}[\tau+1,f_\tau(x,u_\tau)+\vartheta_\tau]\right\} \tag{6.12}$$

If $V^{(\sigma_\tau)}(\tau,x)$ exists, we have

$$V^{(\sigma_\tau)}(\tau,x)=\max_{u_\tau}E_{\vartheta_\tau}\left\{g_\tau\left(x,u_\tau;\theta_\tau^{\sigma_\tau}\right)+\frac{\varpi_\tau}{\sum_{\zeta=\tau}^{T}\varpi_\zeta}q_{\tau+1}[f_\tau(x,u_\tau)+\vartheta_\tau]\right.$$
$$\left.+\frac{\sum_{\hat{T}=\tau+1}^{T}\varpi_{\hat{T}}}{\sum_{\zeta=\tau}^{T}\varpi_\zeta}\sum_{\sigma_{\tau+1}=1}^{\eta_{\tau+1}}\lambda_{\tau+1}^{\sigma_{\tau+1}}V^{(\sigma_{\tau+1})}[\tau+1,f_\tau(x,u_\tau)+\vartheta_\tau]\right\}, \tag{6.13}$$

for $\theta_\tau^{\sigma_\tau}\in\left\{\theta_\tau^1,\theta_\tau^2,\cdots,\theta_\tau^{\eta_\tau}\right\}$ and $\tau\in\{1,2,\cdots,T-2\}$.

Hence Theorem 1.1 follows. ∎

Appendix B. Proof of Proposition 5.1

Consider first the last stage of operation, that is stage 4, when $\theta_4^{\sigma_4}$ occurs. Invoking Proposition 5.1, $V^{(\sigma_4)i}(4,x) = \left[A_4^{(\sigma_4)i}x + C_4^{(\sigma_4)i}\right]\left(\frac{1}{1+r}\right)^3$ and $V^{(\sigma_5)i}(5,x) = q^i x\left(\frac{1}{1+r}\right)^4$, the conditions in Eq. (5.4) become

$$\left[A_4^{(\sigma_4)i}x + C_4^{(\sigma_4)i}\right] = \max_{u_4^{(\sigma_4)i}}\left\{\left[P_4^{(\sigma_4)i}u_4^i - \frac{c_4^{(\sigma_4)i}}{x}\left(u_4^i\right)^2\right]\right.$$
$$+\frac{\varpi_4}{\sum_{\zeta=4}^{4}\varpi_\zeta}\sum_{y=1}^{3}\gamma_4^y q^i\left[x + a - bx - u_4^{(\sigma_4)i} - \phi_4^{(\sigma_4)j*}(x) + \vartheta_4^y\right]\left(\frac{1}{1+r}\right)$$
$$\left.+\frac{\sum_{\mu=4+1}^{4}\varpi_\mu}{\sum_{\zeta=4}^{4}\varpi_\zeta}\sum_{y=1}^{3}\gamma_4^y q^i\left[x + a - bx - u_4^{(\sigma_4)i} - \phi_4^{(\sigma_4)j*}(x) + \vartheta_4^y\right]\left(\frac{1}{1+r}\right)\right\}, \tag{6.14}$$

for $i \in \{1,2\}$ and $\sigma_4 \in \{1,2\}$.

Performing the indicated maximization in (6.14) yields:

$$\left[P_4^{(\sigma_4)i} - \frac{2c_4^{(\sigma_4)i}u_4^{(\sigma_4)i}}{x}\right] - q^i\left(\frac{1}{1+r}\right) = 0, \quad \text{for } i \in \{1,2\} \tag{6.15}$$

The game equilibrium strategies in stage 4 can then be expressed as:

$$\phi_4^{(\sigma_4)i*}(x) = \left[P_4^{(\sigma_4)i} - (1+r)^{-1}q^i\right]\frac{x}{2c_4^{(\sigma_4)i}}, \text{for } i \in \{1,2\} \tag{6.16}$$

Substituting (6.16) into (6.14) yields:

$$\left[A_4^{(\sigma_4)i}x + C_4^{(\sigma_4)i}\right] = P_4^{(\sigma_4)i}\left[P_4^{(\sigma_4)i} - (1+r)^{-1}q^i\right]\frac{x}{2c_4^{(\sigma_4)i}}$$
$$-\left[P_4^{(\sigma_4)i} - (1+r)^{-1}q^i\right]^2\frac{x}{4c_4^{(\sigma_4)i}}$$
$$+q^i\left[x + a - bx - \left[P_4^{(\sigma_4)1} - (1+r)^{-1}q^1\right]\frac{x}{2c_4^{(\sigma_4)1}}\right.$$
$$\left.-\left[P_4^{(\sigma_4)2} - (1+r)^{-1}q^2\right]\frac{x}{2c_4^{(\sigma_4)2}} + \overline{\vartheta}_4\right]\left(\frac{1}{1+r}\right), \text{for } i \in \{1,2\} \text{ and } \sigma_4 \in \{1,2\}. \tag{6.17}$$

Note that both sides of Eq. (6.17) are linear expression of x, the terms $A_4^{(\sigma_4)i}$ and $C_4^{(\sigma_4)i}$, for $i \in \{1,2\}$ and $\sigma_3 \in \{1,2\}$, can be readily obtained as constants:

$$A_4^{(\sigma_4)i} = P_4^{(\sigma_4)i}\left[P_4^{(\sigma_4)i} - (1+r)^{-1}q^i\right]\frac{1}{2c_4^{(\sigma_4)i}}$$

$$-\left[P_4^{(\sigma_4)i} - (1+r)^{-1}q^i\right]^2\frac{1}{4c_4^{(\sigma_4)i}} + q^i\left[1-b\right.$$

$$\left.-\left[P_4^{(\sigma_4)1} - (1+r)^{-1}q^1\right]\frac{1}{2c_4^{(\sigma_4)1}} - \left[P_4^{(\sigma_4)2} - (1+r)^{-1}q^2\right]\frac{1}{2c_4^{(\sigma_4)2}}\right]\left(\frac{1}{1+r}\right),$$

$$\text{and}\, C_4^{(\sigma_4)i} = q^i a\left(\frac{1}{1+r}\right),\ \text{for}\, i \in \{1,2\}\, \text{and}\, \sigma_4 \in \{1,2\}. \tag{6.18}$$

Now we proceed to stage 3, Invoking Proposition 5.1 the conditions in Eq. (6.17) become

$$\left[A_3^{(\sigma_3)i}x + C_3^{(\sigma_3)i}\right] = \max_{u_3^{(\sigma_3)i}}\left\{\left[P_3^{(\sigma_3)i}u_3^i - \frac{c_3^{(\sigma_3)i}}{x}\left(u_3^i\right)^2\right]\right.$$

$$+\frac{\varpi_3}{\sum_{\zeta=3}^{4}\varpi_\zeta}\sum_{y=1}^{3}\gamma_3^y q^i\left[x + a - bx - u_3^{(\sigma_3)i} - \phi_3^{(\sigma_3)j*}(x) + \vartheta_3^y\right]\left(\frac{1}{1+r}\right)$$

$$+\frac{\varpi_4}{\sum_{\zeta=3}^{4}\varpi_\zeta}\sum_{y=1}^{3}\gamma_3^y\sum_{\sigma_4=1}^{2}\lambda_4^{\sigma_4}\left(A_4^{(\sigma_4)i}\left[x + a - bx - u_k^{(\sigma_k)i} - \phi_k^{(\sigma_k)j*}(x) + \vartheta_k^y\right] + C_4^{(\sigma_4)i}\right)$$

$$\left(\frac{1}{1+r}\right),\text{for}\, i \in \{1,2\}\ \text{ and }\ \sigma_3 \in \{1,2\}. \tag{6.19}$$

Performing the indicated maximization in (6.19) yields:

$$\left[P_3^{(\sigma_3)i} - \frac{2c_3^{(\sigma_3)i}u_3^{(\sigma_3)i}}{x}\right] - \frac{\varpi_3}{\sum_{\zeta=3}^{4}\varpi_\zeta}q^i\left(\frac{1}{1+r}\right) - \frac{\varpi_4}{\sum_{\zeta=3}^{4}\varpi_\zeta}\sum_{\sigma_4=1}^{2}\lambda_4^{\sigma_4}A_4^{(\sigma_4)i}\left(\frac{1}{1+r}\right) = 0,$$

$$\text{for}\, i \in \{1,2\}\, \text{and}\, \sigma_3 \in \{1,2\}. \tag{6.20}$$

The game equilibrium strategies in stage 3 can then be expressed as:

$$\phi_3^{(\sigma_3)i*}(x) = \left[P_3^{(\sigma_3)i} - \frac{\varpi_3}{\sum_{\zeta=3}^{4}\varpi_\zeta} q^i\left(\frac{1}{1+r}\right)\right.$$
$$\left. - \frac{\varpi_4}{\sum_{\zeta=3}^{4}\varpi_\zeta}\sum_{\sigma_4=1}^{2}\lambda_4^{\sigma_4}A_4^{(\sigma_4)i}\left(\frac{1}{1+r}\right)\right]\frac{x}{2c_3^{(\sigma_3)i}}, \tag{6.21}$$

for $i \in \{1,2\}$ and $\sigma_3 \in \{1,2\}$.

Substituting (6.21) into (6.19) yields:

$$\left[A_3^{(\sigma_3)i}x + C_3^{(\sigma_3)i}\right] = P_3^{(\sigma_3)i}\left[P_3^{(\sigma_3)i} - \frac{\varpi_3}{\sum_{\zeta=3}^{4}\varpi_\zeta}q^i\left(\frac{1}{1+r}\right)\right.$$
$$\left. - \frac{\varpi_4}{\sum_{\zeta=3}^{4}\varpi_\zeta}\sum_{\sigma_4=1}^{2}\lambda_4^{\sigma_4}A_4^{(\sigma_4)i}\left(\frac{1}{1+r}\right)\right]\frac{x}{2c_3^{(\sigma_3)i}}$$
$$-\left[P_3^{(\sigma_3)i} - \frac{\varpi_3}{\sum_{\zeta=3}^{4}\varpi_\zeta}q^i\left(\frac{1}{1+r}\right)\right.$$
$$\left. - \frac{\varpi_4}{\sum_{\zeta=3}^{4}\varpi_\zeta}\sum_{\sigma_4=1}^{2}\lambda_4^{\sigma_4}A_4^{(\sigma_4)i}\left(\frac{1}{1+r}\right)\right]^2\frac{x}{4c_3^{(\sigma_3)i}}$$
$$+\frac{\varpi_3}{\sum_{\zeta-3}^{4}\varpi_\zeta}\sum_{y=1}^{3}\gamma_3^y q^i\left\{x + a - bx - \left[P_3^{(\sigma_3)1} - \frac{\varpi_3}{\sum_{\zeta=3}^{4}\varpi_\zeta}q^1\left(\frac{1}{1+r}\right)\right.\right.$$
$$\left. - \frac{\varpi_4}{\sum_{\zeta=3}^{4}\varpi_\zeta}\sum_{\sigma_4=1}^{2}\lambda_4^{\sigma_4}A_4^{(\sigma_4)1}\left(\frac{1}{1+r}\right)\right]\frac{x}{2c_3^{(\sigma_3)1}}$$
$$-\left[P_3^{(\sigma_3)2} - \frac{\varpi_3}{\sum_{\zeta=3}^{4}\varpi_\zeta}q^2\left(\frac{1}{1+r}\right) - \frac{\varpi_4}{\sum_{\zeta=3}^{4}\varpi_\zeta}\sum_{\sigma_4=1}^{2}\lambda_4^{\sigma_4}A_4^{(\sigma_4)2}\left(\frac{1}{1+r}\right)\right]\frac{x}{2c_3^{(\sigma_3)}}$$
$$\left.+\vartheta_3^y\right\}\left(\frac{1}{1+r}\right) + \frac{\varpi_4}{\sum_{\zeta=3}^{4}\varpi_\zeta}\sum_{y=1}^{3}\gamma_3^y\sum_{\sigma_4=1}^{2}\lambda_4^{\sigma_4}\left\{A_4^{(\sigma_4)i}\left[x + a - bx\right.\right.$$
$$-\left[P_3^{(\sigma_3)1} - \frac{\varpi_3}{\sum_{\zeta=3}^{4}\varpi_\zeta}q^1\left(\frac{1}{1+r}\right) - \frac{\varpi_4}{\sum_{\zeta=3}^{4}\varpi_\zeta}\sum_{\sigma_4=1}^{2}\lambda_4^{\sigma_4}A_4^{(\sigma_4)1}\left(\frac{1}{1+r}\right)\right]\frac{x}{2c_3^{(\sigma_3)1}}$$
$$-\left[P_3^{(\sigma_3)2} - \frac{\varpi_3}{\sum_{\zeta=3}^{4}\varpi_\zeta}q^2\left(\frac{1}{1+r}\right) - \frac{\varpi_4}{\sum_{\zeta=3}^{4}\varpi_\zeta}\sum_{\sigma_4=1}^{2}\lambda_4^{\sigma_4}A_4^{(\sigma_4)2}\left(\frac{1}{1+r}\right)\right]\frac{x}{2c_3^{(\sigma_3)}}$$
$$\left.\left.+\vartheta_3^y\right] + C_4^{(\sigma_4)i}\right\}\left(\frac{1}{1+r}\right), \quad \text{for } i \in \{1,2\} \text{ and } \sigma_3 \in \{1,2\}. \tag{6.22}$$

Once again, both sides of Eq. (6.22) are linear expression of x. The term $A_3^{(\sigma_3)i}$ equals the coefficient of x on the right-hand-side of (6.22). The term $C_3^{(\sigma_3)i}$ equals the rest of the expression on the right-hand-side of (6.22).

Following the above analysis for stages $k \in \{1,2\}$, one can obtain:

$$\left[A_k^{(\sigma_k)i}x + C_k^{(\sigma_k)i}\right] = P_k^{(\sigma_k)i}\left[P_k^{(\sigma_k)i} - \frac{\varpi_k}{\sum_{\zeta=k}^{4}\varpi_\zeta}q^i\left(\frac{1}{1+r}\right)\right.$$

$$\left. - \frac{\sum_{\mu=\tau+1}^{4}\varpi_\mu}{\sum_{\zeta=\tau}^{4}\varpi_\zeta}\sum_{\sigma_{k+1}=1}^{2}\lambda_{\tau+1}^{\sigma_{k+1}}A_{k+1}^{(\sigma_{k+1})i}\left(\frac{1}{1+r}\right)\right]\frac{x}{2c_k^{(\sigma_k)i}}$$

$$-\left[P_3^{(\sigma_3)i} - \frac{\varpi_3}{\sum_{\zeta=4}^{4}\varpi_\zeta}q^i\left(\frac{1}{1+r}\right) - \frac{\sum_{\mu=\tau+1}^{4}\varpi_\mu}{\sum_{\zeta=\tau}^{4}\varpi_\zeta}\sum_{\sigma_{k+1}=1}^{2}\lambda_{\tau+1}^{\sigma_{k+1}}A_{k+1}^{(\sigma_{k+1})i}\left(\frac{1}{1+r}\right)\right]^2\frac{x}{4c_k^{(\sigma_k)i}}$$

$$+\frac{\varpi_k}{\sum_{\zeta=k}^{4}\varpi_\zeta}\sum_{y=1}^{3}\gamma_k^y q^i\left\{x + a - bx - \sum_{j=1}^{2}\left[P_k^{(\sigma_k)j} - \frac{\varpi_k}{\sum_{\zeta=k}^{4}\varpi_\zeta}q^j\left(\frac{1}{1+r}\right)\right.\right.$$

$$\left.\left. - \frac{\sum_{\mu=\tau+1}^{4}\varpi_\mu}{\sum_{\zeta=\tau}^{4}\varpi_\zeta}\sum_{\sigma_{k+1}=1}^{2}\lambda_{\tau+1}^{\sigma_{k+1}}A_{k+1}^{(\sigma_{k+1})j}\left(\frac{1}{1+r}\right)\right]\frac{x}{2c_k^{(\sigma_k)j}} + \vartheta_k^y\right\}\left(\frac{1}{1+r}\right)$$

$$+\frac{\sum_{\mu=k+1}^{4}\varpi_\mu}{\sum_{\zeta=k}^{4}\varpi_\zeta}\sum_{y=1}^{3}\gamma_k^y\sum_{\sigma_{k+1}=1}^{2}\lambda_{k+1}^{\sigma_{k+1}}\left\{A_{k+1}^{(\sigma_{k+1})i}\left[x + a - bx\right.\right.$$

$$-\sum_{j=1}^{2}\left[P_k^{(\sigma_k)j} - \frac{\varpi_k}{\sum_{\zeta=k}^{4}\varpi_\zeta} qj\left(\frac{1}{1+r}\right)\right.$$

$$\left. - \frac{\sum_{\mu=\tau+1}^{4}\varpi_\mu}{\sum_{\zeta=\tau}^{4}\varpi_\zeta}\sum_{\sigma_{k+1}=1}^{2}\lambda_{\tau+1}^{\sigma_{k+1}}A_{k+1}^{(\sigma_{k+1})j}\left(\frac{1}{1+r}\right)\right]\frac{x}{2c_k^{(\sigma_k)j}} + \vartheta_k^y\Bigg]$$

$$+ C_4^{(\sigma_4)i}\Bigg\}\left(\frac{1}{1+r}\right), \text{for } i$$

$$\in \{1,2\}, \sigma_1 \in \{1\}\sigma_2 \in \{1,2\} \text{ and } k \in \{1,2\}. \tag{6.23}$$

Once again, both sides of Eq. (6.23) are linear expression of x. The term $A_k^{(\sigma_k)i}$ equals the coefficient of x on the right-hand-side of (6.23). The term $C_k^{(\sigma_k)i}$ equals the rest of the expression on the right-hand-side of (6.23). ■

10.7 Chapter Notes

This Chapter considers subgame-consistent cooperative solutions in randomly furcating stochastic dynamic games with random horizon. In the process of obtaining the main results for subgame consistent solution, the forms of Bellman equations and Hamilton-Jacobi-Bellman equations for solving inter-temporal problems with randomly furcating payoffs and random horizon were developed in Yeung and Petrosyan (2014c). They are random horizon versions of the Bellman equations and Hamilton-Jacobi-Bellman equations in Chap. 9. By removing uncertainties in the state dynamics the analysis can be readily applied to randomly furcating dynamic games with random horizon. In particular, the PDP in Theorem 4.1 becomes

$$\xi^{(\sigma_k)i}(k, x_k^*) = B_k^{(\sigma_k)i}(x_k^*) + \frac{\varpi_k}{\sum_{\zeta=k}^{T}\varpi_\zeta} q_{k+1}^i(x_{k+1}^*)$$

$$+ \frac{\sum_{\hat{T}=k+1}^{T}\varpi_{\hat{T}}}{\sum_{\zeta=k}^{T}\varpi_\zeta}\left[\sum_{\tau=k+1}^{\hat{T}}\sum_{\sigma_\tau=1}^{\eta_\tau}\lambda_\tau^{\sigma_\tau}B_\tau^{(\sigma_\tau)i}(x_\tau) + q_{\hat{T}+1}^i(x_{\hat{T}+1})\right], \text{ for } i \in N. \tag{7.1}$$

10.8 Problems

1. Consider an economy endowed with a renewable resource and there are two resource extractors (firms). These firms are given the lease to extract the resource. The lease for resource extraction has to be renewed after each stage (year) for up to a maximum of four stages. At stage 1, it is known that the probabilities that the lease will last up to 1, 2, 3 or 4 years long are respectively 0.1, 0.2, 0.5 and 0.2.

   Let u_k^i denote the resource extracted by firm i at stage k, for $i \in \{1, 2\}$. Let U^i be the set of admissible amount of resource extracted by firm i, and $x_k \in X \subset R^+$ be the size of the resource stock at stage k.

   It is known at each stage there is a random element, θ_k for $k \in \{1, 2, 3, 4\}$, affecting the revenues of these firms and their costs of extraction.

   If θ_k^1 happens at stage $k \in \{2, 3, 4\}$ the profits (in present-value) that firm 1 and firm 2 will obtain at stage k are respectively:

$$\left[3u_k^1 - \frac{2}{x_k}\left(u_k^1\right)^2\right]\left(\frac{1}{1+r}\right)^{k-1} \text{ and } \left[1.5u_k^2 - \frac{1}{x_k}\left(u_k^2\right)^2\right]\left(\frac{1}{1+r}\right)^{k-1},$$

   where $r = 0.05$ is the discount rate.

   If θ_k^2 happens at stage $k \in \{2, 3, 4\}$ the profits (in present-value) that firm 1 and firm 2 will obtain at stage k are respectively:

$$\left[2.5u_k^1 - \frac{1}{x_k}\left(u_k^1\right)^2\right]\left(\frac{1}{1+r}\right)^{k-1} \text{ and } \left[2u_k^2 - \frac{2}{x_k}\left(u_k^2\right)^2\right]\left(\frac{1}{1+r}\right)^{k-1}.$$

   It is known in stage 1 that θ_1^1 has occurred. The probability that θ_k^1 will occur at stage $k \in \{2, 3, 4\}$ is 0.3 and the probability that θ_k^2 will occur at stage $k \in \{2, 3, 4\}$ is 0.7. In stage 5, a terminal payment (again in present-value) equaling $1.5x_5\left(\frac{1}{1+r}\right)^3$ will be paid to firm 1 and a terminal payment (again in present-value) equaling $0.5x_5\left(\frac{1}{1+r}\right)^3$ will be paid to firm 2.

   The growth dynamics of the resource is governed by the stochastic difference equation:

$$x_{k+1} = x_k + 16 - 0.15x_k - \sum_{j=1}^{2} u_k^j + \vartheta_k,$$

   for $k \in \{1, 2, 3, 4\}$ and $x_1 = 12$,

   where ϑ_k is a random variable with non-negative range $\{0, 2, 3\}$ and corresponding probabilities $\{0.2, 0.4, 0.4\}$; moreover ϑ_1, ϑ_2, ϑ_3 are independent. Moreover, we have the constraint $u_k^1 + u_k^2 \le 0.85x_k + 16$.

The objective of extractor $i \in \{1, 2\}$ is to maximize the present value of the expected stream of future profits:

Characterize the feedback Nash equilibrium.

(2) Obtain a group optimal solution that maximizes the joint expected profit.
(3) Consider the case when the extractors agree to share the excess of cooperative gains over their expected noncooperative profits equally. Derive a subgame consistent solution.

Chapter 11
Subgame Consistency in NTU Cooperative Dynamic Games

Cooperative games suggest the possibility of enhancing the participants' well-being in situations involving strategic interactions. Various cooperative solutions have been presented, like the Nash (1950, 1953) bargaining solution, the Shapley (1953) value, and the stable set of von Neumann and Morgenstern (1944). Frequently, the lack of sustainability of the cooperation scheme leads to break-ups of the scheme as the game evolves or even to the outright rejection of the cooperation scheme. One of the ways to uphold sustainability of a cooperation scheme is to maintain the condition of subgame consistency. In non-transferrable utility/payoff (NTU) cooperative dynamic games, the inapplicability of transfer payments makes the derivation of subgame consistent solutions extremely strenuous. In Chap. 6 subgame consistent solution in cooperative differential games with non-transferable payoffs under a constant weight scheme is provided. However, the result is confined to a specific class of games under a very restrictive set of optimality principles. Crucial problems of using constant payoff weights include the possibility of the failure of individual rationality to be fulfilled throughout the cooperative duration and the deviation from the original optimality principle as the game evolves. The use of variable payoff weights provides an effective way in achieving subgame consistency and preserving individual rationality under a wide range of optimality principles.

This Chapter considers subgame consistency in NTU cooperative dynamic games with the use of variable payoff weights. It is based on an elaborated exposition of the analyses in Yeung and Petrosyan (2015a, b). Section 11.1 presents the game formulation and provides the mathematical preliminaries for deriving subgame consistent solutions. The notion of subgame consistency in NTU dynamic games under a variable weights scheme is presented in Sect. 11.2. Derivation of subgame consistent cooperative strategies via variable weights is shown in Sect. 11.3. Section 11.4 gives an illustration in public capital build-up. Section 11.5 provides an extension the analysis to NTU cooperative stochastic dynamic games. Section 11.6 supplies an illustration in stochastic build-up of public capital. Chapter notes are given in Sect. 11.7 and problems in Sect. 11.8.

D.W.K. Yeung, L.A. Petrosyan, *Subgame Consistent Cooperation*,
Theory and Decision Library C 47, DOI 10.1007/978-981-10-1545-8_11

11.1 Game Formulation and Mathematical Preliminaries

Consider the general $T-$ stage $n-$ person nonzero-sum discrete-time dynamic game with initial state x_1^0. The state space of the game is $X \in R^m$ and the state dynamics of the game is characterized by the difference equation:

$$x_{k+1} = f_k\left(x_k, u_k^1, u_k^2, \cdots, u_k^n\right), \tag{1.1}$$

for $k \in \{1, 2, \cdots, T\} \equiv \kappa$ and $x_1 = x_1^0$,

where $u_k^i \in R^{m_i}$ is the control vector of player i at stage k, and $x_k \in X$ is the state of the game. The payoff that player i seeks to maximize is

$$\sum_{k=1}^{T} g_k^i\left(x_k, u_k^1, u_k^2, \cdots, u_k^n\right) + q^i(x_{T+1}), \tag{1.2}$$

for $i \in \{1, 2, \cdots, n\} \equiv N$,

where $q^i(x_{T+1})$ is the terminal payoff that player i will received in stage $T+1$.

The payoffs of the players are not transferable. Let $\left\{\phi_k^i(x), \text{ for } k \in \kappa \text{ and } i \in N\right\}$ denote a set of strategies that provides a feedback Nash equilibrium solution (if it exists) to the game (1.1 and 1.2), and $\left\{V^i(k, x), \text{ for } k \in \kappa \text{ and } i \in N\right\}$ denote the value functions yielding the payoff to player i over the stages from k to T. A way to characterize a feedback Nash equilibrium of the game is given in the Theorem below.

Theorem 1.1 A set of strategies $\left\{\phi_k^i(x), \text{ for } k \in \kappa \text{ and } i \in N\right\}$ provides a feedback Nash equilibrium solution to the game (1.1 and 1.2) if there exist functions $V^i(k, x)$, for $k \in K$ and $i \in N$, satisfying the following recursive relations:

$$\begin{aligned} V^i(T+1, x) &= q^i(x_{T+1}), \\ V^i(k, x) &= \max_{u_k^i} \left\{ g_k^i\left[x, \phi_k^1(x), \phi_k^2(x), \cdots, \phi_k^{i-1}(x), u_k^i, \phi_k^{i+1}(x), \cdots, \phi_k^n(x)\right] \right. \\ &\left. \quad + V^i\left[k+1, f_k\left[x, \phi_k^1(x), \phi_k^2(x), \cdots, \phi_k^{i-1}(x), u_k^i, \phi_k^{i+1}(x), \cdots, \phi_k^n(x)\right]\right] \right\} \end{aligned} \tag{1.3}$$

for $i \in N$ and $k \in \kappa$.

Proof Follow the proof of Theorem 1.1 in Chap. 7. ∎

Since the analysis is on cooperative schemes for improving the non-cooperative outcomes in NTU dynamic games, we would consider games with non-cooperative Nash equilibrium outcomes.

11.1.1 Cooperation Under Constant Weights

To enhance their payoffs the players would consider formulating a cooperative scheme. In particular, the players agree to cooperate and enhance their payoffs according to an agreed-upon optimality principle. Since payoffs are non-transferable the payoffs of individual players are directly determined by the optimal cooperative strategies adopted. Pareto efficient cooperative strategies can be derived from the maximization of the weighted sum of payoffs of the players under a set of agreed-upon payoff weights.

To establish the optimization foundation of the variable weights scheme we consider first the case in which the players adopt a vector of constant payoff weights $\alpha = (\alpha^1, \alpha^2, \cdots, \alpha^n)$ in all stages, where $\sum_{j=1}^{n} \alpha^j = 1$. Conditional upon the vector of weights α, the players' optimal cooperative strategies can be generated by solving the dynamic programming problem of maximizing the weighted sum of payoffs (see Leitmann (1974), Dockner and Jørgensen (1984), Hamalainen et al (1986), Yeung and Petrosyan (2005) and Yeung et al. (2007)):

$$\sum_{j=1}^{n} \left[\sum_{k=1}^{T} \alpha^j g_k^j(x_k, u_k^1, u_k^2, \cdots, u_k^n) + \alpha^j q^j(x_{T+1}) \right] \tag{1.4}$$

subject to (1.1).

An optimal solution to the problem (1.1) and (1.4) can be characterized by the following theorem.

Theorem 1.2 A set of strategies $\left\{ \psi_k^{(\alpha)i}(x), \text{ for } k \in \kappa \text{ and } i \in N \right\}$ provides an optimal solution to the problem (1.1) and (1.4) if there exist functions $W^{(\alpha)}(k, x)$, for $k \in K$, such that the following recursive relations are satisfied:

$$W^{(\alpha)}(T+1, x) = \sum_{j=1}^{n} \alpha^j q^j(x_{T+1}), \tag{1.5}$$

$$\begin{aligned} W^{(\alpha)}(k, x) = \max_{u_k^1, u_k^2, \cdots, u_k^n} & \left\{ \sum_{j=1}^{n} \alpha^j g_k^j(x_k, u_k^1, u_k^2, \cdots, u_k^n) \right. \\ & \left. + W^{(\alpha)} \left[k+1, f_k(x_k, u_k^1, u_k^2, \cdots, u_k^n) \right] \right\} = \\ \sum_{j=1}^{n} \alpha^j g_k^j & \left[x, \psi_k^{(\alpha)1}(x), \psi_k^{(\alpha)2}(x), \cdots, \psi_k^{(\alpha)n}(x) \right] \\ & + W^{(\alpha)} \left[k+1, f_k\left(x, \psi_k^{(\alpha)1}(x), \psi_k^{(\alpha)2}(x), \cdots, \psi_k^{(\alpha)n}(x) \right) \right], \end{aligned} \tag{1.6}$$

Proof The conditions in (1.5 and 1.6) stem directly from the results in discrete-time dynamic programming in Theorem A.5 of the Technical Appendices. ■

Substituting the optimal control $\left\{\psi_k^{(\alpha)i}(x),\ \text{for } k \in \kappa \text{ and } i \in N\right\}$ into the state dynamics (1.1), one can obtain the dynamics of the cooperative trajectory as:

$$x_{k+1} = f_k\left(x_k, \psi_k^{(\alpha)1}(x_k), \psi_k^{(\alpha)2}(x_k), \cdots, \psi_k^{(\alpha)n}(x_k)\right), \tag{1.7}$$

for $k \in \kappa$ and $x_1 = x^0$.

We use $x_k^{(\alpha)} \in X_k^{(\alpha)}$ to denote the value of the state at stage k generated by (1.7). The value function $W^{(\alpha)}(k,x)$ gives the maximized weighted cooperative payoff over the stages from k to T.

11.1.2 Individual Payoffs and Individual Rationality

Given that all players are adopting the cooperative strategies the payoff of player i under cooperation can be obtained as:

$$\begin{aligned} W^{(\alpha)i}(t,x) = \sum_{k=t}^{T} g_k^i\Big[x_k^{(\alpha)}, \psi_k^{(\alpha)1}\left(x_k^{(\alpha)}\right), \psi_k^{(\alpha)2}\left(x_k^{(\alpha)}\right), \cdots \\ \cdots, \psi_k^{(\alpha)n}\left(x_k^{(\alpha)}\right)\Big] + q^i\left(x_{T+1}^{(\alpha)}\right)\bigg|\, x_t^{(\alpha)} = x \end{aligned} \tag{1.8}$$

for $i \in N$ and $t \in \kappa$.

To allow the derivation of the functions $W^{(\alpha)i}(t,K)$ in a more direct way we derive a deterministic counterpart of the analysis in Yeung (2013) and characterize individual player's payoffs under cooperation by the following Theorem.

Theorem 1.3 The payoff of player i over the stages from t to $T+1$ can be characterized as the value function $W^{(\alpha)i}(t,x)$ satisfying the following recursive system of equations:

$$\begin{aligned} &W^{(\alpha)i}(T+1,x) = q^i(x_{T+1}) \\ &W^{(\alpha)i}(t,x) = g_t^i\left[x, \psi_t^{(\alpha)1}(x), \psi_t^{(\alpha)2}(x), \cdots, \psi_t^{(\alpha)n}(x)\right] \\ &\quad + W^{(\alpha)i}\left[t+1, f_t\left(t, \psi_t^{(\alpha)1}(x), \psi_t^{(\alpha)2}(x), \cdots, \psi_t^{(\alpha)n}(x)\right)\right], \\ &\text{for } i \in N \text{ and } t \in \kappa \end{aligned} \tag{1.9}$$

Proof The term $W^{(\alpha)i}(t,x)$ in (1.8) can be expressed as:

$$\begin{aligned} W^{(\alpha)i}(k,x) = g_k^i\left[x, \psi_k^{(\alpha)1}(x), \psi_k^{(\alpha)2}(x), \cdots, \psi_k^{(\alpha)n}(x)\right] \\ + \sum_{\tau=k+1}^{T} g_\tau^i\left[x_\tau^{(\alpha)}, \psi_\tau^{(\alpha)1}\left(x_\tau^{(\alpha)}\right), \psi_\tau^{(\alpha)2}\left(x_\tau^{(\alpha)}\right), \cdots, \psi_\tau^{(\alpha)n}\left(x_\tau^{(\alpha)}\right)\right] + q^i\left(x_{T+1}^{(\alpha)}\right). \end{aligned} \tag{1.10}$$

Invoking (1.8) again, we have:

$$W^{(\alpha)i}\left(k+1,x_{k+1}^{(\alpha)}\right)=\sum_{\tau=k+1}^{T}g_{\tau}^{i}\left[x_{\tau}^{(\alpha)},\psi_{\tau}^{(\alpha)1}\left(x_{\tau}^{(\alpha)}\right),\psi_{\tau}^{(\alpha)2}\left(x_{\tau}^{(\alpha)}\right),\cdots,\psi_{\tau}^{(\alpha)n}\left(x_{\tau}^{(\alpha)}\right)\right]$$
$$+q^{i}\left(x_{T+1}^{(\alpha)}\right). \tag{1.11}$$

Using (1.11) one express (1.10) as:

$$\begin{aligned}W^{(\alpha)i}(k,x)&=g_{k}^{i}\left[x,\psi_{k}^{(\alpha)1}(x),\psi_{k}^{(\alpha)2}(x),\cdots,\psi_{k}^{(\alpha)n}(x)\right]\\&+W^{(\alpha)i}\left[k+1,f_{k}\left(x,\psi_{k}^{(\alpha)1}(x),\psi_{k}^{(\alpha)2}(x),\cdots,\psi_{k}^{(\alpha)n}(x)\right)\right],\end{aligned} \tag{1.12}$$

for $i\in N$ and $k\in\kappa$.

Hence Theorem 1.3 follows. ■

For individual rationality to be maintained throughout all the stages $t\in\kappa$, it is required that:

$$W^{(\alpha)i}\left(t,x_{t}^{(\alpha)}\right)\geq V^{i}\left(t,x_{t}^{(\alpha)}\right),\quad \text{for } i\in N \text{ and } t\in\kappa. \tag{1.13}$$

Let the set of weights α that satisfies (1.13) be denoted by Λ. If Λ is not an empty set, a vector $\hat{\alpha}=\left(\hat{\alpha}^{1},\hat{\alpha}^{2},\cdots\cdots,\hat{\alpha}^{n}\right)\in\Lambda$ agreed upon by all players would yield a cooperative solution which satisfies both individual rationality and Pareto optimality throughout the cooperation duration.

Remark 1.1 The pros of the constant payoff weights scheme is that full Pareto efficiency is satisfied in the sense that there does not exist any strategy path which would enhance the cooperative payoff of a player without lowering the cooperative payoff of at least one of the other players in all stages.

The cons of the constant payoff weights scheme include the inflexibility in accommodating the preferences of the players according to the initial cooperative agreement and the high possibility of the non-existence of the set of weights Λ that satisfies individual rationality throughout the cooperation duration. ■

In the existing literature on NTU cooperative dynamic games only Sorger (2006) and Marin-Solano (2014) adopted a variable payoff weights scheme.

11.2 Subgame Consistent Cooperative Solution Via Variable Weights

Now, we proceed to consider subgame consistent solutions in NTU cooperative dynamic games. A salient property of a subgame consistent solution is that the agreed-upon optimality principle remains in effect at each stage of the game and hence the players do not possess incentives to deviate from the solution plan. Let $\Gamma(t, x_t)$ denote the cooperative game in which the objective of player i is

$$\sum_{k=t}^{T} g_k^i\left(x_k, u_k^1, u_k^2, \cdots, u_k^n\right) + q^i(x_{T+1}), \quad \text{for } i \in N, \tag{2.1}$$

and the state dynamics is

$$x_{k+1} = f_k\left(x_k, u_k^1, u_k^2, \cdots, u_k^n\right), \tag{2.2}$$

for $k \in \{t, t+1, \cdots, T\}$ and the state at stage t is x_t.

Let the agreed-upon optimality principle be denoted by $P(t, x_t)$. For subgame consistency to be maintained the agreed-upon optimality principle $P(t, x_t)$ must be satisfied in the subgame $\Gamma(t, x_t)$ for $t \in \{1, 2, \cdots, T\}$. Hence, when the game proceeds to any stage t, the agreed-upon solution policy remains effective. Examples of optimality principles $P(t, x_t)$ include criteria like the Nash bargaining solution, cooperative gains proportional to non-cooperatives payoffs and the mid-value of feasible payoff weights.

A time-invariant weights scheme is usually hardly applicable for the derivation of a subgame consistent solution in general. As stated in Remark 1.1, the set Λ which satisfies individual rationality throughout the game duration is often empty. In general, typical optimality principles in classical game theory could not be maintained as the game proceeds under a time-invariant payoff weights cooperative scheme. To derive a set of subgame consistent strategies in a cooperative solution with optimality principle $P(t, x_t)$ a variable payoff weight scheme has to be adopted. In particular, at each stage $t \in \kappa$ the players would adopt a vector of payoff weights $\hat{\alpha}_t = \left(\hat{\alpha}_t^1, \hat{\alpha}_t^2, \cdots, \hat{\alpha}_t^n\right)$ for $\sum_{j=1}^{n} \hat{\alpha}_t^j = 1$ which satisfies the agreed-upon optimality principle. The chosen set of weights $\hat{\alpha}_t = \left(\hat{\alpha}_t^1, \hat{\alpha}_t^2, \cdots, \hat{\alpha}_t^n\right)$ must lead to the satisfaction of the optimality principle $P(t, x_t)$ in the subgame $\Gamma(t, x_t)$ for $t \in \{1, 2, \cdots T\}$.

11.3 Derivation of Subgame Consistent Cooperative Strategies

To derive the optimal cooperative strategies in a subgame consistent solution for NTU cooperative dynamic games with variable payoff weights we invoke the principle of optimality in dynamic programming and begin with the final stage of the cooperative game. Section 11.3.1 derived the optimal cooperative strategies in the last two stages and Sect. 11.3.2 presents the analysis for all the preceding stages.

11.3.1 Optimal Cooperative Strategies in Ending Stages

Consider first the last operation stage, that is stage T, with the state $x_T = x \in X$. The players will select a set of payoff weight $\alpha_T = \left(\alpha_T^1, \alpha_T^2, \cdots, \alpha_T^n\right)$ which satisfies the optimality principle $P(T, x)$. The players' optimal cooperative strategies can be generated by solving the dynamic programming problem of maximizing the weighted sum of their payoffs

$$\sum_{j=1}^{n} \left[\alpha_T^j g_T^j\left(x_T, u_T^1, u_T^2, \cdots, u_T^n\right) + \alpha_T^j q^j(x_{T+1}) \right] \tag{3.1}$$

subject to

$$x_{T+1} = f_T\left(x_T, u_T^1, u_T^2, \cdots, u_T^n\right), \; x_T = x. \tag{3.2}$$

Invoking Theorem 1.2, given the payoff weights being α_T the optimal cooperative strategies $\left\{u_T^i = \psi_T^{(\alpha_T)i}, \text{ for } i \in N\right\}$ in stage T are characterized by the conditions

$$\begin{aligned} W^{(\alpha_T)}(T+1, x) &= \sum_{j=1}^{n} \alpha_T^j q^j(x_{T+1}), \\ W(T, x) &= \max_{u_T^1, u_T^2, \cdots, u_T^n} \Bigg\{ \sum_{j=1}^{n} \alpha_T^j g_T^j\left(x_T, u_T^1, u_T^2, \cdots, u_T^n\right) \\ &\quad + W^{(\alpha_T)}\left[T+1, f_T\left(x_T, u_T^1, u_T^2, \cdots, u_T^n\right)\right] \Bigg\} \end{aligned} \tag{3.3}$$

Given that all players are adopting the cooperative strategies the payoff of player i under cooperation covering stages T and $T+1$ can be obtained as:

$$W^{(\alpha_T)i}(T, x) = g_T^i\left[x, \psi_T^{(\alpha_T)1}(x), \psi_T^{(\alpha_T)2}(x), \cdots, \psi_T^{(\alpha_T)n}(x)\right] + q^i\left(x_{T+1}^{(\alpha_T)}\right) \text{ for } i \in N. \tag{3.4}$$

Invoking Theorem 1.3 one can characterize $W^{(\alpha_T)i}(T, x)$ by the following equations

$$\begin{aligned} &W^{(\alpha_T)i}(T+1, x) = q^i(x) \\ &W^{(\alpha_T)i}(T, x) = g_T^i\left[x, \psi_T^{(\alpha_T)1}(x), \psi_T^{(\alpha_T)2}(x), \cdots, \psi_T^{(\alpha_T)n}(x)\right] \\ &\qquad + W^{(\alpha_T)i}\left[T+1, f_T\left(x, \psi_T^{(\alpha_T)1}(x), \psi_T^{(\alpha_T)2}(x), \cdots, \psi_T^{(\alpha_T)n}(x)\right)\right], \text{ for } i \in N. \end{aligned} \tag{3.5}$$

For individual rationality to be maintained, it is required that:

$$W^{(\alpha_T)i}(T,x) \geq V^i(T,x), \text{ for } i \in N. \tag{3.6}$$

We use Λ_T to denote the set of weights α_T that satisfies (3.6). Let $\hat{\alpha}_T = \left(\hat{\alpha}_T^1, \hat{\alpha}_T^2, \cdots, \hat{\alpha}_T^n\right) \in \Lambda_T$ denote the payoff weights in stage T that leads to the satisfaction of the optimality principle $P(T,x)$.

Now we proceed to cooperative scheme in the second last stage. Given that the payoff of player i in stage T is $W^{(\hat{\alpha}_T)i}(T,x)$, his payoff in stage $T-1$ can be expressed as:

$$\begin{aligned}
&g_{T-1}^i\left(x_{T-1}, u_{T-1}^1, u_{T-1}^2, \cdots, u_{T-1}^n\right) \\
&\quad + g_T^i\left[x_T, \psi_T^{(\hat{\alpha}_T)1}(x_T), \psi_T^{(\hat{\alpha}_T)2}(x_T), \cdots, \psi_T^{(\hat{\alpha}_T)n}(x_T)\right] + q^i(x_{T+1}) \\
&= g_{T-1}^i\left(x_{T-1}, u_{T-1}^1, u_{T-1}^2, \cdots, u_{T-1}^n\right) + W^{(\hat{\alpha}_T)i}(T, x_T), \text{ for } i \in N.
\end{aligned} \tag{3.7}$$

In this stage the players will select payoff weights $\alpha_{T-1} = \left(\alpha_{T-1}^1, \alpha_{T-1}^2, \cdots, \alpha_{T-1}^n\right)$ which satisfy optimality principle $P(T-1,x)$. The players' optimal cooperative strategies can be generated by solving the following dynamic programming problem of maximizing the weighted sum of payoffs

$$\sum_{j=1}^n \alpha_{T-1}^j \left[g_{T-1}^j\left(x_{T-1}, u_{T-1}^1, u_{T-1}^2, \cdots, u_{T-1}^n\right) + W^{(\hat{\alpha}_T)j}(T, x_T) \right] \tag{3.8}$$

subject to

$$x_T = f_{T-1}\left(x_{T-1}, u_{T-1}^1, u_{T-1}^2, \cdots, u_{T-1}^n\right), \quad x_{T-1} = x \tag{3.9}$$

Invoking Theorem 1.2, given the payoff weights being α_{T-1} the optimal cooperative strategies $\left\{u_{T-1}^i = \psi_{T-1}^{(\alpha_{T-1})i}, \text{ for } i \in N\right\}$ in stage $T-1$ are characterized by the conditions

$$\begin{aligned}
W^{(\alpha_{T-1})}(T,x) &= \sum_{j=1}^n \alpha_{T-1}^j W^{(\hat{\alpha}_T)j}(T,x) \\
W^{(\alpha_{T-1})}(T-1,x) &= \max_{u_{T-1}^1, u_{T-1}^2, \cdots, u_{T-1}^n} \left\{ \sum_{j=1}^n \alpha_{T-1}^j g_{T-1}^j\left(x_{T-1}, u_{T-1}^1, u_{T-1}^2, \cdots, u_{T-1}^n\right) \right. \\
&\qquad \left. + W^{(\alpha_{T-1})}\left[T, f_{T-1}\left(x_{T-1}, u_{T-1}^1, u_{T-1}^2, \cdots, u_{T-1}^n\right)\right] \right\}
\end{aligned} \tag{3.10}$$

Invoking Theorem 1.3 one can characterize the payoff of player i under cooperation covering the stages $T-1$ to $T+1$ by:

$$W^{(\alpha_{T-1})i}(T,x) = W^{(\hat{\alpha}_T)i}(T,x_T),$$
$$W^{(\alpha_{T-1})i}(T-1,x) = g_{T-1}^{i}\left[x,\psi_{T-1}^{(\alpha_{T-1})1}(x),\psi_{T-1}^{(\alpha_{T-1})2}(x),\cdots,\psi_{T-1}^{(\alpha_{T-1})n}(x)\right]$$
$$+W^{(\alpha_{T-1})i}\left[T,f_{T-1}\left(x,\psi_{T-1}^{(\alpha_{T-1})1}(x),\psi_{T-1}^{(\alpha_{T-1})2}(x),\cdots,\psi_{T-1}^{(\alpha_{T-1})n}(x)\right)\right],\text{ for } i\in N. \quad (3.11)$$

For individual rationality to be maintained, it is required that:

$$W^{(\alpha_{T-1})i}(T-1,x) \geq V^{i}(T-1,x),\text{ for } i\in N. \quad (3.12)$$

We use Λ_{T-1} to denote the set of weights α_{T-1} that satisfies (3.12). Let the vector $\hat{\alpha}_{T-1} = \left(\hat{\alpha}_{T-1}^{1},\hat{\alpha}_{T-1}^{2},\cdots,\hat{\alpha}_{T-1}^{n}\right) \in \Lambda_{T-1}$ be the set of payoff weights that leads to satisfaction of the optimality principle $\Gamma(T-1,x)$.

11.3.2 Optimal Cooperative Strategies in Preceding Stages

Now we proceed to characterize the cooperative scheme in stage $k\in\{1,2,\cdots,T-1\}$. Following the analysis in Sect. 11.3.1, the payoff of player i in stage $k+1$ is $W^{(\hat{\alpha}_{k+1})i}(k+1,x)$ and his payoff in stage k can be expressed as:

$$g_k^i\left(x_k,u_k^1,u_k^2,\cdots,u_k^n\right)$$
$$+\sum_{h=k}^{T} g_h^i\left[x_h,\psi_h^{(\hat{\alpha}_h)1}(x_h),\psi_h^{(\hat{\alpha}_h)2}(x_h),\cdots,\psi_h^{(\hat{\alpha}_h)n}(x_h)\right]+q^i(x_{T+1})$$
$$= g_k^i\left(x_k,u_k^1,u_k^2,\cdots,u_k^n\right)+W^{(\hat{\alpha}_{k+1})i}(k,x_{k+1}),\text{ for } i\in N. \quad (3.13)$$

In this stage the players will select a set of weights $\alpha_k = \left(\alpha_k^1,\alpha_k^2,\cdots,\alpha_k^n\right)$ which satisfies the optimality principle $P(k,x)$. The players' optimal cooperative strategies can be generated by solving the following dynamic programming problem of maximizing the weighted sum of payoffs

$$\sum_{j=1}^{n}\alpha_k^j\left[\,g_k^j\left(x_k,u_k^1,u_k^2,\cdots,u_k^n\right)+W^{(\hat{\alpha}_{k+1})j}(k+1,x_{k+1})\,\right], \quad (3.14)$$

subject to

$$x_{k+1} = f_k\left(x_k,u_k^1,u_k^2,\cdots,u_k^n\right),\quad x_k = x, \quad (3.15)$$

Invoking Theorem 1.2, given the payoff weights being α_k the optimal cooperative strategies $\left\{u_k^i = \psi_k^{(\alpha_k)i},\text{ for } i\in N\right\}$ in stage k are characterized by the conditions

$$W^{(\alpha_k)}(k+1,x) = \sum_{j=1}^{n} \alpha_k^j W^{(\hat{\alpha}_{k+1})j}(k+1,x)$$

$$W^{(\alpha_k)}(k,x) = \max_{u_k^1,u_k^2,\cdots,u_k^n} \left\{ \sum_{j=1}^{n} \alpha_k^j g_k^j(x_k,u_k^1,u_k^2,\cdots,u_k^n) \right.$$
$$\left. +W^{(\alpha_k)}\left[k+1,f_k(x_k,u_k^1,u_k^2,\cdots,u_k^n)\right] \right\}, \tag{3.16}$$

The payoff of player i under cooperation can be obtained as:

$$W^{(\alpha_k)i}(k,x) = g_k^i\left[x,\psi_k^{(\alpha_k)1}(x),\psi_k^{(\alpha_k)2}(x),\cdots,\psi_k^{(\alpha_k)n}(x)\right] + W^{(\hat{\alpha}_{k+1})i}(k+1,x_{k+1}), \tag{3.17}$$

for $i \in N$.

Invoking Theorem 1.3 one can characterize $W^{(\alpha_k)i}(k,x)$ by the following equations

$$W^{(\alpha_k)i}(k+1,x) = W^{(\hat{\alpha}_{k+1})i}(k+1,x),$$
$$W^{(\alpha_k)i}(k,x) = g_k^i\left[x,\psi_k^{(\alpha_k)1}(x),\psi_k^{(\alpha_k)2}(x),\cdots,\psi_k^{(\alpha_k)n}(x)\right]$$
$$+W^{(\alpha_k)i}\left[k+1,f_k\left(x,\psi_k^{(\alpha_k)1}(x),\psi_k^{(\alpha_k)2}(x),\cdots,\psi_k^{(\alpha_k)n}(x)\right)\right], \tag{3.18}$$

for $i \in N$.

For individual rationality to be maintained in stage k, it is required that:

$$W^{(\alpha_k)i}(k,x) \geq V^i(k,x), \text{ for } i \in N. \tag{3.19}$$

We use Λ_k to denote the set of weights α_k that satisfies (3.19). Again, we use $\hat{\alpha}_k = \left(\hat{\alpha}_k^1,\hat{\alpha}_k^2,\cdots,\hat{\alpha}_k^n\right) \in \Lambda_k$ to denote the set of payoff weights that leads to the satisfaction of the optimal principle $P(k,x)$, for $k \in \kappa$.

11.3.3 Subgame Consistent Solution: A Mathematical Theorem

In this subsection, we consider subgame consistents under viable payoff weights. A theorem characterizing a subgame consistent solution of the cooperative dynamic game (1.1–1.2) with the optimality principle $P(k,x_k)$ can be obtained as follows.

Theorem 3.1 A set of payoff weights $\left\{\hat{\alpha}_k = \left(\hat{\alpha}_k^1,\hat{\alpha}_k^2,\cdots,\hat{\alpha}_k^n\right), \text{ for } k \in \kappa\right\}$ and a set of strategies $\left\{\psi_k^{(\hat{\alpha}_k)i}(x), \text{ for } k \in \kappa \text{ and } i \in N\right\}$ provides a subgame consistent solution to the cooperative dynamic game (1.1) and (1.2) with optimality principle $P(k,x)$ if there exist functions $W^{(\hat{\alpha}_k)}(k,x)$ and $W^{(\hat{\alpha}_k)i}(k,x)$, for $i \in N$, $k \in \kappa$, which satisfy the following recursive relations:

$$\begin{aligned}
&W^{(\hat{\alpha}_{T+1})i}(T+1,x) = q^i(x_{T+1}) \\
&W^{(\hat{\alpha}_k)}(k,x) = \max_{u_k^1, u_k^2, \cdots, u_k^n} \left\{ \sum_{j=1}^{n} \hat{\alpha}^j g_k^j(x, u_k^1, u_k^2, \cdots, u_k^n) \right. \\
&\qquad \left. + \sum_{j=1}^{n} \hat{\alpha}_k^j W^{(\hat{\alpha}_{k+1})j}\left[k+1, f_k(x, u_k^1, u_k^2, \cdots, u_k^n)\right] \right\}; \\
&W^{(\hat{\alpha}_k)i}(k,x) = g_k^j\left(x, \psi_k^{(\hat{\alpha}_k)1}(x), \psi_k^{(\hat{\alpha}_k)2}(x), \cdots, \psi_k^{(\hat{\alpha}_k)n}(x)\right) \\
&\qquad + W^{(\hat{\alpha}_{k+1})i}\left[k+1, f_k\left(x, \psi_k^{(\hat{\alpha}_k)1}(x), \psi_k^{(\hat{\alpha}_k)2}(x), \cdots, \psi_k^{(\hat{\alpha}_k)n}(x)\right)\right], \\
&\text{for } i \in N \text{ and } k \in \kappa;
\end{aligned} \tag{3.20}$$

and the optimality principle

$$P(k,x) \text{ in all stages } k \in \kappa. \tag{3.21}$$

Proof See the exposition from equation (3.1) to equation (3.19) in Sects. 11.3.1 and 11.3.2. ■

In the case when the agreed-upon optimality principle requires the proportion each player's cooperative payoff to his non-cooperative payoff being equal, condition (3.21) in Theorem 3.1 becomes

$$\frac{W^{(\hat{\alpha}_t)i}(t,x)}{V^i(t,x)} = \frac{W^{(\hat{\alpha}_t)j}(t,x)}{V^j(t,x)}, \text{ for } i,j \in N \text{ and } t \in \kappa.$$

If the optimality principle requires the satisfaction of the Nash bargaining solution, condition (3.21) in Theorem 3.1 becomes

$$\hat{\alpha}_t = \arg\max_{\alpha_t} \left\{ \prod_{j=1}^{n} \left[W^{(\alpha_t)j}(t,x) - V^j(t,x) \right] \right\}; \quad \text{for } t \in \kappa.$$

In the two-player case, if the optimality principle requires the chosen payoff weights $\hat{\alpha}_t = \{\hat{\alpha}_t^1, \hat{\alpha}_t^2\}$ to be the mid-value of the maximum and minimum of the payoff weight α_t^1 and that of the payoff weights α_t^2 in the set Λ, condition (3.21) in Theorem 3.1 becomes:

$\hat{\alpha}_t^i = \frac{\underline{\alpha}_t^i + \overline{\alpha}_t^i}{2}$, $W^{(\underline{\alpha}_t^i, 1-\underline{\alpha}_t^i)i}(t,x) = V^i(t,x)$ and $W^{(\overline{\alpha}_t^i, 1-\overline{\alpha}_t^i)j}(t,x) = V^j(t,x)$,for $i,j \in \{1,2\}$, $i \neq j$ and $t \in \kappa$.

Substituting the optimal control $\left\{\psi_k^{(\hat{\alpha}_k)i}(x) \text{ for } i \in N \text{ and } k \in \kappa\right\}$ into the state dynamics (1.1), one can obtain the dynamics of the cooperative trajectory as:

$$x_{k+1} = f_k\left(x_k, \psi_k^{(\hat{\alpha}_k)1}(x_k), \psi_k^{(\hat{\alpha}_k)2}(x_k), \cdots, \psi_k^{(\hat{\alpha}_k)n}(x_k)\right)$$
$$x_1 = x_1^0 \text{ and } k \in \kappa \tag{3.22}$$

The cooperative trajectory $\{x_k^* \text{ for } k \in \kappa\}$ is the solution generated by (3.22).

Remark 3.1 The subgame consistent solution presented in Theorem 3.1 is conditional Pareto efficient in the sense that the solution is a Pareto efficient outcome satisfying the condition that the agreed-upon optimality principle is maintained in all stages. ■

In particular, there does not any exist any strategies paths fulfilling the agreed-upon optimality principle in every stage that would lead to the payoff for any player i $W^i(k,x) > W^{(\hat{\alpha}_t)i}(k,x)$, while the payoffs of other players remains no less than $W^{(\hat{\alpha}_t)j}(k,x)$, for $i,j \in N$ and $j \neq i$ and $k \in \kappa$.

Remark 3.2 A subgame consistent solution is fully Pareto efficient only if the optimality principle $P(t,x)$ requires the choice of a set of time-invariant payoff weights. ■

A special example of the optimality principle in a two-player differential game leading to the choice of a set of constant weights for a subgame consistent solution can be found in Yeung and Petrosyan (2005). The optimality principle requires the chosen payoff weights $\hat{\alpha}_t = \{\hat{\alpha}_t^1, \hat{\alpha}_t^2\}$ equals an average value of the maximum and minimum of stage T's payoff weight α_T^1 and that of α_T^2 in the set Λ_T, for $t \in \kappa$, under the pre-condition that $\hat{\alpha}_t \in \Lambda_t$. Note that a subgame consistent solution under this restricted optimality principle exists in very limited situations.

In cooperative dynamic games with transferable payoffs, subgame consistent solution satisfying dynamic consistency and full Pareto efficiency can bė obtained with the use of side-payments (see analysis in Chap. 7). However in dynamic games with non-transferable payoffs, it often not possible to reach a cooperative solution satisfying full Pareto efficiency and individual rationality in all stages because of the absence of side-payments. Since the issue of full Pareto efficiency is of less importance than that of the reaching of a cooperative solution, achieving the latter at the expense of the former is a practical way out.

11.4 An Illustration in Public Capital Build-up

In this section, we provide an illustration of the derivation of subgame consistent solutions of public goods provision in a 2-player cooperative dynamic game with non-transferable payoffs.

11.4.1 Game Formulation and Noncooperative Outcome

Consider an economic region with 2 asymmetric agents. These agents receive benefits from an existing public capital stock x_t at each stage $t \in \{1, 2, 3, 4\}$. The accumulation dynamics of the public capital stock is governed by the difference equation:

$$x_{k+1} = x_k + \sum_{j=1}^{2} u_k^j - \delta x_k \, , \, x{=}x_1^0, \text{ for } t \in \{1, 2, 3\}, \tag{4.1}$$

where u_k^i is the physical amount of investment in the public good and δ is the rate of depreciation.

The objective of agent $i \in \{1, 2\}$ is to maximize the payoff:

$$\sum_{k=1}^{3} \left[a_k^i x_k - c_k^i \left(u_k^i \right)^2 \right] (1+r)^{-(k-1)} + \left(q^i x_4 + m^i \right) (1+r)^{-3}, \tag{4.2}$$

subject to the dynamics (4.1),

where $\alpha_k^i x_k$ gives the gain that agent i derives from the public capital at stage $t \in \{1, 2, 3\}$, $c_k^i (u_k^i)^2$ is the cost of investing u_k^i in the public capital, r is the discount rate and $(q^i x_4 + m^i)$ is the terminal payoff of agent i at stage 4.

The payoffs of the agent are not transferable. We first derive the noncooperative outcome of the game. Invoking Theorem 1.1, one can characterize the noncooperative Nash equilibrium for the game (4.1 and 4.2) as follows. In particular, a set of strategies $\{u_t^{i*} = \phi_t^i(x), \text{ for } t \in \{1, 2, 3\} \text{ and } i \in \{1, 2\}\}$ provides a Nash equilibrium solution to the game (4.1 and 4.2) if there exist functions $V^i(t, x)$, for $i \in \{1, 2\}$ and $t \in \{1, 2, 3\}$, such that the following recursive relations are satisfied:

$$\begin{aligned} V^i(t, x) = \max_{u_t^i} \Bigg\{ & \left[a_t^i x - c_t^i \left(u_t^i \right)^2 \right] (1+r)^{-(t-1)} \\ & + V^i \left[t+1, x + \phi_t^j(x) + u_t^i - \delta x \right] \Bigg\} \text{ for } t \in \{1, 2, 3\}; \end{aligned} \tag{4.3}$$

$$V^i(4, x) = (q^i x + m^i)(1+r)^{-3}, \quad \text{for } i \in \{1, 2\} \, . \tag{4.4}$$

Performing the indicated maximization in (4.3) yields:

$$\phi_t^i(x) = \frac{(1+r)^{t-1}}{2 c_t^i} V_{x_{t+1}}^i \left[t+1, x + \sum_{j=1}^{2} \phi_t^j(x) - \delta x \right], \tag{4.5}$$

for $i \in \{1, 2\}$ and $t \in \{1, 2, 3\}$.

The game equilibrium payoffs of the agents can be obtained as:

Proposition 4.1 The value function which represents the game equilibrium payoff of agent i is:

$$V^i(t,x) = \left[A_t^i x + C_t^i\right](1+r)^{-(t-1)}, \text{ for } i \in \{1,2\} \text{ and } t \in \{1,2,3\}, \tag{4.6}$$

where

$$A_3^i = a_3^i + q^i(1-\delta)(1+r)^{-1}, \text{ and}$$

$$C_3^i = -\frac{(q^i)^2(1+r)^{-2}}{4c_3^i} + \left[q^i \sum_{j=1}^{2} \frac{q^j(1+r)^{-1}}{2c_3^j} + m^i\right](1+r)^{-1};$$

$$A_2^i = a_2^i + A_3^i(1-\delta)(1+r)^{-1}, \text{ and}$$

$$C_2^i = -\frac{1}{4c_2^i}\left(A_3^i(1+r)^{-1}\right)^2$$

$$+\left[A_3^{(\sigma_3)i}\left(\sum_{j=1}^{2}\frac{A_3^j(1+r)^{-1}}{2c_2^j}\right) + C_3^i\right](1+r)^{-1}\Bigg\};$$

$$A_1^i = a_1^i + A_2^i(1-\delta)(1+r)^{-1}, \text{ and}$$

$$C_1^i = -\frac{1}{4c_1^i}\left(A_2^i(1+r)^{-1}\right)^2$$

$$+\left[A_2^i\left(\sum_{j=1}^{2}\frac{A_2^j(1+r)^{-1}}{2c_1^j}\right) + C_2^i\right](1+r)^{-1}\Bigg\}; \text{ for } i \in \{1,2\}. \tag{4.7}$$

Proof Using (4.5) and (4.6) to evaluate the system (4.3 and 4.4) yields the results in (4.6 and 4.7). ■

11.4.2 Cooperative Solution

Now consider first the case when the agents agree to cooperate and maintain an optimality principle $P(t,x_t)$ requiring the adoption of the mid values of the maximum and minimum of the payoff weight α_t^i in the set Λ_t, for $i \in \{1,2\}$ and $t \in \{1,2,3\}$.

In view of Theorem 3.1, to obtain the maximum and minimum values of α_T^i, we first consider deriving the optimal cooperative strategies in stage $T = 3$ by solving the problem:

$$W^{(\alpha_4)}(4,x) = \sum_{j=1}^{2}(q^j x + m^j)(1+r)^{-3}$$

$$W^{(\alpha_3)}(3,x) = \max_{u_3^1, u_3^2}\left\{ \sum_{j=1}^{2}\alpha_3^j\left[a_3^j x - c_3^j\left(u_3^j\right)^2\right](1+r)^{-2} + \sum_{j=1}^{2}\alpha_3^j\left[q^j\left(x + \sum_{j=1}^{2}u_3^j - \delta x\right) + m^j\right](1+r)^{-3}\right\}. \tag{4.8}$$

Performing the indicated maximization in (4.8) yields:

$$\psi_3^{(\alpha_3)i}(x) = \frac{(1+r)^{-1}}{2\alpha_3^i c_3^i}\sum_{j=1}^{n}\alpha_3^j q^j, \text{ for } i \in \{1,2\}. \tag{4.9}$$

The maximized weighted joint payoff under payoff weights α_3 at stage 3 can be obtained as:

Proposition 4.2

$$W^{(\alpha_3)}(3,x) = \left[A_3^{(\alpha_3)}x + C_3^{(\alpha_3)}\right](1+r)^{-2}, \tag{4.10}$$

where

$$A_3^{(\alpha_3)} = \sum_{j=1}^{2}\alpha_3^j\left[a_3^j + q^j(1-\delta)(1+r)^{-1}\right], \text{ and}$$

$$C_3^{(\alpha_3)} = -\sum_{j=1}^{2}\alpha_3^j\left[\frac{(1+r)^{-2}}{4\alpha_3^j c_3^j}\left(\sum_{\ell=1}^{2}\alpha_3^\ell q_3^\ell\right)^2\right] + \sum_{j=1}^{2}\alpha_3^j\left[q^j\left(\sum_{j=1}^{2}\frac{(1+r)^{-1}}{2\alpha_3^j c_3^j}\left(\sum_{\ell=1}^{2}\alpha_3^\ell q_3^\ell\right)\right) + m^j\right](1+r)^{-1}\bigg\}. \tag{4.11}$$

Proof Substitute the cooperative strategies from (4.9) into (4.8) yields the function $W^{(\alpha_3)}(3,x)$ in (4.10). ∎

Invoking Theorem 1.3, the payoff of player i under cooperation can be characterized as:

$$W^{(\alpha_3)i}(3,x) = \left[a_3^i x - \frac{(1+r)^{-2}}{4\alpha_3^i c_3^i} \left(\sum_{\ell=1}^{2} \alpha_3^\ell q_3^\ell \right)^2 \right] (1+r)^{-2}$$
$$+ \left[q^i \left(x + \sum_{j=1}^{2} \frac{(1+r)^{-1}}{2c_3^j} \left(\sum_{\ell=1}^{2} \alpha_3^\ell q_3^\ell \right) \right.\right.$$
$$\left.\left. - \delta x \right) + m^i \right] (1+r)^{-3} \Bigg\}, \tag{4.12}$$

for $i \in \{1,2\}$.

The value function $W^{(\alpha_3)i}(3,x)$ reflecting the cooperative payoff of player i under payoff weights α_3 at stage 3 can be obtained as:

Proposition 4.3

$$W^{(\alpha_3)i}(3,x) = \left[A_3^{(\alpha_3)i} x + C_3^{(\alpha_3)i} \right] (1+r)^{-2}, \text{ for } i \in \{1,2\}, \tag{4.13}$$

where

$$A_3^{(\alpha_3)i} = \left[a_3^j + q^j (1-\delta)(1+r)^{-1} \right], \text{ and}$$

$$C_3^{(\alpha_3)i} = -\left[\frac{(1+r)^{-2}}{4\alpha_3^i c_3^i} \left(\sum_{\ell=1}^{2} \alpha_3^\ell q_3^\ell \right)^2 \right]$$
$$+ \left[q^i \left(\sum_{j=1}^{2} \frac{(1+r)^{-1}}{2\alpha_3^j c_3^j} \left(\sum_{\ell=1}^{2} \alpha_3^\ell q_3^\ell \right) \right) + m^i \right] (1+r)^{-1} \Bigg\} \tag{4.14}$$

Proof The right-hand-side of equation (4.12) is a linear function with coefficients $A_3^{(\alpha_3)i}$ and $A_3^{(\alpha_3)i}$ in (4.14). Hence Proposition 4.3 follows. ■

To identify the range of α_3 that satisfies individual rationality we examine the functions which gives the excess of agent i's cooperative over his non-cooperative payoff:

$$W^{(\alpha_3)i}(3,x) - V^i(3,x) = \left[C_3^{(\alpha_3)i} - C_3^i \right] (1+r)^{-2}, \text{ for } i \in \{1,2\}. \tag{4.15}$$

For individual rationality to be satisfied, it is required that $W^{(\alpha_3)i}(3,x) - V^i(3,x) \geq 0$ for $i \in \{1,2\}$. Using $\alpha_3^j = 1 - \alpha_3^i$ and upon rearranging terms $C_3^{(\alpha_3)i}$ can be expressed as:

$$C_3^{(\alpha_3)i} = q^i\left[\frac{(1+r)^{-2}}{2c_3^i}\left(\frac{\alpha_3^i q^i + (1-\alpha_3^i)q^j}{\alpha_3^1}\right)\right.$$
$$\left.+\frac{(1+r)^{-2}}{2c_3^j}\left(\frac{\alpha_3^i q^i + (1-\alpha_3^i)q^j}{1-\alpha_3^i}\right)\right] + m^i(1+r)^{-1}$$
$$-\frac{(1+r)^{-2}}{4c_3^i}\left(\frac{\alpha_3^i q^i + (1-\alpha_3^i)q^j}{\alpha_3^i}\right)^2, \text{for } i,j \in \{1,2\} \text{ and } i \neq j. \tag{4.16}$$

Differentiating $C_3^{(\alpha_3)i}$ with respect to α_3^i yields

$$\frac{\partial C_3^{(\alpha_3)i}}{\partial \alpha_3^i} = \frac{(1+r)^{-2}}{2c_3^j}\left(\frac{(q^i)^2}{(1-\alpha_3^i)^2}\right)$$
$$+\frac{(1+r)^{-2}}{2c_3^i}\left(\frac{(1-\alpha_3^i)q^j}{\alpha_3^i}\right)\left(\frac{q^j}{(\alpha_3^i)^2}\right), \tag{4.17}$$

which is positive for $\alpha_3^i \in (0,1)$.

One can readily observed that $\lim\limits_{\alpha_3^i \to 0} C_3^{(\alpha_3)i} \to -\infty$ and $\lim\limits_{\alpha_3^i \to 1} C_3^{(\alpha_3)i} \to \infty$. Therefore an $\underline{\alpha}_3^i \in (0,1)$ can be obtained such that $W^{(\underline{\alpha}_3^i, 1-\underline{\alpha}_3^i)i}(3,x) = V^i(3,x)$ and yields agent i's minimum payoff weight value satisfying his own individual rationality. Similarly there exist an $\overline{\alpha}_3^i \in (0,1)$ such that $W^{(\overline{\alpha}_3^i, 1-\overline{\alpha}_3^i)j}(3,x) = V^j(3,x)$ and yields agent i's maximum payoff weight value while maintaining agent j's individual rationality.

Since the maximization of the sum of weighted payoffs in stage 3 yields a Pareto optimum, there exist a non-empty set of α_3 satisfying individual rationality for both agents. Given that the agreed-upon optimality principle $P(t,x_t)$ requires the adoption of the mid values of the maximum and minimum of the payoff weight α_t^i in the set Λ_t, for $t \in \{1,2,3\}$, the cooperative weights in stage 3 is $\hat{\alpha}_3 = \left(\frac{\underline{\alpha}_3^i + \overline{\alpha}_3^i}{2}, 1 - \frac{\underline{\alpha}_3^i + \overline{\alpha}_3^i}{2}\right)$.

Now consider the stage 2 problem, we derive the optimal cooperative strategies in stage 2 by solving the problem:

$$W^{(\alpha_2)}(2,x) = \max_{u_2^1, u_2^2}\left\{\sum_{j=1}^{2}\alpha_2^j\left[a_2^j x - c_2^j\left(u_2^j\right)^2\right](1+r)^{-1}\right.$$
$$\left.+\sum_{j=1}^{2}\alpha_2^j W^{(\hat{\alpha}_3)j}\left[3, x + \sum_{j=1}^{2}u_2^j - \delta x\right]\right\} \tag{4.18}$$

Performing the indicated maximization in (4.18) yields:

$$\psi_2^{(\alpha_2)i}(x) = \frac{(1+r)^{-1}}{2\alpha_2^i c_2^i} \sum_{j=1}^{n} \alpha_2^j A_3^{(\hat{\alpha}_3)j}, \text{ for } i \in \{1,2\}.$$

Following the analysis in stage 3, one can obtain

$$W^{(\alpha_2)}(2,x) = \left[A_2^{(\alpha_2)}x + C_2^{(\alpha_2)}\right](1+r)^{-1},$$
$$W^{(\alpha_2)i}(2,x) = \left[A_2^{(\alpha_2)i}x + C_2^{(\alpha_2)i}\right](1+r)^{-1}, \text{ for } i \in \{1,2\}, \tag{4.19}$$

where $A_2^{(\alpha_2)}$, $C_2^{(\alpha_2)}$, $A_2^{(\alpha_2)i}$ and $C_2^{(\alpha_2)i}$ are functions that depend on α_2.

Similarly, one can readily show that $\frac{\partial C_2^{(\alpha_2)i}}{\partial \alpha_2^i}$ is positive and $\lim_{\alpha_2^i \to 0} C_2^{(\alpha_2)i} \to -\infty$ and $\lim_{\alpha_2^i \to 1} C_2^{(\alpha_2)i} \to \infty$. Agent i's minimum payoff weight is $\underline{\alpha}_2^i \in (0,1)$ which leads to

$W^{(\underline{\alpha}_2^i, 1-\underline{\alpha}_2^i)i}(2,x) = V^i(2,x)$,

and his maximum payoff weight is $\overline{\alpha}_2^i \in (0,1)$ which leads to

$$W^{(\overline{\alpha}_2^i, 1-\overline{\alpha}_2^i)j}(2,x) = V^j(2,x).$$

Invoking the agreed-upon optimality principle $P(t, x_t)$ the cooperative weights in stage 2 is $\hat{\alpha}_2 = \left(\frac{\underline{\alpha}_2^i + \overline{\alpha}_2^i}{2}, 1 - \frac{\underline{\alpha}_2^i + \overline{\alpha}_2^i}{2}\right)$.

Finally, following the analysis in stages 2 and 3, one can obtain the cooperative weights in stage 1 as $\hat{\alpha}_1 = \left(\frac{\underline{\alpha}_1^i + \overline{\alpha}_1^i}{2}, 1 - \frac{\underline{\alpha}_1^i + \overline{\alpha}_1^i}{2}\right)$.

In general, there is no guarantee for the existence of a constant payoff weight such that the basic requirement of individual rationality is satisfied in all subsequent stages. An example is provided below.

Example 4.1 Consider the case in which $q^1 = 3$, $q^2 = 4$, $m^1 = 10$, $m^2 = 20$, $r = 0.05$, $\delta = 0.02$, $c_3^1 = 2$, $c_3^2 = 4$, $a_3^1 = 4$, $a_3^2 = 1$, $c_2^1 = 7$, $c_2^2 = 2$, $a_2^1 = 1$, $a_2^2 = 2$, $c_1^1 = 1$, $c_1^2 = 4$, $a_1^1 = 2$, $a_1^2 = 1$. In stage 1, a constant α_1^1 has to be between $\underline{\alpha}_1^1 = 0.435$ and $\overline{\alpha}_1^1 = 0.545$. In stage 2, a constant α_2^1 has to be between $\underline{\alpha}_2^1 = 0.33$ and $\overline{\alpha}_2^1 = 0.43$. In stage 3, α_3^1 has to be between $\underline{\alpha}_3^1 = 0.55$ and $\overline{\alpha}_3^1 = 0.655$. Therefore there does not exist a constant choice of α_t^1 for $t \in \{1,2,3\}$ such that individual rationality is satisfied in all the subsequent subgame stages. ■

11.4.3 Other Optimality Principles

In this section, we consider deriving subgame consistent solutions for the cooperative dynamic game (4.1 and 4.2) under two other optimality principles.

11.4.3.1 Proportional Cooperative Gains

We first consider the optimality principle $P(t,x_t)$ which requires the proportion of each player's cooperative payoff to his non-cooperative payoff to be equal. In particular, a subgame consistent solution requires payoff weights $\hat{\alpha}_1$, $\hat{\alpha}_2$ and $\hat{\alpha}_3$ leading to

$$\frac{W^{(\hat{\alpha}_t)1}(t,x_t)}{V^1(t,x_t)} = \frac{W^{(\hat{\alpha}_t)2}(t,x_t)}{V^2(t,x_t)}, \text{ for } t \in \{12,3\},$$

along the cooperation trajectory.

Invoking the value functions $V^i(t,x)$ and $W^{(\hat{\alpha}_t)i}(t,x)$, for $i \in \{1,2\}$ and $t \in \{12,3\}$, a subgame consistent solution to the problem can be obtained with the payoff weights $\hat{\alpha}_1$, $\hat{\alpha}_2$ and $\hat{\alpha}_3$ which satisfy:

$$\frac{A_t^{(\hat{\alpha}_t)1}x_t + C_t^{(\hat{\alpha}_t)1}}{A_t^1 x_t + C_t^1} = \frac{A_t^{(\hat{\alpha}_t)2}x_t + C_t^{(\hat{\alpha}_t)2}}{A_t^2 x_t + C_t^2}, \text{ for } t \in \{12,3\}, \quad (4.20)$$

and

$$x_{t+1} = x_t \sum_{j=1}^{2} \left[\frac{(1+r)^{-1}}{2\hat{\alpha}_t^j c_t^i} \sum_{\ell=1}^{2} \hat{\alpha}_t^\ell A_{t+1}^{(\hat{\alpha}_{t+1})\ell} \right] - \delta x_t, \; x_1 = x_1^0, \quad (4.21)$$

for $t \in \{12,3\}$ and $A_4^{(\hat{\alpha}_4)\ell} = q^\ell$.

Note that from (4.14), $A_t^{(\alpha_t)i}$ is independent of α_t. Moreover, $C_t^{(\alpha_t)i}$ is strictly increasing in α_t^i and $C_t^{(\alpha_t)j}$ is strictly decreasing in α_t^i for $\alpha_t^i \in [\underline{\alpha}_t^i, \overline{\alpha}_t^i]$, therefore one can readily identify payoff weights $\hat{\alpha}_t^i$ such that (4.20) is satisfied.

11.4.3.2 Dynamic Nash Bargaining Solution

Now, consider the case where the agents agree with an optimality principle $P(t,x_t)$ that requires the excess of the players' cooperative payoffs over their respective non-cooperative payoffs satisfies the Nash bargaining solution. As Haurie (1976) pointed out that the property of dynamic consistency, which is crucial in maintaining sustainability in cooperation, is absent in the direct application of the Nash bargaining solution in dynamic games. To overcome this problem, a dynamic Nash bargaining solution can be represented by the subgame consistent solution to the cooperative dynamic game problem in which the optimality principle requires the satisfaction of the Nash bargaining solution in every stage of the game. To obtain such a solution, the players would first search for an α_3 in stage 3 to maximize the Nash product

$$\prod_{j=1}^{2}\left[W^{(\alpha_3)j}(3,x) - V^j(3,x)\right].$$

Invoking (4.15) and (4.16), the dynamic Nash bargaining solution has to solve the problem

$$\max_{\alpha_3^i}\prod_{j=1}^{2}\left[C_3^{(\alpha_3)j} - C_3^j\right](1+r)^{-2}, \tag{4.22}$$

in the range of $\alpha_3^i \in \left[\underline{\alpha}_3^i, \overline{\alpha}_3^i\right]$.

Invoking the derivative property of $C_3^{(\alpha_3)i}$ in (4.17) a solution weight $\hat{\alpha}_3^i \in \left[\underline{\alpha}_3^i, \overline{\alpha}_3^i\right]$ can be obtained. Then one can use $W^{(\hat{\alpha}_3)j}(3,x)$ for $j \in \{1,2\}$ to form the terminal payoff $\sum_{j=1}^{2}\alpha_2^j W^{(\hat{\alpha}_3)j}(3,x)$ for the cooperation scheme in stage 2. Repeating the above analysis, one can identify $\hat{\alpha}_2$ which yields the Nash bargaining solution in stage 2. Finally, in a similar manner, $\hat{\alpha}_1$ which yields the Nash bargaining solution in stage 1 can be obtained.

11.5 Stochastic Extension

Consider the general $T-$ stage $n-$ person nonzero-sum discrete-time stochastic dynamic game with initial state x_1^0. The state space of the game is $X \in R^m$ and the state dynamics of the game is characterized by the stochastic difference equation:

$$x_{k+1} = f_k\left(x_k, u_k^1, u_k^2, \cdots, u_k^n\right) + G_k(x_k)\theta_k, \tag{5.1}$$

for $k \in \{1,2,\cdots,T\} \equiv \kappa$ and $x_1 = x_1^0$,

where $u_k^i \in R^{m_i}$ is the control vector of player i at stage k, and $x_k \in X$ is the state of the game and θ_k is a set of independent random variable. The payoff that player i seeks to maximize is

$$E_{\theta_1,\theta_2,\cdots,\theta_T}\left\{\sum_{\zeta=1}^{T} g_\zeta^i\left[x_\zeta, u_\zeta^1, u_\zeta^2, \cdots, u_\zeta^n, x_{\zeta+1}\right] + q^i(x_{T+1})\right\}, \tag{5.2}$$

for $i \in \{1,2,\cdots,n\} \equiv N$,

where $q^i(x_{T+1})$ is the terminal payoff that player i will received in stage $T+1$, and $E_{\theta_1,\theta_2,\cdots,\theta_T}$ is the expectation operation with respect to the statistics of θ_1, θ_2, $\cdots, \theta_T$.

The payoffs of the players are not transferable.

11.5.1 Non-cooperative Outcome and Optima Under Constant Weights

Using Theorem 4.1 in Chap. 7 a feedback Nash equilibrium of the game (5.1 and 5.2) can be characterized as follows. A set of strategies $\{\phi_k^i(x), \text{ for } k \in \kappa \text{ and } i \in N\}$ provides a feedback Nash equilibrium solution to the game (5.1 and 5.2) if there exist functions $V^i(k,x)$, for $k \in K$ and $i \in N$, such that the following recursive relations are satisfied:

$$V^i(k,x) = \max_{u_k^i} E_{\theta_k}\Big\{ g_k^i\big[x, \phi_k^1(x), \phi_k^2(x), \cdots, \phi_k^{i-1}(x), u_k^i, \phi_k^{i+1}(x), \cdots, \phi_k^n(x)\big]$$
$$+V^i\big[k+1, f_k\big[x, \phi_k^1(x), \phi_k^2(x), \cdots, \phi_k^{i-1}(x), u_k^i, \phi_k^{i+1}(x), \cdots, \phi_k^n(x)\big]\Big\}$$

$V^i(T+1,x) = q_{T+1}^i(x)$; for $i \in N$ and $k \in \kappa$.

Again, to establish the optimization foundation of the variable weights scheme we consider first the case in which the players adopt a vector of constant payoff weights $\alpha = (\alpha^1, \alpha^2, \cdots, \alpha^n)$ in all stages, where $\sum_{j=1}^{n} \alpha^j = 1$. Conditional upon the vector of weights α, the players' optimal cooperative strategies can be generated by solving the stochastic dynamic programming problem of maximizing the expected weighted sum of payoffs:

$$E_{\theta_1,\theta_2,\cdots,\theta_T}\left\{ \sum_{j=1}^{n} \left[\sum_{k=1}^{T} \alpha^j g_k^j(x_k, u_k^1, u_k^2, \cdots, u_k^n) + \alpha^j q^j(x_{T+1}) \right] \right\} \tag{5.3}$$

subject to (5.1).

An optimal solution to the problem (5.1) and (5.3) can be characterized by the following theorem.

Theorem 5.1 A set of strategies $\left\{\psi_k^{(\alpha)i}(x), \text{ for } k \in \kappa \text{ and } i \in N\right\}$ provides an optimal solution to the problem (1.1) and (1.4) if there exist functions $W^{(\alpha)}(k,x)$, for $k \in K$, such that the following recursive relations are satisfied:

$$W^{(\alpha)}(T+1,x) = \sum_{j=1}^{n} \alpha^j q^j(x_{T+1}),$$

$$W^{(\alpha)}(k,x) = \max_{u_k^1, u_k^2, \cdots, u_k^n} E_{\theta_k}\Big\{ \sum_{j=1}^{n} \alpha^j g_k^j(x_k, u_k^1, u_k^2, \cdots, u_k^n)$$
$$+ W^{(\alpha)}\big[k+1, f_k(x_k, u_k^1, u_k^2, \cdots, u_k^n)\big]\Big\} \tag{5.4}$$

Proof The conditions in (5.4) stem directly from the results in discrete-time stochastic dynamic programming in Theorem A.6 of the Technical Appendices. ∎

Substituting the optimal control $\left\{\psi_k^{(\alpha)i}(x),\ \text{for } k \in \kappa \text{ and } i \in N\right\}$ into the state dynamics (5.1), one can obtain the dynamics of the cooperative trajectory as:

$$x_{k+1} = f_k\left(x_k, \psi_k^{(\alpha)1}(x_k), \psi_k^{(\alpha)2}(x_k), \cdots, \psi_k^{(\alpha)n}(x_k)\right) + G_k(x_k)\theta_k, \tag{5.5}$$

for $k \in \kappa$ and $x_1 = x^0$.

We use $X_k^{(\alpha)}$ to denote the set of possible values of the state at stage k generated by (5.5). We use $x_k^{(\alpha)} \in X_k^{(\alpha)}$ to denote an element in the set $X_k^{(\alpha)}$. The term $W^{(\alpha)}(k,x)$ gives the weighted cooperative payoff over the stages from k to T.

Given that all players are adopting the cooperative strategies the payoff of player i under cooperation can be obtained as:

$$W^{(\alpha)i}(t,x) = E_{\theta_t,\theta_{t+1},\cdots,\theta_T}\left\{ \sum_{k=t}^{T} g_k^i\left[x_k^{(\alpha)}, \psi_k^{(\alpha)1}\left(x_k^{(\alpha)}\right), \psi_k^{(\alpha)2}\left(x_k^{(\alpha)}\right), \cdots \right.\right.$$
$$\left.\left. \cdots, \psi_k^{(\alpha)n}\left(x_k^{(\alpha)}\right)\right] + q^i\left(x_{T+1}^{(\alpha)}\right) \middle| \, x_t^{(\alpha)} = x \right\}, \tag{5.6}$$

for $i \in N$ and $t \in \kappa$.

To allow the derivation of the functions $W^{(\alpha)i}(t,K)$ in a more direct way we follow the analysis in Yeung (2013) and characterize individual player's payoffs under cooperation by the following Theorem.

Theorem 5.2 The payoff of player i over the stages from t to $T+1$ can be characterized as the value function $W^{(\alpha)i}(t,x)$ satisfying the following recursive system of equations:

$$W^{(\alpha)i}(T+1,x) = q^i(x_{T+1}),$$
$$W^{(\alpha)i}(t,x) = E_{\theta_t}\left\{ g_t^i\left[x, \psi_t^{(\alpha)1}(x), \psi_t^{(\alpha)2}(x), \cdots, \psi_t^{(\alpha)n}(x)\right]\right.$$
$$\left. + W^{(\alpha)i}\left[t+1, f_t\left(t, \psi_t^{(\alpha)1}(x), \psi_t^{(\alpha)2}(x), \cdots, \psi_t^{(\alpha)n}(x)\right) + G(x)\theta_t\right] \right\}, \tag{5.7}$$

for $i \in N$ and $t \in \kappa$.

Proof $W^{(\alpha)i}(t,x)$ in (5.6) can be expressed as:

$$W^{(\alpha)i}(t,x) = E_{\theta_t,\theta_{t+1},\cdots,\theta_T}\left\{ g_t^i\left[x, \psi_t^{(\alpha)1}(x), \psi_t^{(\alpha)2}(x), \cdots, \psi_t^{(\alpha)n}(x)\right]\right.$$
$$\left. + \sum_{k=t+1}^{T} g_k^i\left[x_k^{(\alpha)}, \psi_k^{(\alpha)1}\left(x_k^{(\alpha)}\right), \psi_k^{(\alpha)2}\left(x_k^{(\alpha)}\right), \cdots, \psi_k^{(\alpha)n}\left(x_k^{(\alpha)}\right)\right] + q^i\left(x_{T+1}^{(\alpha)}\right) \right\} \tag{5.8}$$

Invoking (5.6) again, we have:

$$W^{(\alpha)i}\left(t+1,x_{t+1}^{(\alpha)}\right)=E_{\theta_{t+1},\theta_{t+2},\cdots,\theta_T}\left\{\sum_{k=t+1}^{T}g_k^i\left[x_k^{(\alpha)},\psi_k^{(\alpha)1}\left(x_k^{(\alpha)}\right),\psi_k^{(\alpha)2}\left(x_k^{(\alpha)}\right),\cdots\right.\right.$$
$$\left.\left.\cdots,\psi_k^{(\alpha)n}\left(x_k^{(\alpha)}\right)\right]+q^i\left(x_{T+1}^{(\alpha)}\right)\right\}$$
$$(5.9)$$

Using (5.8) and (5.9), one can obtain (5.7). ■

For individual rationality to be maintained throughout all the stages $t\in\kappa$, it is required that:

$$W^{(\alpha)i}\left(t,x_t^{(\alpha)}\right)\geq V^i\left(t,x_t^{(\alpha)}\right),\quad \text{for } i\in N \text{ and } t\in\kappa. \tag{5.10}$$

Let the set of weights α that satisfies (5.10) be denoted by Λ. If Λ is not an empty set, a vector $\hat{\alpha}=\left(\hat{\alpha}^1,\hat{\alpha}^2,\cdots\cdots,\hat{\alpha}^n\right)\in\Lambda$ agreed upon by all players would yield a cooperative solution which satisfies both individual rationality and Pareto optimality throughout the cooperation duration.

11.5.2 Subgame Consistent Cooperative Solution

Now, we proceed to consider subgame consistent solutions in NTU cooperative stochastic dynamic games. Let $\Gamma(t,x_t)$ denote the cooperative game in which the objective of player i is

$$E_{\theta_t,\theta_{t+1},\cdots,\theta_T}\left\{\sum_{k=t}^{T}g_k^i\left(x_k,u_k^1,u_k^2,\cdots,u_k^n\right)+q^i(x_{T+1})\right\},\quad \text{for } i\in N, \tag{5.11}$$

and the state dynamics is

$$x_{k+1}=f_k\left(x_k,u_k^1,u_k^2,\cdots,u_k^n\right)+G_k(x_k)\theta_k, \tag{5.12}$$

for $k\in\{t,t+1,\cdots,T\}$ and the state at stage t is x_t.

Let the agreed-upon optimality principle be denoted by $P(t,x_t)$. For subgame consistency to be maintained the agreed-upon optimality principle $P(t,x_t)$ must be satisfied in the subgame $\Gamma(t,x_t)$ for $t\in\{1,2,\cdots,T\}$. Hence, when the game proceeds to any stage t, the agreed-upon solution policy remains effective. Since in general, typical optimality principles in classical game theory could not be maintained as the game proceeds under a time-invariant payoff weights cooperative scheme we consider deriving a set of subgame consistent strategies in a cooperative solution with a variable payoff weight scheme.

11.5.2.1 Cooperative Strategies in Ending Stages

Invoking the principle of backward induction we begin with the last operation stage, that is stage T, with the state $x_T = x \in X$. The players will select a set of payoff weight $\alpha_T = \left(\alpha_T^1, \alpha_T^2, \cdots, \alpha_T^n\right)$ which satisfies the optimality principle $P(T,x)$. The players' optimal cooperative strategies can be generated by solving the stochastic dynamic programming problem of maximizing the weighted sum of their payoffs

$$E_{\theta_T}\left\{ \sum_{j=1}^{n} \left[\alpha_T^j g_T^j\left(x_T, u_T^1, u_T^2, \cdots, u_T^n\right) + \alpha_T^j q^j(x_{T+1}) \right] \right\} \tag{5.13}$$

subject to

$$x_{T+1} = f_T\left(x_T, u_T^1, u_T^2, \cdots, u_T^n\right) + G_T(x_T)\theta_T, \quad x_T = x\,. \tag{5.14}$$

Let $\left\{ u_T^i = \psi_T^{(\alpha_T)i}, \text{ for } i \in N \right\}$ denote the optimal cooperative strategies in stage T that solves the stochastic dynamic programming problem (5.13 and 5.14). When all players are adopting the cooperative strategies the payoff of player i under cooperation covering stages T and $T+1$ can be obtained as:

$$W^{(\alpha_T)i}(T,x) = E_{\theta_T}\left\{ g_T^i\left[x, \psi_T^{(\alpha_T)1}(x), \psi_T^{(\alpha_T)2}(x), \cdots, \psi_T^{(\alpha_T)n}(x)\right] + q^i\left(x_{T+1}^{(\alpha_T)}\right) \right\} \text{for } i \in N. \tag{5.15}$$

Invoking Theorem 5.2 one can characterize player i's payoff $W^{(\alpha_T)i}(T,x)$ by the following equations

$$\begin{aligned} &W^{(\alpha_T)i}(T+1,x) = q^i(x), \\ &W^{(\alpha_T)i}(T,x) = E_{\theta_T}\left\{ g_T^i\left[x, \psi_T^{(\alpha_T)1}(x), \psi_T^{(\alpha_T)2}(x), \cdots, \psi_T^{(\alpha_T)n}(x)\right] \right. \\ &\qquad \left. + W^{(\alpha_T)i}\left[T+1, f_T\left(x, \psi_T^{(\alpha_T)1}(x), \psi_T^{(\alpha_T)2}(x), \cdots, \psi_T^{(\alpha_T)n}(x)\right) + G_T(x)\theta_T\right] \right\} \\ &\text{for } i \in N. \end{aligned} \tag{5.16}$$

For individual rationality to be maintained, it is required that:

$$W^{(\alpha_T)i}(T,x) \geq V^i(T,x), \text{ for } i \in N. \tag{5.17}$$

Since the maximization problem (5.13 and 5.14) with payoff weight α_T yields a Pareto optimal cooperative solution and the non-cooperative outcome is (in general) suboptimal there always exists a set of weights that satisfies (5.17).

We use Λ_T to denote the set of weights α_T that satisfies (5.17). Then we use $\hat{\alpha}_T = \left(\hat{\alpha}_T^1, \hat{\alpha}_T^2, \cdots, \hat{\alpha}_T^n\right) \in \Lambda_T$ to denote the payoff weights in stage T that leads to the satisfaction of the optimality principle $P(T, x)$.

Now we proceed to cooperative scheme in the second to last stage. Given that the payoff of player i in stage T is $W^{(\hat{\alpha}_T)i}(T, x)$, his payoff covering stages $T-1$ to $T+1$ can be expressed as:

$$E_{\theta_{T-1}}\left\{ g_{T-1}^i\left(x_{T-1}, u_{T-1}^1, u_{T-1}^2, \cdots, u_{T-1}^n\right) + W^{(\hat{\alpha}_T)i}(T, x_T) \right\}, \text{ for } i \in N. \tag{5.18}$$

In this stage the players will select payoff weights $\alpha_{T-1} = \left(\alpha_{T-1}^1, \alpha_{T-1}^2, \cdots, \alpha_{T-1}^n\right)$ which satisfy optimality principle $\Gamma(T-1, x)$. The players' optimal cooperative strategies $\left\{u_{T-1}^i = \psi_{T-1}^{(\alpha_{T-1})i}, \text{ for } i \in N\right\}$ in stage $T-1$ can be generated by solving the stochastic dynamic programming problem of maximizing

$$E_{\theta_{T-1}}\left\{ \sum_{j=1}^n \alpha_{T-1}^j \left[g_{T-1}^j\left(x_{T-1}, u_{T-1}^1, u_{T-1}^2, \cdots, u_{T-1}^n\right) + W^{(\hat{\alpha}_T)j}(T, x_T) \right] \right\} \tag{5.19}$$

subject to

$$x_T = f_{T-1}\left(x_{T-1}, u_{T-1}^1, u_{T-1}^2, \cdots, u_{T-1}^n\right) + G_{T-1}(x_{T-1})\theta_{T-1}, \quad x_{T-1} = x. \tag{5.20}$$

Invoking Theorem 5.2 one can characterize the payoff of player i under cooperation covering the stages $T-1$ to $T+1$ by:

$$\begin{aligned} &W^{(\alpha_{T-1})i}(T, x) = W^{(\hat{\alpha}_T)i}(T, x_T), \\ &W^{(\alpha_{T-1})i}(T-1, x) = E_{\theta_{T-1}}\left\{ g_{T-1}^i\left[x, \psi_{T-1}^{(\alpha_{T-1})1}(x), \psi_{T-1}^{(\alpha_{T-1})2}(x), \cdots, \psi_{T-1}^{(\alpha_{T-1})n}(x)\right] \right. \\ &\left. + W^{(\alpha_{T-1})i}\left[T, f_{T-1}\left(x, \psi_{T-1}^{(\alpha_{T-1})1}(x), \psi_{T-1}^{(\alpha_{T-1})2}(x), \cdots, \psi_{T-1}^{(\alpha_{T-1})n}(x)\right) + G_{T-1}(x)\theta_{T-1}\right] \right\}, \\ &\text{for } i \in N. \end{aligned} \tag{5.21}$$

For individual rationality to be maintained, it is required that:

$$W^{(\alpha_{T-1})i}(T-1, x) \geq V^i(T-1, x), \text{ for } i \in N. \tag{5.22}$$

We use Λ_{T-1} to denote the set of weights α_{T-1} that satisfies (5.22). We use the vector $\hat{\alpha}_{T-1} = \left(\hat{\alpha}_{T-1}^1, \hat{\alpha}_{T-1}^2, \cdots\cdots, \hat{\alpha}_{T-1}^n\right) \in \Lambda_{T-1}$ to denote the set of payoff weights that leads to satisfaction of the optimality principle $\Gamma(T-1, x)$.

11.5.2.2 Cooperative Strategies in Preceding Stages

Now we proceed to characterize the cooperative scheme in stage $k \in \{1, 2, \cdots, T-2\}$. Following the analysis in Sect. 11.4.1, the players will select a set of weights $\alpha_k = \left(\alpha_k^1, \alpha_k^2, \cdots, \alpha_k^n\right)$ which satisfies the optimality principle $P(k, x)$. The players' optimal cooperative strategies $\left\{u_k^i = \psi_k^{(\alpha_k)i}, \text{ for } i \in N\right\}$ in stage k can be generated by solving the following stochastic dynamic programming problem of maximizing

$$E_{\theta_k}\left\{ \sum_{j=1}^{n} \alpha_k^j \left[g_k^j\left(x_k, u_k^1, u_k^2, \cdots, u_k^n\right) + W^{(\hat{\alpha}_{k+1})j}(k+1, x_{k+1}) \right] \right\}, \tag{5.23}$$

subject to

$$x_{k+1} = f_k\left(x_k, u_k^1, u_k^2, \cdots, u_k^n\right) + \theta_k, \quad x_k = x. \tag{5.24}$$

Invoking Theorem 5.2 the payoff of player i under cooperation can be characterized by the following equations

$$\begin{aligned}
&W^{(\alpha_k)i}(k+1, x) = W^{(\hat{\alpha}_{k+1})i}(k+1, x),\\
&W^{(\alpha_k)i}(k, x) = E_{\theta_k}\left\{ g_k^i\left[x, \psi_k^{(\alpha_k)1}(x), \psi_k^{(\alpha_k)2}(x), \cdots, \psi_k^{(\alpha_k)n}(x)\right]\right.\\
&\qquad \left. + W^{(\alpha_k)i}\left[k+1, f_k\left(x, \psi_k^{(\alpha_k)1}(x), \psi_k^{(\alpha_k)2}(x), \cdots, \psi_k^{(\alpha_k)n}(x)\right) + G_k(x)\theta_k\right] \right\}
\end{aligned}$$

for $i \in N$.

$$\tag{5.25}$$

For individual rationality to be maintained in stage k, it is required that:

$$W^{(\alpha_k)i}(k, x) \geq V^i(k, x), \text{ for } i \in N. \tag{5.26}$$

We use Λ_k to denote the set of weights α_k that satisfies (5.26). Again, we use $\hat{\alpha}_k = \left(\hat{\alpha}_k^1, \hat{\alpha}_k^2, \cdots, \hat{\alpha}_k^n\right) \in \Lambda_k$ to denote the set of payoff weights that leads to the satisfaction of the optimal principle $P(k, x)$, for $k \in \kappa$.

11.5.2.3 A Solution Theorem

Similar to the deterministic analysis we consider subgame consistent solutions under variable payoff weights in a stochastic environment. A theorem characterizing a subgame consistent solution of the cooperative stochastic dynamic game (5.1 and 5.2) with the optimality principle $P(k, x_k)$ can be obtained as follows.

Theorem 5.3 A set of payoff weights $\{\hat{\alpha}_k = (\hat{\alpha}_k^1, \hat{\alpha}_k^2, \cdots, \hat{\alpha}_k^n),\ \text{for } k \in \kappa\}$ and a set of strategies $\left\{\psi_k^{(\hat{\alpha}_k)i}(x),\ \text{for } k \in \kappa \text{ and } i \in N\right\}$ provides a subgame consistent solution to the cooperative stochastic dynamic game (5.1 and 5.2) with optimality principle $P(k,x)$ if there exist functions $W^{(\hat{\alpha}_k)}(k,x)$ and $W^{(\hat{\alpha}_k)i}(k,x)$, for $i \in N$, $k \in \kappa$, which satisfy the following recursive relations:

$$W^{(\hat{\alpha}_{T+1})i}(T+1,x) = q^i(x_{T+1}),$$

$$W^{(\hat{\alpha}_k)}(k,x) = \max_{u_k^1, u_k^2, \cdots, u_k^n} \left\{ E_{\theta_k} \left[\sum_{j=1}^{n} \hat{\alpha}^j g_k^j(x, u_k^1, u_k^2, \cdots, u_k^n) \right.\right.$$

$$\left.\left. + \sum_{j=1}^{n} \hat{\alpha}_k^j W^{(\hat{\alpha}_{k+1})j}\left[k+1, f_k(x, u_k^1, u_k^2, \cdots, u_k^n) + G_k(x)\theta_k\right] \right] \right\};$$

$$W^{(\hat{\alpha}_k)i}(k,x) = E_{\theta_k} \left\{ g_k^j\left(x, \psi_k^{(\hat{\alpha}_k)1}(x), \psi_k^{(\hat{\alpha}_k)2}(x), \cdots, \psi_k^{(\hat{\alpha}_k)n}(x)\right) \right.$$

$$\left. + W^{(\hat{\alpha}_{k+1})i}\left[k+1, f_k\left(x, \psi_k^{(\hat{\alpha}_k)1}(x), \psi_k^{(\hat{\alpha}_k)2}(x), \cdots, \psi_k^{(\hat{\alpha}_k)n}(x)\right) + G_k(x)\theta_k\right] \right\},$$

for $i \in N$ and $k \in \kappa$; and the optimality principle $P(k,x)$ in all stages $k \in \kappa$.

(5.27)

Proof Follow the analysis from equation (5.13) to equation (5.25) in Sects. 11.4.1 and 11.4.2. ■

Again, agreed-upon optimality principles may include (i) the proportion each player's cooperative payoff to his non-cooperative payoff being equal, (ii) the satisfaction of the Nash bargaining solution, or (iii) the chosen payoff weights $\hat{\alpha}_k = \{\hat{\alpha}_k^1, \hat{\alpha}_k^2\}$ in a 2-player case to be the mid-value of the maximum and minimum of the payoff weight α_k^1 and that of the payoff weights α_k^2 in the set Λ. Like in the deterministic case the subgame consistent solution presented in Theorem 5.3 is conditional Pareto efficient in the sense that the solution is a Pareto efficient outcome satisfying the condition that the agreed-upon optimality principle is maintained in all stages.

11.6 An Illustration in Stochastic Build-up of Capital

Consider a stochastic version of the example in Sect. 11.4 in which there are 2 asymmetric agents. These agents receive benefits from an existing public capital stock x_t at each stage $t \in \{1, 2, 3, 4\}$. The accumulation dynamics of the public capital stock is governed by the stochastic difference equation:

$$x_{k+1} = x_k + \sum_{j=1}^{2} u_k^j - \delta x_k + \theta_k x_k, x = x_1^0, \text{ for } t \in \{1,2,3\}, \tag{6.1}$$

where u_k^i is the physical amount of investment in the public good and δ is the rate of obsolescence and θ_k is a random variable affecting the rate of obsolescence with range $\{\theta_k^1, \theta_k^2, \theta_k^3, \theta_k^4\}$ and corresponding probabilities $\{\lambda_k^1, \lambda_k^2, \lambda_k^3, \lambda_k^4\}$. In particular, $(\delta - \theta_k^y)$ is non-negative and not greater than one for $y \in \{1,2,3,4\}$ and $t \in \{1,2,3\}$.

The objective of agent $i \in \{1,2\}$ is to maximize the payoff:

$$E_{\theta_1,\theta_2,\theta_3}\left\{ \sum_{k=1}^{3} \left[a_k^i x_k - c_k^i \left(u_k^i\right)^2\right](1+r)^{-(k-1)} + \left(q^i x_4 + m^i\right)(1+r)^{-3} \right\}, \tag{6.2}$$

subject to the dynamics (6.1),

where a_k^i, c_k^i, r, q^i and m^i are positive model parameters.

The payoffs of the agents are not transferable and the non-cooperative payoffs of agent i can be obtained as:

$$V^i(t,x) = \left[A_t^i x + C_t^i\right](1+r)^{-(t-1)}, \\ \text{for } i \in \{1,2\} \text{ and } t \in \{1,2,3\}, \tag{6.3}$$

where

$$A_3^i = a_3^i + q^i\left(1 - \delta + \sum_{\ell=1}^{4} \lambda_3^\ell \theta_3^\ell\right)(1+r)^{-1}, \text{ and}$$

$$C_3^i = -\frac{(q^i)^2(1+r)^{-2}}{4c_3^i} + \left[q^i\left(\sum_{j=1}^{2} \frac{q^j(1+r)^{-1}}{2c_3^j}\right) + m^i\right](1+r)^{-1},$$

$$A_2^i = a_2^i + A_3^i\left(1 - \delta + \sum_{\ell=1}^{4} \lambda_2^\ell \theta_2^\ell\right)(1+r)^{-1}, \text{ and}$$

$$C_2^i = -\frac{1}{4c_2^i}\left(A_3^i(1+r)^{-1}\right)^2 + \left[A_3^i\left(\sum_{j=1}^{2} \frac{A_3^j(1+r)^{-1}}{2c_2^j}\right) + C_3^i\right](1+r)^{-1} \Bigg\}$$

$$A_1^i = a_1^i + A_2^i\left(1 - \delta + \sum_{\ell=1}^{4} \lambda_1^\ell \theta_1^\ell\right)(1+r)^{-1} \quad C_1^i = -\frac{1}{4c_1^i}\left(A_2^i(1+r)^{-1}\right)^2$$

$$+\left[A_2^i\left(\sum_{j=1}^{2} \frac{A_2^j(1+r)^{-1}}{2c_1^j}\right) + C_2^i\right](1+r)^{-1} \Bigg\}; \text{ for } i \in \{1,2\}.$$

11.6.1 Cooperative Solution

Now consider first the case when the agents agree to cooperate and maintain an optimality principle $P(t, x_t)$ requiring the adoption of the mid values of the maximum and minimum of the payoff weight α_t^i in the set Λ_t, for $i \in \{1,2\}$ and $t \in \{1,2,3\}$.

Invoking the technique of stochastic dynamic programming the value function $W^{(\alpha_3)}(3,x)$ in stage 3 can be obtained as:

$$W^{(\alpha_3)}(3,x) = \left[A_3^{(\alpha_3)}x + C_3^{(\alpha_3)}\right](1+r)^{-2}, \tag{6.4}$$

where

$$A_3^{(\alpha_3)} = \sum_{j=1}^{2}\alpha_3^j\left[a_3^j + q^j\left(1 - \delta + \sum_{\ell=1}^{4}\lambda_3^\ell\theta_3^\ell\right)(1+r)^{-1}\right], \text{ and}$$

$$C_3^{(\alpha_3)} = -\sum_{j=1}^{2}\alpha_3^j\left[\frac{(1+r)^{-2}}{4\alpha_3^j c_3^j}\left(\sum_{\ell=1}^{2}\alpha_3^\ell q_3^\ell\right)^2\right]$$
$$+\sum_{j=1}^{2}\alpha_3^j\left[q^j\left(\sum_{j=1}^{2}\frac{(1+r)^{-1}}{2\alpha_3^j c_3^j}\left(\sum_{\ell=1}^{2}\alpha_3^\ell q^\ell\right)\right) + m^j\right](1+r)^{-1}\Bigg\}$$

Invoking Theorem 5.2, the payoff of player i under cooperation can be obtained as:

$$W^{(\alpha_3)i}(3,x) = \left[A_3^{(\alpha_3)i}x + C_3^{(\alpha_3)i}\right](1+r)^{-2}, \tag{6.5}$$

for $i \in \{1,2\}$,

where

$$A_3^{(\alpha_3)i} = \left[a_3^j + q^j\left(1 - \delta + \sum_{\ell=1}^{4}\lambda_3^\ell\theta_3^\ell\right)(1+r)^{-1}\right], \text{ and}$$

$$C_3^{(\alpha_3)i} = -\left[\frac{(1+r)^{-2}}{4\alpha_3^i c_3^i}\left(\sum_{\ell=1}^{2}\alpha_3^\ell q_3^\ell\right)^2\right]$$
$$+\left[q^i\left(\sum_{j=1}^{2}\frac{(1+r)^{-1}}{2\alpha_3^j c_3^j}\left(\sum_{\ell=1}^{2}\alpha_3^\ell q^\ell\right)\right) + m^i\right](1+r)^{-1}\Bigg\}$$

To identify the range of α_3 that satisfies individual rationality we examine the functions which gives the excess of agent i's cooperative over his non-cooperative payoff, that is

$$W^{(\alpha_3)i}(3,x) - V^i(3,x) = \left[C_3^{(\alpha_3)i} - C_3^i\right](1+r)^{-2}, \quad \text{for } i \in \{1,2\}, \tag{6.6}$$

because $A_3^{(\alpha_3)i} = A_3^i$.

For individual rationality to be satisfied, it is required that $W^{(\alpha_3)i}(3,x) - V^i(3,x) \geq 0$ for $i \in \{1,2\}$. Using $\alpha_3^j = 1 - \alpha_3^i$ and upon rearranging terms $C_3^{(\alpha_3)i}$ can be expressed as:

$$C_3^{(\alpha_3)i} = q^i\left[\frac{(1+r)^{-2}}{2c_3^i}\left(\frac{\alpha_3^i q^i + (1-\alpha_3^i)q^j}{\alpha_3^1}\right)\right.$$
$$\left.+\frac{(1+r)^{-2}}{2c_3^j}\left(\frac{\alpha_3^i q^i + (1-\alpha_3^i)q^j}{1-\alpha_3^i}\right)\right] + m^i(1+r)^{-1}$$
$$-\frac{(1+r)^{-2}}{4c_3^i}\left(\frac{\alpha_3^i q^i + (1-\alpha_3^i)q^j}{\alpha_3^i}\right)^2, \quad (6.7)$$

for $i,j \in \{1,2\}$ and $i \neq j$.

Following the analysis in Sect. 11.4 we differentiate $C_3^{(\alpha_3)i}$ with respect to α_3^i to obtain

$$\frac{\partial C_3^{(\alpha_3)i}}{\partial \alpha_3^i} = \frac{(1+r)^{-2}}{2c_3^j}\left(\frac{(q^i)^2}{(1-\alpha_3^i)^2}\right)$$
$$+\frac{(1+r)^{-2}}{2c_3^i}\left(\frac{(1-\alpha_3^i)q^j}{\alpha_3^i}\right)\left(\frac{q^j}{(\alpha_3^i)^2}\right), \quad (6.8)$$

which is positive for $\alpha_3^i \in (0,1)$.

Given that $\lim_{\alpha_3^i \to 0} C_3^{(\alpha_3)i} \to -\infty$ and $\lim_{\alpha_3^i \to 1} C_3^{(\alpha_3)i} \to \infty$, while the cooperative solution is Pareto optimal and the non-cooperative outcome is (in general) suboptimal an $\underline{\alpha}_3^i \in (0,1)$ can be obtained such that

$$W^{(\underline{\alpha}_3^i, 1-\underline{\alpha}_3^i)i}(3,x) = V^i(3,x)$$

and yields agent i's minimum payoff weight value satisfying his own individual rationality. Similarly there exist an $\overline{\alpha}_3^i \in (0,1)$ such that

$$W^{(\overline{\alpha}_3^i, 1-\overline{\alpha}_3^i)j}(3,x) = V^j(3,x)$$

and yields agent i's maximum payoff weight value while maintaining agent j's individual rationality. According to the agreed-upon optimality principle $P(t, x_t)$, the cooperative weights in stage 3 is $\hat{\alpha}_3 = \left(\frac{\underline{\alpha}_3^i + \overline{\alpha}_3^i}{2}, 1 - \frac{\underline{\alpha}_3^i + \overline{\alpha}_3^i}{2}\right)$.

Now consider the stage 2 problem. We use $W^{(\hat{\alpha}_3)j}(3,x)$ for $j \in \{1,2\}$ to form the terminal payoff $\sum_{j=1}^{2} \alpha_2^j W^{(\hat{\alpha}_3)j}(3,x)$ for the cooperation scheme in stage 2. Following the analysis in stage 3, one can obtain

$$W^{(\alpha_2)}(2,x) = \left[A_2^{(\alpha_2)}x + C_2^{(\alpha_2)}\right](1+r)^{-1},$$
$$W^{(\alpha_2)i}(2,x) = \left[A_3^{(\alpha_3)i}x + C_3^{(\alpha_3)i}\right](1+r)^{-2}, \text{ for } i \in \{1,2\},$$

where $A_2^{(\alpha_2)}$, $C_2^{(\alpha_2)}$, $A_3^{(\alpha_3)i}$ and $C_3^{(\alpha_3)i}$ are functions that depend on α_2.

One can readily verified that $A_t^{(\alpha_t)i} = A_t^i$ is independent of α_t and $C_t^{(\alpha_t)i}$ is strictly increasing in α_t^i and $C_t^{(\alpha_t)j}$ is strictly decreasing in α_t^i. Hence agent i's minimum payoff weight is $\underline{\alpha}_2^i \in (0,1)$ which leads to

$$W^{\left(\underline{\alpha}_2^i, 1-\underline{\alpha}_2^i\right)i}(2,x) = V^i(2,x),$$

and his maximum payoff weight is $\overline{\alpha}_2^i \in (0,1)$ which leads to

$$W^{\left(\overline{\alpha}_2^i, 1-\overline{\alpha}_2^i\right)j}(2,x) = V^j(2,x).$$

Invoking the agreed-upon optimality principle $P(t, x_t)$ the cooperative weights in stage 2 is $\hat{\alpha}_2 = \left(\dfrac{\underline{\alpha}_2^i + \overline{\alpha}_2^i}{2}, 1 - \dfrac{\underline{\alpha}_2^i + \overline{\alpha}_2^i}{2}\right)$. Finally, following the analysis in stages 2 and 3, one can obtain the cooperative weights in stage 1 as $\hat{\alpha}_1 = \left(\dfrac{\underline{\alpha}_1^i + \overline{\alpha}_1^i}{2}, 1 - \dfrac{\underline{\alpha}_1^i + \overline{\alpha}_1^i}{2}\right)$.

As shown in Sect. 11.4 for the deterministic case, $\underline{\alpha}_t^i$ and $\overline{\alpha}_t^i$ would change as t changes and in general, there is no guarantee for the existence of a constant payoff weight such that the basic requirement of individual rationality is satisfied in all subsequent stages.

11.6.2 Other Optimality Principles

Consider first the case where the agents agree with an optimality principle $P(t, x_t)$ that requires the satisfaction of the Nash bargaining solution. They would first search for an α_3 in stage 3 to maximize the Nash product. The issue becomes solving the problem

$$\max_{\alpha_3^i} \prod_{j=1}^{2} \left[C_3^{(\alpha_3)j} - C_3^j\right](1+r)^{-2}, \tag{6.9}$$

in the range of $\alpha_3^i \in \left[\underline{\alpha}_3^i, \overline{\alpha}_3^i\right]$.

Invoking the property of $C_3^{(\alpha_3)i}$ a solution weight $\hat{\alpha}_3^i \in \left[\underline{\alpha}_3^i, \overline{\alpha}_3^i\right]$ can be obtained. Then one can use $W^{(\hat{\alpha}_3)j}(3,x)$ for $j \in \{1,2\}$ to form the terminal payoff $\sum_{j=1}^{2} \alpha_2^j$

$W^{(\hat{\alpha}_3)j}(3,x)$ for the cooperation scheme in stage 2. Repeating the above analysis, one can identify $\hat{\alpha}_2$ which yields the Nash bargaining solution in stage 2. Finally, in a similar manner, $\hat{\alpha}_1$ which yields the Nash bargaining solution in stage 1 can be obtained.

Now consider another optimality principle $P(t, x_t)$ which requires the proportion of each player's cooperative payoff to his non-cooperative payoff to be equal. In particular, a subgame consistent solution requires payoff weights $\hat{\alpha}_1$, $\hat{\alpha}_2$ and $\hat{\alpha}_3$ leading to

$$\frac{A_t^{(\hat{\alpha}_t)1}x_t + C_t^{(\hat{\alpha}_t)1}}{A_t^1 x_t + C_t^1} = \frac{A_t^{(\hat{\alpha}_t)2}x_t + C_t^{(\hat{\alpha}_t)2}}{A_t^2 x_t + C_t^2}, \quad \text{for } t \in \{1,2,3\}, \tag{6.10}$$

and

$$x_{t+1} = x_t \sum_{j=1}^{2}\left[\frac{(1+r)^{-1}}{2\hat{\alpha}_t^j c_t^i}\sum_{\ell=1}^{2}\hat{\alpha}_t^\ell A_{t+1}^{(\hat{\alpha}_{t+1})\ell}\right] - \delta x_t, x_1 = x_1^0, \tag{6.11}$$

for $t \in \{12,3\}$ and $A_4^{(\hat{\alpha}_4)\ell} = q^\ell$.

Again with $A_t^{(\alpha_t)i} = A_t^i$ being independent of α_t and $C_t^{(\alpha_t)i}$ being strictly increasing in α_t^i and $C_t^{(\alpha_t)j}$ being strictly decreasing in α_t^i for $\alpha_t^i \in [\underline{\alpha}_t^i, \overline{\alpha}_t^i]$, therefore one can readily identify payoff weights $\hat{\alpha}_t^i$ such that (6.10) is satisfied.

11.7 Chapter Notes

The number of studies in cooperative dynamic games with non-transferrable utility/payoff (NTU) is very scantly. On top of Yeung and Petrosyan (2015a, b), Sorger (2006) presented a recursive Nash bargaining solution for a discrete-time NTU cooperative dynamic game involving a productive asset. Predtetchinski (2007) considered the strong sequential core for stationary NTU cooperative dynamic games. In a two-person dynamic game where the game structures are time invariant and the game horizon approaches infinity the payoff allocations under the set of weights Λ_k that satisfying $W^{(\alpha_k)i}(k,x) \geq V^i(k,x)$ in the steady state converges to the strong sequential core in Predtetchinski (2007).

11.8 Problems

1. Consider an economic region with 2 asymmetric agents. These agents receive benefits from an existing public capital stock x_t at each stage $t \in \{1,2,3,4\}$. The

accumulation dynamics of the public capital stock is governed by the difference equation:

$$x_{k+1} = x_k + \sum_{j=1}^{2} u_k^j - 0.1x_k\,,\ x_1 = {}^1x_1^0,\ \text{for } t \in \{1,2,3\},$$

where u_k^i is the physical amount of investment in the public good and δ is the rate of depreciation.

The objective of agent 1 is to maximize the payoff

$$\sum_{k=1}^{3}\left[2x_k - \left(u_k^i\right)^2\right](1+r)^{-(k-1)} + (x_4+5)(1+r)^{-3},$$

and the objective of agent 1 is to maximize the payoff:

$$\sum_{k=1}^{3}\left[1x_k - \left(u_k^i\right)^2\right](1+r)^{-(k-1)} + (0.5x_4+6)(1+r)^{-3}.$$

The payoffs of the agent are not transferable.

Characterize the feedback Nash equilibrium.

2. Consider the case when the agents agree to cooperate and maintain an optimality principle $P(t, x_t)$ requiring the adoption of the mid values of the maximum and minimum of the payoff weight α_t^i in the set Λ_t, for $i \in \{1,2\}$ and $t \in \{1,2,3\}$.

   Characterize the maximum and minimum values of α_t^i, in the set Λ_t, for $i \in \{1,2\}$ and $t \in \{1,2,3\}$.
3. Derive a subgame consistent solution.

Part III
Applications

Chapter 12
Applications in Cooperative Public Goods Provision

The notion of public goods, which are non-rival and non-excludable, was first introduced by Samuelson (1954). Examples of public goods include clean environment, national security, scientific knowledge, accessible public capital, technical know-how and public information. The non-exclusiveness and positive externalities of public goods constitutes major factors for market failure in their provision. The provision of public goods constitutes a classic case of market failure which calls for cooperative optimization. However, cooperation cannot be sustainable unless there is guarantee that the agreed-upon optimality principle can be maintained throughout the planning duration.

This Chapter presents two sets of applications in subgame consistent cooperative provision of public goods to solve the problem. The first application is based on Yeung and Petrosyan (2013b) in which the analysis is based on a cooperative stochastic differential game framework. The second application is based on Yeung and Petrosyan (2014b) in which the analysis is conducted in a randomly-furcating stochastic dynamic game framework. The continuous-time differential game analysis is provided in Sects. 12.1, 12.2, 12.3 and 12.4. Section 12.1 provides an analytical framework of cooperative public goods provision. An application in multiple asymmetric agents public capital build-up in given in Sect. 12.2. An application in the development of technical knowledge as a public good in an industry is provided in Sect. 12.3. In Sect. 12.4 application in infinite horizon cooperative public capital goods provision is examined. The discrete-time dynamic game analysis is provided in Sects. 12.5 and 12.6. Cooperative public goods provision under accumulation and payoff uncertainties is presented in Sect. 12.5 and an illustration is given in Sect. 12.6. Appendices of the chapter and chapter notes are contained in Sects. 12.7, 12.8 and 12.9 respectively.

D.W.K. Yeung, L.A. Petrosyan, *Subgame Consistent Cooperation*,
Theory and Decision Library C 47, DOI 10.1007/978-981-10-1545-8_12

12.1 Cooperative Public Goods Provision: An Analytical Framework

In this Section we set up an analytical framework to study collaborative public goods provision. In particular, group optimal strategies, subgame consistent cooperative schemes and payoff distribution procedures are investigated.

12.1.1 Game Formulation and Non-cooperative Outcome

Consider the case of the provision of a public good in which a group of n agents carry out a project by making continuous contributions of some inputs or investments to build up a productive stock of a public good. Let $K(s)$ denote the level of the productive stock and $I_i(s)$ denote the contribution or investment by agent i at time s, the stock accumulation dynamics is governed by

$$dK(s) = \left[\sum_{j=1}^{n} I_j(s) - \delta K(s) \right] ds + \sigma K(s) dz(s), \;\; K(0) = K_0, \tag{1.1}$$

where δ is the rate of decay of the productive stock, $z(s)$ is Wiener process and σ is a scaling constant.

The instantaneous payoff to agent i at time instant s is

$$R_i[K(s)] - C_i[I_i(s)], \quad i \in \{1, 2, \cdots, n\} = N, \tag{1.2}$$

where $R_i(K)$ is the revenue/payoff to agent i if the productive stock is K and $C_i[I_i]$ is the cost of investing I_i by agent i. Marginal cost of investment is increasing in I_i. Marginal revenue product of the productive stock is non-negative, that is $R_i'(K) \geq 0$, before a saturation level $\overline{K}$ has been reached; and marginal cost of investment is positive and non-decreasing, that is $C_i'[I_i] > 0$ and $C_i''[I_i] \geq 0$. Moreover, the payoffs of the players are transferable.

The objective of agent $i \in N$ is to maximize its expected net revenue over the planning horizon T, that is

$$E\left\{ \int_0^T \{R_i[K(s)] - C_i[I_i(s)]\} e^{-rs} ds + q_i[K(T)] e^{-rT} \right\} \tag{1.3}$$

subject to the stock accumulation dynamics (1.1), where r is the discount rate, and $q_i[K(T)] \geq 0$ is an amount conditional on the productive stock that agent i would receive at time T.

Acting for individual interests, the agents are involved in a stochastic differential game. In such a framework, a feedback Nash equilibrium has to be sought. Let

$\{\phi_i(s,K) = I_i^*(s) \in I^i, \text{for } i \in N \text{ and } s \in [0,T]\}$ denote a set of feedback strategies that brings about a feedback Nash equilibrium of the game (1.1) and (1.3). Invoking Theorem 1.1 in Chap. 3 for solving stochastic differential games, a feedback solution to the problem (1.1) and (1.3) can characterized by the following set of Hamilton-Jacobi-Bellman equations:

$$-V_t^i(t,K) - \frac{1}{2}V_{KK}^i(t,K)\sigma^2K^2 = \max_{I_i}\left\{[R_i(K) - C_i(I_i)]e^{-rt}\right.$$
$$\left. + V_K^i(t,K)\left[\sum_{\substack{j=1 \\ j\neq i}}^n \phi_j(t,K) + I_i - \delta K\right]\right\}, \tag{1.4}$$

$$V^i(T,K) = q_i(K)e^{-rT}, \quad \text{for } i \in N. \tag{1.5}$$

A Nash equilibrium non-cooperative outcome of public goods provision by the n agents is characterized by the solution of the system of partial differential equations (1.4 and 1.5).

12.1.2 Subgame Consistent Cooperative Scheme

It is well-known problem that noncooperative provision of goods with externalities, in general, would lead to dynamic inefficiency. Cooperative games suggest the possibility of socially optimal and group efficient solutions to decision problems involving strategic action. Now consider the case when the agents agree to cooperate and extract gains from cooperation. In particular, they act cooperatively and agree to distribute the joint payoff among themselves according to an optimality principle. If any agent disagrees and deviates from the cooperation scheme, all agents will revert to the noncooperative framework to counteract the free-rider problem in public goods provision. In particular, free-riding would lead to a lower future payoff due to the loss of cooperative gains. Thus a credible threat is in place. As stated before group optimality, individual rationality and subgame consistency are three crucial properties that sustainable cooperative scheme has to satisfy.

To fulfil group optimality the agents would seek to maximize their expected joint payoff. To maximize their expected joint payoff the agents have to solve the stochastic dynamic programming problem

$$\max_{\{I_1(s),I_2(s),\cdots,I_n(s)\}} E\left\{\sum_{j=1}^n\left[\int_0^T \{R_j[K(s)] - C_j[I_j(s)]\}e^{-rs}ds\right.\right.$$
$$\left.\left. + q_j[K(T)]e^{-rT}\right]\right\} \tag{1.6}$$

subject to the stock dynamics (1.1).

Let $\{\psi_i(s,K)$, for $i \in N$ and denote a set of strategies that brings about an optimal solution to the stochastic control problem (1.1) and (1.6). Invoking the standard stochastic dynamic programming technique in Theorem A.3 of the Technical Appendices an optimal solution to the stochastic control problem (1.1) and (1.6) can characterized by the following set of equations (see also Fleming and Rishel 1975; Ross 1983):

$$-W_t(t,K) - \frac{1}{2}W_{KK}(t,K)\sigma^2K^2$$
$$= \max_{I_1,I_2,\cdots,I_n} \left\{ \sum_{j=1}^{n} \left[[R_j(K) - C_j(I_j)]e^{-rt} \right] + W_K(t,K)\left(\sum_{j=1}^{n} I_j - \delta K \right) \right\}, \quad (1.7)$$

$$W(T,K) = \sum_{j=1}^{n} q_j[K(T)]e^{-rT}. \quad (1.8)$$

A group optimal solution of public goods provision by the n agents is characterized by the solution of the partial differential equation (1.7 and 1.8). In particular, $W(t,K)$ gives the maximized joint payoff of the n players at time $t \in [0,T]$ given that the state is x.

Substituting the optimal strategies $\{\psi_i(s,K)$, for $i \in N$ and $s \in [0,T]\}$ into (1.1) yields the optimal path of productive stock dynamics:

$$dK(s) = \left[\sum_{j=1}^{n} \psi_j(s,K(s)) - \delta K(s) \right] ds + \sigma K(s)dz(s), \ K(0) = K_0. \quad (1.9)$$

We use X_s^* to denote the set of realizable values of $K(s)$ generated by (1.9) at time s. The term $K_s^* \in X_s^*$ is used to denote and element in X_s^*.

Let $\xi(\cdot,\cdot)$ denote the agreed-upon imputation vector guiding the distribution of the total cooperative payoff under the agreed-upon optimality principle along the cooperative trajectory $\{K^*(s)\}_{s\in[0,T]}$. At time s and if the productive stock is K_s^*, the imputation vector according to $\xi(\cdot,\cdot)$ is

$$\xi(s,K_s^*) = \left[\xi^1(s,K_s^*), \xi^2(s,K_s^*), \cdots, \xi^n(s,K_s^*)\right] \text{ for } s \in [0,T]. \quad (1.10)$$

A variety of examples of imputations $\xi(s,K_s^*)$ can be found in Chap. 2. For individual rationality to be maintained throughout all time $s \in [0,T]$, it is required that:

$$\xi^i(s,K_s^*) \geq V^i(s,K_s^*), \text{ for } i \in N \text{ and } s \in [0,T].$$

To satisfy group optimality, the imputation vector has to satisfy

$$W(s, K_s^*) = \sum_{j=1}^{n} \xi^i(s, K_s^*), \text{ for } s \in [0, T].$$

12.1.3 Payoff Distribution Procedure

Following the analysis in Chap. 3, we formulate a Payoff Distribution Procedure so that the agreed-upon imputations (1.10) can be realized. Let $B^i(s, K^*(s))$ for $s \in [0, T)$ denote the payment that agent i will received at time s under the cooperative agreement if $K^*(s)$ is realized at that time.

The payment scheme involving $B^i(s, K^*(s))$ constitutes a PDP in the sense that along the cooperative trajectory $\{K^*(s)\}_{s\in[0,T]}$ the imputation to agent i covering the duration $[\tau, T]$ can be expressed as:

$$\xi^i(\tau, K_\tau^*) = E\left\{\int_\tau^T B^i(s, K^*(s))e^{-rs}ds + q_i[K^*(T)]e^{-rT} \middle| K^*(\tau) = K_\tau^*\right\}, \quad (1.11)$$

for $i \in N$ and $\tau \in [0, T]$.

The values of $B^i(s, K^*(s))$ for $i \in N$ and $s \in [\tau, T)$, which leads to the realization of imputation (1.10) and hence a subgame consistent cooperative solution can be obtained by the following theorem.

Theorem 1.1 A PDP for agent $i \in N$ with a terminal payment $q_i(K_T^*)$ at time T and an instantaneous payment at time $s \in [0, T]$ which present value is:

$$\begin{aligned} B_i(s, K_s^*)e^{-rs} &= -\xi_s^i(s, K_s^*) - \frac{1}{2}\sigma^2(K_s^*)^2 \xi_{K_s K_s}^i(s, K_s^*) \\ &- \xi_{K_s}^i(s, K_s^*)\left[\sum_{j=1}^{n} \psi_j^*(s, K_s^*) - \delta K_s^*\right], \text{ for } i \in N \text{ and } K_s^* \in X_s^*, \end{aligned} \quad (1.12)$$

would lead to the realization of the imputation $\xi(s, K_s^*)$ in (1.10).

Proof See Appendix A. ■

Note that the payoff distribution procedure in Theorem 1.1 would give rise to the agreed-upon imputation in (1.10) and therefore subgame consistency is satisfied.

When all agents are using the cooperative strategies, the payoff that player i will directly receive at time s is

$$R_i(K_s^*) - C_i[\psi_i^*(s, K_s^*)].$$

However, according to the agreed upon imputation, agent i is supposed to receive $B_i^i(s, K_s^*)$. Therefore a transfer payment (which could be positive or negative)

$$\varpi_i(s, K_s^*) = B_i(s, K_s^*) - \{R_i(K_s^*) - C_i[\psi_i^*(s, K_s^*)]\} \tag{1.13}$$

will be imputed to agent $i \in N$ at time $s \in [0, T]$.

12.2 An Application in Asymmetric Agents Public Capital Build-up

In this section, we examine an application of the analysis in the build-up of public capital by multiple asymmetric agents.

12.2.1 Game Model

Consider an economic region with n asymmetric agents. These agents receive benefits from an existing public capital stock $K(s)$. The accumulation dynamics of the public capital stock is governed by

$$dK(s) = \left[\sum_{j=1}^{n} I_j(s) - \delta K(s)\right] ds + \sigma K(s) dz(s), \; K(0) = K_0, \tag{2.1}$$

where δ is the depreciation rate of the public capital and $I_i(s) \in [0, \bar{I}]$ is the investment made by the ith agent in the public capital.

Each agent gains from the existing level of public capital and the ith agent seeks to maximize its expected stream of monetary gains:

$$E\left\{\int_0^T \left\{\alpha_i K(s) - c_i [I_i(s)]^2\right\} e^{-rs} ds + \left[q_1^i K(T) + q_2^i\right] e^{-rT} \middle| K(0) = K_0\right\}, \text{ for } i \in N, \tag{2.2}$$

subject to (2.1);

where α_i, c_i, q_1^i and q_2^i are positive constants, and $\alpha_i \neq \alpha_j, c_i \neq c_j, q_1^i \neq q_1^j$ and $q_2^i \neq q_2^j$, for $i, j \in N$ and $i \neq j$.

In particular, $\alpha_i K(s)$ gives the gain that agent i derives from the public capital, $c_i[I_i(s)]^2$ is the cost of investing $I_i(s)$ in the public capital, and $[q_1^i K(T) + q_2^i]$ is the terminal valuation of the public capital at time T. Invoking the analysis in (1.5 and 1.6) in Sect. 12.1 we obtain the corresponding Hamilton-Jacobi-Bellman equations characterizing a non-cooperative outcome as:

$$-V_t^i(t,K) - \frac{1}{2}V_{KK}^i(t,K)\sigma^2K^2 = \max_{I_i}\left\{\left\{\alpha_i K - c_i(I_i)^2\right\}e^{-rt} + V_K^i(t,K)\left[\sum_{\substack{j=1 \\ j\neq i}}^{n}\phi_j(t,K) + I_i - \delta K\right]\right\}, \tag{2.3}$$

$$V^i(T,K) = \left[q_1^i K(T) + q_2^i\right]e^{-rT}, \quad \text{for } i \in N; \tag{2.4}$$

Performing the maximization operator in (2.4) yields:

$$I_i = \frac{V_K^i(t,K)}{2c_i}e^{rt}, \text{ for } i \in N. \tag{2.5}$$

To solve the game (2.1 and 2.2) we first obtain the value functions indicating the game equilibrium payoffs of the agents as follows.

Proposition 2.1 The value function $V^i(t,K)$ of agent i can be obtained as:

$$V^i(t,K) = [A_i(t)K + C_i(t)]e^{-rt} \text{ for } i \in N; \tag{2.6}$$

where

$$A_i(t) = \left(q_1^i - \frac{\alpha_i}{r+\delta}\right)e^{-(r+\delta)(T-t)} + \frac{\alpha_i}{r+\delta},$$

and the value of $C_i(t)$ is generated by the following first order linear differential equation:

$$\dot{C}_i(t) = rC_i(t) + \frac{[A_i(t)]^2}{4c_i} - \left[\sum_{j=1}^{n}\frac{A_i(t)A_j(t)}{2c_j}\right],$$
$$C_i(T) = q_2^i, \text{ for } i \in N, \tag{2.7}$$

Proof See Appendix B. ■

Using Proposition 2.1 and (2.5) the game equilibrium strategies can be obtained to characterize the market equilibrium. The asymmetry of agents brings about different payoffs and investment levels in public capital investments.

12.2.2 Cooperative Provision of Public Capital

Now we consider the case when the agents agree to act cooperatively and seek higher gains. They agree to maximize their expected joint gain and distribute the cooperative gain proportional to their non-cooperative gains. To maximize their expected joint gains the agents maximize

$$E\left\{\int_0^T \sum_{j=1}^n \left\{\alpha_j K(s) - c_j \left[I_j(s)\right]^2\right\} e^{-rs} ds + \sum_{j=1}^n \left[q_1^j K(T) + q_2^j\right] e^{-rT} \middle| K(0) = K_0 \right\} \tag{2.8}$$

subject to dynamics (2.1).

Following the analysis in (1.7 and 1.8) in Sect. 12.1, the corresponding stochastic dynamic programming equation can be obtained as:

$$-W_t(t,K) - \frac{1}{2} W_{KK}(t,K)\sigma^2 K^2 = \max_{I_1, I_2, \cdots, I_n} \left\{ \sum_{j=1}^n \left[\alpha_j K - c_j (I_j)^2\right] e^{-rt} + W_K(t,K) \left(\sum_{j=1}^n I_j - \delta K \right) \right\}, \tag{2.9}$$

$$W(T,K) = \sum_{j=1}^n \left(q_1^j K + q_2^j\right) e^{-rT}. \tag{2.10}$$

Performing the maximization operator in (2.9) yields:

$$I_i = \frac{W_K(t,K)}{2c_i} e^{rt}, \text{ for } i \in N. \tag{2.11}$$

The maximized expected joint profit of the n participating firms can be obtained as:

Proposition 2.2 The value function $W(t,K)$ indicating the maximized expected joint payoff is

$$W(t,K) = [A(t)K + C(t)]e^{-rt}, \tag{2.12}$$

where

$$A(t) = \left(\sum_{j=1}^n q_1^j - \frac{\sum_{j=1}^n \alpha_j}{r+\delta} \right) e^{-(r+\delta)(T-t)} + \frac{\sum_{j=1}^n \alpha_j}{r+\delta}, \text{ and}$$

and the value of $C(t)$ is generated by the following first order linear differential equation:

$$\dot{C}(t) = rC(t) - \sum_{j=1}^n \frac{[A(t)]^2}{4c_j}, \quad C(T) = \sum_{j=1}^n q_2^j.$$

Proof Follow the proof of Proposition 2.1. ■

Using (2.11) and Proposition 2.2 the optimal trajectory of public capital stock can be expressed as:

$$dK(s) = \left[\sum_{j=1}^{n} \frac{A(s)}{2c_j} - \delta K(s)\right] ds + \sigma K(s) dz(s), \quad K(0) = K_0. \tag{2.13}$$

We use X_s^* to denote the set of realizable values of $K^*(s)$ generated by (2.13) at time s. The term $K_s^* \in X_s^*$ is used to denote and element in X_s^*.

12.2.3 Subgame Consistent Payoff Distribution

Under cooperation every agent will be using the Pareto optimal strategies in (2.11) and the expected payoff that agent i will receive over the cooperative duration $[0, T]$ becomes:

$$E\left\{\int_0^T \left(\alpha_i K^*(s) - \frac{[A(s)]^2}{4c_i} \right) e^{-rs} ds + \sum_{j=1}^{n} [q_1^i K^*(T) + q_2^i] e^{-rT} \right\}, \; i \in N.$$

At initial time 0, the agents agree to distribute the cooperative gain proportional to their non-cooperative gains. Therefore agent i will receive an imputation

$$\begin{aligned}
\zeta^i(0, K_0) &= \frac{V^i(0, K_0)}{\sum_{j=1}^{n} V^j(0, K_0)} W(0, K_0) \\
&= \frac{A_i(0)K_0 + C_i(0)}{\sum_{j=1}^{n} A_j(0)K_0 + C_j(0)} [A(0)K_0 + C(0)], \quad \text{for } i \in N.
\end{aligned}$$

With the agents agreeing to distribute their gains proportional to their non-cooperative gains, the imputation vector becomes

$$\begin{aligned}
\xi^i(s, K_s^*) &= \frac{V^i(s, K_s^*)}{\sum_{j=1}^{n} V^j(s, K_s^*)} W(s, K_s^*) \\
&= \frac{[A_i(t)K + C_i(t)]}{\sum_{j=1}^{n} [A_j(t)K + C_j(t)]} [A(t)K + C(t)] e^{-rt},
\end{aligned} \tag{2.14}$$

for $i \in N$ and $s \in [0, T]$ if the public capital stock is $K_s^* \in X_s^*$.

To guarantee dynamical stability in a dynamic cooperation scheme, the solution has to satisfy the property of subgame consistency which requires the satisfaction of (2.14). Invoking Theorem 1.1, a PDP for agent $i \in N$ with a terminal payment $\left[q_1^i K(T) + q_2^i\right]$ at time T and an instantaneous payment (in present value) at time $s \in [0, T]$

$$
\begin{aligned}
B_i\left(s, K_s^*\right) e^{-rs} = {} & r \frac{\left[A_i(s) K_s^* + C_i(s)\right]}{\sum_{j=1}^{n} \left[A_j(s) K_s^* + C_j(s)\right]} \left[A(s) K_s^* + C(s)\right] e^{-rs} \\
& - \frac{\left[A_i(s) K_s^* + C_i(s)\right]}{\sum_{j=1}^{n} \left[A_j(s) K_s^* + C_j(s)\right]} \left[\dot{A}(s) K_s^* + \dot{C}(s)\right] e^{-rs} \\
& - \frac{\left[A(s) K_s^* + C(s)\right] e^{-rs}}{\left(\sum_{j=1}^{n} \left[A_j(s) K_s^* + C_j(s)\right]\right)^2} \left[\sum_{j=1}^{n} \left[A_j(s) K_s^* + C_j(s)\right] \left[\dot{A}_i(s) K_s^* + \dot{C}_i(s)\right]\right. \\
& \left. - \left[A_i(s) K_s^* + C_i(s)\right] \sum_{j=1}^{n} \left[\dot{A}_j(s) K_{s*} + \dot{C}_j(s)\right]\right] \\
& - \xi_{K_s}^i\left(s, K_s^*\right) \left[\sum_{j=1}^{n} \frac{A(s)}{2c_j} - \delta K_s^*\right] - \frac{1}{2} \sigma^2 \left(K_s^*\right)^2 \xi_{K_s K_s}^i\left(s, K_s^*\right),
\end{aligned}
\tag{2.15}
$$

where

$$
\begin{aligned}
\xi_{K_s}^i\left(s, K_s^*\right) = {} & \left[\frac{\left[A_i(s) K_s^* + C_i(s)\right]}{\sum_{j=1}^{n} \left[A_j(s) K_s^* + C_j(s)\right]} A(s) e^{-rs}\right. \\
& \left. + \frac{A_i(s) \sum_{j=1}^{n} \left[A_j(s) K_s^* + C_j(s)\right] - \left[A_i(s) K_s^* + C_i(s)\right] \sum_{j=1}^{n} A_j(s)}{\left(\sum_{j=1}^{n} \left[A_j(s) K_s^* + C_j(s)\right]\right)^2} \left[A(s) K_s^* + C(s)\right] e^{-rs}\right],
\end{aligned}
$$

and $\xi_{K_s K_s}^i\left(s, K_s^*\right) = \partial \xi_{K_s}^i\left(s, K_s^*\right) / \partial K_s$,

for $i \in N$ and $K_s^* \in X_s^*$,

would lead to the realization of the imputation $\xi(s, K_s^*)$ in (2.14).

The values of the terms $A_j(s), \dot{A}_j(s), C_j(s)$ and $\dot{C}_j(s)$ are given in Proposition 2.2 and its proof.

Finally, when all agents are using the cooperative strategies, the payoff that player i will directly receive at time s is

$$
\alpha_i K_s^* - \frac{[A(s)]^2}{4c_i}.
$$

However, according to the agreed upon imputation, agent i is to receive $B_i(s, K_s^*)$ in (2.15). Therefore a transfer payment

$$\varpi_i^i(s, K_s^*) = B_i(s, K_s^*) - \left[\alpha_i K_s^* - \frac{[A(s)]^2}{4c_i}\right] \tag{2.16}$$

will be imputed to agent $i \in N$ at time $s \in [0, T]$.

12.3 An Application in the Development of Technical Knowledge

In this section, we examine the application of the analysis in the development of technical knowledge as a public good in an industry.

12.3.1 Game Formulation and Noncooperative Market Outcome

Consider an industry with two types of firms using a common type of technology. There are n_1 type 1 firms and n_2 type 2 firms and the planning horizon is $[0, T]$. We use $I_i^{(1)}(s) \in [0, \bar{I}]$ to denote the technology investment of the ith type 1 firm, for $i \in \{1, 2, \cdots, n_1\} \equiv N_1$. Similarly, $I_j^{(2)}(s) \in [0, \bar{I}]$ is used to denote the technology investment of the jth type 2 firm, for $j \in \{n_1 + 1, n_1 + 2, \cdots, n_1 + n_2\} \equiv N_2$. The technology accumulation dynamics is governed by

$$dK(s) = \left[\sum_{i \in N_1} I_i^{(1)}(s) + \sum_{j \in N_2} I_j^{(2)}(s) - \delta K(s)\right] ds + \sigma K(s) dz(s), \ K(0) = K_0, \tag{3.1}$$

where δ is the depreciation rate of technology.

Each firm benefits from the existing level of technology. The ith type 1 firm seeks to maximize its expected stream of profits:

$$E\left\{\int_0^T \left\{\alpha_1 K(s) - b_1 [K(s)]^2 - \rho_1 I_i^{(1)}(s) - (c_1/2)\left[I_i^{(1)}(s)\right]^2\right\} e^{-rs} ds \right.$$
$$\left. + e^{-rT}\left[q_1 (K(T))^2 + q_2 K(T) + q_3\right] \middle| K(0) = K_0 \right\}, \text{ for } i \in N_1, \tag{3.2}$$

subject to (3.1).

In particular, given the technology level $K(s)$, the instantaneous revenue of a type 1 firm is $K(s)[\alpha_1 - b_1 K(s)]$. The cost of investment is $\rho_1 I_i^{(1)}(s) - (1/2)\left[I_i^{(1)}(s)\right]^2$. For each firm, there is a terminal valuation $e^{-rT}\left[q_1(K(T))^2+q_2K(T)+q_3\right]$ with $q_1<0$, $q_2>0$ and $q_3>0$.

The *j*th type 2 firm seeks to maximize its expected stream of profits:

$$E\left\{\int_0^T\left\{\alpha_2K(s)-b_2[K(s)]^2-\rho_2I_i^{(2)}(s)-(c_2/2)\left[I_i^{(2)}(s)\right]^2\right\}e^{-rs}ds\right.$$
$$\left.+e^{-rT}\left[q_1(K(T))^2+q_2K(T)+q_3\right]\left|K(0)=K_0\right.\right\},\quad \text{for } j\in N_2, \tag{3.3}$$

subject to (3.1).

To derive the noncooperative market outcome of the industry we invoke the analysis in (1.4 and 1.5) in Sect. 12.1 and obtain the corresponding Hamilton-Jacobi-Bellman equations

$$-V_t^{(1)i}(t,K)-\frac{1}{2}V_{KK}^{(1)i}(t,K)\sigma^2K^2$$
$$=\max_{I_i^{(1)}}\left\{\left\{\alpha_1K-b_1K^2-\rho_1I_i^{(1)}-(c_1/2)\left(I_i^{(1)}\right)^2\right\}e^{-rt}\right.$$
$$\left.+V_K^{(1)i}(t,K)\left[\sum_{\substack{\ell\in N_1\\ \ell\neq i}}\phi_\ell^{(1)}(t,K)+\sum_{\ell\in N_2}\phi_\ell^{(2)}(t,K)+I_i^{(1)}-\delta K\right]\right\},$$
$$V^{(1)i}(T,K)=e^{-rT}\left(q_1K^2+q_2K+q_3\right),\quad \text{for } i\in N_1; \tag{3.4}$$
$$-V_t^{(2)j}(t,K)-\frac{1}{2}V_{KK}^{(2)j}(t,K)\sigma^2K^2$$
$$=\max_{I_j^{(2)}}\left\{\left\{\alpha_2K-b_2K^2-\rho_2I_j^{(2)}-(c_2/2)\left(I_j^{(2)}\right)^2\right\}e^{-rt}\right.$$
$$\left.+V_K^{(2)j}(t,K)\left[\sum_{\ell\in N_1}\phi_\ell^{(1)}(t,K)+\sum_{\substack{\ell\in N_2\\ \ell\neq j}}\phi_\ell^{(2)}(t,K)+I_j^{(2)}-\delta K\right]s\right\},$$
$$V^{(2)j}(T,K)=e^{-rT}\left(q_1K^2+q_2K+q_3\right),\quad \text{for } j\in N_2. \tag{3.5}$$

Performing the maximization operator in (3.4) and (3.5) yields game equilibrium investment strategies of the type 1 firm and the type 2 firms as:

$$I_i^{(1)}=\frac{V_K^{(1)i}(t,K)e^{rt}-\rho_1}{c_1},\quad \text{for } i\in N_1; \tag{3.6}$$

and

$$I_j^{(2)} = \frac{V_K^{(2)j}(t,K)e^{rt} - \rho_2}{c_2}, \quad \text{for } j \in N_2. \tag{3.7}$$

To solve the game we first obtain the value functions indicating the game equilibrium payoffs of the firms as follows.

Proposition 3.1 The value functions indicating the game equilibrium payoffs of the firms are

$$\begin{aligned} V^{(1)i}(t,K) &= \left[A_1(t)K^2 + B_1(t)K + C_1(t)\right]e^{-rt} \text{ for } i \in N_1; \text{ and} \\ V^{(2)j}(t,K) &= \left[A_2(t)K^2 + B_2(t)K + C_2(t)\right]e^{-rt}, \text{ for } j \in N_2; \end{aligned} \tag{3.8}$$

where the values of $A_1(t), A_2(t), B_1(t), B_2(t), C_1(t)$ and $C_2(t)$ are generated by the following block-recursive ordinary differential equations:

$$\begin{aligned} \dot{A}_1(t) &= \frac{(2-4n_1)}{c_1}[A_1(t)]^2 - \frac{4n_2}{c_2}A_1(t)A_2(t) + \left(r + 2\delta - \sigma^2\right)A_1(t) + b_1, \\ \dot{A}_2(t) &= \frac{(2-4n_2)}{c_2}[A_2(t)]^2 - \frac{4n_1}{c_1}A_1(t)A_2(t) + \left(r + 2\delta - \sigma^2\right)A_2(t) + b_2, \\ A_1(T) &= q_1 \text{ and } A_2(T) = q_1; \end{aligned} \tag{3.9}$$

$$\begin{aligned} \dot{B}_1(t) &= \left[(r+\delta) - \left(\frac{4n_1}{c_1} - \frac{2}{c_1}\right)A_1(t) - 2\frac{n_2}{c_2}A_2(t)\right]B_1(t) - 2\frac{n_2}{c_2}A_1(t)B_2(t) \\ &\quad + 2\left(\frac{n_1\rho_1}{c_1} + \frac{n_2\rho_2}{c_2}\right)A_1(t) - \alpha_1. \\ \dot{B}_2(t) &= \left[(r+\delta) - \left(\frac{4n_2}{c_2} - \frac{2}{c_2}\right)A_2(t) - 2\frac{n_1}{c_1}A_1(t)\right]B_2(t) - 2\frac{n_1}{c_1}A_2(t)B_1(t) \\ &\quad + 2\left(\frac{n_1\rho_1}{c_1} + \frac{n_2\rho_2}{c_2}\right)A_2(t) - \alpha_2. \\ B_1(T) &= q_2 \text{ and } B_2(T) = q_2; \end{aligned} \tag{3.10}$$

$$\begin{aligned} \dot{C}_1(t) &= rC_1(t) - \left(\frac{n_1}{c_1} - \frac{1}{2c_1}\right)[B_1(t)]^2 - \frac{n_2}{c_2}B_1(t)B_2(t) + \left(\frac{n_1\rho_1}{c_1} + \frac{n_2\rho_2}{c_2}\right)B_1(t) - \frac{\rho_1^2}{2c_1}; \\ \dot{C}_2(t) &= rC_2(t) - \left(\frac{n_2}{c_2} - \frac{1}{2c_2}\right)[B_2(t)]^2 - \frac{n_1}{c_1}B_1(t)B_2(t) \\ &\quad + \left(\frac{n_1\rho_1}{c_1} + \frac{n_2\rho_2}{c_2}\right)B_2(t) - \frac{\rho_2^2}{2c_2}; \\ C_1(T) &= q_3 \text{ and } C_2(T) = q_3. \end{aligned} \tag{3.11}$$

Proof See Appendix C. ■

System (3.9, 3.10 and 3.11) is a block-recursive system of ordinary differential equations. In particular, (3.9) is a system which involves $A_1(t)$ and $A_2(t)$; (3.10) is a system which involves $A_1(t), A_2(t), B_1(t)$ and $B_2(t)$; and (3.11) is a system which involves $B_1(t), B_2(t), C_1(t)$ and $C_2(t)$.

A convenient way to solve the problem numerically is to express system (3.9) as an initial value problem with the variables $A_1^*(t) = A_1(T - t)$ and $A_2^*(t) = A_2(T - t)$ where:

$$\begin{aligned}
\dot{A}_1^*(t) &= \frac{(4n_1 - 2)}{c_1}\left[A_1^*(t)\right]^2 + \frac{4n_2}{c_2}A_1^*(t)A_2^*(t) - \left(r + 2\delta - \sigma^2\right)A_1^*(t) - b_1,\\
\dot{A}_2^*(t) &= \frac{(4n_2 - 2)}{c_2}\left[A_2^*(t)\right]^2 + \frac{4n_1}{c_1}A_1^*(t)A_2^*(t) - \left(r + 2\delta - \sigma^2\right)A_2^*(t) - b_2,\\
A_1^*(0) &= q_1 \text{ and } A_2^*(0) = q_1.
\end{aligned} \tag{3.12}$$

Using Euler's method, the numerical solution of (3.12) could be readily evaluated as:

$$\begin{aligned}
A_1^*(t + \Delta t) = A_1^*(t) + \Bigg[&\frac{(4n_1 - 2)}{c_1}\left[A_1^*(t)\right]^2 + \frac{4n_2}{c_2}A_1^*(t)A_2^*(t)\\
&- (r + 2\delta - \sigma^2)A_1^*(t) - b_1\Bigg]\Delta t,\\
A_2^*(t + \Delta t) = A_2^*(t) + \Bigg[&\frac{(4n_2 - 2)}{c_2}\left[A_2^*(t)\right]^2 + \frac{4n_1}{c_1}A_1^*(t)A_2^*(t)\\
&- (r + 2\delta - \sigma^2)A_2^*(t) - b_2\Bigg]\Delta t,
\end{aligned} \tag{3.13}$$

The numerical values generated in (3.13) yields $A_1^*(t) = A_1(T - t)$ and $A_2^*(t) = A_2(T - t)$. Substituting $A_1(t)$ and $A_2(t)$ into (3.10) yields a pair of linear differential equations in $B_1(t)$ and $B_2(t)$ which could readily be solved numerically. Substituting $B_1(t)$ and $B_2(t)$ into (3.11) yields a pair of independent linear differential equations in $C_1(t)$ and $C_2(t)$, which once again is readily solvable numerically.

Using Proposition 3.1 and (3.6 and 3.7) the game equilibrium strategies can be obtained and the market equilibrium be characterized.

12.3.2 Cooperative Development of Technical Knowledge

Now we consider the case when the firms agree to act cooperatively and seek higher expected profits. They agree to maximize their expected joint profit and share the excess of cooperative profits over noncooperative profits equally. To maximize their expected joint profits the firms maximize

$$E\left\{\int_0^T \sum_{h\in N_1}\left\{\alpha_1 K(s) - b_1[K(s)]^2 - \rho_1 I_h^{(1)}(s) - (c_1/2)\left[I_h^{(1)}(s)\right]^2\right\}e^{-rs}ds\right.$$
$$+\int_0^T \sum_{k\in N_2}\left\{\alpha_2 K(s) - b_2[K(s)]^2 - \rho_2 I_k^{(2)}(s) - (c_2/2)\left[I_k^{(2)}(s)\right]^2\right\}e^{-rs}ds$$
$$\left.+(n_1+n_2)e^{-rT}\left[q_1(K(T))^2 + q_2 K(T) + q_3\right]\,\middle|\, K(0) = K_0\right\}, \tag{3.14}$$

subject to dynamics (3.1).

Following the analysis in (1.6, 1.7, 1.8, 1.9 and 1.10) in Sect. 12.1, the corresponding stochastic dynamic programming equation can be obtained as:

$$-W_t(t,K) - \frac{1}{2}W_{KK}(t,K)\sigma^2 K^2$$
$$= \max_{I_1^{(1)},I_2^{(1)},\ldots,I_{n_1}^{(1)};I_{n_1+1}^{(2)},I_{n_1+2}^{(2)},\ldots,I_{n_1+n_2}^{(2)}}\left\{\sum_{h\in N_1}\left[\alpha_1 K - b_1 K^2 - \rho_1 I_h^{(1)} - (c_1/2)\left(I_h^{(1)}\right)^2\right]e^{-rt}\right.$$
$$+\sum_{k\in N_2}\left\{\alpha_2 K - b_2 K^2 - \rho_2 I_k^{(2)} - (c_2/2)\left(I_k^{(2)}\right)^2\right\}e^{-rt}$$
$$\left.+W_K(t,K)\left[\sum_{h\in N_1} I_h^{(1)}(s) + \sum_{k\in N_2} I_k^{(2)}(s) - \delta K(s)\right]\right\}, \tag{3.15}$$
$$W(T,K) = (n_1+n_2)e^{-rT}\left(q_1 K^2 + q_2 K + q_3\right). \tag{3.16}$$

Performing the maximization operator in (3.15) yields:

$$I_i^{(1)} = \frac{W(t,K)e^{rt} - \rho_1}{c_1},\ \text{for}\, i\in N_1;\ \text{and}$$
$$I_j^{(2)} = \frac{W(t,K)e^{rt} - \rho_2}{c_2},\ \text{for}\, j\in N_2. \tag{3.17}$$

The expected joint payoff of the firms can be obtained as:

Proposition 3.2 The value function $W(t,K)$, which reflects the maximized expected joint payoff at time t given the level of technology K is

$$W(t,K) = \left[A(t)K^2 + B(t)K + C(t)\right]e^{-rt}, \tag{3.18}$$

where the values of $A(t), B(t)$ and $C(t)$ are generated by the following block recursive ordinary differential equations:

$$\dot{A}(t) = (r + 2\delta - \sigma^2)A(t) - 2\left(\frac{n_1}{c_1} + \frac{n_2}{c_2}\right)[A(t)]^2 + n_1 b_1 + n_2 b_2,$$
$$A(T) = (n_1+n_2)q_1; \tag{3.19}$$

$$\dot{B}(t) = \left[r + \delta - 2\left(\frac{n_1}{c_1} + \frac{n_2}{c_2}\right)A(t)\right]B(t) + 2\left[\frac{n_1\rho_1}{c_1} + \frac{n_2\rho_2}{c_2}\right]A(t) - n_1\alpha_1 - n_2\alpha_2,$$
$$B(T) = (n_1 + n_2)q_2; \tag{3.20}$$

$$\dot{C}(t) = rC(t) - \frac{n_1}{2c_1}[B(t) - \rho_1]^2 - \frac{n_2}{2c_2}[B(t) - \rho_2]^2,$$
$$C(T) = (n_1 + n_2)q_3. \tag{3.21}$$

Proof Follow the proof of Proposition 3.1. ■

Using (3.17) and Proposition 3.2 the optimal technology accumulation dynamics can be expressed as:

$$dK(s) = \left[\frac{n_1}{c_1}[2A(s)K(s) + B(s) - \rho_1] + \frac{n_2}{c_2}[2A(s)K(s) + B(s) - \rho_2]\right.$$
$$\left. - \delta K(s)\right]ds + \sigma K(s)dz(s), \quad K(0) = K_0 \tag{3.22}$$

We use X_s^* to denote the set of realizable values of $K^*(s)$ generated by (3.22) at time s. The term $K_s^* \in X_s^*$ is used to denote and element in X_s^*.

With the firms agreeing to share the excess of cooperative profits over noncooperative profits equally the imputation vector becomes

$$\xi^{(1)i}(s, K_s^*) = V^{(1)i}(s, K_s^*) + \frac{1}{n_1 + n_2}\left[W(s, K_s^*)\right.$$
$$\left. - \sum_{h \in N_1} V^{(1)h}(s, K_s^*) - \sum_{k \in N_2} V^{(2)k}(s, K_s^*)\right], \text{ for type 1 firm } i \in N_1;$$
$$\xi^{(2)j}(s, K_s^*) = V^{(2)j}(s, K_s^*) + \frac{1}{n_1 + n_2}\left[W(s, K_s^*)\right.$$
$$\left. - \sum_{h \in N_1} V^{(1)h}(s, K_s^*) - \sum_{k \in N_2} V^{(2)k}(s, K_s^*)\right], \text{ for type 2 firm } j \in N_2; \tag{3.23}$$

at time instant $s \in [0, T]$ if the state of technology is $K_s^* \in X_s^*$.

Invoking Theorem 1.1, a PDP for firm $i \in N_1$ and firm $j \in N_2$ with a terminal payment $\left[q_1 (K_T^*)^2 + q_2 K_T^* + q_3\right]$ at time T and an instantaneous payment (in present value) at time $s \in [0, T]$ equalling

$$B_i^{(1)}(s,K_s^*)e^{-rs} = -\xi_s^{(1)i}(s,K_s^*)$$
$$-\xi_{K_s}^{(1)i}(s,K_s^*)\left[\frac{n_1}{c_1}[2A(s)K_s^* + B(s) - \rho_1] + n_2[2A(s)K_s^* + B(s) - \rho_2]\right.$$
$$\left. - \delta K_s^*\right] - \frac{1}{2}\sigma^2(K_s^*)^2\xi_{K_sK_s}^{(1)i}(s,K_s^*), \text{ given to the type 1 firm } i \in N_1;$$

and

$$B_j^{(2)}(s,K_s^*)e^{-rs} = -\xi_s^{(2)j}(s,K_s^*)$$
$$-\xi_{K_s}^{(2)j}(s,K_s^*)\left[\frac{n_1}{c_1}[2A(s)K_s^* + B(s) - \rho_1] + n_2[2A(s)K_s^* + B(s) - \rho_2]\right.$$
$$\left. - \delta K_s^*\right] - \frac{1}{2}\sigma^2(K_s^*)^2\xi_{K_sK_s}^{(2)j}(s,K_s^*), \text{ given to the type 2 firm } j \in N_2;$$

would lead to the realization of the imputation $\xi(s,K_s^*)$ in (3.23) and hence a subgame consistent scheme.

The terms $\xi_s^{(\omega)i_\omega}(s,K_s^*)$, $\xi_{K_s}^{(\omega)i_\omega}(s,K_s^*)$ and $\xi_{K_sK_s}^{(\omega)i_\omega}(s,K_s^*)$, for $\omega \in \{1,2\}$ and $i_\omega \in N_\omega$, can be obtained readily using Proposition 3.1, Proposition 3.2 and (3.23).

Moreover, the game (3.1, 3.2 and 3.3) can be extended to include the case with more than 2 types of firms. Finally, worth-noting is that the payoff structures and state dynamics of the game (3.1, 3.2 and 3.3) encompass those of the existing dynamic games of public goods provision. For instance, Fershtman and Nitzan (1991) is case where $n_1 = n$, $n_2 = 0$, $\rho_1 = \rho_2 = 0$ and $\sigma = 0$. Wirl (1996) is the case where $n_1 = 2$, $n_2 = 0$, $\rho_1 = \rho_2 = 0$ and $\sigma = 0$. Wang and Ewald (2010) is the case where $n_1 = 2$, $n_2 = 0$ and $\rho_1 = \rho_2 = 0$. Dockner et al. (2000) is case where $n_1 = 1$, $n_2 = 1$, $b_1 = b_2 = 1$, $\rho_1 = \rho_2 = \rho$, $c_1 = c_2 = 1$ and $\sigma = 0$.

12.4 Infinite Horizon Analysis

In this section, we consider the case when the planning horizon approaches infinity, that is $T \to \infty$. The objective of agent $i \in N$ is to maximize its expected payoff

$$E\left\{\int_0^\infty \{R_i[K(s)] - C_i[I_i(s)]\}e^{-rs}ds \,\middle|\, K(0) = K_0\right\} \tag{4.1}$$

subject to dynamics (1.1).

The corresponding Hamilton-Jacobi-Bellman equations in current value formulation characterizing a feedback solution of the infinite horizon problem (1.1) and (4.1) are (see Theorem 5.1 in Chap. 3):

$$rV^i(K) - \frac{1}{2}V_{KK}^i(K)\sigma^2K^2 = \max_{I_i}\left\{[R_i(K) - C_i(I_i)]\right.$$
$$\left. + V_K^i(K)\left[\sum_{\substack{j=1 \\ j\neq i}}^{n} \phi_j(K) + I_i - \delta K\right]\right\}, \text{ for } i \in N, \tag{4.2}$$

Performing the maximization operator in (4.2) yields:

$$dC_i(I_i)/dI_i = V_K^i(K),\ \text{for}\, i \in N \tag{4.3}$$

Condition (4.3) reflects that in a non-cooperative equilibrium the marginal cost of investment of agent *i* will be equal to the agent's implicit marginal valuation/benefit of the productive stock in the infinite horizon case.

12.4.1 Subgame Consistent Cooperative Provision

Consider the case when the agents agree to act cooperatively and seek higher gains. They agree to maximize their expected joint gain and distribute the cooperative gain according to the imputation vector

$$\xi(K) = \left[\xi^1(K), \xi^2(K), \cdots, \xi^n(K)\right] \text{ when the state is } K. \tag{4.4}$$

To maximize their expected joint gains the agents maximize

$$\max_{\{I_1(s), I_2(s), \cdots, I_n(s)\}} E\left\{\sum_{j=1}^n \left[\int_0^\infty \{R_j[K(s)] - C_j[I_j(s)]\} e^{-rs} ds\right]\right\} \tag{4.5}$$

subject to dynamics (1.1).

Invoking stochastic dynamic programming techniques an optimal solution to the stochastic control problem (1.1) and (4.5) can characterized by the following set of equations (see Theorem A.4 in the Technical Appendices):

$$\begin{aligned} &rW(K) - \frac{1}{2} W_{KK}(K)\sigma^2 K^2 \\ &= \max_{I_1, I_2, \cdots, I_n} \left\{\sum_{j=1}^n \left[[R_j(K) - C_j(I_j)] + W_K(K)\left(\sum_{j=1}^n I_j - \delta K\right)\right]\right\}. \end{aligned} \tag{4.6}$$

In particular, $W(K)$ gives the maximized expected joint payoff of the n players at time given that the level of technology is K. Let $\psi_j^*(K)$, for $j \in N$, denote the game equilibrium investment strategy of agent i, the optimal trajectory of the public goods can be expressed as:

$$dK(s) = \left[\sum_{j=1}^n \psi_j(K(s)) - \delta K(s)\right] ds + \sigma K(s) dz(s), \tag{4.7}$$

for $K(0) = K_o$,

We use X^* to denote the set of realizable values of K generated by (4.7). The term $K^* \in X^*$ is used to denote an element in X^*.

Following the analysis in Theorem 5.3 in Chap. 3, we formulate a Payoff Distribution Procedure (PDP) so that the agreed-upon imputations (4.4) can be realized. Let $B^i(K^*)$ denote the payment that agent i will received under the cooperative agreement if K^* is realized.

A theorem characterizing a formula for $B^i(K^*)$, for $i \in N$, which yields (4.4) is provided below.

Theorem 4.1 A PDP with an instantaneous payment equaling

$$B_i(K^*) = r\xi^i(K^*) - \xi^i_K(K^*)\left[\sum_{j=1}^{n}\psi^*_j(K^*) - \delta K^*\right]$$
$$-\frac{1}{2}\sigma^2(K^*)^2\xi^i_{KK}(K^*), \quad \text{for } i \in N, \text{ given that the state is } K^* \in X^* \quad (4.8)$$

would lead to the realization of the imputation $\xi(K^*)$ in (4.4).

Proof See Theorem 5.3 in Chap. 3. ■

Note that the payoff distribution procedure in Theorem 4.1 would give rise to the agreed-upon imputation in (4.4) and therefore subgame consistency is satisfied.

When all agents are using the cooperative strategies and the state equals K^*, the payoff that player i will directly receive is

$$R_i(K^*) - C_i[\psi_i(K^*)].$$

However, according to the agreed upon imputation, agent i is to receive $B_i(K^*)$. Therefore a transfer payment

$$\varpi_i(K^*) = B_i(K^*) - \{R_i(K^*) - C_i[\psi_i(K^*)]\}. \quad (4.9)$$

will be imputed to agent $i \in N$.

12.4.2 *Infinite Horizon Public Capital Goods Provision: An Illustration*

In this section we consider the infinite horizon game of public capital goods provision in which the expected payoff to agent $i \in N$ is:

$$E\left\{\int_0^\infty \left\{\alpha_i K(s) - c_i[I_i(s)]^2\right\}e^{-rs}ds \,\middle|\, K(0) = K_0\right\}, \text{ for } i \in N. \quad (4.10)$$

The accumulation dynamics of the public capital stock is governed by (2.1).

Setting up the corresponding Hamilton-Jacobi-Bellman equations according to (4.2) and performing the maximization operator yields:

$$I_i = \frac{V_K^i(K)}{2c_i}, \text{ for } i \in N.$$

The value functions which reflect the expected noncooperative payoffs of the agents can be obtained as:

Proposition 4.1 The value function reflecting the expected noncooperative payoff of agent i is:

$$V^i(K) = (A_i K + C_i), \text{ for } i \in N; \tag{4.11}$$

where $A_i = \dfrac{\alpha_i}{(r+\delta)}$, and

$$C_i = \left[\sum_{j=1}^{n} \frac{A_i A_j}{2c_j r}\right] - \frac{(A_i)^2}{4c_i r}.$$

Proof Following the derivation of Proposition 2.1, one can obtain the value function as in (4.11). ■

Consider the case when the agents agree to act cooperatively and seek higher gains. They agree to maximize their expected joint gain and distribute the cooperative gain proportional to their non-cooperative gains. To maximize their expected joint gains the agents maximize

$$E\left\{\int_0^{\infty} \sum_{j=1}^{n} \left\{\alpha_j K(s) - c_j [I_j(s)]^2\right\} e^{-rs} ds \,\middle|\, K(0) = K_0\right\} \tag{4.12}$$

subject to dynamics (2.1).

Performing the maximization operator in (4.12) yields:

$$I_i = \frac{W_K(K)}{2c_i}, \text{ for } i \in N.$$

The value function $W(K)$ which reflects the maximized expected joint profits of the $n$would lead to the realization of the imputation as:

Proposition 4.2

$$W(K) = [AK + C], \tag{4.13}$$

where $A = \sum_{j=1}^{n} \frac{\alpha_j}{(r+\delta)}$ and $C = \sum_{j=1}^{n} \frac{(A)^2}{4c_j r}$.

Proof Following the derivation of Proposition 2.2, one can obtain the value function as in (4.13). ■

With the agents agreeing to distribute their gains proportional to their non-cooperative gains, the imputation vector becomes

$$\xi^i(K^*) = \frac{V^i(K^*)}{\sum_{j=1}^{n} V^j(K^*)} W(K^*) = \frac{(A_i K + C_i)}{\sum_{j=1}^{n}(A_j K + C_j)}(AK + C), \tag{4.14}$$

for $i \in N$ if the public capital stock is $K^* \in X^*$.

To guarantee dynamical stability in a dynamic cooperation scheme, the solution has to satisfy the property of subgame consistency which requires the satisfaction of (4.14). Following Theorem 4.1 we can obtain the PDP that brings about a subgame consistent solution with instantaneous payments:

$$B_i(K^*) = \frac{r(A_i K^* + C_i)}{\sum_{j=1}^{n}(A_j K^* + C_j)}(AK^* + C) - \xi_K^i(K^*)\left[\sum_{j=1}^{n}\frac{A}{2c_j} - \delta K^*\right] - \frac{1}{2}\sigma^2(K^*)^2\xi_{KK}^i(K^*), \tag{4.15}$$

where

$$\xi_K^i(K^*) = \left\{\frac{[(A_i(AK^* + C) + (A_i K^* + C_i)A]\sum_{j=1}^{n}(A_j K^* + C_j)}{\left[\sum_{j=1}^{n}(A_j K^* + C_j)\right]^2} - \frac{(A_i K^* + C_i)(AK^* + C)\sum_{j=1}^{n}A_j}{\left[\sum_{j=1}^{n}(A_j K^* + C_j)\right]^2}\right\}\left[\sum_{j=1}^{n}\frac{A}{2c_j} - \delta K^*\right], \text{ and}$$

$$\xi_{KK}^i(K^*) = d\xi_K^i(K^*)/dK,$$

for $i \in N$ if the public capital stock is $K^* \in X^*$.

Therefore a transfer payment

$$\varpi_i(K^*) = B_i(K^*) - \left[\alpha_i K^* - (A^2/c_i)\right]$$

will be imputed to agent $i \in N$.

12.5 Public Goods Provision Under Accumulation and Payoff Uncertainties

This Section considers cooperative provision of public goods by asymmetric agents in a discrete-time dynamic game framework with uncertainties in stock accumulation dynamics and future payoff structures. One of the major hindrances for dynamic cooperation in public goods provision is the uncertainty in the future gains. This section resolves the problem with subgame consistent schemes. The analytical framework and the non-cooperative outcome of public goods provision are provided in Sect. 12.5.1. Details of a Pareto optimal cooperative scheme are presented in Sect. 12.5.2. A payment mechanism ensuring subgame consistency is derived in Sect. 12.5.3.

12.5.1 Analytical Framework and Non-cooperative Outcome

Consider the case of the provision of a public good in which a group of n agents carry out a project by making contributions to the building up of the stock of a productive public good. The game involves T stages of operation and after the T stages each agent received a terminal payment in stage $T+1$. We use K_t to denote the level of the productive stock and I_t^i the public capital investment by agent i at stage $t \in \{1, 2, \cdots, T\}$. The stock accumulation dynamics is governed by the stochastic difference equation:

$$K_{t+1} = K_t + \sum_{j=1}^{n} I_t^j - \delta K_t + \vartheta_t, \; K_1 = {}^1K^0, \tag{5.1}$$

for $t \in \{1, 2, \cdots, T\}$,

where δ is the depreciation rate and ϑ_t is a sequence of statistically independent random variables.

The payoff of agent i at stage t is affected by a random variable θ_t. In particular, the payoff to agent i at stage t is

$$R^i(K_t, \theta_t) - C^i\left(I_t^i, \theta_t\right), \; i \in \{1, 2, \cdots, n\} = N, \tag{5.2}$$

where $R^i(K_t, \theta_t)$ is the revenue/payoff to agent i, $C^i(I_t^i, \theta_t)$ is the cost of investing $I_t^i \in X^i$, and θ_t for $t \in \{1, 2, \cdots, T\}$ are independent discrete random variables with range $\{\theta_t^1, \theta_t^2, \cdots, \theta_t^{\eta_t}\}$ and corresponding probabilities $\{\lambda_t^1, \lambda_t^2, \cdots, \lambda_t^{\eta_t}\}$, where η_t is a positive integer for $t \in \{1, 2, \cdots, T\}$. In stage 1, it is known that θ_1 equals θ_1^1 with probability $\lambda_1^1 = 1$.

Marginal revenue product of the productive stock is positive, that is $\partial R^i(K_t, \theta_t)/\partial K_t > 0$, before a saturation level $\overline{K}$ has been reached; and marginal cost of investment is positive and non-decreasing, that is $\partial C^i\left(I_t^i,\theta_t\right)/\partial I_t^i\right) > 0$ and $\partial^2 C^i\left(I_t^i,\theta_t\right)/\partial I_t^{i2} \geq 0$.

The objective of agent $i \in N$ is to maximize its expected net revenue over the planning horizon, that is

$$E_{\theta_1,\theta_2,\cdots,\theta_T;\vartheta_1,\vartheta_2,\cdots,\vartheta_T}\left\{\sum_{s=1}^{T}\left[R^i(K_s,\theta_s)-C^i\left(I_s^i,\theta_s\right)\right](1+r)^{-(s-1)} + q^i(K_{T+1})(1+r)^{-T}\right\} \tag{5.3}$$

subject to the stock accumulation dynamics (5.1),

where $E_{\theta_1,\theta_2,\cdots,\theta_T;\vartheta_1,\vartheta_2,\cdots,\vartheta_T}$ is the expectation operation with respect to the random variables $\theta_1,\theta_2,\cdots,\theta_T$ and $\vartheta_1,\vartheta_2,\cdots,\vartheta_T$; r is the discount rate, and $q^i(K_T) \geq 0$ is an amount conditional on the productive stock that agent i would receive at stage $T+1$. Since there is no uncertainty in stage $T+1$, we use θ_{T+1}^1 to denote the condition in stage $T+1$ with probability $\lambda_{T+1}^1 = 1$.

To solve the game, we follow the analysis in Chap. 9 and begin with the subgame starting at the last operating stage, that is stage T. If $\theta_T^{\sigma_T} \in \{\theta_T^1,\theta_T^2,\cdots,\theta_T^{\eta_T}\}$ has occurred at stage T and the public capital stock is $K_T = K$, the subgame becomes:

$$\max_{I_T^i} E_{\vartheta_T}\left\{\left[R^i(K_T,\theta_T^{\sigma_T})-C^i\left(I_T^i,\theta_T^{\sigma_T}\right)\right](1+r)^{-(T-1)} + q^i(K_{T+1})(1+r)^{-T}\right\},\ \text{for}\, i \in N \tag{5.4}$$

subject to

$$K_{T+1} = K_T + \sum_{j=1}^{n} I_T^j - \delta K_T + \vartheta_T, K_T = K. \tag{5.5}$$

The subgame (5.4 and 5.5) is a stochastic dynamic game. Invoking the standard techniques for solving stochastic dynamic games, a characterization the feedback Nash equilibrium is provided in the following lemma.

Lemma 5.1 A set of strategies $\phi_T^{(\sigma_T)*}(K) = \left\{\phi_T^{(\sigma_T)1*}(K),\phi_T^{(\sigma_T)2*}(K),\cdots\cdots,\phi_T^{(\sigma_T)n*}(K)\right\}$ provides a Nash equilibrium solution to the subgame (5.4 and 5.5) if there exist functions $V^{(\sigma_T)i}(t,K)$, for $i \in N$ and $t \in \{1,2\}$, such that the following conditions are satisfied:

$$V^{(\sigma_T)i}(T,K) = \max_{I_T^i} E_{\vartheta_T}\Bigg\{ \left[R^i(K_T,\theta_T^{\sigma_T}) - C^i\left(I_T^i,\theta_T^{\sigma_T}\right)\right](1+r)^{-(T-1)}$$

$$+ V^{(\sigma_{T+1})i}\left[T+1, K + \sum_{\substack{j=1 \\ j\neq i}}^{n} \phi_T^{(\sigma_T)j*}(K) + I_T^i - \delta K + \vartheta_T\right]\Bigg\},$$

$$V^{(\sigma_{T+1})i}(T+1,K) = q^i(K)(1+r)^{-T};\ \text{for } i\in N \tag{5.6}$$

Proof The system of equations in (5.6) satisfies the standard stochastic dynamic programming property and the Nash property for each agent $i\in N$. Hence a Nash equilibrium of the subgame (5.4 and 5.5) is characterized. Details of the proof of the results can be found in Theorem 4.1 in Chap. 7. ■

Using Lemma 5.1, one can characterize the value functions $V^{(\sigma_T)i}(T,K)$ for all $\sigma_T\in\{1,2,\cdots,\eta_T\}$ if they exist. In particular, $V^{(\sigma_T)i}(T,K)$ yields agent i's expected game equilibrium payoff in the subgame starting at stage T given that $\theta_T^{\sigma_T}$ occurs and $K_T = K$.

Then we proceed to the subgame starting at stage $T-1$ when $\theta_{T-1}^{\sigma_{T-1}}\in\{\theta_{T-1}^1,\theta_{T-1}^2,\cdots,\theta_{T-1}^{\eta_{T-1}}\}$ occurs and $K_{T-1}=K$. In this subgame agent $i\in N$ seeks to maximize his expected payoff

$$E_{\theta_T;\vartheta_{T-1},\vartheta_T}\Bigg\{\sum_{s=T-1}^{T}\left[R^i(K_s,\theta_s) - C^i\left(I_s^i,\theta_s\right)\right](1+r)^{-(s-1)}$$

$$+ q^i(K_{T+1})(1+r)^{-T}\Bigg\}$$

$$= E_{\vartheta_{T-1}}\Bigg\{\left[R^i\left(K_{T-1},\theta_{T-1}^{\sigma_{T-1}}\right) - C^i\left(I_{T-1}^i,\theta_{T-1}^{\sigma_{T-1}}\right)\right](1+r)^{-(T-2)}$$

$$+ \sum_{\sigma_T=1}^{\eta_T}\lambda_T^{\sigma_T}\left[R^i(K_T,\theta_T^{\sigma_T}) - C^i\left(I_T^i,\theta_T^{\sigma_T}\right)\right](1+r)^{-(T-2)}$$

$$+ q^i(K_{T+1})(1+r)^{-T}\Bigg\}, \tag{5.7}$$

subject to the capital accumulation dynamics

$$K_{t+1} = K_t + \sum_{j=1}^{n} I_t^j - \delta K_t + \vartheta_t, K_{T-1} = K,\ \text{for } t\in\{T-1,T\}. \tag{5.8}$$

If the functions $V^{(\sigma_T)i}(T,K)$ for all $\sigma_T \in \{1,2,\cdots,\eta_T\}$ characterized in Lemma 5.1 exist, the subgame (5.7 and 5.8) can be expressed as a game in which agent i seeks to maximize the expected payoff

$$E_{\vartheta_{T-1}}\Bigg\{\left[R^i(K_{T-1},\theta_{T-1}) - C^i\left(I^i_{T-1},\theta_{T-1}\right)\right](1+r)^{-(T-2)} + \sum_{\sigma_T=1}^{\eta_T}\lambda_T^{\sigma_T} V^{(\sigma_T)i}\left[T, K_{T-1} + \sum_{j=1}^{n} I^j_{T-1} - \delta K_{T-1} + \vartheta_{T-1}\right]\Bigg\}, \quad \text{for } i \in N, \tag{5.9}$$

using his control I^i_{T-1}.

A Nash equilibrium of the subgame (5.9) can be characterized by the following lemma.

Lemma 5.2 A set of strategies $\phi_{T-1}^{(\sigma_{T-1})*}(K) = \left\{\phi_{T-1}^{(\sigma_{T-1})1*}(K), \phi_{T-1}^{(\sigma_{T-1})2*}(K), \cdots, \phi_{T-1}^{(\sigma_{T-1})n*}(K)\right\}$ provides a Nash equilibrium solution to the subgame (5.9) if there exist functions $V^{(\sigma_T)i}(T,K_T)$ for $i \in N$ and $\sigma_T = \{1,2,\cdots,\eta_T\}$ characterized in Lemma 5.1, and functions $V^{(\sigma_{T-1})i}(T-1,K)$, for $i \in N$, such that the following conditions are satisfied:

$$V^{(\sigma_{T-1})i}(T-1,K) = \max_{I^i_{T-1}} E_{\vartheta_{T-1}}\Bigg\{\left[R^i\left(K_{T-1},\theta_{T-1}^{\sigma_{T-1}}\right) - C^i\left(I^i_{T-1},\theta_{T-1}^{\sigma_{T-1}}\right)\right](1+r)^{-(T-2)} + \sum_{\sigma_T=1}^{\eta_T}\lambda_T^{\sigma_T} V^{(\sigma_T)i}\left[T, K + \sum_{\substack{j=1 \\ j \neq i}}^{n} \phi_{T-1}^{(\sigma_{T-1})j*}(K) + I^i_{T-1} - \delta K + \vartheta_{T-1}\right]\Bigg\}, \text{ for } i \in N. \tag{5.10}$$

Proof The conditions in Lemma 5.1 and the system of equations in (5.10) satisfies the standard discrete-time stochastic dynamic programming property and the Nash property for each agent $i \in N$. Hence a Nash equilibrium of the subgame (5.9) is characterized. ■

Using Lemma 5.2, one can characterize the functions $V^{(\sigma_T)i}(T-1,K)$ for all $\theta_{T-1}^{\sigma_{T-1}} \in \{\theta^1_{T-1},\theta^2_{T-1},\cdots,\theta^{\eta_{T-1}}_{T-1}\}$, if they exist. In particular, $V^{(\sigma_{T-1})i}(T-1,K)$ yields agent i's expected game equilibrium payoff in the subgame starting at stage $T-1$ given that $\theta_{T-1}^{\sigma_{T-1}}$ occurs and $K_{T-1} = K$.

Consider the subgame starting at stage $t \in \{T-2, T-3, \cdots, 1\}$ when $\theta_t^{\sigma_t} \in \{\theta^1_t, \theta^2_t, \cdots, \theta^{\eta_t}_t\}$ occurs and $K_t = K$, in which agent $i \in N$ maximizes his expected payoff

$$E_{\vartheta_t}\Bigg\{\left[R^i\left(K,\theta_t^{\sigma_t}\right)-C^i\left(I_t^i,\theta_t^{\sigma_t}\right)\right](1+r)^{-(t-1)}$$
$$+\sum_{\sigma_{t+1}=1}^{\eta_{t+1}}\lambda_{t+1}^{\sigma_{t+1}}V^{(\sigma_{t+1})i}\left[t+1,K+\sum_{j=1}^{n}I_t^j-\delta K+\vartheta_t\right]\Bigg\},\ \text{for } i\in N, \tag{5.11}$$

subject to the public capital accumulation dynamics

$$K_{t+1}=K_t+\sum_{j=1}^{n}I_t^j-\delta K_t+\vartheta_t, K_t=K. \tag{5.12}$$

A Nash equilibrium solution for the game (5.1, 5.2 and 5.3) can be characterized by the following theorem.

Theorem 5.1 A set of strategies $\phi_t^{(\sigma_t)*}(K)=\left\{\phi_t^{(\sigma_t)1*}(K),\phi_t^{(\sigma_t)2*}(K),\cdots\cdots,\phi_t^{(\sigma_t)n*}(K)\right\}$,, for $\sigma_t\in\{1,2,\cdots,\eta_t\}$ and $t\in\{1,2,\cdots,T\}$, constitutes a Nash equilibrium solution to the game (5.1, 5.2 and 5.3) if there exist functions $V^{(\sigma_t)i}(t,K)$, for $\sigma_t\in\{1,2,\cdots,\eta_t\}$, $t\in\{1,2,\cdots,T\}$ and $i\in N$, such that the following recursive relations are satisfied:

$$V^{(\sigma_T)i}(T+1,K)=q^i(K_{T+1})(1+r)^{-T},$$
$$V^{(\sigma_t)i}(t,K)=\max_{I_t^i}E_{\vartheta_t}\Bigg\{\left[R^i\left(K_t,\theta_t^{\sigma_t}\right)-C^i\left(I_t^i,\theta_t^{\sigma_t}\right)\right](1+r)^{-(t-1)}$$
$$+\sum_{\sigma_{t+1}=1}^{\eta_{t+1}}\lambda_{t+1}^{\sigma_{t+1}}V^{(\sigma_{t+1})i}\left[t+1,K+\sum_{\substack{j=1\\ j\neq i}}^{n}\phi_t^{(\sigma_t)j*}(K)+I_t^i-\delta K_t+\vartheta_t\right]\Bigg\},$$
$$\text{for } \sigma_t\in\{1,2,\cdots,\eta_t\},\ t\in\{1,2,\cdots,T\}\ \text{and } i\in N. \tag{5.13}$$

Proof The results in (5.13) characterizing the game equilibrium in stage T and stage $T-1$ are proved in Lemma 5.1 and Lemma 5.2. Invoking the subgame in stage $t\in\{1,2,\cdots,T-1\}$ as expressed in (5.11 and 5.12), the results in (5.13) satisfy the optimality conditions in stochastic dynamic programming and the Nash equilibrium property for each agent in each of these subgames. Therefore, a feedback Nash equilibrium of the game (5.1, 5.2 and 5.3) is characterized. ■

Hence, the noncooperative outcome of the public capital provision game (5.1, 5.2 and 5.3) can be obtained.

12.5.2 Optimal Cooperative Scheme

Now consider the case when the agents agree to cooperate and enhance their gains from cooperation. In particular, they act cooperatively to maximize their expected joint payoff and distribute the joint payoff among themselves according to an agreed-upon optimality principle. If any agent deviates from the cooperation scheme, all agents will revert to the noncooperative framework to counteract the free-rider problem in public goods provision. As stated before, group optimality, individual rationality and subgame consistency are three crucial properties that sustainable cooperative scheme has to satisfy.

12.5.2.1 Pareto Optimal Provision

To fulfil group optimality the agents would seek to maximize their expected joint payoff. In particular, they have to solve the discrete-time stochastic dynamic programming problem of maximizing

$$E_{\theta_1,\theta_2,\cdots,\theta_T;\vartheta_1,\vartheta_2,\cdots,\vartheta_T}\left\{\sum_{j=1}^{n}\sum_{s=1}^{T}\left[R^j(K_s,\theta_s)-C^j\left(I_s^j,\theta_s\right)\right](1+r)^{-(s-1)}\right.$$
$$\left.+\sum_{j=1}^{n}q^j(K_{T+1})(1+r)^{-T}\right\} \tag{5.14}$$

subject to dynamics (5.1).

To solve the dynamic programming problem (5.1) and (5.14), we first consider the problem starting at stage T. If $\theta_T^{\sigma_T}\in\left\{\theta_T^1,\theta_T^2,\cdots,\theta_T^{\eta_T}\right\}$ has occurred at stage T and the state $K_T=K$, the problem becomes:

$$\max_{I_T^1,I_T^2,\cdots,I_T^n} E_{\vartheta_T}\left\{\sum_{j=1}^{n}\left[R^j(K,\theta_T^{\sigma_T})-C^j\left(I_T^j,\theta_T^{\sigma_T}\right)\right](1+r)^{-(T-1)}\right.$$
$$\left.+\sum_{j=1}^{n}q^j(K_{T+1})(1+r)^{-T}\right\}, \tag{5.15}$$

$$\text{subject to } K_{T+1}=K_T=\sum_{j=1}^{n}I_T^j-\delta K_T+\vartheta_T, K_T=K. \tag{5.16}$$

A characterization of an optimal solution to the stochastic control problem (5.15 and 5.16) is provided in the following lemma.

Lemma 5.3 A set of controls $I_T^{(\sigma_T)*} = \psi_T^{(\sigma_T)*}(K) = \left\{\psi_T^{(\sigma_T)1*}(K), \psi_T^{(\sigma_T)2*}(K), \cdots\cdots, \psi_T^{(\sigma_T)n*}(K)\right\}$ provides an optimal solution to the stochastic control problem (5.15 and 5.16) if there exist functions $W^{(\sigma_{T+1})}(T,K)$ such that the following conditions are satisfied:

$$
\begin{aligned}
&W^{(\sigma_T)}(T,K) \\
&= \max_{I_T^{(\sigma_T)1}, I_T^{(\sigma_T)2}, \cdots, I_T^{(\sigma_T)n}} E_{\vartheta_T}\left\{\sum_{j=1}^{n}\left[R^j(K,\theta_T^{\sigma_T}) - C^j\left(I_T^j,\theta_T^{\sigma_T}\right)\right](1+r)^{-(T-1)}\right. \\
&\qquad \left. + \sum_{j=1}^{n} q^j\left(K + \sum_{h=1}^{n} I_T^h - \delta K + \vartheta_T\right)(1+r)^{-T}\right\}, \\
&\qquad W^{(\sigma_{T+1})i}(T+1,K) = \sum_{j=1}^{n} q^j(K)(1+r)^{-T}. \qquad (5.17)
\end{aligned}
$$

Proof The system of equations in (5.17) satisfies the standard discrete-time stochastic dynamic programming property. See Theorem A.6 in the Technical Appendices for details of the proof of the results. ■

Using Lemma 5.3, one can characterize the functions $W^{(\sigma_T)}(T,K)$ for all $\theta_T^{\sigma_T} \in \{\theta_T^1, \theta_T^2, \cdots, \theta_T^{\eta_T}\}$, if they exist. In particular, $W^{(\sigma_T)}(T,K)$ yields the expected cooperative payoff starting at stage T given that $\theta_T^{\sigma_T}$ occurs and $K_T = K$.

Following the analysis in Sect. 12.5.1, the control problem starting at stage t when $\theta_t^{\sigma_t} \in \{\theta_t^1, \theta_t^2, \cdots, \theta_t^{\eta_t}\}$ occurs and $K_t = K$ can be expressed as:

$$
\begin{aligned}
&\max_{I_t^{(\sigma_t)1}, I_t^{(\sigma_t)2}, \cdots, I_t^{(\sigma_t)n}} E_{\vartheta_t}\left\{\sum_{j=1}^{n}\left[R^j(K,\theta_t^{\sigma_t}) - C^j\left(I_t^j,\theta_t^{\sigma_t}\right)\right](1+r)^{-(t-1)}\right. \\
&\qquad \left. + \sum_{\sigma_{t+1}=1}^{\eta_{t+1}} \lambda_{t+1}^{\sigma_{t+1}} W^{(\sigma_{t+1})}\left(t+1, K + \sum_{h=1}^{n} I_t^h - \delta K + \vartheta_t\right)\right\}, \qquad (5.18)
\end{aligned}
$$

where $W^{(\sigma_{t+1})}\left(t+1, K + \sum_{h=1}^{n} I_t^h - \delta K + \vartheta_t\right)$ is the expected optimal cooperative payoff in the control problem starting at stage $t+1$ when $\theta_{t+1}^{\sigma_{t+1}} \in \{\theta_{t+1}^1, \theta_{t+1}^2, \cdots, \theta_{t+1}^{\eta_{t+1}}\}$ occurs.

An optimal solution for the stochastic control problem (5.14) can be characterized by the following theorem.

Theorem 5.2 A set of controls $\psi_t^{(\sigma_t)*}(K) = \left\{\psi_t^{(\sigma_t)1*}(K), \psi_t^{(\sigma_t)2*}(K), \cdots\cdots, \psi_t^{(\sigma_t)n*}(K)\right\}$, for $\sigma_t \in \{1,2,\cdots,\eta_t\}$ and $t \in \{1,2,\cdots,T\}$ provides an optimal solution to the stochastic control problem (5.1) and (5.14) if there exist functions $W^{(\sigma_t)}(t,K)$, for

$\sigma_t \in \{1,2,\cdots,\eta_t\}$ and $t \in \{1,2,\cdots,T\}$, such that the following recursive relations are satisfied:

$$W^{(\sigma_T)}(T+1,K) = \sum_{j=1}^{n} q^j(K)(1+r)^{-T},$$

$$W^{(\sigma_T)}(t,K) =$$

$$\max_{I_t^{(\sigma_t)1}, I_t^{(\sigma_t)2}, \cdots, I_t^{(\sigma_t)n}} E_{\vartheta_t}\left\{ \sum_{j=1}^{n} \left[R^j(K,\theta_t^{\sigma_t}) - C^j\left(I_t^j, \theta_t^{\sigma_t}\right) \right] (1+r)^{-(t-1)} \right.$$

$$\left. + \sum_{\sigma_{t+1}=1}^{\eta_{t+1}} \lambda_{t+1}^{\sigma_{t+1}} W^{(\sigma_{t+1})}\left(t+1, K + \sum_{h=1}^{n} I_t^h - \delta K + \vartheta_t \right) \right\}, \quad (5.19)$$

for $\sigma_t \in \{1,2,\cdots,\eta_t\}$ and $t \in \{1,2,\cdots,T\}$.

Proof Invoking Lemma 5.3 and the specification of the control problem starting in stage $t \in \{1,2,\cdots,T-1\}$ as expressed in (5.18), the results in (5.19) satisfy the optimality conditions in discrete-time stochastic dynamic programming. Therefore, an optimal solution of the stochastic control problem is characterized in Theorem 5.2. ■

Substituting the optimal control $\{\psi_t^{(\sigma_t)i*}(K)$, for $t \in \{1,2,\cdots T\}$ and $i \in N\}$ into (5.1), one can obtain the dynamics of the cooperative trajectory of public capital accumulation as:

$$K_{t+1} = K_t + \sum_{j=1}^{n} \psi_t^{(\sigma_t)j*}(K_t) - \delta K_t + \vartheta_t, \; K_1 = {}^1K \text{ if } \theta_t^{\sigma_t} \text{ occurs at stage } t, \quad (5.20)$$

for $t \in \{1,2,\cdots,T\}$, $\sigma_t \in \{1,2,\cdots,\eta_t\}$

We use X_t^* to denote the set of realizable values of K_t at stage t generated by (5.20). The term $K_t^* \in X_t^*$ is used to denote an element in X_t^*.

The term $W^{(\sigma_t)}(t,K_t^*)$ gives the expected total cooperative payoff over the stages from t to T if $\theta_t^{\sigma_t}$ occurs and $K_t^* \in X_t^*$ is realized at stage t.

12.5.2.2 Individually Rational Condition

The agents then have to agree to an optimality principle in distributing the total cooperative payoff among themselves. For individual rationality to be upheld the expected payoffs an agent receives under cooperation have to be no less than his expected noncooperative payoff along the cooperative state trajectory $\{K_t^*\}_{t=1}^{T+1}$. Let $\xi^{(\sigma_t)}(t,K_t^*) = \left[\xi^{(\sigma_t)1}(t,K_t^*), \xi^{(\sigma_t)2}(t,K_t^*), \cdots, \xi^{(\sigma_t)n}(t,K_t^*)\right]$ denote the imputation vector guiding the distribution of the total expected cooperative payoff under the agreed-upon optimality principle along the cooperative trajectory given that $\theta_t^{\sigma_t}$ has occurred in stage t, for $\sigma_t \in \{1, 2, \cdots, \eta_t\}$ and $t \in \{1, 2, \cdots, T\}$.

If for example, the optimality principle specifies that the agents share the expected total cooperative payoff proportional to their non-cooperative payoffs, then the imputation to agent i becomes:

$$\xi^{(\sigma_t)i}(t, K_t^*) = \frac{V^{(\sigma_t)i}(t, K_t^*)}{\sum_{j=1}^{n} V^{(\sigma_t)j}(t, K_t^*)} W^{(\sigma_t)}(t, K_t^*), \tag{5.21}$$

for $i \in N$, $\sigma_t \in \{1, 2, \cdots, \eta_t\}$ and $t \in \{1, 2, \cdots, T\}$.

For individual rationality to be guaranteed in every stage $k \in \{1, 2, \cdots, T\}$, it is required that the imputation satisfies:

$$\xi^{(\sigma_t)i}(t, K_t^*) \geq V^{(\sigma_t)i}(t, K_t^*), \tag{5.22}$$

for $i \in N$, $\sigma_t \in \{1, 2, \cdots, \eta_t\}$ and $t \in \{1, 2, \cdots, T\}$.

To ensure group optimality, the imputation vector has to satisfy

$$W^{(\sigma_t)}(t, K_t^*) = \sum_{j=1}^{n} \xi^{(\sigma_t)j}(t, K_t^*), \tag{5.23}$$

for $\sigma_t \in \{1, 2, \cdots, \eta_t\}$ and $t \in \{1, 2, \cdots, T\}$.

Hence, a valid imputation scheme $\xi^{(\sigma_t)i}(t, K_t^*)$, for $i \in N$ and $\sigma_t \in \{1, 2, \cdots, \eta_t\}$ and $t \in \{1, 2, \cdots, T\}$, has to satisfy conditions (5.22) and (5.23).

12.5.3 Subgame Consistent Payment Mechanism

To guarantee dynamical stability in a stochastic dynamic cooperation scheme, the solution has to satisfy the property of subgame consistency in addition to group optimality and individual rationality. For subgame consistency to be satisfied, the imputation according to the original optimality principle has to be maintained in all the T stages along the cooperative trajectory $\{K_t^*\}_{t=1}^{T}$. In other words, the imputation

$$\xi^{(\sigma_t)}(t, K_t^*) = \left[\xi^{(\sigma_t)1}(t, K_t^*), \xi^{(\sigma_t)2}(t, K_t^*), \cdots, \xi^{(\sigma_t)n}(t, K_t^*)\right] \tag{5.24}$$

has to be upheld for $\sigma_t \in \{1, 2, \cdots, \eta_t\}$ and $t \in \{1, 2, \cdots, T\}$ and $K_t^* \in X_t^*$.

12.5.3.1 Payoff Distribution Procedure

We first formulate a Payoff Distribution Procedure (PDP) so that the agreed-upon imputation (5.24) can be realized. Let $B_t^{(\sigma_t)i}(K_t^*)$ denote the payment that agent i will received at stage t under the cooperative agreement if $\theta_t^{\sigma_t} \in \{\theta_t^1, \theta_t^2, \cdots, \theta_t^{\eta_t}\}$ occurs and $K_t^* \in X_t^*$ is realized at stage $t \in \{1, 2, \cdots, T\}$. The payment scheme $\{B_t^{(\sigma_t)i}(K_t^*)$ for $i \in N$ contingent upon the event $\theta_t^{\sigma_t}$ and state K_t^*, for $t \in \{1, 2, \cdots, T\}\}$ constitutes a PDP in the sense that the imputation to agent i over the stages 1 to T can be expressed as:

$$\xi^{(\sigma_1)i}(1, K^0) = B_1^{(\sigma_1)i}(K^0) + E_{\theta_2, \cdots, \theta_T; \vartheta_1, \vartheta_2, \cdots, \vartheta_T}\left(\sum_{\zeta=2}^{T} B_\zeta^{(\sigma_\zeta)i}\left(K_\zeta^*\right) + q^i(K_{T+1}^*)(1+r)^{-T} \right), \quad (5.25)$$

for $i \in N$.

Moreover, according to the agreed-upon optimality principle in (5.24), if $\theta_t^{\sigma_t}$ occurs and $K_t^* \in X_t^*$ is realized at stage t the imputation to agent i is $\xi^{(\sigma_t)i}(t, K_t^*)$. Therefore the payment scheme $B_t^{(\sigma_t)}(K_t^*)$ has to satisfy the conditions

$$\xi^{(\sigma_t)i}(t, K_t^*) = B_t^{(\sigma_t)i}(K_t^*) + E_{\theta_{t+1}, \theta_{t+2}, \cdots, \theta_T; \vartheta_t, \vartheta_{t+1}, \cdots, \vartheta_T}\left(\sum_{\zeta=t+1}^{T} B_\zeta^{(\sigma_\zeta)i}\left(K_\zeta^*\right) + q^i(K_{T+1}^*)(1+r)^{-T} \right) \quad (5.26)$$

for $i \in N$ and all $t \in \{1, 2, \cdots, T\}$.

For notational convenience the term $\xi^{(\sigma_{T+1})i}(T+1, K_{T+1}^*)$ is used to denote $q^i(K_{T+1}^*)(1+r)^{-T}$. Crucial to the formulation of a subgame consistent solution is the derivation of a payment scheme $\{B_t^{(\sigma_t)i}(K_t^*)$, for $i \in N$, $\sigma_t \in \{1, 2, \cdots, \eta_t\}$, $K_t^* \in X_t^*$ and $t \in \{1, 2, \cdots, T\}$ $\}$ so that the imputation in (5.26) can be realized.

A theorem for the derivation of a subgame consistent payment scheme can be established as follows.

Theorem 5.3 A payment equaling

$$B_t^{(\sigma_t)i}(K_t^*) = (1+r)^{(t-1)}\Bigg\{ \xi^{(\sigma_t)i}(t, K_t^*) - E_{\vartheta_t}\left[\sum_{\sigma_{t+1}=1}^{\eta_{t+1}} \lambda_{t+1}^{\sigma_{t+1}} \left(\xi^{(\sigma_{t+1})i}\left[t+1, K_t^* + \sum_{h=1}^{n} \psi_t^{(\sigma_t)h*}(K_t^*) - \delta K_t^* + \vartheta_t \right] \right) \right] \Bigg\}, \quad (5.27)$$

given to agent $i \in N$ at stage $t \in \{1, 2, \cdots, T\}$, if $\theta_t^{\sigma_t}$ occurs and $K_t^* \in X_t^*$, leads to the realization of the imputation in (5.26).

Proof To construct the proof of Theorem 5.3, we first express the term

$$E_{\theta_{t+1},\theta_{t+2},\cdots,\theta_T;\vartheta_t,\vartheta_{t+1},\cdots,\vartheta_T}\left(\sum_{\zeta=t+1}^{T} B_\zeta^{(\sigma_\zeta)i}\left(K_\zeta^*\right)(1+r)^{-(\zeta-1)}+q^i\left(K_{T+1}^*\right)(1+r)^{-T}\right)$$

$$= E_{\vartheta_{t+1}}\left\{\sum_{\sigma_{t+1}=1}^{\eta_{t+1}} \lambda_{t+1}^{\sigma_{t+1}}\left[B_{t+1}^{(\sigma_{t+1})i}\left(K_{t+1}^*\right)(1+r)^{-(t-1)}\right.\right.$$

$$+ E_{\theta_{t+2},\theta_{t+3},\cdots,\theta_T;\vartheta_{t+2},\vartheta_{t+3},\cdots,\vartheta_T}\left(\sum_{\zeta=t+2}^{T} B_\zeta^{(\sigma_\zeta)i}\left(K_\zeta^*\right)(1+r)^{-(\zeta-1)}\right.$$

$$\left.\left.\left.+ q^i\left(K_{T+1}^*\right)(1+r)^{-T}\right)\right]\right\} \tag{5.28}$$

Then, using (5.26) we can express the term $\xi^{(\sigma_{t+1})i}\left(t+1,K_{t+1}^*\right)$ as

$$\xi^{(\sigma_{t+1})i}\left(t+1,K_{t+1}^*\right) = B_{t+1}^{(\sigma_{t+1})i}\left(K_{t+1}^*\right)(1+r)^{-t}$$
$$+ E_{\theta_{t+2},\theta_{t+3},\cdots,\theta_T;\vartheta_{t+2},\vartheta_{t+3},\cdots,\vartheta_T}\left(\sum_{\zeta=t+2}^{T} B_\zeta^{(\sigma_\zeta)i}\left(K_\zeta^*\right) + q^i\left(K_{T+1}^*\right)(1+r)^{-T}\right). \tag{5.29}$$

The expression on the right-hand-side of equation (5.29) is the same as the expression inside the square brackets of (5.28). Invoking equation (5.29) we can replace the expression inside the square brackets of (5.28) by $\xi^{(\sigma_{t+1})i}\left(t+1,K_{t+1}^*\right)$ and obtain:

$$E_{\theta_{t+1},\theta_{t+2},\cdots,\theta_T;\vartheta_t,\vartheta_{t+1},\cdots,\vartheta_T}\left(\sum_{\zeta=t+1}^{T} B_\zeta^{(\sigma_\zeta)i}\left(K_\zeta^*\right)(1+r)^{-(\zeta-1)}+q^i\left(K_{T+1}^*\right)(1+r)^{-T}\right)$$
$$= E_{\vartheta_t}\left[\sum_{\sigma_{t+1}=1}^{\eta_{t+1}} \lambda_{t+1}^{\sigma_{t+1}}\left(\xi^{(\sigma_{t+1})i}\left(t+1,K_{t+1}^*\right)\right)\right].$$
$$= E_{\vartheta_t}\left[\sum_{\sigma_{t+1}=1}^{\eta_{t+1}} \lambda_{t+1}^{\sigma_{t+1}}\left(\xi^{(\sigma_{t+1})i}\left[t+1,K_t^*+\sum_{h=1}^{n}\psi_t^{(\sigma_t)h*}\left(K_t^*\right)-\delta K_t^*+\vartheta_t\right]\right)\right]$$

Substituting the term

$$E_{\theta_{t+1},\theta_{t+2},\cdots,\theta_T;\vartheta_t,\vartheta_{t+1},\cdots,\vartheta_T}\left(\sum_{\zeta=t+1}^{T} B_\zeta^{(\sigma_\zeta)i}\left(K_\zeta^*\right)(1+r)^{-(\zeta-1)}+q^i\left(K_{T+1}^*\right)(1+r)^{-T}\right)$$

by $E_{\vartheta_t}\left[\sum_{\sigma_{t+1}=1}^{\eta_{t+1}} \lambda_{t+1}^{\sigma_{t+1}}\left(\xi^{(\sigma_{t+1})i}\left[t+1,K_t^*+\sum_{h=1}^{n}\psi_t^{(\sigma_t)h*}\left(K_t^*\right)-\delta K_t^*+\vartheta_t\right]\right)\right]$ in (5.26) we can express (5.26) as:

$$\xi^{(\sigma_t)i}(t, K_t^*) = B_t^{(\sigma_t)i}(K_t^*)(1+r)^{-(t-1)}$$
$$+ E_{\vartheta_t}\left[\sum_{\sigma_{t+1}=1}^{\eta_{t+1}} \lambda_{t+1}^{\sigma_{t+1}} \left(\xi^{(\sigma_{t+1})i}\left[t+1, K_t^* + \sum_{h=1}^{n} \psi_t^{(\sigma_t)h*}(K_t^*) - \delta K_t^* + \vartheta_t\right] \right)\right]. \tag{5.30}$$

For condition (5.30), which is an alternative form of (5.26), to hold it is required that:

$$B_t^{(\sigma_t)i}(K_t^*) = (1+r)^{(t-1)}\Bigg\{ \xi^{(\sigma_t)i}(t, K_t^*)$$
$$- E_{\vartheta_t}\left[\sum_{\sigma_{t+1}=1}^{\eta_{t+1}} \lambda_{t+1}^{\sigma_{t+1}} \left(\xi^{(\sigma_{t+1})i}\left[t+1, K_t^* + \sum_{h=1}^{n} \psi_t^{(\sigma_t)h*}(K_t^*) - \delta K_t^* + \vartheta_t\right] \right)\right]\Bigg\}, \tag{5.31}$$

for $i \in N$ and $t \in \{1, 2, \cdots, T\}$.

Therefore by paying $B_t^{(\sigma_t)i}(K_t^*)$ to agent $i \in N$ at stage $t \in \{1, 2, \cdots, T\}$, if $\theta_t^{\sigma_t}$ occurs and $K_t^* \in X_t^*$ is realized, leads to the realization of the imputation in (5.26). Hence Theorem 5.3 follows. ■

For a given imputation vector

$$\xi^{(\sigma_t)}(t, K_t^*) = \left[\xi^{(\sigma_t)1}(t, K_t^*), \xi^{(\sigma_t)2}(t, K_t^*), \cdots, \xi^{(\sigma_t)n}(t, K_t^*)\right],$$

for $\sigma_t \in \{1, 2, \cdots, \eta_t\}$ and $t \in \{1, 2, \cdots, T\}$, Theorem 5.3 can be used to derive the PDP that leads to the realization this vector.

12.5.3.2 Transfer Payments

When all agents are using the cooperative strategies given that $K_t^* \in X_t^*$, and $\theta_t^{\sigma_t}$ occur, the payoff that agent i will directly receive at stage t becomes

$$\left[R^i(K_t^*, \theta_t^{\sigma_t}) - C^i\left(\psi_t^{(\sigma_t)i*}(K_t^*), \theta_t^{\sigma_t}\right)\right](1+r)^{-(t-1)} \tag{5.32}$$

However, according to the agreed upon imputation, agent i is supposed to receive $B_t^{(\sigma_t)i}(K_t^*)$ at stage t as given in Theorem 5.3. Therefore a transfer payment (which can be positive or negative)

$$\varpi_t^{(\sigma_t)i}(K_t^*) = B_t^{(\sigma_t)i}(K_t^*) - \left[R^i(K_t^*, \theta_t^{\sigma_t}) - C^i\left(\psi_t^{(\sigma_t)i*}(K_t^*), \theta_t^{\sigma_t}\right)\right](1+r)^{-(t-1)}, \tag{5.33}$$

for $t \in \{1, 2, \cdots, T\}$ and $i \in N$,

will be assigned to agent i to yield the cooperative imputation $\xi^{(\sigma_t)}(t, K_t^*)$.

12.6 An Illustration

In this section, we provide an illustration of the derivation of a subgame consistent solution of public goods provision under accumulation and payoff uncertainties in a multiple asymmetric agents situation. The basic game structure is a discrete-time analog of an example in Yeung and Petrosyan (2013b) but with the crucial addition of uncertain future payoff structures to reflect probable changes in preferences, technologies, demographic structures and institutional arrangements.

12.6.1 Public Capital Build-up Amid Uncertainties

We consider an n asymmetric agents economic region in which the agents receive benefits from an existing public capital stock K_t at each stage $t \in \{1, 2, \cdots, T\}$. The accumulation dynamics of the public capital stock is governed by the stochastic difference equation:

$$K_{t+1} = K_t + \sum_{j=1}^{n} I_t^j - \delta K_t + \vartheta_t, \; K_1 = {}^1K^0, \; \text{for}\, t \in \{1, 2, 3\}, \tag{6.1}$$

where ϑ_t is a discrete random variable with non-negative range $\{\vartheta_t^1, \vartheta_t^2, \vartheta_t^3\}$ and corresponding probabilities $\{\gamma_t^1, \gamma_t^2, \gamma_t^3\}$, and $\sum_{j=1}^{3} \gamma_t^j \vartheta_t^j = \varpi_t > 0$.

At stage 1, it is known that $\theta_1^{\sigma_1} = \theta_1^1$ has happened with probability $\lambda_1^1 = 1$, and the payoff of agent i is

$$\alpha_1^{(\sigma_1)i} K_1 - c_1^{(\sigma_1)i} (I_1^i)^2;$$

At stage $t\in\{2,3\}$, the payoff of agent i is

$$\alpha_t^{(\sigma_t)i}K_t - c_t^{(\sigma_t)i}\left(I_t^i\right)^2,$$

if $\theta_t^{\sigma_t}\in\{\theta_t^1,\theta_t^2,\theta_t^3,\theta_t^4\}$ occurs.

In particular, $\alpha_t^{(\sigma_t)i}K_t$ gives the gain that agent i derives from the public capital at stage $t\in\{1,2,3\}$, and $c_t^{(\sigma_t)i}\left(I_t^i\right)^2$ is the cost of investing I_t^i in the public capital.

The probability that $\theta_t^{\sigma_t}\in\{\theta_t^1,\theta_t^2,\theta_t^3,\theta_t^4\}$ will occur at stage $t\in\{2,3\}$ is $\lambda_t^{\sigma_t}\in\{\lambda_t^1,\lambda_t^2,\lambda_t^3,\lambda_t^4\}$. In stage 4, a terminal payment contingent upon the size of the capital stock equaling $(q^iK_4+m^i)(1+r)^{-3}$ will be paid to agent i. Since there is no uncertainty in stage 4, we use θ_4^1 to denote the condition in stage 4 with probability $\lambda_4^1=1$.

The objective of agent $i\in N$ is to maximize the expected payoff:

$$E_{\theta_1,\theta_2,\theta_3;\vartheta_1,\vartheta_2,\vartheta_3}\left\{\sum_{\tau=1}^{3}\left[\alpha_\tau^{(\sigma_\tau)i}K_\tau - c_\tau^{(\sigma_\tau)i}\left(I_\tau^i\right)^2\right](1+r)^{-(\tau-1)} \right.$$
$$\left. +(q^iK_4+m^i)(1+r)^{-3}\right\}, \tag{6.2}$$

subject to the public capital accumulation dynamics (6.1).

The noncooperative outcome will be examined in the next subsection.

12.6.2 Noncooperative Outcome

Invoking Theorem 5.1, one can characterize the noncooperative Nash equilibrium strategies for the game (6.1 and 6.2) as follows. In particular, a set of strategies $\{I_t^{(\sigma_t)i*}=\phi_t^{(\sigma_t)i*}(K)$, for $\sigma_1\in\{1\}$, $\sigma_2,\sigma_3\in\{1,2,3,4\}$, $t\in\{1,2,3\}$ and $i\in N\}$ provides a Nash equilibrium solution to the game (6.1 and 6.2) if there exist functions $V^{(\sigma_t)i}(t,K)$, for $i\in N$ and $t\in\{1,2,3\}$, such that the following recursive relations are satisfied:

$$V^{(\sigma_t)i}(t,K)=\max_{I_t^i}E_{\vartheta_t}\left\{\left[\alpha_t^{(\sigma_t)i}K-c_t^{(\sigma_t)i}\left(I_t^i\right)^2\right](1+r)^{-(t-1)}\right.$$
$$\left.+\sum_{\sigma_{t+1}=1}^{4}\lambda_{t+1}^{\sigma_{t+1}}V^{(\sigma_{t+1})i}\left[t+1,K+\sum_{\substack{j=1\\j\neq i}}^{n}\phi_t^{(\sigma_t)j*}(K)+I_t^i-\delta K+\vartheta_t\right]\right\}$$
$$=\max_{I_t^i}\left\{\left[\alpha_t^{(\sigma_t)i}K-c_t^{(\sigma_t)i}\left(I_t^i\right)^2\right](1+r)^{-(t-1)}\right.$$
$$\left.+\sum_{y=1}^{3}\gamma_t^y\sum_{\sigma_{t+1}=1}^{4}\lambda_{t+1}^{\sigma_{t+1}}V^{(\sigma_{t+1})i}\left[t+1,K+\sum_{\substack{j=1\\j\neq i}}^{n}\phi_t^{(\sigma_t)j*}(K)+I_t^i-\delta K+\vartheta_t^y\right]\right\},$$

for $t\in\{1,2,3\}$; (6.3)

$$V^{(\sigma_4)i}(4,K)=\left(q^iK+m^i\right)(1+r)^{-3}. \quad (6.4)$$

Performing the indicated maximization in (6.3) yields:

$$I_t^i=\phi_t^{(\sigma_t)i*}(K)$$
$$=\frac{(1+r)^{t-1}}{2c_t^{(\sigma_t)i}}\sum_{y=1}^{3}\gamma_t^y\sum_{\sigma_{t+1}=1}^{4}\lambda_{t+1}^{\sigma_{t+1}}V_{K_{t+1}}^{(\sigma_{t+1})i}\left[t+1,K+\sum_{j=1}^{n}\phi_t^{(\sigma_t)j*}(K)-\delta K+\vartheta_t^y\right], \quad (6.5)$$

for $i\in N$, $t\in\{1,2,3\}$, $\sigma_1=1$, and $\sigma_\tau\in\{1,2,3,4\}$ for $\tau\in\{2,3\}$.

The game equilibrium payoffs of the agents can be obtained as:

Proposition 6.1 The value function which represents the expected payoff of agent i is:

$$V^{(\sigma_t)i}(t,K)=\left[A_t^{(\sigma_t)i}K+C_t^{(\sigma_t)i}\right](1+r)^{-(t-1)}, \quad (6.6)$$

for $i\in N, t\in\{1,2,3\}, \sigma_1=1$, and $\sigma_\tau\in\{1,2,3,4\}$ for $\tau\in\{2,3\}$;

where

$A_3^{(\sigma_3)i}=\alpha_3^{(\sigma_3)i}+q^i(1-\delta)(1+r)^{-1}$, and

$$C_3^{(\sigma_3)i} = -\frac{(q^i)^2(1+r)^{-2}}{4c_3^{(\sigma_3)i}} + \left[q^i \sum_{j=1}^{n} \frac{q^j(1+r)^{-1}}{2c_3^{(\sigma_3)j}} + q^i \varpi_3 + m^i \right](1+r)^{-1};$$

$$A_2^{(\sigma_2)i} = \alpha_2^{(\sigma_2)i} + \sum_{\sigma_3=1}^{4} \lambda_3^{\sigma_3} A_3^{(\sigma_3)i}(1-\delta)(1+r)^{-1}, \text{ and}$$

$$C_2^{(\sigma_2)i} = -\frac{1}{4c_2^{(\sigma_2)i}} \left(\sum_{\sigma_3=1}^{4} \lambda_3^{\sigma_3} \left(A_3^{(\sigma_3)i}\right)(1+r)^{-1} \right)^2$$

$$+ \sum_{\sigma_3=1}^{4} \lambda_3^{\sigma_3} \left[A_3^{(\sigma_3)i} \left(\sum_{j=1}^{n} \sum_{\rho_3=1}^{4} \lambda_3^{\rho_3} \frac{A_3^{(\rho_3)j}(1+r)^{-1}}{2c_2^{(\sigma_2)j}} + \varpi_2 \right) + C_3^{(\sigma_3)i} \right](1+r)^{-1} \Bigg\};$$

$$A_1^{(\sigma_1)i} = \alpha_1^{(\sigma_1)i} + \sum_{\sigma_2=1}^{4} \lambda_2^{\sigma_2} A_2^{(\sigma_2)i}(1-\delta)(1+r)^{-1}, \text{ and}$$

$$C_1^{(\sigma_1)i} = -\frac{1}{4c_1^{(\sigma_1)i}} \left(\sum_{\sigma_2=1}^{4} \lambda_2^{\sigma_2} \left(A_2^{(\sigma_2)i}\right)(1+r)^{-1} \right)^2$$

$$+ \sum_{\sigma_2=1}^{4} \lambda_2^{\sigma_2} \left[A_2^{(\sigma_2)i} \left(\sum_{j=1}^{n} \sum_{\rho_2=1}^{4} \lambda_2^{\rho_2} \frac{A_2^{(\rho_2)j}(1+r)^{-1}}{2c_1^{(\sigma_1)j}} + \varpi_1 \right) + C_2^{(\sigma_2)i} \right](1+r)^{-1} \Bigg\};$$

for $i \in N$.

Proof See Appendix D. ■

Substituting the relevant derivatives of the value functions $V^{(\sigma_t)i}(t, K)$ in Proposition 6.1 into the game equilibrium strategies (6.5) yields a noncooperative Nash equilibrium solution of the game (6.1 and 6.2).

12.6.3 Cooperative Provision of Public Capital

Now we consider the case when the agents agree to cooperate and seek to enhance their gains. They agree to maximize their expected joint gain and distribute the cooperative gain proportional to their expected non-cooperative gains. The agents would first maximize their expected joint payoff

$$E_{\theta_1,\theta_2,\theta_3;\vartheta_1,\vartheta_2,\vartheta_3}\left\{\sum_{j=1}^{n}\sum_{\tau=1}^{3}\left[\alpha_\tau^{(\sigma_\tau)j}K_\tau - c_\tau^{(\sigma_\tau)j}\left(I_\tau^j\right)^2\right](1+r)^{-(\tau-1)}\right.$$
$$\left.+\sum_{j=1}^{n}(q^jK_4+m^j)(1+r)^{-3}\right\}, \tag{6.7}$$

subject to the stochastic dynamics (6.1).

Invoking Theorem 5.2, one can characterize the solution of the stochastic dynamic programming problem (6.1) and (6.7) as follows. In particular, a set of control strategies $\{u_t^{(\sigma_t)i*} = \psi_t^{(\sigma_t)i*}(K),$ for $t\in\{1,2,3\}$ and $i\in N, \sigma_1 = 1,\ \sigma_\tau\in\{1,2,3,4\}$ for $\tau\in\{2,3\}\}$, provides an optimal solution to the problem (6.1) and (6.7) if there exist functions $W^{(\sigma_t)}(t,K)$, for $t\in\{1,2,3\}$, such that the following recursive relations are satisfied:

$$W^{(\sigma_t)}(t,K) = \max_{I_t^1,I_t^2\cdots,I_t^n} E_{\vartheta_t}\left\{\sum_{j=1}^{n}\left[\alpha_t^{(\sigma_t)j}K - c_t^{(\sigma_t)j}\left(I_t^j\right)^2\right](1+r)^{-(t-1)}\right.$$
$$\left.+\sum_{\sigma_{t+1}=1}^{4}\lambda_{t+1}^{\sigma_{t+1}}W^{(\sigma_{t+1})}\left[t+1,K+\sum_{j=1}^{n}I_t^j-\delta K+\vartheta_t\right]\right\}$$
$$=\max_{I_t^i}\left\{\sum_{j=1}^{n}\left[\alpha_t^{(\sigma_t)j}K - c_t^{(\sigma_t)j}\left(I_t^j\right)^2\right](1+r)^{-(t-1)}\right.$$
$$\left.+\sum_{y=1}^{3}\gamma_t^y\sum_{\sigma_{t+1}=1}^{4}\lambda_{t+1}^{\sigma_{t+1}}W^{(\sigma_{t+1})i}\left[t+1,K+\sum_{j=1}^{n}I_t^j-\delta K+\vartheta_t^y\right]\right\},$$
for $t\in\{1,2,3\}$; (6.8)

$$W^{(\sigma_4)}(4,K) = \sum_{j=1}^{n}(q^jK+m^j)(1+r)^{-3} \tag{6.9}$$

Performing the indicated maximization in (6.8) yields:

$$I_t^i = \psi_t^{(\sigma_t)i*}(K)$$
$$=\frac{(1+r)^{t-1}}{2c_t^{(\sigma_t)i}}\sum_{y=1}^{3}\gamma_t^y\sum_{\sigma_{t+1}=1}^{4}\lambda_{t+1}^{\sigma_{t+1}}W_{K_{t+1}}^{(\sigma_{t+1})}\left[t+1,K+\sum_{j=1}^{n}\psi_t^{(\sigma_t)j*}(K)-\delta K+\vartheta_t^y\right], \tag{6.10}$$

for $i\in N$, $t\in\{1,2,3\}$, $\sigma_1 = 1$, and $\sigma_\tau\in\{1,2,3,4\}$ for $\tau\in\{2,3\}$.

The expected joint payoff under cooperation can be obtained as:

Proposition 6.2 The value function which represents the expected joint payoff is

$$W^{(\sigma_t)}(t,K) = \left[A_t^{(\sigma_t)}K + C_t^{(\sigma_t)}\right](1+r)^{-(t-1)}, \tag{6.11}$$

for $t\in\{1,2,3\}$, $\sigma_1 = 1$, and $\sigma_\tau\in\{1,2,3,4\}$ for $\tau\in\{2,3\}$;
where

$$A_3^{(\sigma_3)} = \sum_{j=1}^{n}\alpha_3^{(\sigma_3)j} + \sum_{j=1}^{n} q^j(1-\delta)(1+r)^{-1}, \text{ and}$$

$$C_3^{(\sigma_3)} = -\sum_{j=1}^{n}\frac{\left(\sum_{h=1}^{n} q^h(1+r)^{-1}\right)^2}{4c_3^{(\sigma_3)j}}$$
$$+\sum_{j=1}^{n}\left[q^j\left(\sum_{\ell=1}^{n}\frac{\sum_{h=1}^{n} q^h(1+r)^{-1}}{2c_3^{(\sigma_3)\ell}} + \varpi_3\right) + m^j\right](1+r)^{-1};$$

$$A_2^{(\sigma_2)} = \sum_{j=1}^{n}\alpha_2^{(\sigma_2)j} + \sum_{\sigma_3=1}^{4}\lambda_3^{\sigma_3}A_3^{(\sigma_3)}(1-\delta)(1+r)^{-1}, \text{ and}$$

$$C_2^{(\sigma_2)} = -\sum_{j=1}^{n}\frac{1}{4c_2^{(\sigma_2)j}}\left(\sum_{\sigma_3=1}^{4}\lambda_3^{\sigma_3}A_3^{(\sigma_3)}(1+r)^{-1}\right)^2$$
$$+\sum_{\sigma_3=1}^{4}\lambda_3^{\sigma_3}\left[A_3^{(\sigma_3)i}\left(\sum_{j=1}^{n}\sum_{\rho_3=1}^{4}\lambda_3^{\rho_3}\frac{A_3^{(\rho_3)j}(1+r)^{-1}}{2c_2^{(\sigma_2)j}} + \varpi_2\right) + C_3^{(\sigma_3)i}\right](1+r)^{-1}\Bigg\};$$

$$A_1^{(\sigma_1)} = \sum_{j=1}^{n}\alpha_1^{(\sigma_1)j} + \sum_{\sigma_2=1}^{4}\lambda_2^{\sigma_2}A_2^{(\sigma_2)}(1-\delta)(1+r)^{-1}, \text{ and}$$

$$C_1^{(\sigma_1)} = -\sum_{j=1}^{n}\frac{1}{4c_1^{(\sigma_1)j}}\left(\sum_{\sigma_2=1}^{4}\lambda_2^{\sigma_2}A_2^{(\sigma_2)}(1+r)^{-1}\right)^2$$
$$+\sum_{\sigma_2=1}^{4}\lambda_2^{\sigma_2}\left[A_2^{(\sigma_2)}\left(\sum_{j=1}^{n}\sum_{\rho_2=1}^{4}\lambda_2^{\rho_2}\frac{A_2^{(\rho_2)}(1+r)^{-1}}{2c_1^{(\sigma_1)j}} + \varpi_1\right) + C_2^{(\sigma_2)}\right](1+r)^{-1}\Bigg\}.$$

Proof Follow the proof of Proposition 6.1. ■

Using (6.10) and Proposition 6.2, the optimal cooperative strategies of the agents can be obtained as:

$$\psi_3^{(\sigma_3)i*}(K) = \frac{\sum_{h=1}^{n} q^h (1+r)^{-1}}{2c_3^{(\sigma_3)i}},$$

$$\psi_2^{(\sigma_2)i*}(K) = \sum_{\sigma_3=1}^{4} \lambda_3^{\sigma_3} \frac{A_3^{(\sigma_3)} (1+r)^{-1}}{2c_2^{(\sigma_2)i}},$$

$$\psi_1^{(\sigma_1)i*}(K) = \sum_{\sigma_2=1}^{4} \lambda_2^{\sigma_2} \frac{A_2^{(\sigma_2)} (1+r)^{-1}}{2c_1^{(\sigma_1)i}}, \text{ for } i \in N. \tag{6.12}$$

Substituting $\psi_t^{(\sigma_t)i*}(K)$ from (6.12) into (6.1) yields the optimal cooperative accumulation dynamics:

$$K_{t+1} = K_t + \sum_{j=1}^{n} \sum_{\sigma_{t+1}=1}^{4} \lambda_{t+1}^{\sigma_{t+1}} \frac{A_{t+1}^{(\sigma_{t+1})} (1+r)^{-1}}{2c_t^{(\sigma_t)j}} - \delta K_t + \vartheta_t,\ K_1 = {}^1K^0, \tag{6.13}$$

if $\theta_t^{\sigma_t}$ occurs at stage t, for $t \in \{1, 2, 3\}$.

12.6.4 Subgame Consistent Cooperative Solution

Given that the agents agree to share the cooperative gain proportional to their expected non-cooperative payoffs, an imputation

$$\xi^{(\sigma_t)i}(t, K_t^*) = \frac{V^{(\sigma_t)i}(t, K_t^*)}{\sum_{j=1}^{n} V^{(\sigma_t)j}(t, K_t^*)} W^{(\sigma_t)}(t, K_t^*)$$

$$= \frac{\left[A_t^{(\sigma_t)i} K_t^* + C_t^{(\sigma_t)i}\right]}{\sum_{j=1}^{n} \left[A_t^{(\sigma_t)i} K_t^* + C_t^{(\sigma_t)i}\right]} \left[A_t^{(\sigma_t)} K_t^* + C_t^{(\sigma_t)}\right] (1+r)^{-(t-1)}, \text{ for } i \in N, \tag{6.14}$$

if $\theta_t^{\sigma_t}$ occurs at stage t for $t \in \{1, 2, 3\}$ has to be maintained.

Invoking Theorem 5.3, if $\theta_t^{\sigma_t}$ occurs and $K_t^* \in X_t^*$ is realized at stage t a payment equaling

$$B_t^{(\sigma_t)i}(K_t^*) = (1+r)^{(t-1)}\Bigg\{ \xi^{(\sigma_t)i}(t, K_t^*)$$

$$- \left[\sum_{y=1}^{3} \gamma_t^y \sum_{\sigma_{t+1}=1}^{\eta_{t+1}} \lambda_{t+1}^{\sigma_{t+1}} \left(\xi^{(\sigma_{t+1})i} \left[t+1, K_t^* + \sum_{h=1}^{n} \psi_t^{(\sigma_t)h*}(K_t^*) - \delta K_t^* + \vartheta_t^y \right) \right] \right\}$$

$$= \frac{\left[A_t^{(\sigma_t)i} K_t^* + C_t^{(\sigma_t)i}\right]}{\sum_{j=1}^{n} \left[A_t^{(\sigma_t)i} K_t^* + C_t^{(\sigma_t)i}\right]} \left[A_t^{(\sigma_t)} K_t^* + C_t^{(\sigma_t)}\right]$$

$$- \sum_{y=1}^{3} \gamma_t^y \sum_{\sigma_{t+1}=1}^{\eta_{t+1}} \lambda_{t+1}^{\sigma_{t+1}} \frac{\left[A_{t+1}^{(\sigma_{t+1})i} K_{t+1}(\sigma_{t+1}, \vartheta_t^y) + C_{t+1}^{(\sigma_{t+1})i}\right]}{\sum_{j=1}^{n} \left[A_{t+1}^{(\sigma_{t+1})i} K_{t+1}(\sigma_{t+1}, \vartheta_t^y) + C_{t+1}^{(\sigma_{t+1})i}\right]} \left[A_{t+1}^{(\sigma_{t+1})} K_{t+1}(\sigma_{t+1}, \vartheta_t^y)\right.$$

$$\left. + C_{t+1}^{(\sigma_t)}\right] (1+r)^{-1}, \tag{6.15}$$

where $K_{t+1}(\sigma_{t+1}, \vartheta_t^y) = K_t^* + \sum_{j=1}^{n} \sum_{\sigma_{t+1}=1}^{4} \lambda_{t+1}^{\sigma_{t+1}} \frac{A_{t+1}^{(\sigma_{t+1})}(1+r)^{-1}}{2c_t^{(\sigma_t)j}} - \delta K_t^* + \vartheta_t^y$,

given to agent i at stage $t \in \{1, 2, 3\}$ if $\theta_t^{\sigma_t}$ occurs would lead to the realization of the imputation (6.14).

A subgame consistent solution and the corresponding payment schemes can be obtained using Propositions 5.1 and 5.2 and conditions (6.12, 6.13, 6.14 and 6.15).

Finally, since all agents are adopting the cooperative strategies, the payoff that agent i will directly receive at stage t is

$$\alpha_t^{(\sigma_t)i} K_t^* - \frac{1}{4c_t^{(\sigma_t)i}} \left(\sum_{\sigma_{t+1}=1}^{4} \lambda_{t+1}^{\sigma_{t+1}} A_{t+1}^{(\sigma_{t+1})} (1+r)^{-1} \right)^2, \tag{6.16}$$

if $\theta_t^{\sigma_t}$ occurs at stage t.

However, according to the agreed upon imputation, agent i is supposed to receive $\xi^{(\sigma_t)i}(t, K_t^*)$ in (6.15), therefore a transfer payment (which can be positive or negative) equalling

$$\pi^{(\sigma_t)i}(t, K_t^*) = \xi^{(\sigma_t)i}(t, K_t^*) - \alpha_t^{(\sigma_t)i} K_t^*$$

$$+ \frac{1}{4c_t^{(\sigma_t)i}} \left(\sum_{\sigma_{t+1}=1}^{4} \lambda_{t+1}^{\sigma_{t+1}} A_{t+1}^{(\sigma_{t+1})} (1+r)^{-1} \right)^2 \tag{6.17}$$

will be given to agent $i \in N$ at stage t.

12.7 Appendices

Appendix A. Proof of Theorem 1.1 Invoking (1.11), one can obtain

$$\begin{aligned}\xi^i(\tau, K_\tau^*) &= E\left\{ \int_\tau^T B_i(s, K^*(s))e^{-rs}ds + q_i[K^*(T)]e^{-rT} \middle| K^*(\tau) = K_\tau^* \right\}, \\ &= E\left\{ \int_\tau^{\tau+\Delta t} B_i(s, K^*(s))e^{-rs}ds \right. \\ &\quad \left. + \xi^{(\tau+\Delta t)i}(\tau + \Delta t, K_\tau^* + \Delta K_\tau^*) \middle| K^*(\tau) = K_\tau^* \right\},\end{aligned} \tag{7.1}$$

$i \in N$ and $\tau \in [0, T]$,
where

$$\Delta K_\tau^* = \left[\sum_{j=1}^n \psi_j^*(\tau, K_\tau^*) - \delta K_\tau^* \right] \Delta t + \sigma K_\tau^* \Delta z_\tau + o(\Delta t), \text{ and}$$

$\Delta z_\tau = Z(\tau + \Delta t) - z(\tau)$, and $E_\tau[o(\Delta t)]/\Delta t \to 0$ as $\Delta t \to 0$.
Using (7.1), one obtains

$$\begin{aligned}&E\left\{ \int_\tau^{\tau+\Delta t} B_i(s, K^*(s))e^{-rs}ds \middle| K^*(\tau) = K_\tau^* \right\} \\ &= E\left\{ \xi^i(\tau, K_\tau^*) - \xi^{(\tau+\Delta t)i}(\tau + \Delta t, K_\tau^* + \Delta K_\tau^*) \middle| K^*(\tau) = K_\tau^* \right\}, \\ &\text{for all } \tau \in [0, T] \text{ and } i \in N.\end{aligned} \tag{7.2}$$

If the imputations $\xi^i(\tau, K_\tau^*)$ are continuous and differentiable, as $\Delta t \to 0$, one can express condition (7.2) as:

$$\begin{aligned}&E\left\{ B_i(s, K_s^*)e^{-rt}\Delta t + o(\Delta t) \right\} = E\left\{ -\xi_\tau^i(\tau, K_\tau^*)\Delta t \right. \\ &- \xi_{K_\tau}^i(\tau, K_\tau^*) \left[\sum_{j=1}^n \psi_j^*(\tau, K_\tau^*) - \delta K_\tau^* \right] \Delta t \\ &\left. - \frac{1}{2}\xi_{K_\tau}^i(\tau, K_\tau^*)\sigma K_\tau^* \Delta z_\tau - \frac{1}{2}\xi_{K_\tau K_\tau}^i(\tau, K_\tau^*)\sigma^2 (K_\tau^*)^2 \Delta t o(\Delta t) \right\} \\ &\text{for } i \in N.\end{aligned} \tag{7.3}$$

Dividing (7.3) throughout by Δt, with $\Delta t \to 0$, and taking expectation yield (1.12). Thus the payoff distribution procedure in $B_i^i(s, K_s^*)$ in (1.12) would lead to the realization of $\xi(s, K_s^*)$ in (1.10). ■

Appendix B. Proof of Proposition 2.1 Using the value functions in Proposition 2.1 and the optimal strategies in (2.5) the Hamilton-Jacobi-Bellman equations (2.4) reduces to:

$$r[A_i(t)K + C_i(t)] - \left[\dot{A}_i(t)K + \dot{C}_i(t)\right] = \alpha_i K - \frac{[A_i(t)]^2}{4c_i} + A_i(t)\left[\sum_{j=1}^{n}\frac{A_j(t)}{2c_j} - \delta K\right],$$

$$[A_i(T)K + C_i(T)] = q_1^i K + q_2^i, \text{ for } i \in N; \tag{7.4}$$

For (7.4) to hold it is required that

$$\dot{A}_i(t) = (r+\delta)A_i(t) - \alpha_i, \; A_i(T) = q_1^i; \text{ and} \tag{7.5}$$

$$\dot{C}_i(t) = rC_i(t) + \frac{[A_i(t)]^2}{4c_i} - \left[\sum_{j=1}^{n}\frac{A_i(t)A_j(t)}{2c_j}\right], C_i(T) = q_2^i; \quad \text{for } i \in N. \tag{7.6}$$

The differential equation system (7.5 and 7.6) is a block-recursive system with $A_i(t)$ in (7.5) being independent of $A_j(t)$ for $j \neq i$ and all $C_j(t)$ for $j \in N$.

Solving each of the n independent constant-coefficient linear differential equation in (7.5) yields:

$$A_i(t) = \left(q_1^i - \frac{\alpha_i}{r+\delta}\right)e^{-(r+\delta)(T-t)} + \frac{\alpha_i}{r+\delta}, \text{ for } i \in N. \tag{7.7}$$

Substituting the explicit solution of $A_i(t)$ from (7.7) into (7.6) yields:

$$\dot{C}_i(t) = rC_i(t) + \frac{1}{4c_i}\left[\left(q_1^i - \frac{\alpha_i}{r+\delta}\right)e^{-(r+\delta)(T-t)} + \frac{\alpha_i}{r+\delta}\right]^2$$

$$-\sum_{j=1}^{n}\frac{1}{2c_j}\left[\left(q_1^i - \frac{\alpha_i}{r+\delta}\right)e^{-(r+\delta)(T-t)} + \frac{\alpha_i}{r+\delta}\right]$$

$$\left[\left(q_1^j - \frac{\alpha_j}{r+\delta}\right)e^{-(r+\delta)(T-t)} + \frac{\alpha_j}{r+\delta}\right],$$

$$C_i(T) = q_2^i, \text{ for } i \in N, \tag{7.8}$$

which is a system of independent linear differential equations in $C_i(t)$. Note that the coefficients are integrable functions; hence the solution of $C_i(t)$ could be readily obtained. Q.E.D.

Appendix C. Proof of Proposition 3.1 Invoking the fact that firms of the same type are identical, we have $\phi_i^{(1)}(t,K) = \phi_h^{(1)}(t,K)$ and $V^{(1)i}(t,K) = V^{(1)h}(t,K)$ for $i,h \in N_1$; and similarly $\phi_j^{(2)}(t,K) = \phi_\ell^{(2)}(t,K)$ and $V^{(2)j}(t,K) = V^{(2)\ell}(t,K)$ for $j,\ell \in N_2$. Using the value functions in Proposition 3.1 and the optimal strategies in (3.6 and 3.7), one can express Hamilton-Jacobi-Bellman equations (3.4 and 3.5) as:

$$
\begin{aligned}
& r\left[A_1(t)K^2 + B_1(t)K + C_1(t)\right] - \left[\dot{A}_1(t)K^2 + \dot{B}_1(t)K + \dot{C}_1(t)\right] \\
& \quad - A_1(t)\sigma^2K^2 \\
& = \Big\{ \alpha_1K - b_1K^2 - \rho_1[2A_1(t)K + B_1(t) - \rho_1] \\
& \qquad - (c_1/2)[2A_1(t)K + B_1(t) - \rho_1]^2 \Big\} \\
& \qquad + [2A_1(t)K + B_1(t)]\Big[n_1[2A_1(t)K + B_1(t) - \rho_1] \\
& \qquad + n_2[2A_2(t)K + B_2(t) - \rho_2] - \delta K \Big], \left[A_1(T)K^2 + B_1(T)K + C_1(T)\right] \\
& = \left[q_1K^2 + q_2K + q_3\right]; r\left[A_2(t)K^2 + B_2(t)K + C_2(t)\right] \\
& \quad - \left[\dot{A}_2(t)K^2 + \dot{B}_2(t)K + \dot{C}_2(t)\right] - A_2(t)\sigma^2K^2 \\
& = \Big\{ \alpha_2K - b_2K^2 - \rho_2[2A_2(t)K + B_2(t) - \rho_2] \\
& \qquad - (c_2/2)[2A_2(t)K + B_2(t) - \rho_2]^2 \Big\} \\
& \qquad + [2A_2(t)K + B_2(t)]\Big[n_1[2A_1(t)K + B_1(t) - \rho_1] \\
& \qquad + n_2[2A_2(t)K + B_2(t) - \rho_2] - \delta K \Big], \left[A_2(T)K^2 + B_2(T)K + C_2(T)\right] \\
& = \left[q_1K^2 + q_2K + q_3\right].
\end{aligned}
\tag{7.9}
$$

For system (7.9) to hold it is required that

(i) the coefficients multiplying with K^2 and K have to agree with system, and

(ii) the equalities of the other terms as indicated by the system.

These required conditions are given in (3.9, 3.10 and 3.11).
Hence Proposition 3.1 follows. Q.E.D.

Appendix D. Proof of Proposition 6.1 Consider first the last stage, that is stage 3, when $\theta_3^{\sigma_3}$ occurs. Invoking that $V^{(\sigma_3)i}(3, K) = \left[A_3^{(\sigma_3)i}K + C_3^{(\sigma_3)i}\right](1 + r)^{-2}$ and $V^{(\sigma_4)i}(4, K_4) = (q^iK + m^i)(1 + r)^{-3}$ from Proposition 6.1, the condition governing $t = 3$ in equation (6.3) becomes

$$\left[A_3^{(\sigma_3)i}K + C_3^{(\sigma_3)i}\right](1+r)^{-2} = \max_{I_3^i}\left\{ \left[\alpha_3^{(\sigma_3)i}K - c_3^{(\sigma_3)i}\left(I_3^i\right)^2\right](1+r)^{-2}\right.$$

$$+\sum_{y=1}^{3}\gamma_3^y\sum_{\sigma_4=1}^{1}\lambda_4^{\sigma_4}$$

$$\left.\cdot\left[q^i\left(K + \sum_{\substack{j=1\\ j\neq i}}^{n}\phi_3^{(\sigma_3)j*}(K) + I_3^i - \delta K + \vartheta_3^y\right) + m^i\right](1+r)^{-3}\right\}, \text{for } i\in N. \tag{7.10}$$

Performing the indicated maximization in (7.10) yields the game equilibrium strategies in stage 3 as:

$$\phi_3^{(\sigma_3)i*}(K) = \frac{q^i(1+r)^{-1}}{2c_3^{(\sigma_3)i}}, \quad \text{for } i\in N. \tag{7.11}$$

Substituting (7.11) into (7.10) yields:

$$\left[A_3^{(\sigma_3)i}K + C_3^{(\sigma_3)i}\right] = \alpha_3^{(\sigma_3)i}K - \frac{(q^i)^2(1+r)^{-2}}{4c_3^{(\sigma_3)i}}$$

$$+\sum_{y=1}^{3}\gamma_3^y\left[q^i\left(K + \sum_{j=1}^{n}\frac{q^j(1+r)^{-1}}{2c_3^{(\sigma_3)j}} - \delta K + \vartheta_t^y\right) + m^i\right](1+r)^{-1}\Bigg\}, \tag{7.12}$$

for $i\in N$.

Note that both sides of equation (7.12) are linear expressions of K. For (7.12) to hold it is required that:

$$A_3^{(\sigma_3)i} = \alpha_3^{(\sigma_3)i} + q^i(1-\delta)(1+r)^{-1}, \text{ and}$$

$$C_3^{(\sigma_3)i} = -\frac{(q^i)^2(1+r)^{-2}}{4c_3^{(\sigma_3)i}} + \left[q^i\sum_{j=1}^{n}\frac{q^j(1+r)^{-1}}{2c_3^{(\sigma_3)j}} + q^i\varpi_3 + m^i\right](1+r)^{-1}, \tag{7.13}$$

for $i\in N$.

Now we proceed to stage 2, using $V^{(\sigma_3)i}(3,K) = \left[A_3^{(\sigma_3)i}K + C_3^{(\sigma_3)i}\right](1+r)^{-2}$ with $A_3^{(\sigma_3)i}$ and $C_3^{(\sigma_3)i}$ given in (7.13), the conditions in equation (6.3) become

$$\left[A_2^{(\sigma_2)i}K + C_2^{(\sigma_2)i}\right](1+r)^{-1} = \max_{I_2^i}\left\{\left[\alpha_2^{(\sigma_2)i}K - c_2^{(\sigma_2)i}\left(I_2^i\right)^2\right](1+r)^{-1}\right.$$

$$+\sum_{y=1}^{3}\gamma_2^y\sum_{\sigma_3=1}^{4}\lambda_3^{\sigma_3}$$

$$\left.\left[A_3^{(\sigma_3)i}\left(K + \sum_{\substack{j=1\\ j\neq i}}^{n}\phi_2^{(\sigma_2)j*}(K) + I_2^i - \delta K + \vartheta_2^y\right) + C_3^{(\sigma_3)i}\right](1+r)^{-2}\right\},$$

for $i \in N$. (7.14)

Performing the indicated maximization in (7.14) yields the game equilibrium strategies in stage 2 as:

$$\phi_2^{(\sigma_2)i*}(K) = \sum_{\sigma_3=1}^{4}\lambda_3^{\sigma_3}\frac{A_3^{(\sigma_3)i}(1+r)^{-1}}{2c_2^{(\sigma_2)i}},\ \text{for } i \in N. \tag{7.15}$$

Substituting (7.15) into (7.14) yields:

$$\left[A_2^{(\sigma_2)i}K + C_2^{(\sigma_2)i}\right] = \alpha_2^{(\sigma_2)i}K - \frac{1}{4c_2^{(\sigma_2)i}}\left(\sum_{\sigma_3=1}^{4}\lambda_3^{\sigma_3}A_3^{(\sigma_3)i}(1+r)^{-1}\right)^2$$

$$+\sum_{y=1}^{3}\gamma_2^y\sum_{\sigma_3=1}^{4}\lambda_3^{\sigma_3}\left[A_3^{(\sigma_3)i}\left(K + \sum_{j=1}^{n}\sum_{\rho_3=1}^{4}\lambda_3^{\rho_3}\frac{A_3^{(\rho_3)j}(1+r)^{-1}}{2c_2^{(\sigma_2)j}}\right.\right.$$

$$\left.\left.-\delta K + \vartheta_2^y\right) + C_3^{(\sigma_3)i}\right](1+r)^{-1}\Bigg\},\ \text{for } i \in N. \tag{7.16}$$

Both sides of equation (7.16) are linear expressions of K. For (7.16) to hold it is required that:

$$A_2^{(\sigma_2)i} = \alpha_2^{(\sigma_2)i} + \sum_{\sigma_3=1}^{4}\lambda_3^{\sigma_3}A_3^{(\sigma_3)i}(1-\delta)(1+r)^{-1},\ \text{and}$$

$$C_2^{(\sigma_2)i} = -\frac{1}{4c_2^{(\sigma_2)i}}\left(\sum_{\sigma_3=1}^{4}\lambda_3^{\sigma_3}A_3^{(\sigma_3)i}(1+r)^{-1}\right)^2$$

$$+\sum_{\sigma_3=1}^{4}\lambda_3^{\sigma_3}\left[A_3^{(\sigma_3)i}\left(\sum_{j=1}^{n}\sum_{\rho_3=1}^{4}\lambda_3^{\rho_3}\frac{A_3^{(\rho_3)j}(1+r)^{-1}}{2c_2^{(\sigma_2)j}} + \varpi_2\right) + C_3^{(\sigma_3)i}\right](1+r)^{-1}\Bigg\},$$

for $i \in N$. (7.17)

Now we proceed to stage 1, using $V^{(\sigma_2)i}(2,K) = \left[A_2^{(\sigma_2)i}K + C_2^{(\sigma_2)i}\right](1+r)^{-1}$ with $A_2^{(\sigma_2)i}$ and $C_2^{(\sigma_2)i}$ given in (7.17), the conditions in equation (6.3) become

$$\left[A_1^{(\sigma_1)i}K + C_1^{(\sigma_1)i}\right] = \max_{I_1^i}\left\{ \left[\alpha_1^{(\sigma_1)i}K - c_1^{(\sigma_1)i}\left(I_1^i\right)^2\right]\right.$$
$$+\sum_{y=1}^{3}\gamma_1^y\sum_{\sigma_2=1}^{4}\lambda_2^{\sigma_2}$$
$$\left.\left[A_2^{(\sigma_2)i}\left(K + \sum_{\substack{j=1\\ j\neq i}}^{n}\phi_1^{(\sigma_1)j*}(K) + I_1^i - \delta K + \vartheta_1^y\right) + C_2^{(\sigma_2)i}\right](1+r)^{-1}\right\},$$
$$\text{for } i\in N. \tag{7.18}$$

Performing the indicated maximization in (7.18) yields the game equilibrium strategies in stage 1 as:

$$\phi_1^{(\sigma_1)i*}(K) = \sum_{\sigma_2=1}^{4}\lambda_2^{\sigma_2}\frac{A_2^{(\sigma_2)i}(1+r)^{-1}}{2c_1^{(\sigma_1)i}}, \text{ for } i\in N \tag{7.19}$$

Substituting (7.19) into (7.18) yields:

$$\left[A_1^{(\sigma_1)i}K + C_1^{(\sigma_1)i}\right] = \alpha_1^{(\sigma_1)i}K - \frac{1}{4c_1^{(\sigma_1)i}}\left(\sum_{\sigma_2=1}^{4}\lambda_2^{\sigma_2}A_2^{(\sigma_2)i}(1+r)^{-1}\right)^2$$
$$+\sum_{y=1}^{3}\gamma_1^y\sum_{\sigma_2=1}^{4}\lambda_2^{\sigma_2}\left[A_2^{(\sigma_2)i}\left(K + \sum_{j=1}^{n}\sum_{\rho_2=1}^{4}\lambda_2^{\rho_2}\frac{A_2^{(\rho_2)j}(1+r)^{-1}}{2c_1^{(\sigma_1)j}}\right.\right.$$
$$\left.\left.-\delta K + \vartheta_1^y\right) + C_2^{(\sigma_2)i}\right](1+r)^{-1}\Bigg\}, \text{ for } i\in N. \tag{7.20}$$

Both sides of equation (7.20) are linear expressions of K. For (7.20) to hold it is required that:

$$A_1^{(\sigma_1)i} = \alpha_1^{(\sigma_1)i} + \sum_{\sigma_2=1}^{4} \lambda_2^{\sigma_2} A_2^{(\sigma_2)i}(1-\delta)(1+r)^{-1}, \text{ and}$$

$$C_1^{(\sigma_1)i} = -\frac{1}{4c_1^{(\sigma_1)i}}\left(\sum_{\sigma_1=1}^{4} \lambda_2^{\sigma_2} A_2^{(\sigma_2)i}(1+r)^{-1}\right)^2$$

$$+\sum_{\sigma_2=1}^{4} \lambda_2^{\sigma_2}\left[A_2^{(\sigma_2)i}\left(\sum_{j=1}^{n}\sum_{\rho_2=1}^{4} \lambda_2^{\rho_2} \frac{A_2^{(\rho_2)j}(1+r)^{-1}}{2c_1^{(\sigma_1)j}} + \varpi_1\right) + C_2^{(\sigma_2)i}\right](1+r)^{-1}\Bigg\},$$

for $i \in N$. (7.21)

Hence Proposition 6.1 follows. *Q.E.D.*

12.8 Chapter Notes

Though cooperative provision of public goods is the key to a socially optimal solution one may find it hard to be convinced that dynamic cooperation can offer a long-term solution unless the agreed-upon optimality principle can be maintained from the beginning to the end. The notion of public goods, which are non-rival and non-excludable, was first introduced by Samuelson (1954). Problems concerning private provision of public goods are studied in Bergstrom et al. (1986). Static analysis on provision of public goods are found in Chamberlin (1974), McGuire (1974) and Gradstein and Nitzan (1989). In many contexts, the provision and use of public goods are carried out in an intertemporal framework. Fershtman and Nitzan (1991) and Wirl (1996) considered differential games of public goods provision with symmetric agents. Wang and Ewald (2010) introduced stochastic elements into these games. Dockner et al. (2000) presented a game model with two asymmetric agents in which knowledge is a public good. These studies on dynamic game analysis focus on the noncooperative equilibria and the collusive solution that maximizes the joint payoffs of all agents.

This Chapter provides applications of cooperative provision of public goods with a subgame consistent cooperative scheme. The analysis can be readily extended into a multiple public capital goods paradigm. In addition, more complicated stochastic disturbances in the public goods dynamics, like $\sigma[I_1(s), I_2(s), \cdots, I_n(s), K(s)]$, can be adopted.

12.9 Problems

1. Consider a 4-stage 3 asymmetric agents economic game in which the agents receive benefits from an existing public capital stock K_t. The accumulation

dynamics of the public capital stock is governed by the stochastic difference equation:

$$K_{t+1} = K_t + \sum_{j=1}^{5} I_t^j - 0.1K_t + \vartheta_t,\ K_1 {=}^1 20,\ \text{for}\, t \in \{1,2,3,4\},$$

where ϑ_t is a discrete random variable with range $\{1,2,3\}$ and corresponding probabilities $\{0.7, 0.2, 0.1\}$.

At stage 1, it is known that θ_1^1 has happened, and the payoffs of agents 1, 2 and 3 are respectively:

$$5K_1 - 2(I_1)^2,\ 3K_1 - (I_1)^2 \text{ and } 6K_1 - 3(I_1)^2.$$

At stage $t \in \{2,3,4\}$, the payoffs of agent 1, 2 and 3 are respectively

$$5K_1 - 2(I_1)^2,\ 3K_1 - (I_1)^2 \,\text{and}\, 6K_1 - 3(I_1)^2$$

if θ_t^1 occurs; and the payoffs of agent 1, 2 and 3 are respectively

$$6K_1 - 2(I_1)^2,\ 3K_1 - 2(I_1)^2,\ 4K_1 - 2(I_1)^2$$

if θ_t^2 occurs.

The probability that θ_t^1 would occur is 0.6 and the probability that θ_t^2 would occur is 0.4.

In stage 5, the terminal valuations of the agent 1, 2 and 3 are respectively:

$$(2K_5 + 10)(1+r)^{-4},\ (K_5 + 15)(1+r)^{-4} \,\text{and}\, (3K_5 + 5)(1+r)^{-4}.$$

Characterize the feedback Nash equilibrium.

2. Obtain a group optimal solution that maximizes the joint expected profit.
3. Consider the case when the agents agree to share the cooperative gain proportional to their expected non-cooperative payoffs in providing the public good jointly. Derive a subgame consistent solution.

Chapter 13
Collaborative Environmental Management

After several decades of rapid technological advancement and economic growth, alarming levels of pollutions and environmental degradation are emerging all over the world. Due to the geographical diffusion of pollutants, unilateral response on the part of one country or region is often ineffective. Though cooperation in environmental control holds out the best promise of effective action, limited success has been observed. Existing multinational joint initiatives like the Kyoto Protocol or pollution permit trading can hardly be expected to offer a long-term solution because there is no guarantee that participants will always be better off within the entire duration of the agreement. This Chapter presents collaborative schemes in a cooperative differential game framework and derives subgame consistent solutions for the schemes.

Sections 13.1, 13.2, 13.3, and 13.4 of this Chapter give an integrated exposition of the work of Yeung and Petrosyan (2008) on a cooperative stochastic differential game of transboundary industrial pollution. The game formulation is provided in Sect. 13.1 and noncooperative outcomes are characterized in Sect. 13.2. Cooperative arrangements, subgame-consistent imputations and payment distribution mechanism are provided in Sect. 13.3. A numerical example is given in Sect. 13.4. In Sect. 13.5, an extension of the Yeung and Petrosyan (2008) analysis to incorporate uncertainties in future payoffs is presented. Section 13.6 contains the chapter appendices. Chapter notes are given in Sect. 13.7 and problems in Sect. 13.8.

13.1 Game Formulation

In this section we present a stochastic differential game model of environmental with n asymmetric nations or regions.

D.W.K. Yeung, L.A. Petrosyan, *Subgame Consistent Cooperation*,
Theory and Decision Library C 47, DOI 10.1007/978-981-10-1545-8_13

13.1.1 The Industrial Sector

Consider a multinational economy which is comprised of n nations. To allow different degrees of substitutability among the nations' outputs a differentiated products oligopoly model has to be adopted. The differentiated oligopoly model used by Dixit (1979) and Singh and Vives (1984) in industrial organizations is adopted to characterize the interactions in this international market. In particular, the nations' outputs may range from a homogeneous product to n unrelated products. Specifically, the inverse demand function of the output of nation $i \in N \equiv \{1, 2, \cdots, n\}$ at time instant s is

$$P_i(s) = \alpha^i - \sum_{j=1}^{n} \beta_j^i q_j(s), \tag{1.1}$$

where $P_i(s)$ is the price of the output of nation i, $q_j(s)$, is the output of nation j, α^i and β_j^i for $i \in N$ and $j \in N$ are positive constants. The output choice $q_j(s) \in [0, \overline{q}_j]$ is nonnegative and bounded by a maximum output constraint $\overline{q}_j$. Output price equals zero if the right-hand-side of (1.1) becomes negative. The demand system (1.1) shows that the economy is a form of differentiated products oligopoly with substitute goods. In the case when $\alpha^i = \alpha^j$ and $\beta_j^i = \beta_i^j$ for all $i \in N$ and $j \in N$, the industrial outputs resemble a homogeneous good. In the case when $\beta_j^i = 0$ for $i \neq j$, the n nations produce n unrelated products. Moreover, the industry equilibrium generated by this oligopoly model is computable and fully tractable.

Industrial profits of nation i at time s can be expressed as:

$$\pi_i(s) = \left[\alpha^i - \sum_{j=1}^{n} \beta_j^i q_j(s)\right] q_i(s) - c_i q_i(s) - v_i(s) q_i(s), \text{ for } i \in N. \tag{1.2}$$

where $v_i(s) \geq 0$ is the tax rate imposed by government i on its industrial output at time s and c_i is the unit cost of production. At each time instant s, the industrial sector of nation $i \in N$ seeks to maximize (1.2). Note that each industrial sector would consider the information on the demand structure, each other's cost structures and tax policies. In a competitive market equilibrium firms will produce up to a point where marginal cost of production equals marginal revenue and the first order condition for a Nash equilibrium for the n nations economy yields

$$\sum_{j=1}^{n} \beta_j^i q_j(s) + \beta_i^i q_i(s) = \alpha^i - c_i - v_i(s), \text{ for } i \in N. \tag{1.3}$$

With output tax rates $v(s) = \{v_1(s), v_2(s), \cdots, v_n(s)\}$ being regarded as parameters by the industrial sectors (1.3) becomes a system of equations linear in

$q(s) = \{q_1(s), q_2(s), \cdots, q_n(s)\}$. Solving (1.3) yields an industry equilibrium with output in industry i being

$$q_i(s) = \phi_i(v(s)) = \overline{\alpha}^i + \sum_{j \in N} \overline{\beta}_j^i \; v_j(s), \tag{1.4}$$

where $\overline{\alpha}^i$ and $\overline{\beta}_j^i$, for $i \in N$ and $j \in N$, are constants involving the model parameters $\{\beta_1^1, \beta_2^1, \cdots, \beta_n^1; \beta_1^2, \beta_2^2, \cdots, \beta_n^2; \cdots; \beta_1^n, \beta_2^n, \cdots, \beta_n^n\}$, $\{\alpha^1, \alpha^2, \cdots, \alpha^n\}$ and $\{c_1, c_2, \cdots, c_n\}$.

One can readily observe from (1.3) that an increase in the tax rate has the same effect of an increase in cost. Ceteris paribus, an increase in nation i's tax rate would depress the output of industrial sector i and vice versa.

13.1.2 Local and Global Environmental Impacts

Industrial production emits pollutants into the environment. The emitted pollutants cause short term local impacts on neighboring areas of the origin of production in forms like passing-by waste in waterways, wind-driven suspended particles in air, unpleasant odour, noise, dust and heat. For an output of $q_i(s)$ produced by nation i, there will be a short-term local environmental impact (cost) of $\varepsilon_i^i q_i(s)$ on nation i itself and a local impact of $\varepsilon_j^i q_i(s)$ on its neighbor nation j. Nation i will receive short-term local environmental impacts from its adjacent nations measured as $\varepsilon_i^j q_j(s)$ for $j \in \overline{K}^i$. Thus $\overline{K}^i$ is the subset of nations whose outputs produce local environmental impacts to nation i. Moreover, industrial production would also create long-term global environmental impacts by building up existing pollution stocks like Green-house-gas, CFC and atmospheric particulates. Each government adopts its own pollution abatement policy to reduce the pollution stock. Let $x(s) \subset R^+$ denote the level of pollution at time s, the dynamics of pollution stock is governed by the stochastic differential equation:

$$dx(s) = \left[\sum_{j=1}^{n} a_j q_j(s) - \sum_{j=1}^{n} b_j u_j(s) [x(s)]^{1/2} - \delta x(s) \right] ds + \sigma x(s) dz(s), \; x(t_0) = x_{t_0}, \tag{1.5}$$

where σ is a noise parameter and $z(s)$ is a Wiener process, $a_j q_j$ is the amount added to the pollution stock by a unit of nation j's output, $u_j(s)$ is the pollution abatement effort of nation j, $b_j u_j(s)[x(s)]^{1/2}$ is the amount of pollution removed by $u_j(s)$ unit of abatement effort of nation j, and δ is the natural rate of decay of the pollutants.

Short term local impacts are closely related to the level of production activities and hence are characterized by a deterministic scheme. On the other hand, the accumulation of pollution stock like greenhouse gas often involves the interactions between the natural environment and the pollutants emitted and hence stochastic elements would appear. For instance, nature's capability to replenish the environment, the rate of pollution degradation and climate change are subject to certain degrees of uncertainty. Hence a stochastic dynamic game is used to model the evolution of pollution stock (1.5). Finally the damage (cost) of the pollution stock in the environment to nation i at time s is $h_i x(s)$.

13.1.3 The Governments' Objectives

The governments have to promote business interests and at the same time handle the financing of the costs brought about by pollution. In particular, each government maximizes the net gains in the industrial sector minus the sum of expenditures on pollution abatement and damages from pollution. The instantaneous objective of government i at time s can be expressed as:

$$\left[\alpha^i - \sum_{j=1}^{n} \beta_j^i q_j(s)\right] q_i(s) - c_i q_i(s) - c_i^a [u_i(s)]^2 - \sum_{j \in \overline{K}^i} \varepsilon_i^j [q_j(s)] - h_i x(s),\ i \in N, \tag{1.6}$$

where $c_i^a [u_i(s)]^2$ is the cost of employing u_i amount of pollution abatement effort, and $h_i x(s)$ is the value of damage to country i from $x(s)$ amount of pollution.

The governments' planning horizon is $[t_0, T]$. It is possible that T may be very large. At time T, the terminal appraisal associated with the state of pollution is $g^i [\bar{x}^i - x(T)]$ where $g^i \geq 0$ and $\bar{x}^i \geq 0$. The discount rate is r. Each one of the n governments seeks to maximize the integral of its instantaneous objective (1.6) over the planning horizon subject to pollution dynamics (1.5) with controls on the level of abatement effort and output tax.

By substituting $q_i(s)$, for $i \in N$, from (1.4) into (1.5) and (1.6) one obtains a stochastic differential game in which government $i \in N$ seeks to:

$$\max_{v_i(s),\, u_i(s)} E_{t_0}\left\{ \int_{t_0}^{T} \left[\left(\alpha^i - \sum_{j=1}^{n} \beta_j^i \left[\overline{\alpha}^j + \sum_{h\in N} \overline{\beta}_h^j \; v_h(s) \right] \right) \left[\overline{\alpha}^i + \sum_{h\in N} \overline{\beta}_h^i \; v_h(s) \right] \right. \right.$$
$$- c_i \left[\overline{\alpha}^i + \sum_{j\in N} \overline{\beta}_j^i \; v_j(s) \right] - c_i^a [u_i(s)]^2 - \sum_{j\in \overline{K}^i} \varepsilon_i^j \left[\overline{\alpha}^j + \sum_{\ell\in N} \overline{\beta}_\ell^j \; v_\ell(s) \right]$$
$$\left. \left. - h_i x(s) \right] e^{-r(s-t_0)} ds - g^i [x(T) - \overline{x}^i] e^{-r(T-t_0)} \right\} \tag{1.7}$$

subject to

$$dx(s) = \left[\sum_{j=1}^{n} a_j \left[\overline{\alpha}^j + \sum_{h\in N} \overline{\beta}_h^j \; v_h(s) \right] - \sum_{j=1}^{n} b_j u_j(s) [x(s)]^{1/2} - \delta x(s) \right] ds$$
$$+ \sigma x(s) dz(s), \; x(t_0) = x_{t_0}. \tag{1.8}$$

In the game (1.7 and 1.8) one can readily observe that government i's tax policy $v_i(s)$ is not only explicitly reflected in its own output but also on the outputs of other nations. This modeling formulation allows some intriguing scenario to arise. For instance, an increase of $v_i(s)$ may just cause a minor drop in nation i's industrial profit but may cause significant increases in its neighbors' outputs which produce large local negative environmental impacts to nation i. This results in nations' reluctance to increase or impose taxes on industrial outputs.

13.2 Noncooperative Outcomes

In this section we discuss the solution to the noncooperative game (1.7) and (1.8). Since the payoffs of nations are measured in monetary terms, the game is a transferable payoff game. Under a noncooperative framework, a feedback Nash equilibrium solution can be characterized as (see Basar and Olsder (1995)):

Definition 2.1 A set of feedback strategies $\{u_i^*(t) = \mu_i(t,x),\ v_i^*(t) = \phi_i(t,x),$ for $i \in N\}$ provides a Nash equilibrium solution to the game (1.7 and 1.8) if there exist suitably smooth functions $V^i(t,x) : [t_0, T] \times R \to R,\ i \in N$, satisfying the following partial differential equations:

$$-V_t^i(t,x) - \frac{\sigma^2 x^2}{2} V_{xx}^i(t,x) = \max_{v_i, u_i} \Bigg\{$$

$$\Bigg[\Bigg(\alpha^i - \sum_{j=1}^{n} \beta_j^i \Bigg[\overline{\alpha}^j + \sum_{\substack{h \in N \\ h \neq i}}^{n} \overline{\beta}_h^j \ \phi_h(t,x) + \overline{\beta}_i^j v_i\Bigg]\Bigg)\Bigg[\overline{\alpha}^i + \sum_{\substack{h \in N \\ h \neq i}} \overline{\beta}_h^i \ \phi_h(t,x) + \overline{\beta}_i^i v_i\Bigg]$$

$$-c_i \Bigg[\overline{\alpha}^i + \sum_{\substack{j \in N \\ j \neq i}} \overline{\beta}_j^i \ \phi_j(t,x) + \overline{\beta}_i^i v_i\Bigg] - c_i^a [u_i]^2$$

$$-\sum_{j \in \overline{K}^i} \varepsilon_i^j \Bigg[\overline{\alpha}^j + \sum_{\substack{\ell \in N \\ \ell \neq i}} \overline{\beta}_\ell^j \ \phi_\ell(t,x) + \overline{\beta}_i^j v_i\Bigg] - h_i x \Bigg] e^{-r(t-t_0)}$$

$$+V_x^i \Bigg[\sum_{j=1}^{n} a_j \Bigg[\overline{\alpha}^j + \sum_{\substack{h \in N \\ h \neq i}} \overline{\beta}_h^j \ \phi_h(t,x) + \overline{\beta}_i^j v_i\Bigg]$$

$$- \sum_{\substack{j=1 \\ j \neq i}}^{n} b_j \mu_j(t,x) x^{1/2} - b_i u_i x^{1/2} - \delta x\Bigg]\Bigg\}, \tag{2.1}$$

$$V^i(T,x) = -g^i \left[x - \overline{x}^i\right] e^{-r(T-t_0)}. \tag{2.2}$$

Performing the indicated maximization in (2.1) yields:

$$\mu_i(t,x) = -\frac{b_i}{2c_i^a} V_x^i(t,x) e^{r(t-t_0)} x^{1/2}, \tag{2.3}$$

$$\Bigg(\alpha^i - \sum_{j=1}^{n} \beta_j^i \Bigg[\overline{\alpha}^j + \sum_{h \in N}^{n} \overline{\beta}_h^j \ \phi_h(t,x)\Bigg]\Bigg)\overline{\beta}_i^i - \Bigg[\sum_{j=1}^{n} \beta_j^i \overline{\beta}_i^j\Bigg]\Bigg[\overline{\alpha}^i + \sum_{h \in N} \overline{\beta}_h^i \ \phi_h(t,x)\Bigg]$$

$$-c_i \overline{\beta}_i^i - \sum_{j \in \overline{K}^i} \varepsilon_i^j \overline{\beta}_i^j + V_x^i \sum_{j=1}^{n} a_j \overline{\beta}_i^j e^{r(t-t_0)} = 0, \tag{2.4}$$

for $t \in [t_0 < T]$ and $i \in N$.

System (2.4) forms a set of equations linear in $\{\phi_1(t,x), \phi_2(t,x), \cdots, \phi_n(t,x)\}$ with $\{V_x^1(t,x)e^{r(t-t_0)}, V_x^2(t,x)e^{r(t-t_0)}, \cdots, V_x^n(t,x)e^{r(t-t_0)}\}$ being taken as a set of parameters. Solving (2.4) yields:

$$\phi_i(t,x) = \hat{\alpha}^i + \sum_{j\in N} \hat{\beta}_j^i \; V_x^j(t,x)\, e^{r(t-t_0)}, \;\; i \in N, \tag{2.5}$$

where $\hat{\alpha}^i$ and $\hat{\beta}_j^i$, for $i \in N$ and $j \in N$, are constants involving the constant coefficients in (2.4). Substituting the results in (2.3) and (2.5) into (2.1 and 2.2) we obtain game equilibrium expected payoffs of the nations as:

Proposition 2.1

$$V^i(t,x) = [A_i(t)x + C_i(t)] \;\; e^{-r(t-t_0)}, \;\text{for } i \in N, \tag{2.6}$$

where $\{A_1(t), A_2(t), \cdots, A_n(t)\}$ satisfying the following set of constant coefficient quadratic ordinary differential equations:

$$\dot{A}_i(t) = (r+\delta) \; A_i(t) - \frac{b_i^2}{4c_i^a}[A_i(t)]^2 - A_i(t) \sum_{\substack{j=1 \\ j\neq i}}^{n} \frac{b_j^2}{2c_j^a} A_j(t) + h_i,$$

$$A_i(T) = -g^i; \;\; \text{for } i \in N, \tag{2.7}$$

and $\{C_i(t);\;\; i \in N\}$ is given by $C_i(t) = e^{r(\;t-t_0)}\left[\int_{t_0}^{t} F_i(y)e^{-r(y-t_0)}dy + C_i^0\right]$, (2.8)

where $C_i^0 = g^i \bar{x}^i e^{-r\,(T-t_0)} - \int_{t_0}^{T} F_i(y)e^{-r(y-t_0)}dy$

$$\begin{aligned}
F_i(t) = & -\left(\alpha^i - \sum_{j=1}^{n} \beta_j^i \left\{ \overline{\alpha}^j + \sum_{h\in Ni}^{n} \overline{\beta}_h^j \left[\hat{\alpha}^h + \sum_{k\in N} \hat{\beta}_k^h \; A_k(t) \right] \right\} \right) \\
& \left(\overline{\alpha}^i + \sum_{h\in N} \overline{\beta}_h^i \left[\hat{\alpha}^h + \sum_{k\in N} \hat{\beta}_k^h \; A_k(t) \right] \right) \\
& + c_i \left\{ \overline{\alpha}^i - \sum_{j\in N} \overline{\beta}_j^i \left[\hat{\alpha}^j + \sum_{k\in N} \hat{\beta}_k^j \; A_k(t) \right] \right\} \\
& + \sum_{j\in \overline{K}^i} \varepsilon_i^j \left\{ \overline{\alpha}^j + \sum_{\ell\in N} \overline{\beta}_\ell^j \;\; \left[\hat{\alpha}^\ell + \sum_{k\in N} \hat{\beta}_k^\ell \; A_k(t) \right] \right\} \\
& - A_i(t) \left[\sum_{j=1}^{n} a_j \left\{ \overline{\alpha}^j + \sum_{h\in N} \overline{\beta}_h^j \left[\hat{\alpha}^h + \sum_{k\in N} \hat{\beta}_k^h \; A_k(t) \right] \right\} \right].
\end{aligned}$$

Proof See Appendix A. ■

The corresponding feedback Nash equilibrium strategies of the game (1.7 and 1.8) can be obtained as:

$$\mu_i(t,x) = -\frac{b_i}{2c_i^a} A_i(t) x^{1/2} \text{ and } \phi_i(t,x) = \hat{\alpha}^i + \sum_{j \in N} \hat{\beta}_j^i \; A_j(t) \quad \text{for } i \in N \text{ and } t \in [t_0, T]. \tag{2.9}$$

A remark that will be utilized in subsequent analysis is given below.

Remark 2.1 Let $V^{(\tau)i}(t, x_t)$ denote the value function indicating the game equilibrium payoff of nation i in a game with payoffs (1.7) and dynamics (1.8) which starts at time τ. One can readily verify that $V^{(\tau)i}(t, x_t) = V^i(t, x_t) e^{r(\tau - t_0)}$, for $\tau \in [t_0, T]$. ■

13.3 Cooperative Arrangement

Now consider the case when all the nations want to cooperate and agree to act so that an international optimum could be achieved. For the cooperative scheme to be upheld throughout the game horizon both group rationality and individual rationality are required to be satisfied at any time. In addition, to ensure that the cooperative solution is dynamically stable, the agreement must be subgame-consistent. The cooperative plan will dissolve if any of the nations deviates from the agreed-upon plan.

13.3.1 Group Optimality and Cooperative State Trajectory

Consider the cooperative stochastic differential games with payoff structure (1.5) and dynamics (1.3). To secure group optimality the participating nations seek to maximize their joint expected payoff by solving the following stochastic control problem:

$$\max_{v_1, v_2, \cdots, v_n; u_1, u_2, \cdots, u_n} E_{t_0} \Bigg\{ \int_{t_0}^{T} \sum_{\ell=1}^{n} \Bigg[\Bigg(\alpha^\ell - \sum_{j=1}^{n} \beta_j^\ell \Big[\overline{\alpha}^j + \sum_{h \in N} \overline{\beta}_h^j \; v_h(s) \Big] \Bigg) \Big[\overline{\alpha}^\ell + \sum_{h \in N} \overline{\beta}_h^\ell \; v_h(s) \Big]$$
$$- c_\ell \Big[\overline{\alpha}^\ell + \sum_{j \in N} \overline{\beta}_j^\ell \; v_j(s) \Big] - c_\ell^a [u_\ell(s)]^2 - \sum_{j \in \overline{K}^\ell} \varepsilon_\ell^j \Big[\overline{\alpha}^j + \sum_{k \in N} \overline{\beta}_k^j \; v_k(s) \Big]$$
$$- h_\ell x(s) \Bigg] e^{-r(s - t_0)} ds - \sum_{\ell=1}^{n} g^\ell \big[x(T) - \overline{x}^\ell \big] e^{-r(T - t_0)} \Bigg\} \tag{3.1}$$

subject to (1.8).

Invoking Fleming's (1969) technique in stochastic control in Theorem A.3 of the Technical Appendices a set of controls $\{ [v_i^{**}(t), u_i^{**}(t)] = [\psi_i(t,x), \varpi_i(t,x)]$, for $i \in N\}$ constitutes an optimal solution to the stochastic control problem (3.1) and (1.8) if there exists continuously differentiable function $W(t,x) : [t_0, T] \times R \to R$, $i \in N$, satisfying the following partial differential equations:

$$-W_t(t,x) - \frac{\sigma^2 x^2}{2} W_{xx}(t,x) =$$

$$\max_{v_1, v_2, \cdots, v_n; u_1, u_2, \cdots, u_n} \left\{ \sum_{\ell=1}^{n} \left[\left(\alpha^\ell - \sum_{j=1}^{n} \beta_j^\ell \left[\overline{\alpha}^j + \sum_{h \in N} \overline{\beta}_h^j \; v_h \right] \right) \left[\overline{\alpha}^\ell + \sum_{h \in N} \overline{\beta}_h^\ell \; v_h \right] \right. \right.$$

$$\left. - c_\ell \left[\overline{\alpha}^\ell + \sum_{j \in N} \overline{\beta}_j^\ell \; v_j \right] - c_\ell^a [u_\ell]^2 - \sum_{j \in \overline{K}^\ell} \varepsilon_\ell^j \left[\overline{\alpha}^j + \sum_{k \in N} \overline{\beta}_k^j \; v_k \right] - h_\ell x \right] e^{-r(s-t_0)}$$

$$\left. + W_x(t,x) \left[\sum_{j=1}^{n} a_j \left[\overline{\alpha}^j + \sum_{h \in N} \overline{\beta}_h^j \; v_h \right] - \sum_{j=1}^{n} b_j u_j x^{1/2} - \delta x \right] \right\}, \tag{3.2}$$

$$W(T,x) = -\sum_{i=1}^{n} g^i \left[x(T) - \overline{x}^i \right] e^{-r(T-t_0)}. \tag{3.3}$$

Performing the indicated maximization in (3.2) yields the optimal controls under cooperation as:

$$\varpi_i(t,x) = -\frac{b_i}{2c_i^a} W_x(t,x) e^{r(t-t_0)} x^{1/2}, \quad \text{for } i \in N; \tag{3.4}$$

$$\sum_{\ell=1}^{n} \left[\left(\alpha^\ell - \sum_{j=1}^{n} \beta_j^\ell \left[\overline{\alpha}^j + \sum_{h \in N} \overline{\beta}_h^j \; \psi_h(t,x) \right] \right) \overline{\beta}_i^\ell \right.$$

$$\left. - [\sum_{j=1}^{n} \beta_j^\ell \overline{\beta}_i^j] \left[\overline{\alpha}^\ell + \sum_{h \in N} \overline{\beta}_h^\ell \; \psi_h(t,x) \right] \right] \tag{3.5}$$

$$- \sum_{\ell=1}^{n} \left[c_\ell \overline{\beta}_i^\ell + \sum_{j \in \overline{K}^i} \varepsilon_\ell^j \overline{\beta}_i^j \right] + V_x^i \sum_{j=1}^{n} a_j \overline{\beta}_i^j e^{r(t-t_0)} = 0, \text{ for } i \in N.$$

System (3.5) can be viewed as a set of equations linear in $\{\psi_1(t,x), \psi_2(t,x), \cdots, \psi_n(t,x)\}$ with $W_x(t,x) e^{r(t-t_0)}$ being taken as a parameter. Solving (3.5) yields:

$$\psi_i(t,x) = \hat{\hat{\alpha}}^{\,i} + \hat{\hat{\beta}}^{\,i}\; W_x(t,x)e^{r(t-t_0)}, \tag{3.6}$$

where $\hat{\hat{\alpha}}^{\,i}$ and $\hat{\hat{\beta}}^{\,i}$, for $i \in N$, are constants involving the model parameters.

The expected joint payoff of the nations under cooperation can be obtained as:

Proposition 3.1 System (3.2 and 3.3) admits a solution

$$W(t,x) = \left[A^*(t)x + C^*(t)\right]\; e^{-r(t-t_0)}, \tag{3.7}$$

with

$$A^*(t) = A_*^P + \Phi^*(t)\left[\,\overline{C}^* - \int_{t_0}^{t}\sum_{j=1}^{n}\frac{b_j^2}{2c_j^a}\Phi^*(y)dy\,\right]^{-1}, \text{ and}$$

$$C^*(t) = e^{r(\,t-t_0)}\left[\,\int_{t_0}^{t}F^*(y)e^{-r(y-t_0)}dy + C_*^0\,\right],$$

where $\Phi^*(t) = \exp\left\{\int_{t_0}^{t}\left[\,\sum_{j=1}^{n}\frac{b_j^2}{2c_j^a}A_*^P + (r+\delta)\,\right]\,dy\right\}$,

$$\overline{C}^* = \frac{-\Phi^*(T)}{\left(A_*^P + \sum_{j=1}^{n}g^j\right)} + \int_{t_0}^{T}\sum_{j=1}^{n}\frac{b_j^2}{2c_j^a}\Phi^*(y)dy,$$

$$A_*^P(t) = \left\{\,(r+\delta) - \left[\,(r+\delta)^2 + 4\sum_{j=1}^{n}\frac{b_j^2}{2c_j^a}\sum_{j=1}^{n}h_j\,\right]^{1/2}\,\right\}\Big/\sum_{j=1}^{n}\frac{b_j^2}{c_j^a},$$

$$\begin{aligned}F^*(t) = -\sum_{\ell=1}^{n}\Bigg[&\left(\alpha^\ell - \sum_{j=1}^{n}\beta_j^\ell\left\{\overline{\alpha}^j + \sum_{h\in N}\overline{\beta}_h^j\,\left[\hat{\hat{\alpha}}^{\,h} + \hat{\hat{\beta}}^{\,h}\;A^*(t)\right]\right\}\right)\{\overline{\alpha}^\ell\\ &+ \sum_{h\in N}\overline{\beta}_h^\ell\,\left[\hat{\hat{\alpha}}^{\,h} + \hat{\hat{\beta}}^{\,h}\;A^*(t)\right]\Big\} - c_\ell\left\{\overline{\alpha}^\ell + \sum_{j\in N}\overline{\beta}_j^\ell\,\left[\hat{\hat{\alpha}}^{\,j} + \hat{\hat{\beta}}^{\,j}\;A^*(t)\right]\right\}\\ &- \sum_{j\in\overline{K}^\ell}\varepsilon_\ell^j\left\{\overline{\alpha}^j + \sum_{k\in N}\overline{\beta}_k^j\,\left[\hat{\hat{\alpha}}^{\,k} + \hat{\hat{\beta}}^{\,kj}\;A^*(t)\right]\right\}\Bigg]\\ &- A_x^*(t)\left[\,\sum_{j=1}^{n}a_j\left\{\overline{\alpha}^j + \sum_{h\in N}\overline{\beta}_h^j\,\left[\hat{\hat{\alpha}}^{\,h} + \hat{\hat{\beta}}^{\,h}\;A^*(t)\right]\right\}\,\right], \text{ and}\end{aligned}$$

$$C_*^0 = \sum_{j=1}^{n}g^j\overline{x}^je^{-r(\,T-t_0)} - \int_{t_0}^{T}F^*(y)e^{-r(y-t_0)}dy.$$

Proof See Appendix B. ■

Using (3.4), (3.6) and (3.7), the control strategy under cooperation can be obtained as:

$$\psi_i(t,x) = \hat{\hat{\alpha}}^i + \hat{\hat{\beta}}^i A^*(t) \text{ and } \varpi_i(t,x) = -\frac{b_i}{2c_i^a} A^*(t) x^{1/2}, \tag{3.8}$$

for $t \in [t_0 < T]$ and $i = 1, 2, \cdots, n$.

Substituting the optimal control strategy from (3.8) into (1.3) yields the dynamics of pollution accumulation under cooperation. Solving the stochastic cooperative pollution dynamics yields the cooperative state trajectory:

$$x^*(t) = e^{\left[\int_{t_0}^{t} [\sum_{j=1}^{n} \frac{b_j^2}{2c_j^a} A^*(s) - \delta - \frac{\sigma^2}{2}] ds + \int_{t_0}^{t} \sigma dz(s)\right]}$$
$$\left[x_{t_0} + \int_{t_0}^{t} \sum_{j=1}^{n} a_j \left\{ \overline{\alpha}^j + \sum_{h \in N} \overline{\beta}_h^j \left[\hat{\hat{\alpha}}^h + \hat{\hat{\beta}}^h A^*(s) \right] \right\} \right.$$
$$\left. e^{\left[\int_{t_0}^{s} [\frac{\sigma^2}{2} + \delta - \sum_{j=1}^{n} \frac{b_j^2}{2c_j^a} A^*(\tau)] d\tau - \int_{t_0}^{s} \sigma dz(\tau)\right]} ds \right], \tag{3.9}$$

for $t \in [t_0, T]$.

We use X_t^* to denote the set of realizable values of $x^*(t)$ at time t generated by (3.9). The term x_t^* is used to denote an element in the set X_t^*.

A remark that will be utilized in subsequent analysis is given below.

Remark 3.1 Let $W^{(\tau)}(t, x_t)$ denote the value function indicating the maximized joint payoff of the stochastic control problem with objective (3.1) and dynamics (1.8) which starts at time τ. One can readily verify that $W^{(\tau)}(t, x_t^*) = W(t, x_t^*) e^{r(\tau - t_0)}$, for $\tau \in [t_0, T]$. ■

13.3.2 *Individually Rational and Subgame-Consistent Imputation*

An agreed upon optimality principle must be sought to allocate the cooperative payoff. In a dynamic framework individual rationality has to be maintained at every instant of time within the cooperative duration $[t_0, T]$ given any feasible state generated by the cooperative trajectory (3.9). For $\tau \in [t_0, T]$, let $\xi^{(\tau)i}(\tau, x_\tau^*)$ denote the solution imputation (payoff under cooperation) over the period $[\tau, T]$ to player $i \in N$ given that the state is $x_\tau^* \in X_\tau^*$. Individual rationality along the cooperative trajectory requires:

$$\xi^{(\tau)i}\left(\tau,x_{\tau}^{*}\right) \geq V^{(\tau)i}\left(\tau,x_{\tau}^{*}\right),\ \text{for } i \in N,\ x_{\tau}^{*} \in X_{\tau}^{*} \text{ and } \tau \in [t_0,T]. \tag{3.10}$$

Since nations are asymmetric and the number of nations may be large, a reasonable solution optimality principle for gain distribution is to share the expected gain from cooperation proportional to the nations' relative sizes of expected noncooperative payoffs. As mentioned before, a stringent condition – subgame consistency – is required for a credible cooperative solution. In order to satisfy the property of subgame consistency, this optimality principle has to remain in effect throughout the cooperation period. Hence the solution imputation scheme $\left\{\xi^{(\tau)i}\left(\tau,x_{\tau}^{*}\right);\ \text{for } i \in N\right\}$ has to satisfy:

Condition 4.1

$$\xi^{(\tau)i}\left(\tau,x_{\tau}^{*}\right) = V^{(\tau)i}\left(\tau,x_{\tau}^{*}\right) + \frac{V^{(\tau)i}\left(\tau,x_{\tau}^{*}\right)}{\sum_{j=1}^{n} V^{(\tau)j}\left(\tau,x_{\tau}^{*}\right)}\left[W^{(\tau)}\left(\tau,x_{\tau}^{*}\right) - \sum_{j=1}^{n} V^{(\tau)j}\left(\tau,x_{\tau}^{*}\right)\right]$$

$$= \frac{V^{(\tau)i}\left(\tau,x_{\tau}^{*}\right)}{\sum_{j=1}^{n} V^{(\tau)j}\left(\tau,x_{\tau}^{*}\right)} W^{(\tau)}\left(\tau,x_{\tau}^{*}\right), \tag{3.11}$$

for $i \in N, x_{\tau}^{*} \in X_{\tau}^{*}$ and $\tau \in [t_0,T]$. ■

One can easily verify that the imputation scheme in Condition 4.1 satisfies individual rationality. Crucial to the analysis is the formulation of a payment distribution mechanism that would lead to the realization of Condition 4.1. This will be done in the next Section.

13.3.3 Payment Distribution Mechanism

To formulate a payment distribution scheme over time so that the agreed upon imputation (3.11) can be realized for any time instant $\tau \in [t_0,T]$ we apply the techniques developed in Chap. 3. Let the vectors $B\left(s,x_{s}^{*}\right) = \left[B_1\left(s,x_{s}^{*}\right), B_2\left(s,x_{s}^{*}\right), \cdots, B_n\left(s,x_{s}^{*}\right)\right]$ denote the instantaneous payment to the n nations at time instant s when the state is $x_{s}^{*} \in X_{s}^{*}$. A terminal value of $g^i\left[\bar{x}^i - x_{T}^{*}\right]$ is realized by nation i at time T.

To satisfy (3.11) it is required that

$$\xi^{(\tau)i}\left(\tau, x_\tau^*\right) = \frac{V^{(\tau)i}\left(\tau, x_\tau^*\right)}{\sum_{j=1}^{n} V^{(\tau)j}\left(\tau, x_\tau^*\right)} W^{(\tau)}\left(\tau, x_\tau^*\right) =$$

$$E_\tau\left\{\left(\int_\tau^T B_i\left(s, x^*(s)\right) e^{-r(s-\tau)} ds - g^i\left[x_T^* - \overline{x}^i\right] e^{-r(T-\tau)} \middle| x(\tau) = x_\tau^*\right\}, \quad (3.12)$$

for $i \in N, x_\tau^* \in X_\tau^*$ and $\tau \in [t_0, T]$.

To facilitate further exposition, we use the term $\xi^{(\tau)i}(t, x_t^*)$ which equals

$$E_\tau\left\{\left(\int_t^T B_i\left(s, x^*(s)\right) e^{-r(s-\tau)} ds - g^i\left[x_T^* - \overline{x}^i\right] e^{-r(T-\tau)} \middle| x(t) = x_t^*\right\}$$

$$= \frac{V^{(\tau)i}\left(t, x_t^*\right)}{\sum_{j=1}^{n} V^{(\tau)j}\left(t, x_t^*\right)} W^{(\tau)}\left(t, x_t^*\right) = \frac{V^{(t)i}\left(t, x_t^*\right)}{\sum_{j=1}^{n} V^{(t)j}\left(t, x_t^*\right)} W^{(t)}\left(t, x_t^*\right) e^{-r(t-\tau)} \quad (3.13)$$

$$\xi^{(t)i}\left(t, x_t^*\right) e^{-r(t-\tau)},$$

for $x_t^* \in X_t^*$ and $t \in [\tau, T]$,

to denote the expected present value (with initial time set at τ) of nation i's cooperative payoff over the time interval $[t, T]$.

A theorem characterizing a formula for $B_i(\tau, x_\tau^*)$, for $\tau \in [t_0, T]$ and $i \in N$, which yields Condition 4.1 is provided below.

Theorem 3.1 A distribution scheme with a terminal payment $-g^i\left[x_T^* - \overline{x}^i\right]$ at time T and an instantaneous payment at time $\tau \in [t_0, T]$ when $x(\tau) = x_\tau^*$:

$$\begin{aligned} B_i\left(\tau, x_\tau^*\right) = &- \left[\xi_t^{(\tau)i}\left(t, x_t^*\right)\Big|_{t=\tau}\right] - \frac{\sigma^2 x^2}{2}\left[\xi_{x_t^* x_t^*}^{(\tau)i}\left(t, x_t^*\right)\Big|_{t=\tau}\right] \\ &- \left[\xi_{x_t^*}^{(\tau)i}\left(t, x_t^*\right)\Big|_{t=\tau}\right]\left[\sum_{j=1}^{n} a_j\left[\overline{\alpha}^j + \sum_{h \in N} \overline{\beta}_h^j \ \psi_h\left(\tau, x_\tau^*\right)\right]\right. \\ &\left. - \sum_{j=1}^{n} b_j \varpi_j\left(\tau, x_\tau^*\right)\left(x_\tau^*\right)^{1/2} - \delta x_\tau^*\right], \text{for } i \in N, \end{aligned} \quad (3.14)$$

yield Condition 4.1.

Proof Since $\xi^{(\tau)i}(t,x_t^*)$ is continuously differentiable in t and x_t^*, using (3.13) and Remarks 2.1 and 3.1 one can obtain:

$$E_\tau\left\{\int_\tau^{\tau+\Delta t} B_i\big(s,x^*(s)\big)e^{-r(s-\tau)}ds \,\middle|\, x(\tau)=x_\tau^*\right\}$$

$$=E_\tau\left\{\xi^{(\tau)i}\big(\tau,x_\tau^*\big)-e^{-r\Delta t}\xi^{(\tau+\Delta t)i}\big(\tau+\Delta t,x_{\tau+\Delta t}^*\big)\,\middle|\, x(\tau)=x_\tau^*\right\} \quad (3.15)$$

$$=E_\tau\left\{\xi^{(\tau)i}\big(\tau,x_\tau^*\big)-\xi^{(\tau)i}\big(\tau+\Delta t,x_{\tau+\Delta t}^*\big)\,\middle|\, x(\tau)=x_\tau^*\right\},$$

for $i\in N$ and $\tau\in[t_0,T]$,

where

$$\Delta x_\tau=\left[\sum_{j=1}^n a_j\left[\overline{\alpha}^j+\sum_{h\in N}\overline{\beta}_h^j\ \psi_h\big(\tau,x_\tau^*\big)\right]-\sum_{j=1}^n b_j\varpi_j\big(\tau,x_\tau^*\big)\big(x_\tau^*\big)^{1/2}-\delta x_\tau^*\right]\Delta t$$
$$+\sigma x_\tau^*\Delta z_\tau+o(\Delta t),$$

$\Delta z_\tau=z(\tau+\Delta t)-z(\tau)$, and $E_\tau[o(\Delta t)]/\Delta t\to 0$ as $\Delta t\to 0$.

With $\Delta t\to 0$, condition (3.15) can be expressed as:

$$E_\tau\left\{B_i\big(\tau,x_\tau^*\big)\Delta t+o(\Delta t)\right\}=E_\tau\Bigg\{-\left[\xi_t^{(\tau)i}\big(t,x_t^*\big)\Big|_{t=\tau}\right]\Delta t$$
$$-\left[\xi_{x_t^*}^{(\tau)i}\big(t,x_t^*\big)\Big|_{t=\tau}\right]\left[\sum_{j=1}^n a_j\psi_j^q\big(\tau,x_\tau^*\big)\right.$$
$$\left.-\sum_{j=1}^n b_j\psi_j^u\big(\tau,x_\tau^*\big)\big(x_\tau^*\big)^{1/2}-\delta x_\tau^*\right]\Delta t$$
$$-\frac{\sigma^2x^2}{2}\left[\xi_{x_t^*x_t^*}^{(\tau)i}\big(t,x_t^*\big)\Big|_{t=\tau}\right]\Delta t$$
$$-\left[\xi_{x_t^*}^{(\tau)i}\big(t,x_t^*\big)\Big|_{t=\tau}\right]\sigma x\Delta z_\tau-o(\Delta t)\Bigg\}, \quad (3.16)$$

Taking expectation and dividing (3.16) throughout by Δt, with $\Delta t\to 0$, yields (3.14). Hence Theorem 3.1 follows. ■

When all nations are adopting the cooperative strategies the rate of instantaneous payment that nation $\ell \in N$ will realize at time t with the state being x_t^* can be expressed as (see derivation in Appendix II):

$$\begin{aligned}\Re_\ell(t,x_t^*) = &\left(\alpha^\ell - \sum_{j=1}^{n}\beta_j^\ell\left\{\overline{\alpha}^j + \sum_{h\in N}\overline{\beta}_h^j\left[\hat{\alpha}^h + \hat{\beta}^h A^*(t)\right]\right\}\right)\left\{\overline{\alpha}^\ell + \sum_{h\in N}\overline{\beta}_h^\ell\left[\hat{\alpha}^h + \hat{\beta}^h A^*(t)\right]\right\}\\ &-c_\ell\left\{\overline{\alpha}^\ell + \sum_{j\in N}\overline{\beta}_j^\ell\left[\hat{\alpha}^j + \hat{\beta}^j A^*(t)\right]\right\} - c_\ell^a\left[\frac{b_\ell}{2c_\ell^a}A^*(t)\right]^2 x_t^*\\ &-\sum_{j\in\overline{K}^\ell}\varepsilon_\ell^j\left\{\overline{\alpha}^j + \sum_{k\in N}\overline{\beta}_k^j\left[\hat{\alpha}^k + \hat{\beta}^{kj} A^*(t)\right]\right\} - h_\ell x_t^*.\end{aligned} \tag{3.17}$$

Since according to Theorem 3.1 under the cooperative scheme an instantaneous payment to nation ℓ equaling $B_\ell(t,x_t^*)$ at time t with the state being x_t^*, a side payment of the value $B_\ell(t,x_t^*) - \Re_\ell(t,x_t^*)$ will be offered to nation ℓ.

13.4 A Numerical Example

Consider a multinational economy which is comprised of 2 nations. At time instant s the demand functions of the output of nations 1 and 2 are respectively

$$P_1(s) = 50 - q_1(s) - 0.5q_2(s) \text{ and } P_2(s) = 90 - 2q_2(s) - q_1(s). \tag{4.1}$$

The cost of production of a unit of output in nation 1 and nation 2 are respectively 2 and 1. Industrial profits of these nations at time s can be expressed as:

$$\begin{aligned}\pi_1(s) &= [50 - q_1(s) - 0.2q_2(s)]q_1(s) - 2q_1(s) - v_1(s)q_1(s) \text{ and}\\ \pi_2(s) &= [90 - 2q_2(s) - 0.6q_1(s)]q_2(s) - q_2 - v_2(s)q_2(s).\end{aligned} \tag{4.2}$$

where $v_i(s)$ is the tax rate imposed by the government of nation i on its industrial output.

An industry equilibrium can be obtained as:

$$\begin{aligned}q_1(s) &= \frac{4355}{197} - \frac{100}{197}v_1(s) + \frac{5}{197}v_2(s) \text{ and}\\ q_2(s) &= \frac{3730}{197} + \frac{15}{197}v_1(s) - \frac{50}{197}v_2(s).\end{aligned} \tag{4.3}$$

The short-term local environmental impact (cost) of nation 1's output on itself is $0.5q_1(s)$ and that on nation 2 is $0.4q_1(s)$. The short-term local environmental impact (cost) of nation 2's output on itself is $0.8q_2(s)$ and that on nation 1 is $0.6q_2(s)$.

The dynamics of pollution stock is governed by the stochastic differential equation:

$$dx(s) = \left[0.5q_1(s) + q_2(s) - 0.4u_1(s)x(s)^{1/2} - 0.2u_2(s)x(s)^{1/2} - 0.01x(s) \right] ds \\ +0.05x(s)dz(s),\ x(t_0) = 20. \tag{4.4}$$

The damage (cost) of the pollution stock in the environment to nations 1 and 2 are respectively $4x(s)$ and $5x(s)$. The abatement costs are $0.5[u_1(s)]^2$ and $[u_2(s)]^2$ for nations 1 and 2 respectively. The instantaneous objectives of the governments in nations 1 and 2 at time s are respectively:

$$[50 - q_1(s) - 0.2q_2(s)]q_1(s) - 2q_1(s) - 0.5[u_1(s)]^2 - 0.5q_1(s) - 0.6q_2(s) - 4x(s) \tag{4.5}$$

and

$$[90 - 2q_2(s) - 0.6q_1(s)]q_2(s) - q_2(s) - [u_2(s)]^2 - 0.8q_2(s) - 0.4q_1(s) - 5x(s) \tag{4.6}$$

At time $T = 5$ (decades), the terminal value associated with the state of pollution is $2[100 - x(T)]$ for nation 1 and $3[60 - x(T)]$ for nation 2.

Substituting $q_i(s)$, for $i \in \{1, 2\}$, from (4.3) into (4.4, 4.5, and 4.6) one obtains a stochastic differential game in which government 1 seeks to:

$$\max_{v_1(s),\, u_1(s)} E_0 \left\{ \int_0^5 \left[\frac{20668769.5}{38809} - \frac{4988}{38809}v_1(s) + \frac{48967}{38809}v_2(s) \right.\right. \\ -\frac{15}{38809}v_1(s)v_2(s) - \frac{9700}{38809}[v_1(s)]^2 + \frac{25}{38809}[v_2(s)]^2 - 0.5[u_1(s)]^2 \\ \left.\left. -4x(s) \right] e^{-0.05s} ds - 2[x(T) - 100]e^{-0.25} \right\}, \tag{4.7}$$

and government 2 seeks to

$$\max_{v_2(s),\, u_2(s)} E_0 \left\{ \int_0^5 \left[\frac{26894778}{38809} + \frac{229316}{38809}v_1(s) - \frac{3704}{38809}v_2(s) \right.\right. \\ +\frac{450}{38809}[v_1(s)]^2 - \frac{4850}{38809}[v_2(s)]^2 - \frac{45}{38809}v_1(s)v_2(s) - [u_2(s)]^2 \\ \left.\left. -5x(s) \right] e^{-0.05s} ds - 3[x(T) - 60]e^{-0.25} \right\}, \tag{4.8}$$

subject to

$$dx(s) = \left[\frac{5907.5}{197} - \frac{35}{197}v_1(s) - \frac{47.5}{197}v_2(s) - 0.4u_1(s)x(s)^{1/2} - 0.2u_2(s)x(s)^{1/2} - 0.01x(s)\right]ds + 0.05x(s)dz(s), x(t_0) = 10. \tag{4.9}$$

Solving the game yields:

$$V^1(t,x) = [A_1(t)x + C_1(t)]\ e^{-0.05t} \text{ and } V^2(t,x) = [A_2(t)x + C_2(t)]\ e^{-0.05t}, \tag{4.10}$$

where

$$\begin{aligned}
\dot{A}_1(t) &= 0.06A_1(t) - 0.08(A_1(t))^2 - 0.02A_1(t)A_2(t) + 4,\\
\dot{A}_2(t) &= 0.06A_2(t) - 0.01(A_2(t))^2 - 0.16A_1(t)A_2(t) + 5,\\
A_1(5) &= -2,\ A_2(5) = -3;\\
\dot{C}_1(t) &= 0.05C_1(t) - 532.1129418 - 30.12271238A_1(t) + 1.216826461A_2(t)\\
&\quad -0.031957733[A_1(t)]^2 - 0.0005996[A_2(t)]^2 - 0.232474798A_1(t)A_2(t),\\
\dot{C}_2(t) &= 0.05C_2(t) - 691.5051178 + 2.098040543A_1(t) - 30.12885088A_2(t)\\
&\quad -0.0014636068[A_1(t)]^2 - 0.116177969[A_2(t)]^2 - 0.062738812A_1(t)A_2(t),\\
C_1(5) &= 200 \text{ and } C_2(5) = 180.
\end{aligned}$$

The values of $A_1(t), A_2(t), C_1(t)$ and $C_2(t)$ over the time interval $[0, 5]$ are computed and presented in Figs. 13.1a, b.

Now consider the case when all the nations want to cooperate and agree to act so that an international optimum could be achieved. The instantaneous objective of the cooperative scheme is the sum of the individual objectives (4.5) and (4.6). The terminal value associated with the state of pollution is $2[100 - x(T)] + 3[60 - x(T)]$.

To secure group optimality the participating nations seek to maximize their joint expected payoff by solving the following stochastic control problem:

$$\max_{v_1(s), v_2(s), u_1(s), \mu_2(s)} E_0\left\{ \int_0^5 \left\{ \left[\frac{47563547.5}{38809} + \frac{224328}{38809}v_1(s) + \frac{45263}{38809}v_2(s) - \frac{60}{38809}v_1(s)v_2(s) - \frac{9250}{38809}[v_1(s)]^2. - \frac{4825}{38809}[v_2(s)]^2 - 0.5[u_1(s)]^2 - [u_2(s)]^2 - 9x(s) \right] e^{-0.05s} ds - 5[x(T) - 76]e^{-0.25} \right\} \tag{4.11}$$

subject to (4.9).

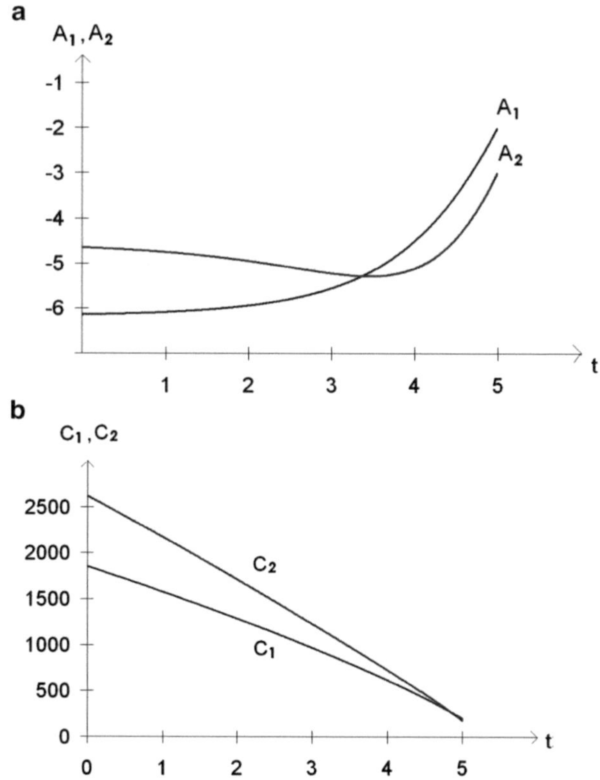

Fig. 13.1 The values of $A_1(t)$, $A_2(t)$, $C_1(t)$ and $C_2(t)$ over the time interval [0,5]

Solving the stochastic control problem (4.11) and (4.9) yields

$$W(t,x) = [A(t)x + C(t)]\ e^{-0.05\,t} \tag{4.12}$$

where

$$\begin{aligned}
&\dot{A}\,(t) = 0.06A(t) - 0.09(A(t))^2 + 9,\\
&\dot{C}\,(t) = 0.05C(t) - 1263.273926 - 26.72283855A(t) - 0.149456522[A(t)]^2,\\
&A(5) = -5,\ \text{and}\ C(5) = 380.
\end{aligned}$$

The values of $A(t)$ and $C(t)$ over the time interval $[0,5]$ are computed and presented in Figs. 13.2a, b.

The cooperative strategies are:

$$\begin{aligned}
&u_1^*(t) = \varpi_1(t,x) = -0.4A(t)x^{1/2}, u_2^*(t) = \varpi_2(t,x) = -0.1A(t)x^{1/2},\\
&v_1^*(t) = \psi_1(t,x) = \frac{216204942}{17852140} - \frac{6597530}{17852140}A(t),\\
&v_2^*(t) = \psi_2(t,x) = \frac{41195291}{8926070} - \frac{8635002.5}{8926070}A(t).
\end{aligned} \tag{4.13}$$

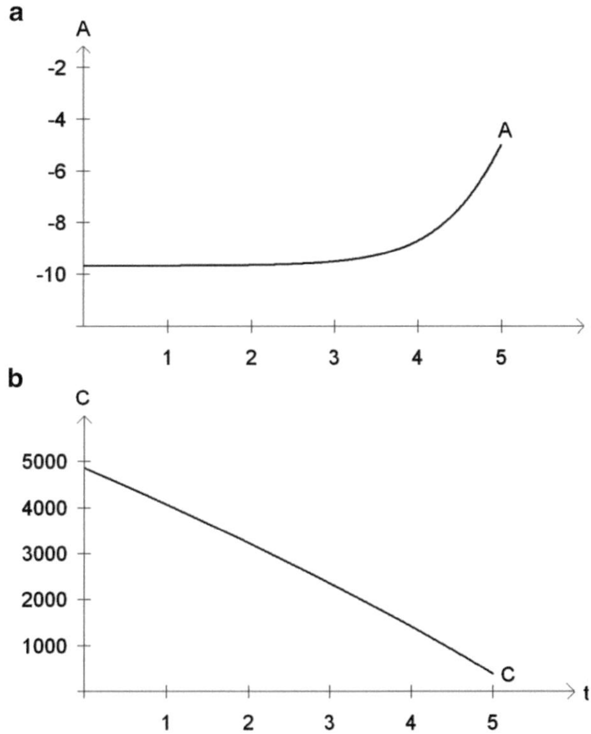

Fig. 13.2 The values of $A(t)$ and $C(t)$ over the time interval [0,5]

Substituting the cooperative strategies into (4.13) yields the dynamics of pollution accumulation under cooperation as:

$$dx(t) = \left[\frac{5907.5}{197} - \frac{35}{197}\left[\frac{216204942}{17852140} - \frac{6597530}{17852140}A(t)\right] - \frac{47.5}{197}\left[\frac{41195291}{8926070} - \frac{8635002.5}{8926070}A(t)\right] + 0.16A(t)x + 0.02A(t)x - 0.01x \right] dt + 0.05x(s)dz(s).$$

Sharing the expected gain from cooperation proportional to the nations' relative sizes of expected noncooperative payoffs yields:

$$\xi^{(\tau)i}(\tau, x_\tau^*) = \frac{\left[A_i(\tau)x_\tau^* + C_i(\tau)\right]}{\sum_{j=1}^{2}\left[A_j(\tau)x_\tau^* + C_j(\tau)\right]}\left[A(\tau)x_\tau^* + C(\tau)\right] \tag{4.14}$$

for $i \in \{1, 2\}, x_\tau^* \in X_\tau^*$ and $\tau \in [t_0, T]$.

Following Theorem 3.1, a subgame consistent payment distribution procedure consists of a terminal payment $2\left[100 - x_T^*\right]$ to nation 1 and a terminal payment $3\left[60 - x_T^*\right]$ to nation 2 at time T and an instantaneous payment at time $\tau \in [t_0, T]$:

$$B_i(\tau, x_\tau^*) = \frac{-\left[A_i(\tau)x_\tau^* + C_i(\tau)\right]}{\left(\sum_{j=1}^{2}\left[A_j(\tau)x_\tau^* + C_j(\tau)\right]\right)}\left\{\left[\dot{A}(\tau)x_\tau^* + \dot{C}(\tau)\right] - 0.05\left[A(\tau)x_\tau^* + C(\tau)\right]\right\}$$

$$- \frac{\left[A(\tau)x_\tau^* + C(\tau)\right]}{\left(\sum_{j=1}^{2}\left[A_j(\tau)x_\tau^* + C_j(\tau)\right]\right)}\left\{\left[\dot{A}_i(\tau)x_\tau^* + \dot{C}_i(\tau)\right] - 0.05\left[A_i(\tau)x_\tau^* + C_i(\tau)\right]\right\}$$

$$+ \frac{\left[A_i(\tau)x_\tau^* + C_i(\tau)\right]\left[A(\tau)x_\tau^* + C(\tau)\right]}{\left(\sum_{j=1}^{2}\left[A_j(\tau)x_\tau^* + C_j(\tau)\right]\right)^2}$$

$$\sum_{j=1}^{2}\left\{\left[\dot{A}_j(\tau)x_\tau^* + \dot{C}_j(\tau)\right] - 0.05\left[A_j(\tau)x_\tau^* + C_j(\tau)\right]\right\}$$

$$+ \left[\frac{\left[A_i(\tau)x_\tau^* + C_i(\tau)\right]\left[A(\tau)x_\tau^* + C(\tau)\right]}{\left(\sum_{j=1}^{2}\left[A_j(\tau)x_\tau^* + C_j(\tau)\right]\right)^2}\left(\sum_{j=1}^{2}A_j(\tau)\right)\right.$$

$$\left. - \frac{\left[A_i(\tau)x_\tau^* + C_i(\tau)\right]A(\tau) + \left[A(\tau)x_\tau^* + C(\tau)\right]A_i(\tau)}{\sum_{j=1}^{2}\left[A_j(\tau)x_\tau^* + C_j(\tau)\right]}\right] \times \left[\frac{5907.5}{197}\right.$$

$$- \frac{35}{197}\left[\frac{216204942}{17852140} - \frac{6597530}{17852140}A(\tau)\right] - \frac{47.5}{197}\left[\frac{41195291}{8926070} - \frac{8635002.5}{8926070}A(\tau)\right]$$

$$\left. + 0.16A(\tau)x_\tau^* + 0.02A(\tau)x_\tau^* - 0.01x_\tau^*\right], \text{for } i \in \{1, 2\}.$$

When both nations are adopting the cooperative strategies the rate of instantaneous payment that nation 1 will realize at time t with the state being x_t^* can be expressed as

$$\mathfrak{R}_1(t, x_t^*) = 763.0983415 - 1.063694026A(t) - 0.115784499[A(t)]^2 - 0.08[A(t)]^2 x_t^* - 4x_t^*. \tag{4.15}$$

Similarly, the rate of instantaneous payment that nation 2 will realize at time t with the state being x_t^* can be expressed as

$$\mathfrak{R}_2\left(t,x_t^*\right) = 500.175621 + 1.063674026A(t) - 0.033672022[A(t)]^2 \\ - 0.01[A(t)]^2 x_t^* - 5x_t^*. \tag{4.16}$$

A side payment of the value $B_\ell\left(t,x_t^*\right) - \mathfrak{R}_\ell\left(t,x_t^*\right)$ will be offered to nation $\ell \in \{1,2\}$. The values of $B_1(t,x_t^*), B_2(t,x_t^*), \mathfrak{R}_1(t,x_t^*), \mathfrak{R}_2(t,x_t^*)$ together with the side payment nation 1 and nation 2 will receive at different time t with given x_t^* are given in Table 13.1 below.

13.5 Extension to Uncertainty in Payoffs

In this section we incorporate uncertainty in future payoffs into the cooperative environmental management presented in the previous sections. Uncertainties in future payoffs are prevalent in fast developing countries. This type of uncertainties often hinders the reaching of cooperative agreements in joint pollution control initiatives. Subgame consistent cooperative schemes provide an effective mean to resolve the problem.

13.5.1 Game Formulation and Non-cooperative Outcome

Consider a randomly furcating counterpart of the stochastic differential game of environmental management in Sect. 13.1 in which the future payoffs are not known with certainty. The game horizon is $[t_0, T]$. When the game commences at t_0, the demand structures, production costs and impacts of the pollution stock of the nations are known. In future instants of time t_k $(k = 1, 2, \cdots, m)$, where $t_0 < t_m < T \equiv t_{m+1}$, the demand structures, production costs and pollution impacts in the time interval $\left[t_k, t_{k+1}\right)$ are affected by a series of random events Θ^k. In particular, Θ^k for $k \in \{1, 2, \cdots, m\}$, are independent and identically distributed random variables with range $\{\theta_1, \theta_2, \ldots, \theta_\eta\}$ and corresponding probabilities

Table 13.1 PDP and transfer payments of nations 1 and 2

t	x_t^*	$B_1(t,x_t^*)$	$\mathfrak{R}_1(t,x_t^*)$	Nation 1 side-pay	$B_2(t,x_t^*)$	$\mathfrak{R}_2(t,x_t^*)$	Nation 2 side-pay
0.5	16.274	417.369	575.71	−158.341	638.585	390.152	248.433
1	14.723	403.075	593.578	−190.503	627.295	399.367	227.928
2	13.825	398.045	604.409	−206.364	616.929	404.801	212.128
3	13.773	404.331	608.059	−203.728	606.139	405.705	200.434
3.5	13.941	409.816	611.371	−201.555	597.160	405.729	191.431
4	14.396	415.961	618.542	−202.581	583.725	405.439	178.286
4.5	15.532	421.425	633.153	−211.728	564.143	404.029	160.114

$\{\lambda_1, \lambda_2, \ldots, \lambda_\eta\}$. Changes in preference, legal arrangements, technology and the physical environments are examples of factors which constitute to these uncertainties.

In the time interval $[t_k, t_{k+1})$ for $k = (1, 2, \cdots, m)$ if the random event θ_{a_k} for $a_k \in \{1, 2, \cdots, \eta\}$ is realized the demand function of the output of nation $i \in N \equiv \{1, 2, \cdots, n\}$ at time instant s is $P_i(s) = \alpha^i_{\theta_{a_k}} - \sum_{j=1}^{n} \beta^i_j q_j(s)$, the unit cost of production is $c_{i(\theta_{a_k})}$, and the value of damage to country i from $x(s)$ amount of pollution is $h_i^{\theta_{a_k}} x(s)$. When the game commences at t_0, the demand structures, production costs and pollution impact in the interval $[t_0, t_1)$ are known to be $P_i(s) = \alpha^i_{\theta_1} - \sum_{j=1}^{n} \beta^i_j q_j(s), c_{i(\theta_1)}$ and $h_i^{\theta_1} x(s)$.

Industrial profits of nation i at time $s \in [t_k, t_{k+1})$ if θ_{a_k} is realized can be expressed as:

$$\pi_i^{\theta_{a_k}}(s) = \left[\alpha^i_{\theta_{a_k}} - \sum_{j=1}^{n} \beta^i_j q_j(s)\right] q_i(s) - c_i^{\theta_{a_k}} q_i(s) - v_i^{\theta_{a_k}}(s) q_i(s), \text{ for } i \in N, \tag{5.1}$$

where $v_i^{\theta_{a_k}}(s) \geq 0$ is the tax rate imposed by government i on its industrial output at time $s \in [t_k, t_{k+1})$.

In a competitive market equilibrium firms will produce up to a point where marginal cost of production equals marginal revenue and the first order condition for a Nash equilibrium for the n nations economy yields

$$\sum_{j=1}^{n} \beta^i_j q_j(s) + \beta^i_i q_i(s) = \alpha^i_{\theta_{a_k}} - c_i^{\theta_{a_k}} - v_i^{\theta_{a_k}}(s), \text{ for } i \in N. \tag{5.2}$$

With output tax rates $v^{\theta_{a_k}}(s) = \left\{v_1^{\theta_{a_k}}(s), v_2^{\theta_{a_k}}(s), \cdots, v_n^{\theta_{a_k}}(s)\right\}$ being regarded as parameters by firms (5.2) becomes a system of equations linear in $q(s) = \{q_1(s), q_2(s), \cdots, q_n(s)\}$. Solving (1.3) yields an industry equilibrium with output in industry i being

$$q_i(s) = \phi_i\left(v^{\theta_{a_k}}(s)\right) = \overline{\alpha}^i_{\theta_{a_k}} + \sum_{j \in N} \overline{\beta}_j^{i(\theta_{a_k})} \; v_j^{\theta_{a_k}}(s), \tag{5.3}$$

where $\overline{\alpha}^i_{\theta_{a_k}}$ and $\overline{\beta}_j^{i(\theta_{a_k})}$, for $i \in N$ and $j \in N$, are constants involving the model parameters
$\{\beta_1^1, \beta_2^1, \cdots, \beta_n^1; \beta_1^2, \beta_2^2, \cdots, \beta_n^2; \cdots; \beta_1^n, \beta_2^n, \cdots, \beta_n^n\}$, $\left\{\alpha^1_{\theta_{a_k}}, \alpha^2_{\theta_{a_k}}, \cdots, \alpha^n_{\theta_{a_k}}\right\}$ and $\left\{c_1^{\theta_{a_k}}, c_2^{\theta_{a_k}}, \cdots, c_n^{\theta_{a_k}}\right\}$

The instantaneous objective of government i at time $s \in [t_k, t_{k+1})$ can be expressed as:

$$\left[\alpha^i_{\theta_{a_k}} - \sum_{j=1}^{n} \beta^i_j q_j(s)\right] q_i(s) - c_i^{\theta_{a_k}} q_i(s) - c_i^a \left[u_i^{\theta_{a_k}}(s)\right]^2 - \sum_{j \in \overline{K}^i} \varepsilon^j_i \left[q_j(s)\right] - h_i^{\theta_{a_k}} x(s),\ i \in N \tag{5.4}$$

By substituting $q_i(s)$, for $i \in N$, from (5.3) into (5.4) and (1.5) one obtains a randomly furcating stochastic differential game in which government $i \in N$ seeks to maximize its payoff:

$$E_{t_0}\left\{\int_{t_0}^{t_1}\left[\left(\alpha^i_{\theta_1} - \sum_{j=1}^{n}\beta^i_j\left[\overline{\alpha}^j_{\theta_1} + \sum_{h\in N}\overline{\beta}_h^{j(\theta_1)} v_h^{\theta_1}(s)\right]\right)\left[\overline{\alpha}^i_{\theta_1} + \sum_{h\in N}\overline{\beta}_h^{i(\theta_1)} v_h^{\theta_1}(s)\right]\right.\right.$$

$$- c_i^{\theta_1}\left[\overline{\alpha}^i_{\theta_1} + \sum_{j\in N}\overline{\beta}_j^{i(\theta_1)} v_j^{\theta_1}(s)\right] - c_i^a\left[u_i^{\theta_1}(s)\right]^2 - \sum_{j\in\overline{K}^i}\varepsilon^j_i\left[\overline{\alpha}^j_{\theta_1} + \sum_{\ell\in N}\overline{\beta}_\ell^{j(\theta_1)} v_\ell^{\theta_1}(s)\right]$$

$$\left. - h_i^{\theta_1} x(s)\right] e^{-r(s-t_0)} ds$$

$$+ \sum_{k=1}^{m}\sum_{a_k=1}^{\eta}\lambda_{a_k}\int_{t_k}^{t_{k+1}}\left[\left(\alpha^i_{\theta_{a_k}} - \sum_{j=1}^{n}\beta^i_j\left[\overline{\alpha}^j_{\theta_{a_k}} + \sum_{h\in N}\overline{\beta}_h^{j(\theta_{a_k})} v_h^{\theta_{a_k}}(s)\right]\right)\left[\overline{\alpha}^i_{\theta_{a_k}} + \sum_{h\in N}\overline{\beta}_h^{i(\theta_{a_k})} v_h^{\theta_{a_k}}(s)\right]\right.$$

$$- c_i^{\theta_{a_k}}\left[\overline{\alpha}^i_{\theta_{a_k}} + \sum_{j\in N}\overline{\beta}_j^{i(\theta_{a_k})} v_j^{\theta_{a_k}}(s)\right] - c_i^a\left[u_i^{\theta_{a_k}}(s)\right]^2 - \sum_{j\in\overline{K}^i}\varepsilon^j_i\left[\overline{\alpha}^j_{\theta_{a_k}} + \sum_{\ell\in N}\overline{\beta}_\ell^{j(\theta_{a_k})} v_\ell^{\theta_{a_k}}(s)\right]$$

$$\left.\left. - h_i^{\theta_{a_k}} x(s)\right] e^{-r(s-t_0)} ds - g^i\left[x(T) - \overline{x}^i\right] e^{-r(T-t_0)}\right\} \tag{5.5}$$

subject to

$$dx(s) = \left[\sum_{j=1}^{n} a_j\left[\overline{\alpha}^j_{\theta_{a_k}} + \sum_{h\in N}\overline{\beta}_h^{j(\theta_{a_k})} v_h^{\theta_{a_k}}(s)\right] - \sum_{j=1}^{n} b_j u_j^{\theta_{a_k}}(s)[x(s)]^{1/2} - \delta x(s)\right] ds$$

$$+ \sigma x(s) dz(s),\ for\, s \in [t_k, t_{k+1})\ \text{if}\ \theta_{a_k}\ \text{occurs in the interval}\ [t_k, t_{k+1}),$$

$$\text{and}\ x(t_0) = x_{t_0} \tag{5.6}$$

Invoking 1.1 in Chap. 4 a Nash equilibrium of the randomly furcating stochastic differential game (5.5 and 5.6) can be characterized by the following theorem.

Theorem 5.1 A set of feedback strategies $\{u_i^{(m)\theta_{am}*}(t) = \mu_i^{(m)\theta_{am}}(t,x), v_i^{(m)\theta_{am}*}(t) = \phi_i^{(m)\theta_{am}}(t,x)$, for $t \in [t_m, T]$; $u_i^{(k)\theta_{a_k}*}(t) = \mu_i^{(k)\theta_{a_k}}(t,x), v_i^{(k)\theta_{a_k}*}(t) = \phi_i^{(k)\theta_{a_k}}(t,x)$, for $t \in [t_k, t_{k+1}), k \in \{0,1,2,\cdots,m-1\}$ and $i \in N\}$, contingent upon the events $\theta_{a_m} \in \{\theta_1, \theta_2, \ldots, \theta_\eta\}$ and $\theta_{a_k} \in \{\theta_1, \theta_2, \ldots, \theta_\eta\}$ for $k \in \{1,2,\cdots,m-1\}$ constitutes a Nash equilibrium solution for the game (5.5 and 5.6), if there exist continuously differentiable functions $V^{i[\theta_{am}](m)}(t,x) : [t_m, T] \times R \to R$ and $V^{i[\theta_{a_k}](k)}(t,x) : [t_k, t_{k+1}] \times R \to R$, for $k \in \{1,2,\cdots,m-1\}$ and $i \in N$, which satisfy the following partial differential equations:

$$
\begin{aligned}
&-V_t^{i[\theta_{am}](m)}(t,x) - \frac{\sigma^2 x^2}{2} V_{xx}^{i[\theta_{am}](m)}(t,x) \\
&= \max_{v_i, u_i} \Bigg\{ \Bigg[\Bigg(\alpha^i_{\theta_{am}} - \sum_{j=1}^n \beta^i_j \Bigg[\overline{\alpha}^j_{\theta_{am}} + \sum_{\substack{h \in N \\ h \neq i}}^n \overline{\beta}_h^{j(\theta_{am})} \phi_h^{m(\theta_{am})}(t,x) + \overline{\beta}_i^{j(\theta_{am})} v_i \Bigg] \Bigg) \\
&\times \Bigg[\overline{\alpha}^i_{\theta_{am}} + \sum_{\substack{h \in N \\ h \neq i}} \overline{\beta}_h^{i(\theta_{am})} \phi_h^{m(\theta_{am})}(t,x) + \overline{\beta}_i^{i(\theta_{am})} v_i \Bigg] \\
&- c_i^{\theta_{am}} \Bigg[\overline{\alpha}^i_{\theta_{am}} + \sum_{\substack{j \in N \\ j \neq i}} \overline{\beta}_j^{i(\theta_{am})} \phi_j^{m(\theta_{am})}(t,x) + \overline{\beta}_i^{i(\theta_{am})} v_i \Bigg] - c_i^a [u_i]^2 \\
&- \sum_{j \in \overline{K}^i} \varepsilon^j_i \Bigg[\overline{\alpha}^j_{\theta_{am}} + \sum_{\substack{\ell \in N \\ \ell \neq i}} \overline{\beta}_\ell^{j(\theta_{am})} \phi_\ell^{m(\theta_{am})}(t,x) + \overline{\beta}_i^{j(\theta_{am})} v_i \Bigg] - h_i^{\theta_{am}} x \Bigg] e^{-r(t-t_0)} \\
&+ V_x^{i[\theta_{am}](m)}(t,x) \Bigg[\sum_{j=1}^n a_j \Bigg[\overline{\alpha}^j_{\theta_{am}} + \sum_{\substack{h \in N \\ h \neq i}} \overline{\beta}_h^{j(\theta_{am})} \phi_h^{m(\theta_{am})}(t,x) + \overline{\beta}_i^{j(\theta_{am})} v_i \Bigg] \\
&- \sum_{\substack{j=1 \\ j \neq i}}^n b_j \mu_j^{m(\theta_{am})}(t,x) x^{1/2} - b_i u_i x^{1/2} - \delta x \Bigg] \Bigg\}, \\
&V^{i[\theta_{am}](m)}(T,x) = -g^i [x - \overline{x}^i] e^{-r(T-t_0)};
\end{aligned}
\tag{5.7}
$$

$$
\begin{aligned}
&-V_t^{i[\theta_{a_k}](k)}(t,x) - \frac{\sigma^2 x^2}{2} V_{xx}^{i[\theta_{a_k}](k)}(t,x) \\
&= \max_{v_i,\, u_i} \left\{ \left[\left(\alpha_{\theta_{a_k}}^i - \sum_{j=1}^{n} \beta_j^i \left[\overline{\alpha}_{\theta_{a_k}}^j + \sum_{\substack{h \in N \\ h \neq i}}^{n} \overline{\beta}_h^{j(\theta_{a_k})} \phi_h^{k(\theta_{a_k})}(t,x) + \overline{\beta}_i^{j(\theta_{a_k})} v_i \right] \right) \right. \right. \\
&\times \left[\overline{\alpha}_{\theta_{a_k}}^i + \sum_{\substack{h \in N \\ h \neq i}} \overline{\beta}_h^{i(\theta_{a_k})} \phi_h^{k(\theta_{a_k})}(t,x) + \overline{\beta}_i^{i(\theta_{a_k})} v_i \right] \\
&-c_i^{\theta_{a_k}} \left[\overline{\alpha}_{\theta_{a_k}}^i + \sum_{\substack{j \in N \\ j \neq i}} \overline{\beta}_j^{i(\theta_{a_k})} \phi_j^{k(\theta_{a_k})}(t,x) + \overline{\beta}_i^{i(\theta_{a_k})} v_i \right] - c_i^a [u_i]^2 \\
&\left. -\sum_{j \in \overline{K}^i} \varepsilon_i^j \left[\overline{\alpha}_{\theta_{a_k}}^j + \sum_{\substack{\ell \in N \\ \ell \neq i}} \overline{\beta}_\ell^{j(\theta_{a_k})} \phi_\ell^{k(\theta_{a_k})}(t,x) + \overline{\beta}_i^{j(\theta_{a_k})} v_i \right] - h_i^{\theta_{am}} x \right] e^{-r(t-t_0)} \\
&+V_x^{i[\theta_{a_k}](k)}(t,x) \left[\sum_{j=1}^{n} a_j \left[\overline{\alpha}_{\theta_{a_k}}^j + \sum_{\substack{h \in N \\ h \neq i}} \overline{\beta}_h^{j(\theta_{a_k})} \phi_h^{k(\theta_{a_k})}(t,x) + \overline{\beta}_i^{j(\theta_{a_k})} v_i \right] \right. \\
&\left. \left. - \sum_{\substack{j=1 \\ j \neq i}}^{n} b_j \mu_j^{k(\theta_{a_k})}(t,x) x^{1/2} - b_i u_i x^{1/2} - \delta x \right] \right\},
\end{aligned}
$$

$$
V^{i[\theta_{a_k}](k)}(t_{k+1},x) = \sum_{a_{k+1}=1}^{\eta} \lambda_{a_{k+1}} V^{i[\theta_{a_{k+1}}](k+1)}(t_{k+1},x), \quad \text{for } \ i \in N \text{ and } k \in \{0,1,2,\cdots,m-1\} \tag{5.8}
$$

Proof Follow the proof of Theorem 1.1 in Chap. 4. ■

Following the analysis in Sect. 13.2 we perform the indicated maximizations in (5.7 and 5.8) to obtain the game equilibrium strategies and the value functions:

$$
V^{i[\theta_{a_k}](k)}(t,x) = \left[A_{k(\theta_{a_k})}^i(t) x + C_{k(\theta_{a_k})}^i(t) \right] e^{-r(t-t_0)}, \tag{5.9}
$$

for $i \in N$ and $k \in \{0,1,2,\cdots,m-1\}$,

where $A^i_{k(\theta_{a_k})}(t)$ and $C^i_{k(\theta_{a_k})}(t)$, for $i \in N$ and $k \in \{0, 1, 2, \cdots, m-1\}$ satisfy a set of constant coefficient quadratic ordinary differential equations similar to that in Proposition 2.1.

13.5.2 Cooperative Arrangement

Now consider the case when all the nations want to cooperate and agree to act so that an international optimum could be achieved. For the cooperative scheme to be upheld throughout the game horizon both group rationality and individual rationality are required to be satisfied at any time. In addition, to ensure that the cooperative solution is dynamically stable, the agreement must be subgame-consistent.

13.5.2.1 Group Optimality and Individual Rationality

To secure group optimality the participating nations seek to maximize their joint expected payoff

$$
\begin{aligned}
E_{t_0}\Bigg\{ & \int_{t_0}^{t_1} \sum_{i=1}^{n} \Bigg[\Bigg(\alpha^i_{\theta_1} - \sum_{j=1}^{n} \beta^i_j \Big[\overline{\alpha}^j_{\theta_1} + \sum_{h\in N} \overline{\beta}^{j(\theta_1)}_h \; v^{\theta_1}_h(s) \Big] \Bigg) \Big[\overline{\alpha}^i_{\theta_1} + \sum_{h\in N} \overline{\beta}^{i(\theta_1)}_h \; v^{\theta_1}_h(s) \Big] \\
& - c^{\theta_1}_i \Big[\overline{\alpha}^i_{\theta_1} + \sum_{j\in N} \overline{\beta}^{i(\theta_1)}_j \; v^{\theta_1}_j(s) \Big] - c^a_i \big[u^{\theta_1}_i(s) \big]^2 - \sum_{j\in \overline{K}^i} \varepsilon^j_i \Big[\overline{\alpha}^j_{\theta_1} + \sum_{\ell\in N} \overline{\beta}^{j(\theta_1)}_\ell \; v^{\theta_1}_\ell(s) \Big] \\
& - h^{\theta_1}_i x(s) \Bigg] e^{-r(s-t_0)} ds \\
& + \sum_{k=1}^{m} \sum_{a_k=1}^{\eta} \lambda_{a_k} \int_{t_h}^{t_{h+1}} \sum_{i=1}^{n} \Bigg[\Bigg(\alpha^i_{\theta_{a_k}} \\
& - \sum_{j=1}^{n} \beta^i_j \Big[\overline{\alpha}^j_{\theta_{a_k}} + \sum_{h\in N} \overline{\beta}^{j(\theta_{a_k})}_h \; v^{\theta_{a_k}}_h(s) \Big] \Bigg) \Big[\overline{\alpha}^i_{\theta_{a_k}} + \sum_{h\in N} \overline{\beta}^{i(\theta_{a_k})}_h \; v^{\theta_{a_k}}_h(s) \Big] \\
& - c^{\theta_{a_k}}_i \Big[\overline{\alpha}^i_{\theta_{a_k}} + \sum_{j\in N} \overline{\beta}^{i(\theta_{a_k})}_j \; v^{\theta_{a_k}}_j(s) \Big] - c^a_i \Big[u^{\theta_{a_k}}_i(s) \Big]^2 \\
& - \sum_{j\in \overline{K}^i} \varepsilon^j_i \Big[\overline{\alpha}^j_{\theta_{a_k}} + \sum_{\ell\in N} \overline{\beta}^{j(\theta_{a_k})}_\ell \; v^{\theta_{a_k}}_\ell(s) \Big] - h^{\theta_{a_k}}_i x(s) \Bigg] e^{-r(s-t_0)} ds \\
& - \sum_{i=1}^{n} g^i \big[x(T) - \overline{x}^i \big] e^{-r(T-t_0)} \Bigg\}
\end{aligned}
\tag{5.10}
$$

subject to (5.6)

Invoking Theorem 2.1 in Chap. 4 an optimal solution to the randomly furcating stochastic control problem (5.6) and (5.10) can be characterized by the theorem below.

Theorem 5.2 A set of control strategies $\{u_i^{(m)\theta_{am}*}(t) = \varpi_i^{(m)\theta_{am}}(t,x), v_i^{(m)\theta_{am}*}(t) = \psi_i^{(m)\theta_{am}}(t,x)$, for $t \in [t_m, T]$; $u_i^{(k)\theta_{a_k}*}(t) = \varpi_i^{(k)\theta_{a_k}}(t,x), v_i^{(k)\theta_{a_k}*}(t) = \psi_i^{(k)\theta_{a_k}}(t,x)$, for $t \in [t_k, t_{k+1})$, $k \in \{0,1,2,\cdots,m-1\}$ and $i \in N\}$, contingent upon the events $\theta_{a_m} \in \{\theta_1, \theta_2, \ldots, \theta_\eta\}$ and $\theta_{a_k} \in \{\theta_1, \theta_2, \ldots, \theta_\eta\}$ for $k \in \{1,2,\cdots,m-1\}$ constitutes a Nash equilibrium solution for the game (5.5 and 5.6), if there exist continuously differentiable functions $W^{[\theta_{am}](m)}(t,x) : [t_m, T] \times R \to R$ and $W^{[\theta_{a_k}](k)}(t,x) : [t_k, t_{k+1}] \times R \to R$, for $k \in \{1,2,\cdots,m-1\}$ and $i \in N$, which satisfy the following partial differential equations:

$$
\begin{aligned}
&-W_t^{[\theta_{am}](m)}(t,x) - \frac{\sigma^2 x^2}{2} W_{xx}^{[\theta_{am}](m)}(t,x) = \\
&\max_{v_1, v_2, \cdots, v_n; u_1, u_2, \cdots, u_n} \Bigg\{ \\
&\sum_{i=1}^n \Bigg[\Bigg(\alpha_{\theta_{am}}^i - \sum_{j=1}^n \beta_j^i \Bigg[\overline{\alpha}_{\theta_{am}}^j + \sum_{h \in N} \overline{\beta}_h^{j(\theta_{am})} v_h \Bigg] \Bigg) \Bigg[\overline{\alpha}_{\theta_{am}}^i + \sum_{h \in N} \overline{\beta}_h^{i(\theta_{am})} v_h \Bigg] \\
&\quad - c_i^{\theta_{am}} \Bigg[\overline{\alpha}_{\theta_{a_k}}^i + \sum_{j \in N} \overline{\beta}_j^{i(\theta_{am})} v_j \Bigg] - c_i^a [u_i]^2 - \sum_{j \in \overline{K}^i} \varepsilon_i^j \Bigg[\overline{\alpha}_{\theta_{am}}^j + \sum_{\ell \in N} \overline{\beta}_\ell^{j(\theta_{am})} v_\ell \Bigg] \\
&\quad - h_i^{\theta_{am}} x \Bigg] \\
&\quad + W_x^{[\theta_{am}](m)}(t,x) \Bigg[\sum_{j=1}^n a_j \Bigg[\overline{\alpha}_{\theta_{am}}^j + \sum_{h \in N} \overline{\beta}_h^{j(\theta_{am})} v_h \Bigg] - \sum_{j=1}^n b_j u_j x^{1/2} - \delta x \Bigg] \Bigg\}, \\
&W^{[\theta_{am}](m)}(T,x) = -\sum_{i=1}^n g^i [x - \overline{x}^i] e^{-r(T-t_0)}; \\
&-W_t^{[\theta_{a_k}](k)}(t,x) - \frac{\sigma^2 x^2}{2} W_{xx}^{[\theta_{a_k}](k)}(t,x) = \\
&\max_{v_1, v_2, \cdots, v_n; u_1, u_2, \cdots, u_n} \\
&\Bigg\{ \sum_{i=1}^n \Bigg[\Bigg(\alpha_{\theta_{a_k}}^i - \sum_{j=1}^n \beta_j^i \Bigg[\overline{\alpha}_{\theta_{a_k}}^j + \sum_{h \in N} \overline{\beta}_h^{j(\theta_{a_k})} v_h \Bigg] \Bigg) \Bigg[\overline{\alpha}_{\theta_{a_k}}^i + \sum_{h \in N} \overline{\beta}_h^{i(\theta_{a_k})} v_h \Bigg] \\
&\quad - c_i^{\theta_{a_k}} \Bigg[\overline{\alpha}_{\theta_{a_k}}^i + \sum_{j \in N} \overline{\beta}_j^{i(\theta_{a_k})} v_j \Bigg] - c_i^a [u_i]^2 - \sum_{j \in \overline{K}^i} \varepsilon_i^j \Bigg[\overline{\alpha}_{\theta_{a_k}}^j + \sum_{\ell \in N} \overline{\beta}_\ell^{j(\theta_{a_k})} v_\ell \Bigg] - h_i^{\theta_{a_k}} x \Bigg] \\
&\quad + W_x^{[\theta_{a_k}](k)}(t,x) \Bigg[\sum_{j=1}^n a_j \Bigg[\overline{\alpha}_{\theta_{a_k}}^j + \sum_{h \in N} \overline{\beta}_h^{j(\theta_{a_k})} v_h \Bigg] - \sum_{\substack{j=1 \\ j \neq i}}^n b_j u_j x^{1/2} - \delta x \Bigg] \Bigg\}, \\
&W^{[\theta_{a_k}](k)}(t_{k+1},x) = \sum_{a_{k+1}=1}^{\eta} \lambda_{a_{k+1}} W^{[\theta_{a_{k+1}}](k+1)}(t_{k+1},x), \text{ for } k \in \{0,1,2,\cdots,m-1\}.
\end{aligned}
$$

Proof Follow the proof of Theorem 2.1 in Chap. 4. ■

Following the analysis in Sect. 13.3 we perform the indicated maximizations in Theorem 5.2 to obtain the game equilibrium strategies and the value functions:

$$W^{[\theta_{a_k}](k)}(t,x) = \left[A_{k(\theta_{a_k})}(t)x + C_{k(\theta_{a_k})}(t)\right]\ e^{-r(t-t_0)}, \tag{5.11}$$

for $k \in \{0,1,2,\cdots,m-1\}$,

where $A_{k(\theta_{a_k})}(t)$ and $C_{k(\theta_{a_k})}(t)$, for $k \in \{0,1,2,\cdots,m-1\}$ satisfy a set of ordinary differential equations similar to that in Proposition 3.1.

Assume that at time t_0 when the initial state is x_0 the agreed upon optimality principle assigns a set of imputation vectors contingent upon the events θ_1 and θ_{a_k} for $\theta_{a_k} \in \{\theta_1,\theta_2,\ \ \ldots,\theta_\eta\}$ and $k \in \{1,2,\cdots,m\}$. We use

$$\left[\xi^{1[\theta_1](0)}(t_0,x_0), \xi^{2[\theta_1^0](0)}(t_0,x_0), \cdots, \xi^{n[\theta_1](0)}(t_0,x_0)\right]$$

to denote an imputation vector of the gains in such a way that the share of the ith player over the time interval $[t_0,T]$ is equal to $\xi^{i[\theta_1](0)}(t_0,x_0)$.

Individual rationality requires that

$$\xi^{i[\theta_1](0)}(t_0,x_0) \geq V^{i[\theta_1](0)}(t_0,x_0),\ \text{for}\ i \in N.$$

In a dynamic framework, individual rationality has to be maintained at every instant of time $t \in [t_0,T]$ along the cooperative trajectory. At time t, for $t \in [t_0,t_1)$, individual rationality requires:

$$\xi^{i[\theta_1](0)}(t,x_t^*) \geq V^{i[\theta_1](0)}(t,x_t^*),\ \text{for}\ i \in N.$$

At time t_k, for $k \in \{1,2,\cdots,m\}$, if $\theta_{a_k} \in \{\theta_1,\theta_2,\ \ \ldots,\theta_\eta\}$ has occurred and the state is $x_{t_k}^*$, the same optimality principle assigns an imputation vector $\left[\xi^{1[\theta_{a_k}](k)}\left(t_k,x_{t_k}^*\right), \xi^{2[\theta_{a_k}](k)}\left(t_k,x_{t_k}^*\right), \cdots, \xi^{n[\theta_{a_k}](k)}\left(t_k,x_{t_k}^*\right)\right]$ (in current value at time t_k). Individual rationality is satisfied if:

$$\xi^{i[\theta_{a_k}](k)}\left(t_k,x_{t_k}^*\right) \geq V^{i[\theta_{a_k}](k)}\left(t_k,x_{t_k}^*\right).\ \text{for}\ i \in N.$$

At time t, for $t \in [t_k,t_{k+1})$, individual rationality requires:

$$\xi^{i[\theta_{a_k}](k)}(t,x_t^*) \geq V^{i[\theta_{a_k}](k)}(t,x_t^*).\ \text{for}\ i \in N.$$

13.5.2.2 Subgame-Consistent Imputation

Finally, we would derive a set of imputation that would like to a subgame consistent solution. Invoking Theorem 3.1 in Chap. 4, a subgame consistent PDP can be derived with the theorem below.

Theorem 5.3 A PDP with a terminal payment $q^i\left(x_T^*\right))$ at time T and an instantaneous payment (in present value) at time $\tau \in [t_k, t_{k+1}]$:

$$B_i^{(\theta_{a_k})k}(\tau) = -\left[\xi_t^{i[\theta_{a_k}](k)}\left(t, x_t^*\right)|_{t=\tau}\right] - \frac{\sigma^2\left(x_\tau^*\right)^2}{2}\left[\xi_{x_t^* x_t^*}^{i[\theta_{a_k}](k)}\left(t, x_t^*\right)|_{t=\tau}\right]$$
$$-\left[\xi_{x_t^*}^{i[\theta_{a_k}](k)}\left(t, x_t^*\right)|_{t=\tau}\right]$$
$$\times\left[\sum_{j=1}^n a_j\left[\overline{\alpha}_{\theta_{a_k}}^j + \sum_{h\in N}\overline{\beta}_h^{j(\theta_{a_k})}\ \psi_h^{(k)\theta_{a_k}}\left(\tau, x_\tau^*\right)\right] - \sum_{j=1}^n b_j \varpi_j^{(k)\theta_{a_k}}\left(\tau, x_\tau^*\right)\left(x_\tau^*\right)^{1/2} - \delta x_\tau^*\right], \tag{5.12}$$

for $i \in N$ and $k \in \{1, 2, \cdots, m\}$,

contingent upon $\theta_{a_k}^k \in \{\theta_1, \theta_2, \ \ldots, \theta_\eta\}$ has occurred at time t_k,

yields a subgame-consistent cooperative solution to the randomly furcating stochastic differential game (5.1 and 5.2).

Proof Follow the proof of Theorem 3.1 in Chap. 4. ■

Thus a subgame consistent cooperative solution is established.

13.6 Appendices

Appendix A: Proof of Proposition 2.1

Using (2.3), (2.5) and (2.6), system (2.1 and 2.2) can be expressed as:

$$r[A_i(t)x + C_i(t)] \ - [\dot{A}_i(t)x + \dot{C}_i(t)]$$
$$= \Bigg[\left(\alpha^i - \sum_{j=1}^n \beta_j^i\left\{\overline{\alpha}^j + \sum_{h\in Ni}^n \overline{\beta}_h^j\left[\hat{\alpha}^h + \sum_{k\in N}\hat{\beta}_k^h\ A_k(t)\right]\right\}\right)$$
$$\left(\overline{\alpha}^i + \sum_{h\in N}\overline{\beta}_h^i\left[\hat{\alpha}^h + \sum_{k\in N}\hat{\beta}_k^h\ A_k(t)\right]\right)$$
$$-c_i\left\{\overline{\alpha}^i + \sum_{j\in N}\overline{\beta}_j^i\left[\hat{\alpha}^j + \sum_{k\in N}\hat{\beta}_k^j\ A_k(t)\right]\right\}$$
$$-c_i^a\left[\frac{b_i}{2c_i^a}A_i(t)\right]^2 x - \sum_{j\in\overline{K}^i}\varepsilon_i^j\left\{\overline{\alpha}^j + \sum_{\ell\in N}\overline{\beta}_\ell^j\left[\hat{\alpha}^\ell + \sum_{k\in N}\hat{\beta}_k^\ell\ A_k(t)\right]\right\} - h_i x\Bigg]$$
$$+A_i(t)\left[\sum_{j=1}^n a_j\left\{\overline{\alpha}^j + \sum_{h\in N}\overline{\beta}_h^j\left[\hat{\alpha}^h + \sum_{k\in N}\hat{\beta}_k^h\ A_k(t)\right]\right\} + \sum_{j=1}^n b_j\frac{b_j}{2c_j^a}A_j(t)x - \delta x\right], \tag{6.1}$$

$$[A_i(T)x + C_i(T)] \ = -g^i\left(x - \overline{x}^i\right), \text{ for } i \in N. \tag{6.2}$$

For (6.1) and (6.2) to hold, it is required that

$$\dot{A}_i(t) = (r+\delta)\ A_i(t) - A_i(t) \sum_{\substack{j=1 \\ j \neq i}}^{n} \frac{b_j^2}{2c_j^a} A_j(t) - \frac{b_i^2}{4c_i^a}[A_i(t)]^2 + h_i, \tag{6.3}$$

$$A_i(T) = -g^i, \tag{6.4}$$

$$\begin{aligned}\dot{C}_i(t) &= rC_i(t) - \left(\alpha^i - \sum_{j=1}^{n} \beta_j^i \left\{ \overline{\alpha}^j + \sum_{h \in Ni}^{n} \overline{\beta}_h^j \left[\hat{\alpha}^h + \sum_{k \in N} \hat{\beta}_k^h\ A_k(t) \right] \right\} \right) \\ &\left(\overline{\alpha}^i + \sum_{h \in N} \overline{\beta}_h^i \left[\hat{\alpha}^h + \sum_{k \in N} \hat{\beta}_k^h\ A_k(t) \right] \right) \\ &+ c_i \left\{ \overline{\alpha}^i - \sum_{j \in N} \overline{\beta}_j^i \left[\hat{\alpha}^j + \sum_{k \in N} \hat{\beta}_k^j\ A_k(t) \right] \right\} \\ &+ \sum_{j \in \overline{K}^i} \varepsilon_i^j \left\{ \overline{\alpha}^j + \sum_{\ell \in N} \overline{\beta}_\ell^j \left[\hat{\alpha}^\ell + \sum_{k \in N} \hat{\beta}_k^\ell\ A_k(t) \right] \right\} \\ &- A_i(t) \left[\sum_{j=1}^{n} a_j \left\{ \overline{\alpha}^j + \sum_{h \in N} \overline{\beta}_h^j \left[\hat{\alpha}^h + \sum_{k \in N} \hat{\beta}_k^h\ A_k(t) \right] \right\} \right] = rC_i(t) + F_i(t),\end{aligned} \tag{6.5}$$

$$C_i(T) = g^i \overline{x}^i. \tag{6.6}$$

Equations (6.3, 6.4, 6.5, and 6.6) forms a block recursive system of differential equations with (6.3) and (6.4) being independent of (6.5) and (6.6).

Solving $\{A_1(t), A_2(t), \cdots, A_n(t)\}$ in (6.3 and 6.4) and upon substituting them into (6.5) and (6.6) yield a system of linear first order differential equations:

$$\dot{C}_i(t) = rC_i(t) + F_i(t), \tag{6.7}$$

$$C_i(T) = g^i \overline{x}^i, \text{ and } i \in N. \tag{6.8}$$

Since $C_i(t)$ is independent of $C_j(t)$ for $i \neq j$, $C_i(t)$ can be solved as:

$$C_i(t) = e^{r\,(t-t_0)} \left[\int_{t_0}^{t} F_i(y) e^{-r(y-t_0)} dy + C_i^0 \right], \tag{6.9}$$

$$\text{where } C_i^0 = g^i \overline{x}^i e^{-r\,(T-t_0)} - \int_{t_0}^{T} F_i(y) e^{-r(y-t_0)} dy. \tag{6.10}$$

Hence Proposition 2.1 follows. Q.E.D.

Appendix B: Proof of Proposition 3.1

Substituting (3.4) and (3.6) into (3.2) and using (3.7) one obtains:

$$
\begin{aligned}
& r\left[A^*(t)x+C^*(t)\right]-\left[\dot{A}^*(t)x+\dot{C}^*(t)\right]= \\
& \sum_{\ell=1}^{n}\Bigg[\left(\alpha^\ell-\sum_{j=1}^{n}\beta_j^\ell\left\{\overline{\alpha}^j+\sum_{h\in N}\overline{\beta}_h^j\left[\hat{\hat{\alpha}}^h+\hat{\hat{\beta}}^h A^*(t)\right]\right\}\right)\left\{\overline{\alpha}^\ell+\sum_{h\in N}\overline{\beta}_h^\ell\left[\hat{\hat{\alpha}}^h+\hat{\hat{\beta}}^h A^*(t)\right]\right\} \\
& -c_\ell\left\{\overline{\alpha}^\ell+\sum_{j\in N}\overline{\beta}_j^\ell\left[\hat{\hat{\alpha}}^j+\hat{\hat{\beta}}^j A^*(t)\right]\right\}-c_\ell^a\left[\frac{b_\ell}{2c_\ell^a}A^*(t)\right]^2 x \\
& -\sum_{j\in\overline{K}^\ell}\varepsilon_\ell^j\left\{\overline{\alpha}^j+\sum_{k\in N}\overline{\beta}_k^j\left[\hat{\hat{\alpha}}^k+\hat{\hat{\beta}}^{kj} A^*(t)\right]\right\}-h_\ell x\Bigg] \\
& +A_x^*(t)\left[\sum_{j=1}^{n}a_j\left\{\overline{\alpha}^j+\sum_{h\in N}\overline{\beta}_h^j\left[\hat{\hat{\alpha}}^h+\hat{\hat{\beta}}^h A^*(t)\right]\right\}+\sum_{j=1}^{n}\frac{b_j^2}{2c_j^a}A^*(t)x-\delta x\right],
\end{aligned}
\tag{6.11}
$$

$$
\left[A^*(T)x+C^*(T)\right]=-\sum_{i=1}^{n}g^i\left[x(T)-\overline{x}^i\right]. \tag{6.12}
$$

For (6.11) and (6.12) to hold, it is required that

$$
\dot{A}^*(t)=(r+\delta)\;A^*(t)-\sum_{j=1}^{n}\frac{b_j^2}{2c_j^a}\left[A^*(t)\right]^2+\sum_{j=1}^{n}h_j, \tag{6.13}
$$

$$
A^*(T)=-\sum_{j=1}^{n}g^j; \tag{6.14}
$$

$$
\begin{aligned}
\dot{C}^*(t)=rC^*(t)\; & -\sum_{\ell=1}^{n}\Bigg[\left(\alpha^\ell-\sum_{j=1}^{n}\beta_j^\ell\left\{\overline{\alpha}^j+\sum_{h\in N}\overline{\beta}_h^j\left[\hat{\hat{\alpha}}^h+\hat{\hat{\beta}}^h A^*(t)\right]\right\}\right)\{\overline{\alpha}^\ell \\
& +\sum_{h\in N}\overline{\beta}_h^\ell\left[\hat{\hat{\alpha}}^h+\hat{\hat{\beta}}^h A^*(t)\right]\}-c_\ell\left\{\overline{\alpha}^\ell+\sum_{j\in N}\overline{\beta}_j^\ell\left[\hat{\hat{\alpha}}^j+\hat{\hat{\beta}}^j A^*(t)\right]\right\} \\
& -\sum_{j\in\overline{K}^\ell}\varepsilon_\ell^j\left\{\overline{\alpha}^j+\sum_{k\in N}\overline{\beta}_k^j\left[\hat{\hat{\alpha}}^k+\hat{\hat{\beta}}^{kj} A^*(t)\right]\right\}\Bigg] \\
& -A_x^*(t)\left[\sum_{j=1}^{n}a_j\left\{\overline{\alpha}^j+\sum_{h\in N}\overline{\beta}_h^j\left[\hat{\hat{\alpha}}^h+\hat{\hat{\beta}}^h A^*(t)\right]\right\}\right]=rC^*(t)+F^*(t),
\end{aligned}
\tag{6.15}
$$

$$
C^*(T)=\sum_{j=1}^{n}g^j\overline{x}^j. \tag{6.16}
$$

Equations (6.13, 6.14, 6.15, and 6.16) forms a block recursive system of differential equations with (6.13 and 6.14) being independent of (6.15 and 6.16). Moreover,

(6.15 and 6.16) is a Riccati equation with constant coefficients which solution can be obtained by standard methods as:

$$A^*(t) = A_*^P + \Phi^*(t)\left[\overline{C}^* - \int_{t_0}^{t} \sum_{j=1}^{n} \frac{b_j^2}{2c_j^a} \Phi^*(y)dy\right]^{-1}, \tag{6.17}$$

where $\Phi^*(t) = \exp\left\{ \int_{t_0}^{t} \left[\sum_{j=1}^{n} \frac{b_j^2}{2c_j^a} A_*^P + (r+\delta) \right] dy \right\}$,

$\overline{C}^* = \dfrac{-\Phi^*(T)}{\left(A_*^P + \sum_{j=1}^{n} g^j\right)} + \int_{t_0}^{T} \sum_{j=1}^{n} \frac{b_j^2}{2c_j^a} \Phi^*(y)dy$, and

$A_*^P(t) = \left\{ (r+\delta) - \left[(r+\delta)^2 + 4\sum_{j=1}^{n} \frac{b_j^2}{2c_j^a} \sum_{j=1}^{n} h_j \right]^{1/2} \right\} / \sum_{j=1}^{n} \frac{b_j^2}{c_j^a}$ is a particular solution of the (6.13).

Upon substituting $A^*(t)$ above into (6.15), the system (6.15 and 6.16) becomes a system of linear first order differential equations:

$$\dot{C}^*(t) = rC^*(t) + +F^*(t), \tag{6.18}$$

$$C^*(T) = \sum_{j=1}^{n} g^j \overline{x}^j. \tag{6.19}$$

Solving (6.18 and 6.19) yields:

$$C^*(t) = e^{r(t-t_0)}\left[\int_{t_0}^{t} F^*(y) e^{-r(y-t_0)} dy + C_*^0 \right], \tag{6.20}$$

where $C_*^0 = \sum_{j=1}^{n} g^j \overline{x}^j e^{-r(T-t_0)} - \int_{t_0}^{T} F^*(y) e^{-r(y-t_0)} dy$.

Hence Proposition 3.1 follows. Q.E.D.

13.7 Chapter Notes

Though cooperation in environmental control holds out the best promise of effective action, limited success has been observed because existing multinational joint initiatives fail to satisfy the property of subgame consistency. In this Chapter we present a cooperative stochastic differential game of transboundary industrial pollution with industries and governments being separate entities. In particular, industrial production creates two types of negative environmental externalities – a short-term local impact and a long-term global impact. Given these impacts the individual government tax policy has to take into consideration the tax policies of other nations and these policies' intricate effects on outputs and environmental effects. A subgame consistent cooperative solution is derived in this stochastic differential game. A payment distribution mechanism is provided to support the subgame consistent solution under which the expected gain from cooperation is shared proportionally to the nations' relative sizes of expected noncooperative payoffs. The incorporation of uncertainties in future payoffs in Sect. 13.5 enriches the analysis with consideration of a realistic concern.

Applications of noncooperative differential games in environmental studies can be found in Yeung (1992); Dockner and Long (1993); Tahvonen (1994); Stimming (1999); Feenstra et al. (2001) and Dockner and Leitmann (2001). Cooperative differential games in environmental control have been presented by Dockner and Long (1993); Jørgensen and Zaccour (2001); Petrosyan and Zaccour (2003); Fredj et al. (2004); Breton et al. (2005, 2006), Yeung (2007a, 2008), Yeung and Petrosyan (2007a, 2012c) and Li (2014).

13.8 Problems

1. Consider an economy which is comprised of 2 nations and the planning horizon is $[0, 4]$. At time instant s the demand functions of the output of nations 1 and 2 are respectively

$$P_1(s) = 60 - 1.5q_1(s) - 0.2q_2(s) \text{ and } P_2(s) = 75 - 3q_2(s) - 0.5q_1(s).$$

The dynamics of pollution stock is governed by the stochastic differential equation:

$$dx(s) = \left[q_1(s) + 0.5q_2(s) - 0.4u_1(s)x(s)^{1/2} - 0.3u_2(s)x(s)^{1/2} - 0.02x(s) \right] ds$$
$$+ 0.04x(s)dz(s), \ x(0) = 25.$$

The damage (cost) of the pollution stock in the environment to nations 1 and 2 are respectively $3x(s)$ and $4x(s)$. The abatement costs are $[u_1(s)]^2$ and $0.4[u_2(s)]^2$ for nations 1 and 2 respectively. The instantaneous objectives of the governments in nations 1 and 2 at time s are respectively:

$$[60 - 1.5q_1(s) - 0.2q_2(s)]q_1(s) - 2q_1(s) - [u_1(s)]^2 - 0.5q_1(s) - 0.6q_2(s) - 3x(s)$$
and

$$[75 - 3q_2(s) - 0.5q_1(s)]q_2(s) - 2q_2(s) - [u_2(s)]^2 - 0.8q_2(s) - 0.4q_1(s) - 4x(s).$$

At terminal time 4, the terminal value associated with the state of pollution is $2[90 - x(T)]$ for nation 1 and $2[70 - x(T)]$ for nation 2.

Characterize a feedback Nash equilibrium solution for this fishery game.

2. If these nations agree to cooperate and maximize their joint payoff, obtain a group optimal cooperative solution.
3. Furthermore, if these nations agree to share their cooperative gain proportional to their expected payoffs, derive a subgame consistent cooperative solution.

Chapter 14
Cooperation with Technology Switching

Under the current situation of environmental degradation, even substantial reduction in industrial production using conventional production technique would only slow down the rate of increase and not be able to reverse the trend of continual pollution accumulation. Adoption of environment-preserving technique plays a central role to solving the problem effectively. Due to the geographical diffusion of pollutants and the globalization of trade, unilateral response on the part of one country or region is often ineffective. Though cooperation in environmental control holds out the best promise of effective action, limited success has been observed. Conventional multinational joint initiatives like the Kyoto Protocol or the Copenhagen Accord can hardly be expected to offer a long-term solution because (i) the plans which focus mainly on emissions reduction would unlikely be able to offer an effective mean to halt the accelerating trend of environmental deterioration, (ii) environment-preserving technique has to be adopted to provide an effective mean in solving the industrial pollution problem, and (iii) there is no guarantee that agreed-upon optimality principle could be maintained throughout the entire duration of cooperation. In this chapter, we present a cooperative dynamic model of collaborative environmental management with production technique choices.

Sections 14.1, 14.2, 14.3 and 14.4 of this Chapter is based on Yeung's (2014) work on subgame consistent collaborative environmental management with the availability of environment-preserving production techniques. Section 14.1 presents a dynamic game model of environmental control with production technique choices. Noncooperative outcomes are characterized in Sect. 14.2. Cooperative arrangements, group optimal actions, solution state trajectories, and time consistent payment distribution mechanism are examined in Sect. 14.3. A numerical illustration is provided in Sect. 14.4. An extension of Yeung's (2014) analysis to multiple types of environment-preserving techniques is provided in Sect. 14.5. Chapter appendices are provided in Sect. 14.6. Chapter notes are given in Sect. 14.7 and problems in Sect. 14.8.

D.W.K. Yeung, L.A. Petrosyan, *Subgame Consistent Cooperation*,
Theory and Decision Library C 47, DOI 10.1007/978-981-10-1545-8_14

14.1 Environmental Model with Production Technique Choices

In this section we present a dynamic game model of transboundary pollution with two production technique choices – a conventional technology and an environment-preserving technology. The game involves $T-$ stages and n asymmetric nations (regions or jurisdictions).

14.1.1 The Industrial Sector

The industrial sectors of the n asymmetric nations form an international economy. The demand for the output of nation $i \in \{1, 2, \cdots, n\} \equiv N$ at stage $t \in \{1, 2, \cdots, T\} \equiv \kappa$ is

$$P_t^i = \alpha_t^i - \sum_{j=1}^{n} \beta_j^i Q_t^j, \tag{1.1}$$

where P_t^i is the price of the output of nation i, Q_t^j is the output of nation j, α_t^i and β_j^i for $i \in N$ and $j \in N$ are positive parameters. The quantity of output $Q_t^j(s) \in \left[0, \overline{Q}^j\right]$ is nonnegative and bounded by a maximum capacity constraint $\overline{Q}^j$. Output price equals zero if the right-hand-side of (1.1) becomes negative. The demand system (1.1) shows that the economy is a form of differentiated products oligopoly. In the case when $\alpha^i = \alpha^j$ and $\beta_j^i = \beta_i^j$ for all $i \in N$ and $j \in N$, the industrial output is a homogeneous good. This type of model was first introduced by Dixit (1979) and later used in analyses in industrial organizations (see for example, Singh and Vives (1984)) and environmental games. In this analysis α_t^i for $i \in N$ is allowed to change over time to reflect different growth rates in different nations.

There are two types of production techniques available to each nation's industrial sector: conventional technique and environment-preserving technique. Industrial sectors pay more for using environment-preserving technique. The amount of pollutants emitted by environment-preserving technique is less than that emitted by conventional technique.

We use q_t^j to denote the output of nation j if it uses conventional technique and $\hat{q}_t^j$ to denote the output of nation j if it uses environment-preserving technique. The average cost of producing a unit of output with conventional technique in nation j is c^j while that of producing a unit of output with environment-preserving technique is $\hat{c}^j$.

Let v_t^i denote the tax rate imposed by government i on industrial output produced by conventional technique in stage t, and $\hat{v}_t^i$ denote the tax rate imposed on output produced by environment-preserving technique. Nation i's industrial sector will choose to use environment-preserving technique if $c^i + v_t^i > \hat{c}^i + \hat{v}_t^i$, otherwise it would choose to use conventional technique. In stage t, let the set of nations using conventional technique be denoted by S_t^1 and the set of nations using environment-preserving technique by S_t^2. The profit of industrial sector $i_t \in S_t^1$ and that of industrial sector $\ell_t \in S_t^1$ in stage t can be expressed respectively as

$$\pi_t^{i_t} = \left[\alpha_t^{i_t} - \sum_{j \in S_t^1} \beta_j^{i_t} q_t^j - \sum_{\zeta \in S_t^2} \beta_\zeta^{i_t} \hat{q}_t^\zeta \right] q_t^{i_t} - c^{i_t} q_t^{i_t} - v_t^{i_t} q_t^{i_t}, \text{ for } i_t \in S_t^1, \tag{1.2}$$

and

$$\hat{\pi}_t^{\ell_t} = \left[\alpha_t^{\ell_t} - \sum_{j \in S_t^1} \beta_j^{\ell_t} q_t^j - \sum_{\zeta \in S_t^2} \beta_\zeta^{\ell_t} \hat{q}_t^\zeta \right] \hat{q}_t^{\ell_t} - c^{\ell_t} \hat{q}_t^{\ell_t} - v_t^{\ell_t} \hat{q}_t^{\ell_t}, \text{ for } \ell_t \in S_t^2 \tag{1.3}$$

In each stage t the industrial sector of nation $i_t \in S_t^1$ seeks to maximize (1.2) and the industrial sector of nation $\ell_t \in S_t^1$ seeks to maximize (1.3). The first order condition for an industry equilibrium in stage t yields

$$\begin{aligned} &\alpha_t^{i_t} - \sum_{j \in S_t^1} \beta_j^{i_t} q_t^j - \sum_{\zeta \in S_t^2} \beta_\zeta^{i_t} \hat{q}_t^\zeta - \beta_{i_t}^{i_t} q_t^{i_t} = c^{i_t} + v_t^{i_t}, \text{ for } i_t \in S_t^1; \text{ and} \\ &\alpha_t^{\ell_t} - \sum_{j \in S_t^1} \beta_j^{\ell_t} q_t^j - \sum_{\zeta \in S_t^2} \beta_\zeta^{\ell_t} \hat{q}_t^\zeta - \beta_{\ell_t}^{\ell_t} \hat{q}_t^{\ell_t} = \hat{c}^{\ell_t} + \hat{v}_t^{\ell_t}, \text{ for } \ell_t \in S_t^2. \end{aligned} \tag{1.4}$$

Condition (1.4) shows that the industrial sectors will produce up to a point where marginal revenue (the left-hand side of the equations) equals the cost plus tax of a unit of output produced (the right-hand-side of the equations).

14.1.2 Pollution Dynamics

Industrial production creates long-term environmental impacts by building up existing pollution stocks like green-house-gas, CFC and atmospheric particulates. Each government adopts its own pollution abatement policy to reduce pollutants in the environment. At the initial stage 1, the level of pollution is $x_1 = x^0$. The dynamics of pollution accumulation is governed by the difference equation:

$$x_{t+1} = x_t + \sum_{i_t \in S_t^1} a^{i_t} q_t^{i_t} + \sum_{\ell_t \in S_t^2} \hat{a}^{\ell_t} \hat{q}_t^{\ell_t} - \sum_{j=1}^{n} b_j u_t^j (x_t)^{1/2} - \delta x_t, \; x_1 = x^0, \tag{1.5}$$

where a^{i_t} is the amount of pollution created by a unit of nation i_t's output using conventional technique,

$\hat{a}^{\ell_t}$ is the amount of pollution created by a unit of nation ℓ_t's output using environment-preserving technique,

u_t^j is the pollution abatement effort of nation j at stage t,

$b_j u_t^j (x_t)^{1/2}$ is the amount of pollution removed by u_t^j units of abatement effort of nation j, δ is the natural rate of decay of the pollutants.

The damage (cost) of x_t amount of pollution to nation is $h^j x_t$. The cost of u_t^j units of abatement effort is $c_i^a (u_t^j)^2$.

14.1.3 The Governments' Objectives

The governments have to promote business interests and at the same time bear the costs brought about by pollution. In particular, each government maximizes the net gains in the industrial sector plus tax revenues minus the sum of expenditures on pollution abatement and damages from pollution. The payoff of government $i_t \in S_t^1$ at stage t can be expressed as:

$$[\alpha_t^{i_t} - \sum_{j \in S_t^1} \beta_j^{i_t} q_t^j - \sum_{\zeta \in S_t^2} \beta_\zeta^{i_t} \hat{q}_t^{\zeta}] q_t^{i_t} - c^{i_t} q_t^{i_t} - c_{i_t}^a \left(u_t^{i_t}\right)^2 - h^{i_t} x_t; \tag{1.6}$$

and the payoff of government $\ell_t \in S_t^2$ at stage t can be expressed as:

$$[\alpha_t^{\ell_t} - \sum_{j \in S_t^1} \beta_j^{\ell_t} q_t^j - \sum_{\zeta \in S_t^2} \beta_\zeta^{\ell_t} \hat{q}_t^{\zeta}] \hat{q}_t^{\ell_t} - \hat{c}^{\ell_t} \hat{q}_t^{\ell_t} - c_{\ell_t}^a \left(u_t^{\ell_t}\right)^2 - h^{\ell_t} x_t. \tag{1.7}$$

The governments' planning horizon is from stage 1 to stage T. It is possible that T may be very large. The discount rate is r. A terminal appraisal of pollution damage is $g^i(\overline{x}^i - x_{T+1})$ will be given to government i at stage $T+1$, where $g^i \geq 0$. In particular, if the level of pollution at stage $T+1$ is higher (lower) than $\overline{x}^i$, government i will receive a bonus (penalty) equaling $g^i(\overline{x}^i - x_{T+1})$. Each one of the n governments seeks to maximize the sum of the discounted payoffs over the T stages plus the terminal appraisal. In particular, government i would seek to maximize the objective

$$\sum_{t=1}^{T} \Bigg[[\alpha_t^i - \sum_{\substack{j \in S_t^1 \\ j \neq i}} \beta_j^i q_t^j - \sum_{\substack{\zeta \in S_t^2 \\ j \neq i}} \beta_\zeta^i \hat{q}_t^{\zeta} - \beta_i^i \overline{q}_t^i] \overline{q}_t^i - \overline{c}^i \overline{q}_t^i - c_i^a \left(u_t^i\right)^2 - h^i x_t \Bigg] \left(\frac{1}{1+r}\right)^{t-1} + g^i(\overline{x}^i - x_{T+1}) \left(\frac{1}{1+r}\right)^T; \; i \in N \tag{1.8}$$

where $\overline{q}_t^i = q_t^i$ and $\overline{q}_t^i = q_t^i$ if industrial sector i uses conventional technique; and $\overline{q}_t^i = \hat{q}_t^{\,i}$ and $\overline{c}_t^i = \hat{c}_t^{\,i}$ if industrial sector i uses environment-preserving technique.

Besides designing an optimal abatement policy, each of the governments also has to design a tax scheme which would lead to the level of output that maximizes its objective. The problem of maximizing objectives (1.8) subject to pollution dynamics (1.5) is a dynamic game between these n governments.

14.2 Noncooperative Outcomes

In this section we discuss the solution to the noncooperative dynamic game (1.5) and (1.8). Since under a noncooperative framework, pre-commitment is not possible, a feedback Nash equilibrium solution is sought. A theorem characterizing a feedback Nash equilibrium solution is presented in the following theorem.

Theorem 2.1 A set of strategies $\Big\{ q_t^{i_t*} = \phi_t^{i_t}(x),\ \hat{q}_t^{\,\hat{i}_t*} = \hat{\phi}_t^{\,\hat{i}_t}(x),\ u_t^{i_t*} = \upsilon_t^{i_t}(x),\ u_t^{\hat{i}_t*} = \hat{\upsilon}_t^{\,\hat{i}_t}(x)$, for $t \in \kappa$ and $i_t \in S_t^1$ and $\hat{i}_t \in S_t^2 \}$ provides a feedback Nash equilibrium solution to the game (1.5) and (1.8) if there exist functions $V^{i_t}(t,x)$ and $\hat{V}^{\hat{i}_t}(t,x)$, for $t \in \kappa$ and $i_t \in S_t^1$ and $\hat{i}_t \in S_t^2$, such that the following recursive relations are satisfied:

$$V^{i_t}(T+1,x) = \hat{V}^{\hat{i}_t}(T+1,x) = V^i(T+1,x) = g^i(\overline{x}^i - x)\left(\tfrac{1}{1+r}\right)^T, \text{ for } i \in N,$$

$$\begin{aligned}
V^{i_t}(t,x) = \max_{q_t^{i_t},\, u_t} \Bigg\{ \Bigg[& \Big[\alpha_t^{i_t} - \sum_{\substack{j \in S_t^1 \\ j \neq i_t}} \beta_j^{i_t} \phi_t^j(x) - \sum_{\zeta \in S_t^2} \beta_\zeta^{i_t} \hat{\phi}_t^{\,\zeta}(x) - \beta_{i_t}^{i_t} q_t^{i_t}\Big] q_t^{i_t} - c^{i_t} q_t^{i_t} \\
& - c_{i_t}^a \left(u_t^{i_t}\right)^2 - h^{i_t} x \Bigg] \left(\tfrac{1}{1+r}\right)^{t-1} \\
& + \overline{V}^{i_t}\Bigg[t+1, x + \sum_{\substack{j \in S_t^1 \\ j \neq i_t}} a^j \phi_t^j(x) + \sum_{\zeta \in S_t^2} \hat{a}^\zeta \hat{\phi}_t^{\,\zeta}(x) + a^{i_t} q_t^{i_t} \\
& - \sum_{\substack{j \in S_t^1 \\ j \neq i_t}} b_j \upsilon_t^j(x) x^{1/2} - \sum_{j \in S_t^2}^{n} b_j \hat{\upsilon}_t^{\,j}(x) x^{1/2} - b_{i_t} u_t^{i_t} x^{1/2} - \delta x \Bigg] \Bigg\},
\end{aligned}$$

for $t \in \kappa$ and $i_t \in S_t^1$; and

$$\hat{V}^{\hat{i}_t}(t,x) = \max_{\hat{q}_t^{\hat{i}_t}, u_t^{\hat{i}_t}} \left\{ \left[\left[\alpha_t^{\hat{i}_t} - \sum_{j\in S_t^1} \beta_j^{\hat{i}_t}\phi_t^j(x) - \sum_{\substack{\zeta \in S_t^2 \\ \zeta \neq \hat{i}_t}} \beta_\zeta^{\hat{i}_t}\hat{\phi}_t^{\zeta}(x) - \beta_{\hat{i}_t}^{\hat{i}_t}\hat{q}_t^{\hat{i}_t} \right] \hat{q}_t^{\hat{i}_t} \right.\right.$$

$$\left. - \hat{c}^{\hat{i}_t}\hat{q}_t^{\hat{i}_t} - c_{\hat{i}_t}^a\left(\hat{u}_t^{\hat{i}_t}\right)^2 - h^{\hat{i}_t}x \right] \left(\frac{1}{1+r}\right)^{t-1}$$

$$+ \overline{V}^{\hat{i}_t}\left[t+1, x + \sum_{j\in S_t^1} a^j\phi_t^j(x) + \sum_{\substack{\zeta \in S_t^2 \\ \zeta \neq \hat{i}_t}} \hat{a}^{\zeta}\hat{\phi}_t^{\zeta}(x) + \hat{a}^{\hat{i}_t}\hat{q}_t^{\hat{i}_t} \right.$$

$$\left.\left. - \sum_{j\in S_t^1} b_j v_t^j(x)x^{1/2} - \sum_{\substack{j \in S_t^2 \\ j \neq \hat{i}_t}}^{n} b_j\hat{v}_t^j(x)x^{1/2} - b_{\hat{i}_t}\hat{u}_t^{\hat{i}_t}x^{1/2} - \delta x \right]\right\},$$

for $t \in \kappa$ and $\hat{i}_t \in S_t^2$ (2.1)

and

$$c^{i_t}\left(\frac{1}{1+r}\right)^{t-1} - \overline{V}_{x_{t+1}}^{i_t}(t+1, x_{t+1})a^{i_t} < \hat{c}^{i_t}\left(\frac{1}{1+r}\right)^{t-1} - \overline{V}_{x_{t+1}}^{i_t}\left(t+1, x_{t+1}^{(i_t)}\right)\hat{a}^{i_t},$$

for $i_t \in S_t^1$;

$$c^{\hat{i}_t}\left(\tfrac{1}{1+r}\right)^{t-1} - \overline{V}_{x_{t+1}}^{\hat{i}_t}\left(t+1, x_{t+1}^{(\hat{i}_t)}\right)a^{\hat{i}_t} \geq \hat{c}^{\hat{i}_t}\left(\tfrac{1}{1+r}\right)^{t-1} - \overline{V}_{x_{t+1}}^{\hat{i}_t}(t+1, x_{t+1})\hat{a}^{\hat{i}_t},$$

for $\hat{i}_t \in S_t^2$ (2.2)

where $\overline{V}^i(t+1, x_{t+1}) = V^i(t+1, x_{t+1})$ if i uses conventional technology in stage $t+1$ and $\overline{V}^i(t+1, x_{t+1}) = \hat{V}^i(t+1, x_{t+1})$ if i uses environment preserving technology in stage $t+1$, and

$$x_{t+1} = x + \sum_{j\in S_t^1} a^j\phi_t^j(x) + \sum_{\zeta\in S_t^2} \hat{a}^{\zeta}\hat{\phi}_t^{\zeta}(x) - \sum_{j\in S_t^1} b_j v_t^j(x)x^{1/2} - \sum_{j\in S_t^2} b_j\hat{v}_t^j(x)x^{1/2} - \delta x,$$

$$x_{t+1}^{(i_t)} = x + \sum_{\substack{j \in S_t^1 \\ j \neq i_t}} a^j\phi_t^j(x) + \hat{a}^{i_t}\hat{q}_t^{i_t} + \sum_{\zeta\in S_t^2} \hat{a}^{\zeta}\hat{\phi}_t^{\zeta}(x)$$

$$- \sum_{\substack{j \in S_t^1 \\ j \neq i_t}} b_j v_t^j(x)x^{1/2} - b_{i_t}u_t^{i_t} - \sum_{j\in S_t^2} b_j\hat{v}_t^j(x)x^{1/2} - \delta x, \text{ and}$$

$$x_{t+1}^{(\hat{i}_t)} = x + \sum_{j\in S_t^1} a^j\phi_t^j(x) + \sum_{\substack{\zeta \in S_t^2 \\ \zeta \neq \hat{i}_t}} \hat{a}^{\zeta}\hat{\phi}_t^{\zeta}(x) + a^{\hat{i}_t}q_t^{\hat{i}_t}$$

$$- \sum_{j\in S_t^1} b_j v_t^j(x)x^{1/2} - \sum_{\substack{j \in S_t^2 \\ j \neq \hat{i}_t}} b_j\hat{v}_t^j(x)x^{1/2} - b_{\hat{i}_t}u_t^{\hat{i}_t} - \delta x$$

Proof If nation $i_t \in S_t^1$ adopts conventional technique and nation $\hat{i}_t \in S_t^2$ adopts environment-preserving technique, the results in (2.1) satisfy the optimality conditions in dynamic programming and the Nash equilibrium.

The inequalities in (2.2) yield the conditions justifying why nation $i_t \in S_t^1$ adopts conventional technique and nation $\hat{i}_t \in S_t^2$ adopts environment-preserving technique. To prove this we perform the indicated maximization in (2.1) and obtain:

$$\begin{aligned}\alpha_t^{i_t} - \sum_{j \in S_t^1} \beta_j^{i_t} \phi_t^j(x) - \sum_{\zeta \in S_t^2} \beta_\zeta^{i_t} \hat{\phi}_t^{\zeta}(x) - \beta_{i_t}^{i_t} \phi_t^{i_t}(x) \\ = c^{i_t} - a^{i_t} \overline{V}_{x_{t+1}}^{i_t}(t+1, x_{t+1})(1+r)^{t-1},\end{aligned} \tag{2.3}$$

for $i_t \in S_t^1$; and

$$\begin{aligned}\alpha_t^{\hat{i}_t} - \sum_{j \in S_t^1} \beta_j^{\hat{i}_t} \phi_t^j(x) - \sum_{\zeta \in S_t^2} \beta_\zeta^{\hat{i}_t} \phi_t^\zeta(x) - \beta_{\hat{i}_t}^{\hat{i}_t} \hat{\phi}_t^{\hat{i}_t}(x) \\ = \hat{c}^{\hat{i}_t} - \hat{a}^{\hat{i}_t} \overline{V}_{x_{t=1}}^{\hat{i}_t}(t+1, x_{t+1})(1+r)^{t-1}, \\ \text{for } \hat{i}_t \in S_t^2\end{aligned} \tag{2.4}$$

In view of (1.4), the left-hand-side of Eqs. (2.3) and that of (2.4) reflect the marginal revenues to the industrial sectors. To motivate the industrial sectors to produce outputs as given in (2.3) government i_t has to impose a tax $v_t^{i_t}$ equaling $-a^{i_t} \overline{V}_{x_{t+1}}^{i_t}$ $(t+1, x_{t+1})(1+r)^{t-1}$ on a unit of output produced with conventional technique. Similarly, government $\hat{\imath}_t$ has to impose a tax $\hat{v}_t^{\hat{i}_t}$ equaling $-\hat{a}^{\hat{i}_t} \overline{V}_{x_{t+1}}^{\hat{i}_t}(t+1, x_{t+1})$ $(1+r)^{t-1}$ on a unit of output produced with environment-preserving technique to arrive at (2.4).

At stage t the unit cost plus unit tax to the industrial sector of nation i for using conventional technique is

$$c^i - a^i \overline{V}_{x_{t+1}}^i(t+1, x_{t+1}, \vartheta_{t+1})(1+r)^{t-1}, \tag{2.5}$$

and the unit cost plus unit tax to the industrial sector of nation i for using environment-preserving technique is

$$\hat{c}^i - \hat{a}^i \overline{V}_{x_{t+1}}^i(t+1, x_{t+1}, \vartheta_{t+1})(1+r)^{t-1}. \tag{2.6}$$

The industrial sector of nation i would adopt the technique which costs (production cost plus tax) less. Therefore for nation $i_t \in S_t^1$ the condition $c^{i_t}\left(\frac{1}{1+r}\right)^{t-1} - \overline{V}_{x_{t+1}}^{i_t}(t+1, x_{t+1})a^{i_t} < \hat{c}^{i_t}\left(\frac{1}{1+r}\right)^{t-1} - \overline{V}_{x_{t+1}}^{i_t}\left(t+1, x_{t+1}^{(i_t)}\right)\hat{a}^{i_t}$ has to be satisfied.

For nation $\hat{i}_t \in S_t^2$ the condition
$c^{\hat{i}_t}\left(\frac{1}{1+r}\right)^{t-1} - \overline{V}^{\hat{i}_t}_{x_{t+1}}\left(t+1, x_{t+1}^{(\hat{i}_t)}\right)a^{\hat{i}_t} \geq \hat{c}^{\hat{i}_t}\left(\frac{1}{1+r}\right)^{t-1} - \overline{V}^{\hat{i}_t}_{x_{t+1}}(t+1, x_{t+1})\hat{a}^{\hat{i}_t}$ has to be satisfied.

Hence a feedback Nash equilibrium is characterized (see Theorem 1.1 in Chap. 7) and Theorem 2.1 follows. ■

The term $-a^{i_t}\overline{V}^{i_t}_{x_{t+1}}(t+1, x_{t+1})(1+r)^{t-1}$ reflects the marginal social cost to nation i_t brought about by a unit of output produced with conventional technique. The term $-\hat{a}^{\hat{i}_t}\overline{V}^{\hat{i}_t}_{x_{t+1}}(t+1, x_{t+1})(1+r)^{t-1}$ reflects the marginal social cost to nation $\hat{\imath}_t$ brought about by a unit of output produced with environment-preserving technique.

Rearranging (2.3) and (2.4) we obtain the system

$$\sum_{j\in S_t^1}\beta_j^{i_t}\phi_t^j(x) + \sum_{\zeta\in S_t^2}\beta_\zeta^{i_t}\hat{\phi}_t^{\zeta}(x) + \beta_{i_t}^{i_t}\phi_t^{i_t}(x)$$
$$= \alpha_t^{i_t} - c^{i_t} + a^{i_t}\overline{V}^{i_t}_{x_{t+1}}(t+1, x_{t+1})(1+r)^{t-1}, \text{ for } i_t \in S_t^1; \quad (2.7)$$

$$\sum_{j\in S_t^1}\beta_j^{\hat{i}_t}\phi_t^j(x) + \sum_{\zeta\in S_t^2}\beta_\zeta^{\hat{i}_t}\hat{\phi}_t^{\zeta}(x) + \beta_{\hat{i}_t}^{\hat{i}_t}\hat{\phi}_t^{\hat{i}_t}(x)$$
$$= \alpha_t^{\hat{i}_t} - \hat{c}^{\hat{i}_t} + \hat{a}^{\hat{i}_t}\overline{V}^{\hat{i}_t}_{x_{t+1}}(t+1, x_{t+1})(1+r)^{t-1}, \text{ for } \hat{i}_t \in S_t^2. \quad (2.8)$$

System (2.7 and 2.8) can be viewed as a set of equations linear in

$$\phi_t^{i_t}(x) \text{ for } i_t \in S_t^1 \text{ and } \hat{\phi}_t^{\hat{i}_t}(x) \text{ for } \hat{i}_t \in S_t^2,$$

with $\overline{V}^{i_t}_{x_{t+1}}(t+1, x_{t+1})(1+r)^{t-1}$ for $i_t \in S_t^1$ and $\overline{V}^{\hat{i}_t}_{x_{t+1}}(t+1, x_{t+1})(1+r)^{t-1}$ for $\hat{i}_t \in S_t^2$ being taken as a set of parameters. Solving (2.7 and 2.8) yields:

$$\phi_t^{i_t}(x) = \overline{\alpha}_t^{i_t} + \sum_{j\in S_t^1}\overline{\beta}_t^{(i_t)j}\; V^j_{x_{t+1}}(t+1, x_{t+1})(1+r)^{t-1}$$
$$+\sum_{j\in S_t^2}\overline{\beta}_t^{(i_t)j}\; \overline{V}^j_{x_{t+1}}(t+1, x_{t+1})(1+r)^{t-1}, \quad \text{for } i_t \in S_t^1;$$

$$\hat{\phi}_t^{\hat{i}_t}(x) = \hat{\overline{\alpha}}_t^{\hat{i}_t} + \sum_{j\in S_t^1}\hat{\overline{\beta}}_t^{(\hat{i}_t)j}\; V^j_{x_{t+1}}(t+1, x_{t+1})(1+r)^{t-1}$$
$$+\sum_{j\in S_t^2}\hat{\overline{\beta}}_t^{(\hat{i}_t)j}\; \overline{V}^j_{x_{t+1}}(t+1, x_{t+1})(1+r)^{t-1}, \quad \text{for } \hat{i}_t \in S_t^2; \quad (2.9)$$

where $\overline{\alpha}_t^{i_t}$ and $\overline{\beta}_t^{(i_t)j}$ for $i_t \in S_t^1$, and $\hat{\overline{\alpha}}_t^{\hat{i}_t}$ and $\hat{\overline{\beta}}_t^{(\hat{i}_t)j}$, $\hat{i}_t \in S_t^2$, are constants involving the model parameters.

In addition, performing the maximization operator in (2.1) with respect to $u_t^{i_t}$ and $u_t^{\hat{i}_t}$ yields

$$\upsilon_t^{i_t}(x) = -\frac{b_{i_t}}{2c_{i_t}^a}\overline{V}_{x_{t+1}}^{i_t}(t+1, x_{t+1})(1+r)^{t-1}x^{1/2}, \text{ for } i_t \in S_t^1; \text{ and}$$
$$\hat{\upsilon}_t^{\hat{i}_t}(x) = -\frac{b_{\hat{i}_t}}{2c_{\hat{i}_t}^a}\overline{V}_{x_{t+1}}^{\hat{i}_t}(t+1, x_{t+1})(1+r)^{t-1}x^{1/2}, \text{ for } \hat{i}_t \in S_t^2. \quad (2.10)$$

The game equilibrium payoffs of the nations can be obtained as:

Proposition 2.1 System (2.1 and 2.2) admits a solution

$$V^{i_t}(t,x) = \left(A_t^{i_t}x + C_t^{i_t}\right)\left(\frac{1}{1+r}\right)^{t-1}, \text{ for } i_t \in S_t^1$$
$$\text{and } \hat{V}^{\hat{i}_t}(t,x) = \left(\hat{A}_t^{\hat{i}_t}x + \hat{C}_t^{\hat{i}_t}\right)\left(\frac{1}{1+r}\right)^{t-1}, \text{ for } \hat{i}_t \in S_t^2, \text{ for } t \in \kappa; \quad (2.11)$$

with $A_t^{i_t}$, $C_t^{i_t}$, $\hat{A}_t^{\hat{i}_t}$ and $\hat{C}_t^{\hat{i}_t}$ being constants involving the model parameters.

Proof See Appendix A. ■

Though conventional technique emits higher level of pollutants, nations have no incentive to switch to environment-preserving technique if the sum of marginal cost of production and the nation's social cost resulted from using conventional technique is lower than that resulted from using environment-preserving technique.

14.3 Cooperative Arrangements in Environmental Control

Now consider the case when all the nations want to collaborate and tackle the pollution problem together. To ensure group optimality, the nations would seek to maximize their joint payoff under cooperation. Since nations are asymmetric and the number of nations may be large, a reasonable optimality principle for gain distribution is to share the gain from cooperation proportional to the nations' relative sizes of noncooperative payoffs. Cooperation will cease if any of the nations refuses to act accordingly at any time within the game horizon.

14.3.1 Group Optimality and Cooperative State Trajectory

Consider the case when all the nations agree to act cooperatively so that the joint payoff will be maximized. Since two technique choices are available they have to determine which nations would use which type of techniques over the T stages. Let M^{γ} be a matrix reflecting the pattern of technique choices by the n nations over the T stages. In particular, according to pattern M^{γ}, the set of nations that use conventional technique is $S_t^{M^{\gamma}[1]}$ and the set of nations that use environment-preserving technique is $S_t^{M^{\gamma}[2]}$ in stage $t \in \kappa$. To select the controls which would maximize joint payoff under pattern M^{γ} the nations have to solve the following optimal control problem which maximizes

$$\sum_{t=1}^{T}\Bigg[\sum_{i_t \in S_t^{M^{\gamma}[1]}}\Bigg([\alpha_t^{i_t} - \sum_{j \in S_t^{M^{\gamma}[1]}}\beta_j^{i_t} q_t^j - \sum_{\zeta \in S_t^{M^{\gamma}[2]}}\beta_\zeta^{i_t}\hat{q}_t^{\zeta}]q_t^{i_t}$$

$$- c^{i_t} q_t^{i_t} - c_{i_t}^a (u_t^{i_t})^2 - h^{i_t} x_t\Bigg)\left(\frac{1}{1+r}\right)^{t-1}$$

$$+ \sum_{\hat{i}_t \in S_t^{M^{\gamma}[2]}}\Bigg(\left[\alpha_t^{\hat{i}_t} - \sum_{j \in S_t^{M^{\gamma}[1]}}\beta_j^{\hat{i}_t} q_t^j - \sum_{\zeta \in S_t^{M^{\gamma}[2]}}\beta_\zeta^{\hat{i}_t}\hat{q}_t^{\zeta}\right]\hat{q}_t^{\hat{i}_t}$$

$$- \hat{c}^{\hat{i}_t}\left(S_t^{M^{\gamma}[2]}\right)\hat{q}_t^{\hat{i}_t} - c_{\hat{i}_t}^a\left(\hat{u}_t^{\hat{i}_t}\right)^2 - h^{\hat{i}_t} x_t\Bigg)\left(\frac{1}{1+r}\right)^{t-1}\Bigg]$$

$$+ \sum_{i=1}^{n} g^i(\bar{x}^i - x_{T+1})\left(\frac{1}{1+r}\right)^{T}\Bigg\}, \tag{3.1}$$

subject to

$$x_{t+1} = x_t + \sum_{\ell_t \in S_t^{M^{\gamma}[1]}} a^{\ell_t} q_t^{\ell_t} + \sum_{\hat{\ell}_t \in S_t^{M^{\gamma}[2]}} \hat{a}^{\hat{\ell}_t}\hat{q}_t^{\hat{\ell}_t} - \sum_{\ell_t \in S_t^{M^{\gamma}[1]}} b_{\ell_t} u_t^{\ell_t}(x_t)^{1/2}$$

$$- \sum_{\hat{\ell}_t \in S_t^{M^{\gamma}[2]}} b_{\hat{\ell}_t}\hat{u}_t^{\hat{\ell}_t}(x_t)^{1/2} - \delta\, x_t, \quad x_1 = x^0. \tag{3.2}$$

The solution to the optimal control problem (3.1 and 3.2) is characterized in the following theorem.

Theorem 3.1 A set of strategies $\Big\{q_t^{\ell_t *} = \psi_t^{(M^{\gamma})\ell_t}(x),\ \hat{q}_t^{\hat{\ell}_t *} = \hat{\psi}_t^{(M^{\gamma})\hat{\ell}_t}(x),\ u_t^{\ell_t *} = \varpi_t^{(M^{\gamma})\ell_t}(x),\ \hat{u}_t^{\hat{\ell}_t *} = \hat{\varpi}_t^{(M^{\gamma})\hat{\ell}_t}(x)$, for $t \in \kappa$ and $\ell_t \in S_t^{M^{\gamma}[1]}$ and $\hat{\ell}_t \in S_t^{M^{\gamma}[2]}\Big\}$ constitutes an optimal solution to the control problem (3.1) and (3.2) if there exist functions $W^{M^{\gamma}}(t,x) : R \to R$, for $t \in \kappa$, such that the following recursive relations are satisfied:

$$W^{M^{\gamma}}(t,x) = \max_{u_t^{i_t}, q_t^{i_t},\ i_t \in S_t^{M^{\gamma}[1]};\ \hat{u}_t^{\hat{i}_t} \hat{q}_t^{\hat{i}_t},\ \hat{i}_t \in S_t^{M^{\lambda}[2]}} \Bigg\{ \sum_{i_t \in S_t^{M^{\gamma}[1]}} \Bigg[$$

$$\Big[\alpha_t^{i_t} - \sum_{j \in S_t^{M^{\gamma}[1]}} \beta_j^{i_t} q_t^j - \sum_{\zeta \in S_t^{M^{\gamma}[2]}} \beta_\zeta^{i_t} \hat{q}_t^{\zeta}\Big] q_t^{i_t} - c^{i_t} q_t^{i_t} - c_{i_t}^a \left(u_t^{i_t}\right)^2 - h^{i_t} x_t \Bigg] \left(\tfrac{1}{1+r}\right)^{t-1} +$$

$$\sum_{\hat{i}_t \in S_t^{M^{\gamma}[2]}} \Bigg[\Big[\alpha_t^{\hat{i}_t} - \sum_{j \in S_t^{M^{\gamma}[1]}} \beta_j^{\hat{i}_t} q_t^j - \sum_{\zeta \in S_t^{M^{\gamma}[2]}} \beta_\zeta^{\hat{i}_t} \hat{q}_t^{\zeta}\Big] \hat{q}_t^{\hat{i}_t}$$

$$-\hat{c}^{\hat{i}_t}\left(S_t^{M^{\gamma}[2]}\right) \hat{q}_t^{\hat{i}_t} - c_{\hat{i}_t}^a \left(\hat{u}_t^{\hat{i}_t}\right)^2 - h^{\hat{i}_t} x_t \Bigg] \left(\tfrac{1}{1+r}\right)^{t-1}$$

$$+ W^{M^{\gamma}} \Bigg[t+1, x + \sum_{\ell_t \in S_t^{M^{\gamma}[1]}} a^{\ell_t} q_t^{\ell_t} + \sum_{\hat{\ell}_t \in S_t^{M^{\gamma}[2]}} \hat{a}^{\hat{\ell}_t} \hat{q}_t^{\hat{\ell}_t} - \sum_{\ell_t \in S_t^{M^{\gamma}[1]}} b_{\ell_t} u_t^{\ell_t}(x_t)^{1/2}$$

$$- \sum_{\hat{\ell}_t \in S_t^{M^{\gamma}[2]}} b_{\hat{\ell}_t} \hat{u}_t^{\hat{\ell}_t}(x)^{1/2} - \delta\, x \Bigg] \Bigg\}. \text{ for } t \in \kappa;$$

$$W^{M^{\gamma}}(T+1,x) = \sum_{i=1}^{n} g^i(\overline{x}^i - x)\left(\tfrac{1}{1+r}\right)^T \tag{3.3}$$

Proof The results in (3.3) satisfy the standard optimality conditions in discrete-time dynamic programming. ■

Performing the indicated maximization in (3.3) yields

$$\varpi_t^{(M^{\gamma})\ell_t}(x) = -\frac{b_{\ell_t}}{2c_{\ell_t}^a} W_{x_{t+1}}^{M^{\gamma}}(t+1, x_{t+1})(1+r)^{t-1} x^{1/2}, \text{ for } \ell_t \in S_t^{M^{\gamma}[1]}, \text{ and}$$

$$\hat{\varpi}_t^{(M^{\gamma})\hat{\ell}_t}(x) = -\frac{b_{\hat{\ell}_t}}{2c_{\hat{\ell}_t}^a} W_{x_{t+1}}^{M^{\gamma}}(t+1, x_{t+1})(1+r)^{t-1} x^{1/2}, \text{ for } \hat{\ell}_t \in S_t^{M^{\gamma}[2]}; \tag{3.4}$$

$$\alpha_t^{\ell_t} - \sum_{j \in S_t^{M^{\gamma}[1]}} \beta_j^{\ell_t} \psi_t^{(M^{\gamma})j}(x) - \sum_{j \in S_t^{M^{\gamma}[2]}} \beta_j^{\ell_t} \hat{\psi}_t^{(M^{\gamma})j}(x) - \sum_{\zeta \in S_t^{M^{\gamma}[1]}} \beta_{\ell_t}^{\zeta} \psi_t^{(M^{\gamma})\zeta}(x)$$
$$- \sum_{\zeta \in S_t^{M^{\gamma}[2]}} \beta_{\ell_t}^{\zeta} \hat{\psi}_t^{(M^{\gamma})\zeta}(x) = c^{\ell_t} - a^{\ell_t} W_{x_{t+1}}^{M^{\gamma}}(t+1, x_{t+1})(1+r)^{t-1},$$
$$\text{for } \ell_t \in S_t^{M^{\gamma}[1]} \tag{3.5}$$

$$\alpha_t^{\hat{\ell}_t} - \sum_{j \in S_t^{M^{\gamma}[1]}} \beta_j^{\hat{\ell}_t} \psi_t^{(M^{\gamma})j}(x) - \sum_{j \in S_t^{M^{\gamma}[2]}} \beta_j^{\hat{\ell}_t} \hat{\psi}_t^{(M^{\gamma})j}(x) - \sum_{\zeta \in S_t^{M^{\gamma}[1]}} \beta_{\hat{\ell}_t}^{\zeta} \psi_t^{(M^{\gamma})\zeta}(x)$$
$$- \sum_{\zeta \in S_t^{M^{\gamma}[2]}} \beta_{\hat{\ell}_t}^{\zeta} \hat{\psi}_t^{(M^{\gamma})\zeta}(x) = c^{\hat{\ell}_t}\left(S_t^{M^{\gamma}[2]}\right) - a^{\hat{\ell}_t} W_{x_{t+1}}^{M^{\gamma}}(t+1, x_{t+1})(1+r)^{t-1},$$

$$\text{for } \hat{\ell}_t \in S_t^{M^{\gamma}[2]}, \tag{3.6}$$

where $W^{M^\gamma}_{x_{t+1}}(t+1,x_{t+1})$ is the short form for

$$W^{M^\gamma}_{x_{t+1}}\Bigg[t+1,x+\sum_{j\in S_t^{M^\gamma[1]}} a^j\psi_t^{(M^\gamma)j}(x)+\sum_{j\in S_t^{M^\gamma[2]}} \hat{a}^j\hat{\psi}_t^{(M^\gamma)j}(x)$$
$$-\sum_{j\in S_t^{M^\gamma[1]}} b_j\varpi_t^{(M^\gamma)j}(x)-\sum_{j\in S_t^{M^\gamma[2]}} b_j\hat{\varpi}_t^{(M^\gamma)j}(x)-\delta x\Bigg].$$

System (3.5) and (3.6) can be viewed as a set of equations linear in $\psi_t^{\ell_t}(x)$ and $\hat{\psi}_t^{\hat{\ell}_t}(x)$ for $\ell_t\in S_t^{M^\gamma[1]}$ and $\hat{\ell}_t\in S_t^{M^\gamma[2]}$ with $W^{M^\gamma}_{x_{t+1}}(t+1,x_{t+1})(1+r)^{t-1}$ being taken a parameter. Solving (3.5) and (3.6) yields:

$$\psi_t^{(M^\gamma)\ell_t}(x)=\tilde{\alpha}_t^{(M^\gamma)\ell_t}+\tilde{\beta}_t^{(M^\gamma)\ell_t}\;W^{M^\gamma}_{x_{t+1}}(t+1,x_{t+1})(1+r)^{t-1},\ \text{for}\ \ell_t\in S_t^{M^\gamma[1]};$$

$$\hat{\psi}_t^{(M^\gamma)\hat{\ell}_t}(x)=\hat{\tilde{\alpha}}_t^{(M^\gamma)\hat{\ell}_t}+\hat{\tilde{\beta}}_t^{(M^\gamma)\hat{\ell}_t}\;W^{M^\gamma}_{x_{t+1}}(t+1,x_{t+1})(1+r)^{t-1},\ \text{for}\ \hat{\ell}_t\in S_t^{M^\gamma[2]};\quad(3.7)$$

where $\tilde{\alpha}_t^{(M^\gamma)\ell_t}$ and $\tilde{\beta}_j^{(M^\gamma)\ell_t}$ for $\ell_t\in S_t^{M^\gamma[1]}$, and $\hat{\tilde{\alpha}}_t^{(M^\gamma)\hat{\ell}_t}$ and $\hat{\tilde{\beta}}_j^{(M^\gamma)\hat{\ell}_t}$, $\hat{\ell}_t\in S_t^{M^\gamma[2]}$ are constants involving the model parameters.

The maximized joint payoff of the nations under technology pattern M^γ can be obtained as:

Proposition 3.1 System (3.3) admits a solution

$$W^{M^\gamma}(t,x)=\left(A_t^{M^\gamma}x+C_t^{M^\gamma}\right)\left(\frac{1}{1+r}\right)^{t-1},\quad t\in\kappa\tag{3.8}$$

where $A_t^{M^\gamma}$ and $C_t^{M^\gamma}$ are constants involving the model parameters.

Proof See Appendix B. ∎

The technique pattern M^γ which yields the highest joint payoff $W^{M^\gamma}(t,x)$ will be adopted in the cooperative scheme. Let us denote the technique pattern that yields the highest joint payoff by $M*$.

Using (3.4), (3.7) and (3.8), the control strategy under cooperation with technique pattern $M*$ can be obtained accordingly. To induce the industrial sector to produce the socially optimal levels of output with the desired technique, we invoke (1.4), (3.5), and (3.6) to obtain the optimal tax rates

$$v_t^{(M^*)\ell_t}=-a^{\ell_t}A_{t+1}^{M^*}(1+r)^{-1}+\sum_{\substack{\zeta\in S_t^{M^*[1]}\\ \zeta\neq\ell_t}}\beta_{\ell_t}^{\zeta}\psi_t^{(M^*)\zeta}(x)+\sum_{\zeta\in S_t^{M^*[2]}}\beta_{\ell_t}^{\zeta}\hat{\psi}_t^{(M^*)\zeta}(x),$$

on conventional technique for $\ell_t\in S_t^{M^*[1]}$; and

$$\hat{v}_t^{(M^*)\hat{\ell}_t} = -a^{\hat{\ell}_t} A_{t+1}^{M^*}(1+r)^{-1} + \sum_{\zeta \in S_t^{M^*[1]}} \beta_{\hat{\ell}_t}^{\zeta} \psi_t^{(M^*)\zeta}(x) + \sum_{\substack{\zeta \in S_t^{M^*[2]} \\ \zeta \neq \hat{\ell}_t}} \beta_{\hat{\ell}_t}^{\zeta} \hat{\psi}_t^{(M^*)\zeta}(x),$$

on environment-preserving techniques for $\hat{\ell}_t \in S_t^{M^*[2]}$.

A lump-sum levy/subsidy will be given to each industrial sector to guarantee that the same profit level as that under a noncooperative equilibrium is maintained.

Substituting the optimal control strategy into (3.2) yields the dynamics of pollution accumulation under cooperation as:

$$\begin{aligned} x_{t+1} = & \sum_{j \in S_t^{M^*[1]}} a^j \left[\tilde{\alpha}_t^{(M^*)j} + \tilde{\beta}_t^{(M^*)j} \, A_{t+1}^{M^*}(1+r)^{-1} \right] \\ & + \sum_{j \in S_t^{M^*[2]}} \hat{a}^j \left[\hat{\tilde{\alpha}}_t^{(M^*)j} + \hat{\tilde{\beta}}_t^{(M^*)j} \, A_{t+1}^{M^*}(1+r)^{-1} \right] \\ & + \left[1 + \sum_{j=1}^{n} \frac{(b_j)^2}{2c_j^a} A_{t+1}^{M^*}(1+r)^{-1} - \delta \right] x_t, \; x_1 = x^0. \end{aligned} \tag{3.9}$$

Equation (3.9) is a linear difference equation with time varying coefficients. We use $\{x_k^*\}_{k=1}^{T+1}$ to denote the solution path satisfying (3.9). Solving (3.9) gives

$$x_t^* = \left(\prod_{\zeta=1}^{t} S_\zeta^1 \right) x^0 + \sum_{k=1}^{t} \left(\prod_{\zeta=k+1}^{t} S_\zeta^1 \right) S_k^2, \tag{3.10}$$

where $S_\zeta^1 = 1 + \sum_{j=1}^{n} \frac{(b_j)^2}{2c_j^a} A_{\zeta+1}^{M^*}(1+r)^{-1} - \delta$, and

$$\begin{aligned} S_\zeta^2 = & \sum_{j \in S_t^{M^*[1]}} a^j \left[\tilde{\alpha}_\zeta^{(M^*)j} + \tilde{\beta}_\zeta^{(M^*)j} \, A_{\zeta+1}^{M^*}(1+r)^{-1} \right] \\ & + \sum_{j \in S_t^{M^*[2]}} \hat{a}^j \left[\hat{\tilde{\alpha}}_t^{(M^*)j} + \hat{\tilde{\beta}}_t^{(M^*)j} \, A_{t+1}^{M^*}(1+r)^{-1} \right] \end{aligned}$$

14.3.2 Subgame Consistent Collaborative Solution

To achieve dynamic consistency the agreed upon optimality principle must be maintained at every stage of collaboration. In particular, the agreed-upon optimality principle requires the nations to share the gain from cooperation proportional to the nations' relative sizes of noncooperative payoffs. Note that this optimality principle satisfies individual rationality because each nation's payoff under cooperation is higher than its payoff under non-cooperation. To achieve subgame consistency the agreed-upon optimality principle must be maintained at every stage of collaboration. Let $\xi^{\ell}(t, x_t^*)$ denote nation ℓ's imputation (payoff under cooperation) covering the stages t to T under the agreed-upon optimality principle along the cooperative trajectory $\left\{x_k^*\right\}_{k=t}^{T+1}$. Hence the solution imputation scheme has to satisfy:

Condition 3.1

$$\xi^{\ell}\left(t, x_t^*\right) = \frac{\overline{V}^{\ell}\left(t, x_t^*\right)}{\sum\limits_{i \in S_t^1} V^i\left(t, x_t^*\right) + \sum\limits_{j \in S_t^2} \hat{V}^j\left(t, x_t^*\right)} W^{M^*}\left(t, x_t^*\right), \tag{3.11}$$

for all $\ell \in N$ and all $t \in \kappa$,

where $\overline{V}^{\ell}\left(t, x_t^*\right) = V^{\ell}\left(t, x_t^*\right)$ if $\ell \in S_t^1$ and $\overline{V}^{\ell}\left(t, x_t^*\right) = \hat{V}^{\ell}\left(t, x_t^*\right)$ if $\ell \in S_t^2$.

Crucial to the analysis is the formulation of a payment distribution mechanism that would lead to the realization of Condition 3.1. This will be done in the next Section.

14.3.3 Payment Distribution Mechanism

To design a payment distribution scheme over time so that the agreed-upon imputation in Condition 3.1 can be realized we apply the techniques developed in Chap. 7. In formulating a Payoff Distribution Procedure (PDP) we let $B_t^{\ell}(x_t^*)$ denote the payment that nation ℓ will received at stage t under the cooperative agreement given the state x_t^* at stage $t \in \kappa$.

The payment scheme involving $B_t^{\ell}(x_t^*)$ constitutes a PDP in the sense that along the optimal state trajectory $\left\{x_k^*\right\}_{k=t}^{T+1}$ the imputation to nation ℓ over the stages from t to T can be expressed as:

$$\xi^{\ell}\left(t, x_t^*\right) = \sum_{\zeta=t}^{T} B_{\zeta}^{\ell}\left(x_{\zeta}^*\right)\left(\frac{1}{1+r}\right)^{\zeta-1} + g^{\ell}\left(\overline{x}^{\ell} - x_{T+1}\right)\left(\frac{1}{1+r}\right)^{T}, \tag{3.12}$$

for $\ell \in N$ and $t \in \kappa$.

Making use of (3.12), one can arrive at:

$$\xi^{\ell}\left(t, x_t^*\right) = \sum_{\zeta=t}^{h-1} B_{\zeta}^{\ell}\left(x_{\zeta}^*\right)\left(\frac{1}{1+r}\right)^{\zeta-1} + \xi^{\ell}\left(h, x_h^*\right), \tag{3.13}$$

for $\ell \in N$ and $t \in \kappa$ and $h \in \{t+1, t+2, \cdots, T\}$.

A theorem characterizing a formula for $B_t^{\ell}(x_t^*)$, for $t \in \{1, 2, \cdots, T-1\}$ and $\ell \in N$, which yields (3.12) is provided below.

Theorem 3.2 A payment

$$B_t^\ell(x_t^*) = (1+r)^{t-1}\left[\xi^\ell(t,x_t^*) - \xi^\ell(t+1,x_{t+1}^*)\right], \text{ for } \ell \in N$$

given to nation $\ell \in N$ at stage $t \in \{1,2,\cdots,T-1\}$, and a payment

$$B_T^\ell(x_T^*) = (1+r)^{T-1}\left[\xi^\ell(T,x_T^*) - g^\ell(\overline{x}^\ell - x_{T+1}^*)\left(\frac{1}{1+r}\right)^T\right], \tag{3.14}$$

given to nation $\ell \in N$ at stage T would lead to the realization of the imputation $\{\xi^\ell(t,x_t^*), \text{for } t \in \kappa \text{ and } \ell \in N\}$.

Proof From (3.13) one can obtain

$$B_t^\ell(x_t^*)\left(\frac{1}{1+r}\right)^{t-1} = \xi^\ell(t,x_t^*) - \xi^\ell(t+1,x_{t+1}^*), \tag{3.15}$$

for $\ell \in N$ and $t \in \kappa$.

Note that $B_t^\ell(x_t^*)\left(\frac{1}{1+r}\right)^{t-1}$ is the present value (as from initial stage 1) of a payment $B_t^\ell(x_t^*)$ that will be given nation ℓ at stage t. Hence if a payment as specified in (3.14) is given to nation ℓ at stage $t \in \kappa$, the imputation $\{\xi^\ell(t,x_t^*)$, for $t \in \kappa$ and $\ell \in N\}$ can be realized by showing that

$$\sum_{\zeta=t}^{T} B_\zeta^\ell(x_\zeta^*)\left(\frac{1}{1+r}\right)^{\zeta-1} + g^\ell(\overline{x}^\ell - x_{T+1})\left(\frac{1}{1+r}\right)^T$$
$$= \sum_{\zeta=t}^{T}\left[\xi^\ell(\zeta,x_\zeta^*) - \xi^\ell(\zeta+1,x_{\zeta+1}^*)\right] = \xi^\ell(t,x_t^*)$$

for $\ell \in N$ and $t \in \kappa$, given that $\xi^\ell(T+1,x_{T+1}^*) = g^\ell(\overline{x}^\ell - x_{T+1}^*)\left(\frac{1}{1+r}\right)^T$. ■

Invoking Condition 3.1 and Theorem 3.2 the payment (in present value terms) to nation ℓ in stage $t \in \kappa$ can be obtained as:

$$B_t^\ell(x_t^*)\left(\frac{1}{1+r}\right)^{t-1} = \xi^\ell(t,x_t^*) - \xi^\ell(t+1,x_{t+1}^*)$$
$$= \frac{\overline{V}^\ell(t,x_t^*)}{\sum_{i\in S_t^1} V^i(t,x_t^*) + \sum_{j\in S_t^2} \hat{V}^j(t,x_t^*)} W^{M^*}(t,x_t^*)$$
$$- \frac{\overline{V}^\ell(t+1,x_{t+1}^*)}{\sum_{i\in S_{t+1}^1} V^i(t+1,x_{t+1}^*) + \sum_{j\in S_{t+1}^2} \hat{V}^j(t+1,x_{t+1}^*)} W^{M^*}(t+1,x_{t+1}^*), \tag{3.16}$$

for $\ell \in N$, $t \in \kappa$,

where $\overline{V}^{\ell}(t, x_t^*) = V^{\ell}(t, x_t^*)$ if $\ell \in S_t^{M^*[1]}$ and $\overline{V}^{\ell}(t, x_t^*) = \hat{V}^{\ell}(t, x_t^*)$ if $\ell \in S_t^{M^*[2]}$.

Formula (3.16) provides a payoff distribution procedure leading to the satisfaction of Condition 3.1 and hence a time-consistent solution will be obtained.

14.4 Numerical Illustration

As a numerical illustration we consider the case where there are 3 nations which have 3 stages of actions. The demand functions of these nations are respectively

$$P_t^1 = 50 - 2Q_t^1 - Q_t^2 - Q_t^3,\ P_t^2 = 72 - Q_t^1 - 4Q_t^2 - 2Q_t^3, \text{ and}$$
$$P_t^3 = 60 - 2Q_t^1 - Q_t^2 - 3Q_t^3.$$

The costs of producing output with conventional technique are $c^1 = 1$, $c^2 = 0.5$, $c^3 = 1$; and those of using environment-preserving technique are $\hat{c}^1 = 2.5, \hat{c}^2 = 2$, $\hat{c}^3 = 2$. The abatement costs are $c_1^a = 2$, $c_2^a = 2$, $c_3^a = 2.5$; and the abatement parameters are $b_1 = 1$, $b_2 = 1$, $b_3 = 1.5$. The pollution dynamics parameters are $a^1 = 2$, $\hat{a}^1 = 0.5$, $a^2 = 2$, $\hat{a}^2 = 0.5$, $a^3 = 2$, $\hat{a}^3 = 1$. The pollution decay rate $\delta = 0.05$ and the pollution damage parameters are $h_1 = 0.7$, $h_2 = 0.8$, $h_3 = 1.8$. The initial pollution stock is $x_1 = 4$ and the discount rate is $r = 0.04$. The terminal bonus (penalty) parameters are $g^1 = 0.5$, $g^2 = 0.4$, $g^3 = 1.7$; $\overline{x}^1 = 200$, $\overline{x}^2 = 500$, $\overline{x}^3 = 100$.

We first compute the outcome under non-cooperation. At stage $T + 1 = 4$, invoking Proposition 3.1 we have

$$A_4^1 = -0.5, A_4^2 = -0.4, A^3 = -1.7;$$
$$C_4^1 = 100; C_4^2 = 200, \text{ and} C_4^3 = 170.$$

Using Condition (2.2), one can show that industrial sectors 1 and 2 will use conventional technique and sector 3 will use environment-preserving technique in stage 3. Industrial outputs can be obtained as:

$$q_3^1 = 9.1169, q_3^2 = 6.3787,\ \hat{q}_3^3 = 5.2921.$$

Invoking Conditions (6.4) and (6.5) in Appendix A we obtain:

$$A_3^1 = -0.727968, A_3^2 = -0.817751, \hat{A}_3^3 = -2.398049,$$
$$C_3^1 = 252.8004,\ C_3^2 = 346.0104 \text{ and} \hat{C}_3^3 = 194.6502.$$

Using Condition (2.2), one can show that industrial sectors 1 and 2 will use conventional technique and sector 3 will use environment-preserving technique in stage 2. Industrial outputs can be obtained as:

$$q_2^1 = 9.0168,\ q_2^2 = 6.3074,\ \hat{q}_2^3 = 5.2255.$$

Invoking (6.4) and (6.5) in Appendix A we obtain:

$$A_2^1 = -0.43983,\ \ A_2^2 = -0.516227,\ \ \hat{A}_2^3 = -1.937482,$$
$$C_2^1 = 393.1945,\ \ C_2^2 = 473.5493,\ \text{and}\ \ \hat{C}_2^3 = 196.2488.$$

According to Condition (2.2), industrial sectors 1 and 2 will use conventional technique and sector 3 will use environment-preserving technique in stage 1. Industrial outputs can be obtained as:

$$q_1^1 = 9.1361,\ q_1^2 = 6.3587,\ q_1^3 = 5.2510.$$

Invoking (6.4) and (6.5) in Appendix A we obtain:

$$A_1^1 = -0.672388,\ A_1^2 = -0.772149,\ \hat{A}_1^3 = -2.360774,$$
$$C_1^1 = 537.4093,\ C_1^2 = 605.3887,\ \text{and}\ \hat{C}_1^3 = 211.4722.$$

The noncooperative state path can be obtained as:

$$x_1 = 4,\ x_2 = 42.0106,\ x_3 = 27.0039,\ x_4 = 41.0968.$$

Now consider the case that the 3 nations agree to collaborate so that they would maximize their joint payoff and share the gain from cooperation proportional to the nations' relative sizes of noncooperative payoffs. The joint payoff maximizing pattern of technique choices is that all 3 nations will adopt environment-preserving technique.

First consider stage 4, from Condition (6.7) in Appendix B we obtain

$$A_{T+1}^{M^*} = -2.6 \text{ and } C_4^{M^*} = 470.$$

Invoking Condition (6.8) in Appendix B we obtain:

$$A_3^{M^*} = -2.70625,\ A_2^{M^*} = -2.555709 \text{ and } A_1^{M^*} = -2.766075.$$

The nations' outputs in the 3 stages under cooperation are

$$\hat{q}_3^1 = 6.2344,\ \hat{q}_3^2 = 5.8281,\ \hat{q}_3^3 = 3.2186;\ \hat{q}_2^1 = 6.2325,\ \hat{q}_2^2 = 5.8275$$
$$\hat{q}_2^3 = 3.2050\ ;\ \hat{q}_1^1 = 6.2363,\ \hat{q}_1^2 = 5.8288,\ \hat{q}_1^3 = 3.2225.$$

Invoking (6.8) in Appendix B again we obtain:

$$C_3^{M^*} = 885.7551,\ C_2^{M^*} = 1284.575 \text{ and } C_1^{M^*} = 1684.066.$$

Solving the optimal cooperative trajectory yields:

$$x_1^* = 4, x_2^* = 3.7169, x_3^* = 3.5777, \; x_4^* = 4.1516.$$

The joint payoffs in stages 1–4 along the optimal cooperative trajectory can be obtained as:

$$W^{M*}(1, x_1^*) = 1673.002, \; W^{M*}(2, x_2^*) = 1226.035,$$
$$W^{M*}(3, x_3^*) = 809.9787 \text{ and } W^{M*}(4, x_4^*) = 408.2323.$$

The individual payoffs for the 3 nations along the optimal cooperative trajectory are

$$V^1(1, x_1^*) = 534.7198, \; V^2(1, x_1^*) = 602.3001, \; V^3(1, x_1^*) = 202.0292,$$
$$V^1(2, x_2^*) = 376.4998, \; V^2(2, x_2^*) = 453.4909, \; V^3(2, x_2^*) = 182.0357,$$
$$V^1(3, x_3^*) = 231.3202, \; V^2(3, x_3^*) = 317.2011, \; V^3(3, x_3^*) = 172.0328,$$
$$V^1(4, x_4^*) = 87.0542, \; V^2(4, x_4^*) = 176.323, \; V^3(4, x_4^*) = 144.8551.$$

To summarize the results we first present the technology patterns, national outputs and levels of pollution stock under non-cooperation and collaboration in Table 14.1. The technique pattern under noncooperation involves nation 3 adopting environment-preserving techniques in stages 2 and 3. While under cooperation all 3 nations adopt environment-preserving techniques in all the 3 stages. The levels of pollution under collaboration are below those with no cooperation.

Then we proceed to compute the imputations to the nations under collaboration using Condition 4.1 and these figure are given in Table 14.2 along with the joint payoff under cooperation.

Finally, payoff distribution procedures leading the realization of the imputations in Table 14.2 are derived using Theorem 3.2 and displayed in Table 14.3. Note that both the current value and present values of these payments in various stages are provided.

Table 14.1 Technology patterns, national outputs and levels of pollution stock under non-cooperation and collaboration

Stage	Non-cooperation				Collaboration			
t	Nation 1	Nation 2	Nation 3	x_t	Nation 1	Nation 2	Nation 3	x_t^*
1	q_1^1	q_1^2	q_1^3	4	$\hat{q}_1^1$	$\hat{q}_1^2$	$\hat{q}_1^3$	4
	9.1361	6.3587	5.2510		6.2363	5.8288	3.2225	
2	q_2^1	q_2^2	$\hat{q}_2^3$	42.0106	$\hat{q}_2^1$	$\hat{q}_2^2$	$\hat{q}_2^3$	3.7169
	9.0168	6.3074	5.2255		6.2325	5.8275	3.2050	
3	q_3^1	q_3^2	$\hat{q}_3^3$	27.0039	$\hat{q}_3^1$	$\hat{q}_3^2$	$\hat{q}_3^3$	3.5777
	9.1169	6.3787	5.2921		6.2344	5.8281	3.2186	
4				41.0968				4.1516

Table 14.2 Total collaborative payoff and nations' imputations

t	Total collaborative payment $W^{M*}(t,x_t^*)$	Nation 1's imputation $\xi^1(t,x_t^*)$	Nation 2's imputation $\xi^2(t,x_t^*)$	Nation 3's imputation $\xi^3(t,x_t^*)$
1	1673.002	668.0765	752.5111	252.4143
2	1226.035	456.1164	549.3885	220.5299
3	809.9787	260.0283	356.5674	193.383
4	408.2323	87.0542	176.323	144.8551

Table 14.3 Payments incurred to nations in each stage – present value and current value

	Stage cooperative payment (in current value)			Stage cooperative payment (in present value)		
t	$B_t^1(x_t^*)$	$B_t^2(x_t^*)$	$B_t^3(x_t^*)$	$R_tB_t^1(x_t^*)$	$R_tB_t^2(x_t^*)$	$R_tB_t^3(x_t^*)$
1	211.9601	203.1226	31.8844	211.9601	203.1226	31.8844
2	203.9317	200.5339	28.2327	196.0882	192.8211	27.1469
3	187.0887	194.9524	52.4879	172.974	180.2445	48.528
4	97.9242	198.3393	162.9422	87.0542	176.323	144.8551

where $R_t = (1+r)^{-(t-1)}$.

14.5 Multi Production Technique Choices

In this Section we consider the case where there are more than one type of environmental-preserving techniques leading to different degrees of pollution. For exposition sake we examine the situation where there are three types of production techniques available to each nation's industrial sector: a conventional technique and two environment-preserving techniques. Industrial sectors pay more for using environment-preserving techniques. The amounts of pollutants emitted by environment-preserving techniques are less than that emitted by conventional technique. The first environment-preserving technique costs less than the second environment-preserving technique but emits more pollutants than that of the second technique.

14.5.1 Game Formulation and Non-cooperative Equilibria

We use q_t^j to denote the output of nation j if it uses conventional technique, $\hat{q}_t^j$ to denote the output of nation j if it uses the first environment-preserving technique and $\hat{\hat{q}}_t^j$ to denote the output of nation j if it uses the second environment-preserving technique. The average cost of producing a unit of output with conventional technique in nation j is c^j, that of producing a unit of output with the first environment-preserving technique is $\hat{c}^j$ and that with the second environment-preserving technique is $\hat{\hat{c}}^j$.

Let v_t^i denote the tax rate imposed by government i on industrial output produced by conventional technique in stage t, $\hat{v}_t^i$ denote the tax rate imposed on output produced by the first environment-preserving technique and $\hat{\hat{v}}_t^i$ denote the tax rate imposed on output produced by the second environment-preserving technique. Nation i's industrial sector will choose the technique that has the lowest value of unit production cost plus unit tax. For instance, if $\hat{\hat{c}}^i + \hat{\hat{v}}_t^i < \hat{c}^i + \hat{v}_t^i < c^i + v_t^i$ then the second environment-preserving technique will be used by nation i. In stage t, let the set of nations using conventional technique be denoted by S_t^1, the set of nations using the first environment-preserving technique by S_t^2, and the set of nations using the second environment-preserving technique by S_t^3. The profits of the industrial sectors in stage t can be expressed as

$$\pi_t^{i_t} = \Big[\alpha_t^{i_t} - \sum_{j\in S_t^1} \beta_j^{i_t} q_t^j - \sum_{\zeta\in S_t^2} \beta_\zeta^{i_t} \hat{q}_t^{\zeta} - \sum_{\zeta\in S_t^3} \beta_\zeta^{i_t} \hat{\hat{q}}_t^{\zeta}\Big] q_t^{i_t} - c^{i_t} q_t^{i_t} - v_t^{i_t} q_t^{i_t},$$

for $i_t \in S_t^1$, (5.1)

$$\hat{\pi}_t^{\ell_t} = \Big[\alpha_t^{\ell_t} - \sum_{j\in S_t^1} \beta_j^{\ell_t} q_t^j - \sum_{\zeta\in S_t^2} \beta_\zeta^{\ell_t} \hat{q}_t^{\zeta} - \sum_{\zeta\in S_t^3} \beta_\zeta^{\ell_t} \hat{\hat{q}}_t^{\zeta}\Big] \hat{q}_t^{\ell_t} - c^{\ell_t} \hat{q}_t^{\ell_t} - v_t^{\ell_t} \hat{q}_t^{\ell_t},$$

for $\ell_t \in S_t^2$, (5.2)

$$\pi_t^{\varpi_t} = \Big[\alpha_t^{\varpi_t} - \sum_{j\in S_t^1} \beta_j^{\varpi_t} q_t^j - \sum_{\zeta\in S_t^2} \beta_\zeta^{\varpi_t} \hat{q}_t^{\zeta} - \sum_{\zeta\in S_t^3} \beta_\zeta^{\varpi_t} \hat{\hat{q}}_t^{\zeta}\Big] q_t^{\varpi_t} - c^{\varpi_t} q_t^{\varpi_t} - v_t^{\varpi_t} q_t^{\varpi_t},$$

for $\varpi_t \in S_t^3$ (5.3)

In each stage t the industrial sector of nation $i_t \in S_t^1$ seeks to maximize (5.1), the industrial sector of nation $\ell_t \in S_t^2$ seeks to maximize (5.2) and the industrial sector of nation $\varpi_t \in S_t^3$ seeks to maximize (5.3). The first order condition for a Nash equilibrium in stage t yields

$$\alpha_t^{i_t} - \sum_{j\in S_t^1} \beta_j^{i_t} q_t^j - \sum_{\zeta\in S_t^2} \beta_\zeta^{i_t} \hat{q}_t^{\zeta} - \sum_{\zeta\in S_t^3} \beta_\zeta^{i_t} \hat{\hat{q}}_t^{\zeta} - \beta_{i_t}^{i_t} q_t^{i_t} = c^{i_t} + v_t^{i_t}, \text{ for } i_t \in S_t^1;$$

$$\alpha_t^{\ell_t} - \sum_{j\in S_t^1} \beta_j^{\ell_t} q_t^j - \sum_{\zeta\in S_t^2} \beta_\zeta^{\ell_t} \hat{q}_t^{\zeta} - \sum_{\zeta\in S_t^3} \beta_\zeta^{\ell_t} \hat{\hat{q}}_t^{\zeta} - \beta_{\ell_t}^{\ell_t} \hat{q}_t^{\ell_t} = \hat{c}^{\ell_t} + \hat{v}_t^{\ell_t}, \text{ for } \ell_t \in S_t^2;$$

and

$$\alpha_t^{\varpi_t} - \sum_{j\in S_t^1} \beta_j^{\varpi_t} q_t^j - \sum_{\zeta\in S_t^2} \beta_\zeta^{\varpi_t} \hat{q}_t^{\zeta} - \sum_{\zeta\in S_t^3} \beta_\zeta^{\varpi_t} \hat{\hat{q}}_t^{\zeta} - \beta_{i_t}^{\varpi_t} \hat{\hat{q}}_t^{\varpi i_t} = \hat{\hat{c}}^{\varpi_t} + \hat{\hat{v}}_t^{i_t}, \text{ for}$$

$\varpi_t \in S_t^3$ (5.4)

Condition (5.4) shows that the industrial sectors will produce up to a point where marginal revenue (the left-hand side of the equations) equals the cost plus tax of a unit of output produced (the right-hand-side of the equations).

The dynamics of pollution accumulation is then governed by the difference equation:

$$x_{t+1} = x_t + \sum_{i_t \in S_t^1} a^{i_t} q_t^{i_t} + \sum_{\ell_t \in S_t^2} \hat{a}^{\ell_t} \hat{q}_t^{\ell_t} + \sum_{\varpi_t \in S_t^3} \hat{a}^{\varpi_t} \hat{\hat{q}}_t^{\varpi_t} - \sum_{j=1}^{n} b_j u_t^j (x_t)^{1/2} - \delta x_t, \quad (5.5)$$

$x_1 = x^0$.

The payoff of government s $i_t \in S_t^1$ at stage t can be expressed as:

$$[\alpha_t^{i_t} - \sum_{j \in S_t^1} \beta_j^{i_t} q_t^j - \sum_{\zeta \in S_t^2} \beta_\zeta^{i_t} \hat{q}_t^\zeta - \sum_{\zeta \in S_t^3} \beta_\zeta^{i_t} \hat{\hat{q}}_t^\zeta] q_t^{i_t} - c^{i_t} q_t^{i_t} - c_{i_t}^a (u_t^{i_t})^2 - h^{i_t} x_t; \quad (5.6)$$

the payoff of government $\ell_t \in S_t^2$ at stage t can be expressed as:

$$[\alpha_t^{\ell_t} - \sum_{j \in S_t^1} \beta_j^{\ell_t} q_t^j - \sum_{\zeta \in S_t^2} \beta_\zeta^{\ell_t} \hat{q}_t^\zeta - \sum_{\zeta \in S_t^3} \beta_\zeta^{\ell_t} \hat{\hat{q}}_t^\zeta] \hat{q}_t^{\ell_t} - \hat{c}^{\ell_t} \hat{q}_t^{\ell_t} - c_{\ell_t}^a (u_t^{\ell_t})^2 - h^{\ell_t} x_t, \quad (5.7)$$

and the payoff of government $\varpi_t \in S_t^3$ at stage t can be expressed as:

$$[\alpha_t^{\varpi_t} - \sum_{j \in S_t^1} \beta_j^{\varpi_t} q_t^j - \sum_{\zeta \in S_t^2} \beta_\zeta^{\varpi_t} \hat{q}_t^\zeta - \sum_{\zeta \in S_t^3} \beta_\zeta^{\varpi_t} \hat{\hat{q}}_t^\zeta] \hat{q}_t^{\varpi_t} - \hat{\hat{c}}^{\varpi_t} \hat{\hat{q}}_t^{\varpi_t} - c_{\varpi_t}^a (u_t^{\varpi_t})^2 - h^{\varpi_t} x_t. \quad (5.8)$$

The governments' planning horizon is from stage 1 to stage T. It is possible that T may be very large. The discount rate is r. A terminal appraisal of pollution damage is $g^i(\overline{x}^i - x_{T+1})$ will be given to government i at stage $T+1$, where $g^i \geq 0$. Each one of the n governments seeks to maximize the sum of the discounted payoffs over the T stages plus the terminal appraisal. In particular, government i would seek to maximize the objective

$$\sum_{t=1}^{T} \Bigg[[\alpha_t^i - \sum_{\substack{j \in S_t^1 \\ j \neq i}} \beta_j^i q_t^j - \sum_{\substack{\zeta \in S_t^2 \\ j \neq i}} \beta_\zeta^i \hat{q}_t^\zeta - \sum_{\substack{\zeta \in S_t^3 \\ j \neq i}} \beta_\zeta^i \hat{\hat{q}}_t^\zeta - \beta_i^i \overline{q}_t^i] \overline{q}_t^i - \overline{c}^i \overline{q}_t^i - c_i^a (u_t^i)^2$$

$$- h^i x_t \Bigg] \left(\frac{1}{1+r}\right)^{t-1} + g^i(\overline{x}^i - x_{T+1}) \left(\frac{1}{1+r}\right)^T; \text{ for } i \in N; \quad (5.9)$$

where $\overline{q}_t^i = q_t^i$ and $\overline{q}_t^i = q_t^i$ if industrial sector i uses conventional technique; and $\overline{q}_t^i = \hat{q}_t^i$ and $\overline{c}_t^i = \hat{c}_t^i$ if industrial sector i uses the first environment-preserving technique; and $\overline{q}_t^i = \hat{\hat{q}}_t^i$ and $\overline{c}_t^i = \hat{\hat{c}}_t^i$ if industrial sector i uses the second environment-preserving technique.

Following the proof of Theorem 2.1 a solution to the noncooperative dynamic game (5.5) and (5.9) can be characterized by the following theorem.

Theorem 5.1 A set of strategies $\{q_t^{i_t*} = \phi_t^{i_t}(x), \hat{q}_t^{\hat{i}_t*} = \hat{\phi}_t^{\hat{i}_t}(x), \hat{\hat{q}}_t^{\hat{\hat{i}}_t*} = \hat{\hat{\phi}}_t^{\hat{\hat{i}}_t}(x),$ $u_t^{i_t*} = \upsilon_t^{i_t}(x), u_t^{\hat{i}_t*} = \hat{\upsilon}_t^{\hat{i}_t}(x), u_t^{\hat{\hat{i}}_t*} = \hat{\hat{\upsilon}}_t^{\hat{\hat{i}}_t}(x),$ for $t \in \kappa i_t \in S_t^1$, $\hat{i}_t \in S_t^2$ and $\hat{\hat{i}}_t \in S_t^3\}$ provides a feedback Nash equilibrium solution to the game (5.5) and (5.9) if there exist functions $V^{i_t}(t,x)$, $\hat{V}^{\hat{i}_t}(t,x) : R \to R$, and $\overleftarrow{\hat{V}}^{\hat{\hat{i}}_t}(t,x)$, for $t \in \kappa$, $i_t \in S_t^1$, $\hat{i}_t \in S_t^2$ and $\hat{\hat{i}}_t \in S_t^3$, such that the following recursive relations are satisfied:

$$V^{i_t}(T+1,x) = V^{\hat{i}_t}(T+1,x) = V^{\hat{\hat{i}}_t}(T+1,x) = V^i(T+1,x)$$
$$= g^i\left(\bar{x}^i - x\right)\left(\frac{1}{1+r}\right)^T, \text{ for } i \in N,$$

$$V^{i_t}(t,x) = \max_{q_t^{i_t}, u_t}\Bigg\{\Bigg[\Big[\alpha_t^{i_t} - \sum_{\substack{j\in S_t^1\\ j\neq i_t}}\beta_j^{i_t}\phi_t^j(x) - \sum_{\zeta\in S_t^2}\beta_\zeta^{i_t}\hat{\phi}_t^{\zeta}(x) - \sum_{\zeta\in S_t^3}\beta_\zeta^{i_t}\hat{\hat{\phi}}_t^{\zeta}(x) - \beta_{i_t}^{i_t}q_t^{i_t}\Big]q_t^{i_t}$$
$$-c^{i_t}q_t^{i_t} - c_{i_t}^a\left(u_t^{i_t}\right)^2 - h^{i_t}x\Bigg]\left(\tfrac{1}{1+r}\right)^{t-1}$$
$$+\overline{V}^{i_t}\Bigg[t+1, x+\sum_{\substack{j\in S_t^1\\ j\neq i_t}}a^j\phi_t^j(x) + \sum_{\zeta\in S_t^2}\hat{a}^{\zeta}\hat{\phi}_t^{\zeta}(x) + \sum_{\zeta\in S_t^3}\hat{\hat{a}}^{\zeta}\hat{\hat{\phi}}_t^{\zeta}(x) + a^{i_t}q_t^{i_t}$$
$$-\sum_{\substack{j\in S_t^1\\ j\neq i_t}}b_j\upsilon_t^j(x)x^{1/2}$$
$$-\sum_{j\in S_t^2}^{n}b_j\hat{\upsilon}_t^j(x)x^{1/2} - \sum_{j\in S_t^3}^{n}b_j\hat{\hat{\upsilon}}_t^j(x)x^{1/2} - b_{i_t}u_t^{i_t}x^{1/2} - \delta x\Bigg]\Bigg\},$$

for $t \in \kappa$ and $i_t \in S_t^1$;

$$\overline{V}^{\hat{i}_t}(t,x) = \max_{\hat{q}_t^{\hat{i}_t}, u_t^{\hat{i}_t}}\Bigg\{$$
$$\Bigg[\Big[\alpha_t^{\hat{i}_t} - \sum_{j\in S_t^1}\beta_j^{\hat{i}_t}\phi_t^j(x) - \sum_{\substack{\zeta\in S_t^2\\ \zeta\neq\hat{i}_t}}\beta_\zeta^{\hat{i}_t}\hat{\phi}_t^{\zeta}(x) - \sum_{\zeta\in S_t^3}\beta_\zeta^{\hat{i}_t}\hat{\hat{\phi}}_t^{\zeta}(x) - \beta_{\hat{i}_t}^{\hat{i}_t}\hat{q}_t^{\hat{i}_t}\Big]\hat{q}_t^{\hat{i}_t}$$
$$-\hat{c}^{\hat{i}_t}\hat{q}_t^{\hat{i}_t} - c_{\hat{i}_t}^a\left(u_t^{\hat{i}_t}\right)^2 - h^{\hat{i}_t}x\Bigg]\left(\tfrac{1}{1+r}\right)^{t-1}$$
$$+\overline{V}^{\hat{i}_t}\Bigg[t+1, x+\sum_{j\in S_t^1}a^j\phi_t^j(x) + \sum_{\substack{\zeta\in S_t^2\\ \zeta\neq\hat{i}_t}}\hat{a}^{\zeta}\hat{\phi}_t^{\zeta}(x) + \sum_{\zeta\in S_t^3}\hat{\hat{a}}^{\zeta}\hat{\hat{\phi}}_t^{\zeta}(x) + \hat{a}^{\hat{i}_t}\hat{q}_t^{\hat{i}_t}$$
$$-\sum_{j\in S_t^1}b_j\upsilon_t^j(x)x^{1/2}$$
$$-\sum_{\substack{j\in S_t^2\\ j\neq\hat{i}_t}}^{n}b_j\hat{\upsilon}_t^j(x)x^{1/2} - \sum_{j\in S_t^3}^{n}b_j\hat{\hat{\upsilon}}_t^j(x)x^{1/2} - b_{\hat{i}_t}u_t^{\hat{i}_t}x^{1/2} - \delta x\Bigg]\Bigg\},$$

for $t \in \kappa$ and $\hat{i}_t \in S_t^2$; and

$$\hat{\hat{V}}^{\hat{\hat{i}}_t}(t,x) = \max_{\hat{\hat{q}}_t^{\hat{\hat{i}}_t}, u_t^{\hat{\hat{i}}_t}} \Bigg\{$$

$$\Bigg[\; [\alpha_t^{\hat{\hat{i}}_t} - \sum_{j \in S_t^1} \beta_j^{\hat{\hat{i}}_t} \phi_t^j(x) - \sum_{\zeta \in S_t^2} \beta_\zeta^{\hat{\hat{i}}_t} \hat{\phi}_t^\zeta(x) - \sum_{\substack{\zeta \in S_t^3 \\ \zeta \neq \hat{\hat{i}}_t}} \beta_\zeta^{\hat{\hat{i}}_t} \hat{\hat{\phi}}_t^\zeta(x) - \beta_{\hat{\hat{i}}_t}^{\hat{\hat{i}}_t} \hat{\hat{q}}_t^{\hat{\hat{i}}_t}] \hat{\hat{q}}_t^{\hat{\hat{i}}_t}$$

$$- \hat{\hat{c}}^{\hat{\hat{i}}_t} \hat{\hat{q}}_t^{\hat{\hat{i}}_t} - c_{\hat{\hat{i}}_t}^a \left(u_t^{\hat{\hat{i}}_t}\right)^2 - h^{\hat{\hat{i}}_t} x \; \Bigg] \left(\frac{1}{1+r}\right)^{t-1}$$

$$+ \overline{V}^{\hat{\hat{i}}_t} \Bigg[\; t+1, x + \sum_{j \in S_t^1} a^j \phi_t^j(x) + \sum_{\zeta \in S_t^2} \hat{a}^\zeta \hat{\phi}_t^\zeta(x) + \sum_{\substack{\zeta \in S_t^3 \\ \zeta \neq \hat{\hat{i}}_t}} \hat{\hat{a}}^\zeta \hat{\hat{\phi}}_t^\zeta(x) + \hat{\hat{a}}^{\hat{\hat{i}}_t} \hat{\hat{q}}_t^{\hat{\hat{i}}_t}$$

$$- \sum_{j \in S_t^1} b_j \upsilon_t^j(x) x^{1/2}$$

$$- \sum_{j \in S_t^2}^{n} b_j \hat{\upsilon}_t^j(x) x^{1/2} - \sum_{\substack{j \in S_t^3 \\ j \neq \hat{\hat{i}}_t}}^{n} b_j \hat{\hat{\upsilon}}_t^j(x) x^{1/2} - b_{\hat{\hat{i}}_t} u_t^{\hat{\hat{i}}_t} x^{1/2} - \delta x \; \Bigg] \Bigg\},$$

for $t \in \kappa$ and $\hat{\hat{i}}_t \in S_t^3$;

and

$$c^{i_t}\left(\frac{1}{1+r}\right)^{t-1} - \overline{V}_{x_{t+1}}^{i_t}(t+1, x_{t+1}) a^{i_t} < \hat{c}^{i_t}\left(\frac{1}{1+r}\right)^{t-1} - \overline{V}_{x_{t+1}}^{i_t}\left(t+1, x_{t+1}^{(i_t)1}\right) \hat{a}^{i_t},$$

$$c^{i_t}\left(\frac{1}{1+r}\right)^{t-1} - \overline{V}_{x_{t+1}}^{i_t}(t+1, x_{t+1}) a^{i_t} < \hat{\hat{c}}^{i_t}\left(\frac{1}{1+r}\right)^{t-1} - \overline{V}_{x_{t+1}}^{i_t}\left(t+1, x_{t+1}^{(i_t)2}\right) \hat{\hat{a}}^{i_t},$$

for $i_t \in S_t^1$;

$$\hat{c}^{\hat{i}_t}\left(\frac{1}{1+r}\right)^{t-1} - \overline{V}_{x_{t+1}}^{\hat{i}_t}(t+1, x_{t+1}) \hat{a}^{\hat{i}_t} \le c^{\hat{i}_t}\left(\frac{1}{1+r}\right)^{t-1} - \overline{V}_{x_{t+1}}^{\hat{i}_t}\left(t+1, x_{t+1}^{(\hat{i}_t)0}\right) a^{\hat{i}_t}$$

$$\hat{c}^{\hat{i}_t}\left(\frac{1}{1+r}\right)^{t-1} - \overline{V}_{x_{t+1}}^{\hat{i}_t}(t+1, x_{t+1}) \hat{a}^{\hat{i}_t} < \hat{\hat{c}}^{\hat{i}_t}\left(\frac{1}{1+r}\right)^{t-1} - \overline{V}_{x_{t+1}}^{\hat{i}_t}\left(t+1, x_{t+1}^{(\hat{i}_t)2}\right) \hat{\hat{a}}^{\hat{i}_t},$$

for $\hat{i}_t \in S_t^2$;

$$c^{\hat{\hat{i}}_t}\left(\frac{1}{1+r}\right)^{t-1} - \overline{V}_{x_{t+1}}^{\hat{\hat{i}}_t}(t+1, x_{t+1}) \hat{\hat{a}}^{\hat{\hat{i}}_t} \le c^{\hat{\hat{i}}_t}\left(\frac{1}{1+r}\right)^{t-1} - \overline{V}_{x_{t+1}}^{\hat{\hat{i}}_t}\left(t+1, x_{t+1}^{(\hat{\hat{i}}_t)0}\right) a^{\hat{\hat{i}}_t},$$

$$c^{\hat{\hat{i}}_t}\left(\frac{1}{1+r}\right)^{t-1} - \overline{V}_{x_{t+1}}^{\hat{\hat{i}}_t}(t+1, x_{t+1}) \hat{\hat{a}}^{\hat{\hat{i}}_t} \le \hat{c}^{\hat{\hat{i}}_t}\left(\frac{1}{1+r}\right)^{t-1} - \overline{V}_{x_{t+1}}^{\hat{\hat{i}}_t}\left(t+1, x_{t+1}^{(\hat{\hat{i}}_t)1}\right) \hat{a}^{\hat{\hat{i}}_t},$$

for $\hat{\hat{i}}_t \in S_t^3$;

where $\overline{V}^i(t+1, x_{t+1}) = V^i(t+1, x_{t+1})$ if i uses conventional technology in stage $t+1$ and $\overline{V}^i(t+1, x_{t+1}) = \hat{V}^i(t+1, x_{t+1})$ if i uses the first environment preserving technology in stage $t+1$, $\overline{V}^i(t+1, x_{t+1}) = \hat{\hat{V}}^i(t+1, x_{t+1})$ if i uses the second environment preserving technology in stage $t+1$, and

$$x_{t+1} = x + \sum_{j\in S_t^1} a^j\phi_t^j(x) + \sum_{\zeta\in S_t^2} \hat{a}^\zeta\hat{\phi}_t^\zeta(x) + \sum_{\zeta\in S_t^3} \hat{\hat{a}}^\zeta\hat{\hat{\phi}}_t^\zeta(x) - \sum_{j\in S_t^1} b_j\upsilon_t^j(x)x^{1/2}$$
$$-\sum_{j\in S_t^2} b_j\hat{\upsilon}_t^j(x)x^{1/2} - \sum_{j\in S_t^3} b_j\hat{\hat{\upsilon}}_t^j(x)x^{1/2} - \delta x,$$

$$x_{t+1}^{(i_t)1} = x + \sum_{\substack{j\in S_t^1\\ j\neq i_t}} a^j\phi_t^j(x) + \sum_{\zeta\in S_t^2} \hat{a}^\zeta\hat{\phi}_t^\zeta(x) + \sum_{\zeta\in S_t^3} \hat{\hat{a}}^\zeta\hat{\hat{\phi}}_t^\zeta(x) + \hat{a}^{i_t}\hat{q}_t^{i_t} - \sum_{\substack{j\in S_t^1\\ j\neq i_t}} b_j\upsilon_t^j(x)x^{1/2}$$
$$-\sum_{j\in S_t^2} b_j\hat{\upsilon}_t^j(x)x^{1/2} - \sum_{j\in S_t^3} b_j\hat{\hat{\upsilon}}_t^j(x)x^{1/2} - b_{i_t}u_t^{i_t} - \delta x,$$

$$x_{t+1}^{(i_t)2} = x + \sum_{\substack{j\in S_t^1\\ j\neq i_t}} a^j\phi_t^j(x) + \sum_{\zeta\in S_t^2} \hat{a}^\zeta\hat{\phi}_t^\zeta(x) + \sum_{\zeta\in S_t^3} \hat{\hat{a}}^\zeta\hat{\hat{\phi}}_t^\zeta(x) + \hat{\hat{a}}^{i_t}\hat{\hat{q}}_t^{i_t} - \sum_{\substack{j\in S_t^1\\ j\neq i_t}} b_j\upsilon_t^j(x)x^{1/2}$$
$$-\sum_{j\in S_t^2} b_j\hat{\upsilon}_t^j(x)x^{1/2} - \sum_{j\in S_t^3} b_j\hat{\hat{\upsilon}}_t^j(x)x^{1/2} - b_{i_t}u_t^{i_t} - \delta x,$$

$$x_{t+1}^{(\hat{i}_t)0} = x + \sum_{j\in S_t^1} a^j\phi_t^j(x) + \sum_{\substack{\zeta\in S_t^2\\ \zeta\neq \hat{i}_t}} \hat{a}^\zeta\hat{\phi}_t^\zeta(x) + \sum_{\zeta\in S_t^3} \hat{\hat{a}}^\zeta\hat{\hat{\phi}}_t^\zeta(x) + a^{\hat{i}_t}q_t^{\hat{i}_t} - \sum_{j\in S_t^1} b_j\upsilon_t^j(x)x^{1/2}$$
$$-\sum_{\substack{j\in S_t^2\\ j\neq \hat{i}_t}} b_j\hat{\upsilon}_t^j(x)x^{1/2} - \sum_{j\in S_t^3} b_j\hat{\hat{\upsilon}}_t^j(x)x^{1/2} - b_{\hat{i}_t}u_t^{\hat{i}_t} - \delta x,$$

$$x_{t+1}^{(\hat{i}_t)2} = x + \sum_{j\in S_t^1} a^j\phi_t^j(x) + \sum_{\substack{\zeta\in S_t^2\\ \zeta\neq \hat{i}_t}} \hat{a}^\zeta\hat{\phi}_t^\zeta(x) + \sum_{\zeta\in S_t^3} \hat{\hat{a}}^\zeta\hat{\hat{\phi}}_t^\zeta(x) + \hat{\hat{a}}^{\hat{i}_t}\hat{\hat{q}}_t^{\hat{i}_t}$$
$$-\sum_{j\in S_t^1} b_j\upsilon_t^j(x)x^{1/2} - \sum_{\substack{j\in S_t^2\\ j\neq \hat{i}_t}} b_j\hat{\upsilon}_t^j(x)x^{1/2} - \sum_{j\in S_t^3} b_j\hat{\hat{\upsilon}}_t^j(x)x^{1/2} - b_{\hat{i}_t}u_t^{\hat{i}_t} - \delta x,$$

$$x_{t+1}^{(\hat{\hat{i}}_t)0} = x + \sum_{j \in S_t^1} a^j \phi_t^j(x) + \sum_{\zeta \in S_t^2} \hat{a}^\zeta \hat{\phi}_t^\zeta(x) + \sum_{\substack{\zeta \in S_t^3 \\ \zeta \neq \hat{\hat{i}}_t}} \hat{\hat{a}}^\zeta \hat{\hat{\phi}}_t^\zeta(x) + a^{\hat{\hat{i}}_t} q_t^{\hat{\hat{i}}_t}$$

$$- \sum_{j \in S_t^1} b_j v_t^j(x) x^{1/2} - \sum_{j \in S_t^2} b_j \hat{v}_t^j(x) x^{1/2} - \sum_{\substack{j \in S_t^3 \\ j \neq \hat{\hat{i}}_t}} b_j \hat{\hat{v}}_t^j(x) x^{1/2} - b_{\hat{\hat{i}}_t} u_t^{\hat{\hat{i}}_t} - \delta x,$$

$$x_{t+1}^{(\hat{\hat{i}}_t)1} = x + \sum_{j \in S_t^1} a^j \phi_t^j(x) + \sum_{\zeta \in S_t^2} \hat{a}^\zeta \hat{\phi}_t^\zeta(x) + \sum_{\substack{\zeta \in S_t^3 \\ \zeta \neq \hat{\hat{i}}_t}} \hat{\hat{a}}^\zeta \hat{\hat{\phi}}_t^\zeta(x) + \hat{a}^{\hat{\hat{i}}_t} \hat{q}_t^{\hat{\hat{i}}_t}$$

$$- \sum_{j \in S_t^1} b_j v_t^j(x) x^{1/2} - \sum_{j \in S_t^2} b_j \hat{v}_t^j(x) x^{1/2} - \sum_{\substack{j \in S_t^3 \\ j \neq \hat{\hat{i}}_t}} b_j \hat{\hat{v}}_t^j(x) x^{1/2} - b_{\hat{\hat{i}}_t} u_t^{\hat{\hat{i}}_t} - \delta x.$$

Proof Follow the proof of Theorem 2.1. ■

Following the analysis in Sect. 14.2 the value functions in Theorem 5.1 can be obtained as:

$$V^{i_t}(t,x) = \left(A_t^{i_t} x + C_t^{i_t}\right)\left(\frac{1}{1+r}\right)^{t-1}, \text{ for } i_t \in S_t^1 \text{ and } t \in \kappa;$$

$$\hat{V}^{\hat{i}_t}(t,x) = \left(\hat{A}_t^{\hat{i}_t} x + \hat{C}_t^{\hat{i}_t}\right)\left(\frac{1}{1+r}\right)^{t-1}, \text{ for } \hat{i}_t \in S_t^2 \text{ and } t \in \kappa;$$

$$\hat{\hat{V}}^{\hat{\hat{i}}_t}(t,x) = \left(\hat{\hat{A}}_t^{\hat{\hat{i}}_t} x + \hat{\hat{C}}_t^{\hat{\hat{i}}_t}\right)\left(\frac{1}{1+r}\right)^{t-1}, \text{ for } \hat{\hat{i}}_t \in S_t^3 \text{ and } t \in \kappa; \quad (5.10)$$

with $A_t^{i_t}, C_t^{i_t}, \hat{A}_t^{\hat{i}_t}, \hat{C}_t^{\hat{i}_t}, \hat{\hat{A}}_t^{\hat{\hat{i}}_t}$ and $\hat{\hat{C}}_t^{\hat{\hat{i}}_t}$ being constants involving the model parameters.

14.5.2 Group Optimality and Subgame Consistent Payment Scheme

Now consider the case when all the nations want to collaborate and tackle the pollution problem together. To achieve group optimality all the nations would act cooperatively so that the joint payoff will be maximized. Since three technique choices are available they have to determine which nations would use which type of techniques over the T stages. Let M^γ be a matrix reflecting the pattern of technique choices by the n nations over the T stages. In particular, according to pattern M^γ, the set of nations that use conventional technique is $S_t^{M^\gamma[1]}$, the set of nations that use the first environment-preserving technique is $S_t^{M^\gamma[2]}$ and the set of nations that use the second environment-preserving technique is $S_t^{M^\gamma[3]}$ in stage $t \in \kappa$. To select the controls which would maximize joint payoff under pattern M^γ the nations have to solve the following optimal control problem of maximizing

$$\sum_{t=1}^{T}\Bigg[\sum_{i_t\in S_t^{M^\gamma[1]}}\Bigg(\Big[\alpha_t^{i_t}-\sum_{j\in S_t^{M^\gamma[1]}}\beta_j^{i_t}q_t^j-\sum_{\zeta\in S_t^{M^\gamma[2]}}\beta_\zeta^{i_t}\hat{q}_t^{\zeta}-\sum_{\zeta\in S_t^{M^\gamma[3]}}\beta_\zeta^{i_t}\hat{\hat{q}}_t^{\zeta}\Big]q_t^{i_t}$$
$$-c^{i_t}q_t^{i_t}-c_{i_t}^a\left(u_t^{i_t}\right)^2-h^{i_t}x_t\Bigg)\left(\tfrac{1}{1+r}\right)^{t-1} \qquad -\hat{c}^{\,\hat{i}_t}\left(S_t^{M^\gamma[2]}\right)\hat{q}_t^{\hat{i}_t}$$
$$+\sum_{\hat{i}_t\in S_t^{M^\gamma[2]}}\Bigg(\Big[\alpha_t^{\hat{i}_t}-\sum_{j\in S_t^{M^\gamma[1]}}\beta_j^{\hat{i}_t}q_t^j-\sum_{\zeta\in S_t^{M^\gamma[2]}}\beta_\zeta^{\hat{i}_t}\hat{q}_t^{\zeta}-\sum_{\zeta\in S_t^{M^\gamma[3]}}\beta_\zeta^{\hat{i}_t}\hat{\hat{q}}_t^{\zeta}\Big]\hat{q}_t^{\hat{i}_t}$$
$$-c_{\hat{i}_t}^a\left(u_t^{\hat{i}_t}\right)^2-h^{\hat{i}_t}x_t\Bigg)\left(\frac{1}{1+r}\right)^{t-1}$$
$$+\sum_{\hat{\hat{i}}_t\in S_t^{M^\gamma[3]}}\Bigg(\Big[\alpha_t^{\hat{\hat{i}}_t}-\sum_{j\in S_t^{M^\gamma[1]}}\beta_j^{\hat{\hat{i}}_t}q_t^j-\sum_{\zeta\in S_t^{M^\gamma[2]}}\beta_\zeta^{\hat{\hat{i}}_t}\hat{q}_t^{\zeta}-\sum_{\zeta\in S_t^{M^\gamma[3]}}\beta_\zeta^{\hat{\hat{i}}_t}\hat{\hat{q}}_t^{\zeta}\Big]\hat{\hat{q}}_t^{\hat{\hat{i}}_t}-\hat{\hat{c}}^{\,\hat{\hat{i}}_t}\left(S_t^{M^\gamma[2]}\right)\hat{\hat{q}}_t^{\hat{\hat{i}}_t}$$
$$-c_{\hat{\hat{i}}_t}^a\left(u_t^{\hat{\hat{i}}_t}\right)^2-h^{\hat{\hat{i}}_t}x_t\Bigg)\left(\frac{1}{1+r}\right)^{t-1}\Bigg]+\sum_{i=1}^{n}g^i\left(\bar{x}^i-x_{T+1}\right)\left(\frac{1}{1+r}\right)^T \tag{5.11}$$

subject to

$$x_{t+1}=x_t+\sum_{\ell_t\in S_t^{M^\gamma[1]}}a^{\ell_t}q_t^{\ell_t}+\sum_{\hat{\ell}_t\in S_t^{M^\gamma[2]}}\hat{a}^{\hat{\ell}_t}\hat{q}_t^{\hat{\ell}_t}+\sum_{\hat{\hat{\ell}}_t\in S_t^{M^\gamma[3]}}\hat{\hat{a}}^{\hat{\hat{\ell}}_t}\hat{\hat{q}}_t^{\hat{\hat{\ell}}_t}-\sum_{\ell_t\in S_t^{M^\gamma[1]}}b_{\ell_t}u_t^{\ell_t}(x_t)^{1/2}$$
$$-\sum_{\hat{\ell}_t\in S_t^{M^\gamma[2]}}b_{\hat{\ell}_t}u_t^{\hat{\ell}_t}(x_t)^{1/2}-\sum_{\hat{\hat{\ell}}_t\in S_t^{M^\gamma[3]}}b_{\hat{\hat{\ell}}_t}u_t^{\hat{\hat{\ell}}_t}(x_t)^{1/2}-\delta\,x_t,$$
$$x_1=x^0. \tag{5.12}$$

The solution to the optimal control problem (5.11 and 5.12) can be characterized by the following theorem.

Theorem 5.2 A set of strategies $\{q_t^{\ell_t*}=\psi_t^{(M^\gamma)\ell_t}(x),\ \hat{q}_t^{\hat{\ell}_t*}=\hat{\psi}_t^{(M^\gamma)\hat{\ell}_t}(x),\ \hat{\hat{q}}_t^{\hat{\hat{\ell}}_t*}=\hat{\hat{\psi}}_t^{(M^\gamma)\hat{\hat{\ell}}_t}(x),\ u_t^{\ell_t*}=\varpi_t^{(M^\gamma)\ell_t}(x),\ u_t^{\hat{\ell}_t*}=\hat{\varpi}_t^{(M^\gamma)\hat{\ell}_t}(x),\ u_t^{\hat{\hat{\ell}}_t*}=\hat{\hat{\varpi}}_t^{(M^\gamma)\hat{\ell}_t}(x)$ for $t\in\kappa$ and ℓ_t

$\in S_t^{M^\gamma[1]}$, $\hat{\ell}_t \in S_t^{M^\gamma[2]}$ and $\hat{\hat{\ell}}_t \in S_t^{M^\gamma[3]}\}$ constitutes an optimal solution to the control problem (5.11) and (5.12) if there exist functions $W^{M^\gamma}(t,x)$, for $t \in \kappa$, such that the following recursive relations are satisfied:

$$W^{M^\gamma}(t,x) = \max_{\substack{u_t^{i_t}, q_t^{i_t}, i_t \in S_t^{M^\gamma[1]}; \\ \hat{u}_t^{\hat{i}_t} \hat{q}_t^{\hat{i}_t}, \hat{i}_t \in S_t^{M^\lambda[2]}; \\ \hat{\hat{u}}_t^{\hat{\hat{i}}_t} \hat{\hat{q}}_t^{\hat{\hat{i}}_t}, \hat{\hat{i}}_t \in S_t^{M^\lambda[3]},}} \left\{ \sum_{i_t \in S_t^{M^\gamma[1]}} \left(\left[\alpha_t^{i_t} - \sum_{j \in S_t^{M^\gamma[1]}} \beta_j^{i_t} q_t^j - \sum_{\zeta \in S_t^{M^\gamma[2]}} \beta_\zeta^{i_t} \hat{q}_t^\zeta - \sum_{\zeta \in S_t^{M^\gamma[3]}} \beta_\zeta^{i_t} \hat{\hat{q}}_t^\zeta \right] q_t^{i_t} \right.\right.$$

$$\left. - c^{i_t} q_t^{i_t} - c_{i_t}^a \left(u_t^{i_t}\right)^2 - h^{i_t} x_t \right) \left(\frac{1}{1+r}\right)^{t-1}$$

$$+ \sum_{\hat{i}_t \in S_t^{M^\gamma[2]}} \left(\left[\alpha_t^{\hat{i}_t} - \sum_{j \in S_t^{M^\gamma[1]}} \beta_j^{\hat{i}_t} q_t^j - \sum_{\zeta \in S_t^{M^\gamma[2]}} \beta_\zeta^{\hat{i}_t} \hat{q}_t^\zeta - \sum_{\zeta \in S_t^{M^\gamma[3]}} \beta_\zeta^{\hat{i}_t} \hat{\hat{q}}_t^\zeta \right] \hat{q}_t^{\hat{i}_t} \right.$$

$$\left. - \hat{c}^{\hat{i}_t}\left(S_t^{M^\gamma[2]}\right) \hat{q}_t^{\hat{i}_t} - c_{\hat{i}_t}^a \left(\hat{u}_t^{\hat{i}_t}\right)^2 - h^{\hat{i}_t} x_t \right) \left(\frac{1}{1+r}\right)^{t-1}$$

$$+ \sum_{\hat{\hat{i}}_t \in S_t^{M^\gamma[3]}} \left(\left[\alpha_t^{\hat{\hat{i}}_t} - \sum_{j \in S_t^{M^\gamma[1]}} \beta_j^{\hat{\hat{i}}_t} q_t^j - \sum_{\zeta \in S_t^{M^\gamma[2]}} \beta_\zeta^{\hat{\hat{i}}_t} \hat{q}_t^\zeta - \sum_{\zeta \in S_t^{M^\gamma[3]}} \beta_\zeta^{\hat{\hat{i}}_t} \hat{\hat{q}}_t^\zeta \right] \hat{\hat{q}}_t^{\hat{\hat{i}}_t} - \hat{\hat{c}}^{\hat{\hat{i}}_t}\left(S_t^{M^\gamma[2]}\right) \hat{\hat{q}}_t^{\hat{\hat{i}}_t} \right.$$

$$\left.\left. - c_{\hat{\hat{i}}_t}^a \left(\hat{\hat{u}}_t^{\hat{\hat{i}}_t}\right)^2 - h^{\hat{\hat{i}}_t} x_t \right) \left(\frac{1}{1+r}\right)^{t-1} \right]$$

$$+ W^{M^\gamma}\left[t+1, x + \sum_{\ell_t \in S_t^{M^\gamma[1]}} a^{\ell_t} q_t^{\ell_t} + \sum_{\hat{\ell}_t \in S_t^{M^\gamma[2]}} \hat{a}^{\hat{\ell}_t} \hat{q}_t^{\hat{\ell}_t} + \sum_{\hat{\hat{\ell}}_t \in S_t^{M^\gamma[3]}} \hat{\hat{a}}^{\hat{\hat{\ell}}_t} \hat{\hat{q}}_t^{\hat{\hat{\ell}}_t} - \sum_{\ell_t \in S_t^{M^\gamma[1]}} b_{\ell_t} u_t^{\ell_t} (x_t)^{1/2} \right.$$

$$\left.\left. - \sum_{\hat{\ell}_t \in S_t^{M^\gamma[2]}} b_{\hat{\ell}_t} \hat{u}_t^{\hat{\ell}_t} (x)^{1/2} - \sum_{\hat{\hat{\ell}}_t \in S_t^{M^\gamma[3]}} b_{\hat{\hat{\ell}}_t} \hat{\hat{u}}_t^{\hat{\hat{\ell}}_t} (x)^{1/2} - \delta x \right]\right\}, \text{ for } t \in \kappa;$$

$$W^{M^\gamma}(T+1,x) = \sum_{i=1}^{n} g^i \left(\bar{x}^i - x\right) \left(\frac{1}{1+r}\right)^T \tag{5.13}$$

Proof The results in (3.3) satisfy the standard optimality conditions in discrete-time dynamic programming. ■

Following the analysis in Sect. 14.3 with the indicated maximization and solving of (5.13) yields the value function indicating the maximized joint payoff under pattern M^γ as

$$W^{M^{\gamma}}(t,x) = \left(A_t^{M^{\gamma}} x + C_t^{M^{\gamma}}\right)\left(\frac{1}{1+r}\right)^{t-1}, t \in \kappa \tag{5.14}$$

where $A_t^{M^{\gamma}}$ and $C_t^{M^{\gamma}}$ are constants involving the model parameters.

The technique pattern M^{γ} which yields the highest joint payoff $W^{M^{\gamma}}(t,x)$ will be adopted in the cooperative scheme. Let us denote the technique pattern that yields the highest joint payoff by M^*. To achieve dynamic consistency the agreed upon optimality principle must be maintained at every stage of collaboration.

The agreed-upon optimality principle requires the nations to share the gain from cooperation proportional to the nations' relative sizes of noncooperative payoffs. In a dynamic framework this condition has to be maintained at every stage. Let $\xi^{\ell}(t, x_t^*)$ denote nation ℓ's imputation (payoff under cooperation) covering the stages t to T under the agreed-upon optimality principle along the cooperative trajectory $\{x_k^*\}_{k=t}^{T}$. Following Theorem 3.2, a payment

$$B_t^{\ell}(x_t^*) = (1+r)^{t-1}\left[\xi^{\ell}(t,x_t^*) - \xi^{\ell}(t+1,x_{t+1}^*)\right], \text{ for } \ell \in N$$

given to nation $\ell \in N$ at stage $t \in \{1,2,\cdots,T-1\}$, and a payment

$$B_T^{\ell}(x_T^*) = (1+r)^{T-1}\left[\xi^{\ell}(T,x_T^*) - g^{\ell}(\bar{x}^{\ell} - x_{T+1}^*)\left(\frac{1}{1+r}\right)^{T}\right], \tag{5.15}$$

given to nation $\ell \in N$ at stage T would lead to the realization of the imputation $\{\xi^{\ell}(t,x_t^*), \text{ for } t \in \kappa \text{ and } \ell \in N\}$.

If the agreed-upon optimality principle requires the nations to share the gain from cooperation proportional to the nations' relative sizes of noncooperative payoffs the payment (in present value terms) to nation ℓ in stage $t \in \kappa$ can be obtained as:

$$\begin{aligned} B_t^{\ell}(x_t^*)\left(\tfrac{1}{1+r}\right)^{t-1} &= \xi^{\ell}(t,x_t^*) - \xi^{\ell}(t+1,x_{t+1}^*) \\ &= \frac{\overline{V}^{\ell}(t,x_t^*)}{\sum_{j=1}^{n} V^j(t,x_t^*)} W^{M^*}(t,x_t^*) - \frac{\overline{V}^{\ell}(t+1,x_{t+1}^*)}{\sum_{j=1}^{n} V^j(t+1,x_{t+1}^*)} W^{M^*}(t+1,x_{t+1}^*), \end{aligned} \tag{5.16}$$

for $\ell \in N$, $t \in \kappa$,

where $\overline{V}^{\ell}(t,x_t^*) = V^{\ell}(t,x_t^*)$ if $\ell \in S_t^1$, $\overline{V}^{\ell}(t,x_t^*) = \hat{V}^{\ell}(t,x_t^*)$ if $\ell \in S_t^2$ and $\overline{V}^{\ell}(t,x_t^*) = \hat{\hat{V}}^{\ell}(t,x_t^*)$ if $\ell \in S_t^3$.

Finally, the analysis can readily be extended to the case where there are three or more environmental-preserving techniques in a similar manner.

14.6 Appendices

Appendix A: Proof of Proposition 2.1

From (2.11) we can obtain $V^{i_t}_{x_{t+1}}(t+1, x_{t+1})(1+r)^{t-1}$ as $A^{i_t}_{t+1}(1+r)^{-1}$ and $\overline{V}^{\hat{i}_t}_{x_{t+1}}(t+1, x_{t+1})(1+r)^{t-1}$ as $\overline{A}^{\hat{i}_t}_{t+1}(1+r)^{-1}$. Substituting these results into the game equilibrium strategies (2.9) and (2.10), and then into (2.1) yield:

$$
\begin{aligned}
&A^i_{T+1}x + C^i_{T+1} = g^i(\overline{x}^i - x),\ i \in N;\\
&A^{i_t}_t x + C^{i_t}_t = \Bigg[\Bigg(\alpha^{i_t}_t - \sum_{j\in S^1_t}\beta^{i_t}_j\phi^j_t(x) - \sum_{j\in S^2_t}\beta^{i_t}_j\hat{\phi}^j_t(x)\Bigg)\phi^{i_t}_t(x)\\
&-c^{i_t}\phi^{i_t}_t(x) - \frac{(b_{i_t})^2}{4c^a_{i_t}}\left[\overline{A}^{i_t}_{t+1}(1+r)^{-1}\right]^2 x - h^{i_t}x\Bigg]\\
&+(1+r)^{-1}\Bigg[\overline{A}^{i_t}_{t+1}\Bigg(x + \sum_{j\in S^1_t}a^j\phi^j_t(x) + \sum_{j\in S^2_t}\hat{a}^j\hat{\phi}^j_t(x)\\
&+\sum_{j\in S^1_t}\frac{(b_j)^2}{2c^a_j}\overline{A}^j_{t+1}(1+r)^{-1}x + \sum_{j\in S^2_t}\frac{(b_j)^2}{2c^a_j}\overline{A}^j_{t+1}(1+r)^{-1}x - \delta x\Bigg) + \overline{C}^{i_t}_{t+1}\Bigg],
\end{aligned}
$$

for $t \in \kappa$ and $i_t \in S^1_t$,

$$
\begin{aligned}
&\hat{A}^{\hat{i}_t}_t x + \hat{C}^{\hat{i}_t}_t = \Bigg[\Bigg(\alpha^{\hat{i}_t}_t - \sum_{j\in S^1_t}\beta^{\hat{i}_t}_j\phi^j_t(x) - \sum_{j\in S^2_t}\beta^{\hat{i}_t}_j\hat{\phi}^j_t(x)\Bigg)\hat{\phi}^{\hat{i}_t}_t(x)\\
&-\hat{c}^{\hat{i}_t}\hat{\phi}^{\hat{i}_t}_t(x) - \frac{\left(b_{\hat{i}_t}\right)^2}{4c^a_{\hat{i}_t}}\left[\overline{A}^{\hat{i}_t}_{t+1}(1+r)^{-1}\right]^2 x - h^{\hat{i}_t}x\Bigg]\\
&+(1+r)^{-1}\Bigg[\overline{A}^{\hat{i}_t}_{t+1}\Bigg(x + \sum_{j\in S^1_t}a^j\phi^j_t(x) + \sum_{j\in S^2_t}\hat{a}^j\hat{\phi}^j_t(x)\\
&+\sum_{j\in S^1_t}\frac{(b_j)^2}{2c^a_j}\overline{A}^j_{t+1}(1+r)^{-1}x + \sum_{j\in S^2_t}\frac{(b_j)^2}{2c^a_j}\overline{A}^j_{t+1}(1+r)^{-1}x - \delta x\Bigg) + \overline{C}^{\hat{i}_t}_{t+1}\Bigg],
\end{aligned}
\tag{6.1}
$$

for $t \in \kappa$ and $\hat{i}_t \in S^2_t$,

$$\overline{A}_{t+1}^{i_t}(1+r)^{-1} > \frac{\left(\hat{c}^{i_t} - c^{i_t}\right)}{\left(\hat{a}^{i_t} - a^{i_t}\right)}, \text{ for } i_t \in S_t^1 \text{ and}$$

$$\overline{A}_{t+1}^{\hat{i}_t}(1+r)^{-1} \le \frac{\left(\hat{c}^{\hat{i}_t} - c^{\hat{i}_t}\right)}{\left(\hat{a}^{\hat{i}_t} - a^{\hat{i}_t}\right)}, \text{ for } \hat{i}_t \in S_t^2. \tag{6.2}$$

Where

$$\phi_t^j(x) = \left[\overline{\alpha}_t^j + \sum_{\zeta \in S_t^1} \overline{\beta}_t^{(j)\zeta}\, \overline{A}_{t+1}^{\zeta}(1+r)^{-1} + \sum_{\zeta \in S_t^2} \overline{\beta}_t^{(j)\zeta}\, \hat{A}_{t+1}^{\zeta}(1+r)^{-1}\right] \text{ and}$$

$$\hat{\phi}_t^{k}(x) = \left[\hat{\overline{\alpha}}_t^{k} + \sum_{\zeta \in S_t^1} \hat{\overline{\beta}}_t^{(k)\zeta}\, \overline{A}_{t+1}^{\zeta}(1+r)^{-1} + \sum_{\zeta \in S_t^2} \hat{\overline{\beta}}_t^{(k)\zeta}\, \hat{A}_{t+1}^{\zeta}(1+r)^{-1}\right].$$

First consider the stage $T+1$, from (6.1) we obtain

$$A_{T+1}^i = -g^i,\; C_{T+1}^i = g^i \overline{x}^i, \text{ for } i \in N. \tag{6.3}$$

At stage T, invoking (6.2), industrial sector i which has $A_{T+1}^i(1+r)^{-1} > \frac{\left(\hat{c}^i - c^i\right)}{\left(\hat{a}^i - a^i\right)}$ would use conventional technique, otherwise it would use environment-preserving technique.

Note that on the left-hand-side of (6.1) the expressions are $A_t^{i_t}x + C_t^{i_t}$ and $\hat{A}_t^{\hat{i}_t}x + \hat{C}_t^{\hat{i}_t}$. On the right-hand-side there are expressions which are linear in x with coefficients involving the terms $A_{t+1}^{i_t}$, $C_{t+1}^{i_t}$, $\hat{A}_{t+1}^{\hat{i}_t}$, and $\hat{C}_{t+1}^{\hat{i}_t}$. The values of $A_T^{i_t}$, $C_T^{i_t}$, $\hat{A}_T^{\hat{i}_t}$, and $\hat{C}_T^{\hat{i}_t}$ for $i_t \in S_t^1$ and $\hat{i}_t \in S_t^2$ can be obtained using the values of $A_{T+1}^{i_t} = \hat{A}_{T+1}^{\hat{i}_t} = -g^i$ and $C_{T+1}^{i_t} = \hat{C}_{T+1}^{\hat{i}_t} = g^i \overline{x}^i$ in (6.3).

In stage $\tau \in \{t, t+1, \cdots, T\}$ if nation i_t chooses to adopt conventional technique at stage τ, one can invoke (6.1) to obtain the explicit solutions of $A_\tau^{i_t}$ and $C_\tau^{i_t}$ as:

$$C_\tau^{i_t} = \left(\alpha_\tau^{i_t} - \sum_{j \in S_\tau^1} \beta_j^{i_\tau} \phi_\tau^j(x) - \sum_{j \in S_\tau^2} \beta_j^{i_t} \hat{\phi}_\tau^{j}(x)\right) \phi_\tau^{i_t}(x) - c^{i_t} \phi_\tau^{i_t}(x)$$
$$+ (1+r)^{-1}\left[\overline{A}_{\tau+1}^{i_t}\left(\sum_{j \in S_\tau^1} a^j \phi_\tau^j(x) + \sum_{j \in S_\tau^2} \hat{a}^j \hat{\phi}_\tau^{j}(x)\right) + C_{\tau+1}^{i_t}\right], \text{ and}$$
$$A_\tau^{i_t} = -\frac{(b_{i_t})^2}{4c_{i_t}^a}\left[\overline{A}_{\tau+1}^{i_t}(1+r)^{-1}\right]^2 - h^{i_t}$$
$$+ (1+r)^{-1}\overline{A}_{\tau+1}^{i_t}\left(1 + \sum_{j \in S_\tau^1} \frac{(b_j)^2}{2c_j^a} \overline{A}_{\tau+1}^j (1+r)^{-1} + \sum_{j \in S_\tau^2} \frac{(b_j)^2}{2c_j^a} \overline{A}_{\tau+1}^j (1+r)^{-1} - \delta\right). \tag{6.4}$$

In stage $\tau \in \{t, t+1, \cdots, T\}$ if nation i_t chooses to adopt environment-preserving technique at stage τ, one can invoke (6.1) to obtain the explicit solutions of $\hat{A}_{\tau}^{\hat{i}_t}$ and $\hat{C}_{\tau}^{\hat{i}_t}$ as:

$$\hat{C}_{\tau}^{\hat{i}_t} = \left(\alpha_{\tau}^{\hat{i}_t} - \sum_{j \in S_{\tau}^1} \beta_j^{\hat{i}_t} \phi_{\tau}^j(x) - \sum_{j \in S_{\tau}^2} \beta_j^{\hat{i}_t} \hat{\phi}_{\tau}^j \right) \hat{\phi}_{\tau}^{i_t}(x) - \hat{c}^{\hat{i}_t} \hat{\phi}_{\tau}^{i_t}(x)$$
$$+ (1+r)^{-1} \left[\overline{A}_{\tau+1}^{\hat{i}_t} \left(\sum_{j \in S_{\tau}^1} a^j \phi_{\tau}^j(x) + \sum_{j \in S_{\tau}^2} \hat{a}^j \hat{\phi}_{\tau}^j(x) \right) + \overline{C}_{\tau+1}^{\hat{i}_t} \right], \text{ and}$$
$$\hat{A}_{\tau}^{\hat{i}_t} = - \frac{\left(b_{\hat{i}_t}\right)^2}{4c_{\hat{i}_t}^a} \left[\overline{A}_{\tau+1}^{\hat{i}_t} (1+r)^{-1} \right]^2 - h^{\hat{i}_t}$$
$$+ (1+r)^{-1} \overline{A}_{\tau+1}^{\hat{i}_t} \left(1 + \sum_{j \in S_{\tau}^1} \frac{(b_j)^2}{2c_j^a} \overline{A}_{\tau+1}^j (1+r)^{-1} + \sum_{j \in S_{\tau}^2} \frac{(b_j)^2}{2c_j^a} \overline{A}_{\tau+1}^j (1+r)^{-1} - \delta \right). \tag{6.5}$$

Repeating the process for τ from T to t, one can obtain $A_t^{i_t}, C_t^{i_t}, \hat{A}_t^{\hat{i}_t}$ and $\hat{C}_t^{\hat{i}_t}$. A_t^i and C_t^i, for $i_t \in S_t^1$ and $\hat{i}_t \in S_t^2$ explicitly as constants from the model parameters. *Q.E.D.*

Appendix B: Proof of Proposition 3.1

From (3.8) we can obtain $W_{x_{t+1}}^{M^\gamma}(t+1, x_{t+1})(1+r)^{t-1}$ as $A_{t+1}^{M^\gamma}(1+r)^{-1}$. Substituting this result into the optimal controls in (3.4) and (3.7), and then into (3.3) yields

$$A_t^{M^\gamma} x + C_t^{M^\gamma} = \sum_{i \in S_t^{M^\gamma[1]}} \left[\left(\alpha_t^i - \sum_{j \in S_t^{M^\gamma[1]}} \beta_j^i \psi_t^{(M^\gamma)j}(x) \right. \right.$$
$$\left. - \sum_{\zeta \in S_t^{M^\gamma[2]}} \beta_\zeta^i \hat{\psi}_t^{(M^\gamma)\zeta} \right) \psi_t^{(M^\gamma)i}(x)$$
$$\left. - c^i \psi_t^{(M^\gamma)i}(x) - \frac{(\hat{b}_i)^2}{4c_i^a} \left[A_{t+1}^{M^\gamma} (1+r)^{-1} \right]^2 x - h^i x \right]$$
$$+ \sum_{i \in S_t^{M^\gamma[2]}} \left[\left(\alpha_t^i - \sum_{j \in S_t^{M^\gamma[1]}} \beta_j^i \psi_t^{(M^\gamma)j}(x) - \sum_{\zeta \in S_t^{M^\gamma[2]}} \beta_\zeta^i \hat{\psi}_t^{(M^\gamma)\zeta}(x) \right) \hat{\psi}_t^{(M^\gamma)i}(x) \right.$$
$$\left. - \hat{c}^{i\left(S_t^{M^\gamma[2]}\right)} \hat{\psi}_t^{(M^\gamma)i}(x) - \frac{(\hat{b}_i)^2}{4c_i^a} \left[A_{t+1}^{M^\gamma} (1+r)^{-1} \right]^2 x - h^i x \right]$$
$$+ (1+r)^{-1} \left[A_{t+1}^{M^\gamma} \left(x + \sum_{j \in S_t^{M^\gamma[1]}} a^j \psi_t^{(M^\gamma)j}(x) + \sum_{j \in S_t^{M^\gamma[2]}} \hat{a}^j \hat{\psi}_t^{(M^\gamma)j}(x) \right. \right.$$
$$\left. \left. + \sum_{j=1}^n \frac{(\hat{b}_i)^2}{2c_i^a} A_{t+1}^{M^\gamma} (1+r)^{-1} x - \delta\, x \right) + C_{t+1}^{M^\gamma} \right], \text{ for } t \in \kappa;$$
$$A_{T+1}^{M^\gamma} x + C_{T+1}^{M^\gamma} = \sum_{i=1}^n g^i (\overline{x}^i - x). \tag{6.6}$$

Where

$$\psi_t^{(M^\gamma)j}(x) = \tilde{\alpha}_t^{(M^\gamma)j} + \tilde{\beta}_t^{(M^\gamma)j} \; A_{t+1}^{M^\gamma}(1+r)^{-1} \text{ and}$$
$$\hat{\psi}_t^{(M^\gamma)\zeta}(x) = \hat{\tilde{\alpha}}_t^{(M^\gamma)j} + \hat{\tilde{\beta}}_t^{(M^\gamma)j} \; A_{t+1}^{M^\gamma}(1+r)^{-1}.$$

First consider the stage $T+1$, from (6.6) we obtain

$$A_{T+1}^{M^\gamma} = \sum_{i=1}^{n} -g^i \text{ and } C_{T+1}^{M^\gamma} = \sum_{i=1}^{n} g^i \bar{x}^i. \tag{6.7}$$

Now we consider the stage T. Note that the left-hand-side of (6.6) consists of the expression $A_T^{M^\gamma} x + C_T^{M^\gamma}$. On the right-hand-side there is an expression which is linear in x with coefficients involving $A_{T+1}^{M^\gamma}$ and $C_{T+1}^{M^\gamma}$. The values of $A_T^{M^\gamma}$ and $C_T^{M^\gamma}$ can be obtained using $A_{T+1}^{M^\gamma}$ and $C_{T+1}^{M^\gamma}$ in (6.7). Using (6.6) yields the explicit solution of $C_t^{M^\gamma}$ and $A_t^{M^\gamma}$ for $t = T$ as:

$$\begin{aligned}
C_t^{M^\gamma} = & \sum_{i \in S_t^{M^\gamma[1]}} \left[\left(\alpha_t^i - \sum_{j \in S_t^{M^\gamma[1]}} \beta_j^i \, \psi_t^{(M^\gamma)j}(x) - \sum_{\zeta \in S_t^{M^\gamma[2]}} \beta_\zeta^i \hat{\psi}_t^{(M^\gamma)\zeta} \right) \psi_t^{(M^\gamma)i}(x) \right. \\
& \left. - c^i \psi_t^{(M^\gamma)i}(x) \right] + \sum_{i \in S_t^{M^\gamma[2]}} \left[\left(\alpha_t^i - \sum_{j \in S_t^{M^\gamma[1]}} \beta_j^i \, \psi_t^{(M^\gamma)j}(x) \right. \right. \\
& \left. \left. - \sum_{\zeta \in S_t^{M^\gamma[2]}} \beta_\zeta^i \hat{\psi}_t^{(M^\gamma)\zeta}(x) \right) \hat{\psi}_t^{(M^\gamma)i}(x) - \hat{c}^{\,i}\left(S_t^{M^\gamma[2]}\right) \hat{\psi}_t^{(M^\gamma)i}(x) \right] \\
& + (1+r)^{-1} \left[A_{t+1}^{M^\gamma} \left(\sum_{j \in S_t^{M^\gamma[1]}} a^j \psi_t^{(M^\gamma)j}(x) + \sum_{j \in S_t^{M^\gamma[2]}} \hat{a}^j \, \hat{\psi}_t^{(M^\gamma)j}(x) \right) + C_{t+1}^{M^\gamma} \right],
\end{aligned}$$

and

$$\begin{aligned}
A_t^{M^\gamma} = & \sum_{i=1}^{n} \left[-\frac{(b_i)^2}{4c_i^a} \left[A_{t+1}^{M^\gamma}(1+r)^{-1} \right]^2 - h^i \right] \\
& + (1+r)^{-1} \left[A_{t+1}^{M^\gamma} \left(1 + \sum_{j=1}^{n} \frac{(b_i)^2}{2c_i^a} A_{t+1}^{M^\gamma}(1+r)^{-1} - \delta \right) \right].
\end{aligned} \tag{6.8}$$

Now consider stage $T-1$. One can obtain $A_{T-1}^{M^\gamma}$ and $C_{T-1}^{M^\gamma}$ as in (6.8) by setting $t = T-1$. Repeating the process, $A_t^{M^\gamma}$ and $C_t^{M^\gamma}$ for $t \in \{1, 2, \cdots, T-2\}$ can be explicitly obtained. *Q.E.D.*

14.7 Chapter Notes

Adoption of environment-preserving production technique plays a key role to effectively solving the continual worsening global industrial pollution problem. In this Chapter a dynamic game of collaborative pollution management with production technique choices is presented. Various extensions can also be incorporated into the analysis readily. First, one may introduce costs of technique switching. Second, the natural rate of decay may be related to the pattern of technique choice. Finally, the number of industrial products produced by a nation could be more than one.

14.8 Problems

1. Consider a 2-nation version of the dynamic game model of transboundary pollution with two production technique choices in Sect. 14.1. The demand functions of these nations are respectively

$$P_t^1 = 40 - 2Q_t^1 - Q_t^2 \text{ and } P_t^2 = 65 - 0.5Q_t^1 - 4Q_t^2.$$

The costs of producing output with conventional technique are $c^1 = 1.2$, $c^2 = 0.8$; and those of using environment-preserving technique are $\hat{c}^{\,1} = 2.5$, $\hat{c}^{\,2} = 2$. The abatement costs are $c_1^a = 2$, $c_2^a = 3$; and the abatement parameters are $b_1 = 1.5$, $b_2 = 0.5$. The pollution dynamics parameters are $a^1 = 2$, $\hat{a}^{\,1} = 0.5$, $a^2 = 3$, $\hat{a}^{\,2} = 1$. The pollution decay rate $\delta = 0.05$ and the pollution damage parameters are $h_1 = 0.9$, $h_2 = 0.5$. The initial pollution stock is $x_1 = 3$ and the discount rate is $r = 0.05$. The terminal bonus (penalty) parameters are $g^1 = 0.8$, $g^2 = 0.2$; $\bar{x}^1 = 400$, $\bar{x}^2 = 450$.

   Characterize a feedback Nash equilibrium solution and show the pattern of technology used.
2. If these nations agree to cooperate and maximize their joint payoff, obtain a group optimal cooperative solution.
3. Furthermore, if these nations agree to share their cooperative gain proportional to their expected payoffs, derive a subgame consistent cooperative solution.

Chapter 15
Applications in Business Collaboration

In this Chapter, we present two applications in business collaboration. The first one is on corporate joint venture and the second one is on cartel. The joint venture analysis is from Yeung and Petrosyan (2006a), Yeung (2010) and Chapter 9 of Yeung and Petrosyan (2012a). The Cartel analysis is extracted from Yeung (2005) and Chapter 11 of Yeung and Petrosyan (2012a). Sections 15.1, 15.2 and 15.3 contain the analysis of a corporate joint venture in which gains can be obtained from cost saving cooperation. In section 15.1, a dynamic corporate joint venture under uncertainty is formulated. The expected venture profit maximization, subgame consistent PDP and an illustration are provided. In Sect. 15.2, the Shapley Value Solution for the joint venture is derived. An analysis on joint venture under an infinite horizon is given in Sect. 15.3. Sections 15.4, 15.5 and 15.6 present a cartel analysis which contains dormant firms. Section 15.4 presents a stochastic dynamic dormant-firm cartel. The basic settings, market outcome, optimal cartel output and subgame-consistent cartel profit sharing are investigated. An illustration with explicit functional forms is given in Sect. 15.5. An analysis on infinite horizon cartel is provided in Sect. 15.6. An Appendix of the Chapter is given in Sect. 15.7. Chapter notes are given in Sect. 15.8 and problems in Sect. 15.9.

15.1 Dynamic Corporate Joint Venture Under Uncertainty

As markets become increasingly globalized and firms become more multinational, corporate joint ventures are likely to yield opportunities to quickly create economies of scale and critical mass, incorporate new skills and technology, and facilitate rational resource sharing (see Bleeke and Ernst (1993)). With joint ventures becoming a powerful force shaping global corporate strategy, partnerships between firms have significantly increased. Despite their purported benefits, however, joint ventures are highly unstable and have a consistently high rate of failure (Blodgett

D.W.K. Yeung, L.A. Petrosyan, *Subgame Consistent Cooperation*,
Theory and Decision Library C 47, DOI 10.1007/978-981-10-1545-8_15

1992; Parkhe 1993). Subgame consistent solution for joint ventures would provide a solution to the problem.

In this section, we present a general framework of a dynamic joint venture in which there are n firms. The venture horizon is $[t_0, T]$. The state dynamics of the ith firm is characterized by the set of vector-valued stochastic differential equations:

$$dx^i(s) = f^i\left[s, x^i(s), u_i(s)\right] ds + \sigma_i\left[s, x^i(s)\right] dz_i(s), \; x^i(t_0) = x_0^i, \text{ for } i \in N, \quad (1.1)$$

where $x^i(s) \in X^i \subset R^{m_i}$ denotes the technology state of firm i, $u_i \in U_i \subset compR^\lambda$ is the control vector of firm i, $\sigma_i[s, x^i(s)]$ is a $m_i \times \Theta_i$ and $z_i(s)$ is a Θ_i-dimensional Wiener process and the initial state x_0^i is given. Let $\Omega_i[s, x^i(s)] = \sigma_i[s, x^i(s)]\, \sigma_i[s, x^i(s)]^T$ denote the covariance matrix with its element in row h and column ζ denoted by $\Omega_i^{h\zeta}[s, x^i(s)]$. For $i \neq j$, $x^i \cap x^j = \varnothing$, and $z_i(s)$ and $z_j(s)$ are independent Wiener processes. We also used $x^N(s)$ to denote the vector $[x^1(s), x^2(s), \ldots, x^n(s)]$ and x_0^N the vector $[x_0^1, x_0^2, \ldots, x_0^n]$.

The expected profit of firm i is:

$$E_{t_0}\left\{\int_{t_0}^{T} \left(g^i\left[s, x^i(s)\right] - c_i^{\{i\}} u_i(s)\right]\right) \exp\left[-\int_{t_0}^{s} r(y) dy\right] ds,$$
$$+ \exp\left[-\int_{t_0}^{T} r(y) dy\right] q^i(x^i(T))\right\}, \text{ for } i \in [1, 2, \ldots, n] \equiv N, \quad (1.2)$$

where $\exp\left[-\int_{t_0}^{t} r(y) dy\right]$ is the discount factor, and $q^i(x_i(T))$ the terminal payoff. In particular, $g^i[s, x_i, u_i]$ and $q^i(x_i)$ are positively related to x_i, reflecting the earning potent of the technology.

A set of investment strategies $\phi_i^*(t, x^i)$ for firm i constitutes a Nash equilibrium to the stochastic differential game (1.1 and 1.2) if there exist continuously twice differentiable function $V^{(t_0)i}(t, x^i) : [t_0, T] \times R^{m_i} \to R$ satisfying the following equations:

$$-V_t^{(t_0)i}(t, x^i) - \frac{1}{2}\sum_{h,\zeta=1}^{m_i} \Omega_i^{h\zeta}(t, x^i) V_{x^{i(h)} x^{i(\zeta)}}^{(t_0)i}(t, x^i) =$$
$$\max_{u_i}\left\{\left[g(t, x^i) - c_i^{\{i\}}(u_i)\right] \exp\left[-\int_{t_0}^{t} r(y) dy\right] + V_x^{(t_0)i}(t, x) f^i\left[t, x^i, u_i\right]\right\},$$
$$V^{(t_0)i}(T, x^i) = q^i(x^i) \exp\left[-\int_{t_0}^{T} r(y) dy\right], \text{ for } i \in N.$$

Let $V^{(\tau)i}(t, x^i)$ denote the game equilibrium payoff of firm i in a game with dynamics (1.1) and payoff (1.2) which starts at time τ for $\tau \in [t_0, T)$. One can readily obtain

$$\exp\left[\int_{t_0}^{\tau} r(y)dy\right] V^{(t_0)i}(t,x^i) = V^{(\tau)i}(t,x^i),$$

for $\tau \in [t_0, T]$ and $i \in N$.

For the sake of clarity in exposition, we consider the case where $m_i = 1$, for $i \in N$.

15.1.1 Expected Venture Profit Maximization

Consider a joint venture consisting of all these n companies. The participating firms can gain core skills and technology that would be impossible for them to obtain on their own individually. Cost saving opportunities are created under joint venture, for instance, savings in joint R&D, administration, marketing, customer services, purchasing, financing, and economy of scales and scope. The cost of control of firm j under the joint venture becomes $c_j^N[u_j(s)]$. With absolute joint venture cost advantage we have

$$c_j^N(u_j) \leq c_j^{\{j\}}(u_j), \text{ for } j \in N, \tag{1.3}$$

Moreover, marginal cost advantages lead to:

$$\partial c_j^N(u_j)/\partial u_j \leq \partial c_j^{\{j\}}(u_j)/\partial u_j\,, \text{ for } j \in N.$$

At time t_0, the joint venture would maximize the expected joint venture profit:

$$E_{t_0}\left\{\int_{t_0}^{T} \sum_{j=1}^{n} (g^j[s, x^j(s)] - c_j^N[u_j(s)]) \exp\left[-\int_{t_0}^{s} r(y)dy\right] ds \right.$$
$$\left. + \sum_{j=1}^{n} \exp\left[-\int_{t_0}^{T} r(y)dy\right] q^j(x^j(T))\right\} \tag{1.4}$$

subject to (1.3).

Invoking Fleming's techniques of stochastic optimal control, the solution to the problem (1.3 and 1.4) can be characterized as follows.

Corollary 1.1 A set of controls $\{\psi_i^*(t,x), \text{ for } i \in N \text{ and } t \in [t_0, T]\}$ provides an optimal solution to the control problem (1.3 and 1.4) if there exists continuously twice differentiable function $W^{(t_0)}(t,x) : [t_0, T] \times R^n \to R$ satisfying the following Bellman equation:

$$-W_t^{(t_0)}(t,x) - \frac{1}{2}\sum_{h,\zeta=1}^{n} \Omega^{h\zeta}(t,x) W_{x^h x^\zeta}^{(t_0)}(t,x) =$$

$$= \max_{u_1,u_2,\cdots,u_n} \left\{ \sum_{j=1}^{n} \left[g^j(t,x^j) - c_j^N(u_j) \right] \exp\left[-\int_{t_0}^{t} r(y)dy \right] \right.$$

$$\left. + \sum_{j=1}^{n} W_{x_j}^{(t_0)}(t,x) f^j(t,x^j,u_j) \right\},$$

$$W^{(t_0)}(T,x) = \exp\left[-\int_{t_0}^{T} r(y)dy \right] \sum_{j=1}^{n} q^j(x^j), \tag{1.5}$$

where $x = \{x^1, x^2, \ldots, x^n\}$. ■

Hence the firms will adopt the cooperative control $\{\psi_i^*(t,x)$, for $i \in N$ and $t \in [t_0, T]\}$ to obtain the maximized level of expected joint profit. Substituting this set of control into (1.3) yields the dynamics of technology advancement under cooperation as:

$$dx^i(s) = f^i\left[s, x^i(s), \psi_i^*(s, x(s))\right] ds + \sigma_i[s, x^i(s)]\, dz_i(s),$$
$$x^i(t_0) = x_0^i, \text{ for } i \in N. \tag{1.6}$$

Let $x^*(t) = \{x^{1*}(t), x^{2*}(t), \ldots, x^{n*}(t)\}$ denote the solution to (1.6). The optimal cooperative trajectory can be expressed as:

$$x^{i*}(t) = x_0^i + \int_{t_0}^{t} f^i\left[s, x^{i*}(s), \psi_i^*(s, x^*(s))\right] ds + \int_{t_0}^{t} \sigma_i\left[s, x^{i*}(s)\right] dz_i(s), \tag{1.7}$$

for $i \in N$.

We use X_t^* to denote the set of realizable values of $x^{**}(t)$ at time t generated by (1.6). The term $x_t^* \in X_t^*$ is used to denote an element in X_t^*.

The cooperative investment strategies for the joint venture with dynamics (1.3) and expected joint venture profit (1.4) over the time interval $[t_0, T]$ can be expressed more precisely as

$$\{\psi_i^*(t, x^*(t)), \text{ for } i \in N \text{ and } t \in [t_0, T]\}. \tag{1.8}$$

Note that for group optimality to be achievable, the cooperative investment strategies $\{\psi_i^*(t, x^*(t))$, for $i \in N$ and $t \in [t_0, T]\}$ must be exercised throughout time interval $[t_0, T]$.

Along the cooperative investment path $\{x^*(t)\}_{t=t_0}^{T}$ the present value of total expected joint venture profit over the interval $[t, T]$, for $t \in [t_0, T)$, can be expressed as:

$$W^{(t_0)}\left(t,x_t^*\right) =$$
$$E_{t_0}\left\{\int_t^T \sum_{j=1}^n \left(g^j\left[s,x^{j*}(s)\right] - c_j^N\left[\psi_j^*\left(s,x^*(s)\right)\right]\right)\exp\left[-\int_{t_0}^s r(y)dy\right]ds\right.$$
$$\left. + \exp\left[-\int_{t_0}^T r(y)dy\right]\sum_{j=1}^n q^j\left(x^{j*}(T)\right) \,\middle|\, x^*(t) = x_t^* \in X_t^*\right\}. \tag{1.9}$$

Let $W^{(\tau)}(t,x_t^*)$ denote the total venture profit from the control problem with dynamics (1.3) and payoff (1.4) which begins at time $\tau \in [t_0,T]$ with initial state x_τ^*. We can readily have $\exp\left[\int_{t_0}^{\tau} r(y)dy\right] W^{(t_0)}\left(t,x_t^*\right) = W^{(\tau)}\left(t,x_t^*\right)$, for $\tau \in [t_0,T]$ and $t \in \left[\tau,T\right)$.

15.1.2 Subgame Consistent Venture PDP

Since the sizes and earning potentials of the firms in a corporate joint venture may vary significantly, we consider the case when the venture agrees to share the excess of the expected total cooperative payoff over the sum of expected individual noncooperative payoffs proportionally to the firms' expected noncooperative payoffs.

The imputation scheme has to fulfil:

Condition 1.1 An imputation

$$\xi^{(t_0)i}(t_0,x_0) = \frac{V^{(t_0)i}\left(t_0,x_0^i\right)}{\sum_{j=1}^n V^{(t_0)j}\left(t_0,x_0^j\right)} W^{(t_0)}(t_0,x_0)$$

is assigned to firm i, for $i \in N$ at the outset;
and an imputation

$$\xi^{(\tau)i}\left(\tau,x_\tau^*\right) = \frac{V^{(\tau)i}\left(\tau,x_\tau^{i*}\right)}{\sum_{j=1}^n V^{(\tau)j}\left(\tau,x_\tau^{j*}\right)} W^{(\tau)}\left(\tau,x_\tau^*\right) \tag{1.10}$$

is assigned to firm i, for $i \in N$ at time $\tau \in \left(t_0,T\right]$. ■

The imputation (1.10) satisfies

(i) $\xi^{(\tau)i}\left(\tau,x_\tau^*\right) \geq V^{(\tau)i}\left(\tau,x_\tau^{i*}\right)$, for $i \in N$ and $\tau \in [t_0,T]$; and

(ii) $\sum_{j=1}^n \xi^{(\tau)j}\left(\tau,x_\tau^*\right) = W^{(\tau)}\left(\tau,x_\tau^*\right)$, for $\tau \in [t_0,T]$.

Hence the imputation vector $\xi^{(\tau)}(\tau, x_\tau^*)$ in (1.10) satisfies individual rationality and group optimality throughout the game horizon $[t_0, T]$. Invoking Theorem 3.1 in Chap. 3, a PDP with a terminal payment $q^i(x_T^*)$ at time T and an instantaneous payment at time $s \in [\tau, T]$:

$$\begin{aligned}
B_i(s, x_s^*) = &-\left[\xi_t^{(s)i}(t, x_t^*)\Big|_{t=s}\right] - \frac{1}{2}\sum_{h,\zeta=1}^{n} \Omega^{h\zeta}(s, x_s^*)\left[\xi_{x_t^h x_t^\zeta}^{(s)i}(t, x_t^*)\Big|_{t=s}\right] \\
&- \sum_{h=1}^{n}\left[\xi_{x_t^{h*}}^{(s)i}(t, x_t^*)\Big|_{t=s}\right] f^h\left[s, x_s^{h*}, \psi_h^*(s, x_s^*)\right] \\
= &-\frac{\partial}{\partial t}\left[\frac{V^{(s)i}(t, x_t^{i*})}{\sum_{j=1}^{n} V^{(s)j}\left(t, x_t^{j*}\right)} W^{(s)}(t, x_t^*)\Bigg|_{t=s}\right] \\
&- \frac{1}{2}\sum_{h,\zeta=1}^{n} \Omega^{h\zeta}(s, x_s^*)\frac{\partial^2}{\partial x_t^{h*}\partial x_t^{\zeta*}}\left[\frac{V^{(s)i}(t, x_t^{i*})}{\sum_{j=1}^{n} V^{(s)j}\left(t, x_t^{j*}\right)} W^{(s)}(t, x_t^*)\Bigg|_{t=s}\right] \\
&- \sum_{h=1}^{n}\frac{\partial}{\partial x_t^{h*}}\left[\frac{V^{(s)i}(t, x_t^{i*})}{\sum_{j=1}^{n} V^{(t)j}\left(t, x_t^{j*}\right)} W^{(s)}(t, x_t^*)\Bigg|_{t=s}\right] f^h\left[s, x_s^{h*}, \psi_h^*(s, x_s^*)\right],
\end{aligned} \tag{1.11}$$

for $i \in N$ and $x_s^* \in X_s^*$;

yields imputation vectors which satisfy (1.10) and is hence subgame consistent. ■

With firms using the cooperative investment strategies $\{\psi_i^*(\tau, x_\tau^*)$, for $\tau \in [t_0, T]$ and $i \in N\}$, the instantaneous receipt of firm i at time instant τ is:

$$\zeta_i(\tau, x_\tau^*) = g^i(\tau, x_\tau^{i*}) - c_i^N\left[\psi_i^*(\tau, x_\tau^*)\right], \tag{1.12}$$

for $\tau \in [t_0, T]$ and $i \in N$.

According to (1.1), the instantaneous payment that firm i should receive under the agreed-upon optimality principle is $B_i(\tau, x_\tau^*)$ as stated in (1.11). Hence an instantaneous transfer payment

$$\chi^i(\tau, x_\tau^*) = B_i(\tau, x_\tau^*) - \zeta_i(\tau, x_\tau^*) \tag{1.13}$$

has to be given or charged to firm i at time τ, for $i \in N$ and $\tau \in [t_0, T]$.

15.1.3 An Illustration

Consider the case where there are three companies involved in a joint venture. The planning period is $[t_0, T]$. Company i's expected profit is

$$E_{t_0}\left\{\int_{t_0}^{T}\left[P_i\left[x^i(s)\right]^{1/2}-c_i^{\{i\}}u_i(s)\right]\exp[-r(s-t_0)]\,ds\right.$$
$$\left.+\exp[-r(T-t_0)]q_i[x^i(T)]^{1/2}\right\},\ \text{for } i\in\{1,2,3\},\tag{1.14}$$

where P_i, c_i^i and q_i are positive constants, r is the discount rate, $x_i(s)\subset R^+$ is the level of technology of company i at time s, and $u_i(s)\subset R^+$ is its physical investment in technological advancement. The term $P_i[x^i(s)]^{1/2}$ reflects the net operating revenue of company i at technology level $x_i(s)$, and $c_i^i u_i$ is the cost of investment if firm i operates on its own. The term $q_i[x^i(T)]^{1/2}$ gives the salvage value of company i's technology at time T.

The dynamics of the technology level of company i follows the stochastic differential equation:

$$dx^i(s)=\left[\alpha_i[u_i(s)x^i(s)]^{1/2}-\delta\,x^i(s)\right]ds+\sigma_i x^i(s)dz_i(s),\ x^i(t_0)=x_0^i\in X^i,$$
$$\text{for } i\in\{1,2,3\},\tag{1.15}$$

where $\alpha_i[u_i(s)x^i(s)]^{1/2}$ is the addition to the technology brought about by $u_i(s)$ amount of physical investment, δ is the rate of obsolescence, and $z_1(s)$, $z_2(s)$ and $z_3(s)$ are independent Wiener processes.

In the case when each of these three firms acts independently we obtain the corresponding partial differential equations characterizing a non-cooperative equilibrium as:

$$-V_t^{(t_0)i}(t,x^i)-\frac{(\sigma_i x^i)^2}{2}V_{x^ix^i}^{(t_0)i}(t,x^i)$$
$$=\max_{u_i}\left\{\left[P_i(x^i)^{1/2}-c_i^{\{i\}}u_i\right]\exp[-r(t-t_0)]\right.$$
$$\left.+V_{x^i}^{(t_0)i}(t,x^i)\left[\alpha_i(u^ix^i)^{1/2}-\delta\,x^i\right]\right\},$$
$$V^{(t_0)i}(T,x^i)=\exp[-r(T-t_0)]\,q_i(x^i)^{1/2},\ \text{for } i\in\{1,2,3\}.$$

Performing the indicated maximization yields

$$u_i=\frac{\alpha_i^2}{4\left(c_i^{\{i\}}\right)^2}\left[V_{x^i}^{(t_0)i}(t,x^i)\exp[r(t-t_0)]\right]^2x^i,\ \text{for } i\in\{1,2,3\}.$$

Substituting u_i into the above partial differential equations yields:

$$-V_t^{(t_0)i}(t,x^i) - \frac{(\sigma_i x^i)^2}{2} V_{x^i x^i}^{(t_0)i}(t,x^i)$$

$$= P_i(x^i)^{1/2} \exp[-r(t-t_0)] - \frac{\alpha_i^2}{4c_i^{\{i\}}} \left[V_{x^i}^{(t_0)i}(t,x^i)\right]^2 \exp[r(t-t_0)]\, x^i$$

$$+ \frac{\alpha_i^2}{2c_i^{\{i\}}} \left[V_{x^i}^{(t_0)i}(t,x^i)\right]^2 \exp[r(t-\tau)]\, x^i - \delta\, V_{x^i}^{(t_0)i}(t,x^i) x_i, \text{ for } i \in [1,2,3].$$

Solving the above system of partial differential equations yields

$$V^{(t_0)i}(t,x^i) = \left[A_i^{\{i\}}(t)(x^i)^{1/2} + C_i^{\{i\}}(t)\right] \exp[-r(\tau - t_0)], \text{ for } i \in \{1,2,3\}, \quad (1.16)$$

where

$$\dot{A}_i^{\{i\}}(t) = \left(r + \frac{\delta}{2} + \frac{\sigma_i^2}{8}\right) A_i^{\{i\}}(t) - P_i,\ \dot{C}_i^{\{i\}}(t) = rC_i^{\{i\}}(t) - \frac{\alpha_i^2}{16c_i^{\{i\}}} \left[A_i^{\{i\}}(t)\right]^2,$$
$$A_i^{\{i\}}(T) = q_i \text{ and } C_i^{\{i\}}(T) = 0. \quad (1.17)$$

The first equation in the block-recursive system (1.17) is a first-order linear differential equation in $A_i^i(t)$ which can be solved independently by standard techniques. Upon substituting the solution of $A_i^i(t)$ into the second equation of (1.17) yields a first-order linear differential equation in $C_i^i(t)$. The solution of $C_i^i(t)$ can be readily obtained by standard techniques.

Moreover, one can easily derive for $\tau \in [t_0, T]$

$$V^{(\tau)i}(t,x^i) = \left[A_i^{\{i\}}(t)(x^i)^{1/2} + C_i^{\{i\}}(t)\right] \exp[-r(t-\tau)], \quad \text{for } i \in \{1,2,3\} \text{ and } \tau \in [t_0, T].$$

15.1.3.1 Expected Venture Profit and Cost Savings

Consider the case when all these three firms agree to form a joint venture and share their expected joint profit proportionally to their expected noncooperative profits. Cost saving opportunities are created under joint venture from joint R&D, administration, purchasing, financing, and economy of scales and scope. The cost of control of firm j under the joint venture becomes $c_j^{1,2,3}[u_j(s)]$. With joint venture cost advantage

$$c_j^{\{1,2,3\}} \le c_j^{\{j\}}, \text{ for } j \in N, \quad (1.18)$$

The expected profit of the joint venture is the sum of the participating firms' expected profits:

$$E_{t_0}\left\{\int_{t_0}^{T}\sum_{j=1}^{3}\left[P_j\left[x^j(s)\right]^{1/2}-c_j^{\{1,2,3\}}u_j(s)\right]\ \exp[-r(s-t_0)]\,ds\right.$$
$$\left.+\sum_{j=1}^{3}\exp[-r(T-t_0)]q_j\left[x^j(T)\right]^{1/2}\right\}. \tag{1.19}$$

The firms in the joint venture then act cooperatively to maximize (1.19) subject to (1.18). Invoking Theorem A.3 in the Technical Appendices, we obtain the equations characterizing an optimal solution of the stochastic control problem (1.18 and 1.19) as:

$$-W_t^{(t_0)\{1,2,3\}}(t,x^1,x^2,x^3)-\sum_{h,\zeta=1}^{3}\frac{(\sigma_h x^h)(\sigma_\zeta x^\zeta)}{2}W_{x^h x^\zeta}^{(t_0)\{1,2,3\}}(t,x^1,x^2,x^3)$$
$$=\max_{u_1,u_2,u_3}\left\{\sum_{i=1}^{3}\left[P_i\left(x^i\right)^{1/2}-c_i^{\{1,2,3\}}u_i\right]\exp[-r(t-t_0)]\right.$$
$$\left.+\sum_{i=1}^{3}W_{x^i}^{(t_0)\{1,2,3\}}(t,x^1,x^2,x^3)\left[\ \alpha_i(u_ix^i)^{1/2}-\delta\,x^i\ \right]\right\},$$
$$W^{(t_0)\{1,2,3\}}(T,x^1,x^2,x^3)=\sum_{j=1}^{3}\exp[-r(T-t_0)]q_j(x^j)^{1/2}. \tag{1.20}$$

Performing the indicated maximization yields

$$u_i=\frac{\alpha_i^2}{4\left(c_i^{\{1,2,3\}}\right)^2}\left[W_{x^i}^{(t_0)\{1,2,3\}}(t,x^1,x^2,x^3)\exp[r(t-t_0)]\ \right]^2x^i,\ \text{for}\ i\in\{1,2,3\}. \tag{1.21}$$

Substituting (1.21) into (1.20) and solving yields the value function reflecting the expected joint maximized payoffs:

$$W^{(t_0)\{1,2,3\}}(t,x^1,x^2,x^3)$$
$$=\left[A_1^{\{1,2,3\}}(t)(x^1)^{1/2}+A_2^{\{1,2,3\}}(t)(x^2)^{1/2}+A_3^{\{1,2,3\}}(t)(x^3)^{1/2}+C^{\{1,2,3\}}(t)\right]$$
$$\exp[-r(t-t_0)], \tag{1.22}$$

where $A_1^{1,2,3}(t), A_2^{1,2,3}(t), A_3^{1,2,3}(t)$ and $x_3, C^{\{1,2,3\}}(t)$ satisfy

$\dot{A}_i^{\{1,2,3\}}(t)=\left(r+\frac{\delta}{2}+\frac{\sigma_i^{2i}}{8}\right)A_i^{\{1,2,3\}}(t)-\frac{b_i^{[i,j]}}{2}A_j^{\{1,2,3\}}(t)-\frac{b_i^{[i,h]}}{2}A_h^{\{1,2,3\}}(t)-P_i$ for $i,j,h\in\{1,2,3\}$ and $i\neq j\neq h$,

$$\dot{C}^{\{1,2,3\}}(t) = rC^{\{1,2,3\}}(t) - \sum_{i=1}^{3} \frac{\alpha_i^2}{16c_i^{\{1,2,3\}}} \left[A_i^{\{1,2,3\}}(t)\right]^2,$$
$$A_i^{\{1,2,3\}}(T) = q_i \text{ for } i \in \{1,2,3\}, \text{ and } C^{\{1,2,3\}}(T) = 0. \tag{1.23}$$

The first three equations in the block recursive system (1.23) is a system of three linear differential equations which can be solved explicitly by standard techniques. Upon solving $A_i^{1,2,3}(t)$ for $i \in \{1,2,3\}$, and substituting them into the fourth equation of (1.23), one has a linear differential equation in $C^{\{1,2,3\}}(t)$.

The investment strategies of the grand coalition joint venture can be derived as:

$$\psi_i^{\{1,2,3\}}(t,x) = \frac{\alpha_i^2}{16\left(c_i^{\{1,2,3\}}\right)^2} \left[A_i^{\{1,2,3\}}(t)\right]^2, \text{ for } i \in \{1,2,3\}. \tag{1.24}$$

The dynamics of technological progress of the joint venture over the time interval $s \in [t_0, T]$ can be expressed as:

$$dx^i(s) = \left(\frac{\alpha_i^2}{4c_i^{\{1,2,3\}}} A_i^{\{1,2,3\}}(t) \left[x^i(s)\right]^{1/2} - \delta\, x^i(s)\right) ds + \sigma_i x^i(s) dz_i(s), \tag{1.25}$$
$$x^i(t_0) = x_0^i,$$

for $i \in \{1,2,3\}$.

Taking the transforming $y^i(s) = x^i(s)^{1/2}$, for $i \in \{1,2,3\}$, equation system (1.25) can be expressed as:

$$dy^i(s) = \left(\frac{\alpha_i^2}{8c_i} A_i^{\{1,2,3\}}(t) - \frac{\delta}{2} y^i(s) - \frac{\sigma_i^2}{8} A_i^{\{1,2,3\}}(s) y^i(s)\right) ds + \frac{1}{2}\sigma_i y^i(s) dz_i(s),$$
$$y^i(t_0) = \left(x_0^i\right)^{1/2}, \tag{1.26}$$

for $i \in \{1,2,3\}$

Equation (1.26) is a system of linear stochastic differential equations which can be solved by standard techniques. Solving (1.26) yields the joint venture's state trajectory. Let $\{y^{1*}(t), y^{2*}(t), y^{3*}(t)\}$ denote the solution to (1.26). Transforming $x^i = (y^i)^2$, we obtain the state trajectories of the joint venture over the time interval $s \in [t_0, T]$ as

$$x^*(s) = \{x^{1*}(t), x^{2*}(t), x^{3*}(t)\}_{t=t_0}^{T} = \left\{\left[y^{1*}(t)\right]^2, \left[y^{2*}(t)\right]^2, \left[y^{3*}(t)\right]^2\right\}_{t=t_0}^{T}. \tag{1.27}$$

We use X_t^* to denote the set of realizable values of $x^*(t)$ at time t and the term $x_t^* \in X_t^*$ is used to denote an element in X_t^*.

Remark 1.1 One can readily verify that:

$W^{(t_0)\{1,2,3\}}\left(t,x^{1^*},x^{2^*},x^{3^*}\right) = W^{(t)\{1,2,3\}}\left(t,x^{1^*},x^{2^*},x^{3^*}\right)\exp[-r(t-t_0)]$, for $i\in\{1,2,3\}$. ■

15.1.3.2 Subgame-Consistent Venture Profit Sharing

Since the firms agree to share their expected joint profit proportionally to their expected noncooperative profits, the imputation scheme has to fulfill:

Condition 1.2 In the game $\Gamma_c(x_0, T-t_0)$, an imputation

$$\xi^{(t_0)i}\left(t_0,x_0\right)^0 = \frac{V^{(t_0)i}\left(t_0,x_0^i\right)}{\sum_{j=1}^{n} V^{(t_0)j}\left(t_0,x_0^i\right)} W^{(t_0)\{1,2,3\}}\left(t_0,x_0^1,x_0^2,x_0^3\right)$$

is assigned to firm i, for $i\in\{1,2,3\}$;
and in the subgame $\Gamma_c\left(x_\tau^*, T-\tau\right)$, for $\tau\in\left(t_0,T\right]$, an imputation

$$\xi^{(\tau)i}\left(\tau,x_\tau^*\right) = \frac{V^{(\tau)i}\left(\tau,x_\tau^{i*}\right)}{\sum_{j=1}^{n} V^{(\tau)j}\left(\tau,x_\tau^{i*}\right)} W^{(\tau)\{1,2,3\}}\left(\tau,x_\tau^{1*},x_\tau^{2*},x_\tau^{3*}\right) \tag{1.28}$$

is assigned to firm i, for $i\in\{1,2,3\}$. ■

To formulate a payoff distribution procedure over time so that the agreed-upon imputation in Condition 1.2 is satisfied we invoke (1.11) to obtain:

$$B_i\left(\tau,x_\tau^{1*},x_\tau^{2*},x_\tau^{3*}\right) = -\frac{\partial}{\partial t}\left[\frac{V^{(\tau)i}\left(t,x_t^{i*}\right)}{\sum_{j=1}^{3} V^{(\tau)j}\left(t,x_t^{j*}\right)} W^{(\tau)\{1,2,3\}}\left(t,x_t^{1*},x_t^{2*},x_t^{3*}\right)\Bigg|_{t=\tau}\right]$$

$$-\frac{1}{2}\sum_{h,\zeta=1}^{3}\sigma_h x^h \sigma_\zeta x^\zeta \frac{\partial^2}{\partial x_\tau^{h*}\partial x_\tau^{\zeta*}}\left[\frac{V^{(\tau)i}\left(\tau,x_\tau^{i*}\right)}{\sum_{j=1}^{3} V^{(\tau)j}\left(\tau,x_\tau^{j*}\right)} W^{(\tau)\{1,2,3\}}\left(\tau,x_\tau^{1*},x_\tau^{2*},x_\tau^{3*}\right)\right]$$

$$-\sum_{h=1}^{3}\frac{\partial}{\partial x_\tau^{h*}}\left[\frac{V^{(\tau)i}\left(\tau,x_\tau^{i*}\right)}{\sum_{j=1}^{3} V^{(\tau)j}\left(\tau,x_\tau^{j*}\right)} W^{(\tau)\{1,2,3\}}\left(\tau,x_\tau^{1*},x_\tau^{2*},x_\tau^{3*}\right)\right]$$

$$\times\left[\frac{\alpha_h^2}{4c_h^{\{1,2,3\}}} A_h^{\{1,2,3\}}(\tau)\left(x_\tau^{h*}\right)^{1/2} - \delta\, x_\tau^{h*}(s)\right],$$

$$\text{for } i\in\{1,2,3\},\ x_\tau^*\in X_\tau^*\ \tau\in[t_0,T]. \tag{1.29}$$

Finally, with firms using the cooperative investment strategies the instantaneous receipt of firm i at time instant τ is:

$$\zeta_i(\tau, x_\tau^*) = P_i(x_\tau^{i*})^{1/2} - \frac{\alpha_i^2}{16\left(c_i^{\{1,2,3\}}\right)}\left[A_i^{\{1,2,3\}}(\tau)\right]^2,$$
$$\text{for } i \in \{1,2,3\},\ x_\tau^* \in X_\tau^* \text{ and } \tau \in [t_0, T]. \tag{1.30}$$

Under cooperation, the instantaneous payment that firm i should receive under the agreed-upon optimality principle is $B_i(\tau, x_\tau^*)$ as stated in (1.29). Hence an instantaneous transfer payment

$$\chi^i(\tau, x_\tau^*) = B_i(\tau, x_\tau^*) - \zeta_i(\tau, x_\tau^*) \tag{1.31}$$

has to be given or charged to firm i at time τ, for $i \in \{1,2,3\}$, $x_\tau^* \in X_\tau^*$ and $\tau \in [t_0, T]$.

15.2 Shapley Value Solution for a Joint Venture

Consider again the stochastic dynamic venture model (1.1 and 1.2). If firms are allowed to form different coalitions consisting of a subset of companies $K \subseteq N$. There are k firms in the subset K. The participating firms in a coalition can obtain cost savings from cooperation. In particular, they can obtain cost reduction and with joint venture cost advantage as below

$$c_j^K[u_j(s)] \le c_j^L[u_j(s)], \text{ for } j \in L \subseteq K, \tag{2.1}$$

where $c_j^K[u_j(s)]$ represents the costs of the controls of the firm j in the subset K and $c_j^L[u_j(s)]$ represents the costs of the controls of the firm j in the subset L.

Moreover, marginal cost advantages lead to:

$$\partial c_j^K[u_j(s)]/\partial u_j(s) \le \partial c_j^L[u_j(s)]/\partial u_j(s), \text{ for } j \in L \subseteq K.$$

At time t_0, the expected profit to the joint venture K becomes:

$$E_{t_0}\left\{\int_{t_0}^{T} \sum_{j \in K} \left(g^j[s, x^j(s)] - c_j^K[u_j(s)]\right) \exp\left[-\int_{t_0}^{s} r(y)dy\right] ds + \sum_{j \in K} \exp\left[-\int_{t_0}^{T} r(y)dy\right] q^j(x^j(T))\right\}, \text{ for } K \subseteq N. \tag{2.2}$$

15.2.1 Expected Joint Venture Profits

To compute the expected profit of the joint venture K we have to consider the stochastic control problem $\varpi[K; t_0, x_0^K]$ which maximizes expected joint venture profit (2.2) subject to technology accumulation dynamics (2.1).

Invoking Theorem A.3 in the Technical Appendices, the solution to the control problem $\varpi[K; t_0, x_0^K]$ can be characterized as follows.

Corollary 2.1 A set of controls $\{\psi_i^{K*}(t, x^K)$, for $i \in K$ and $t \in [t_0, T]\}$ provides an optimal solution to the stochastic control problem $\varpi[K; t_0, x_0^K]$ if there exists continuously twice differentiable function $W^{(t_0)K}(t, x) : [t_0, T] \times R^k \to R$ satisfying the following partial differential equation:

$$-W_t^{(t_0)K}(t, x^K) - \frac{1}{2}\sum_{h,\zeta \in K} \Omega^{h\zeta}(t, x^K) W_{x^h x^\zeta}^{(t_0)K}(t, x^K) =$$

$$\max_{u_K}\left\{\sum_{j\in K}\left[g^j(t, x^j) - c_j^K(u_j)\right]\exp\left[-\int_{t_0}^{t} r(y)dy\right] + \sum_{j\in K} W_{x_j}^{(t_0)K}(t, x^K) f^j(s, x^j, u_j)\right\},$$

$$W^{(t_0)K}(T, x^K) = \exp\left[-\int_{t_0}^{T} r(y)dy\right]\sum_{j\in K} q^j(x^j). \quad (2.3)\blacksquare$$

Following Corollary 2.1, one can characterize the maximized expected payoff $W^{(\tau)K}(t, x^K)$ to the optimal control problem $\varpi[K; \tau, x_\tau^K]$ which maximizes

$$E_\tau\left\{\int_\tau^T \sum_{j\in K}\left(g^j\left[(s, x^j(s)\right] - c_j^K\left[u_j(s)\right]\right)\exp\left[-\int_\tau^s r(y)dy\right]ds + \sum_{j\in K}\exp\left[-\int_\tau^T r(y)dy\right] q^j(x^j(T))\right\}$$

subject to $dx^j(s) = f^j\left[s, x^j(s), u_j(s)\right] ds + \sigma_j[s, x^j(s)]\, dz_j(s)$, $x^j(\tau) = x_\tau^j$; for $j \in K$.

Superadditivity of the expected coalition payoffs is demonstrated in the condition below.

Condition 2.1 The expected coalition profits $W^{(\tau)K}(t, x^K)$ is superadditive, that is

$$W^{(\tau)K}(\tau, x^K) \ge W^{(\tau)L}(\tau, x^L) + W^{(\tau)K\backslash L}(\tau, x^{K\backslash L}), \text{ for } L \subset K \subseteq N,$$

where $K\backslash L$ is the relative complement of L in K.

Proof See Appendix. ∎

Now consider the case of a grand coalition N in which all the n firms are in the coalition. Following Corollary 2.1, the solution to the stochastic control problem $\varpi[N; t_0, x_0^N]$ can be characterized as in Corollary 1.1. The cooperative state dynamics is (1.6), the optimal stochastic trajectory is (1.7) and the optimal cooperative strategies are in (1.8). Along the cooperative investment path $\{x^*(t)\}_{t=t_0}^T$ the expected total venture profit over the interval $[t, T]$, for $t \in [t_0, T)$, can be expressed as (1.9).

15.2.2 The PDP for Shapley Value

Consider the case where the participating firms agree to share their expected cooperative profits according to the Shapley Value (1953). The imputation has to satisfy

Condition 2.1 In the game $\Gamma_c(x_0, T - t_0)$, an imputation

$$\xi^{(t_0)i}(t_0, x_0^N) = \sum_{K \subseteq N} \frac{(k-1)!(n-k)!}{n!} \left[W^{(t_0)K}(t_0, x_0^K) - W^{(t_0)K \backslash i}(t_0, x_0^{K \backslash i}) \right],$$

is assigned to firm i, for $i \in N$;
and in the subgame $\Gamma_c(x_\tau^*, T - \tau)$, for $\tau \in (t_0, T]$, an imputation

$$\xi^{(\tau)i}(\tau, x_\tau^{N*}) = \sum_{K \subseteq N} \frac{(k-1)!(n-k)!}{n!} \left[W^{(\tau)K}(\tau, x_\tau^{K*}) - W^{(\tau)K \backslash i}(\tau, x_\tau^{K \backslash i *}) \right] \quad (2.4)$$

is assigned to firm i, for $i \in N$. ■

To formulate a payoff distribution procedure over time so that the agreed imputations satisfy the Shapley Value in Condition 2.1 we invoke Theorem 3.1 in Chap. 3 and obtain:

Corollary 2.2 A PDP with a terminal payment $q^i(x_T^*))$ at time T and an instantaneous payment at time $\tau \in [t_0, T]$ when $x^*(\tau) = x_\tau^* \in X_\tau^*$:

$$\begin{aligned}
B_i(\tau, x_\tau^*) = & -\sum_{K \subseteq N} \frac{(k-1)!(n-k)!}{n!} \Bigg\{ \left[W_t^{(\tau)K}(t, x_t^{K*})|_{t=\tau} \right] - \left[W_t^{(\tau)K \backslash i}(t, x_t^{K \backslash i*})|_{t=\tau} \right] \\
& + \sum_{h \in K} \left[\frac{\partial}{\partial x_\tau^{h*}} W^{(\tau)K}(\tau, x_\tau^{K*}) \right] f^h\left[\tau, x_\tau^{h*}, \psi_h^*(\tau, x_\tau^*)\right] \\
& - \sum_{h \in K \backslash i} \left[\frac{\partial}{\partial x_\tau^{h*}} W^{(\tau)K \backslash i}(\tau, x_\tau^{K \backslash i*}) \right] f^h\left[\tau, x_\tau^{h*}, \psi_h^*(\tau, x_\tau^*)\right] \\
& + \frac{1}{2} \sum_{h, \zeta \in K} \Omega^{h\zeta}(\tau, x_\tau^*) \frac{\partial^2}{\partial x_\tau^{h*} \partial x_\tau^{\zeta*}} \left[W^{(\tau)K}(\tau, x_\tau^{K*}) \right] \\
& - \frac{1}{2} \sum_{h, \zeta \in K \backslash i} \Omega^{h\zeta}(\tau, x_\tau^*) \frac{\partial^2}{\partial x_\tau^{h*} \partial x_\tau^{\zeta*}} \left[W^{(\tau)K \backslash i}(\tau, x_\tau^{K \backslash i*}) \right] \Bigg\}, \text{ for } i \in N,
\end{aligned} \quad (2.5)$$

would lead to the realization of the Shapley Value imputations $\xi^{(\tau)i}(\tau, x_\tau^{N*})$ in Condition 2.1. ■

Finally, with firms using the cooperative investment strategies $\{\psi_i^*(\tau, x_\tau^*)$, for $\tau \in [t_0, T]$ and $i \in N\}$, the instantaneous receipt of firm i at time instant τ when $x^*(\tau) = x_\tau^* \in X_\tau^*$ is:

$$\zeta_i(\tau, x_\tau^*) = g^i(\tau, x_\tau^{i*}) - c_i^N[\psi_i^*(\tau, x_\tau^*)],$$
$$\text{for } \tau \in [t_0, T] \text{ and } i \in N. \tag{2.6}$$

According to Corollary 2.2, the instantaneous payment that firm i should receive under the agreed-upon optimality principle is $B_i(\tau, x_\tau^*)$ as stated in (2.5). Hence an instantaneous transfer payment

$$\chi^i(\tau, x_\tau^*) = B_i(\tau, x_\tau^*) - \zeta_i(\tau, x_\tau^*) \tag{2.7}$$

would be given or charged to firm i at time τ, for $i \in N$, $x_\tau^* \in X_\tau^*$ and $\tau \in [t_0, T]$.

15.2.3 Shapley Value Profit Sharing: An Illustration

Consider the venture in Sect. 15.1.3. When the firms act independently, their expected profits and state dynamics are respectively (1.14) and (1.15). The expected profits of firm $i \in \{1, 2, 3\}$ are given in (1.16). However, the participating firms would like to share their expected cooperative profits according to the Shapley Value.

15.2.3.1 Expected Coalition Payoffs

Cost saving opportunities are created under joint venture. In particular, the cost savings in joint venture is depicted as follows

$$c_i^{\{i\}} \leq c_i^{\{i,j\}}, \text{ for } i, j \in \{1, 2, 3\} \text{ and } i \neq j,$$
$$c_i^{\{i,j\}} \leq c_i^{\{i,j,k\}}, \text{ for } i, j, k \in \{1, 2, 3\} \text{ and } i \neq j \neq k. \tag{2.8}$$

The firms in the joint venture maximize the sum of their expected profits:

$$E_{t_0}\left\{\int_{t_0}^{T} \sum_{j=1}^{3} \left[P_j [x^j(s)]^{1/2} - c_j u_j^{\{1,2,3\}}(s)\right] \exp[-r(s - t_0)]\, ds \right.$$
$$\left. + \sum_{j=1}^{3} \exp[-r(T - t_0)] q_j [x^j(T)]^{1/2} \right\} \tag{2.9}$$

subject to (1.15).

Following the analysis in Sect. 15.2, one can obtain the maximized expected venture profit $W^{(t_0)\{1,2,3\}}(t,x^1,x^2,x^3)$ as in (1.22), and the investment strategies:

$$\psi_i^{\{1,2,3\}}(t,x)=\frac{\alpha_i^2}{16(c_i)^2}\left[A_i^{\{1,2,3\}}(t)\right]^2,\ \text{ for } i\in\{1,2,3\}. \tag{2.10}$$

The cooperative state dynamics of the joint venture over the time interval $s\in[t_0,T]$ is in (1.25).

For the computation of the dynamic the Shapley Value, we consider cases when two of the firms form a coalition $\{i,j\}\subset\{1,2,3\}$ to maximize expected joint profit:

$$E_{t_0}\left\{\int_{t_0}^{T}\left[P_i\left[x^i(s)\right]^{1/2}-c_i^{\{i,j\}}u_i(s)+P_j\left[x^j(s)\right]^{1/2}-c_j^{\{i,j\}}u_j(s)\right]\ \exp[-r(s-t_0)]\,ds\right.$$
$$\left.+\exp[-r(T-t_0)]\left\{q_i[x^i(T)]^{1/2}+q_j[x^j(T)]^{1/2}\right\}\right\} \tag{2.11}$$

subject to

$$dx^i(s)=\left[\alpha_i[u_i(s)x^i(s)]^{1/2}-\delta\,x_i(s)\right]ds+\sigma_i x^i(s)dz_i(s),$$
$$x^i(t_0)=x_0^i\in X^i,\ \text{ for } i,j,\in\{1,2,3\}\ \text{ and }\ i\neq j. \tag{2.12}$$

Following the analysis in Sect. 15.1, we obtain the following value functions:

$$W^{(t_0)\{i,j\}}(t,x^i,x^j)=\left[A_i^{\{i,j\}}(t)\left(x^i\right)^{1/2}+A_j^{\{i,j\}}(t)\left(x^j\right)^{1/2}+C^{\{i,j\}}(t)\right]\exp[-r(t-t_0)], \tag{2.13}$$

for $i,j,\in\{1,2,3\}$ and $i\neq j$,
where $A_i^{i,j}(t)$, $A_j^{i,j}(t)$ and $C^{\{i,j\}}(t)$ satisfy

$$\dot{A}_i^{\{i,j\}}(t)=\left(r+\frac{\delta}{2}+\frac{\sigma_i^2}{8}\right)A_i^{\{i,j\}}(t)-P_i,\ \text{and } A_i^{\{i,j\}}(T)=q_i$$

for $i,j,\in\{1,2,3\}$ and $i\neq j$;

$$\dot{C}^{\{i,j\}}(t)=rC^{\{i,j\}}(t)-\sum_{h\in\{i,j\}}\frac{\alpha_h^2}{16c_h^{\{i,j\}}}\left[A_h^{\{i,j\}}(t)\right]^2,$$
$$C^{\{i,j\}}(T)=0.$$

Moreover, one can easily derive for $\tau\in[t_0,T]$

$$W^{(t_0)\{i,j\}}(t,x^i,x^j)=\exp[-r(\tau-t_0)]\ W^{(\tau)\{i,j\}}(t,x^i,x^j),\ \text{for } i,j,\in\{1,2,3\}\text{ and } i\neq j$$

15.2.3.2 Subgame Consistent Shapley Value PDP

To formulate a payoff distribution procedure over time so that the agreed imputations satisfy the Shapley Value we invoke Corollary 2.2 and obtain:

A PDP with a terminal payment $q^i(x_T^*))$ at time T and an instantaneous payment at time $\tau \in [t_0, T]$:

$$
\begin{aligned}
B_i(\tau, x_\tau^*) = &-\sum_{K \subseteq \{1,2,3\}} \frac{(k-1)!(3-k)!}{3!} \Bigg\{ \left[W_t^{(\tau)K}(t, x_t^{K*})|_{t=\tau} \right] \\
&- \left[W_t^{(\tau)K \backslash i}(t, x_t^{K \backslash i*})|_{t=\tau} \right] \\
&+ \sum_{h \in K} \left[\frac{\partial}{\partial x_\tau^{h*}} W^{(\tau)K}(\tau, x_\tau^{K*}) \right] \left[\frac{\alpha_h^2}{4c_h^{\{1,2,3\}}} A_h^{\{1,2,3\}}(\tau)(x_\tau^{i*})^{1/2} - \delta x_\tau^{h*} \right] \\
&- \sum_{h \in K \backslash i} \left[\frac{\partial}{\partial x_\tau^{h*}} W^{(\tau)K \backslash i}(\tau, x_\tau^{K \backslash i*}) \right] \left[\frac{\alpha_h^2}{4c_h^{\{1,2,3\}}} A_h^{\{1,2,3\}}(\tau)(x_\tau^{i*})^{1/2} - \delta x_\tau^{h*} \right] \\
&+ \frac{1}{2} \sum_{h, \zeta \in K} (\sigma_h x_\tau^{h*})(\sigma_\zeta x_\tau^{\zeta*}) \frac{\partial^2}{\partial x_\tau^{h*} \partial x_\tau^{\zeta*}} \left[W^{(\tau)K}(\tau, x_\tau^{K*}) \right] \\
&- \frac{1}{2} \sum_{h, \zeta \in K \backslash i} (\sigma_h x_\tau^{h*})(\sigma_\zeta x_\tau^{\zeta*}) \frac{\partial^2}{\partial x_\tau^{h*} \partial x_\tau^{\zeta*}} \left[W^{(\tau)K \backslash i}(\tau, x_\tau^{K \backslash i*}) \right] \Bigg\},
\end{aligned}
$$

for $i \in \{1, 2, 3\}$, (2.14)

would lead to the realization of the Shapley Value imputations in Condition 2.1.

Using (1.22) and (2.13),

$$
\left[W_t^{(\tau)i}(t, x_t^{i*})|_{t=\tau} \right] = r\left[A_i^{\{i\}}(\tau)(x_\tau^{i*})^{1/2} + C_i^{\{i\}}(\tau) \right] + \left[\dot{A}_i^{\{i\}}(\tau)(x_\tau^{i*})^{1/2} + \dot{C}_i^{\{i\}}(\tau) \right],
$$

for $i \in \{1, 2, 3\}$;

$$
\begin{aligned}
\left[W_t^{(\tau)\{i,j\}}(t, x_t^{i*})|_{t=\tau} \right] &= r\left[A_i^{\{i,j\}}(\tau)(x_\tau^{i*})^{1/2} + A_j^{\{i,j\}}(\tau)(x_\tau^{j*})^{1/2} + C^{\{i,j\}}(\tau) \right] \\
&+ \left[\dot{A}_i^{\{i,j\}}(\tau)(x_\tau^{i*})^{1/2} + \dot{A}_j^{\{i,j\}}(\tau)(x_\tau^{j*})^{1/2} + \dot{C}^{\{i,j\}}(\tau) \right],
\end{aligned}
$$

for $i, j \in \{1, 2, 3\}$ and $i \neq j$;

$$
\begin{aligned}
&\left[W_t^{(\tau)\{1,2,3\}}(t, x_t^{i*})|_{t=\tau} \right] = \\
&r\left[A_1^{\{1,2,3\}}(\tau)(x_\tau^{1*})^{1/2} + A_2^{\{1,2,3\}}(\tau)(x_\tau^{2*})^{1/2} + A_3^{\{1,2,3\}}(\tau)(x_\tau^{3*})^{1/2} + C^{\{1,2,3\}}(\tau) \right] \\
&+ \left[\dot{A}_1^{\{1,2,3\}}(\tau)(x_\tau^{1*})^{1/2} + \dot{A}_2^{\{1,2,3\}}(\tau)(x_\tau^{2*})^{1/2} + \dot{A}_3^{\{1,2,3\}}(\tau)(x_\tau^{3*})^{1/2} + \dot{C}^{\{1,2,3\}}(\tau) \right];
\end{aligned}
$$

$$
\left[\frac{\partial}{\partial x_\tau^{h*}} W^{(\tau)K}(\tau, x_\tau^{K*}) \right] = \frac{1}{2} A_h^K(\tau)(x_\tau^{h*})^{-1/2}, \text{ for } h \in K \subseteq \{1, 2, 3\},
$$

$$
\frac{\partial^2}{\partial (x_\tau^{h*})^2} \left[W^{(\tau)K}(\tau, x_\tau^{K*}) \right] = \frac{-1}{4} A_h^K(\tau)(x_\tau^{h*})^{-3/2}, \text{ and}
$$

$$\frac{\partial^2}{\partial x_\tau^{h*} \partial x_\tau^{\zeta*}} \left[W^{(\tau)K}\left(\tau, x_\tau^{K*}\right) \right] = 0, \text{ for } h \neq \zeta.$$

Finally, with firms using the cooperative investment strategies the instantaneous receipt of firm i at time instant τ is:

$$\zeta_i\left(\tau, x_\tau^*\right) = P_i\left(x_\tau^{i*}\right)^{1/2} - \frac{\alpha_i^2}{16\left(c_i^{\{1,2,3\}}\right)} \left[A_i^{\{1,2,3\}}(\tau) \right]^2,$$
$$\text{for } i \in \{1,2,3\},\ x_\tau^* \in X_\tau^* \text{ and } \tau \in [t_0, T]. \tag{2.15}$$

According to (2.14), the instantaneous payment that firm i should receive under the agreed-upon optimality principle is $B_i(\tau, x_\tau^*)$. Hence an instantaneous transfer payment

$$\chi^i\left(\tau, x_\tau^*\right) = B_i\left(\tau, x_\tau^*\right) - \zeta_i\left(\tau, x_\tau^*\right) \tag{2.16}$$

has to be given or charged to firm i at time τ, for $i \in \{1,2,3\}$, $x_\tau^* \in X_\tau^*$ and $\tau \in [t_0, T]$.

15.3 Infinite Horizon Analysis

Consider the case when the horizon of the analysis approaches infinity. The state dynamics of the ith firm is characterized by the set of vector-valued differential equations:

$$dx^i(s) = f^i\left[x^i(s), u_i(s)\right] ds + \sigma_i\left[x^i(s)\right] dz_i(s),\ x^i(t_0) = x_0^i, \text{ for } i \in N, \tag{3.1}$$

where $\sigma_i[x^i(s)]$ is a $m_i \times \Theta_i$ and $z_i(s)$ is a Θ_i-dimensional Wiener process and the initial state x_0^i is given. Let $\Omega_i[x^i(s)] = \sigma_i[x^i(s)]\sigma_i[x^i(s)]^T$ denote the covariance matrix with its element in row h and column ζ denoted by $\Omega_i^{h\zeta}[x^i(s)]$.

The objective of firm i to be maximized is:

$$E_{t_0}\left\{ \int_{t_0}^{\infty} \left(g^i\left[x^i(s)\right] - c_i^{\{i\}}[u_i(s)]\right) \exp[-r(s-t_0)] ds \right\}, \text{ for } i \in N. \tag{3.2}$$

Consider the alternative formulation of (3.1 and 3.2) as:

$$\max_{u_i} E_t\left\{ \int_{t}^{\infty} \left(g^i\left[x^i(s)\right] - c_i^{\{i\}}[u_i(s)]\right) \exp[-r(s-t)] ds \right\}, \text{ for } i \in N, \tag{3.3}$$

subject to

$$dx^i(s) = f_i\left[x^i(s), u_i(s)\right] ds + \sigma_i\left[x^i(s)\right] dz_i(s),\ x^i(t) = x^i, \text{ for } i \in N. \tag{3.4}$$

The infinite-horizon problem (3.3 and 3.4) is independent of the choice of t and dependent only upon the state at the starting time.

A noncooperative equilibrium of problem (3.3 and 3.4) can be characterized by a set of strategies $\{ u_i = \phi_i^*(x) \text{ for } i \in N \}$ and value functions $\hat{V}^i(x^i) : R^{m_i} \rightarrow R$ for $i \in N$, satisfying the following set of partial differential equations:

$$r\hat{V}^i(x^i) - \frac{1}{2}\sum_{h,\zeta=1}^{m_i} \Omega^{h\zeta}(x^i)\hat{V}^i_{x^{i(h)}x^{i(\zeta)}}(x^i)$$
$$= \max_{u_i}\left\{ g(x^i) - c_i^{\{i\}}(u_i) + \hat{V}^i_x(x) f\left[x^i, u_i\right] \right\} \tag{3.5}$$

Once again, for the sake of clarity in exposition, we consider the case where $m_i=1$, for $i \in N$.

15.3.1 Infinite-horizon Dynamic Joint Venture

These n companies agree to form a joint venture to enhance their profits. Cost-saving opportunities are created under joint venture. The cost of control of firm j under the joint venture becomes $c_j^{1,2,3}[u_j(s)]$. With joint venture cost advantage

$$c_j^{\{1,2,3\}}(u_j) \leq c_j^{\{j\}}(u_j), \text{ for } j \in N, \tag{3.6}$$

The joint venture would maximize the expected joint venture profit:

$$E_t\left\{ \int_t^{\infty} \sum_{j=1}^{n} \left(g^j\left[x^j(s)\right] - c_j^N\left[u_j(s)\right]\right) \exp[-r(s-t)]ds \right\}, \tag{3.7}$$

subject to (3.4).

An optimal solution of the control problem (3.4) and (3.7) can be characterized as:

Corollary 3.1 A set of control strategies $\{ \psi_i^*(x) \text{ for } i \in N_1 \}$ provides a solution to the control problem (3.4) and (3.7), if there exist continuously twice differentiable functions $W(x) : R^n \rightarrow R$, satisfying the following partial differential equation:

$$rW(x) - \frac{1}{2}\sum_{h,\zeta=1}^{n} \Omega^{h\zeta}(x) W_{x^h x^\zeta}(x)$$
$$= \max_{u_1, u_2, \cdots, u_n}\left\{ \sum_{j=1}^{n}\left[g^j(x^j) - c_j^N(u_j)\right] + \sum_{j=1}^{n} W_{x_j}(x) f^j(x^j, u_j) \right\}, \tag{3.8}$$

where $x = \{x^1, x^2, \cdots, x^n\}$. ■

Hence the firms will adopt the cooperative control $\{\psi_i^*(x), \text{ for } i \in N\}$ to obtain the maximized level of expected joint profit. Substituting this set of control into (3.4) yields the dynamics of technology advancement under cooperation as:

$$dx^i(s) = f^i\left[x^i(s), \psi_i^*(x(s))\right] ds + \sigma_i[x^i(s)]\, dz_i(s),$$
$$x^i(t_0) = x_0^i, \text{ for } i \in N. \tag{3.9}$$

Let $x^*(t) = \{x^{1*}(t), x^{2*}(t), \ldots, x^{n*}(t)\}$ denote the solution to (3.9). The optimal trajectory $\{x^*(t)\}_{t=t_0}^{\infty}$ can be expressed as:

$$x^{i*}(t) = x_0^i + \int_{t_0}^{t} f^i\left[x^{i*}(s), \psi_i^*\left(x^*(s)\right)\right] ds + \int_{t_0}^{t} \sigma_i\left[s, x^{i*}(s)\right] dz_i(s),$$
$$\text{for } i \in N. \tag{3.10}$$

We use X_t^* to denote the set of realizable values of $x^*(t)$ at time t generated by (3.10). The term $x_t^* \in X_t^*$ is used to denote an element in X_t^*.

Substituting the optimal extraction strategies in $\{\psi_i^*(x), \text{ for } i \in N\}$ into (3.7) yields the expected venture profit as:

$$W\left(x_t^*\right) = E_t\left\{\int_t^{\infty} \sum_{j=1}^{n} \left(g^j\left[x^{j*}(s)\right] - c_j^N\left[\psi_j^*\left(x^*(s)\right)\right]\right) \exp[-r(s-t)] ds\right\}, \tag{3.11}$$

15.3.2 Subgame-Consistent Venture Profit Sharing

In this Section we consider deriving the PDP for two commonly used sharing optimality principles – (i) sharing venture profits proportional to the participating firms' noncooperative profit, and (ii) sharing venture profits according to the Shapley Value.

(i) Consider the case when the firms in the venture share the excess of the total expected cooperative payoff over the sum of expected individual noncooperative payoffs proportionally to the firms' expected noncooperative payoffs.

The imputation scheme has to fulfil:

Condition 3.1 An imputation

$$\xi^{(\tau)i}\left(\tau, x_\tau^*\right) = \frac{\hat{V}^i\left(x_\tau^*\right)}{\sum_{i=1}^{n} \hat{V}^i\left(x_\tau^*\right)} W\left(x_\tau^*\right) \tag{3.12}$$

is assigned to firm i, for $i \in N$ at time $\tau \in \left[t_0, \infty\right)$ if $x^*(\tau) = x_\tau^* \in X_\tau^*$. ■

To formulate a payoff distribution procedure over time so that the agreed imputations satisfy Condition 3.1 we invoke Theorem 5.3 of Chap. 3 and obtain:

Corollary 3.2 A PDP with an instantaneous payment at time $\tau \in [t_0, \infty)$:

$$B_i(\tau, x_\tau^*) = r\left[\frac{\hat{V}^i(x_\tau^{i*})}{\sum_{j=1}^n \hat{V}^j(x_\tau^{j*})} W(x_\tau^*)\right]$$
$$-\frac{1}{2}\sum_{h,\zeta=1}^n \Omega^{h\zeta}(x_\tau^*)\frac{\partial^2}{\partial x_\tau^{h*}\partial x_\tau^{\zeta*}}\left[\frac{\hat{V}^i(x_\tau^{i*})}{\sum_{j=1}^n \hat{V}^j(x_\tau^{j*})} W(x_\tau^*)\right]$$
$$-\sum_{h=1}^n \frac{\partial}{\partial x_\tau^{h*}}\left[\frac{\hat{V}^i(x_\tau^{i*})}{\sum_{j=1}^n \hat{V}^j(x_\tau^{j*})} W(x_\tau^*)\right] f^h\left[x_\tau^{h*}, \psi_h^*(x_\tau^*)\right], \quad (3.13)$$

for $i \in N$ and $x^*(\tau) = x_\tau^* \in X_\tau^*$.

would lead to realization of the solution imputations in Condition 3.1. ■

With (3.13) a subgame consistent solution can be obtained. Note that while firms are using the cooperative investment strategies $\{\psi_i^*(x_\tau^*)$, and $i \in N\}$, the instantaneous receipt of firm i at time instant τ is:

$$\zeta_i(\tau, x_\tau^*) = g^i(x_\tau^{i*}) - c_i^N[\psi_i^*(x_\tau^*)],$$
$$\text{for } i \in N,\ x^*(\tau) = x_\tau^* \in X_\tau^* \text{ and } \tau \in [t_0, \infty).$$

According to Corollary 3.2, the instantaneous payment that firm i should receive under the agreed-upon optimality principle is $B_i(\tau, x_\tau^*)$, for $i \in N$, as stated in (3.13). Hence an instantaneous transfer payment

$$\chi^i(\tau, x_\tau^*) = B_i(\tau, x_\tau^*) - \zeta_i(\tau, x_\tau^*)$$

has to be given or charged to firm i at time τ, for $i \in N$ if $x^*(\tau) = x_\tau^* \in X_\tau^*$.

(ii) Consider again the infinite horizon dynamic venture model (3.3 and 3.4). The member firms would maximize their expected joint profit and share their expected cooperative profits according to the Shapley Value.

The expected profit to the joint venture K becomes:

$$E_t\left\{\int_t^\infty \sum_{j \in K} \left(g^j[s, x^j(s)] - c_j^K[u_j(s)]\right)\exp[-r(s-t)]ds\right\}, \text{ for } K \subseteq N. \quad (3.14)$$

To compute the profit of the joint venture K we have to consider the optimal control problem (3.4) and (3.14). Invoking Theorem A.4 of the Technical Appendices, the solution to the stochastic control problem can be characterized as follows.

Corollary 3.3 A set of controls $\{\psi_i^{K*}(x^K),\ \text{for}\ i\in K\ \text{and}\ t\in[t_0,\infty)\}$ provides an optimal solution to the stochastic control problem (3.4) and (3.14) if there exists continuously twice differentiable function $W^K(x^K)\ R^k \to R$ satisfying the following equation:

$$rW^K(x^K)-\frac{1}{2}\sum_{h,\zeta\in K}\Omega^{h\zeta}(x^K)W^K_{x^h x^\zeta}(x^K)$$
$$=\max_{u_K}\left\{\sum_{j\in K}\left[g^j(x^j)-c_j^K(u_j)\right]+\sum_{j\in K}W^K_{x_j}(x^K)f^j(x^j,u_j)\right\} \tag{3.15}$$

Now consider the case of a grand coalition N in which all the n firms are in the coalition. Using the result in Corollary 3.1, the cooperative state trajectory can be obtained as in (3.10).

To share the venture profit among participating firms according to the Shapley Value, the imputation has to satisfy

Condition 3.2 An imputation

$$\xi^{(\tau)i}(\tau,x_\tau^{N*})=\sum_{K\subseteq N}\frac{(k-1)!(n-k)!}{n!}\left[W^K(x_\tau^{K*})-W^{K\backslash i}(x_\tau^{K\backslash i*})\right] \tag{3.16}$$

is assigned to firm i, for $i\in N$ at time τ when the state is x_τ^*. ■

To formulate a payoff distribution procedure over time so that the agreed imputations satisfy the Shapley Value in Condition 3.2 we invoke Theorem 5.3 in Chap. 3 and obtain:

Corollary 3.4 A PDP with an instantaneous payment at time $\tau\in[t_0,\infty)$:

$$B_i(\tau,x_\tau^*)=-\sum_{K\subseteq N}\frac{(k-1)!(n-k)!}{n!}\Bigg\{r\ W^{K\backslash i}(x_\tau^{K\backslash i*})-r\ W^K(x_\tau^{K*})$$
$$+\sum_{h\in K}\left[\frac{\partial}{\partial x_\tau^{h*}}W^K(x_\tau^{K*})\right]f^h\left[x_\tau^{h*},\psi_h^*(x_\tau^*)\right]$$
$$-\sum_{h\in K\backslash i}\left[\frac{\partial}{\partial x_\tau^{h*}}W^{K\backslash i}(x_\tau^{K\backslash i*})\right]f^h\left[x_\tau^{h*},\psi_h^*(x_\tau^*)\right]$$
$$+\frac{1}{2}\sum_{h,\zeta\in K}\Omega^{h\zeta}(x_\tau^*)\frac{\partial^2}{\partial x_\tau^{h*}\partial x_\tau^{\zeta*}}\left[W^K(x_\tau^{K*})\right]$$
$$-\frac{1}{2}\sum_{h,\zeta\in K\backslash i}\Omega^{h\zeta}(x_\tau^*)\frac{\partial^2}{\partial x_\tau^{h*}\partial x_\tau^{\zeta*}}\left[W^{K\backslash i}(x_\tau^{K\backslash i*})\right]\Bigg\},\ \text{for}\ i\in N, \tag{3.17}$$

would lead to the realization of the Shapley Value in Condition 3.2. ■

A subgame consistent solution can be constructed with the optimal cooperative strategies and the PDP in (3.17).

15.3.3 An Example of Infinite Horizon Venture

Consider an infinite horizon version 3-company joint venture. The planning period is $[t_0, \infty)$. Company i's expected profit is

$$E_{t_0}\left\{ \int_{t_0}^{\infty} \left[P_i\left[x^i(s)\right]^{1/2} - c_i^{\{i\}} u_i(s)\right] \exp[-r(s-t_0)]\, ds \right\}, \tag{3.18}$$

for $i \in \{1, 2, 3\}$,

The evolution of the technology level of company i follows the dynamics:

$$\begin{aligned} dx^i(s) &= \left[\alpha_i[u_i(s)x^i(s)]^{1/2} - \delta\, x^i(s)\right] ds + \sigma_i x^i(s)\, dz_i(s), \\ x^i(t_0) &= x_0^i \in X^i, \text{ for } i \in \{1, 2, 3\}, \end{aligned} \tag{3.19}$$

In the case when each of these three firms acts independently. The conditions characterizing the non-cooperative payoff of firm i can be obtained as:

$$\begin{aligned} rW^i(x^i) &- \frac{(\sigma_i x^i)^2}{2} W^i_{x^i x^i}(x^i) \\ &= \max_{u_i} \left\{ \left[P_i(x^i)^{1/2} - c_i^{\{i\}} u_i\right] + W^i_{x^i}(x^i)\left[\alpha_i (u^i x_i)^{1/2} - \delta\, x^i\right] \right\}, \\ &\text{for } i \in \{1, 2, 3\}. \end{aligned} \tag{3.20}$$

Solving (3.20) yields

$$W^i(x^i) = \left[A_i^{\{i\}} (x^i)^{1/2} + C_i^{\{i\}}\right], \text{ for } i \in \{1, 2, 3\}, \tag{3.21}$$

where

$$0 = \left(r + \frac{\delta}{2} + \frac{\sigma_i^2}{8}\right) A_i^{\{i\}} - P_i, rC_i^{\{i\}} = \frac{\alpha_i^2}{16 c_i^{\{i\}}} \left(A_i^{\{i\}}\right)^2.$$

15.3.3.1 Cost-Saving Joint Venture

Consider the case when all these three firms agree to form a joint venture and share their expected joint profit proportionally to their expected noncooperative profits.

The cost of control of firm j under the joint venture becomes $c_j^{1,2,3} u_j(s)$. With joint venture cost advantage

$$c_j^{\{1,2,3\}} \leq c_j^{\{j\}}, \text{ for } j \in N. \tag{3.22}$$

The expected profit of the joint venture is the sum of the participating firms' profits:

$$E_t \left\{ \int_t^\infty \sum_{j=1}^3 \left[P_j \left[x^j(s)\right]^{1/2} - c_j^{[1,2,3]} u_j(s) \right] \exp[-r(s-t)]\, ds. \tag{3.23}$$

The optimal solution to the problem of maximizing (3.23) subject to (3.19) can be characterized by

$$\begin{aligned} & rW^{\{1,2,3\}}(x^1,x^2,x^3) - \sum_{h,\zeta=1}^{3} \frac{(\sigma_h x^h)(\sigma_\zeta x^\zeta)}{2} W^{\{1,2,3\}}_{x^h x^\zeta}(x^1,x^2,x^3) \\ & = \max_{u_1,u_2,u_3} \left\{ \sum_{i=1}^{3} \left[P_i (x^i)^{1/2} - c_i^{\{1,2,3\}} u_i \right] \right. \\ & \left. + \sum_{i=1}^{3} W^{\{1,2,3\}}_{x^i}(x^1,x^2,x^3) \left[\alpha_i [u_i x^i]^{1/2} - \delta x^i \right] \right\}. \end{aligned} \tag{3.24}$$

Solving (3.24) yields

$$W^{\{1,2,3\}}(x^1,x^2,x^3) = \left[A_1^{\{1,2,3\}} (x^1)^{1/2} + A_2^{\{1,2,3\}} (x^2)^{1/2} + A_3^{\{1,2,3\}} (x^3)^{1/2} + C^{\{1,2,3\}} \right], \tag{3.25}$$

where $A_1^{1,2,3}, A_2^{1,2,3}, A_3^{1,2,3}$ and $C^{\{1,2,3\}}$ satisfy

$$0 = \left(r + \frac{\delta}{2} + \frac{\sigma_i^2}{8} \right) A_i^{\{1,2,3\}} - P_i$$

for $i,j,h \in \{1,2,3\}$ and $i \neq j \neq h$,

$$rC^{\{1,2,3\}} = \sum_{i=1}^{3} \frac{\alpha_i^2}{16 c_i^{\{1,2,3\}}} \left(A_i^{\{1,2,3\}} \right)^2 \tag{3.26}$$

The investment strategies of the grand coalition joint venture can be derived as:

$$\psi_i^{\{1,2,3\}}(x) = \frac{\alpha_i^2}{16 \left(c_i^{\{1,2,3\}} \right)^2} \left[A_i^{\{1,2,3\}} \right]^2, \text{ for } i \in \{1,2,3\}. \tag{3.27}$$

The dynamics of technological progress of the joint venture over the time interval $s \in [t_0, \infty)$ can be expressed as:

$$dx^i(s) = \left(\frac{\alpha_i^2}{4c_i^{\{1,2,3\}}} A_i^{\{1,2,3\}} [x^i(s)]^{1/2} - \delta\, x^i(s) \right) ds + +\sigma_i x^i(s)\, dz_i(s),$$
$$x^i(t_0) = x_0^i, \tag{3.28}$$

for $i \in \{1, 2, 3\}$.

Taking the transforming $y^i(s) = x^i(s)^{1/2}$, for $i \in \{1, 2, 3\}$, equation system (3.28) can be expressed as:

$$dy^i(s) = \left(\frac{\alpha_i^2}{8c_i^{\{1,2,3\}}} A_i^{\{1,2,3\}} - \frac{\delta}{2} y^i(s) - \frac{\sigma_i^2}{8} A_i^{\{1,2,3\}} y^i(s) \right) ds + \frac{1}{2}\sigma_i y^i(s) dz_i(s),$$
$$y^i(t_0) = \left(x_0^i\right)^{1/2},$$
$$\text{for } i \in \{1, 2, 3\}. \tag{3.29}$$

(3.29) is a system of linear stochastic differential equations which can be solved by standard techniques. Solving (3.29) yields the joint venture's state trajectory. Let $\{y^{1\,*}(t), y^{2\,*}(t), y^{3\,*}(t)\}$ denote the solution to (3.29). Transforming $x^i = (y^i)^2$, we obtain the state trajectories of the joint venture over the time interval $s \in [t_0, \infty)$ as

$$\{x^*(t)\}_{t=t_0}^{\infty} \equiv \{x^{1*}(t), x^{2*}(t), x^{3*}(t)\}_{t=t_0}^{T} = \left\{ [y^{1*}(t)]^2, [y^{2*}(t)]^2, [y^{3*}(t)]^2 \right\}_{t=t_0}^{T}. \tag{3.30}$$

We use X_t^* to denote the set of realizable values of $x^*(t)$ at time t and the term $x_t^* \in X_t^*$ is used to denote an element in X_t^*.

15.3.3.2 Subgame-Consistent Venture Profit Sharing

We consider deriving the PDP for two commonly used sharing optimality principles – (i) sharing venture profits proportional to the participating firms' noncooperative profit, and (ii) sharing venture profits according to the Shapley Value.

(i) If the firms agree to share their expected joint profit proportionally to their expected noncooperative profits, the imputation scheme has to fulfill:

$$\xi^{(\tau)i}(\tau, x_\tau^*) = \frac{\hat{V}^i(x_\tau^{i*})}{\sum_{j=1}^{3} \hat{V}^j(x_\tau^{i*})} W^{\{1,2,3\}}(x_\tau^{1*}, x_\tau^{2*}, x_\tau^{3*}) \tag{3.31}$$

is assigned to firm i, for $i \in \{1, 2, 3\}$ at time τ when the state is $x_\tau^* \in X_\tau^*$. ■

A PDP with and an instantaneous payment at time $\tau \in [t_0, \infty)$ when $x^*(\tau) = x_\tau^* \in X_\tau^*$:

$$\begin{aligned}
B_i(\tau, x_\tau^*) &= r \frac{\hat{V}^i(x_\tau^{i*})}{\sum_{j=1}^{3} \hat{V}^j(x_\tau^{i*})} W^{\{1,2,3\}}(x_\tau^{1*}, x_\tau^{2*}, x_\tau^{3*}) \\
&- \frac{1}{2} \sum_{h,\zeta=1}^{3} \sigma_h x^h \sigma_\zeta x^\zeta \frac{\partial^2}{\partial x_\tau^{h*} \partial x_\tau^{\zeta*}} \left[\frac{\hat{V}^i(x_\tau^{i*})}{\sum_{j=1}^{3} \hat{V}^j(x_\tau^{j*})} W^{\{1,2,3\}}(x_\tau^{1*}, x_\tau^{2*}, x_\tau^{3*}) \right] \\
&- \sum_{h=1}^{3} \frac{\partial}{\partial x_\tau^{h*}} \frac{\hat{V}^i(x_\tau^{i*})}{\sum_{j=1}^{3} \hat{V}^j(x_\tau^{i*})} W^{\{1,2,3\}}(x_\tau^{1*}, x_\tau^{2*}, x_\tau^{3*}) \\
&\times \left[\frac{\alpha_h^2}{4c_h^{\{1,2,3\}}} A_h^{\{1,2,3\}} (x_\tau^{h*})^{1/2} - \delta\, x_\tau^{h*}(s) \right], \text{for } i \in \{1,2,3\},
\end{aligned} \tag{3.32}$$

will lead to the realization of the Imputation in (3.31). ■

A subgame consistent solution can be readily obtained using (3.27) and (3.32). Using the cooperative strategies the instantaneous receipt of firm i at time instant τ given $x^*(\tau) = x_\tau^* \in X_\tau^*$ is:

$$\zeta_i(\tau, x_\tau^*) = P_i (x_\tau^{i*})^{1/2} - \frac{\alpha_i^2}{16\left(c_i^{\{1,2,3\}}\right)} \left[A_i^{\{1,2,3\}}\right]^2, \tag{3.33}$$

for $i \in \{1,2,3\}$ along the cooperative path $\{x^*(t)\}_{t=t_0}^{\infty}$.

(ii) Consider the case when the participating firms agree to share their expected cooperative profits according to the Shapley Value. For the computation of the dynamic the Shapley Value, we consider cases when two of the firms form a coalition $\{i,j\} \subset \{1,2,3\}$. The cost savings in joint venture is depicted as follows

$$\begin{aligned}
c_i^{\{i\}} &\le c_i^{\{i,j\}}, \text{ for } i,j \in \{1,2,3\} \text{ and } i \ne j, \\
c_i^{\{i,j\}} &\le c_i^{\{i,j,k\}}, \text{ for } i,j,k \in \{1,2,3\} \text{ and } i,j,k \in \{1,2,3\}.
\end{aligned} \tag{3.34}$$

Coalition $\{i,j\}$ would maximize expected joint profit:

$$E_t \left\{ \int_t^\infty \left[P_i [x^i(s)]^{1/2} - c_i u_i(s) + P_j [x^j(s)]^{1/2} - c_j u_j(s) \right] \exp[-r(s-t)]\, ds \right\} \tag{3.35}$$

subject to (3.19).

Following the above analysis, we obtain the following value functions:

$$W^{\{i,j\}}\left(x^i,x^j\right)=\left[A_i^{\{i,j\}}\left(x^i\right)^{1/2}+A_j^{\{i,j\}}\left(x^j\right)^{1/2}+C^{\{i,j\}}\right], \tag{3.36}$$

for $i,j,\in\{1,2,3\}$ and $i\neq j$,

where $A_i^{i,j}$, $A_j^{i,j}$ and $C^{\{i,j\}}$ satisfy

$0=\left(r+\dfrac{\delta}{2}+\dfrac{\sigma_i^2}{8}\right)A_i^{\{1,2\}}-P_i$, for $i,j,\in\{1,2,3\}$ and $i\neq j$ and ..

and $rC^{\{i,j\}}=\sum\limits_{h\in\{i,j\}}\dfrac{\alpha_h^2}{16c_h^{\{i,j\}}}\left(A_h^{\{i,j\}}\right)^2$

Invoke Corollary 3.4 and obtain:

A PDP with an instantaneous payment at time $\tau\in\left[t_0,\infty\right)$:

$$\begin{aligned}
B_i\left(\tau,x_\tau^*\right) &= -\sum_{K\subseteq\{1,2,3\}}\frac{(k-1)!(3-k)!}{3!}\Bigg\{r\ W^{K\backslash i}\left(x_\tau^{K\backslash i*}\right)-r\ W^K\left(x_\tau^{K*}\right)\\
&+\sum_{h\in K}\left[\frac{\partial}{\partial x_\tau^{h*}}W^K\left(x_\tau^{K*}\right)\right]\left[\frac{\alpha_h^2}{4c_h^{\{i,j\}}}A_h^{\{1,2,3\}}\left(x_\tau^{i*}\right)^{1/2}-\delta\,x_\tau^{h*}\right]\\
&-\sum_{h\in K\backslash i}\left[\frac{\partial}{\partial x_\tau^{h*}}W^{K\backslash i}\left(x_\tau^{K\backslash i*}\right)\right]\left[\frac{\alpha_h^2}{4c_h^{\{i,j\}}}A_h^{\{1,2,3\}}\left(x_\tau^{i*}\right)^{1/2}-\delta\,x_\tau^{h*}\right]\\
&+\frac{1}{2}\sum_{h,\zeta\in K}\left(\sigma_h x_\tau^{h*}\right)\left(\sigma_\zeta x_\tau^{\zeta*}\right)\frac{\partial^2}{\partial x_\tau^{h*}\partial x_\tau^{\zeta*}}\left[W^K\left(x_\tau^{K*}\right)\right]\\
&-\frac{1}{2}\sum_{h,\zeta\in K\backslash i}\left(\sigma_h x_\tau^{h*}\right)\left(\sigma_\zeta x_\tau^{\zeta*}\right)\frac{\partial^2}{\partial x_\tau^{h*}\partial x_\tau^{\zeta*}}\left[W^{K\backslash i}\left(x_\tau^{K\backslash i*}\right)\right]\Bigg\},
\end{aligned} \tag{3.37}$$

for $i\in\{1,2,3\}$,

would lead to the realization of the Shapley Value;

where

$W^K(x_\tau^{K*})$ is given in (3.21), (3.25) and (3.36); and

$$\left[\frac{\partial}{\partial x_\tau^{h*}}W^K\left(x_\tau^{K*}\right)\right]=\frac{1}{2}A_h^K\left(x_\tau^{h*}\right)^{-1/2},\ \text{for } h\in K\subseteq\{1,2,3\};$$

$$\frac{\partial^2}{\partial\left(x_\tau^{h*}\right)^2}\left[W^K\left(x_\tau^{K*}\right)\right]=\frac{-1}{4}A_h^K\left(x_\tau^{h*}\right)^{-3/2},\ \text{and}$$

$$\frac{\partial^2}{\partial x_\tau^{h*}\partial x_\tau^{\zeta*}}\left[W^K\left(x_\tau^{K*}\right)\right]=0,\ \text{for } h\neq\zeta.$$

15.4 A Stochastic Dynamic Dormant-Firm Cartel

It is well known that cartels restrict their outputs to enhance their joint profit. In this Section, we consider oligopolies in which firms agree to form a cartel to restraint output and enhance their profits. Some firms have cost disadvantages and would elect to become dormant partners.

15.4.1 Basic Settings and Market Outcome

Consider an oligopoly in which n firms are allowed to extract a renewable resource within the duration $[t_0, T]$. Among the n firms, n_1 of them have absolute and marginal cost disadvantages over the other $n_2 = n - n_1$ firms. For notational convenience, the firms with cost advantages are numbered from 1 to n_1 and the firms with cost disadvantages are numbered from $n_1 + 1$ to n. The subset of firms with cost advantages is denoted by N_1 and that of firms with cost disadvantages is denoted by N_2. The firms with cost advantages are identical and so are the firms with cost disadvantages.

The dynamics of the resource is characterized by the stochastic differential equations:

$$dx(s) = \left(f\left[s, x(s), \sum_{j^1 \in N_1} u_{j^1}(s) + \sum_{j^2 \in N_2} u_{j^2}(s))\right] \right) ds + +\sigma[s, x(s)]dz(s),$$

$$x(t_0) = x_0 \in X, \tag{4.1}$$

where $u_j \in U_j$ is the (nonnegative) amount of resource extracted by firm i, for $i \in N$, and $x(s)$ is the resource stock, $\sigma[s, x(s)]$ is a $m \times \Theta$ matrix and $z(s)$ is a Θ-dimensional Wiener process and the initial state x_0 is given. Let $\Omega[s, x(s)] = \sigma[s, x(s)]\sigma[s, x(s)]'$ denote the covariance matrix with its element in row h and column ζ denoted by $\Omega^{h\varsigma}[s, x(s)]$.

The extraction cost depends on the quantity of resource extracted $u^i(s)$ and the resource stock size $x(s)$. In particular,

The extraction cost for the n_1 firms with cost advantages is:

$$c^{j^1}\left[u_{j^1}(s), x(s)\right], \text{ for } j^1 \in N_1; \text{ and}$$

the extraction cost for the n_1 firms with cost advantages can is:

$$c^{j^2}\left[u_{j^2}(s), x(s)\right], \text{ for } j^2 \in N_2.$$

This formulation of cost follows from two assumptions: (i) the cost of extraction is positively related to extraction effort, and (ii) the amount of resource extracted, seen as the output of a production function of two inputs (effort and stock level), is increasing in both inputs (see Clark 1976). In particular, firm $j^1 \in N_1$ has cost advantage so that

$\partial c^{j^1}\left[u_{j^1}(s), x(s)\right] / \partial u_{j^1}(s) < \partial c^{j^2}\left[u_{j^2}(s), x(s)\right] / \partial u_{j^2}(s)$, for all levels of $u_{j^1} \in U_{j^1}$ and $u_{j^2} \in U_{j^2}$ at any $x \in X$.

The market price of the resource depends on the total amount extracted and supplied to the market. The price-output relationship at time s is given by the following downward sloping inverse demand curve $P(s) = g[Q(s)]$, where $Q(s) = \sum_{j^1 \in N_1} u_{j^1}(s) + \sum_{j^2 \in N_2} u_{j^2}(s))$ is the total amount of resource extracted and marketed at time s. At time T, firm $j^1 \in N_1$ will receive a termination bonus $q^{j^1}[x(T)]$ and firm $j^2 \in N_2$ will receive a termination bonus $q^{j^2}[x(T)]$. There exists a discount rate r, and profits received at time t has to be discounted by the factor $\exp[-r(t - t_0)]$.

At time t_0, firm $j^1 \in N_1$ which has cost advantages seeks to maximize its expected profit

$$E_{t_0}\left\{ \int_{t_0}^{T} \left(g\left[\sum_{h \in N_i} u_h(s) + \sum_{\ell \in N_2} u_\ell(s) \right] u_{j^1}(s) - c^{j^1}\left[u_{j^1}(s), x(s)\right] \right) \exp[-r(s - t_0)]\, ds + \exp[-r(T - t_0)] q^{j^1}[x(T)] \right\} \tag{4.2}$$

subject to (4.1).

At time t_0, firm $j^2 \in N_2$ which has cost disadvantages seeks to maximize expected profit

$$E_{t_0}\left\{ \int_{t_0}^{T} \left(g\left[\sum_{h \in N_i} u_h(s) + \sum_{\ell \in N_2} u_\ell(s) \right] u_{j^2}(s) - c^{j^2}\left[u_{j^2}(s), x(s)\right] \right) \exp[-r(s - t_0)]\, ds + \exp[-r(T - t_0)] q^{j^2}[x(T)] \right\} \tag{4.3}$$

subject to (4.1).

We use $\Gamma(x_0, T - t_0)$ to denote the game (4.1, 4.2, and 4.3) and $\Gamma(x_\tau, T - \tau)$ to denote an alternative game with state dynamics (4.1) and payoff structures (4.2 and 4.3), which starts at time $\tau \in [t_0, T]$ with initial state $x_\tau \in X$. A non-cooperative Nash equilibrium solution of the game $\Gamma(x_\tau, T - \tau)$ can be characterized as:

Corollary 4.1 A set of feedback strategies $\left\{ \phi^{*}_{j^1}(t,x) \text{ for } j^1 \in N_1 \text{ and } \phi^{*}_{j^2}(t,x) \text{ for } j^2 \in N_2 \right\}$ provides a Nash equilibrium solution to the game $\Gamma(x_\tau, T-\tau)$, if there exist continuously twice differentiable functions $V^{(\tau)j^1}(t,x)[\tau,T] \times R \to R$ for $j^1 \in N_1$ and $V^{(\tau)j^2}(t,x) : [\tau,T] \times R \to R$ for $j^2 \in N_2$, satisfying the following partial differential equations:

$$
\begin{aligned}
&-V^{(\tau)j^1}_t(t,x) - \frac{1}{2}\sum_{h,\zeta=1}^{m} \Omega^{h\zeta}(t,x) V^{(\tau)j^1}_{x^h x^\zeta}(t,x) = \\
&\max_{u_{j^1}} \left\{ \left(g\left[\sum_{\substack{h\in N_i \\ h\neq j^1}} \phi^{*}_{h}(t,x) + u_{j^1} + \sum_{\ell\in N_2} \phi^{*}_{\ell}(t,x) \right] u_{j^1}, - c^{j^1}\left(u_{j^1},x\right) \right) \exp[-r(t-\tau)] \right. \\
&\left. + V^{(\tau)j^1}_x(t,x) f\left[t,x, \sum_{\substack{h\in N_i \\ h\neq j^1}} \phi^{*}_{h}(t,x) + u_{j^1} + \sum_{\ell\in N_2} \phi^{*}_{\ell}(t,x) \right] \right\}, \text{ and} \\
&V^{(\tau)j^1}(T,x) = \exp[-r(T-t_0)] q^{j^1}(x), \text{ for } j^1 \in N_1; \\
&-V^{(\tau)j^2}_t(t,x) - \frac{1}{2}\sum_{h,\zeta=1}^{m} \Omega^{h\zeta}(t,x) V^{(\tau)j^2}_{x^h x^\zeta}(t,x) = \\
&\max_{u_{j^2}} \left\{ \left(g\left[\sum_{h\in N_i} \phi^{*}_{h}(t,x) + \sum_{\substack{\ell\in N_2 \\ \ell\neq j^2}} \phi^{*}_{\ell}(t,x) + u_{j^2} \right] u_{j^2} - c^{j^2}\left(u_{j^2},x\right) \right) \exp[-r(t-\tau)] \right. \\
&\left. + V^{(\tau)j^2}_x(t,x) f\left[t,x, \sum_{h\in N_i} \phi^{*}_{h}(t,x) + \sum_{\substack{\ell\in N_2 \\ \ell\neq j^2}} \phi^{*}_{\ell}(t,x) + u_{j^2} \right] \right\}, \text{ and} \\
&V^{(\tau)j^2}(T,x) = \exp[-r(T-t_0)] q^{j^2}(x), \text{ for } j^2 \in N_2.
\end{aligned}
\tag{4.4}
$$

First order conditions satisfying the indicated maximization in (4.4) yields:

$$
\left\{ g\left(\sum_{\substack{h\in N_i \\ h\neq j^1}} \phi^{*}_{h}(t,x) + u_{j^1} + \sum_{\ell\in N_2} \phi^{*}_{\ell}(t,x) \right) + g'\left(\sum_{\substack{h\in N_i \\ h\neq j^1}} \phi^{*}_{h}(t,x) + u_{j^1} + \sum_{\ell\in N_2} \phi^{*}_{\ell}(t,x) \right) u_{j^1} \right.
$$

$$-\frac{\partial}{\partial u_{j^1}} c^{j^1}\left(u_{j^1}, x\right) \Bigg\} \exp[-r(t-\tau)]$$

$$+V_x^{(\tau)j^1}(t,x) \frac{\partial}{\partial u_{j^1}} f\left[t, x, \sum_{\substack{h \in N_i \\ h \neq j^1}} \phi_h^*(t,x) + u_{j^1} + \sum_{\ell \in N_2} \phi_\ell^*(t,x)\right] = 0,$$

for $j^1 \in N_1$;

$$\Bigg\{ g\left(\sum_{h \in N_i} \phi_h^*(t,x) + \sum_{\substack{\ell \in N_2 \\ \ell \neq j^2}} \phi_\ell^*(t,x) + u_{j^2}\right) + g'\left(\sum_{h \in N_i} \phi_h^*(t,x) + \sum_{\substack{\ell \in N_2 \\ \ell \neq j^2}} \phi_\ell^*(t,x) + u_{j^2}\right) u_{j^2}$$

$$-\frac{\partial}{\partial u_{j^2}} c^{j^2}\left(u_{j^1}, x\right) \Bigg\} \exp[-r(t-\tau)]$$

$$+V_x^{(\tau)j^2}(t,x) \frac{\partial}{\partial u_{j^2}} f\left[t, x, \sum_{h \in N_i} \phi_h^*(t,x) + \sum_{\substack{\ell \in N_2 \\ \ell \neq j^2}} \phi_\ell^*(t,x) + u_{j^2}\right] = 0, \tag{4.5}$$

for $j^2 \in N_2$.

The expected profits of firm $j^1 \in N_1$ which has cost advantages can be expressed as:

$$V^{(\tau)j^1}(t, x_\tau) = E_\tau \Bigg\{ \int_\tau^T \Bigg(g\left[\sum_{h \in N_i} \phi_h^*[s, x(s)] + \sum_{\ell \in N_2} \phi_\ell^*[s, x(s)]\right] \phi_{j^1}^*[s, x(s)]$$

$$- c^{j^1}\left[\phi_{j^1}^*(s, x(s)), x(s)\right] \Bigg) \exp[-r(s-\tau)]\, ds$$

$$+ \exp[-r(T-\tau)] q^{j^1}[x(T)] \Bigg\},$$

for $j^1 \in N_1$; and

the expected profits of firm $j^2 \in N_2$ which has cost disadvantages can be expressed as:

$$V^{(\tau)j^2}(t,x_\tau) = E_\tau\Bigg\{ \int_\tau^T \Bigg(g\Bigg[\sum_{h\in N_i} \phi_h^*[s,x(s)] + \sum_{\ell\in N_2} \phi_\ell^*[s,x(s)]\Bigg] \phi_{j^2}^*[s,x(s)]$$
$$-c^{j^2}\Big[\phi_{j^2}^*(s,x(s)),x(s)\Big] \Bigg) \exp[-r(s-\tau)]\,ds$$
$$+\exp[-r(T-\tau)]q^{j^2}[x(T)] \Bigg\},$$

for $j^2\in N_2$;
where

$$dx(s) = f\Bigg[s,x(s),\sum_{j^1\in N_1}\phi_{j^1}^*(s,x(s)) + \sum_{j^2\in N_2}\phi_{j^2}^*(s,x(s))\Bigg]ds + \sigma[s,x(s)]dz(s)$$
$$x(\tau) = x_\tau\in X.$$

15.4.2 Optimal Cartel Output

Assume that the firms in the oligopoly agree to form a cartel to restraint output and enhance their expected profits. To achieve a group optimum, these firms are required to solve the following expected joint profit maximization problem:

$$\max_{u_1,u_2,\cdots,u_n} E_{t_0}\Bigg\{ \int_{t_0}^T \Bigg(g\Bigg[\sum_{h\in N_i} u_h(s) + \sum_{\ell\in N_2} u_\ell(s)\Bigg] \Bigg[\sum_{h\in N_i} u_h(s) + \sum_{\ell\in N_2} u_\ell(s)\Bigg]$$
$$-\Bigg[\sum_{h\in N_i} c^h[u_h(s),x(s)] + \sum_{\ell\in N_2} c^\ell[u_\ell(s),x(s)]\Bigg]\Bigg)\exp[-r(s-t_0)]\,ds$$
$$+\exp[-r(T-t_0)]\Bigg[\sum_{h\in N_i} q^h[x(T)] + \sum_{\ell\in N_2} q^\ell[x(T)]\Bigg]\Bigg\} \tag{4.6}$$

subject to (4.1).

An optimal solution of the stochastic control problem (4.1) and (4.6) can be characterized using Theorem A.3 in the Technical Appendices as:

Corollary 4.2 A set of control strategies $\left\{ \psi_{j^1}^*(t,x) \text{ for } j^1\in N_1 \text{ and } \psi_{j^2}^*(t,x) \text{ for } j^2 \in N_2\right\}$ provides a solution to the control problem (4.1) and (4.6), if there exist continuously twice differentiable functions $W^{(t_0)}(t,x) : [\tau,T]\times R^m \to R$, satisfying the following partial differential equation:

$$-W_t^{(t_0)}(t,x) - \frac{1}{2}\sum_{h,\zeta=1}^{m} \Omega^{h\zeta}(t,x) W_{x^h x^\zeta}^{(t_0)}(t,x) =$$

$$\max_{u_1,u_2,\cdots,u_n} \left\{ \left(g\left[\sum_{h\in N_i} u_h + \sum_{\ell\in N_2} u_\ell\right]\left[\sum_{h\in N_i} u_h + \sum_{\ell\in N_2} u_\ell\right] \right.\right.$$

$$\left. - \left[\sum_{h\in N_i} c^h[u_h,x] + \sum_{\ell\in N_2} c^\ell[u_\ell,x]\right]\right) \exp[-r(t-t_0)]$$

$$\left. + W_x^{(t_0)}(t,x) f\left[t,x,\sum_{h\in N_1} u_h + \sum_{\ell\in N_2} u_\ell\right]\right\}, \text{ and}$$

$$W^{(t_0)}(T,x) = \exp[-r(T-t_0)]\left[\sum_{h\in N_i} q^h x + \sum_{\ell\in N_2} q^\ell x\right]. \tag{4.7}$$

Conditions satisfying the indicated maximization in (4.7) include:

$$\left\{ g\left[\sum_{h\in N_i} u_h + \sum_{\ell\in N_2} u_\ell\right] + g'\left[\sum_{h\in N_i} u_h + \sum_{\ell\in N_2} u_\ell\right]\left[\sum_{h\in N_i} u_h + \sum_{\ell\in N_2} u_\ell\right]\right.$$

$$\left. - \frac{\partial}{\partial u_{j^1}} c^{j^1}\left(u_{j^1},x\right) \right\} \exp[-r(t-t_0)]$$

$$+ W_x^{(t_0)}(t,x) \frac{\partial}{\partial u_{j^1}} f\left[t,x,\sum_{h\in N_1} u_h + \sum_{\ell\in N_2} u_\ell\right] \leq 0, u_{j^1} \geq 0,$$

and if $u_{j^1} > 0$, the equality sign must hold, for $j^1 \in N_1$;

$$\left\{ g\left[\sum_{h\in N_i} u_h + \sum_{\ell\in N_2} u_\ell\right] + g'\left[\sum_{h\in N_i} u_h + \sum_{\ell\in N_2} u_\ell\right]\left[\sum_{h\in N_i} u_h + \sum_{\ell\in N_2} u_\ell\right]\right.$$

$$\left. - \frac{\partial}{\partial u_{j^2}} c^{j^2}\left(u_{j^1},x\right) \right\} \exp[-r(t-t_0)]$$

$$+ W_x^{(t_0)}(t,x) \frac{\partial}{\partial u_{j^2}} f\left[t,x,\sum_{h\in N_1} u_h + \sum_{\ell\in N_2} u_\ell\right] \leq 0, u_{j^2} \geq 0, \tag{4.8}$$

and if $u_{j^2} > 0$, the equality sign must hold, for $j^2 \in N_2$.

Since $\frac{\partial}{\partial u_{j^1}} c^{j^1}\left(u_{j^1},x\right) < \frac{\partial}{\partial u_{j^2}} c^{j^2}\left(u_{j^1},x\right)$, all the firm which has cost disadvantages would refrain from extraction. The optimal extraction strategies under cooperation become:

$$u_{j^1}^*(t) = \psi_{j^1}^*(t,x) \text{ for } j^1 \in N_1 \text{ and } u_{j^2}^*(t) = 0 \text{ for } j^2 \in N_2. \tag{4.9}$$

The optimal cooperative state dynamics follows:

$$dx(s) = f\left[s, x(s), \sum_{j^{i}\in N_1} \psi^{*}_{j^{1}}(s, x(s))\right] ds + \sigma[s, x(s)]dz(s),\ x(t_0) = x_0. \tag{4.10}$$

The solution to (4.10) yields a group optimal trajectory, which can be expressed as:

$$x^{*}(t) = x_0 + \int_{t_0}^{t} f\left[s, x^{*}(s), \sum_{j^{i}\in N_1} \psi^{*}_{j^{1}}\left(s, x^{*}(s)\right)\right] ds + \int_{t_0}^{t} \sigma[s, x(s)]dz(s). \tag{4.11}$$

We use X_t^{*} to denote the set of realizable values of $x^{*}(t)$ at time t generated by (4.11). The term $x_t^{*} \in X_t^{*}$ is used to denote an element in X_t^{*}.

Substituting the optimal extraction strategies in (4.9) into (4.6) yields the expected cartel profit as:

$$\begin{aligned} W^{(t_0)}(t_0, x_0) = E_{t_0}\Bigg\{ \int_{t_0}^{T} \Bigg(& g\left[\sum_{h\in N_i} \psi^{*}_{h}\left[s, x^{*}(s)\right]\right] \left[\sum_{h\in N_i} \psi^{*}_{h}\left[s, x^{*}(s)\right]\right] \\ & - \left[\sum_{h\in N_i} c^{h}\left[\psi^{*}_{h}\left(s, x^{*}(s)\right), x^{*}(s)\right]\right] \Bigg) \exp[-r(s-t_0)] \\ & + \exp[-r(T-t_0)] \left[\sum_{h\in N_i} q^{h}\left[x^{*}(T)\right] + \sum_{\ell\in N_2} q^{\ell}\left[x^{*}(T)\right]\right] \Bigg\}. \end{aligned} \tag{4.12}$$

Let $W^{(\tau)}(t, x_t^{*})$ denote the expected total venture profit from the control problem with dynamics (4.1) and payoff (4.6) which begins at time $\tau \in [t_0, T]$ with initial state $x_\tau^{*} \in X_\tau^{*}$. One can readily obtain

$$\exp\left[\int_{t_0}^{\tau} r(y)dy\right] W^{(t_0)}\left(t, x_t^{*}\right) = W^{(\tau)}\left(t, x_t^{*}\right),$$

for $\tau \in [t_0, T]$ and $t \in \left[\tau, T\right)$ and $x_t^{*} \in X_t^{*}$.

Next we consider subgame-consistent profit sharing schemes for the cartel along the optimal output path.

15.4.3 Subgame-Consistent Cartel Profit Sharing

In a dormant firm cartel firms having cost disadvantages will refrain from extraction in order to enhance the cartel's expected profit to a group optimum. Compensation must be made to the dormant firms for stopping their production activities. Consider the case when the firms in the cartel agree to share the expected total cooperative payoff proportional to the firms' expected noncooperative payoffs.

The imputation scheme has to fulfill:

Condition 4.1 An imputation

$$\xi^{(t_0)j^1}(t_0,x_0)=\frac{V^{(t_0)j^1}(t_0,x_0)}{\sum_{h\in N_1}V^{(t_0)h}(t_0,x_0)+\sum_{\ell\in N_2}V^{(t_0)\ell}(t_0,x_0)}W^{(t_0)}(t_0,x_0),$$

is assigned to firm j^1, for $j^1\in N_1$ at the outset; and an imputation

$$\xi^{(t_0)j^2}(t_0,x_0)=\frac{V^{(t_0)j^2}(t_0,x_0)}{\sum_{h\in N_1}V^{(t_0)h}(t_0,x_0)+\sum_{\ell\in N_2}V^{(t_0)\ell}(t_0,x_0)}W^{(t_0)}(t_0,x_0), \qquad (4.13)\blacksquare$$

is assigned to firm j^2, for $j^2\in N_2$ at the outset;
and an imputation

$$\xi^{(\tau)j^1}(\tau,x_\tau^*)=\frac{V^{(\tau)j^1}(\tau,x_\tau^*)}{\sum_{h\in N_1}V^{(\tau)h}(\tau,x_\tau^*)+\sum_{\ell\in N_2}V^{(\tau)\ell}(\tau,x_\tau^*)}W^{(\tau)}(\tau,x_\tau^*)$$

is assigned to firm j^1, for $j^1\in N_1$ at time $\tau\in(t_0,T]$; and an imputation

$$\xi^{(\tau)j^2}(\tau,x_\tau^*)=\frac{V^{(\tau)j^2}(\tau,x_\tau^*)}{\sum_{h\in N_1}V^{(\tau)h}(\tau,x_\tau^*)+\sum_{\ell\in N_2}V^{(\tau)\ell}(\tau,x_\tau^*)}W^{(\tau)}(\tau,x_\tau^*)$$

is assigned to firm j^2 , for $j^2\in N_2$ at time $\tau\in(t_0,T]$

Invoking Theorem 3.1 in Chap. 3, a subgame consistent PDP for the cartel can then be expressed as:

$$\begin{aligned}B_{j^1}(s,x_s^*) &= -\frac{\partial}{\partial t}\left[\left.\frac{V^{(s)j^1}(t,x_t^*)}{\sum_{h\in N_1}V^{(s)h}(t,x_t^*)+\sum_{\ell\in N_2}V^{(s)\ell}(t,x_t^*)}W^{(s)}(t,x_t^*)\right|_{t=s}\right]\\
&\quad-\frac{\partial}{\partial x_s^*}\left[\frac{V^{(s)j^1}(s,x_s^*)}{\sum_{h\in N_1}V^{(s)h}(s,x_s^*)+\sum_{\ell\in N_2}V^{(s)\ell}(s,x_s^*)}W^{(s)}(s,x_s^*)\right]\\
&\quad\times f\left[s,x^*(s),\sum_{j^1\in N_1}\psi_{j^1}^*(s,x^*(s))\right]\end{aligned}$$

$$-\frac{1}{2}\sum_{h,\zeta=1}^{m}\Omega^{h\zeta}\left(s,x_s^*\right)\frac{\partial^2}{\partial x_s^{h*}\partial x_s^{\zeta*}}\left[\frac{V^{(s)j^1}\left(s,x_s^*\right)}{\sum_{h\in N_1}V^{(s)h}\left(s,x_s^*\right)+\sum_{\ell\in N_2}V^{(s)\ell}\left(s,x_s^*\right)}W^{(s)}\left(s,x_s^*\right)\right],$$

for $j^1 \in N_1$;

$$B_{j^2}\left(s,x_s^*\right) = -\frac{\partial}{\partial t}\left[\frac{V^{(s)j^2}\left(t,x_t^*\right)}{\sum_{h\in N_1}V^{(s)h}\left(t,x_t^*\right)+\sum_{\ell\in N_2}V^{(s)\ell}\left(t,x_t^*\right)}W^{(s)}\left(t,x_t^*\right)\Bigg|_{t=s}\right]$$

$$-\frac{\partial}{\partial x_s^*}\left[\frac{V^{(s)j^2}\left(s,x_s^*\right)}{\sum_{h\in N_1}V^{(s)h}\left(s,x_s^*\right)+\sum_{\ell\in N_2}V^{(s)\ell}\left(s,x_s^*\right)}W^{(s)}\left(s,x_s^*\right)\right]$$

$$\times f\left[s,x^*(s),\sum_{j^i\in N_1}\psi_{j^1}^*\left(s,x^*(s)\right)\right]$$

$$-\frac{1}{2}\sum_{h,\zeta=1}^{m}\Omega^{h\zeta}\left(s,x_s^*\right)\frac{\partial^2}{\partial x_s^{h*}\partial x_s^{\zeta*}}\left[\frac{V^{(s)j^2}\left(s,x_s^*\right)}{\sum_{h\in N_1}V^{(s)h}\left(s,x_s^*\right)+\sum_{\ell\in N_2}V^{(s)\ell}\left(s,x_s^*\right)}W^{(s)}\left(s,x_s^*\right)\right],$$

for $j^2 \in N_1$. (4.14)

With firms having cost advantages producing an output $\psi_{j^1}^*(t,x)$ for $j^1 \subset N_1$ and firms having cost disadvantages refraining from production, the instantaneous receipt of firm i at time instant τ when $x_\tau^*(\tau) = x_\tau^* \in X_\tau^*$ is:

$$\zeta_{j^1}\left(\tau,x_\tau^*\right) = g\left[\sum_{h\in N_i}\psi_h^*\left(\tau,x_\tau^*\right)\right]\psi_{j^1}^*\left(\tau,x_\tau^*\right) - c^{j^1}\left[\psi_{j^1}^*\left(\tau,x_\tau^*\right),x_\tau^*(s)\right],$$

for $\tau \in [t_0,T]$ and $j^1 \in N_1$, and

$$\zeta_{j^2}\left(\tau,x_\tau^*\right) = 0 \text{ for } \tau \in [t_0,T] \text{ and } .$$

According to (4.14), the instantaneous payment that firm i should receive under the agreed-upon optimality principle is $B_{j^1}\left(\tau,x_\tau^*\right)$ for $j^1 \in N_1$ and $B_{j^2}\left(\tau,x_\tau^*\right)$ for $j^2 \in N_2$. Hence an instantaneous transfer payment

$$\chi^{j^1}\left(\tau,x_\tau^*\right) = \zeta_{j^1}\left(\tau,x_\tau^*\right) - B_{j^1}\left(\tau,x_\tau^*\right), \text{ for firm } j^1 \in N_1 \text{ and } \tau \in [t_0,T];$$
$$\chi^{j^2}\left(\tau,x_\tau^*\right) = B_{j^1}\left(\tau,x_\tau^*\right), \text{ for firm } j^2 \in N_2 \text{ and } \tau \in [t_0,T]; \quad (4.15)$$

would have to be arranged.

15.5 An Illustration

Consider a dormant firm duopoly in which two firms are allowed to extract a renewable resource within the duration $[t_0, T]$. The dynamics of the resource is characterized by the differential equations:

$$dx(s) = \left[ax(s)^{1/2} - bx(s) - u_1(s) - u_2(s)\right] ds + \sigma x(s) dz(s),$$
$$x(t_0) = x_0 \in X, \tag{5.1}$$

where $u_i \in U_i$ is the (nonnegative) amount of resource extracted by firm i, for $i \in \{1, 2\}$, a and b are positive constants.

The extraction cost for firm $i \in \{1, 2\}$ depends on the quantity of resource extracted $u_i(s)$, the resource stock size $x(s)$, and a parameter c_i. In particular, firm i's extraction cost can be specified as $c_i u^i(s) x(s)^{-1/2}$. In particular, firm 1 has absolute and marginal cost advantages with $c_1 < c_2$.

The market price of the resource depends on the total amount extracted and supplied to the market. The price-output relationship at time s is given by the following downward sloping inverse demand curve $P(s) = Q(s)^{-1/2}$, where $Q(s) = u_1(s) + u_2(s)$ is the total amount of resource extracted and marketed at time s. At time T, firm i will receive a termination bonus with satisfaction $q_i x(T)^{1/2}$, where q_i is nonnegative. There exists a discount rate r, and profits received at time t has to be discounted by the factor $\exp[-r(t - t_0)]$.

At time t_0 the expected profit of firm $i \in \{1, 2\}$ is:

$$E_{t_0}\left\{ \int_{t_0}^{T} \left[\frac{u_i(s)}{[u_1(s) + u_2(s)]^{1/2}} - \frac{c_i}{x(s)^{1/2}} u_i(s) \right] \exp[-r(s - t_0)]\, ds \right.$$
$$\left. + \exp[-r(T - t_0)] q_i x(T)^{\frac{1}{2}} \right\}. \tag{5.2}$$

A set of strategies $\{ u_i^*(t) = \varphi_i^*(t, x), \text{ for } i \in \{1, 2\}\}$ provides a Nash equilibrium solution to the stochastic differential game (5.1 and 5.2), if there exist continuously twice differentiable functions $V^{(t_0)i}(t, x) : [t_0, T] \times R \to R$, $i \in \{1, 2\}$, satisfying the following partial differential equations:

$$-V_t^{(t_0)i}(t, x) - \frac{1}{2}\sigma^2 x^2 V_{xx}^{(t_0)i}(t, x)$$
$$= \max_{u_i} \left\{ \left[\frac{u_i}{\left(u_i + \phi_j^*(t, x)\right)^{1/2}} - \frac{c_i}{x^{1/2}} u_i \right] \exp[-r(t - t_0)] \right.$$
$$\left. + V_x^{(t_0)i}(t, x) \left[ax^{1/2} - bx - u_i - \phi_j^*(t, x)\right] \right\}, \text{ and}$$
$$V^{(t_0)i}(T, x) = q_i x^{1/2} \exp[-r(T - t_0)], \text{ for } i \in \{1, 2\}, j \in \{1, 2\} \text{ and } j \neq i. \tag{5.3}$$

Performing the indicated maximization and solving (5.3) yields:

$$\phi_1^*(t,x) = \frac{x}{4\left[c_1 + V_x^{(t_0)1}\exp[r(t-t_0)]\,x^{1/2}\right]^2} \text{ and}$$
$$\phi_2^*(t,x) = \frac{x}{4\left[c_2 + V_x^{(t_0)2}\exp[r(t-t_0)]\,x^{1/2}\right]^2}. \tag{5.4}$$

The value function reflecting the expected game equilibrium payoffs of the firms in the game (5.1 and 5.2) can be obtained as:

Proposition 5.1 The game equilibrium value function of firm i is

$$V^{(t_0)i}(t,x) = \exp[-r(t-t_0)]\,\left[A_i(t)x^{1/2} + C_i(t)\right]$$
$$\text{for } i\in\{1,2\} \text{ and } t\in[t_0,T], \tag{5.5}$$

where $A_i(t)$, $C_i(t)$, $A_j(t)$ and $C_j(t)$, for $i\in\{1,2\}$ and $j\in\{1,2\}$ and $i\neq j$, satisfy:

$$\begin{aligned}\dot{A}_i(t) &= \left[r + \frac{b}{2} + \frac{\sigma^2}{8}\right]A_i(t) - \left(\frac{3}{2}\right)\frac{\left[2c_j - c_i + A_j(t) - A_i(t)/2\right]}{\left[c_1 + c_2 + A_1(t)/2 + A_2(t)/2\right]^2}\\ &\quad + \left(\frac{3}{2}\right)^2 \frac{c_i\left[2c_j - c_i + A_j(t) - A_i(t)/2\right]}{\left[c_1 + c_2 + A_1(t)/2 + A_2(t)/2\right]^3}\\ &\quad + \left(\frac{9}{8}\right)\frac{A_i(t)}{\left[c_1 + c_2 + A_1(t)/2 + A_2(t)/2\right]^2},\\ A_i(T) &= q_i;\\ \dot{C}_i(t) &= rC_i(t) - \frac{a}{2}A_i(t), \text{ and } C_i(T) = 0.\end{aligned}$$

Proof First substitute the results in (5.4), and $V^{(t_0)1}(t,x)$, $V_x^{(t_0)1}(t,x)$, $V^{(t_0)2}(t,x)$ and $V_x^{(t_0)2}(t,x)$ obtained via (5.5) into the set of partial differential equations (5.3). One can readily show that for this set of equations to be satisfied, Proposition 5.1 has to hold. ■

One can readily verify that

$$V^{(\tau)i}(t,x) = \exp[-r(t-\tau)]\,\left[A_i(t)x^{1/2} + C_i(t)\right], \text{ for } i\in\{1,2\} \text{ and } t\in[\tau,T].$$

Assume that the firms agree to form a cartel and seek to solve the following expected joint profit maximization problem to achieve a group optimum:

$$\max_{u_1,u_2} E_{t_0}\left\{\int_{t_0}^{T}\left[\left[u_1(s)+u_2(s)\right]^{1/2} - \frac{c_1u_1(s)+c_2u_2(s)}{x(s)^{1/2}}\right]\exp[-r(s-t_0)]\,ds \right.$$
$$\left. +\exp[-r(T-t_0)][q_1+q_2]x(T)^{1/2}\right\}, \tag{5.6}$$

subject to dynamics (5.1).

A set of strategies $\left[\psi_1^*(s,x), \psi_2^*(s,x)\right]$, for $s \in [t_0, T]$ provides an optimal solution to the stochastic control problem (5.1) and (5.6), if there exist a continuously twice differentiable function $W^{(t_0)}(t,x) : [t_0, T] \times R \to R$ satisfying the following partial differential equations:

$$\begin{aligned}
&-W_t^{(t_0)}(t,x) - \frac{1}{2}\sigma^2 x^2 W_{xx}^{(t_0)}(t,x) \\
&= \max_{u_1, u_2} \left\{ \left[(u_1 + u_2)^{1/2} - (c_1 u_1 + c_2 u_2)x^{-1/2}\right] \exp[-r(t - t_0)] \right. \\
&\left. + W_x^{(t_0)}(t,x)\left[ax^{1/2} - bx - u_1 - u_2\right] \right\}, \text{ and} \\
&W^{(t_0)}(T,x) = (q_1 + q_2)x^{1/2} \exp[-r(T - t_0)] .
\end{aligned} \tag{5.7}$$

Performing the indicated maximization operation in (5.7) yields:

$$\psi_1^*(t,x) = \frac{x}{4\left[c_1 + W_x \exp[r(t - t_0)]\, x^{1/2}\right]^2} \text{ and } \psi_2^*(t,x) = 0. \tag{5.8}$$

Firm 2 has to refrain from extraction. The more efficient firm (firm 1) would buy the less efficient firm (firm 2) out from the resource extraction process. Firm 2 becomes a dormant firm under cooperation. The value function indicating the maximized expected joint payoff of firms in the control problem (5.1) and (5.6) can be obtained as:

Proposition 5.2

$$W^{(t_0)}(t,x) = \exp[-r(t - t_0)] \left[A(t)x^{1/2} + C(t)\right], \tag{5.9}$$

where $A(t)$ and $B(t)$ satisfy:

$$\begin{aligned}
&\dot{A}(t) = \left[r + \frac{b}{2} + \frac{\sigma^2}{8}\right] A(t) - \frac{1}{4[c_1 + A(t)/2]}, \\
&A(T) = q_1 + q_2 \\
&\dot{C}(t) = rC(t) - \frac{a}{2}A(t), \text{ and } B(T) = 0.
\end{aligned}$$

Proof First substitute the results in (5.8), and $W^{(t_0)}(t,x)$, and $W_x^{(t_0)}(t,x)$ obtained via (5.9) into the set of partial differential equations (5.7). One can readily show that for this set of equations to be satisfied, Proposition 5.2 has to hold. ■

Again, one can readily verify that $W^{(\tau)}(t,x) = \exp[-r(t - \tau)]\left[A(t)x^{1/2} + B(t)\right]$

Upon substituting $\psi_1^*(t,x)$ and $\psi_2^*(t,x)$ into (5.1) yields the optimal cooperative state dynamics as:

$$\begin{aligned}
&dx(s) = \left[ax(s)^{1/2} - bx(s) - \frac{x(s)}{4[c_1 + A(s)/2]^2} \right] ds + \sigma\, x(s) dz(s), \\
&x(t_0) = x_0 \in X.
\end{aligned} \tag{5.10}$$

The solution to (5.10) yields a Pareto optimal trajectory, which can be expressed as:

$$x^*(t) = \left\{ \Phi(t, t_0) \left[x_0^{1/2} + \int_{t_0}^{t} \Phi^{-1}(s, t_0) \frac{a}{2} ds \right] \right\}^2 \tag{5.11}$$

where

$$\Phi(t, t_0) = \exp\left[\int_{t_0}^{t} \left(\frac{-b}{2} - \frac{1}{8[c_1 + A(s)/2]^2} - \frac{3\sigma^2}{8} \right) ds + \int_{t_0}^{t} \frac{\sigma}{2} dz(s) \right].$$

We denote the set containing realizable values of $x^*(t)$ by X_t, for $t \in (t_0, T]$.

Consider the case when the firms in the cartel agree to share the total expected cooperative payoff proportional to the firms' expected noncooperative payoffs. The imputation scheme has to fulfill:

$$\xi^{(\tau)i}(\tau, x_\tau^*) = \frac{V^{(\tau)i}(\tau, x_\tau^*)}{\sum_{j=1}^{2} V^{(\tau)j}(\tau, x_\tau^*)} W^{(\tau)}(\tau, x_\tau^*), \text{ for } i \in \{1, 2\} \text{ and } \tau \in [t_0, T].$$

Invoking the results in (4.14), a subgame consistent PDP for the cartel can then be obtained as

$$B_i(s, x_s^*) = -\frac{\partial}{\partial t}\left[\frac{V^{(s)i}(t, x_t^*)}{\sum_{j=1}^{2} V^{(s)j}(t, x_t^*)} W^{(s)}(t, x_t^*) \bigg|_{t=s} \right]$$

$$-\frac{\partial}{\partial x_s^*}\left[\frac{V^{(s)i}(s, x_s^*)}{\sum_{j=1}^{2} V^{(s)j}(s, x_s^*)} W^{(s)}(s, x_s^*) \right] \left[a(x_s^*)^{1/2} - bx_s^* - \frac{x_s^*}{4[c_1 + A(s)/2]^2} \right]$$

$$-\frac{\sigma^2 (x_s^*)^2}{2} \frac{\partial^2}{\partial (x_s^*)^2}\left[\frac{V^{(s)i}(s, x_s^*)}{\sum_{j=1}^{2} V^{(s)j}(s, x_s^*)} W^{(s)}(s, x_s^*) \right], \text{for } i \in \{1, 2\}, \tag{5.12}$$

Under cooperation, firm 1 would derive an expected payoff:

$$W^{(t_0)1}(t_0, x_0) = E_{t_0}\Bigg\{ \int_{t_0}^{T} \left[[\psi_1^*(s, x^*(s))]^{1/2} - \frac{c_1}{x^*(s)^{1/2}} \psi_1^*(s, x^*(s)) \right] \exp[-r(s - t_0)]\, ds + \exp[-r(T - t_0)] q_1 x^*(T)^{\frac{1}{2}} \Bigg\},$$

where $\psi_1^*\left(s, x^*(s)\right) = \dfrac{x^*(s)}{4[c_1 + A(s)/2]^2}$, and firm 2 would derive an expected payoff:

$$W^{(t_0)2}(t_0, x_0) = 0 \text{ for being dormant.} \tag{5.13}$$

The instantaneous receipt of firm 1 at time instant τ is:

$$\zeta_1\left(\tau, x_\tau^*\right) = \frac{\left(x_\tau^*\right)^{1/2}}{2[c_1 + A(\tau)/2]} - \frac{c_1\left(x_\tau^*\right)^{1/2}}{4[c_1 + A(\tau)/2]^2}$$

for $\tau \in [t_0, T]$ when $x^*(\tau) = x_\tau^* \in X_\tau^*$.

The instantaneous receipt of firm 2 at time instant τ is

$$\zeta_2\left(\tau, x_\tau^*\right) = 0, \quad \text{for } \tau \in [t_0, T].$$

According to (5.12), the instantaneous payment that firm i should receive under the agreed-upon optimality principle is $B_i(\tau, x_\tau^*)$. Hence an instantaneous transfer payment

$$\begin{aligned} &\chi^1\left(\tau, x_\tau^*\right) = \zeta_1\left(\tau, x_\tau^*\right) - B_1\left(\tau, x_\tau^*\right), \text{for firm 1, and} \\ &\chi^2\left(\tau, x_\tau^*\right) = B_2\left(\tau, x_\tau^*\right), \text{ for firm 2 at time } \tau \in [t_0, T] \text{ when } x^*(\tau) = x_\tau^* \in X_\tau^*; \end{aligned} \tag{5.14}$$

would be arranged.

15.6 Infinite Horizon Cartel

In this Section we consider the dormant firm cartel in Sect. 15.5 with an infinite horizon. An oligopoly in which n firms are given perpetual rights to extract a renewable resource.

The dynamics of the resource is characterized by the differential equations:

$$\begin{aligned} &dx(s) = f\left[x(s), \sum_{j^1 \in N_1} u_{j^1}(s) + \sum_{j^2 \in N_2} u_{j^2}(s)\right) \Bigg] ds + \sigma[x(s)]dz(s), \\ &x(t_0) = x_0 \in X, \end{aligned} \tag{6.1}$$

where $u_j \in U_j$ is the (nonnegative) amount of resource extracted by firm i, for $i \in N$, and $x(s)$ is the resource stock, $\sigma[x(s)]$ is a $m \times \Theta$ matrix and $z(s)$ is a $\Theta_i$dimensional Wiener process and the initial state x_0 is given. Let $\Omega[x(s)] = \sigma[x(s)] \; \sigma[x(s)]'$ denote the covariance matrix with its element in row h and column ζ denoted by $\Omega_i^{h\zeta}[x^i(s)]$.

At time t_0, firm $j^1 \in N_1$ which has cost advantages seeks to maximize its expected profit

$$E_{t_0}\left\{ \int_{t_0}^{\infty} \left(g\left[\sum_{h \in N_i} u_h(s) + \sum_{\ell \in N_2} u_\ell(s) \right] u_{j^1}(s) - c^{j^1}\left[u_{j^1}(s), x(s) \right] \right) \exp[-r(s-t_0)]\, ds \right\}, \tag{6.2}$$

subject to (6.1).

At time t_0, firm $j^2 \in N_2$ which has cost disadvantages seeks to maximize its expected profit

$$E_{t_0}\left\{ \int_{t_0}^{\infty} \left(g\left[\sum_{h \in N_i} u_h(s) + \sum_{\ell \in N_2} u_\ell(s) \right] u_{j^2}(s) - c^{j^2}\left[u_{j^2}(s), x(s) \right] \right) \exp[-r(s-t_0)]\, ds \right\}, \tag{6.3}$$

subject to (6.1).

Consider the alternative formulation of (6.1, 6.2, and 6.3) as:

$$\max_{u_{j^1}} E_t\left\{ \int_{t}^{\infty} \left(g\left[\sum_{h \in N_i} u_h(s) + \sum_{\ell \in N_2} u_\ell(s) \right] u_{j^1}(s) - c^{j^1}\left[u_{j^1}(s), x(s) \right] \right) \exp[-r(s-t)]\, ds \right\}, \tag{6.4}$$

for $j^1 \in N_1$, and

$$\max_{u_{j^2}} E_t\left\{ \int_{t}^{\infty} \left(g\left[\sum_{h \in N_i} u_h(s) + \sum_{\ell \in N_2} u_\ell(s) \right] u_{j^2}(s) - c^{j^2}\left[u_{j^2}(s), x(s) \right] \right) \exp[-r(s-t)]\, ds \right\}, \tag{6.5}$$

subject to the state dynamics

$$dx(s) = f\left[x(s), \sum_{j^i \in N_1} u_{j^1}(s) + \sum_{j^2 \in N_2} u_{j^2}(s)) \right] ds + \sigma[x(s)]dz(s), \;\; x(t) = x. \tag{6.6}$$

The infinite-horizon autonomous game (6.4, 6.5, and 6.6) is independent of the choice of t and dependent only upon the state at the starting time, that is x.

Invoking Theorem 5.1 in Chap. 3 a noncooperative feedback Nash equilibrium solution can be characterized by a set of strategies $\left\{ \phi^*_{j^1}(x) \text{ for } j^1 \in N_1 \text{ and } \phi^*_{j^2}(x) \text{ for } j^2 \in N_2 \right\}$ constitutes a Nash equilibrium solution to the game (6.4, 6.5, and 6.6), if there exist functionals $\hat{V}^{j^1}(x) : R^m \to R$ for $j^1 \in N_1$ and $\hat{V}^{j^2}(x) : R^m \to R$ for $j^2 \in N_2$, satisfying the following set of partial differential equations:

$$
\begin{aligned}
& r\hat{V}^{j^1}(x) - \frac{1}{2}\sum_{h,\zeta=1}^{m} \Omega^{h\zeta}(x)\hat{V}^{j^1}_{x^h x^\zeta}(x) = \\
& \max_{u_{j^1}} \left\{ \left(g\left[\sum_{\substack{h\in N_i \\ h\neq j^1}} \phi^*_h(x) + u_{j^1} + \sum_{\ell\in N_2} \phi^*_\ell(x) \right] u_{j^1}, - c^{j^1}\left(u_{j^1}, x\right) \right) \right. \\
& \left. + \hat{V}^{j^1}_x(x) f\left[x, \sum_{\substack{h\in N_i \\ h\neq j^1}} \phi^*_h(x) + u_{j^1} + \sum_{\ell\in N_2} \phi^*_\ell(x) \right] \right\}, \text{ for } j^1 \in N_1 \text{ and} \\
& r\hat{V}^{j^2}(x) - \frac{1}{2}\sum_{h,\zeta=1}^{m} \Omega^{h\zeta}(x)\hat{V}^{j^2}_{x^h x^\zeta}(x) = \\
& \max_{u_{j^2}} \left\{ \left(g\left[\sum_{h\in N_i} \phi^*_h(x) + \sum_{\substack{\ell\in N_2 \\ \ell\neq j^2}} \phi^*_\ell(x) + u_{j^2} \right] u_{j^2} - c^{j^2}\left(u_{j^2}, x\right) \right) \right. \\
& \left. + \hat{V}^{j^2}_x(x) f\left[x, \sum_{h\in N_i} \phi^*_h(x) + \sum_{\substack{\ell\in N_2 \\ \ell\neq j^2}} \phi^*_\ell(x) + u_{j^2} \right] \right\}, \text{ for } j^2 \in N_2.
\end{aligned}
\tag{6.7}
$$

After characterizing the noncooperative market we proceed to consider the Pareto optimal output trajectory if a cartel of these firms is formed.

15.6.1 Pareto Optimal Trajectory

Assume that the firms agree to form a cartel to restraint output and enhance their expected profits. To achieve a group optimum, these firms are required to solve the following expected joint profit maximization problem:

$$\max_{u_1, u_2, \cdots, u_n} E_t \left\{ \int_t^{\infty} \left(g\left[\sum_{h\in N_i} u_h(s) + \sum_{\ell\in N_2} u_\ell(s)\right] \left[\sum_{h\in N_i} u_h(s) + \sum_{\ell\in N_2} u_\ell(s)\right] \right.\right.$$
$$\left.\left. - \left[\sum_{h\in N_i} c^h[u_h(s), x(s)] + \sum_{\ell\in N_2} c^\ell[u_\ell(s), x(s)]\right]\right) \exp[-r(s-t)]\, ds \right\} \tag{6.8}$$

subject to (6.6).

An optimal solution of the stochastic control problem (6.6) and (6.8) can be characterized using Theorem A.4 in the Technical Appendices as:

Corollary 6.1 A set of control strategies $\left\{ \psi^*_{j^1}(x) \text{ for } j^1 \in N_1 \text{ and } \psi^*_{j^2}(x) \text{ for } j^2 \in N_2 \right\}$ provides a solution to the stochastic control problem (6.6) and (6.8), if there exist continuously twice differentiable functions $W(x) : R^m \to R$, satisfying the following partial differential equation:

$$rW(x) - \frac{1}{2}\sum_{h,\zeta=1}^{m} \Omega^{h\zeta}(x) W_{x^h x^\zeta}(x) =$$
$$\max_{u_1, u_2, \cdots, u_n} \left\{ \left(g\left[\sum_{h\in N_i} u_h + \sum_{\ell\in N_2} u_\ell\right] \left[\sum_{h\in N_i} u_h + \sum_{\ell\in N_2} u_\ell\right] \right.\right.$$
$$\left.\left. - \left[\sum_{h\in N_i} c^h[u_h, x] + \sum_{\ell\in N_2} c^\ell[u_\ell, x]\right]\right) + W_x(x) f\left[t, x, \sum_{h\in N_1} u_h + \sum_{\ell\in N_2} u_\ell\right] \right\}. \tag{6.9} \blacksquare$$

Conditions satisfying the indicated maximization in (6.9) include:

$$\left\{ g\left[\sum_{h\in N_i} u_h + \sum_{\ell\in N_2} u_\ell\right] + g'\left[\sum_{h\in N_i} u_h + \sum_{\ell\in N_2} u_\ell\right] \left[\sum_{h\in N_i} u_h + \sum_{\ell\in N_2} u_\ell\right] \right.$$
$$\left. - \frac{\partial}{\partial u_{j^1}} c^{j^1}\left(u_{j^1}, x\right) \right\} + W_x(x) \frac{\partial}{\partial u_{j^1}} f\left[x, \sum_{h\in N_1} u_h + \sum_{\ell\in N_2} u_\ell\right] \le 0,$$
$$u_{j^1} \ge 0,$$

and if $u_{j^1} > 0$, the equality sign must hold, for $j^1 \in N_1$;

$$\left\{ g\left[\sum_{h\in N_i} u_h + \sum_{\ell\in N_2} u_\ell\right] + g'\left[\sum_{h\in N_i} u_h + \sum_{\ell\in N_2} u_\ell\right]\left[\sum_{h\in N_i} u_h + \sum_{\ell\in N_2} u_\ell\right]\right.$$

$$\left. - \frac{\partial}{\partial u_{j^2}} c^{j^2}\left(u_{j^1}, x\right) \right\} + W_x(x)\frac{\partial}{\partial u_{j^2}} f\left[x, \sum_{h\in N_1} u_h + \sum_{\ell\in N_2} u_\ell\right] \le 0, \qquad (6.10)$$

$u_{j^2} \ge 0,$

and if $u_{j^2} > 0$, the equality sign must hold, for $j^2 \in N_2$

Since $\frac{\partial}{\partial u_{j^1}} c^{j^1}\left(u_{j^1}, x\right) < \frac{\partial}{\partial u_{j^2}} c^{j^2}\left(u_{j^1}, x\right)$, all the firm which has cost disadvantages would refrain from extraction. The optimal extraction strategies under cooperation become:

$$u_{j^1}^*(t) = \psi_{j^1}^*(x) \text{ for } j^1 \in N_1 \text{ and } u_{j^2}^*(t) = 0 \text{ for } j^2 \in N_2. \qquad (6.11)$$

The optimal cooperative state dynamics follows:

$$dx(s) = f\left[x(s), \sum_{j^i\in N_1} \psi_{j^1}^*(x(s))\right] ds + +\sigma[x(s)]dz(s),\ x(t_0) = x_0. \qquad (6.12)$$

The solution to (6.12) yields a group optimal trajectory, which can be expressed as:

$$x^*(t) = x_0 + \int_{t_0}^t f\left[x^*(s), \sum_{j^i\in N_1} \psi_{j^1}^*\left(x^*(s)\right)\right] ds + \int_{t_0}^t \sigma[x(s)]dz(s). \qquad (6.13)$$

Substituting the optimal extraction strategies in (6.11) into (6.6) yields the expected cartel profit as:

$$W(x) = E_t\left\{ \int_t^\infty \left(g\left[\sum_{h\in N_i} \psi_h^*[x^*(s)]\right]\left[\sum_{h\in N_i} \psi_h^*[x^*(s)]\right] \right.\right.$$
$$\left.\left. - \left[\sum_{h\in N_i} c^h\left[\psi_h^*(x^*(s)), x^*(s)\right]\right]\right) \exp[-r(s-t)] \right| x(t) = x \Bigg\}. \qquad (6.14)$$

We then examine subgame consistent cartel profit sharing mechanisms in the next subsection.

15.6.2 Subgame Consistent Cartel Profit Sharing

In a dormant firm cartel firms having cost disadvantages will refrain from extraction in order to enhance the cartel's expected profit to a group optimum. Consider the case when the firms in the cartel agree to share the excess of the total expected cooperative payoff proportional to the firms' expected noncooperative payoffs.

The imputation scheme has to fulfill:

Condition 6.1 An imputation

$$\xi^{(\tau)j^1}\left(\tau, x_\tau^*\right) = \frac{\hat{V}^{j^1}\left(x_\tau^*\right)}{\sum\limits_{h \in N_1} \hat{V}^h\left(x_\tau^*\right) + \sum\limits_{\ell \in N_2} \hat{V}^\ell\left(x_\tau^*\right)} W\left(x_\tau^*\right)$$

is assigned to firm j^1, for $j^1 \in N_1$ at time $\tau \in \left[t_0, \infty\right)$ when $x^*(\tau) = x_\tau^* \in X_\tau^*$; and an imputation

$$\xi^{(\tau)j^2}\left(\tau, x_\tau^*\right) = \frac{\hat{V}^{j^2}\left(x_\tau^*\right)}{\sum\limits_{h \in N_1} \hat{V}^h\left(x_\tau^*\right) + \sum\limits_{\ell \in N_2} \hat{V}^\ell\left(x_\tau^*\right)} W\left(x_\tau^*\right)$$

is assigned to firm j^2, for $j^2 \in N_2$ at time $\tau \in \left[t_0, \infty\right)$ when $x^*(\tau) = x_\tau^* \in X_\tau^*$ (6.15)

■

To formulate a set of subgame consistent payoff distribution procedure we invoke Theorem 5.3 in Chap. 3 and obtain:

Corollary 6.2 A PDP with instantaneous payments at time $\tau \in \left[t_0, \infty\right)$ when $x^*(\tau) = x_\tau^* \in X_\tau^*$ equaling

$$\begin{aligned} B_{j^1}\left(x_\tau^*\right) = {} & \frac{r\hat{V}^{j^1}\left(x_\tau^*\right)}{\sum\limits_{h \in N_1} \hat{V}^h\left(x_\tau^*\right) + \sum\limits_{\ell \in N_2} \hat{V}^\ell\left(x_\tau^*\right)} W\left(x_\tau^*\right) \\ & - \frac{\partial}{\partial x_\tau^*}\left[\frac{\hat{V}^{j^1}\left(x_\tau^*\right)}{\sum\limits_{h \in N_1} \hat{V}^h\left(x_\tau^*\right) + \sum\limits_{\ell \in N_2} \hat{V}^\ell\left(x_\tau^*\right)} W\left(x_\tau^*\right)\right] f\left[x_\tau^*, \sum_{j^1 \in N_1} \psi_{j^1}^*\left(x_\tau^*\right)\right] \\ & - \frac{1}{2}\sum_{h,\zeta=1}^{m} \Omega^{h\zeta}\left(x_\tau^*\right) \frac{\partial^2}{\partial x_\tau^{h*} \partial x_\tau^{\zeta*}}\left[\frac{\hat{V}^{j^1}\left(x_\tau^*\right)}{\sum\limits_{h \in N_1} \hat{V}^h\left(x_\tau^*\right) + \sum\limits_{\ell \in N_2} \hat{V}^\ell\left(x_\tau^*\right)} W\left(x_\tau^*\right)\right], \end{aligned}$$

for $j^1 \in N_1$;

$$B_{j^2}\left(x_\tau^*\right) = \frac{r\hat{V}^{j^2}\left(x_\tau^*\right)}{\sum_{h\in N_1}\hat{V}^h\left(x_\tau^*\right) + \sum_{\ell\in N_2}\hat{V}^\ell\left(x_\tau^*\right)} W\left(x_\tau^*\right)$$

$$-\frac{\partial}{\partial x_\tau^*}\left[\frac{\hat{V}^{j^2}\left(x_\tau^*\right)}{\sum_{h\in N_1}\hat{V}^h\left(x_\tau^*\right) + \sum_{\ell\in N_2}\hat{V}^\ell\left(x_\tau^*\right)} W\left(x_\tau^*\right)\right] f\left[x_\tau^*, \sum_{j^1\in N_1}\psi_{j^1}^*\left(x_\tau^*\right)\right]$$

$$-\frac{1}{2}\sum_{h,\zeta=1}^m \Omega^{h\zeta}\left(x_\tau^*\right)\frac{\partial^2}{\partial x_\tau^{h*}\partial x_\tau^{\zeta*}}\left[\frac{\hat{V}^{j^2}\left(x_\tau^*\right)}{\sum_{h\in N_1}\hat{V}^h\left(x_\tau^*\right) + \sum_{\ell\in N_2}\hat{V}^\ell\left(x_\tau^*\right)} W\left(x_\tau^*\right)\right],$$

$$\text{for } j^2 \in N_2; \tag{6.16}$$

yields a subgame consistent payoff distribution procedure to the cooperative game $\Gamma_c(x_0)$ with imputation as specified in Condition 6.1. ■

With firms having cost advantages producing an output $\psi_{j^1}^*(x)$ for $j^1 \in N_1$ and firms having cost disadvantages refraining from production, the instantaneous receipt of firm i at time instant τ when $x^*(\tau) = x_\tau^* \in X_\tau^*$ is:

$$\zeta_{j^1}\left(\tau, x_\tau^*\right) = g\left[\sum_{h\in N_i}\psi_h^*\left(x_\tau^*\right)\right]\psi_{j^1}^*\left(x_\tau^*\right) - c^{j^1}\left[\psi_{j^1}^*\left(x_\tau^*\right), x_\tau^*(s)\right],$$

for $\tau \in \left[t_0, \infty\right)$ and $j^1 \in N_1$, and

$$\zeta_{j^2}\left(\tau, x_\tau^*\right) = 0\,,\ \text{for } \tau \in \left[t_0, \infty\right) \text{ and } j^2 \in N_2.$$

According Corollary 6.2, the instantaneous payment that firm i should receive under the agreed-upon optimality principle is $B_{j^1}\left(\tau, x_\tau^*\right)$ for $j^1 \in N_1$ and $B_{j^2}\left(\tau, x_\tau^*\right)$ for $j^2 \in N_2$ as stated in (6.16). Hence an instantaneous transfer payment

$$\begin{aligned} &\chi^{j^1}\left(\tau, x_\tau^*\right) = \zeta_{j^1}\left(\tau, x_\tau^*\right) - B_{j^1}\left(\tau, x_\tau^*\right), \text{ for firm } j^1 \in N_1 \text{ and } \tau \in [t_0, T]; \\ &\chi^{j^2}\left(\tau, x_\tau^*\right) = B_{j^2}\left(\tau, x_\tau^*\right), \text{ for firm} j^2 \in N_2 \text{ and } \tau \in [t_0, T]; \end{aligned} \tag{6.17}$$

would have to be arranged.

15.6.3 Infinite-Horizon Dormant-Firm Cartel: An Illustration

Consider an infinite horizon version of the game in Sect. 15.3. The dynamics of the resource is characterized by the stochastic differential equations:

$$dx(s) = \left[ax(s)^{1/2} - bx(s) - u_1(s) - u_2(s)\right] ds + \sigma\, x(s)dz(s),$$
$$x(t_0) = x_0 \in X, \tag{6.18}$$

where $u_i \in U_i$ is the (nonnegative) amount of resource extracted by firm i, for $i \in \{1, 2\}$, a and b are positive constants.

At time t_0 the expected profit of firm $i \in \{1, 2\}$ is:

$$E_{t_0}\left\{ \int_{t_0}^{\infty} \left[\frac{u_i(s)}{\left[u_1(s) + u_2(s)\right]^{1/2}} - \frac{c_i}{x(s)^{1/2}} u_i(s) \right] \exp[-r(s - t_0)]\, ds \right\}, \tag{6.19}$$

wehere $c_1 < c_2$.

Consider the alternative game problem which starts at time $t \in [t_0, \infty)$ with initial state $x(t) = x$:

$$\max_{u_i} E_{t_0}\left\{ \int_{t}^{\infty} \left[\frac{u_i(s)}{\left[u_1(s) + u_2(s)\right]^{1/2}} - \frac{c_i}{x(s)^{1/2}} u_i(s) \right] \exp[-r(s - t)]\, ds \right\} \tag{6.20}$$

subject to

$$dx(s) = \left[ax(s)^{1/2} - bx(s) - u_1(s) - u_2(s)\right] ds$$
$$x(t) = x \in X, \tag{6.21}$$

A set of strategies $\{\ \phi_i^*(x)$, for $i \in \{1, 2\}$ provides a Nash equilibrium solution to the game (6.20 and 6.21), if there exist continuously twice differentiable functions $\hat{V}^i(x) : R \to R$, $i \in \{1, 2\}$, satisfying the following partial differential equations:

$$r\hat{V}^i(x) - \frac{1}{2}\sigma^2 x^2 \hat{V}^i_{xx}(x) = \max_{u_i} \left\{ \left[\frac{u_i}{\left(u_i + \phi_j^*(x)\right)^{1/2}} - \frac{c_i}{x^{1/2}} u_i \right] \right.$$
$$\left. + \hat{V}^i_x(t, x)\left[ax^{1/2} - bx - u_i - \phi_j^*(x)\right] \right\}, \tag{6.22}$$

for $i \in \{1, 2\}$, $j \in \{1, 2\}$ and $j \neq i$.

Performing the indicated maximization yields:

$$\phi_1^*(x) = \frac{x}{4\left[c_1 + \hat{V}_x^1 x^{1/2}\right]^2} \text{ and } \phi_2^*(t,x) = \frac{x}{4\left[c_2 + \hat{V}_x^2 x^{1/2}\right]^2}. \tag{6.23}$$

The game equilibrium value functions reflecting the expected payoffs of the firms in the game (6.20 and 6.21) can be obtained as:

Proposition 6.1

$$\hat{V}^i(x) = \left[A_i x^{1/2} + C_i\right], \text{ for } i \in \{1,2\}, \tag{6.24}$$

where A_i, C_i, A_j and C_j, for $i \in \{1,2\}$ and $j \in \{1,2\}$ and $i \neq j$, satisfy:

$$0 = \left[r + \frac{b}{2} + \frac{\sigma^2}{8}\right] A_i - \left(\frac{3}{2}\right) \frac{\left[2c_j - c_i + A_j - A_i/2\right]}{\left[c_1 + c_2 + A_1/2 + A_2/2\right]^2}$$
$$+ \left(\frac{3}{2}\right)^2 \frac{c_i\left[2c_j - c_i + A_j - A_i/2\right]}{\left[c_1 + c_2 + A_1/2 + A_2/2\right]^3} + \left(\frac{9}{8}\right) \frac{A_i}{\left[c_1 + c_2 + A_1/2 + A_2/2\right]^2},$$
$$rC_i = \frac{a}{2} A_i.$$

Proof First substitute the results in (6.23), and $\hat{V}^1(x)$, $\hat{V}_x^1(x)$, $\hat{V}^2(x)$ and $\hat{V}_x^2(x)$ obtained via (6.24) into the set of partial differential equations (6.22). One can readily show that for this set of equations to be satisfied, Proposition 6.1 has to hold. ■

A noncooperative market equilibrium can be explicitly obtained from (6.23) and (6.24).

15.6.3.1 Cartel Output

Assume that the firms agree to form a cartel and seek to solve the following expected joint profit maximization problem to achieve a group optimum:

$$\max_{u_1, u_2} E_{t_0} \left\{ \int_{t_0}^{\infty} \left[[u_1(s) + u_2(s)]^{1/2} - \frac{c_1 u_1(s) + c_2 u_2(s)}{x(s)^{1/2}} \right] \exp[-r(s - t_0)]\, ds \right\} \tag{6.25}$$

subject to dynamics (6.18).

Consider the alternative stochastic control problem which starts at time $t \in [t_0, \infty)$ with initial state $x(t) = x$:

$$\max_{u_1, u_2} E_t \left\{ \int_t^\infty \left[[u_1(s) + u_2(s)]^{1/2} - \frac{c_1 u_1(s) + c_2 u_2(s)}{x(s)^{1/2}} \right] \exp[-r(s-t)]\, ds \right\} \tag{6.26}$$

subject to (6.21)

A set of strategies $[\psi_1^*(x), \psi_2^*(x)]$ provides an optimal solution to the problem (6.26) and (6.21), if there exist a continuously twice differentiable function $W(x)$ $: R \to R$ satisfying the following partial differential equations:

$$\begin{aligned} &rW(x) - \frac{1}{2}\sigma^2 x^2 W_{xx}(x) \\ &= \max_{u_1, u_2} \left\{ \left[(u_1 + u_2)^{1/2} - (c_1 u_1 + c_2 u_2) x^{-1/2} \right] + W_x(x) \left[a x^{1/2} - bx - u_1 - u_2 \right] \right\}. \end{aligned} \tag{6.27}$$

Performing the indicated maximization operation in (6.27) yields:

$$\psi_1^*(x) = \frac{x}{4[c_1 + W_x\, x^{1/2}]^2} \text{ and } \psi_2^*(x) = 0. \tag{6.28}$$

Firm 2 has to refrain from extraction. The more efficient firm (firm 1) would buy the less efficient firm (firm 2) out from the resource extraction process. Firm 2 becomes a dormant firm under cooperation. The maximized cooperative payoff in the control problem (6.26) and (6.21) can be obtained as:

Proposition 6.2

$$W(x) = \left[A x^{1/2} + C \right], \tag{6.29}$$

where A and B satisfy:

$$0 = \left[r + \frac{b}{2} + \frac{\sigma^2}{8} \right] A - \frac{1}{4[c_1 + A/2]} \text{ and } rC = \frac{a}{2} A.$$

Proof First substitute the results in (6.28), and $W(x)$, and $W_x(x)$ obtained via (6.29) into the partial differential Eq. (6.27). One can readily show that for this equation to be satisfied, Proposition 6.2 has to hold. ■

Upon substituting $\psi_1^*(x)$ and $\psi_2^*(x)$ into (6.16) yields the optimal cooperative state dynamics as:

$$\begin{aligned} dx(s) &= \left[a x(s)^{1/2} - b x(s) - \frac{x(s)}{4[c_1 + A/2]^2} \right] + \sigma\, x(s) dz(s), \\ x(t_0) &= x_0 \in X. \end{aligned} \tag{6.30}$$

The solution to (5.10) yields a Pareto optimal trajectory, which can be expressed as:

$$x^*(t) = \left\{ \Phi(t,t_0)\left[x_0^{1/2} + \int_{t_0}^{t} \Phi^{-1}(s,t_0)\frac{a}{2}ds \right] \right\}^2, \tag{6.31}$$

where $\Phi(t,t_0) = \exp\left[\int_{t_0}^{t} \left(\frac{-b}{2} - \frac{1}{8[c_1 + A/2]^2} - \frac{3\sigma^2}{8} \right) ds + \int_{t_0}^{t} \frac{\sigma}{2} dz(s) \right]$.

We denote the set containing realizable values of $x^*(t)$ by X_t, for $t \in (t_0, T]$.

15.6.3.2 Subgame Consistent Cartel Profits Sharing

Consider the case when the firms in the cartel agree to share the expected cooperative payoff proportional to the firms' expected noncooperative payoffs.

The imputation scheme has to fulfill Condition 6.1, that is

$$\xi^{(\tau)i}(\tau, x_\tau^*) = \frac{\hat{V}^i(x_\tau^*)}{\sum_{j=1}^{2} \hat{V}^j(x_\tau^*)} W(x_\tau^*), \text{ for } i \in \{1,2\} \text{ along the cooperative path } \{x_\tau^*\}_{\tau=t_0}^{\infty}. \tag{6.32}$$

To formulate a set of subgame consistent payoff distribution procedure we invoke Corollary 6.2 and obtain:

Corollary 6.3 A PDP with an instantaneous payment at time $\tau \in [t_0, \infty)$ when $x^*(\tau) = x_\tau^* \in X_\tau^*$:

$$B_i(\tau, x_\tau^*) = r \frac{\hat{V}^i(x_\tau^*)}{\sum_{j=1}^{2} \hat{V}^j(x_\tau^*)} W(x_\tau^*) - \frac{\sigma^2 (x_\tau^*)^2}{2} \frac{\partial^2}{\partial (x_\tau^*)^2} \left[\frac{\hat{V}^i(x_\tau^*)}{\sum_{j=1}^{2} \hat{V}^j(x_\tau^*)} W(x_\tau^*) \right],$$
$$- \frac{\partial}{\partial x_\tau^*} \left[\frac{\hat{V}^i(x_\tau^*)}{\sum_{j=1}^{2} \hat{V}^j(x_\tau^*)} W(x_\tau^*) \right] \left[a(x_\tau^*)^{1/2} - b x_\tau^* - \frac{x_\tau^*}{4[c_1 + A/2]^2} \right],$$

for $i \in \{1,2\}$,

$$\tag{6.33}$$

yields a subgame consistent PDP to the cooperative game with payoff (6.19), dynamics (6.18) and imputation as specified in (6.32). ■

The instantaneous receipt of firm 1 at time instant τ with $x^*(\tau) = x_\tau^* \in X_\tau^*$ is:

$$\zeta_1(\tau, x_\tau^*) = \frac{(x_\tau^*)^{1/2}}{2[c_1 + A/2]} - \frac{c_1 (x_\tau^*)^{1/2}}{4[c_1 + A/2]^2}, \text{ for } \tau \in [t_0, \infty).$$

The instantaneous receipt of firm 2 at time instant τ is

$$\zeta_2(\tau, x_\tau^*) = 0, \text{ for } \tau \in [t_0, \infty) \text{ with } x^*(\tau) = x_\tau^* \in X_\tau^*.$$

According Corollary 6.3, the instantaneous payment that firm i should receive under the agreed-upon optimality principle is $B_1(\tau, x_\tau^*)$ and $B_2(\tau, x_\tau^*)$ as stated in (6.33). Hence when $x^*(\tau) = x_\tau^* \in X_\tau^*$ an instantaneous transfer payment

$$\chi^1(\tau, x_\tau^*) = \zeta_1(\tau, x_\tau^*) - B_1(\tau, x_\tau^*), \text{ for firm 1 and}$$
$$\chi^2(\tau, x_\tau^*) = B_2(\tau, x_\tau^*), \text{ for firm 2}$$

would be arranged.

15.7 Appendix: Proof of Condition 2.1

To prove that

$$W^{(\tau)K}(\tau, x^K) \geq W^{(\tau)L}(\tau, x^L) + W^{(\tau)K\backslash L}(\tau, x^{K\backslash L}), \text{ for } L \subset K \subseteq N,$$

we first use $\hat{x}^{j(L)}$, for $j \in L$, to denote the optimal trajectory of the stochastic optimal control problem $\varpi[L; \tau, x_\tau^L]$ which maximizes

$$E_\tau \Bigg\{ \int_\tau^T \sum_{j \in L} \{ g^j[s, x^j(s)] - c_j^L[u_j(s)] \} \exp\left[-\int_\tau^s r(y) dy \right] ds + \sum_{j \in L} \exp\left[-\int_{t_0}^T r(y) dy \right] q^j(x^j(T)) \Bigg\}$$

subject to

$$dx^j(s) = f^j[s, x^j(s), u_j(s)]\, ds + \sigma_j[s, x^j(s)]\, dz_j(s), \; x^j(\tau) = x_\tau^j,$$

for $j \in L$.

Note that

$$W^{(\tau)L}\left(\tau, x_\tau^L\right) =$$
$$E_\tau\left\{ \int_\tau^T \sum_{j\in L} \{g^j\left[s, \hat{x}^{j(L)}(s)\right] - c_j^L\left[\psi_j^{(\tau)L*}\left(s, \hat{x}^{L(L)}(s)\right)\right]\} \exp\left[-\int_\tau^s r(y)dy\right] ds \right.$$
$$\left. + \sum_{j\in L} \exp\left[-\int_\tau^T r(y)dy\right] q^j\left(\hat{x}^{j(L)}(T)\right) \right\}$$
$$\leq E_\tau\Bigg\{ \tag{7.1}$$
$$\int_\tau^T \sum_{j\in L} \{g^j\left[s, \hat{x}^{j(L)}(s)\right] - c_j^K\left[\psi_j^{(\tau)L*}\left(s, \hat{x}^{L(L)}(s)\right)\right]\} \exp\left[-\int_\tau^s r(y)dy\right] ds$$
$$+ \sum_{j\in L} \exp\left[-\int_\tau^T r(y)dy\right] q^j\left(\hat{x}^{j(L)}(T)\right) \Bigg\},$$
because $c_j^K\left[u_j(s)\right] \leq c_j^L\left[u_j(s)\right]$, for $j \in L \subseteq K$.

Similarly, for the optimal control problem $\varpi[K\backslash L; \tau, x_\tau^{K\ L}]$, we have

$$W^{(\tau)K\backslash L}\left(\tau, x_\tau^{K\backslash L}\right) = E_\tau\Bigg\{$$
$$\int_\tau^T \sum_{j\in K\backslash L} \{g^j\left[s, \hat{x}^{j(K\backslash L)}(s)\right] - c_j^{K\backslash L}\left[\psi_j^{(\tau)K\backslash L*}\left(s, \hat{x}^{K\backslash L(K\backslash L)}(s)\right)\right]\} \exp\left[-\int_\tau^s r(y)dy\right] ds$$
$$+ \sum_{j\in K\backslash L} \exp\left[-\int_\tau^T r(y)dy\right] q^j\left(\hat{x}^{j(K\backslash L)}(T)\right) \Bigg\}$$
$$\leq E_\tau\Bigg\{$$
$$\int_\tau^T \sum_{j\in K\backslash L} \{g^j\left[s, \hat{x}^{j(K\backslash L)}(s)\right] - c_j^K\left[\psi_j^{(\tau)K\backslash L*}\left(s, \hat{x}^{K\backslash L(K\backslash L)}(s)\right)\right]\} \exp\left[-\int_\tau^s r(y)dy\right] ds$$
$$+ \sum_{j\in K\backslash L} \exp\left[-\int_\tau^T r(y)dy\right] q^j\left(\hat{x}^{j(K\backslash L)}(T)\right) \Bigg\},$$
because $c_j^K\left[u_j(s)\right] \leq c_j^{K\backslash L}\left[u_j(s)\right]$, for $j \in K\backslash L \subseteq K$.

$$\tag{7.2}$$

Now consider the optimal control problem $\varpi[K; \tau, x_\tau^K]$ which maximizes

$$E_\tau\left\{ \int_\tau^T \sum_{j\in K} \{g^j\left[s, x^j(s)\right] - c_j^K\left[u_j(s)\right]\} \exp\left[-\int_\tau^s r(y)dy\right] ds \right.$$
$$\left. + \sum_{j\in K} \exp\left[-\int_{t_0}^T r(y)dy\right] q^j(x^j(T)) \right\}$$

subject to

$$dx^j(s) = f^j\left[s, x^j(s), u_j(s)\right] ds + \sigma_j\left[s, x^j(s)\right] dz_j(s), \quad x^j(\tau) = x_\tau^j,$$

for $j \in K$.

Since $\psi_j^{(\tau)K*}\left(s, \hat{x}^{K(K)}(s)\right)$ and $\hat{x}^{K(K)}(s)$ are respectively the optimal control and optimal state trajectory of the control problem $\varpi[K; \tau, x_\tau^K]$,

$$\begin{aligned}
&W^{(\tau)K}\left(\tau, x_\tau^K\right) = \\
&\quad E_\tau\left\{ \int_\tau^T \sum_{j\in K} \{g^j\left[s, \hat{x}^{j(K)}(s)\right] - c_j^K\left[\psi_j^{(\tau)K*}\left(s, \hat{x}^{K(K)}(s)\right)\right]\} \exp\left[-\int_\tau^s r(y)dy\right] ds \right. \\
&\qquad\qquad \left. + \sum_{j\in K} \exp\left[-\int_\tau^T r(y)dy\right] q^j\left(\hat{x}^{j(K)}(T)\right) \right\} \\
&\geq E_\tau\left\{ \int_\tau^T \sum_{j\in L} \{g^j\left[s, \hat{x}^{j(L)}(s)\right] - c_j^K\left[\psi_j^{(\tau)L*}\left(s, \hat{x}^{L(L)}(s)\right)\right]\} \exp\left[-\int_\tau^s r(y)dy\right] ds \right. \\
&\qquad\qquad \left. + \sum_{j\in L} \exp\left[-\int_\tau^T r(y)dy\right] q^j\left(\hat{x}^{j(L)}(T)\right) \right\} \\
&+E_\tau\left\{ \right. \\
&\int_\tau^T \sum_{j\in K\backslash L} \{g^j\left[s, \hat{x}^{j(K\backslash L)}(s)\right] - c_j^K\left[\psi_j^{(\tau)K\backslash L*}\left(s, \hat{x}^{K\backslash L(K\backslash L)}(s)\right)\right]\} \exp\left[-\int_\tau^s r(y)dy\right] ds \\
&\qquad \left. + \sum_{j\in K\backslash L} \exp\left[-\int_\tau^T r(y)dy\right] q^j\left(\hat{x}^{j(K\backslash L)}(T)\right) \right\}
\end{aligned} \tag{7.3}$$

Invoking (7.1), (7.2) and (7.3), we have $W^{(\tau)K}\left(\tau, x_\tau^K\right) \geq W^{(\tau)L}\left(\tau, x_\tau^L\right) + W^{(\tau)K\backslash L}\left(\tau, x_\tau^{K\backslash L}\right)$. Hence Condition 2.1 follows. ■

15.8 Chapter Notes

In this Chapter two applications in business cooperation are considered – cost saving joint venture and dormant firm cartel. Despite all their purported benefits, joint ventures are highly unstable because of the lack of dynamical stable profit sharing schemes. Subgame consistent scheme can be perceived as a solution. The analysis can be used as a foundation for future research to develop solutions to joint ventures other than cost saving venture. For instance, other adverse effects, such as uncompensated transfers of technology, operational difficulties, disagreements and anxiety over the loss of proprietary information have been found (Hamel et al. (1989) and Gomes-Casseres (1987)). D'Aspremont and Jacquemin (1988), Kamien et al (1992) and Suzumura (1992) have studied cooperative R&D with spillovers in joint ventures under a static framework. Cellini and Lambertini (2002, 2004) considered cooperative solutions to investment in product differentiation in a dynamic approach.

15.9 Problems

1. Consider the case when there are three companies involved in a joint venture. The planning period is $[0, 3]$. We use $x_i(s)$ to denote the level of technology of company i at time $s \in [0, 3]$, and $u_i(s) \subset R^+$ is its physical investment in technological advancement. The increments of the levels of technology are subject to stochastic disturbances. The discount rate is 0.05. The salvage values of the firms' technologies are $2[x^1(2)]^{1/2}$, $[x^2(2)]^{1/2}$ and $1.5[x^3(2)]^{1/2}$ at time 3. If the companies act independently, the costs of physical investment of these three firms are respectively $3u_1(s)$, $2.5u_2(s)$ and $2u_3(s)$.

   The expected profits for companies 1, 2 and 3 are respectively:

$$E\left\{ \int_0^3 \left[9\left[x^1(s)\right]^{1/2} - 3u_1(s)\right] \exp(-0.05s)\, ds + \exp[-0.05(3)]2[x^1(3)]^{1/2} \right\},$$
$$E\left\{ \int_0^3 \left[6\left[x^2(s)\right]^{1/2} - 2.5u_2(s)\right] \exp(-0.05s)\, ds + \exp[-0.05(3)][x^2(3)]^{1/2} \right\},$$

   and

$$E\left\{ \int_0^3 \left[10\left[x^3(s)\right]^{1/2} - 2u_3(s)\right] \exp(-0.05s)\, ds + \exp[-0.05(3)]\, 1.5\left[x^3(3)\right]^{1/2} \right\}.$$

   The evolution of the technology level of company $i \in \{1, 2, 3\}$ follows a system of stochastic dynamics:

$$dx^1(s) = \left[3[u_1(s)x^1(s)]^{1/2} - 0.2x^1(s)\right]ds + 0.1x^1(s)dz_1(s), x^1(0) = 35,$$
$$dx^2(s) = \left[4[u_2(s)x^2(s)]^{1/2} - 0.1x^2(s)\right]ds + 0.2x^2(s)dz_2(s),\ x^2(0) = 22, \text{ and}$$
$$dx^3(s) = \left[2[u_3(s)x^3(s)]^{1/2} - 0.05x^3(s)\right]ds + 0.1x^i(s)dz_3(s),\ x^3(0) = 20,$$

   where $z_1(s)$, $z_2(s)$ and $z_3(s)$ are independent Wiener processes.

   Characterize a feedback Nash equilibrium solution when these three firms act independently.
2. Consider the case when these three companies form a joint venture. The participating firms in a coalition can gain core skills and technology from each other. In particular, they can obtain cost reduction and with absolute joint venture cost advantage.

With joint venture cost advantage, the cost of investment of firm $j \in \{1,2,3\}$ under the joint venture become $c_j^{1,2,3} u_j(s)$ where $c_1^{\{1,2,3\}} = 1.5$, $c_2^{\{1,2,3\}} = 1$ and $c_3^{\{1,2,3\}} = 1$.

If the joint venture firms agree to maximize their expected joint profit and share the excess gain equally, characterize a subgame consistent solution.

3. Consider a duopoly in which two firms are allowed to extract a renewable resource within the duration [0, 4]. The dynamics of the resource is characterized by the stochastic dynamics

$$dx(s) = \left[4x(s)^{1/2} - 0.3x(s) - u_1(s) - u_2(s)\right]ds \ + 0.1x(s)dz(s),\ x(0) = 100,$$

where $z(s)$ is a Wiener process and $x(s)$ is the resource biomass and $u_i(s)$ is the amount of resource extracted by firm i at time $s \in [0,4]$, for $i \in \{1,2\}$.

The extraction cost for firm 1 and firm 2 are respectively $2u_1(s)x(s)^{-1/2}$ and $3u_1(s)x(s)^{-1/2}$. The market price of the resource depends on the total amount extracted and supplied to the market. The price-output relationship at time s is given by the following downward sloping inverse demand curve $P(s) = Q(s)^{-1/2}$, where $Q(s) = u_1(s) + u_2(s)$ is the total amount of resource extracted and marketed at time s. At terminal time 4, firm 1 will receive a termination bonus $3x(3)^{1/2}$ and firm 2 a bonus $1.5x(3)^{1/2}$. The discount factor is 0.05.

Characterize a feedback Nash equilibrium solution when these firms act independently.

4. If these two firms form a cartel, show that firm 2 has to be dormant. Derive the optimal output strategies of the cartel.
5. Consider the case when the firms in the cartel agree to share the excess of the total expected cooperative profits proportional to the firms' expected noncooperative profits. Characterize a subgame consistent solution.

Technical Appendices

Continuous-Time Dynamic Programming

Consider the dynamic optimization problem in which the single decision-maker maximizes the payoff

$$\int_{t_0}^{T} g[s, x(s), u(s)] \exp\left[-\int_{t_0}^{s} r(y)dy\right] ds + \exp\left[-\int_{t_0}^{T} r(y)dy\right] q(x(T)), \tag{A.1}$$

subject to the vector-valued differential equation:

$$\dot{x}(s) = f[s, x(s), u(s)]ds, \quad x(t_0) = x_0, \tag{A.2}$$

where $x(s) \in X \subset R^m$ denotes the state variables of game, and $u \in U$ is the control.

A frequently adopted approach to dynamic optimization problems is the technique of dynamic programming. The technique was developed by Bellman (1957). The technique is given in Theorem A.1 below.

Theorem A.1 (Bellman's Dynamic Programming) A set of controls $u^*(t) = \phi^*(t, x)$ constitutes an optimal solution to the control problem (A.1 and A.2) if there exist continuously differentiable function $V(t, x)$ defined on $[t_0, T] \times R^m \to R$ and satisfying the following Bellman equation:

$$\begin{aligned} -V_t(t, x) &= \max_u \left\{ g[t, x, u] \exp\left[-\int_{t_0}^{t} r(y)dy\right] + V_x(t, x) f[t, x, u] \right\} \\ &= \left\{ g[t, x, \phi^*(t, x)] \exp\left[-\int_{t_0}^{t} r(y)dy\right] + V_x(t, x) f[t, x, \phi^*(t, x)] \right\}, \end{aligned}$$

D.W.K. Yeung, L.A. Petrosyan, *Subgame Consistent Cooperation*,
Theory and Decision Library C 47, DOI 10.1007/978-981-10-1545-8

$$V(T,x) = q(x)\exp\left[-\int_{t_0}^{T} r(y)dy\right]. \tag{A.3}$$

Proof Define the maximized payoff at time t with current state x as a *value function* in the form:

$$\begin{aligned} V(t,x) &= \max_u \left\{ \int_t^T g(s,x(s),u(s))\exp\left[-\int_{t_0}^{s} r(y)dy\right] ds \right. \\ &\quad \left. + \exp\left[-\int_{t_0}^{T} r(y)dy\right] q(x(T)) \right\} \\ &= \int_t^T g\left[s,x^*(s),\phi^*\left(s,x^*(s)\right)\right]\exp\left[-\int_{t_0}^{s} r(y)dy\right] ds \\ &\quad + \exp\left[-\int_{t_0}^{T} r(y)dy\right] q\left(x^*(T)\right)\right] \end{aligned}$$

satisfying the boundary condition

$$\begin{aligned} &V\left(T,x^*(T)\right) = q\left(x^*(T)\right)\exp\left[-\int_{t_0}^{T} r(y)dy\right], \text{ and} \\ &\dot{x}^*(s) = f\left[s,x^*(s),\phi^*\left(s,x^*(s)\right)\right], \quad x^*(t_0) = x_0 \end{aligned} \tag{A.4}$$

One can express $V(t,x_t^*)$ as:

$$\begin{aligned} V\left(t,x_t^*\right) &= \max_u \left\{ \int_t^T g^i[s,x(s),u(s)]\exp\left[-\int_{t_0}^{s} r(y)dy\right] ds \right. \\ &\quad \left. + q(x(T))\exp\left[-\int_{t_0}^{T} r(y)dy\right] \right\} \\ &= \max_u \left\{ \int_t^{t+\Delta t} g^i[s,x(s),u(s)]\exp\left[-\int_{t_0}^{s} r(y)dy\right] ds. \right. \\ &\quad \left. + V\left(t+\Delta t, x_t^* + \Delta x_t^*\right) \right\} \end{aligned} \tag{A.5}$$

where $\Delta x_t^* = f\left[t,x_t^*,\phi^*\left(t,x_t^*\right)\right]\Delta t + o(\Delta t)$, and $o(\Delta t)/\Delta t \to 0$ as $\Delta t \to 0$.

With $\Delta t \to 0$, Eq. (A.5) can be expressed as:

$$\begin{aligned} V\left(t,x_t^*\right) &= \max_u \left\{ g^i\left[t,x_t^*,u\right]\exp\left[-\int_{t_0}^{t} r(y)dy\right]\Delta t + V\left(t,x_t^*\right) + V_t\left(t,x_t^*\right)\Delta t \right. \\ &\quad \left. + V_{x_t}\left(t,x_t^*\right) f\left[t,x_t^*,\phi^*\left(t,x_t^*\right)\right]\Delta t + o(\Delta t) \right\}. \end{aligned} \tag{A.6}$$

Dividing (A.6) throughout by Δt, with $\Delta t \to 0$, yields

$$-V_t(t,x_t^*) = \max_u \left\{ g^i[t,x_t^*,u]\exp\left[-\int_{t_0}^{t} r(y)dy\right] + V_{x_t}(t,x_t^*)f[t,x_t^*,\phi^*(t,x_t^*)] \right\}, \tag{A.7}$$

with boundary condition

$$V(T,x^*(T)) = q(x^*(T))\exp\left[-\int_{t_0}^{T} r(y)dy\right].$$

Hence Theorem A.1 follows. ■

Infinite-Horizon Continuous-Time Dynamic Programming

Consider the infinite-horizon dynamic optimization problem with a constant discount rate:

$$\max_u \left\{ \int_{t_0}^{\infty} g[x(s),u(s)]\ \exp[-r(s-t_0)]ds \right\}, \tag{A.8}$$

subject to the vector-valued differential equation:

$$\dot{x}(s) = f[x(s),u(s)]ds, \quad x(t_0) = x_0. \tag{A.9}$$

Since s does not appear in $g[x(s),u(s)]$ and the state dynamics explicitly, the problem (A.8 and A.9) is an autonomous problem.

Consider the alternative problem which starts at time $t \in [t_0,\infty)$ with initial state $x(t) = x$:

$$\max_u \int_{t}^{\infty} g[x(s),u(s)]\exp[-r(s-t)]ds, \tag{A.10}$$

subject to

$$\dot{x}(s) = f[x(s),u(s)], x(t) = x. \tag{A.11}$$

Theorem A.2 A set of controls $u = \phi^*(x)$ constitutes an optimal solution to the infinite horizon control problem (A.10 and A.11) if there exists continuously differentiable function $W(x)$ defined on $R^m \to R$ which satisfies the following equation:

$$\begin{aligned} rW(x) &= \max_u \{ g[x,u] + W_x(x) f[x,u] \} \\ &= \{ g[x, \phi^*(x)] + W_x(x) f[x, \phi^*(x)] \}. \end{aligned}$$

Proof The infinite-horizon autonomous problem (A.10 and A.11) is independent of the choice of t and dependent only upon the state at the starting time, that is x. Define the value function to the problem (A.8 and A.9) by

$$\widetilde{W}(t,x) = \max_u \left\{ \int_t^\infty g[x(s),u(s)] \; \exp[-r(s-t_0)]ds \middle| \, x(t) = x = x_t^* \right\},$$

where x_t^* is the state at time t along the optimal trajectory. Moreover, we can write

$$\widetilde{W}(t,x) = \exp[-r(t-t_0)] \max_u \left\{ \int_t^\infty g[x(s),u(s)] \; \exp[-r(s-t)]ds \middle| \, x(t) = x = x_t^* \right\}.$$

Since the problem $\max_u \left\{ \int_t^\infty g[x(s),u(s)] \; \exp[-r(s-t)]ds \middle| \, x(t) = x = x_t^* \right\}$ depends on the current state x only, we can write:

$$W(x) = \max_u \left\{ \int_t^\infty g[x(s),u(s)] \; \exp[-r(s-t)]ds \middle| \, x(t) = x = x_t^* \right\}.$$

It follows that:

$$\begin{aligned} \widetilde{W}(t,x) &= \exp[-r(t-t_0)] \, W(x), \\ \widetilde{W}_t(t,x) &= -r\exp[-r(t-t_0)] \, W(x), \text{ and} \\ \widetilde{W}_x(t,x) &= -r\exp[-r(t-t_0)] \, W_x(x). \end{aligned} \quad \text{(A.12)}$$

Substituting the results from (A.12) into Theorem A.1 yields

$$rW(x) = \max_u \{ g[x,u] + W_x(x) f[x,u] \}. \quad \text{(A.13)}$$

Hence Theorem A.2 follows. ■

Stochastic Control

Consider the dynamic optimization problem in which the single decision-maker:

$$\max_u E_{t_0}\left\{\int_{t_0}^{T} g^i[s,x(s),u(s)]\exp\left[-\int_{t_0}^{s} r(y)dy\right]ds + q(x(T))\exp\left[-\int_{t_0}^{T} r(y)dy\right]\right\}, \tag{A.14}$$

subject to the vector-valued stochastic differential equation:

$$dx(s) = f[s,x(s),u(s)]ds + \sigma[s,x(s)]dz(s), \qquad x(t_0) = x_0, \tag{A.15}$$

where E_{t_0} denotes the expectation operator performed at time t_0, and $\sigma[s,x(s)]$ is a $m \times \Theta$ matrix and $z(s)$ is a Θ-dimensional Wiener process and the initial state x_0 is given. Let $\Omega[s,x(s)] = \sigma[s,x(s)]\,\sigma[s,x(s)]^T$ denote the covariance matrix with its element in row h and column ζ denoted by $\Omega^{h\zeta}[s,x(s)]$.

The technique of stochastic control developed by Fleming (1969) can be applied to solve the problem.

Theorem A.3 A set of controls $u^*(t) = \phi^*(t,x)$ constitutes an optimal solution to the problem (A.14 and A.15), if there exist continuously twice differentiable functions $V(t,x) : [t_0,T] \times R^m \to R$, satisfying the following partial differential equation:

$$-V_t(t,x) - \frac{1}{2}\sum_{h,\zeta=1}^{m} \Omega^{h\zeta}(t,x)V_{x^h x^\zeta}(t,x) =$$

$$\max_u \left\{ g^i[t,x,u]\exp\left[-\int_{t_0}^{t} r(y)dy\right] + V_x(t,x)f[t,x,u] \right\}, \text{ and}$$

$$V(T,x) = q(x)\exp\left[-\int_{t_0}^{T} r(y)dy\right].$$

Proof Substitute the optimal control $\phi^*(t,x)$ into the (A.15) to obtain the optimal state dynamics as

$$dx(s) = f\left[s,x(s),\phi^*(s,x(s))\right]ds + \sigma[s,x(s)]dz(s), \quad x(t_0) = x_0. \tag{A.16}$$

The solution to (A.16), denoted by $x^*(t)$ can be expressed as:

$$x^*(t) = x_0 + \int_{t_0}^{t} f\left[s,x^*(s),\psi_1^{(t_0)*}\left(s,x^*(s)\right),\psi_2^{(t_0)*}\left(s,x^*(s)\right)\right]ds$$
$$+\int_{t_0}^{t} \sigma\left[s,x^*(s)\right]dz(s). \tag{A.17}$$

We use X_t^* to denote the set of realizable values of $x^*(t)$ at time t generated by (A.17). The term x_t^* is used to denote an element in the set X_t^*.

Define the maximized payoff at time t with current state x_t^* as a *value function* in the form:

$$\begin{aligned}
V(t,x_t^*) &= \max_u E_{t_0}\left\{\int_t^T g^i[s,x(s),u(s)]\exp\left[-\int_{t_0}^s r(y)dy\right]ds\right.\\
&\quad\left.+q(x(T))\exp\left[-\int_{t_0}^T r(y)dy\right]\middle|x(t)=x_t^*\right\}\\
&= E_{t_0}\left\{\int_t^T g\left[s,x^*(s),\phi^*(s,x^*(s))\right]\exp\left[-\int_{t_0}^s r(y)dy\right]ds\right.\\
&\quad\left.+q(x^*(T))\exp\left[-\int_{t_0}^T r(y)dy\right]\right\},
\end{aligned}$$

satisfying the boundary condition

$$V(T,x^*(T)) = q(x^*(T))\exp\left[-\int_{t_0}^T r(y)dy\right].$$

One can express $V(t,x_t^*)$ as:

$$\begin{aligned}
V(t,x_t^*) &= \max_u E_{t_0}\left\{\int_t^T g^i[s,x(s),u(s)]\exp\left[-\int_{t_0}^s r(y)dy\right]ds\right.\\
&\quad\left.+q(x(T))\exp\left[-\int_{t_0}^T r(y)dy\right]\middle|x(t)=x_t^*\right\}\\
&= \max_u E_{t_0}\left\{\int_t^{t+\Delta t} g^i[s,x(s),u(s)]\exp\left[-\int_{t_0}^s r(y)dy\right]ds.\right.\\
&\quad\left.+V(t+\Delta t,x_t^*+\Delta x_t^*)\middle|x(t)=x_t^*\right\}.
\end{aligned}\tag{A.18}$$

where

$$\Delta x_t^* = f\left[t,x_t^*,\phi^*(t,x_t^*)\right]\Delta t+\sigma\left[t,x_t^*\right]\Delta z_t+o(\Delta t),$$

$\Delta z_t = z(t+\Delta t)-z(t)$, and $E_t[o(\Delta t)]/\Delta t \to 0$ as $\Delta t \to 0$.

With $\Delta t \to 0$, applying Ito's lemma Eq. (A.18) can be expressed as:

$$\begin{aligned}
V(t,x_t^*) = {} & \max_u E_{t_0}\left\{g^i\left[t,x_t^*,u\right]\exp\left[-\int_{t_0}^t r(y)dy\right]\Delta t+V(t,x_t^*)+V_t(t,x_t^*)\Delta t\right.\\
&+V_{x_t}(t,x_t^*)f\left[t,x_t^*,\phi^*(t,x_t^*)\right]\Delta t+V_{x_t}(t,x_t^*)\sigma\left[t,x_t^*\right]\Delta z_t\\
&\left.+\frac{1}{2}\sum_{h,\zeta=1}^m \Omega^{h\zeta}(t,x)V_{x^h x^\zeta}(t,x)\Delta t+o(\Delta t)\right\}.
\end{aligned}\tag{A.19}$$

Dividing (A.19) throughout by Δt, with $\Delta t \to 0$, and taking expectation yield

$$-V_t(t, x_t^*) - \frac{1}{2}\sum_{h,\zeta=1}^{m} \Omega^{h\zeta}(t,x) V_{x^h x^\zeta}(t,x)$$
$$= \max_u \left\{ g^i[t, x_t^*, u] \exp\left[-\int_{t_0}^{t} r(y)dy\right] + V_{x_t}(t, x_t^*) f[t, x_t^*, \phi^*(t, x_t^*)] \right\}, \qquad \text{(A.20)}$$

with boundary condition $V(T, x^*(T)) = q(x^*(T))\exp\left[-\int_{t_0}^{T} r(y)dy\right]$.

Hence Theorem A.3 follows. ■

Infinite Horizon Stochastic Control

Consider the infinite-horizon stochastic control problem with a constant discount rate:

$$\max_u E_{t_0}\left\{\int_{t_0}^{\infty} g^i[x(s), u(s)] \exp[-r(s - t_0)]ds\right\}, \qquad \text{(A.21)}$$

subject to the vector-valued stochastic differential equation:

$$dx(s) = f[x(s), u(s)]ds + \sigma[x(s)]dz(s), \qquad x(t_0) = x_0, \qquad \text{(A.22)}$$

Let $\Omega[x(s)] = \sigma[x(s)]\sigma[x(s)]^T$ denote the covariance matrix with its element in row h and column ζ denoted by $\Omega^{h\zeta}[x(s)]$.

Since s does not appear in $g[x(s), u(s)]$ and the state dynamics explicitly, the problem (A.21 and A.22) is an autonomous problem.

Consider the alternative problem which starts at time $t \in [t_0, \infty)$ with initial state $x(t) = x$:

$$\max_u E_t\left\{\int_{t}^{\infty} g^i[x(s), u(s)] \exp[-r(s - t)]ds\right\}, \qquad \text{(A.23)}$$

subject to the vector-valued stochastic differential equation:

$$dx(s) = f[x(s), u(s)]ds + \sigma[x(s)]dz(s), \qquad x(t) = x_t. \qquad \text{(A.24)}$$

Theorem A.4 A set of controls $u = \phi^*(x)$ constitutes an optimal solution to the infinite horizon stochastic control problem (A.21 and A.22) if there exists continuously twice differentiable function $W(x)$ defined on $R^m \to R$ which satisfies the following equation:

$$rW(x) - \frac{1}{2}\sum_{h,\zeta=1}^{m} \Omega^{h\zeta}(x)W_{x^h x^\zeta}(x) = \max_u \{g[x,u] + W_x(x)f[x,u]\}$$
$$= \{g[x,\phi^*(x)] + W_x(x)f[x,\phi^*(x)]\}.$$

Proof The infinite-horizon autonomous problem (A.23 and A.24) is independent of the choice of t and dependent only upon the state at the starting time, that is x_t.

Define the value function to the problem (A.23 and A.24) by

$$V(t,x_t^*) = \max_u E_{t_0}\left\{\int_t^\infty g[x(s),u(s)]\ \exp[-r(s-t_0)]ds\,\middle|\,x(t) = x_t^*\right\},$$

where x_t^* is an element belonging to the set of realizable values along the optimal state trajectory at time t. Moreover, we can write

$$V(t,x_t^*) = \exp[-r(t-t_0)]\max_u E_{t_0}\left\{\int_t^\infty g[x(s),u(s)]\ \exp[-r(s-t)]ds\,\middle|\,x(t) = x_t^*\right\}.$$

Since the problem $\max_u E_{t_0}\left\{\int_t^\infty g[x(s),u(s)]\ \exp[-r(s-t)]ds\,\middle|\,x(t) = x_t^*\right\}$ depends on the current state x_t^* only, we can write:

$$W(x_t^*) = \max_u E_{t_0}\left\{\int_t^\infty g[x(s),u(s)]\ \exp[-r(s-t)]ds\,\middle|\,x(t) = x_t^*\right\}.$$

It follows that:

$$\begin{aligned} V(t,x_t^*) &= \exp[-r(t-t_0)]\,W(x_t^*)\\ V_t(t,x_t^*) &= -r\exp[-r(t-t_0)]\,W(x_t^*),\\ V_{x_t}(t,x_t^*) &= -r\exp[-r(t-t_0)]\,W_{x_t}(x_t^*), \text{and}\\ V_{x_t x_t}(t,x_t^*) &= -r\exp[-r(t-t_0)]\,W_{x_t x_t}(x_t^*). \end{aligned} \tag{A.25}$$

Substituting the results from (A.25) into Theorem A.3 yields

$$rW(x) - \frac{1}{2}\sum_{h,\zeta=1}^{m} \Omega^{h\zeta}(x)W_{x^h x^\zeta}(x) = \max_u \{g[x,u] + W_x(x)f[x,u]\}. \tag{A.26}$$

Since time is not explicitly involved (A.26), the derived control u will be a function of x only. Hence Theorem A.4 follows. ■

Discrete-Time Dynamic programming

Consider the general $T-$ stage discrete-time dynamic programming problem with initial state x^0. The state space of the game is $X \in R^m$ and the state dynamics is characterized by the difference equation:

$$x_{k+1} = f_k(x_k, u_k), \qquad \text{(A.27)}$$
$$\text{for } k \in \{1, 2, \cdots, T\} \equiv \kappa \text{ and } x_1 = x^0,$$

where $u_k \in U \subset R^m$ is the control vector of the decision-maker at stage $k, x_k \in X$ is the state. The payoff to be maximized is

$$\sum_{\zeta=1}^{T} g_\zeta^i(x_\zeta, u_\zeta) \left(\frac{1}{1+r}\right)^{\zeta-1} + q_{T+1}(x_{T+1}) \left(\frac{1}{1+r}\right)^{T}, \qquad \text{(A.28)}$$

where r is the discount rate, and $q_{T+1}(x_{T+1})$ is the terminal benefit that the decision-maker will receive at stage $T+1$.

Invoking Bellman's (1957) technique of dynamic programming, an optimal solution can be characterized as follows:

Theorem A.5 A set of controls $\{u_k^* = \psi_k(x), \text{ for } k \in \{1, 2, \cdots, T\}\}$ provides an optimal solution to the dynamic problem (A.27 and A.28) if there exist functions $V(k, x)$, for $k \in \{1, 2, \cdots, T\}\}$, such that the following recursive relations are satisfied:

$$\begin{aligned} V(k,x) &= \max_{u_k} \{g_k(x, u_k) + V[k+1, f_k(x, u_k)]\} \\ &= g_k[x, \psi_k(x)] + V[k+1, f_k(x, \psi_k(x))], \end{aligned} \quad , \quad \text{for} \quad k \in \{1, 2, \cdots, T\}\}; \qquad \text{(A.29)}$$

$$V(T+1, x) = q_{T+1}(x). \qquad \text{(A.30)}$$

Proof Invoking the technique of backward induction we first consider the last operating stage, that is stage T. In that stage one has an optimization problem which maximizes

$$g_T(x, u_T) + q_{T+1}(x_{T+1}) \qquad \text{(A.31)}$$

subject to

$$x_{T+1} = f_T(x, u_T), \qquad x \in X. \qquad \text{(A.32)}$$

Substituting (A.32) into (A.31), the problem (A.31 and A.32) becomes a single-period maximization problem:

$$\max_{u_T} \left\{ g_T(x, u_T) + q_{T+1}[f_T(x, u_T)] \right\}. \tag{A.33}$$

Invoking $V(T+1, x) = q(x)$ in (A.30), the maximized payoff of problem (A.33) can be expressed as:

$$\begin{aligned} V(T, x) &= \max_{u_T} \{ g_T(x, u_T) + V[T+1, f_T(x, u_T)] \} \\ &= g_T[x, \psi_T(x)] + V\left[T+1, f_T\left(x, \psi_T(x)\right)\right]. \end{aligned} \tag{A.34}$$

Now consider the problem in stage $T-1$. The problem becomes

$$\max_{u_{T-1}, u_T} \left\{ \sum_{\zeta = T-1}^{T} g_\zeta(x_\zeta, u_\zeta) + q(x_{T+1}) \right\}, \tag{A.35}$$

subject to $x_T = f_{T-1}(x_{T-1}, u_{T-1})$ and $x_{T+1} = f_T(x_T, u_T)$, $x_{T-1} = x \in X$.

Invoking the result from the analysis in stage T, the stage $T-1$ problem in (A.35) can be expressed as:

$$\max_{u_{T-1}} \left\{ g_{T-1}(x, u_{T-1}) + V^i(T, x_T) \right\} \tag{A.36}$$

$$\text{subject to} x_T = f_{T-1}(x_{T-1}, u_{T-1}) \text{ and } x_{T-1} = x \in X. \tag{A.37}$$

Substituting (A.37) into (A.36), the problem in stage $T-1$ becomes a single-stage problem

$$\max_{u_{T-1}} \{ g_{T-1}(x, u_{T-1}) + V[T, f_{T-1}(x_{T-1}, u_{T-1})] \}. \tag{A.38}$$

The maximized payoff of the stage $T-1$ problem can be expressed as:

$$\begin{aligned} V(T-1, x) &= \max_{u_{T-1}} \{ g_{T-1}(x, u_{T-1}) + V[T, f_{T-1}(x, u_{T-1})] \} \\ &= g_{T-1}[x, \psi_{T-1}(x)] + V[T, f_{T-1}(x, \psi_{T-1}(x))]. \end{aligned} \tag{A.39}$$

Proceeding recursively onwards for stage $k \in \{T-2, T-1, \cdots, 1\}$, the stage k problem can be expressed as a single-stage problem:

$$\max_{u_k} \{ g_k(x, u_k) + V[k+1, f_k(x, u_k)] \}, \tag{A.40}$$

where $x_k = x \in X$.

The maximized payoff of the stage k problem can be expressed as:

$$V(k,x) = \max_{u_k} \{g_k(x,u_k) + V[k+1, f_k(x,u_k)]\}$$
$$= g_k[x, \psi_k(x)] + V[k+1, f_k(x, \psi_k(x))]. \qquad (A.41a)$$

Hence Theorem A.5 follows. ■

Discrete-Time Stochastic Dynamic Programming

Consider the general $T-$ stage discrete-time stochastic dynamic programming problem with initial state x^0. The state space of the game is $X \in R^m$ and the state dynamics of the game is characterized by the stochastic difference equation:

$$x_{k+1} = f_k(x_k, u_k) + \theta_k, \qquad (A.41b)$$
$$\text{for } k \in \{1, 2, \cdots, T\} \equiv \kappa \text{ and } x_1 = x^0,$$

where $u_k \in R^{\varpi}$ is the control vector of agent i at stage k, $x_k \in X$ is the state, and θ_k is a set of statistically independent random variables.

The expected payoff to be maximized is

$$E_{\theta_1, \theta_2, \cdots, \theta_T} \left\{ \sum_{\zeta=1}^{T} g_\zeta[x_\zeta, u_\zeta] \left(\frac{1}{1+r} \right)^{\zeta-1} \right\}, \qquad (A.42)$$

where r is the discount rate and $E_{\theta_1, \theta_2, \cdots, \theta_T}$ is the expectation operation with respect to the statistics of $\theta_1, \theta_2, \cdots, \theta_T$.

Invoking the technique of stochastic dynamic programming, an optimal solution can be characterized as follows:

Theorem A.6 A set of controls $\{u_k^* = \psi_k(x), \text{ for } k \in \{1, 2, \cdots, T\}\}$ provides an optimal solution to the stochastic dynamic problem (A.41 and A.42) if there exist functions $V(k, x)$, for $k \in \{1, 2, \cdots, T\}\}$, such that the following recursive relations are satisfied:

$$V(k,x) = \max_{u_k} E_{\theta_k} \{g_k(x,u_k) + V[k+1, f_k(x,u_k) + \theta_k]\}$$
$$= E_{\theta_k} \{g_k[x, \psi_k(x)] + V[k+1, f_k(x, \psi_k(x)) + \theta_k]\},$$
$$\text{for } k \in \{1, 2, \cdots, T\}\}, \qquad (A.43)$$
$$V(T+1, x) = q(x). \qquad (A.44)$$

Proof We first consider the last stage of operation, that is stage T. One has an optimization problem which maximizes

$$E_{\theta_T}\{g_T(x,u_T) + q_{T+1}(x_{T+1})\} \tag{A.45}$$

subject to

$$x_{T+1} = f_T(x,u_T) + \theta_T, \qquad x \in X. \tag{A.46}$$

Substituting (A.46) into (A.45), the problem (A.45 and A.46) becomes a single-period stochastic maximization problem:

$$\max_{u_T} E_{\theta_T}\{g_T(x,u_T) + q_{T+1}[f_T(x,u_T) + \theta_T]\}. \tag{A.47}$$

Invoking $V(T+1,x) = q(x)$ in (A.44), the maximized expected payoff of problem (A.47) can be expressed as:

$$\begin{aligned} V(T,x) &= \max_{u_T} E_{\theta_T}\{g_T(x,u_T) + V[T+1, f_T(x,u_T) + \theta_T]\} \\ &= E_{\theta_T}\{g_T[x,\psi_T(x)] + V[T+1, f_T(x,\psi_T(x)) + \theta_T]\}. \end{aligned} \tag{A.48}$$

Now consider the problem in stage $T-1$. The problem becomes

$$\max_{u_{T-1},u_T} E_{\theta_{T-1},\theta_T}\left\{\sum_{\zeta=T-1}^{T} g_\zeta(x_\zeta,u_\zeta) + q_{T+1}(x_{T+1})\right\}, \tag{A.49}$$

subject to

$$x_T = f_{T-1}(x_{T-1},u_{T-1}) + \theta_{T-1} \text{ and } x_{T+1} = f_T(x_T,u_T) + \theta_T,\ x_{T-1} = x \in X.$$

Invoking the result from the analysis in stage T,, the stage $T-1$ problem in (A.49) can be expressed as:

$$\max_{u_{T-1}} E_{\theta_{T-1}}\{g_{T-1}(x,u_{T-1}) + V^i(T,x_T)\} \tag{A.50}$$

$$\text{subject to } x_T = f_{T-1}(x_{T-1},u_{T-1}) + \theta_{T-1} \text{ and } x_{T-1} = x \in X. \tag{A.51}$$

Substituting (A.51) into (A.50), the problem in stage $T-1$ becomes a single-stage problem

$$\max_{u_{T-1}} E_{\theta_{T-1}}\{g_{T-1}(x,u_{T-1}) + V[T, f_{T-1}(x_{T-1},u_{T-1}) + \theta_{T-1}]\}. \tag{A.52}$$

The maximized expected payoff of the stage $T-1$ problem can be expressed as:

$$\begin{aligned} V(T-1,x) &= \max_{u_{T-1}} E_{\theta_{T-1}}\{g_{T-1}(x,u_{T-1}) + V[T, f_{T-1}(x,u_{T-1}) + \theta_{T-1}]\} \\ &= E_{\theta_{T-1}}\{g_{T-1}[x,\psi_{T-1}(x)] + V[T, f_{T-1}(x,\psi_{T-1}(x)) + \theta_{T-1}]\}. \end{aligned} \tag{A.53}$$

Proceeding recursively onwards for stage $k \in \{T-2, T-1, \cdots, 1\}$, the stage k problem can be expressed as a single-stage problem:

$$\max_{u_k} E_{\theta_k}\{g_k(x, u_k) + V[k+1, f_k(x, u_k) + \theta_k], \quad (A.54)$$

where $x_k = x \in X$.

The maximized expected payoff of the stage k problem can be expressed as:

$$\begin{aligned} V(k, x) &= \max_{u_k} E_{\theta_k}\{g_k(x, u_k) + V[k+1, f_k(x, u_k) + \theta_k]\} \\ &= E_{\theta_k}\{g_k[x, \psi_k(x)] + V[k+1, f_k(x, \psi_k(x)) + \theta_k]\}. \end{aligned} \quad (A.55)$$

Hence Theorem A.6 follows. ■

References

Amir, R., Nannerup, N.: Information structure and the tragedy of the commons in resource extraction. J. Bioecon. **8**, 147–165 (2006)

Basar, T.: A counter example in linear-quadratic games: existence of non-linear Nash solutions. J. Optim. Theory Appl. **14**, 425–430 (1974)

Basar, T.: Nash strategies for M-person differential games with mixed information structures. Automatica **11**, 547–551 (1975)

Basar, T.: On the uniqueness of the Nash solution in linear-quadratic differential games. Int. J. Game Theory **5**, 65–90 (1976)

Basar, T.: Two general properties of the saddle-point solutions of dynamic games. IEEE Trans. Autom. Control. **AC-22**, 124–126 (1977a)

Basar, T.: Existence of unique equilibrium solutions in nonzero-sum stochastic differential games. In: Roxin, E.O., Liu, P.T., Sternberg, R. (eds.) Differential Games and Control Theory II, pp. 201–228. Marcel Dekker, Inc., New York (1977b)

Basar, T.: Decentralized multicriteria optimization of linear stochastic systems. IEEE Trans. Autom. Control **AC-23**, 233–243 (1978)

Basar, T.: Information structures and equilibria in dynamic games. In: Aoki, M., Marzollo, A. (eds.) New Trends in Dynamic System Theory and Economics, pp. 3–55. Academic Press, New York/London (1979)

Basar, T.: Time consistency and robustness of equilibria in noncooperative dynamic games. In: Van der Ploeg, F., de Zeeuw, A. (eds.) Dynamic Policy Games in Economics, pp. 9–54. Elsevier Science Pub Co, Amsterdam/North-Holland (1989)

Basar, T., Ho, Y.C.: Informational properties of the nash solutions of two stochastic nonzero-sum games. J. Econ. Theory **7**, 370–387 (1974)

Basar, T., Mintz, M.: On the existence of linear saddle-point strategies for a two-person zero-sum stochastic game. In: Proceedings of the IEEE 11th Conference on Decision and Control, New Orleans, LA, 1972, pp. 188–192, IEEE Computer Society Press, Los Alamitos, CA (1972)

Basar, T., Mintz, M.: A multistage pursuit-evasion game that admits a Gaussian random process as a maximum control policy. Stochastics **1**, 25–69 (1973)

Basar, T., Olsder, G.J.: Dynamic Noncooperative Game Theory, 2nd edn. Academic Press, London (1995)

Basar, T., Olsder, G.J.: Dynamic Noncooperative Game Theory, SIAM Edition, SIAm's Classics in Applied mathematics. Society for Industrial and Applied Mathematics, Philadelphia (1999)

Bauso, D., Timmer, J.: Robust dynamic cooperative games. Int. J. Game Theory **38**, 23–36 (2009)

Beard, R., McDonald, S.: Time-consistent fair water sharing agreements. Ann. Int. Soc. Dyn. Games **9**, 393–410 (2007)

Bellman, R.: Dynamic Programming. Princeton University Press, Princeton (1957)

D.W.K. Yeung, L.A. Petrosyan, *Subgame Consistent Cooperation*,
Theory and Decision Library C 47, DOI 10.1007/978-981-10-1545-8

Bergstrom, T., Blume, C., Varian, H.: On the private provision of public goods. J. Public Econ. **29**, 25–49 (1986)
Berkovitz, L.D.: A variational approach to differential games. In: Dresher, M., Shapley, L.S., Tucker, A.W. (eds.) Advances in Game Theory, pp. 127–174. Princeton University Press, Princeton (1964)
Bertsekas, D., Shreve, S.: Stochastic Optimal Control: The Discrete-Time Case. Athena Scientific, Belmont (1996)
Bleeke, J., Ernst, D.: Collaborating to Compete. Wiley, New York (1993)
Blodgett, L.L.: Factors in the instability of international joint ventures: an event history analysis. Strateg. Manag. J. **13**, 475–481 (1992)
Breton, M., Zaccour, G., Zahaf, M.: A differential game of joint implementation of environmental projects. Automatica **41**, 1737–1749 (2005)
Breton, M., Zaccour, G., Zahaf, M.: A game-theoretic formulation of joint implementation of environmental projects. Eur. J. Oper. Res. **168**, 221–239 (2006)
Bylka, S., Ambroszkiewicz, S., Komar, J.: Discrete time dynamic game model for price competition in an oligopoly. Ann. Oper. Res. **97**, 69–89 (2000)
Cellini, R., Lambertini, L.: A differential game approach to investment in product differentiation. J. Econ. Dyn. Control. **27**, 51–62 (2002)
Cellini, R., Lambertini, L.: Private and social incentives towards investment in product differentiation. Int. Game Theory Rev. **6**(4), 493–508 (2004)
Chamberlin, J.: Provision of collective goods as a function of group size. Am. Polit. Sci. Rev. **65**, 707–716 (1974)
Clark, C.W.: Mathematical Bioeconomics: The Optimal Management of Renewable Resources. Wiley, New York (1976)
D'Aspremont, C., Jacquemin, A.: Cooperative and noncooperative R&D in duopoly with spillovers. Am. Econ. Rev. **78**(5), 1133–1137 (1988)
De-Paz, A., Marín-Solano, J., Navas, J.: Time-consistent equilibria in common access resource games with asymmetric players under partial cooperation. Environ. Model. Assess. **18**(2), 171–184 (2013)
Dixit, A.K.: A model of duopoly suggesting a theory of entry barriers. Bell J. Econ. **10**, 20–32 (1979)
Dockner, E.J., Jørgensen, S.: Cooperative and non-cooperative differential games solutions to an investment and pricing problem. J. Int. Oper. Res. Soc. **35**, 731–739 (1984)
Dockner, E.J., Leitmann, G.: Coordinate transformation and derivation of open-loop nash equilibria. J. Econ. Dyn. Control **110**, 1–15 (2001)
Dockner, E.J., Long, N.V.: International pollution control: cooperative versus noncooperative strategies. J. Environ. Econ. Manag. **25**, 13–29 (1993)
Dockner, E.J., Nishimura, K.: Transboundary pollution in a dynamic game model. Jpn. Econ. Rev. **50**(4), 443–456 (1999)
Dockner, E.J., Jørgensen, S., Long, N.V., Sorger, G.: Differential Games in Economics and Management Science. Cambridge University Press, Cambridge (2000)
Dutta, P.-K., Radner, R.: Population growth and technological change in a global warming model. Economic Theory **29**(2), 251–270 (2006)
Ehtamo, H., Hamalainen, R.: A cooperative incentive equilibrium for a resource management. J. Econ. Dyn. Control **17**(4), 659–678 (1993)
Esteban-Bravo, M., Nogales, F.J.: Solving dynamic stochastic economic models by mathematical programming decomposition methods. Comput. Oper. Res. **35**, 226–240 (2008)
Feenstra, T., Kort, P.M., De Zeeuw, A.: Environmental policy instruments in an international duopoly with feedback investment strategies. J. Econ. Dyn. Control **25**, 1665–1687 (2001)
Fershtman, C., Nitzan, S.: Dynamic voluntary provision of public goods. Eur. Econ. Rev. **35**, 1057–1067 (1991)
Filar, J.A., Petrosyan, L.A.: Dynamic cooperative games. Int. Game Theory Rev. **2**, 47–65 (2000)

Fleming, W.H.: Optimal continuous-parameter stochastic control. SIAM Rev. **11**, 470–509 (1969)

Fleming, W.H., Rishel, R.W.: Deterministic and Stochastic Optimal Control. Applications of mathematics, vol. 1. Springer, New York/Heidelberg/Berlin (1975)

Fredj, K., Martín-Herrán, G., Zaccour, G.: Slowing deforestation pace through subsidies: a differential game. Automatica **40**, 301–309 (2004)

Gomes-Casseres, B.: Joint venture instability: is it a problem? Columbia J. World Bus. **XXII**(2), 97–102 (1987)

Gradstein, M., Nitzan, S.: Binary participation and incremental provision of public goods. Soc. Choice Welf. **7**, 171–192 (1989)

Hamalainen, R., Haurie, A., Kaitala, V.: Equilibria and threats in a fishery management game. Optim. Control Appl. Methods **6**, 315–333 (1986)

Hamel, G., Doz, Y., Prahalad, C.K.: Collaborate with your competitors – and win. Harv. Bus. Rev. **67**(1), 133–139 (1989)

Hartwick, J., Yeung, D.W.K.: The tragedy of the commons revisited. Pac. Econ. Rev. **2**(1), 45–62 (1997)

Haurie, A.: A note on nonzero-sum differential games with bargaining solutions. J. Optim. Theory Appl. **18**, 31–39 (1976)

Haurie, A., Zaccour, G.: A differential game model of power exchange between interconnected utilizes. Proceedings of the 25th IEEE conference on decision and control, Athens, Greece, pp. 262–266 (1986)

Haurie, A., Zaccour, G.: A game programming approach to efficient management of interconnected power networks. In: Hämäläinen, R.P., Ehtamo, H. (eds.) Differential Game-Developments in Modeling and Computation. Springer, Berlin (1991)

Huang, X., He, P., Zhang, W.: A cooperative differential game of transboundary industrial pollution between two regions. J. Clean. Prod. **120**, 43–52 (2016)

Isaacs, R.: Differential Games. Wiley, New York (1965)

Jørgensen, S., Yeung, D.W.K.: Stochastic differential game model of a common property fishery. J. Optim. Theory Appl. **90**, 391–403 (1996)

Jørgensen, S., Yeung, D.W.K.: Inter- and intragenerational renewable resource extraction. Ann. Oper. Res. **88**, 275–289 (1999)

Jørgensen, S., Yeung, D.W.K.: Overlapping generations stochastic differential games. Automatica **41**, 69–74 (2005)

Jørgensen, S., Zaccour, G.: Time consistent side payments in a dynamic game of downstream pollution. J. Econ. Dyn. Control. **25**, 1973–1987 (2001)

Jørgensen, S., Zaccour, G.: Differential Games in Marketing. Kluwer Academic Publishers, Boston (2004)

Kaitala, V., Pohjola, M.: Optimal recovery of a shared resource stock: a differential game with efficient memory equilibria. Nat. Resour. Model. **3**, 91–118 (1988)

Kaitala, V., Pohjola, M.: Economic development and agreeable redistribution in capitalism: Efficient game equilibria in a two-class neoclassical growth model. Int. Econ. Rev. **31**, 421–427 (1990)

Kaitala, V., Pohjola, M.: Sustainable international agreements on greenhouse warming: a game theory study. Ann. Int. Soc. Dyn. Games **2**, 67–87 (1995)

Kaitala, V., Mäler, K.G., Tulkens, H.: The acid rain game as a resource allocation process with an application to the international cooperation among Finland, Russia and Estonia. Scand. J. Econ. **97**, 325–343 (1995)

Kamien, M.I., Muller, E., Zang, I.: Research joint ventures and R&D cartels. Am. Econ. Rev. **82** (5), 1293–1306 (1992)

Krawczyk, J.B., Tidball, M.: A discrete-time dynamic game of seasonal water allocation. J. Optim. Theory Appl. **128**(2), 411–429 (2006)

Lehrer, E., Scarsini, M.: On the core of dynamic cooperative games. Dyn. Games Appl. **3**, 359–373 (2013)

Leitmann, G.: Cooperative and Non-Cooperative any Players Differential Games. Springer, New York (1974)

Leitmann, G., Mon, G.: Some geometric aspects of differential games. J. Astronaut. Sci. **14**, 56–65 (1967)

Li, S.: A differential game of transboundary industrial pollution with emission permits trading. J. Optim. Theory Appl. **163**, 642–659 (2014)

Lin, F., Liu, Q., Zhou, X., Chen, Y.: Cooperative differential game for model energy-bandwidth efficiency tradeoff in the internet of things. China Commun. **11**, 92–102 (2014)

Long, N.V.: A Survey of Dynamic Games in Economics. World Scientific, Singapore (2010)

Mäler, K.-G., de Zeeuw, A.: The acid rain differential game. Environ. Resour. Econ. **12**, 167–184 (1998)

Marin-Solano, J.: Time-consistent equilibria in a differential game model with time inconsistent preferences and partial cooperation. In: Haunschmied, J., Veliov, V., Wrzaczek, S. (eds.) Dynamic Games in Economics, pp. 219–238. Springer, New York (2014)

McGuire, M.: Group size, group homogeneity, and the aggregate provision of a pure public good under Cournot behavior. Public Choice **18**, 107–126 (1974)

Miao, X.N., Zhou, X.W., Wu, H.Y.: A cooperative differential game model based on transmission rate in wireless networks. Oper. Res. Lett. **38**, 292–295 (2010)

Nash, J.F.: The bargaining problem. Econometrica **18**, 155–162 (1950)

Nash, J.F.: Non-cooperative games. Ann. Math. **54**, 286–295 (1951)

Nash, J.F.: Two-person cooperative games. Econometrica **21**, 128–140 (1953)

Nie, P., Chen, L., Fukushima, M.: Dynamic programming approach to discrete time dynamic feedback stackelberg games with independent and dependent followers. Eur. J. Oper. Res. **169**, 310–328 (2006)

Parkhe, A.: "Messy" research, methodological predispositions and theory development in international joint ventures. Acad. Manage. Rev. **18**(2), 227–268 (1993)

Petrosyan, L.A.: Stable solutions of differential games with many participants. Viestnik Leningrad Univ. **19**, 46–52 (1977)

Petrosyan, L.A.: Agreeable solutions in differential games. Int. J. Math. Game Theory Algebra **7** (2/3), 165–177 (1997)

Petrosyan, L.A., Danilov, N.N.: Stability of solutions in non-zero sum differential games with transferable payoffs. Viestnik Leningrad Univ. **1**, 52–59 (1979)

Petrosyan, L.A., Danilov, N.N.: Cooperative Differential Games and Their Applications. Izd. Tomskogo University, Tomsk (1982)

Petrosyan, L.A., Murzov, N.V.: Game theoretic problems in mechanics. Litvsk. Math. Sb. **6**, 423–433 (1966)

Petrosyan, L.A., Shevkoplyas, E.V.: Cooperative solutions for games with random duration. Game Theory Appl. **IX**, 125–139 (2003)

Petrosyan, L.A., Yeung, D.W.K.: Dynamically stable solutions in randomly-furcating differential games. Trans. Steklov Inst. Math. **253**(Supplement 1), S208–S220 (2006)

Petrosyan, L.A., Yeung, D.W.K.: Subgame-consistent cooperative solutions in randomly-furcating stochastic differential games. Int. J. Math. Comput. Model. (Special Issue on Lyapunov's Methods in Stability and Control), **45**, 1294–1307 (2007)

Petrosyan, L.A., Zaccour, G.: Time-consistent Shapley value allocation of pollution cost reduction. J. Econ. Dyn. Control. **27**, 381–398 (2003)

Petrosyan, L.A., Zenkevich, N.A.: Game Theory. World Scientific Publishing Co. Pte. Ltd., Republic of Singapore (1996)

Pontryagin, L.S.: On the theory of differential games. Uspekhi Mat. Nauk **21**, 219–274 (1966)

Pontryagin, L.S., Boltyanskii, V.G., Gamkrelidze, R.V., Mishchenko, E.F.: The Mathematical Theory of Optimal Processes. Interscience Publishers, New York (1962)

Predtetchinski, A.: The strong sequential core for stationary cooperative games. Games Econ. Behav. **61**, 50–66 (2007)

Puterman, M.L.: Markov Decision Processes: Discrete Stochastic Dynamic Programming. Wiley, New York (1994)

Ross, S.: Introduction to Stochastic Dynamic Programming. Academic Press, New York (1983)

Rubio, S.J., Casino, B.: A note on cooperative versus non-cooperative strategies in international pollution control. Resour. Energy Econ. **24**, 251–261 (2002)

Rubio, S.J., Ulph, A.: An infinite-horizon model of dynamic membership of international environmental agreements. J. Environ. Econ. Manag. **54**(3), 296–310 (2007)

Samuelson, P.A.: The pure theory of public expenditure. Rev. Econ. Stat. **36**, 350–356 (1954)

Shapley, L.S.: A Value for N-Person Games, Contributions to the Theory of Games. In: Kuhn, H. W., Tucker, A.W. (eds.) Princeton University Press, Princeton, **2**, 307–317 (1953)

Shevkoplyas, E.V.: The Shapley value in cooperative differential games with random duration. Ann. Dyn. Games (2011), Adv. Dyn. Games **11**, 359–373 (2011)

Shubik, M.: Edgeworth market games. In: Contributions to the Theory of Games. Princeton University Press, Princeton (1959a)

Shubik, M.: Strategy and Market Structure. Wiley, New York (1959b)

Singh, N., Vives, X.: Price and quantity competition in a differentiated duopoly. Rand J. Econ. **15**, 546–554 (1984)

Smith, A.E., Zenou, Y.: A discrete-time stochastic model of job matching. Rev. Econ. Dyn. **6**, 54–79 (2003)

Sorger, G.: Recursive bargaining over a productive asset. J. Econ. Dyn. Control. **30**, 2637–2659 (2006)

Stimming, M.: Capital accumulation subject to pollution control: open-loop versus feedback investment strategies. Ann. Oper. Res. **88**, 309–336 (1999)

Suzumura, K.: Cooperative and noncooperative R&D in an oligopoly with spillovers. Am. Econ. Rev. **82**(5), 1307–1320 (1992)

Tahvonen, O.: Carbon dioxide abatement as a differential game. Eur. J. Polit. Econ. **10**, 685–705 (1994)

Tolwinski, B., Haurie, A., Leitmann, G.: Cooperative equilibria in differential games. J. Math. Anal. Appl. **119**, 182–202 (1986)

von Neumann, J., Morgenstern, O.: Theory of Games and Economic Behavior. Wiley, New York (1944)

Vorob'ev, N.N.: Foundations of Game Theory – Noncooperative Games. Birkhäuser Verlag, Boston (1972)

Wang, W.-K., Ewald, C.-O.: Dynamic voluntary provision of public goods with uncertainty: a stochastic differential game model. Decisions Econ. Finan. **3**, 97–116 (2010)

Wie, B.W., Choi, K.: The computation of dynamic Cournot-Nask traffic network equilibria in discrete time. KSCE J. Civ. Eng. **4**(4), 239–248 (2000)

Wirl, F.: Dynamic voluntary provision of public goods: extension to nonlinear strategies. Eur. J. Polit. Econ. **12**, 555–560 (1996)

Yeung, D.W.K.: A differential game of industrial pollution management. Ann. Oper. Res. **37**, 297–311 (1992)

Yeung, D.W.K.: Infinite horizon stochastic differential games with branching payoffs. J. Optim. Theory Appl. **111**(2), 445–460 (2001)

Yeung, D.W.K.: Randomly-furcating stochastic differential games. In: Petrosyan, L., Yeung, D. (eds.) ICM Millennium Lectures on Games, pp. 107–126. Springer, Berlin (2003)

Yeung, D.W.K.: Nontransferable individual payoff functions under stochastic dynamic cooperation. Int. Game Theory Rev. **6**, 281–289 (2004)

Yeung, D.W.K.: Subgame consistent dormant-firm cartel. In: Haurie, A., Zaccour, G. (eds.) Dynamic Games and Applications, pp. 255–271. Springer, Berlin (2005)

Yeung, D.W.K.: An irrational-behavior-proof condition in cooperative differential games. Int. Game Theory Rev. **8**, 739–744 (2006a)

Yeung, D.W.K.: Solution mechanisms for cooperative stochastic differential games. Int. Game Theory Rev. **8**(2), 309–326 (2006b)

Yeung, D.W.K.: Dynamically consistent cooperative solution in a differential game of transboundary industrial pollution. J. Optim. Theory Appl., **134**, 143–160, (2007a)

Yeung, D.W.K.: Noncooperative-equivalent imputation formula in cooperative differential games. Int. J. Algebra, 393–398, (2007b)
Yeung, D.W.K.: Dynamically consistent solution for a pollution management game in collaborative abatement with uncertain future payoffs. In: Yeung, D.W.K., Petrosyan L.A. (Guest eds.) Special issue on frontiers in game theory: In honour of John F. Nash. Int. Game Theory Rev. **10** (4), 517–538 (2008)
Yeung, D.W.K.: Subgame consistent Shapley value imputation for cost-saving joint ventures. Math. Game Theory Appl **2**(3), 137–149 (2010)
Yeung, D.W.K.: Dynamically consistent cooperative solutions in differential games with asynchronous players' horizons. Ann. Int. Soc. Dyn. Games **11**, 375–395 (2011)
Yeung, D.W.K.: Subgame consistent cooperative solutions in stochastic differential games with asynchronous horizons and uncertain types of players. Contrib. Game Theory Manag. **5**, 334–355 (2012)
Yeung, D.W.K.: Nontransferable individual payoffs in cooperative stochastic dynamic games. Int. J. Algebra **7**(13), 597–606 (2013)
Yeung, D.W.K.: Dynamically consistent collaborative environmental management with production technique choices. Ann. Oper. Res. **220**(1), 181–204 (2014)
Yeung, D.W.K., Petrosyan, L.A.: Subgame consistent cooperative solution in stochastic differential games. J. Optim. Theory Appl. **120**(3), 651–666 (2004)
Yeung, D.W.K., Petrosyan, L.A.: Subgame consistent solution of a cooperative stochastic differential game with nontransferable payoffs. J. Optim. Theory Appl. **124**(3), 701–724 (2005)
Yeung, D.W.K., Petrosyan, L.A.: Dynamically stable corporate joint ventures, Automatica, **42**, 365–370 (2006a)
Yeung, D.W.K., Petrosyan, L.A.: Cooperative Stochastic Differential Games. Springer-Verlag, New York, 242pp (2006b)
Yeung, D.W.K., Petrosyan, L.A.: Managing catastrophe-bound Industrial pollution with game-theoretic algorithm: The St Petersburg initiative. In: Petrosyan, L.A., Zenkevich, N.A. (eds.) Contributions to Game Theory and Managemen, pp. 524–538. St Petersburg State University, St Petersburg (2007a)
Yeung, D.W.K., Petrosyan, L.A.: The crux of dynamic economic cooperation: Subgame consistency and equilibrating transitory compensation. Game Theory Appl. **11**, 207–221 (2007b)
Yeung, D.W.K., Petrosyan, L.A.: The tenet of transitory compensation in dynamically stable cooperation. Int. J. Tomogr. Stat. **7**, 60–65 (2007c)
Yeung, D.W.K., Petrosyan, L.A.: A cooperative stochastic differential game of transboundary industrial pollution. Automatica **44**(6), 1532–1544 (2008)
Yeung, D.W.K., Petrosyan, L.A.: Subgame consistent solutions for cooperative stochastic dynamic games. J. Optim. Theory Appl. **145**(3), 579–596 (2010)
Yeung, D.W.K., Petrosyan, L.A.: Subgame consistent cooperative solution of dynamic games with random horizon. J. Optim. Theory Appl. **150**, 78–97 (2011)
Yeung, D.W.K., Petrosyan, L.A.: Subgame Consistent Economic Optimization: An Advanced Cooperative Dynamic Game Analysis. Birkhäuser, Boston, 395 pp (2012a)
Yeung, D.W.K., Petrosyan, L.A.: Subgame consistent solution for cooperative stochastic dynamic games with random horizon. Int. Game Theory Rev., 14(2), 1440012.01- 1440012.29 (2012b)
Yeung, D.W.K., Petrosyan, L.A.: Subgame consistent solution for a cooperative differential game of climate change control. Contrib. Game Theory Manag., **5**, 356–386 (2012c)
Yeung, D.W.K., Petrosyan, L.A.: A notational note on agreeable solutions in cooperative differential games. Game Theory Appl., **15**, 203–206 (2012d)
Yeung, D.W.K., Petrosyan, L.A.: Subgame-consistent cooperative solutions in randomly furcating stochastic dynamic games. Math. Comput. Model. **57**(3–4), 976–991 (2013a)
Yeung, D.W.K., Petrosyan, L.A.: Subgame consistent cooperative provision of public goods. Dyn. Games Appl. **3**(3), 419–442 (2013b)
Yeung, D.W.K., Petrosyan, L.A.: Subgame consistent solution for a dynamic game of pollution management. In: Petrosyan, L.A., Romanovsky, J.V., Yeung, D.W.K. (eds.) Advances in

Economics and Optimization: Collected Scientific Studies Dedicated to the Memory of L.V. Kantorovich, pp. 95–112. New Nova Science Publishers, Hauppauge (2014a)

Yeung, D.W.K., Petrosyan, L.A.: Subgame consistent cooperative provision of public goods under accumulation and payoff uncertainties. In: Haunschmied, J., Veliov, V.M., Wrzaczek, S. (eds.) Dynamic Games in Economics, pp. 289–315. Springer, Berlin/Heidelberg (2014b)

Yeung, D.W.K., Petrosyan, L.A.: Subgame consistent cooperative solutions for randomly furcating stochastic dynamic games with uncertain horizon. Int. Game Theory Rev. **16**(2), 1440012.01–1440012.29 (2014c)

Yeung, D.W.K. Petrosyan, L.A.: Subgame consistent cooperative solution for NTU dynamic games via variable weights. Automatica, **50**, 84–89 (2015a)

Yeung, D.W.K., Petrosyan, L.A.: On subgame consistent solution for NTU cooperative stochastic dynamic games. Paper presented at European meeting on game theory (SING11-GTM2015) at St Petersburg, 8–10 July (2015b)

Yeung, D.W.K., Petrosyan, L.A., Yeung, P.M.: Subgame consistent solutions for a class of cooperative stochastic differential games with nontransferable payoffs. Ann. Int. Soc. Dyn. Games. **9**, 153–170 (2007)

Index

D.W.K. Yeung, L.A. Petrosyan, *Subgame Consistent Cooperation*,
Theory and Decision Library C 47, DOI 10.1007/978-981-10-1545-8

MIX
Papier aus verantwortungsvollen Quellen
Paper from responsible sources
FSC® C105338

If you have any concerns about our products,
you can contact us on
ProductSafety@springernature.com

In case Publisher is established outside the EU,
the EU authorized representative is:
Springer Nature Customer Service Center GmbH
Europaplatz 3, 69115 Heidelberg, Germany

Printed by Libri Plureos GmbH
in Hamburg, Germany